ADVANCES IN
MOLECULAR AND CELL BIOLOGY
VOLUME 33

Lipobiology

ADVANCES IN
MOLECULAR AND CELL BIOLOGY
VOLUME 33

Lipobiology

Series Editor:

E. Edward Bittar
University of Wisconsin - Madison
Madison, Wisconsin,
USA

Volume Editor:

Ger van der Vusse
Maastricht,
The Netherlands

2004

ELSEVIER

Amsterdam - Boston - Heidelberg - London - New York - Oxford
Paris - San Diego - San Francisco - Singapore - Sydney - Tokyo

ELSEVIER B.V.
Sara Burgerhartstraat 25
P.O. Box 211, 1000 AE
Amsterdam, The Netherlands

ELSEVIER Inc.
525 B Street, Suite 1900
San Diego, CA 92101-4495
USA

ELSEVIER Ltd
The Boulevard, Langford Lane
Kidlington, Oxford OX5 1GB
UK

ELSEVIER Ltd
84 Theobalds Road
London WC1X 8RR
UK

First edition 2004

Library of Congress Cataloging in Publication Data
A catalog record is available from the Library of Congress.

British Library Cataloguing in Publication Data
A catalogue record is available from the British Library.

ISBN: 0-444-51496-1
ISSN: 1569-2558 (series)

∞ The paper used in this publication meets the requirements of ANSI/NISO Z39.48-1992 (Permanence of Paper).
Printed in The Netherlands.

TABLE OF CONTENTS

PREFACE

The evolutionary process resulted in living creatures composed of organs, integrated cellular networks within organs and specialized compartments inside cells. These structures require the presence of biological membranes, transport routes for nutrients, storage sites of oxidizable substrates, transporting and metabolising proteins, short term and long term regulation of the activity and production of the substrate handling proteins, and signalling pathways ensuring proper communication between and within organs and cells. Since lipids are playing an essential role in the creation, development and maintenance of living systems, detailed insight into lipobiology is a prequisite to understand the basics of life.

The present volume reflects the state of the art of a selected number of topics of lipobiology.

The editor is fully responsible for the selection, which has to be made due to shortage of space. This selection, in great part, reflects my personal fascination for some specific roles of lipids in the living organism. The chapters of this volume focus on absorption of lipids via the intestinal tract and their transport in blood, storage of lipid nutrients in and release from adipose tissue. Attention has also been paid to uptake, transport and metabolism of fatty acids in parenchymal cells under normal and pathological conditions, and the role of lipids in the regulation of gene expression. Moreover, synthesis, composition, and physical properties of biological membranes and the significance of phospholipids as precursors of compounds involved in signalling cascades are discussed. Finally, various aspects of the relationship between lipids and atherosclerosis, and the involvement of cholesterol-carrying proteins in steroid production have been described.

I gratefully thank Prof. E. Edward Bittar for providing me the opportunity to act as editor of this volume and my colleagues and friends for sharing with us their latest insights in the fascinating field of lipobiology. Their expertise and enthusiasm resulted in twenty-seven chapters summarizing the diverse roles of lipids in biological systems.

GER J. VAN DER VUSSE
Dept. of Physiology
Cardiovascular Research Institute Maastricht (CARIM)
Maastricht, the Netherlands

Brief overview on lipobiology

Ger J. van der Vusse*

Department of Physiology, Cardiovascular Research Institute Maastricht, Maastricht University, P.O. Box 616, 6200 MD Maastricht, The Netherlands
**Correspondence address: Tel.: +31-43-3881086; fax: +31-43-3884166*
E-mail: vandervusse@fys.unimaas.nl(G.J.van der Vusse)

1. Introduction

Lipids are a diverse group of molecules characterized by a predominantly hydrophobic nature. In aqueous environments, such as blood and intracellular compartments, lipids show the propensity to clump together forming either particles like micelles and triacylglycerol-containing lipid droplets or biological membranes such as phospholipid bilayers. Lipid clustering can be modulated by specific lipid binding proteins, enhancing the solubility of lipid molecules in the aqueous compartment and facilitating transport of lipid molecules through water phase.

A major class of biologically important lipids are fatty acids and their derivatives. A second class of lipids that play a crucial role in biological systems are cholesterol and its derivatives and metabolic products. Lipids are involved in a plethora of processes required to maintain cellular structures and to execute cellular functions. Among others, fatty acids are substrates for energy conversion, starting materials for intracellular storage of fat in triacylglycerol-rich droplets and serve as substrates for biologically active substances such as prostaglandins and leukotrienes. Moreover, fatty acids are required for the formation of phospholipids, the building blocks of cellular membranes, and the coating of lipoproteins, the lipid transporting particles in blood plasma. Finally, fatty acids themselves and their derivatives are information carriers interacting with intracellular signaling pathways and nuclear factors involved in the regulation of DNA transcription.

Cholesterol is an important lipid constituent of biological membranes, specifically influencing membrane fluidity. Cholesterol is also the substrate for the formation of bile acids in liver cells. After storage in the gall bladder, bile acids are of paramount importance to facilitate the digestion of dietary lipids and the transport of the digestion products to the epithelial cells of the intestinal wall. Equally important is the fact that cholesterol is the precursor of various classes of steroid hormones, including mineralocorticoids and glucocorticoids produced in the adrenal cortex and sex steroid hormones synthetized in the female and male gonads.

Advances in Molecular and Cell Biology, Vol. 33, pages 1–7

ISSN: 1569-2558 / DOI: 10.1016/S1569-2558(03)33001-2

2. Dietary fatty acids, fat cells and transport in blood

The majority of fatty acyl moieties in the human body is derived from dietary fat. Prior to absorption, the fatty acyl moieties are liberated from neutral dietary fat (triacylglycerols) and transported to the luminal membrane of the enterocytes of the intestinal wall. Evidence is accumulating that transport of fatty acids across the luminal membrane is mediated by membrane-associated lipid binding proteins. Candidates are fatty acid transport protein-4, caveolin-1 and the fatty acid transporter FAT/CD36. Inside the enterocyte, cytoplasmic transport of fatty acids is facilitated by soluble lipid binding proteins, belonging to a multigenic family of 14–15 kDa proteins. In the enterocyte, two types are abundantly expressed: the intestinal type (I-FABP) and the liver type (L-FABP) fatty acid-binding protein. After conversion into acyl CoA, the fatty acyl chain is incorporated into triacylglycerols and released into the intestinal lymphatic system as chylomicron.

During the post-prandial phase, circulating chylomicrons reach the adipose cells, where the neutral fat content of the lipid particle is hydrolyzed by lipoprotein lipase. The activity of lipoprotein lipase is regulated by hormones such as insulin. The fatty acids released are taken up by the adipocytes and intracellularly stored as triacylglycerols. Besides, fatty acyl moieties derived from very low density lipoproteins, produced by the liver, and de novo synthetized fatty acyl chains by the adipocytes also serve as substrates for the formation of triacylglycerol inside the fat cells. The mechanism of uptake of fatty acids by the adipocyte is still debated although increasing evidence points to a substantial role of fatty acid translocase (FAT/CD36) in facilitating transmembrane transport of exogenous fatty acids.

Mobilization of fatty acids from the adipocyte triacylglycerol pool is governed by hormones. The key regulating enzyme is hormone-sensitive lipase. Whether FAT/CD36 is involved in the release of fatty acids from the adipocytes remains to be established. After release from fat cells, fatty acids are bound to plasma albumin and transported to organs and cells in need of these hydrophobic substrates. Albumin is present in blood plasma in relatively high concentrations (~0.6 mM) and can carry at least 6 molecules of fatty acids. This ensures that more than 99% of the fatty acid present in plasma is protein-bound. Fatty acids are released from albumin prior to transport across the endothelium of the organ in question. There are indications that the albumin–fatty acid complex interacts with the luminal endothelial cell membrane (either with phospholipids or with membrane-associated proteins) to facilitate the release of fatty acid from albumin. However, recent observations showing that the glycocalix of the endothelial cell might be a constraint for albumin diffusion to the luminal membrane are not in favor of this mechanism. At present, the kinetics of fatty acid binding to albumin and the position of the binding sites have been partly elucidated.

3. Uptake and transport of fatty acids in tissues

After release from albumin, fatty acids cross the endothelium prior to be taken up by the parenchymal cells. A computer model specifically designed to investigate fatty acid uptake predicts that transendothelial fatty acid transport represents a major constraint for overall fatty acid utilization in cardiac tissue. Although part of the transmembrane trafficking of fatty acids occurs via simple diffusion (absorption into the outer leaflet of the phospholipid bilayer, flip to the inner layer and desorption from the outer layer into the cytosolic compartment), evidence is accumulating that membrane-associated proteins are required to facilitate the movement of fatty acids from the extracellular compartment into the cytoplasm. At present at least three proteins have been identified as putative fatty acids "transporters": plasma membrane fatty acid-binding proteins ($FABP_{pm}$), fatty acid translocase (FAT/CD36) and a family of fatty acid transporting proteins (FATPs). $FABP_{pm}$, now known to be identical to mitochondrial aspartate aminotransferase (AsAT), was found to play an important role in fatty acid uptake by adipocytes and liver cells. Recent studies revealed that prior to reaching the plasma membrane of the cell, $FABP_{pm}$/AsAT has to pass through the mitochondrion, where mitochondrial endopeptidase converts the inactive protein into its active form.

Until now, at least six members of the membrane-associated FATP family have been described. FATPs possess enzymatic activity, catalyzing the conversion of very long chain fatty acids into their acyl CoA esters. It is of interest to note that very long chain fatty acids are essential components of sphingolipids, which are needed to maintain the structure of caveolae in cellular membranes. Since caveolae are involved in cellular fatty acid uptake, FATPs may indirectly facilitate the uptake of fatty acids by maintaining or augmenting caveolar vehicles required for transmembrane fatty acid transport.

Overwhelming evidence has been provided that FAT/CD36 plays a quantitatively important role in fatty acid transport across the plasmalemma of, among others, heart and skeletal muscle cells. The molecular mechanisms by which FAT/CD36 enhances transmembrane trafficking of fatty acids is incompletely understood, but may relate to facilitation of absorption of fatty acids into the outer leaflet of the phospholipid bilayer and/or desorption from the inner leaflet into the cytoplasm. A recently developed computer model predicts the first option. Recent studies revealed that, at least in cardiac and skeletal muscle, FAT/CD36 is also stored in an intracellular site from which it can be recruited by hormones such as insulin and by contractile activity of the myocyte. This feature enables the muscle cells to adjust the uptake of extracellular fatty acids to the intracellular need of oxidizable substrates for energy conversion or to replenish the intracellular triacylglycerol stores depleted during strenuous exercise. The precise mechanism by which the combined activities of $FABP_{pm}$, FATPs and FAT/CD36 are fine tuned and synchronized remains to be elucidated.

4. Intracellular transport and metabolism of fatty acids

Inside the parenchymal cell, diffusion of hydrophobic fatty acids from the plasmalemma to the sites of conversion is facilitated by cytosolic fatty acid-binding proteins ($FABCP_c$). These soluble lipid binding proteins possess a high affinity for fatty acids, are present in relatively high concentrations in the cytoplasma and show a high degree of tissue and cell-specific expression. Besides facilitation of diffusion, they may also facilitate desorption of fatty acids from the outer leaflet of the plasmalemma and might be instrumental in intracellular targeting of the fatty acyl moieties. A pivotal role for intracellular conversion of fatty acids is played by the enzyme fatty acyl CoA synthetase.

The bulk of the enzyme activity is localized at the outer membrane of the mitochondrion, whereas activity is also present in the peroxisomes, and, to a minor extent, in the plasmalemma. Fatty acyl CoA synthetase catalyses the formation of fatty acyl CoAs, substrates for mitochondrial β-oxidation, cytoplasmic formation of triacylglycerols, phospholipids and the covalent binding of fatty acyl chains to proteins, i.e. protein acylation. Hence, fatty acyl CoA synthetase resides at a cross-road of multiple intracellular metabolic pathways.

Fatty acyl CoA, once formed, strongly binds to acyl CoA binding protein (ACBP). This cytoplasmic protein (~ 10 kDa) is present in all eukaryotic species tested so far. ACBP is able to create a cytoplasmic protected pool of acyl CoA esters and, hence, is likely to play a key role in acyl CoA regulated cellular functions. It has been shown that acyl CoA bound to ACPB is the preferred substrate of carnitine acyl transferase, the enzyme catalyzing the formation of acyl carnitine from acyl CoA, a rate-governing step in overall mitochondrial fatty acid oxidation. ACBP may also play a role in cellular Ca^{2+} handling and cell growth.

In addition to the availability of acyl CoA, the rate of conversion of acyl CoA into acyl carnitine, catalyzed by carnitine acyltransferase 1, is modulated by malonyl CoA. This compound inhibits carnitine acyltransferase 1, and hence governs the rate of mitochondrial fatty acid oxidation. Malonyl CoA is produced primarily by carboxylation of acetyl CoA, catalyzed by acetyl CoA carboxylase. Degradation of malonyl CoA is facilitated by malonyl CoA decarboxylase. Alterations in the concentration of malonyl CoA in the cardiac muscle cell can be achieved by alterations in AMP kinase activity, resulting in changes in acetyl CoA carboxylase and malonyl CoA decarboxylase activity.

5. Impaired fatty acid handling and gene expression

Under normal, healthy conditions, cellular fatty acid uptake and metabolism are finely tuned to the cellular need of these substrates. Because of the lipotoxic effects of elevated cellular fatty acid concentrations and accumulation of derivatives such as triacylglycerols, any disturbance in uptake and utilization may affect proper cellular functioning. During the past decade, evidence is accumulating that, for instance in the diabetic state, besides carbohydrate handling fatty acid homeostasis is severely disturbed. In the obese Zucker rat, a model for type-2 diabetes mellitus, the elevated abundance of FAT/CD36 was found to be caused by a permanent reallocation of this membrane-associated fatty acid transporter from intracellular stores to sarcolemma. As a consequence, fatty acid uptake may surpass fatty acid oxidation, which most likely results in the accumulation of triacylglycerols in the affected muscle cells. This event may, in turn, add to insulin resistance as readily seen in type-2 diabetes.

It has been reported that impaired mitochondrial and peroxisomal fatty acid oxidation can also be caused by genetic defects in the expression and synthesis of proteins involved in the mitochondrial and peroxisomal metabolic pathways. Besides defects in energy conversion, impaired fatty acid handling may also affect proper control of the expression of a multitude of genes encoding for fatty acid and glucose handling proteins.

Recent studies have revealed that the eukaryotic cellular genome is able to respond to alterations in the lipid environment. The three peroxisome proliferator-activator receptor (PPAR) isotypes PPARα, PPARβ/δ and PPARγ may act as "liposensors" by using fatty acids and their derivatives as natural ligands. In addition to ligand availability, PPAR activity is modulated by phosphorylation, which, in turn, modulates the ability of PPARs to interact with co-repressor and co-activator proteins. This multilevel control system most likely helps the cell to adjust its enzyme content and activity to meet the metabolic needs in a chronic fashion.

Since fatty acids and their CoA esters are known to exert lipotoxic effects on a multitude of cellular functions, including mitochondrial energy conversion, the mitochondrial content of fatty acids and their derivatives should be kept as low as possible. It has therefore been postulated that the presence of mitochondrial uncoupling protein-3 (UCP-3) is instrumental in transporting fatty acids from the mitochondrial matrix back into the cytoplasmic compartments. Since the outward transport of fatty acid anions lowers the mitochondrial proton gradient, the activity of UCP-3 might result in a mild mitochondrial uncoupling as a side effect.

6. Essential fatty acids and membrane phospholipids

Experimental findings have shown that essential fatty acids and their longer chain, more unsaturated products, are major structural and functional components of cell membrane phospholipids. In particular, under conditions that accretion of tissue is high, e.g. during pregnancy, unrestricted availability of this class of fatty acids is of paramount importance. Evidence is accumulating that the bioavailability of essential fatty acids in pregnant women may be at risk when the diet is low in these substances.

Phospholipids, being the most abundant class of lipids in all eukaryotic membranes, provide the cell with a barrier where selective permeability guarantees proper cellular functioning. Phospholipids represent a class of chemically and functionally different glycerolipids consisting of, among others, phosphatidylcholine, phosphatidylethanolamine, phosphatidylinositol, phosphatidylserine and cardiolipin. Most eukaryotic cells are equipped with a set of enzymes to synthesize all phospholipids required for the diverse array of cellular functions. Following biosynthesis inside the cell, the phospholipid moieties are incorporated in the phospholipid bilayer of the cellular membranes. Moreover, these phospholipids are subjected to a continuous acylation–reacylation cycle. The mechanisms underlying the processes required for the cells to adjust the molecular lipid composition of their membranes to alterations in intra- and extracellular conditions are as yet incompletely understood.

Detailed studies have revealed that the phospholipid composition of the bilayer is asymmetric. In plasma membranes the choline-containing phospholipids, phosphatidylcholine and sphingomyoline, are enriched in the outer leaflet, whereas the bulk of phosphatidylethanolamine and phosphatidylinositol is present in the inner leaflet. Virtually all phosphatidylserine is also located in the inner membrane leaflet. Membrane asymmetry is considered to be of utmost importance for proper functioning of the plasmalemma. However, under certain conditions, i.e. cell activation or differentiation, and during programmed cell death bilayer asymmetry readily collapses. A number of membrane-associated proteins have been identified to be involved in the regulation of membrane phospholipid sidedness. At present three different proteins can be distinguished: flippase facilitating the inward-directed transport of lipids, floppase, which promotes migration in the outward direction and scramblase catalyzing the mixing of lipids between the two membrane leaflets. The precise mechanism of action of these proteins and the way how their activity is regulated are, however, incompletely elucidated.

A special group of membrane-associated lipids are sphingolipids. Besides acting as building blocks of cellular membranes, sphingolipids and their metabolic products play an important role in cellular processes such as apoptosis, proliferation, angiogenesis and stress responses. The complexity of spingolipid-mediated signaling requires a strict regulation of the enzymes involved in their biosynthesis and catabolism.

Since most phospholipids are rich in arachidonic acid, and this particular fatty acid is the precursor of biologically active substances such as prostaglandins and leukotrienes, phospholipids are not only building blocks of cellular membranes, but suppliers of substances that play a regulatory role in processes such as cell growth, inflammation, platelet activation and cytotoxicity as well. The release of arachidonic acid from phospholipids is catalyzed by a variety of phospholipases A_2. Since cytosolic phospholipase A_2 ($cPlA_2$) distinguishes itself by its arachidonic acid preference, this particular enzyme is thought to play a crucial role in phospholipid-mediated cellular signaling pathways.

In addition to $cPlA_2$, other phospholipid hydrolyzing enzymes are identified to be involved in the generation of biologically active substances. For instance, activation of phospholipase C (PLC), catalyzing the hydrolysis of phosphatidylinositol-4,5,biphosphate, is a major signal transduction pathway of a variety of cell surface receptors. At present, 11 mammalian PLC isoforms have been identified. They possess family-specific regulating domains and were found to take differential positions in receptor signaling.

Finally, mammalian cells contain phospholipase D (PLD), an enzyme that catalyses the hydrolysis of the choline head group from phosphatidylcholine. The resulting phosphatidic acid is generally considered the signaling molecule produced by PLD. It has been shown that the catalytic activity is regulated by a variety of factors, including the activity of protein kinase C, various GTPases and tyrosine kinase.

7. Transport of esterified lipids in blood and atherosclerosis

Transport of lipids between organs, other than albumin-mediated transport of fatty acids, is accomplished by lipoprotein particles in blood plasma. These particles comprise among others chylomicrons, released by the intestines, very low density lipoproteins, produced and secreted by hepatocytes, and low density and high density lipoproteins. The latter two are involved in the transport of cholesterol and its fatty acyl esters. During the past decade, a family of proteins has been identified, possessing the ability to transfer lipids, i.e. phospholipids and cholesterol esters, between the lipoprotein particles and tissues. Particular attention attracted one of the members of lipid transfer protein family, i.e. LTP-II or PLTP, phospholipid transfer protein. The putative physiological function of PLTP is to promote the reverse cholesterol transport pathway, and hence, to stimulate the transport of cholesterol from extrahepatic tissue to the liver prior to excretion of cholesterol. Therefore, PLTP may play an important role in mitigating the atherosclerotic process. Since atherosclerosis is a complex and multifactoral disease, research is not only directed to events in the blood compartment but also in the vascular wall. Recent findings strongly suggest that members of the PPAR family are involved in the atherosclerotic process. PPARα and PPARγ have been shown to slow down the progression of atherosclerosis. Agonists of these nuclear receptors are therefore novel pharmaceutical targets to combat life-threatening atherosclerotic lesions of the vascular wall.

8. Cholesterol and steroidogenesis

Besides its unfortunate role in atherosclerosis, cholesterol is of paramount importance for the maintenance of a host of physiological processes. In addition to phospholipid membrane stabilization, cholesterol serves as the substrate for the biosynthesis of steroid hormones. The first rate-governing step in the conversion of cholesterol in biologically active steroid hormones, produced in either the adrenal cortex or the female and male gonadal glands, is the intramitochondrial production of pregnenolone by the catalytic activity of the cholesterol side chain cleaving (CSCC)–enzyme complex. This step is regulated by hormones released by the pituitary gland, i.e. ACTH and LH, and requires increased availability of cholesterol for the CSCC–enzyme complex.

Recent findings indicate that a cholesterol transporting protein, designated StAR (steroidogenic acute regulatory) protein, is involved in the hormone-stimulated increase of pregnenolone synthesis. The precise mechanisms of action of StAR in mediating cholesterol transfer in steroidogenic mitochondria, however, are incompletely understood. The same holds for several recently discovered new members of the StAR-gene family in non-steroidogenic tissues.

This volume on lipobiology contains 27 chapters dealing with a selected number of subjects of lipid homeostasis in the eukaryotic cells and mammalian body. Because of limited space, it was not possible to cover all aspects of lipobiology. Attempts were made to focus on subjects in the field of lipobiology characterized by new insights in underlying regulatory mechanisms. Special attention was paid to proteins involved in lipid transfer and enzymes in lipid metabolism and the regulation of expression of genes encoding for lipid handling proteins. The editor is greatly indebted to his colleagues who enthusiastically accepted his request to contribute to this volume and for their willingness to share with us their thoughtful insights and latest results in the intriguing complexity of lipid homeostasis in biological structures.

Intestinal uptake and transport of fatty acids

Isabelle Niot and Philippe Besnard*

Physiologie de la Nutrition, Ecole Nationale Supérieure de Biologie Appliquée à la Nutrition et à l'Alimentation (ENSBANA), UMR5170 CESG CNRS/INRA/Université de Bourgogne, 1, Esplanade Erasme, F-21000 Dijon, France
**Correspondence address: Tel./fax: +33-03-80-39-66-91*
E-mail: pbesnard@u-bourgogne.fr(P.B.)

Abbreviations

LCFA: long-chain fatty acids; LCA: long-chain acyl-CoA esters; TG: triglycerides; LBP: lipid-binding proteins; FABPpm: plasma membrane fatty acid-binding protein; mAspAT: mitochondrial aspartate aminotransferase; FAT/CD36: fatty acid transporter; FATP: fatty acid transport protein; I-FABP: intestinal fatty acid-binding protein; L-FABP: liver fatty acid-binding protein; ACBP: acyl-CoA-binding protein; ACS: acyl-CoA synthetase; EGF: epidermal growth factor; PPAR: peroxisome proliferator-activated receptor.

1. Introduction

Long-chain fatty acids (LCFA) exert crucial functions in the cell as energetic source, membrane components and precursors of lipid mediators (prostaglandins, leukotrienes, thromboxanes, prostacyclin) known to exert a large variety of regulatory effects. They also modulate cellular activity through the activation of ion-channels, myristoylation or palmitoylation of membrane proteins and regulation of gene expression through the activation or inhibition of nuclear receptors [peroxisome proliferator-activated receptors (PPARs) and LXRs] considered as cellular lipid sensors. These pleiotropic functions suggest that the bio-availability of LCFA is critical, especially for peripheral organs exhibiting a high lipid requirement (i.e. adipose tissue, skeletal muscles, heart and liver). Therefore, the intestinal fat absorption must be a very efficient phenomenon. It is apparently the case since the fecal fat loss is usually below 5% in healthy humans, diagnosis of steatorrhea being considered as soon as this value reaches 7% [1]. Nevertheless, LCFA being hydrophobic molecules, their intestinal absorption remains complex. For didactic reasons, it is classically divided into three successive steps: uptake,

Advances in Molecular and Cell Biology, Vol. 33, pages 9–28

ISSN: 1569-2558/DOI: 10.1016/S1569-2558(03)33002-4

trafficking and lipoprotein synthesis (Fig. 1). This chapter is focused on events and players, which influence LCFA permeation and transport into the intestinal absorptive cells. Readers interested in a review about the new insights on lipoproteins synthesis can refer to Ref. [2].

2. Fat absorption and delivery: a short overview

Lipids found in intestinal lumen originate from feeding and, in much lesser extent, from bile and intestine itself. In humans, around 95% of dietary lipids are triglycerides (TG) mainly composed of LCFA (number of carbons >16). The remaining dietary fats are phospholipids (>4.5%) and cholesterol. Because TG cannot cross the membranes, they must be hydrolyzed before their cellular use. During the post-prandial period, digestion of dietary TG occurs essentially in the duodeno-jejunal lumen through the action of a colipase-dependent pancreatic lipase and leads to the release of 2-monoglycerides and LCFA. In contrast to other energetic nutrients, LCFA are poorly soluble in aqueous solution. Moreover, they exhibit detergent properties potentially harmful for the cellular integrity. To overcome these limitations, LCFA are successively dispersed into mixed

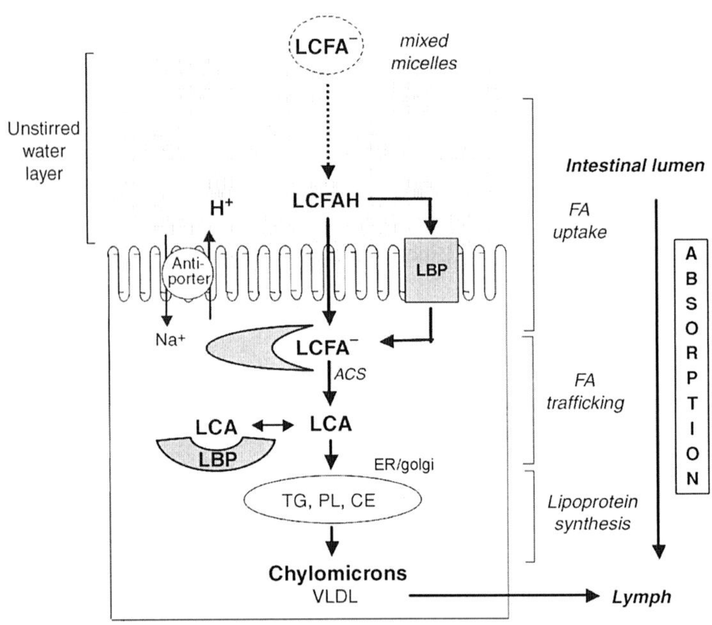

Fig. 1. Intestinal fat absorption. The main steps and players involved in LCFA absorption are depicted: micellar dissociation secondary to the acidic microclimate lining the brush border membrane of enterocytes, cellular uptake through passive and protein-mediated transports, intracellular trafficking involving soluble LBP, TG-rich lipoprotein synthesis and exocytosis into the lymph. $LCFA^-$, ionized long-chain fatty acids; LCFAH, protonated long-chain fatty acids; LBP, lipid-binding proteins; ACS, acyl-CoA synthetases; LCA, long-chain acyl-CoA; TG: triglycerides; PL, phospholipids; CE, cholesterol esters; ER, endoplasmic reticulum; VLDL, very low density lipoproteins.

micelles in the intestinal lumen, bound to soluble lipid-binding proteins (LBP) in the absorptive cells and, after re-esterification in TG, are secreted into lymph as TG-rich lipoproteins (Fig. 1). In blood, the action of endothelial lipoprotein lipase provides LCFA to peripheral tissues converting progressively TG-rich lipoproteins to remnants. The TG remaining in remnants are further hydrolyzed by the hepatic lipase in the liver, and then the delipidated remnants are cleared from the circulation by hepatocytes. LCFA are used as fuel by muscles and liver, while they are stored as TG in adipocytes constituting an easily recruited energetic source for the organism. Therefore, the existence of an efficient transport system in tissues characterized by a high lipid metabolism is essential to facilitate the bio-availability of LCFA and to adjust their cellular uptake to the metabolic needs of the organism.

3. Cellular LCFA uptake: simple diffusion and/or protein-facilitated transport?

How do LCFA cross cell membranes is a highly controversial question that has led to several recent, and sometimes, conflicting reviews [3–9]. Since they have a lipophilic character, the passive diffusion through the phospholipid bilayer was thought, for a long time, to be the exclusive mechanism involved in cellular LCFA uptake. However, this concept has been challenged over the last two decades by the progressive isolation and characterization of several membrane proteins exhibiting a high binding affinity for LCFA in different cell types including the enterocyte.

Simplified models, as protein-free phospholipid vesicles, have been largely used to study the transbilayer movement of native or fluorescent fatty acids in vitro. This transport can be viewed as three successive steps: adsorption of LCFA upon the membrane surface, "flip–flop" movement from external hemileaflet of the bilayer to internal hemileaflet and desorption from the internal bilayer into the inner of the vesicle. The adsorption step is thermodynamically favorable [4] and is extremely fast. Similarly, desorption seems to be a spontaneous phenomenon, at least, in model membranes [4,7]. The most controversial step is the transbilayer (flip–flop) transport. To determine the time course of native LCFA movement through the phospholipid bilayer, Kleinfeld et al. have used the fluorescent probe ADIFAB trapped within lipid vesicles [10]. This technique allows to accurately determine the concentration of LCFA in an aqueous phase. Times found for fatty acid transbilayer transfer were within a 70 ms to 10 s range, in function of the fatty acid type, vesicle size and temperature. Therefore, these authors conclude that the flip–flop step is rate limiting. By extrapolation, they suggest that the passive transport alone might be insufficient to support the metabolic activity of cells known to have high requirements in LCFA, as cardiomyocytes. Kamp and collaborators brought the demonstration that the kinetic of fatty acid translocation across phospholipid bilayers is, in fact, variable in function of the charge of the molecule. Using the pH sensitive fluorophore pyranin, they have shown that the presence of unionized fatty acid in the medium induced a short-term decrease of the pH in the internal compartment [11,12]. This finding is consistent with a rapid movement of protonated LCFA across the lipid bilayer, followed by their ionization in the inner of the vesicle and the diffusion of protons to pyranin. A flip–flop rate shorter than 20 ms independent of fatty acid chain length was reported. By contrast, the transmembrane movement of negatively charged LCFA was several orders of magnitude slower. Taken together, these last data strongly suggest that LCFA can spontaneously cross the cell surface by passive diffusion, when they are protonated. According to this conclusion, LCFA were found to distribute across model membranes [13] and plasma membrane from intact 3T3-L1 adipocytes [14] in response to imposed pH gradients.

If artificial membrane models are useful to dissect the fatty acid fluxes through phospholipid bilayers in a controlled environment, the data obtained, being independent of subsequent metabolic use, are not easily transposable to biological systems. In various cultured cells exhibiting a rapid fatty acid influx, kinetic studies of LCFA uptake support the concept of the existence of a facilitated uptake process in parallel to the passive diffusion. This conclusion is in good agreement with previously published in vivo data [15]. This observation has led to the progressive identification of several plasma membrane LBP in different cell types including the intestinal absorptive cells.

4. In what way does the small intestine differ from the other lipid-utilizing tissues?

Enterocytes are polarized cells with an apical domain facing the intestinal lumen and a basolateral side in contact with lymph and blood. Luminal lipid content is highly variable as compared to what is found in blood. Therefore, the intestinal absorptive cells are daily subjected to dramatic changes in fat supply, in contrast to the other lipid-utilizing cells (i.e. adipocytes, myocytes and hepatocytes). Adequate morphology and luminal environment explain why intestinal fat absorption remains efficient despite this challenge. First, the presence of apical microvilli increases dramatically the cellular absorption area. Secondly, the unstirred water layer lining the cellular apical side promotes the LCFA permeation mainly by passive diffusion in the enterocyte (Fig. 1). With a thickness from 50 to 500 μm [16], the unstirred water layer is a low renewal area produced by the trapping of water molecules into a complex glycoprotein network constituted by both mucus and glycocalyx lining the intestinal epithelium. It is a low renewal compartment characterized by a low pH gradient generated by an efficient Na^+/H^+ antiport exchange system located in the brush border membrane (Fig. 1) [17,18]. The first step of fat absorption is the shuttling of LCFA from the luminal water phase to the vicinity of microvilli by diffusion through the unstirred water layer. By reason of their hydrophobicity, LCFA cross this aqueous diffusion barrier into mixed micelles associated with bile salts and phospholipids (Fig. 1). The micellar solubilization increases the aqueous concentration of LCFA 100–1000-fold [19]. In aqueous solution, more than 99% of LCFA are ionized at physiological pH [8]. The low pH microclimate found near the brush border membrane induces a massive protonation of LCFA when the local pH becomes lower than their pKa (about 5). This event is essential for fat absorption. Indeed, by reducing the LCFA solubility in mixed micelles, it induces their release near the microvilli of absorptive cells [20]. Moreover, it facilitates their subsequent cellular uptake by passive diffusion since protonated LCFA are known to have a greater membrane permeation than their corresponding ionized species [11,12]. The fact that amiloride-mediated disruption of the Na^+/H^+ antiporter decreases, in dose-dependent manner, the LCFA uptake by rat jejunal sheets [21] demonstrates the fundamental role exerted by this acidic microclimate.

It is noteworthy that a similar extracellular environment, especially well adapted to the intestinal post-prandial challenge, is not reproduced in the other lipid-utilizing tissues. Therefore, the direct extrapolation of the fatty acid uptake data obtained in muscles, adipose tissue or liver to intestine appears to be physiologically irrelevant. The same functional restriction must be applied to intestinal cell lines which do not fully reproduce this complex extracellular microclimate.

5. Plasma membrane LBP: what function(s) in the small intestine?

Despite this limitation, cultured intestinal cell lines remain useful models to better understand the complexity of LCFA uptake. In isolated rodent enterocytes as well as in human enterocyte like Caco-2 cell line, transport of LCFA in both apical and basolateral membranes appears to be saturable, at least when low LCFA concentrations are used [22,23]. In these conditions, a competitive inhibition by structurally related LCFA has been reported [24]. Taken together, these data strongly suggest that intestinal LCFA permeation can also take place through a protein-facilitated process (Fig. 1).

The identification of several unrelated proteins exhibiting a high affinity for LCFA in the brush border membrane of the enterocyte correlates quite well with this assumption. To date, four plausible candidate membrane LBP have been identified: the fatty acid transport protein 4 (FATP4) [25], the caveolin-1 (Cav1) [26], the fatty acid transporter (FAT/CD36) [27] and the plasma membrane fatty acid-binding protein (FABPpm) [28]. It has been shown that each of these LBP can enhance LCFA uptake when they are overexpressed in various cell lines.

5.1. FATP4: a fatty acid transporter or an acyl-CoA synthetase (ACS)?

FATP are 63 kDa proteins first identified in 3T3-L1 pre-adipocytes by expression cloning strategy on the basis of their facilitation to uptake LCFA [25]. Therefore, it is not surprising that FATP are undetectable in tissues characterized by a low LCFA uptake (e.g. colon, spleen). Five and six different isoforms of FATP have been found in rodents and humans, respectively [29]. Each isoform exhibits a specific pattern of tissue expression [29]. For example, FATP5 and FATP2 are the major isoforms expressed in the liver [30], while FATP1 is essentially found in the adipose tissue and FATP4 in the small intestine [31]. This latter isoform appears to be a good candidate for a protein-mediated transport of LCFA in the small intestine. Indeed, FATP4 expression correlates quite well with the main site of fat intestinal absorption since its expression level is high in the jejunum, as compared to the duodenum, but lacking in the colon [31]. Moreover, the expression of FATP4 is especially sustained in the brush border membrane of mature enterocytes located in the top of villi, while it is low or lacking in undifferentiated cryptic cells [31]. Finally, the induced depletion of FATP4 protein by antisense strategy leads to a significant decrease in the LCFA uptake by isolated enterocytes [31].

By reference to the only known predictive structure (FATP1), it appears that FATP are essentially oriented toward the cytosol with only a short N-terminus extracellular sequence [30]. Since this extracellular segment does not contain any putative

LCFA-binding site, a role of FATP as fatty acid pool former on the surface of the plasma membrane seems unlikely. The high amino acid sequence identity between FATP1 or FATP4 and ACS, as well as a comparable predictive structure, strongly suggests that these FATP isoforms might be a plasma membrane ACS exhibiting preferential substrate specificity for very long-chain fatty acids [32]. In agreement with this assumption, it was demonstrated that membrane extracts of Cos1 cells transfected with a murine FATP4 expression vector exhibit elevated ACS activity [32]. By contrast, murine FATP1 over-expression enhances cellular uptake of LCFA without any significant change in ACS activity [33]. Analysis of FATP amino acid sequences has revealed a highly conserved ATP-binding motif, which seems to be essential for the protein function. Indeed, mutations within this sequence abolish fatty acid transport [34]. Therefore, if the mechanism by which FATP facilitate the cellular fatty acid uptake is not yet fully understood, it seems to be energy dependent, similarly to ACS. In brief, it is presently unclear if FATP are ACS-associated LBP proteins or plasma membrane ACS. Generation of transgenic mice for FATP4 might allow to determine the physiological function(s) of this membrane LBP.

5.2. Caveolin-1 and FAT/CD36: caveolae-associated LBP

Caveolae are plasma membrane microdomains rich in cholesterol and sphingolipids found in many cell types including enterocytes (for a review, see Ref. [35]). They are proposed to be involved in signal transduction and constitute a non-clathrin route for endocytosis. Being ligand-triggered, this transport system involves a complex signaling which is starting to emerge. Two LBP, the caveolin 1 (Cav-1) and the fatty acid transporter (FAT/CD36), are found in caveolae suggesting that LCFA uptake might also take place through a vesicular transport system.

Cav-1 has been first identified as a putative LBP in 3T3-L1 adipocyte plasma membrane by using a photoreactive LCFA [36]. Its binding affinity for LCFA is in nanomolar range ($K_d = 200$ nM) [36,37]. Its belongs to a small family of 21–24 kDa membrane proteins essential for the formation and stability of caveolae. To date, three distinct caveolin isotypes have been identified. Cav-1 is found associated with Cav-2 to form stable hetero-oligomers [38]. They are especially expressed in well-differentiated cells (e.g. mature enterocyte), while Cav-3 is mainly found in the muscle [39]. Cav-1 is a hairpin-like protein facing the cytosol and anchored in the plasma membrane through a unique peptide sequence [40]. Cav-1 seems to exert pleiotropic functions [41]. A role in the signal transduction is strongly suggested by the demonstration of direct interactions of Cav-1 with resident caveolar signaling molecules [G-protein subunits, Ha-Ras, Src tyrosine kinases or epidermal growth factor (EGF) receptor] [42]. They are also implicated in the cellular trafficking, more particularly in the targeted delivery of specific molecules to organelles [43]. For example, a part of cholesterol destined to lipoprotein synthesis in the enterocyte seems to be targeted to the endoplasmic reticulum via cav-1 [44,45]. An involvement of this protein in intestinal fatty acid uptake and trafficking is likely since Cav-1 expression has been reported in both human intestine biopsies and in well-differentiated enterocyte-like Caco-2 cells [44] in which it appears to be mostly confined to the apical membrane. It is noteworthy that Cav-1 null mice exhibit a hypertriglyceridemia phenotype, which becomes especially dramatic during post-prandial period. The analysis of the lipoprotein profile of fed Cav-1 null mice reveals a strong rise in the lipoprotein fraction corresponding to TG-rich lipoproteins (chylomicrons and very low density lipoproteins) [46] suggesting that Cav-1 play a role in the metabolic fate of LCFA in the enterocyte.

The second LBP found in caveolae is the fatty acid transporter (FAT). It is a 88 kDa transmembrane glycoprotein first identified in rat adipocytes by labeling with sulfo-*N*-succinimidyl derivates of LCFA under conditions where LCFA uptake was significantly inhibited [27]. Rat FAT is homologous to the human scavenger receptor CD36 highly expressed in platelets, monocytes/macrophages and endothelial cells [47]. It is especially found in various organs characterized by a high FA uptake [27,47], as well as in tissues which display a low LCFA requirement (e.g. colon, spleen, tongue). Despite a large spectrum of binding specificity, FAT/CD36 is considered a LBP since it can bind ionized LCFA with an affinity in the nanomolar range and a stoichiometry of 3 mol FA by 1 mol protein [48,49]. Amino acid sequence analysis of FAT/CD36 predicts two transmembrane domains located near the N- and C-terminal tails resulting in a hairpin configuration with a large extracellular hydrophobic domain which might constitute an LCFA binding site [27,47]. A dual palmitoylation might target FAT/CD36 into caveolae, where it has been shown to be co-localized with Cav-1 [50–52]. Concordant data strongly suggest that FAT/CD36 plays a significant role in LCFA uptake by adipose tissue and muscles (reviewed in Ref. [53]). In the small intestine, its expression level (mRNA and protein) is especially high in the major site of fat absorption, i.e. duodeno-jejunum [54]. Immunocytochemical studies in humans and rats demonstrate that FAT/CD36 is strictly localized in the brush border membrane of well-differentiated enterocytes [54,55]. The intestinal FAT/CD36 gene expression has been shown to parallel the lipid contents of the diets. Indeed, jejunal mRNA levels are significantly up-regulated, when rats are chronically submitted to a high fat diet [54], and down-regulated, when they are fed a low fat chow [56]. This positive relationship and its location correlate quite well with an implication of FAT/CD36 in intestinal absorption of dietary fat. However, this involvement remains to be demonstrated.

FAT/CD36 function in caveolae is also unclear. According to its high binding affinity, it might concentrate LCFA in the cholesterol/sphingolipids-rich segments of intestinal plasma membrane leading to a selective fatty acid uptake by a vesicular transport. In agreement with this assumption, we have recently found that re-feeding starved rats a standard laboratory chow containing 3% lipids (wt/wt) leads to a short-term decrease of FAT/CD36 protein from the brush border membrane of enterocytes (I.N. and P.B., unpublished data). Interestingly, this phenomenon appears to be strictly lipid dependent, since it is not reproduced when animals were re-fed an alipidic meal. According to this in vivo finding, it is tempting to speculate that FAT/CD36 might be involved in a vesicular-mediated transport of LCFA in the enterocyte via caveolae. Such a mechanism might explain why a decrease in V_{max} is found in isolated enterocytes from a rat previously subjected to an intraduodenal oleate load [57]. Moreover, intracellular movement of FAT/CD36 has recently been reported in other cell types. In muscles, FAT/CD36 is able to translocate rapidly from an intracellular pool to the sarcolemma during contraction leading to an increase in the cellular FA uptake [58]. This membrane recruitment is an insulin and leptin-dependent phenomenon [59,60]. Taken together these new data indicate that FAT/CD36 likely contributes to a physiological adaptive response of skeletal and cardiac muscles to environmental challenges (e.g. acute exercise). Likewise, a FAT/CD36 translocation from cytosol to plasma membrane has also been demonstrated in the type II pneumocytes in response to an increase of cellular cholesterol, providing a short-term regulation of fatty acids uptake [61]. In the small intestine, the physiological impact of intracellular FAT/CD36 movement remains to be determined. Nevertheless, the lack of dramatic intestinal phenotype in FAT/CD36 null mice as well as in patients with type I CD36 deficiency seems to preclude a fundamental role of this LBP in LCFA intestinal absorption, at least in the standard dietary condition [62,63]. According to this assumption, no difference in lipid absorption is found between wild type and FAT/CD36 null mice subjected to a lipid bolus containing ^{3}H-labeled triolein and ^{14}C-labeled palmitate in a condition in which peripheral lipolysis is inhibited [64].

5.3. FABPpm/mAspAT: a protein in search of a function

FABPpm is a 43 kDa peripheral-associated membrane protein expressed in organs with high lipid needs [65]. It is thought to adhere on the plasma membrane through a specific N-terminal peptide [66]. Immunofluorescence studies realized on rat intestine have shown that FABPpm is expressed in the jejunum and in a lesser extent in the ileum. FABPpm protein is found both in the apical and baso-lateral membranes of enterocytes located in villi and crypts [28]. Binding studies revealed that FABPpm binds various LCFA with an apparent dissociation constant (K_d) of 80 nM [22]. The protein also exhibits a high binding affinity for lysophosphatidylcholine, monoglycerides and cholesterol [22]. FABPpm is identical to the mitochondrial aspartate aminotransferase (mAspAT), an enzyme found in the cytoplasm and mitochondrial matrix, in which it catalyses the transamination reaction linking the urea and the Krebs cycles [67]. FABPpm/mAspAT isolated from plasma membranes lacks the leader sequence with which pre-mAspAT is initially synthesized. Therefore, the different subcellular targeting of the enzyme (i.e. mitochondria, cytoplasm or plasma membrane) could be explained by a post-translational maturation [68]. Overexpression of the mAspAT expression vector in 3T3 fibroblasts triggers the appearance of a FABPpm-like protein in the cell surface and induces an increase in the cellular fatty acid uptake [68]. According to this data, pre-incubation of jejunal explants with a monospecific FABPpm antibody results in a partial, but significant, inhibition of [^{3}H]-oleate uptake suggesting that FABPpm might also play a role in the intestinal absorption of LCFA [22]. Nevertheless, large amounts of antiserum were required to achieve such an inhibition. Moreover, conflicting reports regarding the ability of FABPpm to induce LCFA transport in *Xenopus laevis* oocytes as well as its expression in intestinal zone not involved in fat absorption (i.e. crypt) raise a doubt about the physiological role of this protein in the cellular LCFA uptake [68,69]. Even if results obtained during fasting in red skeletal muscles [70], diabetes in adipocytes from Zucker rats [71] or ethanol load in human hepatoma HepG2 cells [72] are consistent with an involvement of FABPpm/mAspAT in the FA metabolism, the precise role of this protein in the intestinal fat absorption remains questioned.

In brief, the precise contribution to fatty acid uptake of different plasma membrane LBP found in the enterocyte remains to be fully established. To date, it is not clear if their impact is direct or indirect. The fact that some of these LBP (e.g. Cav-1 and FAT/CD36) are multifunctional proteins with a large board of binding specificity, as well as the lack of precise information about the tertiary structure of membrane LBP, adds to the confusion. However, the predictive molecular organization of Cav-1 and FATP suggesting that most of the molecule is facing the cytosol is not in favor of a translocase-like action.

6. Soluble LBPs: do I-FABP and L-FABP exert different function(s) in the enterocyte?

Once into the enterocyte, LCFA are reversibly bound to soluble LBP (Fig. 1). These fatty acid-binding proteins (FABPs) belong to a multigenic family of 14–15 kDa proteins exhibiting a high affinity for various hydrophobic molecules (LCFA, bile acids or retinoids) [73]. Two different FABPs are abundantly expressed in the small intestine: the intestinal type (I-FABP), which is strictly confined to this organ, and the liver type (L-FABP), also found in the liver and kidney. There is an extensive inter-species conservation of the FABP peptide sequence. For instance, more than 80% homology exists between human and rat L-FABP. By contrast, two different FABPs are poorly related within a species. For example, there is lesser than 30% homology between I-FABP and L-FABP in humans [74]. Despite different amino acid sequences, the members of FABP family exhibit a similar tertiary structure that consists of two α helices (αI, αII) and ten antiparallel β strands (βA–βJ) organized in two almost orthogonal β sheets forming an hydrophobic pocket [75]. A "portal" region constituted by the α helices connected to the βC/βD and βE/βF strands allows the entry and exit of LCFA [75,76].

The enterocyte constitutes a unique example of cell in which two different FABPs are simultaneously and abundantly expressed. It is likely that these FABPs exert both common and specific physiological functions in this organ. Transfection studies carried out in cell lines in which either I-FABP or L-FABP is lacking, as in fibroblastic L cells [77] and pluripotent mouse embryonic ES cells [78], or in which only one FABP type is constitutively expressed, as in hBRIE380i (+I-FABP/−L-FABP) [79] and HepG2 (+L-FABP/−I-FABP) [80], demonstrate that I-FABP or L-FABP overexpression leads to an increased LCFA uptake and diffusion.

However, these soluble LBP exhibit some specific features, which lead to a metabolic specialization. First, they exhibit strong differences in their binding properties (stoichiometry, specificity and affinity). Indeed, I-FABP is characterized by a tight binding specificity, since it binds only LCFA with a ratio of one mole for one mole of protein. Because L-FABP has a hydrophobic pocket much larger (440 vs. 234 Å^3) than I-FABP − [74], one molecule of L-FABP can bind two LCFA. This high volume cavity also allows the binding of various bulky hydrophobic molecules such as bile acids [81] or heme [82] but with a stoichiometry of 1/1. FABP binding affinities for various LCFA have accurately been determined by Richieri and collaborators. The dissociation constants (K_d) found ranged from 1 to 1000 nM. The two FABPs have similar K_d values for saturated LCFA. By contrast, one of the two binding sites of L-FABP exhibits a higher avidity for unsaturated LCFA than I-FABP [83]. Second, differences in the regulation of FABP genes are also found. For example, L-FABP expression is transcriptionally induced by LCFA in the small intestine [84] while a similar regulation has not been highlighted for I-FABP as well as in vivo than in vitro [84–86]. Conversely, the peptide YY specifically induces I-FABP gene expression in enterocyte-like hBRIE380i cells [87]. This gut regulatory peptide is secreted mainly by ileal endocrine cells, especially when dietary fat reaches the distal part of the gut, for instance after a dietary overload. The fact that the peptide YY acts as a paracrine agent [88] might explain why I-FABP induction found in rats subjected to a high fat diet occurs only in the ileum [89]. More recently, a down-regulation of I-FABP by the EGF has been reported in highly differentiated Caco-2 cells without change in L-FABP expression levels [90].

The highlighting of a human polymorphism in the gene coding for I-FABP (FABP2 gene) has greatly contributed to a better understanding of the I-FABP role in intestinal lipid metabolism. One base substitution in the codon 54 of FABP2 gene leads to the change of the Ala^{54} by a Thr^{54} in the protein sequence. Initially found in the Pima Indians, this substitution is associated with a high TG plasma level, an insulin resistance [91] and an increase in the body mass index [92]. The fact that the Thr^{54} I-FABP has a twofold greater affinity for LCFA than the Ala^{54} wild protein may explain these metabolic disturbances [91]. Indeed, a greater avidity of the mutant I-FABP for LCFA could lead to an increase in both cellular fatty acid uptake and TG-rich lipoproteins synthesis [91]. According to this assumption, a dramatic rise in LCFA transport and TG secretion is found in Thr^{54} I-FABP-transfected Caco-2 cells as compared to cells transfected with the wild isoform [93]. Similar data have been recently reported in the human jejunal organ culture model in which the Thr^{54} allele appears to be associated with a strong rise in TG synthesis and chylomicron output [94]. Interestingly, the EGF-mediated inhibition of I-FABP is also associated with a decrease in [^{14}C]-palmitate uptake, TG synthesis and secretion in Caco-2 cells [90]. Collectively, these data show that a change in I-FABP binding properties and/or expression level influences the intestinal fat absorption. They strongly suggest that I-FABP is involved in the targeting of dietary LCFA towards the endoplasmic reticulum, where they participate in the synthesis of TG-rich lipoproteins.

Functional differences between I-FABP and L-FABP can be explained by structural specificities. Using a resonance energy transfer assay, Hsu and Storch have demonstrated that the transfer of fluorescent anthroyloxy-FA (AOFA) from I-FABP or L-FABP to acceptor membrane vesicles occurred by different molecular mechanisms [95]. For I-FABP, the transfer of AOFA requires a direct collisional interaction with the phospholipid bilayer, while the FA exchanges between L-FABP and membranes are carried out by an aqueous diffusion. Collisional exchanges occur by ionic interactions between few amino acid residues of the helicoidal domain of I-FABP with anionic phospholipids of the membranes. Indeed, in helix-less mutant I-FABP obtained by site-directed mutagenesis, the collisional exchanges are totally suppressed. In these conditions, the transfer of the AOFA occurs by aqueous diffusion as for L-FABP [96]. It is thought that the collision with a target membrane yields a conformational change in the flexible region of the I-FABP backbone constituted by the distal half of α-II helix and βC–βD turns that triggers the "hinged opening" of the portal domain and the release of the FA [97]. This dynamic domain seems to play a critical role in the binding characteristics of I-FABP.

Therefore, a single modification of its amino acid sequence can have significant physiological consequences. A good illustration of this assumption is the $Ala^{54}Thr$ mutation that affects the binding affinity of I-FABP leading to a dramatic alteration in the lipid metabolism. It is noteworthy that I-FABP may also utilize the membrane–protein interaction for FA acquisition [98]. Since the LCFA delivered on the apical side of the enterocyte are preferentially bound to I-FABP [99], this dynamic exchange system appears to be especially well adapted to facilitate the vectorial transport of dietary lipids from the brush border membrane to the cellular compartment devoted to lipoprotein synthesis (i.e. endoplasmic reticulum).

In the intestinal absorptive cells, the predicted ratio of binding of LCFA to L-FABP/I-FABP is 3.3 [99]. Therefore, L-FABP can be considered as a cytosolic reservoir for fatty acids in the enterocyte. By this action, L-FABP might support lipid metabolism when the lipid supply is low or conversely, protect the cell against the harmful effect of an excess of free LCFA. The fact that L-FABP gene expression is modulated by the lipid content of the diet is in keeping with the latter assumption [84]. From a functional point of view, such a positive regulatory loop seems to be essential to maintain the intestinal mucosa integrity required for an efficient lipid absorption. L-FABP seems also to exert an active role as a partner of the peroxisome proliferator-activated receptor (PPAR), a nuclear receptor considered as a cellular lipid sensor [100]. Indeed, it has been recently demonstrated that L-FABP, through a direct protein–protein interaction with PPARα or PPARγ is able to ensure LCFA transfer to the ligand-binding domain of these nuclear receptors leading to the regulation of specific target genes. In the small intestine, the L-FABP-mediated PPAR activation might lead to the induction of several PPAR target genes encoding for LBP or enzymes (FAT/CD36, FATP, L-FABP, ACS) expected to play a role in fat absorption.

In summary, the physiological functions of gut FABP seem more complex than those which are generally assigned, i.e. ligand desorption from plasma membrane and facilitation of their intracellular diffusion. Their respective structural features speak for a specialization, at least, in standard dietary fat supply: I-FABP being predominantly devoted to the TG-rich lipoprotein synthesis, while L-FABP might be preferentially involved in membrane protection and gene regulation. Because I-FABP and L-FABP gene disruption does not lead to a severe phenotype in mice fed a standard laboratory chow [101,102] the existence of a functional redundancy to ensure a correct fat absorption, especially during the post-prandial period, is likely.

7. ACBP: a housekeeping protein

Thio-esterification of LCFA in long-chain fatty acyl-CoA (LCA) is an obligatory step initiating the lipid metabolism. It is catalyzed by a set of membrane-associated ACS. The newly synthesized LCAs are bound to a specific carrier protein, the acyl-CoA-binding protein (ACBP) (Fig. 1). This LBP is a ubiquitous 10 kDa soluble LCA transporter conserved from the yeast to the mammal (for a review see Ref. [103]). The 86 amino acid residues of ACBP are folded in four α-helix, forming a boomerang structure [104,105]. The acyl chain of LCA is buried in a hydrophobic groove of the binding pocket and is totally protected from the aqueous solvent by its acyl-CoA head [106,107]. ACBP binds with a stoichiometry of 1:1 and an affinity in the nanomolar range with both medium and long-chain LCA [107,108]. In the small intestine from rat and mice, ACBP is co-expressed with FABP in absorptive cells [109]. In rat, a similar ACBP mRNA level is found along the cephalo-caudal axis of the small intestine (I.N. and P.B., unpublished data) in contrast to what is observed for L-FABP and I-FABP. The precise function of ACBP in this organ is not yet determined. However, it is likely that ACBP contributes to the lipoprotein synthesis through the facilitation of LCA unloading from apical membrane (e.g. from FATP4) and their subsequent disposal for microsomal TG and phospholipid synthesis [106]. The fact that an abrogation of ACBP gene expression in pre-adipocytes leads to a drastic reduction in TG accumulation [110] is in keeping with this hypothesis. The detection of ACBP and LCA in the nucleus of different cells [111] suggests a possible interference of ACBP expression and/or LCA disposal with fundamental regulatory pathways [110].

8. A working model for intestinal LCFA uptake and trafficking

Different concepts of cellular fatty acid uptake have been recently proposed [30,112, 113]. These integrative models are especially valid to explain the FA uptake in myocytes or hepatocytes. Nevertheless, they do not take into account the specificities of other cell types, such as the enterocyte. Therefore, the following working model more adapted to the small intestine is proposed (Fig. 2).

Intestinal LCFA uptake is a complex phenomenon which involves diffusion and protein-mediated processes. In contrast to myocytes, adipocytes or hepatocytes, the intestinal absorptive cells are polarized cells with a specific apical environment, which facilitates the LCFA membrane permeation by passive diffusion. The physiological role of the small intestine being to ensure the most efficient absorption, the predominance of such a low affinity/high capacity transport avoids the LCFA uptake becoming a rate-limiting step in the fat absorption during the post-prandial time. Compelling evidences suggest that some of the membrane and soluble LBP play a permissive role improving the diffusional uptake of dietary LCFA. To be efficient, membrane permeation requires a rapid partitioning of the molecule into the lipid bilayer of the plasma membrane and the maintenance of a favorable concentration gradient. The low pH microclimate lining the brush border membrane, which leads to a large LCFA protonation, satisfies the first condition. The second one seems to be realized by the concomitant presence of FATP4 and soluble LBP (I-FABP, L-FABP and ACBP). Indeed, the plasma membrane-associated ACS activity of FATP4 promoting the rapid esterification of LCFA in acyl-CoA esters creates a driving force, which contributes to the maintenance of a low cytoplasmic LCFA concentration. Moreover, the fact that the plasma membrane is impermeable to LCA esters prevents their cellular efflux. The transmembrane trafficking of LCFA is reinforced by the presence of soluble LBP. Indeed, I-FABP, L-FABP and ACBP act as intracellular acceptors and carrier proteins promoting FA and acyl-CoA esters desorption from apical membrane and facilitating their subsequent metabolism. Moreover, by limiting the intracellular free LCFA and LCA levels, L-FABP and ACBP protect the intestinal absorptive cells against the detergent effect of these molecules, while I-FABP targets LCFA to endoplasmic reticulum. Through this cooperative system, the small intestine can

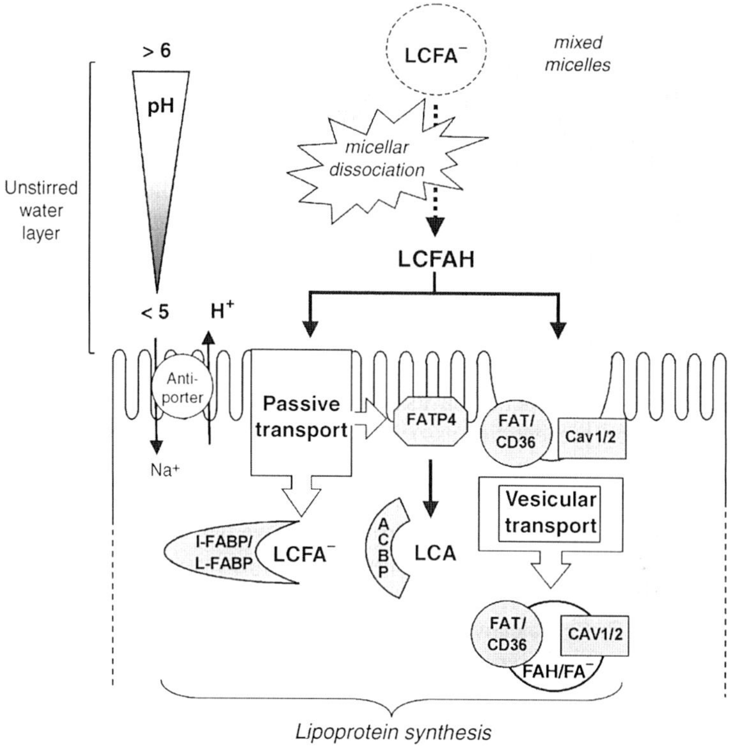

Fig. 2. Intestinal absorption of LCFA: a working model. For detailed explanation, see Section 8. LCFA, long-chain fatty acids; LCA, long-chain acyl-CoA; TG, triglycerides; PL, phospholipids; CE, cholesterol esters; FATP4, fatty acid transport protein 4; FAT/CD36, fatty acid transporter; Cav-1, caveolin-1; I-FABP, intestinal fatty acid-binding protein; L-FABP, liver fatty acid-binding protein; ACBP, acyl-CoA-binding protein; ACS, acyl-CoA synthetases; PPAR, peroxisome proliferator-activated receptor; VLDL, very low density lipoprotein; HDL, high density lipoprotein.

fully assume its physiological role, which is to ensure an optimal uptake and delivery of especially high energetic density molecules.

During the post-prandial period, other routes of lipid uptake are also used by the small intestine. By analogy with the caveolae-dependent endocytosis of cholesterol [44], it is tempting to hypothesize that a vesicle-mediated transport involving FAT/CD36 and/or Cav-1 might contribute to dietary LCFA uptake in complement to the diffusional process. Several arguments suggest that such a mode of uptake is plausible in the small intestine. First, caveolae-associated LBP, Cav-1 and FAT/CD36 are co-expressed in intestinal microvilli. Second, a circulation of FAT/CD36 between plasma membrane and intracellular pools has been described in other tissues [113]. Third, a Cav-1-mediated vesicular uptake of LCFA has been found in hepatoma and endothelial cells [114,115]. Finally, a lipid-dependent decrease in the brush border content of FAT/CD36 protein occurs in re-fed rat as compared to fasted animals (I.N. and P.B., unpublished data). Such a vesicular transport might direct LCFA from the apical membrane to the endoplasmic reticulum facilitating their re-esterification in TG and the subsequent lipoprotein synthesis. Nevertheless, the existence of this route of uptake remains to be fully demonstrated in the small intestine. If this hypothesis is correct, this pathway does not seem to be quantitatively crucial since intestinal fat absorption is not affected in FAT/CD36 null mice fed a standard chow. Simultaneously or alternatively, membrane LBP might constitute a low capacity, high affinity system, which might constitute a local concentration system favoring the LCFA uptake when the lipid level is low in the intestinal lumen (inter-prandial periods, fasting).

9. Conclusions and future directions

FA uptake by a cell depends on its physico-chemical environment (i.e. extracellular LCFA concentration, presentation associated with bile acids in mixed micelles or bound to albumin, pH microclimate) and its physiological specialization (i.e. genotypic expression, metabolic requirements). Among the cells exhibiting a high LCFA trafficking, only the enterocytes have an extracellular environment (low pH microclimate and high lipid supply), which privileges LCFA permeation mainly by passive diffusion. Such a feature is unique and crucial to maintain an efficient fat absorption especially during the post-prandial period. The efficiency of fat absorption is reinforced by the simultaneous expression of several membrane and soluble LBP, which, all together but differently, favor the cellular LCFA influx into the enterocyte. This protein-mediated system might also constitute a local high affinity, low capacity transport system absolutely required for the maintenance of lipoprotein synthesis, when the lipid concentration is low in the intestinal lumen (inter-prandial periods, fasting or distal part of the intestine).

Recent generation of LBP-knockout mice has provided a promising tool to investigate the role of LBP in intestinal fat absorption. According to the putative functions of both plasma membrane and soluble LBP, one would expect that disruption of genes encoding for a plasma membrane or soluble LBP might lead to a severe phenotype. However, FAT/CD36, Cav-1, I-FABP or L-FABP null mice are all viable and apparently healthy when they are fed a normolipidic diet. This observation, not fully surprising, is probably due to compensatory or redundant mechanisms. It provides a new illustration of the remarkable capacity of adaptation which characterizes the small intestine, which continues, for instance, to absorb dietary fat even in the absence of key enzymes for TG synthesis [116]. Future in vivo investigations realized in transgenic animals with intestinal-targeted modifications and subjected to drastic nutritional conditions (e.g. high fat diet, long fasting) are required to determine the real functions of membrane and soluble LBP in the gut.

A relatively abundant literature is devoted to the FA-mediated regulation of genes involved in intestinal lipid metabolism (including LBPs). It is also well established that nutritional manipulations induce modifications in the proliferation rate of the mucosa leading to rapid changes in the absorptive capacity of the small intestine [117,118]. Dietary lipids are responsible for the proliferation of enterocytes through a neuro-endocrine pathway involving the entero-endocrine hormone, glucagon-like peptide-2 and enteric nerves [119,120]. Therefore, the small intestine appears to be a highly versatile organ, which is able to adapt its absorptive capacity (not only gene expression, but also absorptive area) to the lipid content of the diet. This property is probably the consequence of evolutionary pressure, which has led to select a set of very efficient and redundant mechanisms to maximize the absorption of energy-rich nutrients. If this intestinal feature constitutes a vital advantage when the organism is subjected to seasonal changes in the dietary supplies, conversely, it becomes a risk factor for health in situations of constant plethora. In the western diet, dietary fat represents around 40% of the daily caloric intake, while the nutritional advices are 10% lower. This high fat supply associated with a qualitative imbalance (excess of saturated fatty acids, high polyunsaturated fatty acids ω6/ω3 ratio) greatly contributes to the appearance of diseases of plethora (i.e. obesity, atherosclerosis, non-insulin-dependent diabetes, cancers), of which the human and social costs are dramatic. A better knowledge of the mechanisms underlying the intestinal fat absorption might provide new therapeutic strategies allowing the correction of lipid disorders due to the excessive lipid content of the western diet.

References

[1] Ross, A.C., 1993. Overview of retinoid metabolism. J. Nutr. 123, 346–350.

[2] Hussain, M.M., 2001. Structural, biochemical and signaling properties of the low-density lipoprotein receptor gene family. Front. Biosci. 6, D417–D428.

[3] Berk, P.D., 1996. How do long-chain free fatty acids cross cell membranes? Proc. Soc. Exp. Biol. Med. 212, 1–4.

[4] Zakim, D., 1996. Fatty acids enter cells by simple diffusion. Proc. Soc. Exp. Biol. Med. 212, 5–14.

[5] Glatz, J.F., Luiken, J.J., van Nieuwenhoven, F.A., van der Vusse, G.J., 1997. Molecular mechanism of cellular uptake and intracellular translocation of fatty acids. Prostaglandins Leukot. Essent. Fatty Acids 57, 3–9.

[6] Abumrad, N., Harmon, C., Ibrahimi, A., 1998. Membrane transport of long-chain fatty acids: evidence for a facilitated process. J. Lipid Res. 39, 2309–2318.

[7] Hamilton, J.A., 1998. Fatty acid transport: difficult or easy? J. Lipid Res. 39, 467–481.

[8] McArthur, M.J., Atshaves, B.P., Frolov, A., Foxworth, W.D., Kier, A.B., Schroeder, F., 1999. Cellular uptake and intracellular trafficking of long chain fatty acids. J. Lipid Res. 40, 1371–1383.

[9] Hamilton, J.A., Guo, W., Kamp, F., 2002. Mechanism of cellular uptake of long-chain fatty acids: do we need cellular proteins? Mol. Cell. Biochem. 239, 17–23.

[10] Kleinfeld, A.M., Chu, P., Romero, C., 1997. Transport of long-chain native fatty acids across lipid bilayer membranes indicates that transbilayer flip–flop is rate limiting. Biochemistry 36, 14146–14158.

[11] Kamp, F., Westterhoff, H.V., Hamilton, J.A., 1993. Movement of fatty acids, fatty acid analogues and bile acids across phospholipid bilayers. Biochemistry 32, 11074–11086.

[12] Kamp, F., Zakim, D., Zhang, F., Noy, N., Hamilton, J.A., 1995. Fatty acid flip–flop in phospholipid bilayers is extremely fast. Biochemistry 34, 11928–11937.

[13] Hope, M.J., Cullis, P.R., 1987. Lipid asymmetry induced by transmembrane pH gradients in large unilamellar vesicles. J. Biol. Chem. 262, 4360–4366.

[14] Trigatti, B.L., Gerber, G.E., 1996. The effect of intracellular pH on long-chain fatty acid uptake in 3T3-L1 adipocytes: evidence that uptake involves the passive diffusion of protonated long-chain fatty acids across the plasma membrane. Biochem. J. 313, 487–494.

[15] Chow, S.L., Hollander, D., 1979. Linoleic acid absorption in the unanesthetized rat: mechanism of transport and influence of luminal factors on absorption. Lipids 14, 378–385.

[16] Wilson, F.A., Sallee, V.L., Dietschy, J.M., 1971. Unstirred water layers in intestine: rate determinant of fatty acid absorption from micellar solutions. Science 174, 1031–1033.

[17] Shiau, Y.F., Fernandez, P., Jackson, M.J., McMonagle, S., 1985. Mechanisms maintaining a low-pH microclimate in the intestine. Am. J. Physiol. 248, G608–G617.
[18] Orsenigo, M.N., Tosco, M., Zoppi, S., Faelli, A., 1990. Characterization of basolateral membrane Na/H antiport in rat jejunum. Biochim. Biophys. Acta 1026, 64–68.
[19] Tso, P., 1994. Intestinal Lipid Absorption, 3rd ed., Physiology of the Gastrointestinal Tract, 3rd ed, pp. 1867–1907, L.R. Johnson, editor, New York, Raven Press.
[20] Shiau, Y.F., 1981. Mechanism of fat absorption. Am. J. Physiol. 240, G1–G9.
[21] Schoeller, C., Keelan, M., Mulvey, G., Stremmel, W., Thomson, A.B., 1995. Oleic acid uptake into rat and rabbit jejunal brush border membrane. Biochim. Biophys. Acta 1236, 51–64.
[22] Stremmel, W., 1988. Uptake of fatty acids by jejunal mucosal cells is mediated by a fatty acid binding membrane protein. J. Clin. Invest. 82, 2001–2010.
[23] Trotter, P.J., Ho, S.Y., Storch, J., 1996. Fatty acid uptake by Caco-2 human intestinal cells. J. Lipid Res. 37, 336–346.
[24] Tranchant, T., Besson, P., Hoinard, C., Pinault, M., Alessandri, J.M., Delarue, J., Couet, C., Gore, J., 1998. Long-term supplementation of culture medium with essential fatty acids alters alpha-linolenic acid uptake in Caco-2 clone TC7. Can. J. Physiol. Pharmacol. 76, 621–629.
[25] Schaffer, J.E., Lodish, H.F., 1994. Expression, cloning and characterization of a novel adipocyte long-chain fatty acid transport protein. Cell 79, 427–436.
[26] Trigatti, B.L., Mangroo, D., Gerber, G.E., 1991. Photoaffinity labeling and fatty acid permeation in 3T3-L1 adipocytes. J. Biol. Chem. 266, 22621–22625.
[27] Abumrad, N.A., el-Maghrabi, M.R., Amri, E.Z., Lopez, E., Grimaldi, P.A., 1993. Cloning of a rat adipocyte membrane protein implicated in binding or transport of long-chain fatty acids that is induced during preadipocyte differentiation. Homology with human CD36. J. Biol. Chem. 268, 17665–17668.
[28] Stremmel, W., Lotz, G., Strohmeyer, G., Berk, P.D., 1985. Identification, isolation, and partial characterization of a fatty acid binding protein from rat jejunal microvillous membranes. J. Clin. Invest. 75, 1068–1076.
[29] Hirsch, D., Stahl, A., Lodish, H.F., 1998. A family of fatty acid transporters conserved from mycobacterium to man. Proc. Natl Acad. Sci. USA 95, 8625–8629.
[30] Stahl, A., Gimeno, R.E., Tartaglia, L.A., Lodish, H.F., 2001. Fatty acid transport proteins: a current view of a growing family. Trends Endocrinol. Metab. 12, 266–273.
[31] Stahl, A., Hirsch, D.J., Gimeno, R.E., Punreddy, S., Ge, P., Watson, N., Patel, S., Kotler, M., Raimondi, A., Tartaglia, L.A., Lodish, H.F., 1999. Identification of the major intestinal fatty acid transport protein. Mol. Cell 4, 299–308.
[32] Herrmann, T., Buchkremer, F., Gosch, I., Hall, A.M., Bernlohr, D.A., Stremmel, W., 2001. Mouse fatty acid transport protein 4 (FATP4): characterization of the gene and functional assessment as a very long chain acyl-CoA synthetase. Gene 270, 31–40.
[33] Coe, N.R., Smith, A.J., Frohnert, B.I., Watkins, P.A., Bernlohr, D.A., 1999. The fatty acid transport protein (FATP1) is a very long chain acyl-CoA synthetase. J. Biol. Chem. 274, 36300–36304.
[34] Stuhlsatz-Krouper, S.M., Bennett, N.E., Schaffer, J.E., 1999. Molecular aspects of fatty acid transport: mutations in the IYTSGTTGXPK motif impair fatty acid transport protein function. Prostaglandins Leukot. Essent. Fatty Acids 60, 285–289.
[35] Pelkmans, L., Helenius, A., 2002. Endocytosis via caveolae. Traffic 3, 311–320.
[36] Gerber, G.E., Mangroo, D., Trigatti, B.L., 1993. Identification of high affinity membrane-bound fatty acid-binding proteins using a photoreactive fatty acid. Mol. Cell. Biochem. 123, 39–44.
[37] Trigatti, B.L., Anderson, R.G., Gerber, G.E., 1999. Identification of caveolin-1 as a fatty acid binding protein. Biochem. Biophys. Res. Commun. 255, 34–39.
[38] Scherer, P.E., Lewis, R.Y., Volonte, D., Engelman, J.A., Galbiati, F., Couet, J., Kohtz, D.S., van Donselaar, E., Peters, P., Lisanti, M.P., 1997. Cell-type and tissue-specific expression of caveolin-2. Caveolins 1 and 2 co-localize and form a stable hetero-oligomeric complex in vivo. J. Biol. Chem. 272, 29337–29346.
[39] Parton, R.G., 1996. Caveolae and caveolins. Curr. Opin. Cell. Biol. 8, 542–548.
[40] Liu, P., Li, W.P., Machleidt, T., Anderson, R.G., 1999. Identification of caveolin-1 in lipoprotein particles secreted by exocrine cells. Nat. Cell. Biol. 1, 369–375.
[41] Liu, P., Rudick, M., Anderson, R.G., 2002. Multiple functions of caveolin-1. J. Biol. Chem. 277, 41295–41298.

[42] Couet, J., Li, S., Okamoto, T., Ikezu, T., Lisanti, M.P., 1997. Identification of peptide and protein ligands for the caveolin-scaffolding domain. Implications for the interaction of caveolin with caveolae-associated proteins. J. Biol. Chem. 272, 6525–6533.
[43] Conrad, P.A., Smart, E.J., Ying, Y.S., Anderson, R.G., Bloom, G.S., 1995. Caveolin cycles between plasma membrane caveolae and the Golgi complex by microtubule-dependent and microtubule-independent steps. J. Cell. Biol. 131, 1421–1433.
[44] Field, F.J., Born, E., Murthy, S., Mathur, S.N., 1998. Caveolin is present in intestinal cells: role in cholesterol trafficking? J. Lipid Res. 39, 1938–1950.
[45] Schroeder, F., Gallegos, A.M., Atshaves, B.P., Storey, S.M., McIntosh, A.L., Petrescu, A.D., Huang, H., Starodub, O., Chao, H., Yang, H., Frolov, A., Kier, A.B., 2001. Recent advances in membrane microdomains: rafts, caveolae, and intracellular cholesterol trafficking. Exp. Biol. Med. (Maywood) 226, 873–890.
[46] Razani, B., Combs, T.P., Wang, X.B., Frank, P.G., Park, D.S., Russell, R.G., Li, M., Tang, B., Jelicks, L.A., Scherer, P.E., Lisanti, M.P., 2002. Caveolin-1-deficient mice are lean, resistant to diet-induced obesity, and show hypertriglyceridemia with adipocyte abnormalities. J. Biol. Chem. 277, 8635–8647.
[47] Greenwalt, D.E., Lipsky, R.H., Ockenhouse, C.F., Ikeda, H., Tandon, N.N., Jamieson, G.A., 1992. Membrane glycoprotein CD36: a review of its roles in adherence, signal transduction, and transfusion medicine. Blood 80, 1105–1115.
[48] Baillie, A.G., Coburn, C.T., Abumrad, N.A., 1996. Reversible binding of long-chain fatty acids to purified FAT, the adipose CD36 homolog. J. Membr. Biol. 153, 75–81.
[49] Ibrahimi, A., Sfeir, Z., Magharaie, H., Amri, E.Z., Grimaldi, P., Abumrad, N.A., 1996. Expression of the CD36 homolog (FAT) in fibroblast cells: effects on fatty acid transport. Proc. Natl Acad. Sci. USA 93, 2646–2651.
[50] Lisanti, M.P., Scherer, P.E., Vidugiriene, J., Tang, Z., Hermanowski-Vosatka, A., Tu, Y.H., Cook, R.F., Sargiacomo, M., 1994. Characterization of caveolin-rich membrane domains isolated from an endothelial-rich source: implications for human disease. J. Cell. Biol. 126, 111–126.
[51] Tao, N., Wagner, S.J., Lublin, D.M., 1996. CD36 is palmitoylated on both N- and C-terminal cytoplasmic tails. J. Biol. Chem. 271, 22315–22320.
[52] Galbiati, F., Volonté, D., Minetti, C., Chu, J.B., Lisanti, M.P., 1999. Phenotypic behavior of caveolin-3 mutations that cause autosomal dominant limb girdle muscular dystrophy (LGMD-1C). J. Biol. Chem. 274, 25632–25641.
[53] Ibrahimi, A., Abumrad, N.A., 2002. Role of CD36 in membrane transport of long-chain fatty acids. Curr. Opin. Clin. Nutr. Metab. Care 5, 139–145.
[54] Poirier, H., Degrace, P., Niot, I., Bernard, A., Besnard, P., 1996. Localization and regulation of the putative membrane fatty-acid transporter (FAT) in the small intestine. Comparison with fatty acid-binding proteins (FABP). Eur. J. Biochem. 238, 368–373.
[55] Lobo, M.V., Huerta, L., Ruiz-Velasco, N., Teixeiro, E., de la Cueva, P., Celdran, A., Martin-Hidalgo, A., Vega, M.A., Bragado, R., 2001. Localization of the lipid receptors CD36 and CLA-1/SR-BI in the human gastrointestinal tract: towards the identification of receptors mediating the intestinal absorption of dietary lipids. J. Histochem. Cytochem. 49, 1253–1260.
[56] Sukhotnik, I., Gork, A.S., Chen, M., Drongowski, R.A., Coran, A.G., Harmon, C.M., 2001. Effect of low fat diet on lipid absorption and fatty-acid transport following bowel resection. Pediatr. Surg. Int. 17, 259–264.
[57] Chen, M., Yang, Y., Braunstein, E., Georgeson, K.E., Harmon, C.M., 2001. Gut expression and regulation of FAT/CD36: possible role in fatty acid transport in rat enterocytes. Am. J. Physiol. Endocrinol. Metab. 281, E916–E923.
[58] Bonen, A., Luiken, J.J., Arumugam, Y., Glatz, J.F., Tandon, N.N., 2000. Acute regulation of fatty acid uptake involves the cellular redistribution of fatty acid translocase. J. Biol. Chem. 275, 14501–14508.
[59] Luiken, J.J., Dyck, D.J., Han, X.X., Tandon, N.N., Arumugam, Y., Glatz, J.F., Bonen, A., 2002. Insulin induces the translocation of the fatty acid transporter FAT/CD36 to the plasma membrane. Am. J. Physiol. Endocrinol. Metab. 282, E491–E495.
[60] Steinberg, G.R., Dyck, D.J., Calles-Escandon, J., Tandon, N.N., Luiken, J.J., Glatz, J.F., Bonen, A., 2002. Chronic leptin administration decreases fatty acid uptake and fatty acid transporters in rat skeletal muscle. J. Biol. Chem. 277, 8854–8860.

[61] Kolleck, I., Guthmann, F., Ladhoff, A.M., Tandon, N.N., Schlame, M., Rustow, B., 2002. Cellular cholesterol stimulates acute uptake of palmitate by redistribution of fatty acid translocase in type II pneumocytes. Biochemistry 41, 6369–6375.

[62] Febbraio, M., Abumrad, N.A., Hajjar, D.P., Sharma, K., Cheng, W., Pearce, S.F., Silverstein, R.L., 1999. A null mutation in murine CD36 reveals an important role in fatty acid and lipoprotein metabolism. J. Biol. Chem. 274, 19055–19062.

[63] Kashiwagi, H., Tomiyama, Y., Nozaki, S., Honda, S., Kosugi, S., Shiraga, M., Nakagawa, T., Nagao, N., Kanakura, Y., Kurata, Y., Matsuzawa, Y., 1996. A single nucleotide insertion in codon 317 of the CD36 gene leads to CD36 deficiency. Arterioscler. Thromb. Vasc. Biol. 16, 1026–1032.

[64] Goudriaan, J.R., Dahlmans, V.E., Febbraio, M., Teusink, B., Romijn, J.A., Havekes, L.M., Voshol, P.J., 2002. Intestinal lipid absorption is not affected in CD36 deficient mice. Mol. Cell. Biochem. 239, 199–202.

[65] Potter, B.J., Stump, D., Schwieterman, W., Sorrentino, D., Jacobs, L.N., Kiang, C.L., Rand, J.H., Berk, P.D., 1987. Isolation and partial characterization of plasma membrane fatty acid binding proteins from myocardium and adipose tissue and their relationship to analogous proteins in liver and gut. Biochem. Biophys. Res. Commun. 148, 1370–1376.

[66] Berk, P.D., Wada, H., Horio, Y., Potter, B.J., Sorrentino, D., Zhou, S.L., Isola, L.M., Stump, D., Kiang, C.L., Thung, S., 1990. Plasma membrane fatty acid-binding protein and mitochondrial glutamic-oxaloacetic transaminase of rat liver are related. Proc. Natl Acad. Sci. USA 87, 3484–3488.

[67] Stump, D.D., Zhou, S.L., Berk, P.D., 1993. Comparison of plasma membrane FABP and mitochondrial isoform of aspartate aminotransferase from rat liver. Am. J. Physiol. 265, G894–G902.

[68] Isola, L.M., Zhou, S.L., Kiang, C.L., Stump, D.D., Bradbury, M.W., Berk, P.D., 1995. 3T3 fibroblasts transfected with a cDNA for mitochondrial aspartate aminotransferase express plasma membrane fatty acid-binding protein and saturable fatty acid uptake. Proc. Natl Acad. Sci. USA 92, 9866–9870.

[69] Zhou, S.L., Stump, D., Isola, L., Berk, P.D., 1994. Constitutive expression of a saturable transport system for non-esterified fatty acids in *Xenopus laevis* oocytes. Biochem. J. 297, 315–319.

[70] Turcotte, L.P., Srivastava, A.K., Chiasson, J.L., 1997. Fasting increases plasma membrane fatty acid-binding protein (FABP(PM)) in red skeletal muscle. Mol. Cell. Biochem. 166, 153–158.

[71] Berk, P.D., Zhou, S.L., Kiang, C.L., Stump, D., Bradbury, M., Isola, L.M., 1997. Uptake of long chain free fatty acids is selectively up-regulated in adipocytes of Zucker rats with genetic obesity and non-insulin-dependent diabetes mellitus. J. Biol. Chem. 272, 8830–8835.

[72] Zhou, S.L., Gordon, R.E., Bradbury, M., Stump, D., Kiang, C.L., Berk, P.D., 1998. Ethanol up-regulates fatty acid uptake and plasma membrane expression and export of mitochondrial aspartate aminotransferase in HepG2 cells. Hepatology 19998, 1064–1074.

[73] Bernlohr, D.A., Simpson, M.A., Hertzel, A.V., Banaszak, L.J., 1997. Intracellular lipid-binding proteins and their genes. Annu. Rev. Nutr. 17, 277–303.

[74] Thompson, J., Winter, N., Terwey, D., Bratt, J., Banaszak, L., 1997. The crystal structure of the liver fatty acid-binding protein. A complex with two bound oleates. J. Biol. Chem. 272, 7140–7150.

[75] Sacchettini, J.C., Gordon, J.I., 1993. Rat intestinal fatty acid binding protein. A model system for analyzing the forces that can bind fatty acids to proteins. J. Biol. Chem. 268, 18399–18402.

[76] Thompson, H., Zhu, Z., Banni, S., Darcy, K., Loftus, T., Ip, C., 1997. Morphological and biochemical status of the mammary gland as influenced by conjugated linoleic acid: implication for a reduction in mammary cancer risk. Cancer Res. 57, 5067–5072.

[77] Prows, D.R., Schroeder, F., 1997. Metallothionein-IIA promoter induction alters rat intestinal fatty acid binding protein expression, fatty acid uptake, and lipid metabolism in transfected L-cells. Arch. Biochem. Biophys. 340, 135–143.

[78] Atshaves, B.P., Foxworth, W.B., Frolov, A., Roths, J.B., Kier, A.B., Oetama, B.K., Piedrahita, J.A., Schroeder, F., 1998. Cellular differentiation and I-FABP protein expression modulate fatty acid uptake and diffusion. Am. J. Physiol. 274, C633–C644.

[79] Holehouse, E.L., Liu, M.L., Aponte, G.W., 1998. Oleic acid distribution in small intestinal epithelial cells expressing intestinal-fatty acid binding protein. Biochim. Biophys. Acta 1390, 52–64.

[80] Wolfrum, C., Ellinghaus, P., Fobker, M., Seedorf, U., Assmann, G., Börchers, T., Spener, F., 1999. Phytanic acid is ligand and transcriptional activator of murine liver fatty acid binding protein. J. Lipid Res. 40, 708–714.

[81] Thumser, A.E., Wilton, D.C., 1996. The binding of cholesterol and bile salts to recombinant rat liver fatty acid-binding protein. Biochem. J. 320, 729–733.
[82] Thompson, J., Ory, J., Reese-Wagoner, A., Banaszak, L., 1999. The liver fatty acid binding protein – comparison of cavity properties of intracellular lipid-binding proteins. Mol. Cell. Biochem. 192, 9–16.
[83] Richieri, G.V., Ogata, R.T., Kleinfeld, A.M., 1994. Equilibrium constants for the binding of fatty acids with fatty acid-binding proteins from adipocyte, intestine, heart, and liver measured with the fluorescent probe ADIFAB. J. Biol. Chem. 269, 23918–23930.
[84] Poirier, H., Niot, I., Degrace, P., Monnot, M.C., Bernard, A., Besnard, P., 1997. Fatty acid regulation of fatty acid-binding protein expression in the small intestine. Am. J. Physiol. 273, G289–G295.
[85] Hallden, G., Holehouse, E.L., Dong, X., Aponte, G.W., 1994. Expression of intestinal fatty acid binding protein in intestinal epithelial cell lines, hBRIE, 380 cells. Am. J. Physiol. 267, G730–G743.
[86] Le Beyec, J., Delers, F., Jourdant, F., Schreider, C., Chambaz, J., Cardot, P., Pincon-Raymond, M., 1997. A complete epithelial organization of Caco-2 cells induces I-FABP and potentializes apolipoprotein gene expression. Exp. Cell Res. 236, 311–320.
[87] Hallden, G., Aponte, G.W., 1997. Evidence for a role of the gut hormone PYY in the regulation of intestinal fatty acid-binding protein transcripts in differentiated subpopulations of intestinal epithelial cell hybrids. J. Biol. Chem. 272, 12591–12600.
[88] Laburthe, M., Chenut, B., Rouyer-Fessard, C., Tatemoto, K., Couvineau, A., Servin, A., Amiranoff, B., 1986. Interaction of peptide YY with rat intestinal epithelial plasma membranes: binding of the radioiodinated peptide. Endocrinology 118, 1910–1917.
[89] Bass, N.M., 1988. The cellular fatty acid binding proteins: aspects of structure, regulation, and function. Int. Rev. Cytol. 111, 143–184.
[90] Darimont, C., Gradoux, N., de Pover, A., 1999. Epidermal growth factor regulates fatty acid uptake and metabolism in Caco-2 cells. Am. J. Physiol. 276, G606–G612.
[91] Baier, L.J., Sacchettini, J.C., Knowler, W.C., Eads, J., Paolisso, G., Tataranni, P.A., Mochizuki, H., Bennett, P.H., Bogardus, C., Prochazka, M., 1995. An amino acid substitution in the human intestinal fatty acid binding protein is associated with increased fatty acid binding, increased fat oxidation, and insulin resistance. J. Clin. Invest. 95, 1281–1287.
[92] Hegele, R.A., Harris, S.B., Hanley, A.J., Sadikian, S., Connelly, P.W., Zinman, B., 1996. Genetic variation of intestinal fatty acid-binding protein associated with variation in body mass in aboriginal Canadians. J. Clin. Endocrinol. Metab. 81, 4334–4337.
[93] Baier, L.J., Bogardus, C., Sacchettini, J.C., 1996. A polymorphism in the human intestinal fatty acid binding protein alters fatty acid transport across Caco-2 cells. J. Biol. Chem. 271, 10892–10896.
[94] Levy, E., Menard, D., Delvin, E., Stan, S., Mitchell, G., Lambert, M., Ziv, E., Feoli-Fonseca, J.C., Seidman, E., 2001. The polymorphism at codon 54 of the FABP2 gene increases fat absorption in human intestinal explants. J. Biol. Chem. 276, 39679–39684.
[95] Hsu, K.T., Storch, J., 1996. Fatty acid transfer from liver and intestinal fatty acid-binding proteins to membranes occurs by different mechanisms. J. Biol. Chem. 271, 13317–13323.
[96] Corsico, B., Cistola, D.P., Frieden, C., Storch, J., 1998. The helical domain of intestinal fatty acid binding protein is critical for collisional transfer of fatty acids to phospholipid membranes. Proc. Natl Acad. Sci. USA 95, 12174–12178.
[97] Hodsdon, M.E., Cistola, D.P., 1997. Discrete backbone disorder in the nuclear magnetic resonance structure of apo intestinal fatty acid-binding protein: implications for the mechanism of ligand entry. Biochemistry 36, 1450–1460.
[98] Storch, J., Thumser, A.E., 2000. The fatty acid transport function of fatty acid-binding proteins. Biochim. Biophys. Acta 1486, 28–44.
[99] Alpers, D.H., Bass, N.M., Engle, M.J., DeSchryver-Kecskemeti, K., 2000. Intestinal fatty acid binding protein may favor differential apical fatty acid binding in the intestine. Biochim. Biophys. Acta 1483, 352–362.
[100] Wolfrum, C., Borrmann, C.M., Franke, W.W., Gorski, J., Spener, F., 1999. Fatty acids and drugs interacting with FABP and PPAR in hepatocytes: a signaling path to the nucleus. Chem. Phys. Lipids 101, 149.
[101] Vassileva, G., Huwyler, L., Poirier, K., Agellon, L.B., Toth, M.J., 2000. The intestinal fatty acid binding protein is not essential for dietary fat absorption in mice. Faseb J. 14, 2040–2046.

[102] Martin, G.G., Danneberg, H., Kumar, L.S., Atshaves, B.P., Erol, E., Bader, M., Schroeder, F., Binas, B., 2003. Decreased liver fatty acid binding capacity and altered liver lipid distribution in mice lacking the liver fatty acid binding protein (L-FABP) gene. J. Biol. Chem. 1, 1.

[103] Knudsen, J., Neergaard, T.B., Gaigg, B., Jensen, M.V., Hansen, J.K., 2000. Role of acyl-CoA binding protein in acyl-CoA metabolism and acyl-CoA-mediated cell signaling. J. Nutr. 130, 294S–298S.

[104] Knudsen, J., Mandrup, S., Rasmussen, J.T., Andreasen, P.H., Poulsen, F., Kristiansen, K., 1993. The function of acyl-CoA-binding protein (ACBP)/diazepam binding inhibitor (DBI). Mol. Cell. Biochem. 123, 129–138.

[105] Kragelund, B.B., Andersen, K.V., Madsen, J.C., Knudsen, J., Poulsen, F.M., 1993. Three-dimensional structure of the complex between acyl-coenzyme A binding protein and palmitoyl-coenzyme A. J. Mol. Biol. 230, 1260–1277.

[106] Knudsen, J., Jensen, M.V., Hansen, J.K., Faergeman, N.J., Neergaard, T.B., Gaigg, B., 1999. Role of acylCoA binding protein in acylCoA transport, metabolism and cell signaling. Mol. Cell. Biochem. 192, 95–103.

[107] Frolov, A., Schroeder, F., 1998. Acyl coenzyme A binding protein. Conformational sensitivity to long chain fatty acyl-CoA. J. Biol. Chem. 273, 11049–11055.

[108] Rosendal, J., Ertbjerg, P., Knudsen, J., 1993. Characterization of ligand binding to acyl-CoA-binding protein. Biochem. J. 290, 321–326.

[109] Yanase, H., Shimizu, H., Kanda, T., Fujii, H., Iwanaga, T., 2001. Cellular localization of the diazepam binding inhibitor (DBI) in the gastrointestinal tract of mice and its coexistence with the fatty acid binding protein (FABP). Arch. Histol. Cytol. 64, 449–460.

[110] Mandrup, S., Sorensen, R.V., Helledie, T., Nohr, J., Baldursson, T., Gram, C., Knudsen, J., Kristiansen, K., 1998. Inhibition of 3T3-L1 adipocyte differentiation by expression of acyl-CoA-binding protein antisense RNA. J. Biol. Chem. 273, 23897–23903.

[111] Helledie, T., Antonius, M., Sorensen, R.V., Hertzel, A.V., Bernlohr, D.A., Kolvraa, S., Kristiansen, K., Mandrup, S., 2000. Lipid-binding proteins modulate ligand-dependent trans-activation by peroxisome proliferator-activated receptors and localize to the nucleus as well as the cytoplasm. J. Lipid Res. 41, 1740–1751.

[112] Stremmel, W., Pohl, L., Ring, A., Herrmann, T., 2001. A new concept of cellular uptake and intracellular trafficking of long-chain fatty acids. Lipids 36, 981–989.

[113] Glatz, J.F., Storch, J., 2001. Unravelling the significance of cellular fatty acid-binding proteins. Curr. Opin. Lipidol. 12, 267–274.

[114] Pohl, J., Ring, A., Stremmel, W., 2002. Uptake of long-chain fatty acids in HepG2 cells involves caveolae: analysis of a novel pathway. J. Lipid Res. 43, 1390–1399.

[115] Ring, A., Pohl, J., Volkl, A., Stremmel, W., 2002. Evidence for vesicles that mediate long-chain fatty acid uptake by human microvascular endothelial cells. J. Lipid Res. 43, 2095–2104.

[116] Buhman, K.K., Smith, S.J., Stone, S.J., Repa, J.J., Wong, J.S., Knapp, F.F. Jr., Burri, B.J., Hamilton, R.L., Farese, R.V. Jr., 2002. DGAT1 is not essential for intestinal triacylglycerol absorption or chylomicron synthesis. J. Biol. Chem. 16, 16.

[117] Kiba, T., Tanaka, K., Hoshino, M., Numata, K., Okano, K., Inoue, S., 1995. Ventromedial hypothalamic lesions induce the proliferation of gastrointestinal mucosal cells in the rat. Life Sci. 57, 827–832.

[118] Dunel-Erb, S., Chevalier, C., Laurent, P., Bach, A., Decrock, F., Le Maho, Y., 2001. Restoration of the jejunal mucosa in rats refed after prolonged fasting. Comp. Biochem. Physiol. A Mol. Integr. Physiol. 129, 933–947.

[119] Lovshin, J., Yusta, B., Iliopoulos, I., Migirdicyan, A., Dableh, L., Brubaker, P.L., Drucker, D.J., 2000. Ontogeny of the glucagon-like peptide-2 receptor axis in the developing rat intestine. Endocrinology 141, 4194–4201.

[120] Bjerknes, M., Cheng, H., 2001. Modulation of specific intestinal epithelial progenitors by enteric neurons. Proc. Natl Acad. Sci. USA 98, 12497–12502.

Plasma albumin as a fatty acid carrier

Stephen Curry*

Department of Biological Sciences, Imperial College London, Room 746, Huxley Building, South Kensington Campus, London SW7 2AZ, UK
**Correspondence address: Tel.: +44-20-7594-7632; fax: +44-20-7589-0191*
E-mail: s.curry@imperial.ac.uk(S.C.)

1. Introduction

Fatty acids are important components in the synthesis of complex lipids and provide a vital source of metabolic energy. Weight for weight they contain more energy than other fuel reserves such as glycogen or muscle protein because they are more highly reduced and, being hydrophobic, are anhydrous. In mammals 5–25% of the body weight is fat, most of it stored as long-chain fatty acids (>C16) esterified into triacylglycerols in adipose tissue. However, the hydrophobicity that enables the dense packing of these high-energy molecules in adipocytes presents a problem when they are released in the non-esterified form to be delivered to other tissues of the body because their aqueous solubilities are known to be vanishingly small [1]. Albumin, a highly soluble protein, greatly augments the transport capacity of serum, since it is present at a high concentration (~640 μM) and can carry at least six molecules of fatty acid. As the major transporter of non-esterified fatty acids, it plays an important role in the lipid economy of the body.

The natural abundance of albumin, coupled with the development during the second world war of cost-effective means for purifying the protein from human serum, helped to establish it as an attractive target for biochemists in the post war years. Interest in the molecule grew as the extent of its astonishing binding capacity was gradually revealed. In addition to fatty acids, albumin binds bilirubin, heme, thyroxine, steroid hormones, bile acids, metal ions and a wide array of drugs [2]. The protein exerts a significant effect on the kinetics of distribution and elimination of all these compounds. The drug-binding capabilities of albumin, in particular, have stimulated a great deal of pharmaceutical interest in the protein. More recently, the natural propensity of albumin for binding small molecule ligands has begun to be exploited in the development of novel diagnostic and therapeutic reagents. In the last 10 years, high resolution crystallographic investigations, coupled with the development of methods for expressing human serum albumin (HSA) in yeast, have yielded more detailed insights into the binding properties of this remarkable protein and provided a renewed impetus to functional investigations.

Advances in Molecular and Cell Biology, Vol. 33, pages 29–46

ISSN: 1569-2558/DOI: 10.1016/S1569-2558(03)33003-6

The quantity of literature on albumin is truly staggering: a search on Pubmed for papers containing "serum albumin" in the title revealed over 5300 entries. Fortunately, the albumin field has benefited from a number of excellent reviews which give a comprehensive overview of all aspects of the protein [2–5]. However, it is now several years since these were published and the aim of this chapter is therefore to focus on the most recent findings that shed light on albumin's role as a fatty acid transporter.

2. Distribution of albumin

Synthesised in the liver for secretion into the bloodstream, serum albumin is continuously exchanged between plasma and the fluid occupying much of the extravascular space. Albumin expression begins early in foetal development reaching normal levels by birth. The protein is continuously turned over and has a half-life in circulation of around 19 days. Though by name a serum or plasma protein, most of the albumin in the body (~60%) actually resides in tissue fluids. Both passive and active mechanisms appear to facilitate transfer of albumin (and its cargo) between the intravascular and extravascular compartments. In organs with fenestrated capillaries (e.g. liver, spleen) or sinusoids (e.g. pancreas, bowel), passive transport through large pores or fenestrations would appear to dominate. However, in the rest of the body there is a continuous capillary endothelium and albumin appears to traverse this barrier largely via receptor-mediated transcytosis [6]. By this process, albumin binds to glycoprotein receptors on the luminal side of the endothelial cell membrane and is drawn into non-coated plasmalemmal vesicles (caveolae). These vesicles migrate to the opposite side of the cell and fuse with the abluminal membrane to deliver the protein to the tissue fluid in the interstitial space.

Several glycoprotein receptors on endothelial cells with significant affinity for albumin have been characterised to date: gp18, gp31 and gp60 [7–11], each named for its apparent molecular weight in kilodaltons. Of these, gp60, also known as albondin, appears to be primarily implicated in mediating transcytosis of albumin [9,12]. It is expressed specifically on the surface of continuous endothelial cells [13], binds albumin specifically and with sub-micromolar affinity and co-localises with the protein in caveolae [14]. The circulatory protein SPARC (secreted protein, acidic and rich in cysteines) is immunologically and functionally related to gp60. Both proteins bind albumin with high affinity via domains that appear to be structurally related since antibodies raised against SPARC that block albumin binding to the protein also prevent binding of albumin to gp60 [15]; equally αgp60 antibodies can block albumin binding to SPARC [9,12]. The functional significance of the SPARC–albumin interaction is not well understood; one possibility is that binding of albumin inhibits proteolytic degradation of SPARC, prolonging its lifetime in solution [15]; however, there is no evidence as yet that SPARC plays a role in albumin distribution, despite its similarity to gp60.

In contrast to gp60, gp18 and gp31 appear to be expressed in a pattern that does not correlate with the need for transcytosis and have a particular affinity for modified albumins [8,12,16]. They have thus been identified as possible scavenger receptors for albumin, mediating removal of aged proteins from the circulation.

The mechanisms of albumin distribution in the body remain active and interesting areas of research [6,14,17,18]. A key goal must surely be to obtain clones of the albumin receptors which should illuminate the relationship between gp18, gp31, gp60 and SPARC and permit a finer dissection of their properties and their particular contributions to albumin distribution.

While on the subject of albumin-binding proteins, it is worth noting that a number of important bacterial pathogens of humans carry surface proteins that possess high-affinity albumin-binding domains. These include protein G from group C and G streptococci [19], the albumin-binding protein (PAB) from *Peptostreptococcus magnus* [20], and M proteins and protein H from *Streptococcus pyogenes* [21]. Protein G and PAB share a homologous albumin-binding domain known as a GA module (protein G related Albumin binding), which contains just 46 amino acids folded into a three-helix bundle [22,23]. M proteins and protein H contain modules known as C repeats which bind albumin with the same sort of affinity as GA modules. Although C repeats and GA modules are unrelated in sequence, they appear to bind to overlapping sites on the protein [24]. These albumin-binding surface proteins occur in more virulent bacterial strains [20,21] and may operate as virulence factors by helping to shield the bacterium from immune surveillance. In addition, at least one study suggests that the ability to bind albumin stimulates growth, perhaps by providing access to fatty acids [20].

3. Distribution of fatty acids

Although transcytosis is clearly important for the movement of albumin through the endothelial layer and undoubtedly results in transport of albumin-associated fatty acids to the interstitum, this accounts for less than 1% of the fatty acids delivered to tissue fluids [5]. The bulk of fatty acid transfer from serum to the interstitial fluid is achieved by diffusion through the endothelial cytoplasm, aided by intracellular fatty acid binding proteins [17]. Fatty acids released into the interstitum re-associate with albumin prior to delivery to the cells where they will ultimately be metabolised.

The presence of high concentrations of albumin in human plasma (42 g/L; 640 μM) and tissue fluids (10–30 g/L; 150–450 μM), coupled with its high avidity for fatty acids, ensures that more than 99% of the fatty acid present is bound to the protein. Though the total concentration of fatty acids in serum is in the range 90–1200 μM under normal conditions [5], the unbound concentration of fatty acids is only around 7.5 nM [25]; the free concentration in tissue fluid is likely to be similar. Although the fatty acids found most commonly in the circulation [palmitic (C16:0), palmitoleic (C16:1), stearic (C18:0), oleic (C18:1) and linoleic (C18:2)] [26] all bind albumin with high affinities ($K_d \sim 1-10$ nM), they associate and dissociate on a rapid timescale (observed $k_{off} \sim 1.5-8\ s^{-1}$) [27]. Albumin therefore acts as an effective buffer against short-term fluctuations in fatty acid concentrations while at the same time providing a readily accessible supply of fatty acids for energy production and synthesis of more complex lipids. By drastically limiting the free fatty acid concentration, the short-term buffering role of albumin probably helps to protect the body from the toxic effects of high levels of free fatty acid [5]; longer term buffering appears to be provided by regulation of their uptake by adipose cells [28].

Although the basal mole ratio of fatty acid to albumin in human plasma is around 0.7 (implying a somewhat higher mole ratio in tissue fluids where the albumin concentration is lower), the protein has the capacity to absorb wide variations in fatty acid levels in these body fluids; the buffering capacity of albumin is only significantly diminished when the fatty acid:albumin mole ratio exceeds 5 [29], a limit that is not usually encountered in normal physiological conditions, even after prolonged exercise or fasting [5,30]. However, higher fatty acid:albumin ratios, sometimes exceeding 6, may occur in conjunction with conditions such as obesity or diabetes [31]. Under these conditions, the albumin molecule would be expected to be more or less saturated with fatty acid and the free concentration of the lipid would be expected to rise, though redistribution to other sinks such as membranes or lipoproteins also occurs [31].

Within tissue fluids fatty acids dissociate from albumin prior to uptake by the cells that are their final destination. There has been some indication that interactions of albumin–fatty acid complexes with cell membranes may facilitate off-loading their fatty acid cargo at the cell surface [32–34] but such mechanisms are not widely accepted. Presently, the majority view is that the observed rates of association and dissociation would appear to be fast enough to allow passage of adequate quantities of fatty acid to cells under normal conditions [17,27,35]. The molecular mechanisms of cellular uptake of fatty acids will be treated in subsequent chapters. For now we would like to turn to consider the albumin molecule itself to examine those properties that account for its capabilities as a fatty acid transporter.

4. Structure of albumin

Albumin was sequenced in the 1970s, cloned in the 1980s and its three-dimensional structure was solved in the 1990s: three decades of work have yielded a highly detailed view of the albumin molecule and many of its most important properties.

HSA is synthesised in the liver as preproalbumin, a single polypeptide chain of 609 amino acids. The 18-amino acid signal peptide is removed upon extrusion into the endoplasmic reticulum. The remaining 6-amino acid N-terminal propeptide is cleaved at a late stage, either in the *trans*-Golgi or the secretory vesicle by furin. The precise function of the propeptide is not known but it may be needed to direct intracellular trafficking since deletion of this feature from the albumin gene of rats results in accumulation of the protein in the endoplasmic reticulum and a lower secretion rate [36].

The mature protein (585 amino acids) is monomeric and divided into three homologous domains, first identified when the amino acid sequence of the protein was elucidated [37]. Determination of the crystal structure of HSA revealed that the secondary structure content of the protein is purely α-helical [38]. The three homologous domains (I–III) each contain 10 helices connected by turns or extended segments of polypeptide and are structurally very similar to one another [3]. They can be further divided into two subdomains (A and B) which have a common four-helix core; each of the A subdomains contains an additional pair of helices at the C-terminal end followed by an extended polypeptide that links to the B subdomain (Fig. 1a). For domains I and II the last helix of subdomain B is continuous with the first helix of the succeeding A subdomain. These long

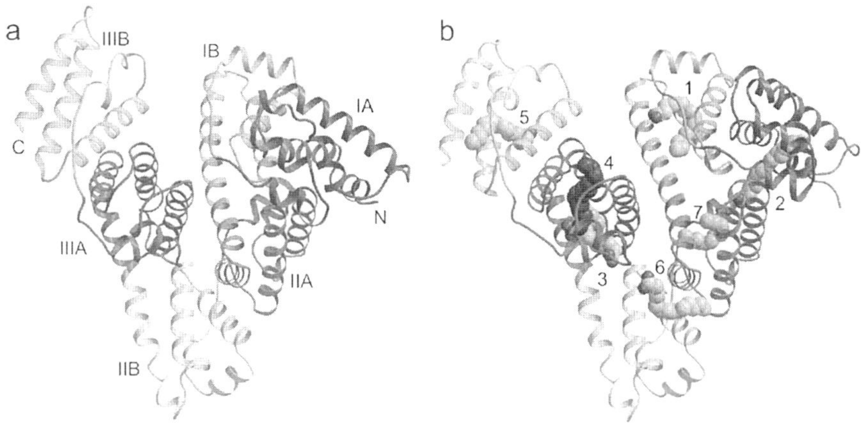

Fig. 1. Domain structure and fatty acid binding sites on HSA. (a) The secondary structure of defatted HSA (PDB ID 1e78 [46]) is shown in a ribbon representation with subdomains coloured as follows: IA, red; IB, light red; IIA, green; IIB, light green; IIIA, blue; IIIB, light blue. (b) Structure of the HSA–oleic acid complex (PDB ID 1gni; [61]). Ligand atoms are shown in a space-filling representation colour-coded by atom type: carbon – grey; oxygen – red. The seven binding sites common to medium- and long-chain fatty acids are numbered 1–7. The fatty acid in site 4 is coloured dark grey to distinguish it from the ligand in site 3. *(For a colored version of this figure, see plate section, page 580.)*

inter-domain helices help to fix the relative orientations of the three domains in HSA. The protein contains 35 cysteine residues, 34 of which pair up within subdomains to form disulphide bridges that undoubtedly contribute to the remarkable thermostability of the protein [39,40].

The three domains are arranged to give the protein a heart-shaped configuration that is at odds with the elongated oblate ellipsoid shape (axial dimensions of 40 Å × 140 Å) indicated by initial solution hydrodynamic studies [2]. This early model exerted a powerful grip on the imagination of albumin investigators prior to the initiation of detailed structural analyses [41,42] and probably contributed to the misinterpretation of the first low-resolution crystallographic electron density maps of the protein [43]. Given this particular history and the known flexibility of the protein, some doubts have persisted regarding validity of the heart-shaped configuration revealed by crystallography as a model for the structure of the protein in solution [2,44]. However, recent work has largely laid these concerns to rest. There are now three independently determined structures of defatted HSA in the protein databank [38,45,46], all in different packing environments. Although there are some small variations between the three structures, particularly in the disposition of domain III (Fig. 2a), the heart shape is maintained, indicating that it is unlikely to be an artefact of crystallisation. Moreover, the crystal structure of equine serum albumin also revealed a heart-shaped molecule [47]. More recently, investigations of the conformation of bovine serum albumin (BSA) in solution using either phosphorescence depolarisation techniques [44] or small-angle X-ray scattering [48] have provided robust support that the crystallographic structures are a proper indication of the protein conformation in solution. Insights derived from these structures may therefore be applied with some confidence to the interpretation of functional data acquired in solution.

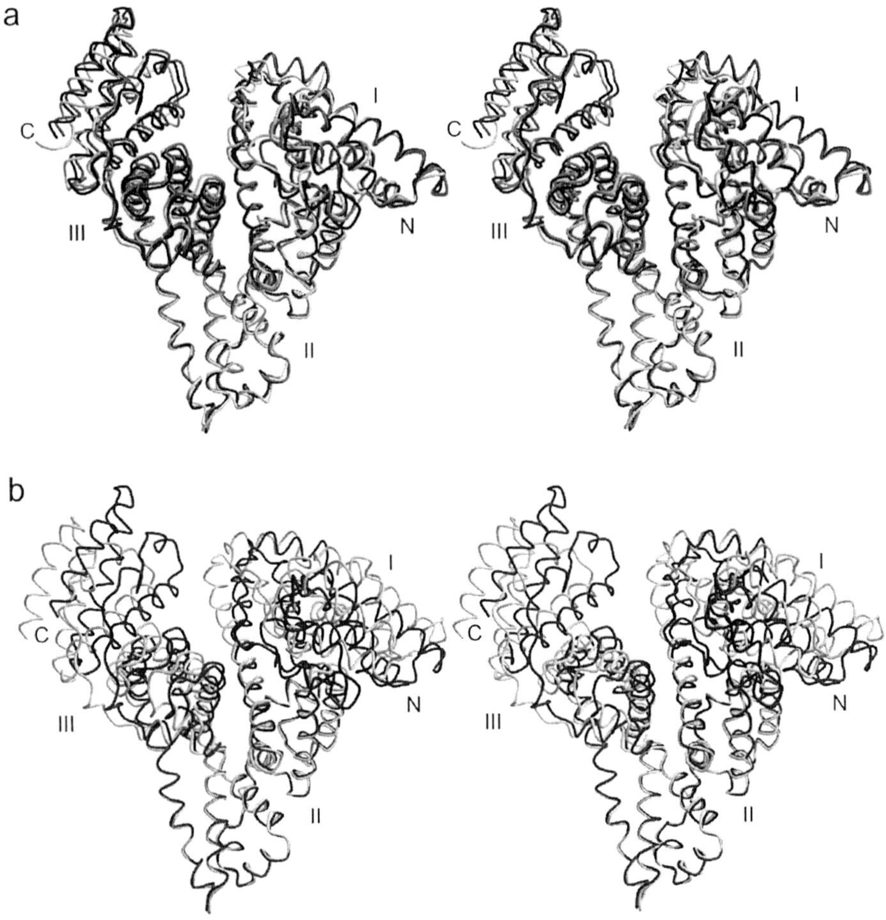

Fig. 2. Stereoviews of conformational variations observed in crystal structures of HSA. (a) Backbone trace of three independently determined crystal structures of defatted HSA. The three structures, PDB ID codes 1uor [38], 1ao6 [45] and 1e78 [46] are in light, medium and dark shades of grey, respectively, and were superposed using just the C_α atoms of domain II. Most of the structural differences occur in domain III. (b) Backbone traces illustrating the structural changes associated with fatty acid binding. The structures of defatted HSA (PDB ID 1e78; [46]) and the HSA–myristate complex (PDB ID 1e7g; [60]), superposed using just the C_α atoms of domain II, are shaded light and dark grey, respectively. Note that domains I and III appear to swing out from the centre upon fatty acid binding.

Nevertheless, the preservation of the heart-shaped configuration of albumin in solution should not be taken to imply that the protein lacks flexibility. As mentioned above, the three crystal structures of defatted HSA give some indication of the conformational flexibility inherent in the molecule. Moreover, albumin has long been known to undergo significant conformational changes associated with variations in pH or binding of fatty acids [2,3] (Figs. 1b and 2b). Distinct conformations of albumin are believed to occur in acidic and basic conditions. The N ("normal" or "neutral") form, which is likely to correspond to the crystal structures of defatted HSA that were determined at pH 7–7.5, converts to the F ("fast") form as the pH drops below 4.0 and then to the E ("expanded") form below 3.5. In the E form, the protein reversibly unfolds into its constituent domains or subdomains and readily releases hydrophobic ligands; this is demonstrated by the need to reduce the pH to 3.0 to achieve effective delipidation of the protein [49]. However, the in vivo function of structural changes dependent of such acidic conditions has yet to be demonstrated. Of more physiological relevance may be the transition to the B ("basic") form, initially characterised as occurring within the range pH 6–9 [2]. This transition has been monitored by observing changes in the circular dichroism spectrum or the fluorescence of bound marker ligands as a function of pH but the conformational changes involved have yet to be characterised structurally. The presence of plasma concentrations of Ca^{2+} (2 mM) depresses the mid-point of the N–B transition to around pH 7 suggesting that the B form may actually be prevalent in plasma or tissue fluids (pH 7.4) [50,51]. Work with proteolytic fragments indicates that the N–B transition primarily involves changes in domains I and II, a view that is consistent with observations that drugs that bind to subdomain IIA but not subdomain IIIA are affected by the transition [52]. A nuclear magnetic resonance (NMR) study also suggested that deprotonation of histidine residues located in domain I may be part of the driving mechanism for the transition [53]. More recent work has emphasised the possible parallel between the N–B transition and the structural changes associated with fatty acid binding (see below).

5. Comparison of albumin with related proteins

Since albumin sequences are highly conserved among mammals, sharing around 75% amino acid sequence identity, their three-dimensional structures are likely to be very similar to one another. Indeed, the close structural similarity between human and equine serum albumins has already been demonstrated [47]. Thus, in mammals, not only are the three internal domains conserved, but also their relative dispositions are likely to be the same. This may reflect conservation of fatty acid binding capabilities since at least two of the fatty acid binding sites traverse domain boundaries (see below).

Comparison of HSA with albumins from more distantly related species including chicken, Chinese cobra, frog, salmon and Atlantic lamprey reveals sequence identities declining from 47 to 21%. Like HSA these albumin orthologues all have three homologous domains, with the sole exception of the Atlantic lamprey, which has extended its complement to 7. Whether these species also maintain the same relative disposition of domains as found in HSA is still unknown.

Humans appear to possess several paralogues of HSA, the most closely related of which is α-fetoprotein. This protein is around 40% identical in amino acid sequence to HSA and is expressed predominantly in the foetal stage of development, rising to a maximum level of 3 g/L in foetal plasma before falling to a level of less than 5 μg/L in adult life. HSA is also expressed in the foetal stage, beginning in the fist few weeks and rising steadily to reach adult levels (around 40 g/L) at birth. Since albumin and α-fetoprotein are both present at substantial levels, particularly after the early weeks of gestation, the particular role of α-fetoprotein seems unclear. It is known that α-fetoprotein binds saturated fatty acids relatively poorly and seems to have a distinct preference for polyunsaturated fatty acids such as arachidonic acid (reviewed in Ref. [54]), perhaps reflecting particular metabolic requirements of the developing foetus.

A second albumin-like protein – α-albumin, also known as afamin – is 36% identical in sequence to HSA but is only present in plasma at around 30 mg/L. It has recently been implicated in the transport of vitamin E [55].

The most distantly related albumin paralogue is vitamin D binding protein (DBP) which has 19% sequence identity with HSA but lacks the C-terminal subdomain (IIIB). DBP occurs in plasma at 0.5 g/L and is involved in the transport of vitamin D and its metabolites. Intriguingly, this protein has also acquired the ability to bind actin monomers. Actin filaments released following tissue damage are disassembled by gelsolin and bound by DBP to prevent repolymerisation and aid disposal. Recent structural work shows that the protein has a single vitamin D binding site in its N-terminal subdomain (IA) [56] and that actin binds with a large interface of contact with DBP involving all three domains [57]. The structure shows strikingly that the relative dispositions of the three domains are very different from that observed in HSA and possibly reflects an adaptation required for binding of actin since the conformational changes in DBP upon actin binding are relatively modest [57].

6. Fatty acid binding

The structure of defatted HSA gave few clues about the nature of fatty acid binding sites on the protein. Crystals of HSA–fatty acid complexes were reported by Carter in 1994 [58] and some preliminary structural data were mentioned in a review article published in the same year [3]. However, the first full report of a crystal structure of an HSA–fatty acid complex was only published in 1998 [59] and was followed by comparative investigations of the binding of medium and long-chain fatty acids [60] and of saturated and unsaturated fatty acids [61]. These analyses revealed the presence of seven binding sites on the protein that are common to the range of fatty acids studied and provide a detailed characterisation of structural aspects of the binding sites. A number of additional sites have been observed for medium chain fatty acids, most likely because the greater solubility of shorter chain fatty acids allowed higher concentrations to be used in co-crystallisation experiments. The additional sites are likely to be physiologically relevant only at very high molar ratios of fatty acid to albumin [60].

The seven common sites are distributed throughout the protein in an asymmetric manner with regard to the internal homology of the domains, as anticipated by previous NMR studies [62]. According to the numbering scheme that we have adopted, site 1 is located in subdomain IB, site 2 is at the interface between IA and IIA, sites 3 and 4 are both contained within IIIA, site 5 is in IIIB, site 6 lies along the interface between IIA and IIB and site 7, finally, is contained within subdomain IIA (Fig. 1b). Thus, each of the domains has been adapted differently for fatty acid binding. Sites 1, 4, 5 and 7 are each wholly contained within a single subdomain whereas sites 2, 3 and 6 are formed by residues from two subdomains. In fact, site 2 is formed from residues from domains I and II, while site 3 is made up of contributions from domains II and III. Thus, the integrity of both these sites is dependent on maintenance of the overall heart-shaped configuration of the protein; interestingly, binding of fatty acids to sites 2 and 3 is implicated in driving adjustments of the relative domain orientation (see below).

Although there are variations in the modes of binding of fatty acids at the different sites, some common aspects are nevertheless evident. With one notable exception (site 2), the binding sites generally consist of preformed hydrophobic cavities which appear to expand upon fatty acid binding as the result of mainchain and sidechain adjustments. Perhaps, the most striking example of this occurs in fatty acid site 1 in subdomain IB where a pair of stacked tyrosine residues (Tyr 138 and Tyr 161) which partially fill the pocket in the defatted protein rotate by about 90° to allow them to contact opposite sides of the methylene tail of the fatty acid ligand (see Fig. 3 of Ref. [63]). More modest sidechain movements are observed at other sites (Fig. 3). The formation of fatty acid site 2 as a result of a fatty acid-induced conformational change is discussed in more detail below. For sites 1–5 the carboxylate moieties of the fatty acids are anchored in the same position via electrostatic interactions with basic or polar sidechains, irrespective of the chain length or the degree of unsaturation. The coordinating amino acids are conserved in mammalian albumins, underlining their functional importance [59]. The methylene tails are accommodated within deep hydrophobic pockets, extending further into these pockets with increasing chain length. This binding configuration occurs irrespective of whether the binding site is a short, wide cavity (sites 1 and 3) or a long, narrow tunnel (sites 2, 4, 5); both electrostatic and hydrophobic interactions thus contribute to the free energy of binding and the structure plausibly accounts for the observed increases in binding affinity with methylene chain length [29,64] (Fig. 3). The mode of binding observed in the crystal structure is consistent with NMR data which not only indicated the presence of specific electrostatic interactions of the fatty acid carboxylates with basic sidechains at several sites [41] but also demonstrated that the carboxylate end was the least mobile portion of the bound fatty acids [65].

In contrast, sites 6 and 7 do not appear to have defined basic or polar sidechains to coordinate the carboxylate group of the bound fatty acids and the orientation of the fatty acids in these sites is less certain at present. Lack of a clear electrostatic interaction between the protein and the fatty acid ligand in these sites also suggests that they may have a relatively low binding affinity, at least for long-chain fatty acids [41,60,66].

The flexibility of the fatty acid tails allows the ligand to follow the contours of the binding pocket. Thus, in sites 2, 4 and 5 which may be considered as linear tunnels, or in site 6 which is more of a surface groove, the different fatty acids bind in similar, extended configurations (Fig. 3). The presence of a single *cis*-double bond in oleic acid (C18:1) makes little detectable difference to the binding configuration; clearly, there is sufficient room in these channel-like sites to easily accommodate the kink introduced in the middle of the methylene tail [61]. Even arachidonic acid (C20:4) that has four *cis*-double bonds and is more restricted in its conformational freedom extends to a more-or-less linear configuration in these narrow sites. In broader cavities (sites 1, 3, 7), the fatty acids bind in bent configurations; for example, site 1 is D shaped and the fatty acids simply follow the curve of the interior wall of the cavity. Site 3 is significantly narrower and fatty acids with chains longer than C14 bind in a relatively tight U-bend configuration (Fig. 3). As before the presence of single or multiple double bonds apparently makes little difference to the binding configuration.

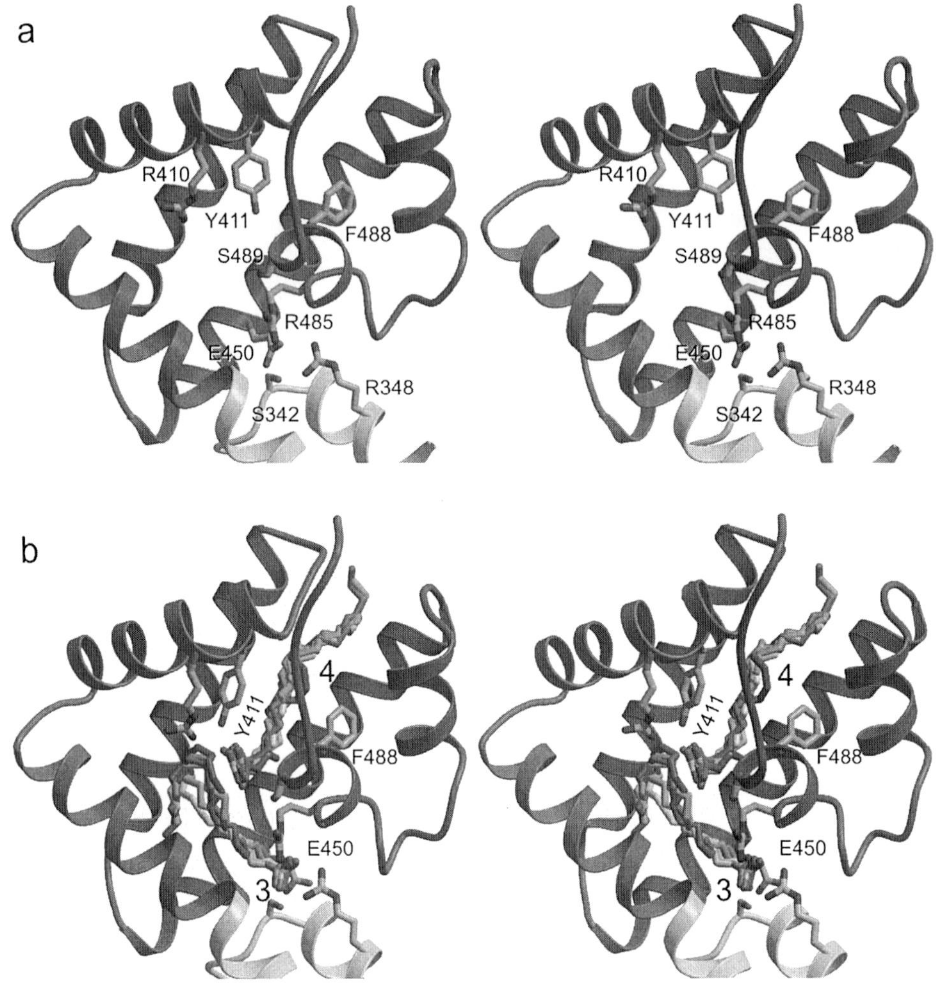

Fig. 3. Stereoviews of fatty acid binding to sites 3 and 4 in subdomain IIIA. (a) Subdomain IIIA structure in defatted HSA [PDB ID 1e78]. The secondary structure coloured as in Fig. 2. A subset of amino acids involved in interactions with fatty acid is labelled. (b) Comparative binding of medium- and long-chain fatty acids to subdomain IIIA structure in HSA [60,61]. Carbon atoms in the different fatty acids are coloured as follows: C10:0, red; C12:0, orange; C14:0, yellow; C16:0, green; C18:0, blue; C18:1, violet; C20:4, brown). Although the carboxylate atoms of the ligands in sites 3 and 4 are tightly anchored, the disposition of the methylene tail depends on the dimensions of the hydrophobic cavity. Note also how certain amino acid sidechains are displaced to make way for the fatty acid ligands, especially Glu 450, Tyr 411 and Phe 488. (*For a colored version of this figure, see plate section, page 581.*)

By accommodating non-esterified fatty acids in discrete, hydrophobic pockets with polar residues at their entrances to coordinate the carboxylate moiety, albumin exhibits a mode of interaction that is common to many other proteins that have evolved to bind this type of ligand, including intracellular lipid binding proteins [67], peroxisome proliferator-activated receptors [68], β-lactoglobulin [69] and maize lipid-transfer protein [70].

The proteins that bind non-esterified fatty acids are nevertheless a structurally diverse set; for example, while albumin is composed of three α-helical domains, intracellular lipid binding proteins consist of a single β-barrel capped by two short helices. This contrast serves to emphasise the variety of ways in which proteins may be configured to bind fatty acids. The interaction of albumin and other proteins with non-esterified fatty acids is quite distinct from the way that apolipoproteins coordinate lipid aggregates within lipoproteins. Lipoproteins are large, micellar structures (5–500 nm) containing around 50–98% by weight of lipids composed variously of phospholipids, triacylglygerol and cholesterol. Apolipoprotein A-1, for example, appears to form a "belt" that constrains a micellar core formed mainly of cholesterol, cholesterol esters and phospholipids [71].

Although medium- and short-chain fatty acids will certainly bind to albumin, the molecule appears to be adapted to accommodate long-chain fatty acids (≥C16) which are the predominant species circulating in vivo [26]. Structural analysis shows that the binding sites are essentially completely filled by fatty acids of length C18–C20 [60,61]. The protein is nevertheless capable of binding even longer fatty acids, though the number of detectable sites appears to decline steadily with increasing length; for C26 fatty acids, there appears to be just a single binding site [72]. Modelling experiments indicate that the apparent loss of binding for such very long-chain fatty acids arises because the additional methylene groups in the hydrophobic tail cannot be fully accommodated in the binding site and protrude into the solvent. The extension of the methylene tail thereby contributes unfavourably to the free energy of binding [72].

7. Conformational changes associated with fatty acid binding

Strikingly, fatty acid binding induces a significant alteration of the relative dispositions of the three domains of HSA [59,63]; the overall conformational change can be considered essentially as rigid body rotations of domains I and III relative to domain II: there are only relatively modest distortions of the backbone of individual domains upon binding [59,60, 63]. According to the view shown in Fig. 2b, fatty acid binding causes domains I and III to rotate in opposite directions about an approximately vertical axis through the centre of the protein. Notably, the long helix connecting domains I and II are significantly bent during this conformational transition, though the distortion in the II–III helix is much less. The rotation of domain I relative to domain II is stabilised by the binding of fatty acid to site 2 which straddles the interface between subdomains IA and IIA. This site is composed of two parts contributed by the two subdomains which are separated by about 10 Å in the defatted structure. The hydrophobic cavity in subdomain IA is occupied by the sidechain of Leu 251 (see Fig. 2 of Ref. [63]) but when fatty acids bind the subdomain detaches from this residue and realigns with subdomain IIA to form a contiguous channel to accommodate the long hydrophobic ligand. Insertion of the fatty acid in this site serves to lock the altered conformation of these two domains. Fatty acids as short as C10:0 appear to be able to drive the reconfiguration of domains I and II but modelling experiments suggest that C8:0 may be too short for its methylene tail to make the stabilising contacts with subdomain IA that are necessary for the conformational change [60].

The mechanism driving the rotation of domain III relative to domain II appears more complex. It is likely to involve the binding of fatty acid to site 3 in subdomain IIIA where the carboxylate moiety of the fatty acid replaces a salt-bridge interaction to Arg 348 made by the sidechain of Glu 450 in the defatted structure (Fig. 3). The displaced Glu sidechain makes a new interaction with the backbone amide of Val 343 in subdomain IIB, and alters the II–III domain interface. Additionally, the helix containing Glu 450 packs against the long interdomain helix connecting domains I and II that is distorted by the conformational adjustment in the relative positioning of these domains due to fatty acid binding. Thus, the rotation of domain III observed upon fatty acid binding may also be due to the propagation via these helix–helix contacts of structural changes occurring between domains I and II.

The same fatty acid-induced conformational change in HSA is observed irrespective of the length or degree of saturation of the bound lipid (subject to the caveats noted above). At first sight this might be taken to indicate that the protein can switch between two relatively well-defined conformational states, depending on whether it is defatted or has fatty acid bound. However, it should be borne in mind that the crystallised HSA–fatty acid complexes were invariably prepared using a high mole ratio of fatty acid to HSA (10–40) [59–61]. More recent work on the structure of the HSA–hemin–myristic acid complex, in which fatty acids were complexed with HSA at a 4:1 mole ratio, revealed a conformation that is intermediate between defatted HSA and the fully loaded HSA–fatty acid complex [73]. The structural basis for this intermediate state is being investigated but the observation nevertheless suggests that the conformational adaptability of HSA is perhaps even greater than was previously appreciated.

Binding experiments that monitored the impact of fatty acids on the binding of other albumin ligands suggest that structural changes become evident at fatty acid to albumin mole ratios of less than 1 and are essentially complete when the ratio reaches a level of 3–4 [2,74,75]. Given that under normal conditions, the fatty acid to HSA mole ratio in plasma is less than 1, thermodynamic considerations would seem to dictate that only a minority of albumin molecules is likely to be in the conformationally altered state and this casts some doubt on the physiological relevance of the fatty acid-induced structural change. Although this change may be responsible for the observation that fatty acids enhance the affinity of albumin for the endothelial cell surface and accelerate internalisation and presumably transfer of the protein and its associated fatty acids to the tissue fluids [76,77], this is not a significant pathway for import of fatty acids [5]. Furthermore, there is no evidence for a cellular receptor which might specifically recognise the conformational state of loaded albumin and thereby facilitate fatty acid uptake. The altered conformation is likely to predominate in diseased states that either increase levels of circulating fatty acid or reduce the available albumin and can raise the fatty acid to HSA mole ratio as high as 6 [31]. It is tempting to speculate that the fatty acid-induced structural change in albumin may serve as an indicator for conditions in which the transport system is becoming overloaded.

The structural changes associated with fatty acid binding also appear to open up the surface recess where Cys 34 is located and thereby increase its activity [78,79]. Through this single unpaired cysteine residue, which is conserved in mammals and other species, HSA provides significant antioxidant activity in plasma [2]. The residue reacts with exogenous thiol-containing compounds such as cysteine and glutathione and various metal ions, including Cd^{2+}, Au^{3+}, Hg^{2+} and Ag^{+}. It also binds nitric oxide (NO) though very recent evidence suggests that in HSA Trp 214 is the primary site of NO binding [80]. It has been postulated that fatty acid binding may serve as an allosteric regulator of the activity of Cys 34 in HSA [78].

A further aspect of the structural changes wrought in albumin by fatty acid binding is that they can impact the binding of other ligands such as bilirubin, thyroxine and drugs [74, 75,81,82]. A recent study revealed the impact of fatty acid-induced structural changes on the binding of the hormone thyroxine to HSA [82]. Thyroxine was observed to bind at four sites in the defatted protein, one in subdomain IIA, one in IIIA and two in IIIB, all of which overlap with fatty acid binding sites. Intriguingly, the structure of an HSA–myristate–thyroxine complex showed that, while fatty acids prevented thyroxine binding at these four sites, the conformational changes due to fatty acid binding opened up a new site for the hormone in the cleft between domains I and III. This site appears to have an affinity for thyroxine that is comparable to the affinity of the primary thyroxine site – most likely subdomain IIA – in defatted HSA [82].

It has long been known that fatty acids can enhance the binding affinity of bilirubin and drugs such as warfarin that bind to drug site 1 (subdomain IIA) [74,75,81]. Though bilirubin is currently presumed to bind in IIA, this has yet to be definitively established. Structural studies on binding of the drug warfarin indicate that the affinity enhancement is primarily due to rearrangement of a pair of sidechains – Tyr 150 and Arg 257 – which adjust their positions to co-ordinate the carboxylate moiety of the fatty acid that binds between subdomains IA and IIA (fatty acid site 2) [59,60]. The hydroxyl group on Tyr 150 is shifted out of the pocket and the positively charged guanidinium on Arg 257 is partially neutralised by interaction with the negatively charged carboxylate on the fatty acid; as a result the deepest part of the pocket becomes more hydrophobic and is better able to bind the apolar coumarin group of the drug [83]. Whether all other drugs bound at this site experience a similar fatty acid-induced enhancement is an interesting question for future investigations.

There are intriguing parallels between the conformational changes due to fatty acid binding and those involved in the N-B transition that occur as the pH is raised. Both transitions primarily involve domains I and II [52] and appear to enhance the binding of site 1 drugs (subdomain IIA) while having little or no effect on the binding of site 2 drugs (subdomain IIIA) [75,81]. The connection between them is further underscored by the finding that binding of fatty acids depresses the mid-point of the N–B transition. A recent study found that mutations of residues Phe 211 and His 242 in subdomain IIA reduced the mid-point of the N–B transition by about 0.6 pH units [84]. This pair of residues packs in a cluster of sidechains that includes Tyr 150, one of the amino acids that is reoriented to coordinate fatty acid bound to fatty acid site 2 (at the IA–IIA interface; see above). The mutagenesis results thus suggest that the mechanistic basis of the N–B transition may be closely related to the fatty acid-induced transition and it is an appealing possibility that the structural changes associated with the two transitions are similar. As a counter to this view, it has been argued that, since fatty acid binding merely lowers the mid-point of the N–B transition rather than abolishing it altogether, the structural changes induced by fatty acid binding and increasing the pH must be distinct [85]. However, this view is based on observations of the effect of pH on the fluorescence of a drug site 1 marker compound (5-dimethylaminonaphthalene-1-sulfonamide) in the presence and absence of fatty acids [85], rather than a more direct probe of changes in protein structure. Proper resolution of this issue must await the determination of the structure of HSA in the B form. This should be an interesting structure in its own right since the B form of the protein may be prevalent in plasma in vivo.

8. Fatty acid binding affinities

Efforts to determine affinities of fatty acid binding sites on albumin have occupied researchers for many years. It is a testament to the technical difficulty of the endeavour that although studies were initiated 40 years ago [86], they are still ongoing [1,29,64]. Binding experiments with fatty acids are problematical because albumin has several non-homogenous binding sites of different affinities, which complicates the data analysis, and because fatty acids themselves are barely soluble, which makes it difficult to perform the necessary measurements of the unbound fatty acid concentration. This latter difficulty is exacerbated particularly in the case of long-chain fatty acids which have a high propensity to form aggregates in aqueous media [87]. The problems with experimental measurements of fatty acid interactions with albumin have been discussed in detail by Vorum et al. [1], who have proposed new methods for determining fatty acid binding affinities. Despite all the technical difficulties, most recent studies appear to concur that there are 2–3 high-affinity binding sites on the protein along with a similar number of secondary sites [1,29, 62,64,88,89]. Long-chain saturated fatty acids bind to the primary high-affinity sites with dissociation constants in the range 1–10 nM; mono- and polyunsaturated long-chain fatty acids bind more weakly ($K_d \sim 10$–100 nM), though they are found to occupy the same sites on the protein [61]. The secondary sites bind fatty acids with affinities that are only around 5–10 times lower than the primary sites [29]. The implication of the relatively narrow range of binding affinities for a given species of fatty acid is that even at a low mole ratio of fatty acid to albumin, as is found under normal physiological conditions, the population of albumin molecules is predicted to be heterogeneous with respect to fatty acid content and to contain proteins with 0, 1, 2 or more bound fatty acid molecules [2,5]. This is supported by the recent determination of the crystal structure of the HSA–myristate–hemin complex, prepared with a myristic acid to HSA mole ratio of 4 [73]. The crystal structure effectively shows the average structure of the population in the crystals and fatty acids were observed to be distributed across at least six binding sites.

The location of the highest affinity fatty acid binding sites on albumin is an enduring problem. The issue would be rather academic if the sole function of albumin were to transport fatty acids. But albumin binds many different types of ligand, both endogenous and exogenous, in sites that overlap with fatty acid binding pockets and it is important to try to determine the nature of the interplay between fatty acids and other ligands for the protein. As discussed above, both competitive and cooperative interactions appear to operate. The issue of ligand–ligand interactions is likely to be particularly complicated in individuals suffering from diseases affecting the levels of albumin or of fatty acids or other albumin ligands (e.g. bilirubin) who may also require treatment using drugs that happen to bind to the protein [31].

Evidence collected in several laboratories suggests that the primary binding sites for medium- and long-chain fatty acids may be distinct (reviewed in Ref. [60]). Competition binding studies with drugs indicate that subdomain IIIA contains the primary site for medium-chain fatty acids such as octanoic acid [90]. Subdomain IIA (fatty acid site 7) also appears to bind medium-chain but not long-chain fatty acids with high affinity [42]. However, the primary sites for long-chain fatty acids have yet to be clearly assigned. The current data on the locations of the highest affinity fatty acid binding sites have been reviewed in some detail [63]. Perhaps, the most telling investigation used NMR to examine the titration of fatty acids into binding sites in proteolytic fragments of BSA [62], which binds fatty acids in a very similar manner to HSA [29]. This study found one high-affinity site in a fragment corresponding to subdomains IA–IB–IIA and two further sites in domain III. NMR analyses also showed that high-affinity sites possessed basic ligands which make salt-bridge interactions with the carboxylate moiety of the bound fatty acid [66]. In the crystal structures of HSA–fatty acid complexes sites 1–5 all have basic sidechains coordinating the fatty acid carboxylate and therefore fit the high-affinity profile identified by NMR [60] but which of these five sites have the highest binding affinities for fatty acid? We argued previously that, on the weight of current evidence, fatty acid site 5 in subdomain IIIB is likely to be a high-affinity site [63] but that it was not yet possible to identify other high-affinity sites with great confidence. It seems likely that the second high-affinity fatty acid site in domain III is located in subdomain IIIA, and therefore overlaps with drug site 2, but whether this is fatty acid site 3 or 4 remains to be seen. Preliminary experiments in which crystals of HSA–myristate were back-soaked in solutions lacking myristate would seem to indicate that fatty acid site 1 is most readily depopulated (Ghuman, Zunszain and Curry, unpublished data), suggesting that fatty acid site 2 is the high-affinity site identified in the IA–IB–IIA fragment of albumin. However, a definitive assignment will require further work, though it is now clear that carefully designed mutagenesis experiments based on the crystal structures of HSA–fatty acid complexes should provide the requisite answers.

Acknowledgements

I am very grateful to all my co-workers on the albumin project at Imperial College: Hendrik Mandelkow, Ananyo Bhattacharya, Tim Grüne, Isabelle Petitpas, Jamie Ghuman, Patricia Zunszain, Peter Brick and Nick Franks. I would like to thank Prof. Theodore Peters, Prof. Ulrich-Kragh-Hansen, Dr Anders Petersen and Dr Patricia Zunszain for critical reading of the manuscript. I would also like to acknowledge the contribution of Delta Biotechnology Ltd. who provided much of the recombinant albumin (Recombumin®) used in our work. I thank the BBSRC, the MRC and the Wellcome Trust for funding support and the staff at Daresbury SRS and EMBL/DESY for assistance with data collection.

References

[1] Vorum, H., Fisker, K., Honore, B., 1997. J. Pept. Res. 49, 347–354.

[2] Peters, T., 1995. All About Albumin: Biochemistry, Genetics and Medical Applications. Academic Press, San Diego, CA.

[3] Carter, D.C., Ho, J.X., 1994. Adv. Protein Chem. 45, 152–203.
[4] Kragh-Hansen, U., 1990. Dan. Med. Bull. 37, 57–84.
[5] Spector, A.A., 1986. In: Scanu, A.M., Spector, A.A. (Eds.), Biochemistry and Biology of Plasma Lipoproteins. Dekker, New York, pp. 247–279.
[6] Simionescu, M., Gafencu, A., Antohe, F., 2002. Microsc. Res. Tech. 57, 269–288.
[7] Ghinea, N., Eskenasy, M., Simionescu, M., Simionescu, N., 1989. J. Biol. Chem. 264, 4755–4758.
[8] Schnitzer, J.E., Sung, A., Horvat, R., Bravo, J., 1992. J. Biol. Chem. 267, 24544–24553.
[9] Tiruppathi, C., Finnegan, A., Malik, A.B., 1996. Proc. Natl Acad. Sci. USA 93, 250–254.
[10] Siflinger-Birnboim, A., Schnitzer, J., Lum, H., Blumenstock, F.A., Shen, C.P., Del Vecchio, P.J., Malik, A.B., 1991. J. Cell. Physiol. 149, 575–584.
[11] Schnitzer, J.E., Carley, W.W., Palade, G.E., 1988. Proc. Natl Acad. Sci. USA 85, 6773–6777.
[12] Schnitzer, J.E., Oh, P., 1994. J. Biol. Chem. 269, 6072–6082.
[13] Schnitzer, J.E., 1992. Am. J. Physiol. 262, H246–H254.
[14] John, T.A., Vogel, S.M., Tiruppathi, C., Malik, A.B., Minshall, R.D., 2003. Am. J. Physiol. Lung Cell. Mol. Physiol. 284, L187–L196.
[15] Schnitzer, J.E., Oh, P., 1992. Am. J. Physiol. 263, H1872–H1879.
[16] Schnitzer, J.E., Bravo, J., 1993. J. Biol. Chem. 268, 7562–7570.
[17] van der Vusse, G.J., van Bilsen, M., Glatz, J.F., Hasselbaink, D.M., Luiken, J.J., 2002. Mol. Cell. Biochem. 239, 9–15.
[18] Predescu, D., Predescu, S., Malik, A.B., 2002. Proc. Natl Acad. Sci. USA 99, 13932–13937.
[19] Kraulis, P., Jonasson, P., Nygren, P.-Å., Uhlén, M., Jendeberg, L., Nilsson, B., Kördel, J., 1996. FEBS Lett. 378, 190–194.
[20] de Château, M., Holst, E., Björck, L., 1996. J. Biol. Chem. 271, 26609–26615.
[21] Åkesson, P., Schmidt, K.-H., Cooney, J., Björck, L., 1994. Biochem. J. 300, 877–886.
[22] Johansson, M.J., de Château, M., Wikström, M., Forsén, S., Drakenberg, T., Björck, L., 1997. J. Mol. Biol. 266, 859–865.
[23] Johansson, M.U., Frick, I.M., Nilsson, H., Kraulis, P.J., Hober, S., Jonasson, P., Linhult, M., Nygren, P.A., Uhlen, M., Bjorck, L., Drakenberg, T., Forsen, S., Wikstrom, M., 2002. J. Biol. Chem. 277, 8114–8120.
[24] Frick, I.M., Åkesson, P., Cooney, J., Sjöbring, U., Schmidt, K.-H., Gomi, H., Hattori, S., Tagawa, C., Kishimoto, F., Björck, L., 1994. Mol. Microbiol. 12, 143–151.
[25] Richieri, G.V., Kleinfeld, A.M., 1995. J. Lipid Res. 36, 229–240.
[26] Saifer, A., Goldman, L., 1961. J. Lipid Res. 2, 268–270.
[27] Demant, E.J., Richieri, G.V., Kleinfeld, A.M., 2002. Biochem. J. 363, 809–815.
[28] Frayn, K.N., 2002. Diabetologia 45, 1201–1210.
[29] Richieri, G.V., Anel, A., Kleinfeld, A.M., 1993. Biochemistry 32, 7574–7580.
[30] Brodersen, R., Andersen, S., Vorum, H., Nielsen, S.U., Pedersen, A.O., 1990. Eur. J. Biochem. 189, 343–349.
[31] Cistola, D.P., Small, D.M., 1991. J. Clin. Invest. 87, 1431–1441.
[32] Weisiger, R., Gollan, J., Ockner, R., 1981. Science 211, 1048–1051.
[33] Trigatti, B.L., Gerber, G.E., 1995. Biochem. J. 308(Pt 1), 155–159.
[34] Hutter, J.F., Piper, H.M., Spieckermann, P.G., 1984. J. Mol. Cell. Cardiol. 16, 219–226.
[35] Hamilton, J.A., Guo, W., Kamp, F., 2002. Mol. Cell. Biochem. 239, 17–23.
[36] McCracken, A.A., Kruse, K.B., 1989. J. Biol. Chem. 264, 20843–20846.
[37] Behrens, P.Q., Spiekerman, A.M., Brown, J.R., 1975. Fed. Proc. 34, 591.
[38] He, X.M., Carter, D.C., 1992. Nature 358, 209–215.
[39] Shrake, A., Finlayson, J.S., Ross, P.D., 1984. Vox Sang. 47, 7–18.
[40] Gumpen, S., Hegg, P.O., Martens, H., 1979. Biochim. Biophys. Acta 574, 189–196.
[41] Cistola, D.P., Small, D.M., Hamilton, J.A., 1987. J. Biol. Chem. 262, 10971–10979.
[42] Brown, J.R., Shockley, P., 1982. In: Jost, P., Griffith, O.H. (Eds.), Lipid–Protein Interactions. Wiley, New York, pp. 26–68.
[43] Carter, D.C., He, X.M., Munson, S.H., Twigg, P.D., Gernert, K.M., Broom, M.B., Miller, T.Y., 1989. Science 244, 1195–1198.

[44] Ferrer, M.L., Duchowicz, R., Carrasco, B., de la Torre, J.G., Acuna, A.U., 2001. Biophys. J. 80, 2422–2430.
[45] Sugio, S., Kashima, A., Mochizuki, S., Noda, M., Kobayashi, K., 1999. Protein Eng. 12, 439–446.
[46] Bhattacharya, A.A., Curry, S., Franks, N.P., 2000. J. Biol. Chem. 275, 38731–38738.
[47] Ho, J.X., Holowachuk, E.W., Norton, E.J., Twigg, P.D., Carter, D.C., 1993. Eur. J. Biochem. 215, 205–212.
[48] Svergun, D.I., Petoukhov, M.V., Koch, M.H., 2001. Biophys. J. 80, 2946–2953.
[49] Kragh-Hansen, U., 1993. Anal. Biochem. 210, 318–327.
[50] Wilting, J., Weideman, M.M., Roomer, A.C., Perrin, J.H., 1979. Biochim. Biophys. Acta 579, 469–473.
[51] Wilting, J., van der Giesen, W.F., Janssen, L.H., Weideman, M.M., Otagiri, M., Perrin, J.H., 1980. J. Biol. Chem. 255, 3032–3037.
[52] Bos, O.J., Remijn, J.P., Fischer, M.J., Wilting, J., Janssen, L.H., 1988. Biochem. Pharmacol. 37, 3905–3909.
[53] Bos, O.J., Fischer, M.J., Wilting, J., Janssen, L.H., 1989. Biochem. Pharmacol. 38, 1979–1984.
[54] Deutsch, H.F., 1991. Adv. Cancer Res. 56, 253–312.
[55] Voegele, A.F., Jerkovic, L., Wellenzohn, B., Eller, P., Kronenberg, F., Liedl, K.R., Dieplinger, H., 2002. Biochemistry 41, 14532–14538.
[56] Verboven, C., Rabijns, A., De Maeyer, M., Van Baelen, H., Bouillon, R., De Ranter, C., 2002. Nat. Struct. Biol. 9, 131–136.
[57] Otterbein, L.R., Cosio, C., Graceffa, P., Dominguez, R., 2002. Proc. Natl Acad. Sci. USA 99, 8003–8008.
[58] Carter, D.C., Chang, B., Ho, J.X., Keeling, K., Krishnasami, Z., 1994. Eur. J. Biochem. 226, 1049–1052.
[59] Curry, S., Mandelkow, H., Brick, P., Franks, N., 1998. Nat. Struct. Biol. 5, 827–835.
[60] Bhattacharya, A.A., Grüne, T., Curry, S., 2000. J. Mol. Biol. 303, 721–732.
[61] Petitpas, I., Grüne, T., Bhattacharya, A.A., Curry, S., 2001. J. Mol. Biol. 314, 955–960.
[62] Hamilton, J.A., Era, S., Bhamidipati, S.P., Reed, R.G., 1991. Proc. Natl Acad. Sci. USA 88, 2051–2054.
[63] Curry, S., Brick, P., Franks, N.P., 1999. Biochim. Biophys. Acta 1441, 131–140.
[64] Rose, H., Conventz, M., Fischer, Y., Jungling, E., Hennecke, T., Kammermeier, H., 1994. Biochim. Biophys. Acta 1215, 321–326.
[65] Hamilton, J.A., Cistola, D.P., Morrisett, J.D., Sparrow, J.T., Small, D.M., 1984. Proc. Natl Acad. Sci. USA 81, 3718–3722.
[66] Cistola, D.P., Small, D.M., Hamilton, J.A., 1987. J. Biol. Chem. 262, 10980–10985.
[67] Banaszak, L., Winter, N., Xu, Z., Bernlohr, D.A., Cowan, S., Jones, T.A., 1994. Adv. Protein Chem. 45, 89–151.
[68] Xu, H.E., Lambert, M.H., Montana, V.G., Parks, D.J., Blanchard, S.G., Brown, P.J., Sternbach, D.D., Lehmann, J.M., Wisely, G.B., Willson, T.M., Kliewer, S.A., Milburn, M.V., 1999. Mol. Cell 3, 397–403.
[69] Wu, S.Y., Perez, M.D., Puyol, P., Sawyer, L., 1999. J. Biol. Chem. 274, 170–174.
[70] Han, G.W., Lee, J.Y., Song, H.K., Chang, C., Min, K., Moon, J., Shin, D.H., Kopka, M.L., Sawaya, M.R., Yuan, H.S., Kim, T.D., Choe, J., Lim, D., Moon, H.J., Suh, S.W., 2001. J. Mol. Biol. 308, 263–278.
[71] Borhani, D.W., Rogers, D.P., Engler, J.A., Brouillette, C.G., 1997. Proc. Natl Acad. Sci. USA 94, 12291–12296.
[72] Choi, J.K., Ho, J., Curry, S., Qin, D., Bittman, R., Hamilton, J.A., 2002. J. Lipid Res. 43, 1000–1010.
[73] Zunszain, P.A., Ghuman, J., Komatsu, T., Tsuchida, E., Curry, S., 2003. BMC Struct. Biol. 3, 6.
[74] Reed, R., 1977. J. Biol. Chem. 252, 7483–7487.
[75] Vorum, H., Honoré, B., 1996. J. Pharm. Pharmacol. 48, 870–875.
[76] Galis, Z., Ghitescu, L., Simionescu, M., 1988. Eur. J. Cell Biol. 47, 358–365.
[77] Antohe, F., Dobrila, L., Heltianu, C., Simionescu, N., Simionescu, M., 1993. Eur. J. Cell Biol. 60, 268–275.
[78] Narazaki, R., Maruyama, T., Otagiri, M., 1997. Biochim. Biophys. Acta 1338, 275–281.
[79] Takabayashi, K., Imada, T., Saito, Y., Inada, Y., 1983. Eur. J. Biochem. 136, 291–295.
[80] Harohalli, K., Petersen, C.E., Ha, C.E., Feix, J.B., Bhagavan, N.V., 2002. J. Biomed. Sci. 9, 47–58.
[81] Birkett, D.J., Myers, S.P., Sudlow, G., 1977. Mol. Pharmacol. 13, 987–992.
[82] Petitpas, I., Petersen, C.E., Ha, C.E., Bhattacharya, A.A., Zunszain, P.A., Ghuman, J., Bhagavan, N.V., Curry, S., 2003. Proc. Natl Acad. Sci. USA 100, 6440–6445.
[83] Petitpas, I., Bhattacharya, A.A., Twine, S., East, M., Curry, S., 2001. J. Biol. Chem. 276, 22804–22809.
[84] Petersen, C.E., Ha, C.E., Curry, S., Bhagavan, N.V., 2002. Proteins 47, 116–125.

[85] Wanwimolruk, S., Birkett, D.J., 1982. Biochim. Biophys. Acta 709, 247–255.
[86] Goodman, D.S., 1958. J. Am. Chem. Soc. 80, 3892–3898.
[87] Vorum, H., Brodersen, R., Kragh-Hansen, U., Pedersen, A.O., 1992. Biochim. Biophys. Acta 1126, 135–142.
[88] Bojesen, I.N., Bojesen, E., 1992. J. Lipid Res. 33, 1327–1334.
[89] Bojesen, I.N., Bojesen, E., 1994. J. Lipid Res. 35, 770–778.
[90] Kragh-Hansen, U., 1991. Biochem. J. 273, 641–644.

Cellular uptake of long chain free fatty acids: the structure and function of plasma membrane fatty acid binding protein

Michael W. Bradbury and Paul D. Berk*

Departments of Medicine (Division of Liver Disease) and Developmental, Cell, and Molecular Biology, Mount Sinai School of Medicine, New York, NY 10029, USA

**Correspondence address: Division of Liver Disease (Box 1039), Mount Sinai School of Medicine, 1 Gustave L. Levy Place, New York, NY 10029, USA. Tel.: +1-212-659-8370; fax: +1-212-348-3517*

E-mail: paul.berk@msnyuhealth.org(P.D.B.)

1. Introduction

Non-esterified long chain fatty acids (LCFA) are important energy substrates, building blocks for the complex lipid components of cellular membranes, and precursors of biological mediators such as the prostaglandins and leukotrienes. They are also increasingly recognized as important intracellular modulators of gene expression (e.g. Refs. [18,39,119]). Given these roles, control of their intracellular concentrations through regulation of their cellular uptake and efflux would be of great value. However, cellular uptake of LCFA was long considered an entirely passive process, which was therefore unregulated and of relatively little intrinsic interest.

Our interest in fatty acid uptake began with the 1981 report by Weisiger et al. [175] suggesting that hepatocellular uptake of LCFA in the isolated, perfused rat liver was mediated by an albumin receptor on the surface of hepatocytes. A prior abstract by the same group [173] suggested that this receptor also mediated the uptake of bilirubin. This latter process had for many years been a major focus of work in our laboratory. Reanalysis of the published 1981 data suggested to us that, rather than indicating the presence of an albumin receptor, the data were consistent with LCFA uptake being a saturable function of the unbound LCFA concentration (reviews by us: [11,17,128]). Identical results and conclusions were obtained when a similar experimental design was applied to studies with isolated rat hepatocytes [93].

These provocative conclusions led to a major redirection of effort in our laboratory, which has focused for the subsequent 20 years on uptake processes for LCFA. Accordingly, in the sections that follow, we will first examine the evidence indicating that there are specific, regulatable, protein-mediated processes for cellular LCFA uptake.

Advances in Molecular and Cell Biology, Vol. 33, pages 47–80

ISSN: 1569-2558/DOI: 10.1016/S1569-2558(03)33004-8

We will then examine evidence that plasma membrane fatty acid binding protein ($FABP_{pm}$) plays a role in these facilitated LCFA uptake process(es), and finally, examine situations in which altered regulation of LCFA uptake contributes to the pathophysiology of specific diseases.

2. Mechanisms of cellular uptake of long chain fatty acids

The apparent saturability of LCFA uptake challenged the long-held view that LCFA uptake occurred solely by passive diffusion. Saturability is a necessary, but not sufficient, condition to establish the existence of facilitated transport. Abumrad et al. had reported as early as 1981 [3,4] that LCFA uptake by adipocytes was saturable. Subsequent studies showed that it exhibited additional kinetic features of facilitated transport, including stereo-specificity, trans-stimulation, and competitive inhibition [1,3,129,131,154]. Subsequently, several plasma membrane proteins were identified as putative LCFA transporters [2,42,115,150,162]. Despite a very large body of evidence favoring the existence in several cell types of facilitated LCFA uptake process(es), the nature of cellular LCFA uptake remains in dispute to this day [9,51,54], with a small but vocal group continuing to insist that LCFA enter cells by purely passive "flip-flop" across the plasma membrane. The controversy reflects fundamental disagreement about three issues: (i) the nature and (ii) the rate of the rate-limiting step, and (iii) whether membrane proteins play a role in transmembrane LCFA movement (reviewed in Refs. [12,51,71,78,92]).

2.1. Studies with synthetic membrane vesicles

Most data on the first two of these issues come from studies of LCFA entry into synthetic vesicles, which typically used indirect measures of transmembrane LCFA movement [36,72]. Recent studies, for example, used the rate of vesicle acidification as a measure of the rate of inwardly directed flip-flop and subsequent ionization of protonated LCFA [25,67–69]. Since flip-flop of LCFA *anions* is ~3 orders of magnitude slower than protonated acids [49], such studies largely ignored *their* role in LCFA uptake although, with p*K*a's in the range of 4.5–6.5, most extracellular LCFA are in their anionic form. Studies using rates of change in the intracellular pH (pH_I) have reported transmembrane transfer rates for natural LCFA into small (~200 Å) unilamellar vesicles that were extremely rapid ($k_{ff} > 200\ s^{-1}$) [69]. Studies in large (~1000 Å) [69,72] or giant (>2000 Å) unilamellar vesicles (GUVs) [72] found progressively smaller k_{ff} values. Flip-flop rates in erythrocyte ghosts were similar to those in GUVs [74]. The magnitude of k_{ff} varied with vesicle composition and decreased with increasing vesicle size (reviewed in Ref. [12]). For vesicles closest in size and lipid composition to cell membranes, flip-flop $t_{1/2}$'s for, e.g. oleate and palmitate are ~1–10 s [72,74]. The development of ADIFAB, a fluorescent fatty acid binding protein that exhibits a spectral shift upon binding LCFA, fostered experiments that clarified important aspects of this approach [72,74,106,109,111], proving, e.g. that flip-flop is the rate-limiting step in the movement of protonated LCFA into vesicles [72,73].

Changes in pH_i also have been used to estimate rates of LCFA entry into cells [25,32, 52,107]. The data all reveal *intracellular* acidification rates orders of magnitude slower than those reported in vesicles or RBC ghosts. However, pH_i studies are insensitive, requiring the use of un-physiologically high unbound LCFA concentrations ($[LCFA_u]$) to obtain quantifiable data. They are, therefore, conducted without albumin or at un-physiologically low albumin levels. The resulting data are therefore of uncertain relevance to what occurs in vivo. LCFA entering cells in such studies do not, for example, enter normal metabolic pathways, remaining available for efflux after ≥4 min [25,32], whereas 75% of LCFA taken up under physiologic conditions are esterified and *unavailable* for efflux by 15 s [153]. Thus, the rate of change of pH_i is unlikely to accurately reflect the rate of physiologic LCFA uptake.

Furthermore, cellular uptake of albumin-bound ligands is strongly influenced by experimental conditions. At very low albumin levels, non-membrane events, e.g. dissociation from albumin or diffusion across the pericellular unstirred water layer may become rate limiting [129,132,134,174], with effects on uptake kinetics that are often incorrectly attributed to membrane transport processes. To avoid such artifacts, we have consistently studied LCFA uptake at physiologic temperature, albumin, and LCFA concentrations. Under such conditions, the major LCFA uptake component in hepatocytes [142,148,154], adipocytes [3,4,118], cardiac and skeletal muscle [21,78,130,140,167], ileal and jejunal epithelia [103,141], 3T3-L1 adipocytes [183,184], and other cells [23,24, 80,163,164] exhibits saturability and other kinetic features of facilitated transport: e.g. cis-inhibition, trans-stimulation, and counter-transport [1,3,128,129,153]. Competitive inhibition [131] and selective inhibition of LCFA uptake by antibodies to putative LCFA transporters [118,130,140,141,148,151,184] (see below) indicate that this process is specific for LCFA. Although >90% of uptake by hepatocytes and adipocytes is normally saturable, each also has a non-saturable component. By contrast, uptake in 3T3-L1 preadipocytes is almost exclusively non-saturable [153,184].

2.2. *Mechanisms of cellular LCFA uptake*

We studied oleic acid (OA) uptake by isolated hepatocytes and its binding to hepatocyte plasma membranes over a wide range of LCFA:BSA molar ratios (ν), which determine the unbound OA concentration $[OA_u]$ [154]. At physiologic albumin concentrations, hepatocellular OA uptake (Fig. 1) and OA binding to hepatocyte plasma membranes exhibit both saturable and non-saturable components [129,154], i.e.

$$\mathrm{UT}([\mathrm{OA_u}]) = (V_{\max}\cdot[\mathrm{OA_u}])/(K_\mathrm{m} + [\mathrm{OA_u}]) + k_\mathrm{u}\cdot[\mathrm{OA_u}], \text{ and} \tag{1}$$

$$B([\mathrm{OA_u}]) = (B_{\max}\cdot[\mathrm{OA_u}])/(K_\mathrm{d} + [\mathrm{OA_u}]) + k_\mathrm{b}\cdot[\mathrm{OA_u}] \tag{2}$$

where $\mathrm{UT}([\mathrm{OA_u}])$ and $B([\mathrm{OA_u}])$ represent experimental values for uptake and binding, respectively, at the unbound oleic acid concentration $[OA_u]$; V_{max} and B_{max} are the maximal uptake velocity and maximal binding capacity; K_m and K_d the unbound oleate concentrations at 50% of maximal uptake and maximal binding capacities, respectively; and k_u and k_b the respective rate constants for non-saturable uptake and non-saturable binding. By plotting the experimental values for saturable uptake as a function of saturable

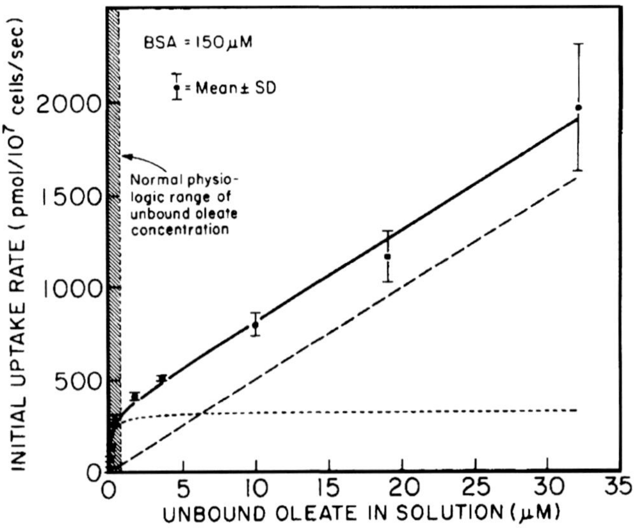

Fig. 1. Cellular uptake of LCFA is the sum of both saturable and non-saturable components. Initial oleate uptake rates were determined at various oleate/albumin ratios, which varies with the unbound oleate concentration in the solution (plotted on the *X*-axis). Means and standard errors of triplicate determinations are plotted, and the solid line represents the computer fit using Eq. (1). The curve is the sum of the saturable (- - -) and non-saturable (– – –) components.

binding (Fig. 2) we asked, in effect, if the saturable component of OA uptake bore some functional relationship to the saturable component of OA binding, i.e.

$$\text{Does } (V_{\text{max}}\cdot[O_{\text{u}}])/(K_{\text{m}} + [O_{\text{u}}]) = f((B_{\text{max}}\cdot[O_{\text{u}}])/(K_{\text{d}} + [O_{\text{u}}]))? \qquad (3)$$

and if so, what is the nature of the functional relationship *f*. Comparison of the computer-fitted binding and uptake functions revealed a very high degree of linear correlation (Fig. 2), suggesting that the saturable LCFA uptake component derives from the pool of LCFA saturably bound to the plasma membrane. A similar analysis suggested that the non-saturable component is derived principally from the separate, non-saturably bound substrate pool. We then calculated rate constants for the saturable (k_{s}) and non-saturable (k_{ns}) transmembrane movement of OA into adipocytes from the general relationship:

$$k_x(\text{s}^{-1}) = \text{UT}_x(\text{pmol}/50,000 \text{ cells s}^{-1})/B_x(\text{pmol}/50,000 \text{ cells}), \qquad (4)$$

where k_x is the rate constant for component *x*, and UT_x and B_x are the relationships describing that component (i.e. saturable or non-saturable) of uptake and binding as a function of $[\text{OA}_{\text{u}}]$. In hepatocytes, the rate of transmembrane OA movement via the saturable pathway ($k_{\text{s}} = 0.7 \text{ s}^{-1}$, $t_{1/2} = 1$ s) was nearly 15 times faster than via the non-saturable pathway ($k_{\text{ns}} = 0.04 \text{ s}^{-1}$, $t_{1/2} = 14$ s) [154]. We have also reported similar studies in isolated rat adipocytes [153]. Although the overall uptake velocity in these cells was significantly faster than in hepatocytes, the transmembrane movement of LCFA via the saturable pathway was again an order of magnitude faster that that via the non-saturable route ($k_{\text{s}} = 2.9 \text{ s}^{-1}$, $t_{1/2} = 0.24$ s; $k_{\text{ns}} = 0.26 \pm 0.04 \text{ s}^{-1}$, $t_{1/2} = 2.7 \pm 0.4$ s).

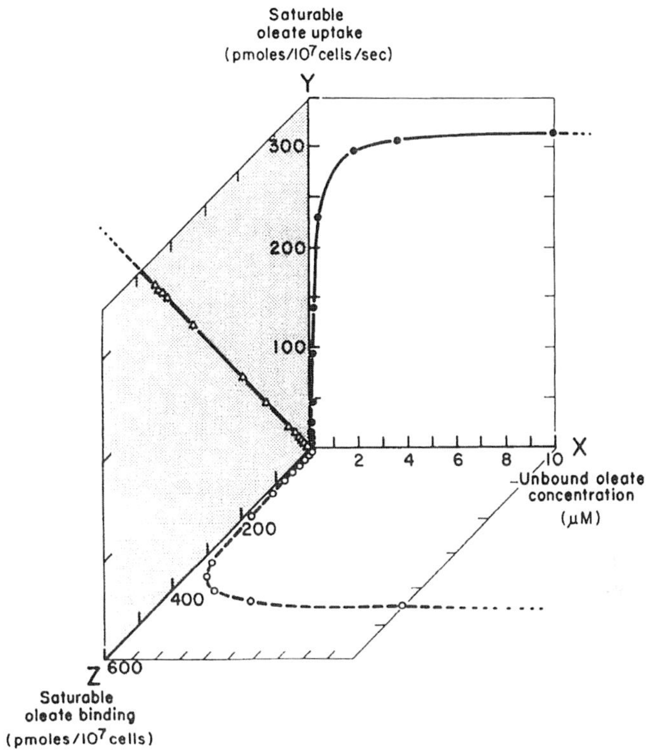

Fig. 2. Relationships between the unbound oleate concentration [OA_u] (*X*), saturable oleate uptake (*Y*) (• — •) and the saturable binding to liver plasma membranes (*Z*) (○---○). Values for *Y* and *Z* represent computer fits from replicate experimental measurements of total uptake and total binding using Eqs. (1) and (2). Data points represent values of [OA_u] used in experiments for determinations of uptake and binding. The relationship between saturable uptake and binding is shown on the *YZ* plane (△—△). This indicates that the saturable component of oleate uptake is a linear function of the saturable binding component ($r = 0.997$).

These analyses used the 1971 OA:BSA binding constants of Spector et al. [135] to calculate the required values of [OA_u]. Three new sets of binding constants, determined by different methods, were reported in 1993/94 [20,106,114], all of which suggested that the Spector constants overestimated [OA_u] by $\geq$10-fold. Since the most detailed of the new studies was by Richieri et al. [106], we compared the effect on the analysis of the adipocyte [^{3}H]-OA binding and uptake studies of calculating [OA_u] from *their* OA:BSA association constants with [OA_u] values obtained with the 1971 Spector constants [135]. Using Richieri's constants, adipocyte OA uptake was still best described as a sum of saturable and non-saturable component, with the saturable process accounting for >90% of LCFA uptake at physiologic BSA concentrations and values of ν. Rate constants for both the saturable (k_s) and non-saturable (k_{ns}) uptake components were calculated from the two sets of computed [OA_u]. Within the physiologic range ($\nu \sim 1.5$), calculated values for k_s in adipocytes were identical ($k_s = 2.9\ s^{-1}$, $t_{1/2} = 0.24$ s). k_{ns} was 0.26 ± 0.04 and $0.10 \pm 0.03\ s^{-1}$, based on the constants of Spector and Richieri, respectively

($t_{1/2} = 2.7 \pm 0.4$ and 6.6 ± 1.7 s, $p > 0.1$). As in hepatocytes [154], k_s in adipocytes was $\geq$10-fold greater than k_{ns}. Thus, demonstration of both saturable and non-saturable OA uptake processes, as well as the magnitudes of their associated rate constants, does not depend on the use of any particular set of association constants to calculate [OA_u] [153].

2.2.1. *Nature of the saturable uptake pathway*

The hepatocellular uptake kinetics of [^{3}H]-α_2,β_2,ω_3-heptafluoro-stearate (HFS), a strongly acidic ($pK_a = 0.5$) LCFA analog existing almost entirely as an anion at physiologic pH, are virtually identical to those of stearate [139]. Uptake of stearate and [^{3}H]-HFS were similarly inhibited by photoaffinity labeling of the cell surface by 11,11-azistearate [117], and by antibodies to the putative LCFA transporter $FABP_{pm}$ [148] (Berk, P.D. et al., unpublished). These data indicate that the saturable LCFA uptake component reflects transport of fatty acid anions. This idea is supported by reports that saturable LCFA uptake is inhibited by anion transport inhibitors such as DIDS (reviewed in Ref. [1]), and by recognition that uncoupling protein facilitates H^+ import into the mitochondrial matrix by transporting LCFA anions across the inner mitochondrial membrane from matrix to intra-membranous space, where they associate with a proton. The resulting protonated acids diffuse down a concentration gradient back to the matrix [44,64,120].

2.2.2. *The non-saturable uptake pathway*

In both hepatocytes and adipocytes, when the LCFA:albumin molar ratio is within the physiologic range ($\nu = 0.5$–3.0), >90% of cellular LCFA uptake occurs via the saturable pathway. That LCFA sequestered under these conditions enter physiologic intracellular compartments in both cell types is indicated by the fact that $\geq$90% of the radiolabeled LCFA taken up are identified in esterified compounds, principally triglycerides, within 2–5 min [133,142,153]. As ν increases progressively beyond 5:1, the proportion of saturable uptake diminishes, eventually becoming vanishingly small. However, most studies of LCFA flip-flop (e.g. Refs. [32,52,67–69]) were conducted at very high, non-physiological values of ν. This suggests (i) that the different uptake processes seen in various studies might simply reflect which process predominates under particular conditions and (ii) that the non-saturable uptake in our hepatocyte and adipocyte studies is the same process described as flip-flop in liposomes and certain cells.

To test this hypothesis we compared the values for k_{ns}, calculated as above, with the rate constant for OA flip-flop (k_{ff}) into rat adipocytes from albumin-free incubation mixtures, which we derived from the data of Civelek et al. [32]. In the best-described study in that paper, 130 nmol of OA were added to 1.3 mL of albumin-free incubation mixture containing isolated rat adipocytes (putative [OA_u] = 100 μM) at 30 °C, and subsequent deviations from the basal pH_i of 7.0, detected with an internalized indicator dye, were recorded. As values for k_{ff} were not reported, the published pH_i data were used to generate curves of the actual intracellular [H^+] and [OH^-]. Changes in these concentrations over the initial 240 s were fitted to hypothetical statistical models. The observed changes were well described by a sum of two exponential components with $t_{1/2}$'s of 6 and ~100 s. Even the rapid component ($k_1 = 0.114\ s^{-1}$, $t_{1/2} = 6.1$ s) is far slower than the computed $t_{1/2}$ for saturable transfer of

LCFA across hepatocyte ($k_s = 0.7\ s^{-1}$) [154] and adipocyte ($k_s = 2.9\ s^{-1}$, *vide supra*) plasma membranes. Data describing the time course of the effects on pH_i of inhibiting Na^+/H^+ exchange [32] suggest that the rapid component reflects the initial entry rate of protonated LCFA (k_{ff}), while the slower exponential reflects modification of the rate of LCFA-induced changes in pH_i due to activation of Na^+/H^+ exchange [32]. The value for the rapid exponential rate constant (k_1), measured at 30 °C, is equivalent to $0.16 \pm 0.02\ s^{-1}$ at 37 °C, and is thus very similar to k_{ns} in adipocytes (see above). That k_1 and k_{ns} are both measures of LCFA flip-flop is suggested by their correlation with flip-flop rates in synthetic vesicles of various sizes and in red cell ghosts ($r = 0.96$) (Fig. 3).

These studies clarify many important aspects of LCFA disposition. Not only are there two distinct pathways for cellular LCFA uptake, but also they move different molecular species. The saturable process predominates at LCFA concentrations found in the bulk plasma of mammals between meals. However, under particular circumstances, such as during the hydrolysis of lipoprotein triglycerides by lipoprotein lipase on the luminal surface of vascular endothelium, or during hormone-sensitive lipase-mediated lipolysis of intracellular triglycerides in adipocytes [153], local concentrations of LCFA may become high enough so that non-saturable transmembrane fluxes transiently exceed those of the saturable process. In effect, the non-saturable pathway allows cells to deal virtually

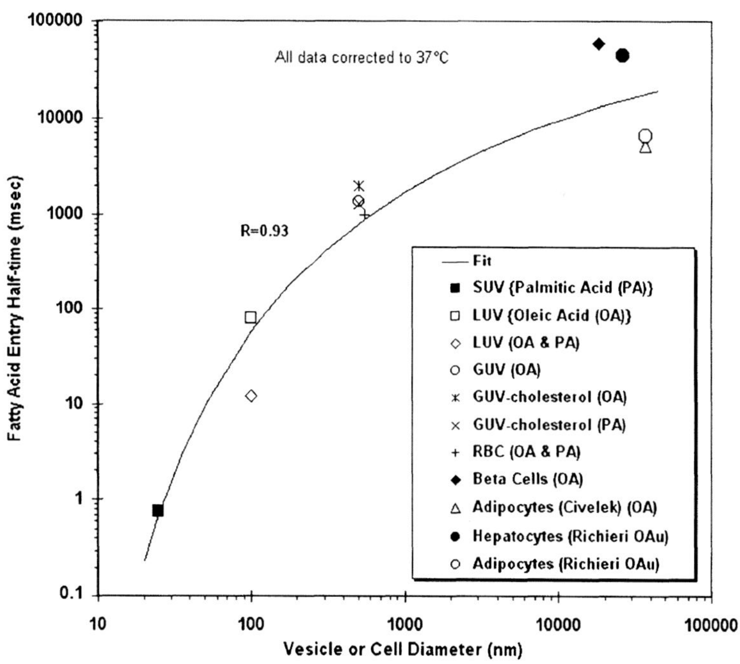

Fig. 3. Relationship of vesicle diameter to fatty acid flip-flop rate. Half-time of fatty acids entry into synthetic vesicles of various sizes, red cell ghosts, and cells was plotted as a function of diameter. Data were compiled from the literature normalized to the rate at 37 °C. Rapid flip-flop is seen in the smallest vesicles and decreases with increasing diameter, possibly reflecting the change in the radius of curvature of the membrane.

instantaneously with high [$LCFA_u$] without the need either to synthesize new transporter molecules or to redistribute pre-existing transporter to the plasma membrane. At the very high [$LCFA_u$] that occur in albumin-free media, non-saturable uptake may dwarf the saturable component to a degree that makes the latter virtually undetectable. Thus, like the proverbial blind men and the elephant, what is seen to be the dominant LCFA uptake process depends largely on which end of the spectrum of [$LCFA_u$] was studied.

Based on these observations, our working model of cellular LCFA uptake is one in which LCFA uptake occurs by two distinct processes: a rapid, facilitated process that transports LCFA anions from saturable sites on the plasma membrane and a slower process representing passive flip-flop of non-saturably bound, protonated LCFA across the lipid bilayer. While conceptually simple, these findings are of fundamental importance to the field of LCFA transport.

2.3. *LCFA transporters*

Support for facilitated LCFA transport also comes from identification of several putative plasma membrane LCFA transporters. The first, a 43 kDa protein designated $FABP_{pm}$, was identified by our laboratory in 1985 in rat hepatocyte plasma membranes [150]. It unexpectedly proved identical to the mitochondrial form of aspartate aminotransferase (mAspAT) [11,13,150]. Despite early skepticism (e.g. Ref. [143]), identity of the two proteins is well established [13,155,183], as is a role for mAspAT in LCFA uptake. The role of $FABP_{pm}$/mAspAT in cellular uptake of LCFA will be reviewed in detail below.

Although $FABP_{pm}$ was the first protein to be identified as a putative LCFA transporter, other putative transporters have subsequently been identified, and will be discussed in detail in other chapters. In addition to $FABP_{pm}$/mAspAT, fatty acid translocase (FAT) [2, 55], the fatty acid transport protein (FATP) family [57,115,137], and caveolin-1 [161,162] are additional putative LCFA transporters whose role in transport is supported by considerable evidence. FAT, an 88-kDa protein identified in adipocyte plasma membranes [55], is identical to the platelet membrane protein CD36. FAT/CD36 reversibly binds LCFA [6] and expression of its cDNA in non-expressing *ob*1771 fibroblasts leads to cell surface expression of the protein and appearance of high affinity, saturable LCFA uptake [77]. It is expressed in adipocytes, intestinal epithelium, and cardiac and skeletal muscle [1]. Its expression is selectively up-regulated in cardiac muscle and adipose tissue in murine models of obesity and/or diabetes [14,48,119], and in the jejunum, where most LCFA absorption occurs, after an LCFA load [100]. It is deficient in hereditary hypertrophic cardiomyopathy [157]. Thus, evidence that FAT/CD36 plays a role in LCFA uptake is strong.

The first member of the FATP family, now designated FATP1, was identified in a 3T3-L1 adipocyte cDNA library by a novel cloning strategy that also isolated a cDNA for fatty acid CoA-synthase (FACS) [115]. A gene family was later identified that encodes additional mouse and human FATPs, with distinct patterns of cellular distribution, and homologs in lower organisms [57]. All FATPs tested increased LCFA uptake after transfection into COS cells or *Escherichia coli* [57]. Mouse and human FATP4 is reportedly the major transporter mediating LCFA uptake on the apical surface of intestinal enterocytes [137]. mAspAT also participates in LCFA uptake at the apical PM of jejunal and ileal enterocytes [103,141,145], and it has been suggested that these proteins and FAT/CD36 may work together to translocate LCFA from intestinal lumen to the circulation. Although FATPs participate in LCFA uptake, it is unclear if they are *transporters* that facilitate influx or *enzymes* that increase cellular LCFA accumulation, inhibiting efflux via esterification [33,40,152,172]. The assays used in most FATP studies may not distinguish between these processes, as shown by isolation of both FATP and FACS clones during the initial screening process [115]. There are no homologies among mAspAT, FAT, and the FATPs. However, it would not be surprising if several transporters, with different tissue distributions and K_m's, and operating under different regulatory control, were involved in LCFA disposition, as occurs with glucose uptake (e.g. Ref. [66]). The hypothesis that $mAspAT_{pm}$, FAT, and FATP cooperate in LCFA uptake into skeletal and cardiac myocytes is intriguing [78], but transfection with cDNAs for each of these proteins separately result in increased LCFA uptake. Furthermore, expression of these transporters under various stimuli often does not change in a coordinated fashion [14,16,85]. Factors regulating their expression include *inter alia* LCFA themselves (e.g. Ref. [5,85]); endotoxin, TNFα, and IL-1 [85]; insulin [61,79]; PPARα and PPARγ agonists [81,88]; and ethanol [181].

A possible role for caveolin-1 in fatty acid transport has also been identified. Photaffinity labeling of 3T3-L1 plasma membrane fractions with a fatty acid analog demonstrated that a 22 kDa protein was labeled efficiently [161]. Further study showed protein to be caveolin-1, a major protein component of caveolae. The role of caveolin in hepatocellular LCFA uptake was suggested by studies in HepG2 cells [99]. Inhibition of caveolae formation, but not clathrin-coated pit formation, inhibited LCFA uptake. Further, the fluorescent fatty acid NBD-stearate co-localized with α-caveolin-1/cyanofluorescent protein in transfected cells. These studies suggested a new pathway for cellular LCFA uptake in which caveolin works in concert with other transporters [146].

3. Identification and characterization of $FABP_{pm}$

Research on LCFA uptake in our laboratory was influenced by the albumin receptor hypothesis [94,173,175,176]. In the course of investigating whether the binding of [^{125}-I] albumin to rat liver plasma membranes suggested the presence of specific binding sites that might represent the putative albumin receptor [147], an affinity chromatography study was performed. When solubilized plasma membrane proteins were passed through an albumin–agarose affinity column, no proteins showed specific binding to the column. In contrast, both BSP–agarose and oleate–agarose columns retained specific proteins from the plasma membrane extracts. These results suggested that hepatocyte plasma membranes contained high affinity binding proteins for BSP, an analog for bilirubin, and oleate, a fatty acid, but not for albumin. The protein eluted from the oleate–agarose column, eventually designated $FABP_{pm}$, had a MW of approximately 40 kDa.

The possible existence of a protein specific for LCFA binding was further examined by standard biochemical methods. Studies in which plasma membrane preparations were incubated with [^{14}C]-oleate under various conditions [144] demonstrated that the binding of oleate was saturable and inhibited by excess cold ligand. Studies in the presence of albumin demonstrated that oleate binding was proportional to the unbound fraction. Binding was inhibited by prior heat denaturation or trypsin treatment, confirmed that binding was due to protein(s). Using the oleate–agarose chromatography method, the 40 kDa protein was isolated [149,150]. Gel filtration co-chromatography studies demonstrated that binding to the protein was specific for LCFA with a free, ionizable carboxyl group; labeled oleate migrated with the protein, but other ligands such as bilirubin, BSP, taurocholate, phosphatidylcholine, and cholestyrloleate did not. Once sufficient purified protein was available, a rabbit polyclonal antibody was raised. It proved to be monospecific for $FABP_{pm}$, showing no reactivity with cytosolic proteins or other targets such as albumin. Immunofluorescence and histochemistry revealed that the protein was present on plasma membranes in rat and mouse liver sections (Fig. 4). The antibody also inhibited binding of oleate to isolated plasma membranes [145]. The data indicated that we had isolated a protein from rat liver plasma membranes responsible for fatty acid binding, and possibly involved in LCFA uptake into hepatocytes.

Similar experiments were conducted on jejunal microvillus membranes [145]. Oleate binding was shown to be saturable, inhibited by heat or trypsin treatment. A ~40 kDa protein, isolated with oleate–agarose chromatography, showed immunological identity with liver $FABP_{pm}$ and binding to various labeled LCFAs, but not to other ligands. Binding of oleate to native, but not heat-denatured, membrane preparations was inhibited by antibody to liver $FABP_{pm}$, further indicating identity of the proteins and implying a similar function. The protein was located by immunofluorescence on the plasma membrane of villus cells and crypt cells, but not in other areas of the gastrointestinal mucosa, where LCFA absorption was not likely to occur. Analogous studies identified a protein of similar molecular weight in plasma membranes of tissue adipocytes and cardiac myocytes [102,118,130], differentiated 3T3-L1 cells [183,184], and vascular endothelial cells [12] that was, in each case, immunologically identical to that isolated from liver.

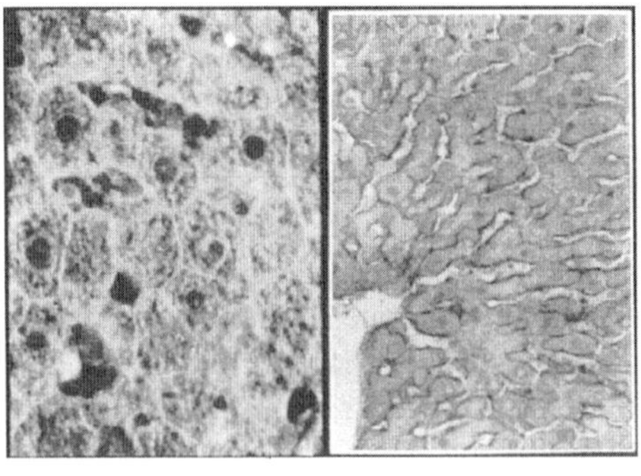

Fig. 4. Immunofluorescent and immunohistochemical identification of $FABP_{pm}$ on liver plasma membranes. Sections of rat liver were treated with anti-$FABP_{pm}$ followed by second antibody conjugated to FITC (left) or horseradish peroxidase (right). Intense membrane staining is evident in both cases. (*For a colored version of this figure, see plate section, page 582.*)

3.1. $FABP_{pm}$ as an LCFA transporter

To explore the potential role of $FABP_{pm}$ in hepatocellular LCFA uptake, a rapid filtration uptake assay initially employed by Abumrad et al. in adipocytes [3,4] was modified and extensively validated [142]. Uptake was found to be linear with time over the first 30 s, consistent with zero-trans conditions, and suggesting that efflux and metabolism had little effect on net LCFA sequestration during this period. The subsequent gradual flattening of the cumulative uptake curve with time was shown to result from efflux and metabolism of the tracer, not saturation of the uptake mechanism. This was established with a dual-label experiment in which cells were first incubated with [^{3}H]-oleate throughout and beyond the linear portion of the uptake curve, after which they were further incubated with [^{14}C]-oleate. The slope of the initial, linear portion of the [^{14}C]-oleate uptake curve was identical to that of the [^{3}H]-oleate curve. Uptake of both tracers showed identical kinetics, with the same calculated V_{max}, indicating that the uptake system was not saturating with time. Although the addition of glucose to the incubation medium enhanced the rate of oleate esterification, it did not increase initial uptake velocity. These and other observations established that the initial, linear portion of the uptake curves reflected membrane transport, largely independent of intracellular binding or metabolism, whereas the latter portions of the curve reflected the net effects of influx, efflux, and cellular metabolism [142]. Subsequent studies [148] showed that hepatocellular oleate uptake was dependent on the Na^+ concentration in the medium, and sensitive to ouabain and various metabolic inhibitors. It was not inhibited by BSP, bile acids, or their analogs, but was competitively inhibited by other LCFA [131], and was also selectively inhibited by a monospecific, polyclonal antibody to $FABP_{pm}$ that had no effect on the uptake of BSP, taurocholate, or cholate [148]. The data thus suggested that the saturable component of LCFA uptake reflected an LCFA-specific transport mechanism that was mediated by $FABP_{pm}$. Although LCFA uptake rates varied in different cell types, those in adipocytes and cardiac myocytes being significantly faster than that in hepatocytes, LCFA uptake processes in adipocytes [118], jejunal enterocytes [141], and cardiac myocytes [130,140] exhibited kinetics similar to those in hepatocytes, were inhibited by phloretin and/or proteases, and were selectively inhibited by anti-$FABP_{pm}$.

$FABP_{pm}$ proved to have an unusual pI of 9.1, with a characteristic pattern of charge isoforms. It was possible to fractionate membrane proteins from adipocytes and cardiac myocytes by preparative isoelectric focusing (IEF), and to apply the fraction with a p$I > 9$ onto oleate–agarose for purification of $FABP_{pm}$ from both tissues [102]. Co-chromatography and immunologic tests showed that the proteins thus isolated bound oleate with high affinity and were identical to $FABP_{pm}$ from liver, arguing for a single protein being present on the plasma membrane in all tissues with high transmembrane fatty acid fluxes and having a function in LCFA uptake.

3.2. Structural and immunological comparison of $FABP_{pm}$ and mAspAT

To further characterize $FABP_{pm}$, highly purified samples were analyzed by protein microsequencing [10,13]. The results were initially disturbing, as the amino-terminal sequence of 22 identifiable amino acids out of the first 24 showed 100% concordance with the well-studied protein variously named mitochondrial glutamate-oxaloacetate transaminase (mGOT) or mAspAT. This observation was met with considerable skepticism [143], not the least in our own laboratory. However, in addition to the N-terminal sequence data, amino acid analyses of purified $FABP_{pm}$ were concordant with published analyses of mAspAT [10,13], and enzymatic assays, performed on the various plasma membrane protein fractions produced during the purification of $FABP_{pm}$, documented that mAspAT activity co-purified with $FABP_{pm}$. mAspAT enzymatic activity was enriched in plasma membrane fractions in parallel with the classical plasma membrane enzymes Mg^{++}- and Na^{+}, K^{+}-ATPases, whereas mitochondrial markers other than mAspAT, such as succinate dehydrogenase, were depleted. Samples of $FABP_{pm}$ isolated from highly purified rat liver plasma membrane preparations and mGOT (mAspAT) from equally purified rat liver mitochondria were examined for similarity. Additional samples of mAspAT were prepared using a novel rapid HPLC purification scheme developed in our laboratory [156]. $FABP_{pm}$ and mAspAT had the same apparent molecular weight by SDS-PAGE, the same basic pI and pattern of charge isoforms, identical behavior on four different HPLC separations, and similar absorption spectra which showed a peak at 435 nm at an acid pH, shifting to 345 nm in basic conditions [10,13]. Both proteins had a similar mAspAT enzymatic specific activity of 140–180 U mg^{-1} protein, and mGOT was shown to bind oleate and to be purifiable from protein mixtures, using oleate–agarose chromatography [10]. Antibodies to $FABP_{pm}$ and mAspAT reacted against the other protein on Western blots, and both antibodies inhibited cellular LCFA uptake. Both antibodies identified proteins in hepatocyte and adipocyte plasma membranes by immunofluorescence. Immunologic activity of either antiserum in each of these situations was abolished by pre-incubation with the other protein [10,13,155]. All of the evidence acquired with the purified proteins and their respective antisera indicated that the two proteins are identical.

3.3. Expression of mAspAT and LCFA uptake

One advantage of the apparent identity of mAspAT and $FABP_{pm}$ was that, as an extremely well-studied enzyme, genomic and cDNA clones of mAspAT had already been produced [58,82,166]. A rat mAspAT cDNA clone was used to produce an expression vector, linking it to a mouse metallothionein-I (MT) promoter. This construct (pMAAT2) was co-transfected into 3T3 fibroblasts with a plasmid bearing an antibiotic selectable marker [62] and stably-transfected cell lines were selected. Cloned cell lines were tested for the presence of the MT-AAT construct by Southern blotting. Cell lines with both the pMAAT2 plasmid and the resistance plasmid, or the resistance plasmid alone, were assayed for LCFA uptake following incubation in the presence and absence of Zn^{++}. In addition, plasma membranes were prepared from the cell lines before and after treatment with Zn^{++}, and Western blotting was used to estimate the amount of $FABP_{pm}$/mAspAT in the membranes. As the MT promoter is responsive to heavy metals, higher expression of the gene was expected in the presence of Zn^{++}. Cells bearing only the resistance plasmid showed very low LCFA uptake, similar to untransfected 3T3 cells, as well as little or no detectable $FABP_{pm}$/mAspAT protein in the membranes. Cell lines bearing the pMAAT2 construct showed de novo $FABP_{pm}$/mAspAT expression in the plasma membranes and a 10-fold increase in the LCFA uptake V_{max} [62]. The increase in both plasma membrane protein expression and LCFA uptake V_{max} was significantly greater after treatment with Zn^{++}.

A similar increase in LCFA uptake followed injection of a capped mAspAT mRNA into *Xenopus* oocytes [182]. LCFA uptake by the oocytes was substantially increased over baseline 2–4 days after injection of the mAspAT mRNA.

3.4. Liposome reconstitution

In preliminary studies, purified $FABP_{pm}$ or mAspAT was incorporated into the lipid bilayer of lecithin:cholesterol liposomes by sonication [101]. LCFA uptake was markedly enhanced in liposomes carrying one of the purified proteins vs. protein-free liposomes. Thus, insertion of $FABP_{pm}$ or mAspAT into liposomes reconstitutes LCFA uptake [101].

3.5. Regulation of $FABP_{pm}$/mAspAT expression and LCFA uptake into cells

3.5.1. Studies in differentiating 3T3-L1 cells

Model systems also demonstrate the relationship between mAspAT expression and LCFA uptake. Differentiation of 3T3-L1 cells to an adipocyte phenotype involves increased expression of mAspAT mRNA and cell surface mAspAT protein (Fig. 5) in parallel with an increased LCFA uptake V_{max}. Anti-$FABP_{pm}$ or -mAspAT antibodies inhibit LCFA uptake in differentiated 3T3-L1 adipocytes, but not in fibroblasts, which express little mAspAT on their plasma membranes [184].

3.5.2. Effects of ethanol (EtOH) in HepG2 cells

In HepG2 cells cultured in 0, 20, 40, or 80 mM EtOH [181], mAspAT mRNA increased progressively with increasing EtOH ($r = 0.98$), to 920% of control at 80 mM (Fig. 6). There was no corresponding increase in total cellular mAspAT enzymatic

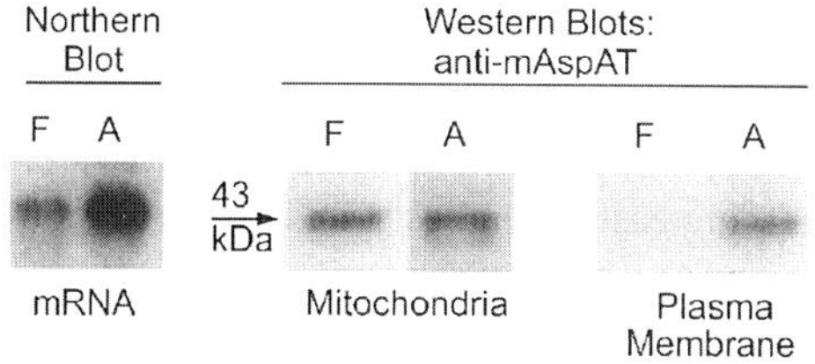

Fig. 5. Expression of mAspAT mRNA and protein during 3T3-L1 preadipocyte differentiation. Cells were harvested at either the proliferating fibroblastic stage (F) or after 8 days of differentiation into adipocytes (A). The Northern blot demonstrates an increase in mAspAT mRNA during differentiation. After isolation of mitochondria and plasma membrane fractions, homogenates were analyzed by Western blotting. No significant change is seen in mitochondria, while plasma membrane protein, which is almost undetectable at first, increases substantially.

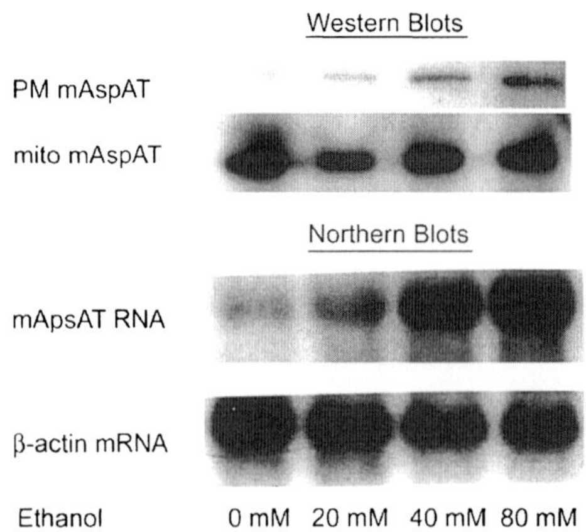

Fig. 6. Northern and Western blot analysis of HepG2 cells cultured in EtOH. Plasma membrane and mitochondrial fractions were isolated from HepG2 cells cultured in 0, 20, 40, or 80 mM EtOH. Western blot analysis showed that plasma membrane mAspAT protein increased with EtOH concentration, while mitochondrial protein did not. Northern blot analysis showed an increase in mAspAT mRNA with increasing EtOH concentration compared to β-actin.

activity, or in the reference mitochondrial matrix enzyme citrate synthase (CS), the cytoplasmic enzymes ALT, cAspAT, and LDH, or the plasma membrane enzyme GGT. Recoveries in the medium of CS, cAspAT, ALT, LDH, and GGT were all $\leq 17\%$ of cellular enzyme content/24 h at 0 mM EtOH, and were not altered by EtOH. mAspAT recovery in the medium was similar at 0 mM ($16 \pm 7\%$/24 h), but showed a striking, concentration dependent increase, to 808% of cellular content/24 h, at 80 mM. OA uptake V_{max} ($r = 0.96$), mAspAT recovered in the medium ($r = 0.90$) and *mAspAT in the plasma membrane* ($r = 0.96$), determined by Western blotting (Fig. 6), were all highly correlated with mAspAT mRNA. mAspAT levels in the *mitochondria*, in contrast, were unchanged. A similar increase in hepatocellular mAspAT mRNA levels and LCFA uptake V_{max} was found in EtOH-fed Wistar rats and C56BL/6J mice (Berk, P.D. et al., unpublished). In alcoholic liver disease (ALD), a specific increase in plasma mAspAT leads to the typical increase in the ratio of aspartate to alanine transaminases [90], and has been widely attributed to mitochondrial injury. In our HepG2 cell studies, we saw no mitochondrial morphologic abnormalities by electron microscopy at any EtOH concentration examined [181]. Nevertheless, computer modeling indicated that increased synthesis and export of mAspAT from the liver in vivo would be sufficient to explain the increase seen clinically in ALD. This study has two important clinical implications: (1) The increased plasma mAspAT in alcoholics may reflect pharmacologic up-regulation of mAspAT mRNA and of mAspAT synthesis by EtOH and (2) the resultant increased mAspAT-mediated fatty acid uptake may contribute to alcoholic fatty liver. The study made two other important findings: that mAspAT is exported from HepG2 cells in large amounts, which can be regulated by EtOH; and that regulation of mAspAT mRNA has different effects on plasma membrane and mitochondrial mAspAT levels.

4. Physiologic regulation of $FABP_{pm}/mAspAT_{pm}$ and LCFA uptake

Evidence that saturable LCFA uptake is physiologically regulated would support the concept that it reflects facilitated, protein-mediated transport. If regulation could be shown to occur in models of human diseases, the findings could also have clinical implications. As just discussed, we have shown that saturable LCFA uptake is greatly enhanced during adipocyte differentiation of mouse 3T3-L1 pre-adipocytes [184]. This led us to examine LCFA uptake in tissues derived from animals with models of obesity and type II diabetes. These studies have important implications for both basic physiology and human disease.

4.1. LCFA uptake in rodent obesity models

4.1.1. Single gene obesity models

Studies in adult male Wistar (+/+), Zucker lean (*fa*/+) and fatty (*fa*/*fa*) [70,165], and Zucker diabetic fatty (*ZDF*) rats [97], found a striking, tissue-specific increase in [^{3}H]OA uptake in *fa*/*fa* and *ZDF adipocytes*, in which V_{max} was increased ninefold ($p < 0.005$) and 13-fold ($p < 0.001$), respectively, and was highly correlated with mAspAT mRNA levels ($p < 0.01$) [16]. The increase denoted up-regulation of a membrane transport process: it greatly exceeded the 2.1-fold increase in the surface area of adipocytes from obese animals [38], and did not result from trans-stimulation due to increased lipolysis. A much smaller increase was seen in heterozygote (*fa*/+) adipocytes. Small increases in V_{max} of <2-fold were also observed in *cardiac myocytes* from *fa*/*fa* and *ZDF* animals. There were no changes in V_{max} in *hepatocytes* (Fig. 7) [12,15,16]. [^{3}H]-OA uptake was also studied in

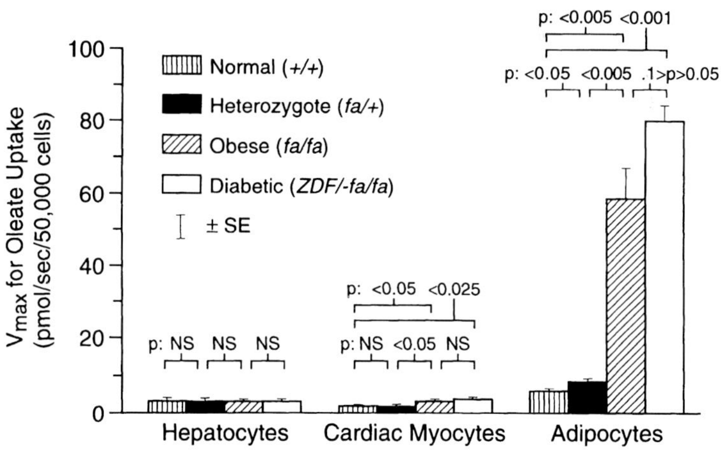

Fig. 7. Comparison of hepatocyte, cardiac myocyte, and adipocyte LCFA uptake rates in Zucker *fatty* rats and controls. Oleate uptake studies were performed using all three cell types from control Wistar (+/+), lean Zucker heterozygote (fa/+), Zucker *fatty* obese (*fa*/*fa*), and ZDF (*fa*/*fa*) rats. Hepatocytes are similar in all genotypes, and a small but significant change is seen in cardiac myocytes of *fa*/*fa* rats. However, the V_{max} for uptake is much larger in adipocyte from *fa*/*fa* rats compared to controls. Similar changes in the V_{max} for adipocyte LCFA uptake is seen in other genetic obesity models.

adipocytes from 20 to 24-day-old male +/+, *fa*/+, and *fa*/*fa* weanlings. The *fa*/*fa* animals were not yet overtly obese, and had plasma fatty acid and glucose levels equivalent to those of the control groups. While adipocyte TNFα mRNA levels, a measure of fat accumulation [59], increased in adult animals in the predicted order +/+ < *fa*/*fa* < *ZDF*, levels in weanling +/+ and *fa*/*fa* adipocytes were similar. Nevertheless, V_{max} was increased 2.9-fold ($p < 0.005$) in *fa*/*fa* compared +/+ cells. These studies show (1) that regulation of LCFA uptake is tissue specific and (2) that up-regulation of adipocyte LCFA uptake is an early event in evolution of the obese phenotype in Zucker *fa*/*fa* rats.

Adipocyte-specific up-regulation of LCFA uptake in Zucker fatty rats alters nutrient partitioning, diverting fatty acids from sites of oxidation toward storage in adipose tissue. To determine if this is a general feature of obesity, we studied [^{3}H]-OA uptake by adipocytes and hepatocytes from (1) homozygous male obese (*ob*), diabetic (*db*), fat (*fat*), and tubby (*tub*) mice [29,30,76,89,91,180] and (2) male Sprague–Dawley rats fed for 7 weeks a diet containing 55% of calories from fat. V_{max} and K_m were compared with controls of appropriate background strain (C57Bl/6J or C57BL/KS) or diet (13% fat calories). Adipocyte LCFA uptake V_{max} was increased 5–6 fold in *ob*, *db*, *fat*, and *tub* mice vs. controls ($p < 0.01$), whereas no differences were seen in the corresponding hepatocytes. Similar changes in V_{max} occurred in fat-fed Sprague–Dawley rats. In mice, the increases in V_{max} were closely correlated with mAspAT mRNA levels in *ob* and *db* animals [16]. Although *fat* and *tub* mice exhibited increases in saturable LCFA uptake equivalent to those in *ob*, *db*, and Zucker animals, mAspAT was not as consistently or extensively up-regulated in these strains. Thus, the strong correlation of mAspAT mRNA levels and saturable LCFA uptake was seen in all rodent strains with defective leptin signaling due to mutations in either leptin itself (*ob* mouse) or the leptin receptor, OB-R (*db* mouse, Zucker rat), but not in at least some strains with normal leptin signaling capability (*fat*, *tub*). Although the expression in adipocytes of the known LCFA transporter genes was somewhat variable in the different obesity models, adipocyte LCFA uptake was uniformly up-regulated in all of the rodent obesity models studied.

4.1.2. Dietary obesity models

Osborne-Mendel (OM) rats are very sensitive and S5B/Pl (S) are resistant to becoming obese when fed with a high fat diet [116]. To explore the basis for this finding, adult male S or OM rats were randomized to a diet of 35% lard (55% of calories from fat) or lab chow (13% fat calories) for up to 11 weeks. Baseline [^{3}H]-OA uptake studies in chow-fed OM and S adipocytes were similar (V_{max}: 6.5 ± 0.4 vs. 6.1 ± 0.5 pmol/s/50,000 cells; K_m: 21 ± 2 vs. 22 ± 1 nM). In the OM rats, V_{max} increased by fourfold after only 2 days of high fat diet, and then gradually decreased over time. The S5B/PL rats briefly exhibited a small but non-significant increase in V_{max} on day 2, which rapidly returned to baseline. At 11 weeks, weight gain was greater in fat-fed vs. chow-fed OM animals ($p < 0.025$) but not in S animals ($0.1 > p > 0.05$). Moreover, the high fat diet produced a small but significant increase over baseline in V_{max} for saturable OA uptake in OM animals (to 8.5 ± 0.6 pmol/s/50,000 cells, $p < 0.025$), but no change in S animals (6.3 ± 0.3 pmol/s/50,000 cells, $p < 0.6$). Hepatocyte OA uptake kinetics were unaltered by the diet in either strain. Thus, differences between OM and S rats in the tendency to gain weight on a high fat diet parallel the regulated responses of their saturable adipocyte LCFA uptake systems.

4.2. Leptin and the regulation of LCFA uptake in obesity

Leptin is an appetite suppressant, and also has an as yet poorly defined role in decreasing the efficiency of energy expenditure. Increased utilization of substrates to meet basal energy requirements results in less being left for storage as fat. The changes in LCFA partitioning resulting from up-regulation of adipocyte LCFA uptake [14,16] produce the opposite effect, favoring storage, rather than consumption, of energy-dense LCFA. To test leptin's role in regulating adipocyte LCFA uptake, we are studying its effects in leptin-deficient *ob/ob* mice [180], in which the leptin receptor and leptin receptor signaling appear to be intact, and leptin treatment results in weight loss [50]. Hyperinsulinemic (50 ± 6 IU mL^{-1}) adult (55–65 g) male *ob/ob* mice are infused with leptin (500 ng h^{-1} s.c.) for 21 days (Group 1). Group 2 receives no leptin, but is pair fed with Group 1. Group 3 mice are saline-treated *ob/ob* controls. At the leptin dose employed, insulin fell by $>50\%$ within 8 h and declined to normal thereafter, and the LCFA uptake V_{max} was significantly reduced by 24 h of treatment. Reductions in food intake or body weight did not occur until days 2–3, respectively. Changes in both V_{max} and body weight occurred more slowly in pair-fed than in leptin-treated animals. These data are consistent with our hypothesis [14,16] that insulin may up-regulate and leptin down-regulate adipocyte LCFA uptake. Our findings that selective up-regulation of adipocyte LCFA uptake precedes weight gain in weanling Zucker rat pups, and down-regulation precedes weight loss during leptin treatment of *ob/ob* mice suggest that regulation of adipocyte LCFA uptake may be an important aspect of the control of body adiposity.

4.3. Human obesity

LCFA uptake by omental adipocytes obtained during laparoscopic surgery from morbidly obese patients undergoing bariatric operations is markedly up-regulated compared to the results in adipocytes from non-obese patients undergoing a variety of clinically indicated procedures. As in the various rodent models studied, adipocytes from obese individuals are appreciably larger that those from non-obese subjects. However, V_{max} remains significantly up-regulated in obese subjects even when expressed per unit of adipocyte surface area. The increase therefore reflects up-regulation of a membrane transport process, and is not simple a reflection of larger cell size. The results of these studies confirm the relevance of data obtained in animal models to humans.

5. Is mAspAT plausible as an LCFA transporter?

Although we have presented considerable evidence that mAspAT functions at the plasma membrane as an LCFA transporter, there remains considerable skepticism about this new and novel function for a long known and well studied protein. Much of the skepticism revolves around four specific questions. In the sections that follow, we present those questions and summarize the known, relevant information on which our responses are based.

1. Does mAspAT really reside on the plasma membrane, or is this finding an artifact?

As a prototypical mitochondrial enzyme, reports that mAspAT existed in other sub-cellular compartments were instantly controversial. The first direct observations of mAspAT on plasma membranes were from immunofluorescent studies on liver [13], jejunum [150], and adipocytes [118]. Immunohistochemistry was also used to demonstrate that mAspAT was present on the surface of differentiated 3T3-L1 adipocytes, but not on fibroblast-like preadipocytes [184]. These reports are supported by higher resolution studies using immuno-electron microscopy (IEM).

5.1. Studies in rat liver and HepG2 cells

In collaboration with Dr Ronald Gordon, we studied the distribution of mAspAT in rat liver sections and in HepG2 cells cultured in both the presence and absence of EtOH by IEM, using monospecific rabbit anti-mAspAT as primary antibody and colloidal-gold conjugated goat anti-rabbit IgG for detection [181]. By IEM, gold particles localized mAspAT within mitochondria, as expected, and in linear arrays contiguous with the outer surface of the outer mitochondrial membrane. In addition [181], small amounts of mAspAT were seen within the ER; in the Golgi stack and vesicles; in cytoplasmic vesicles of various sizes; and on the plasma membrane. Of particular interest was the identification of mAspAT antigen (1) in and around coated pits on the plasma membrane and (2) in adjacent coated vesicles and in small vesicular structures just beneath the plasma membrane. These observations, coupled with evidence of extensive mAspAT export, raise the possibility of an exocytic/endocytic cycle for mAspAT.

5.2. Studies in other tissues

Prof. RS Gupta's studies of the sub-cellular distribution of mitochondrial proteins establish that many are consistently found in particular extra-mitochondrial sites, performing functions different from those conducted in mitochondria [26,28,63,122, 124–127]. We have collaborated with Prof. Gupta to determine by IEM if the presence of mAspAT on plasma membranes was a general phenomenon or was restricted to particular cells. IEM surveys were conducted in other rat tissues, using the same antibodies as in our studies in liver and HepG2 cells [181]. Intense mitochondrial labeling was observed in all tissues examined. No extra-mitochondrial mAspAT was detected in pancreas, spleen, pituitary, or sub-mandibular gland, although other mitochondrial proteins had been found in extra-mitochondrial sites in these tissues. There was substantial plasma membrane labeling at several sites within the kidney, including the thick distal tubule and the cortical collecting duct. Labeling was particularly intense along the boundaries of apposed epithelial cells (Fig. 8). Surface labeling was also seen in plasma membranes of arteriolar endothelial cells from several tissues [27]. It is noteworthy that a facilitated LCFA transport mechanism has been reported in rat kidney plasma membrane vesicles [163]. In addition, the need for a mechanism to transport LCFA from the lumenal surface of the

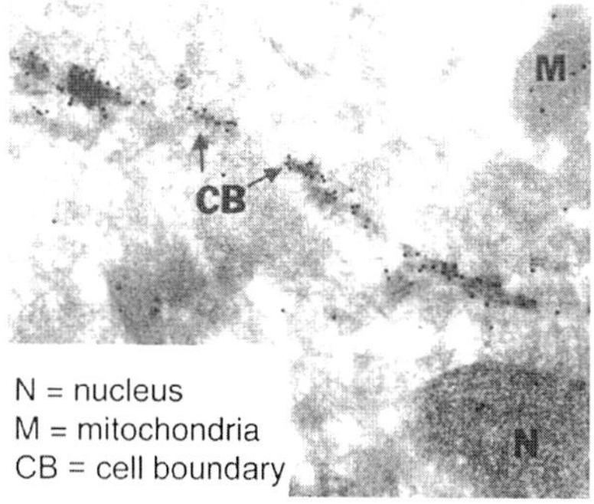

Fig. 8. IEM of mAspAT. Kidney sections treated with anti-mAspAT and second antibody conjugated to colloidal gold demonstrates that mAspAT is found outside of mitochondria. Colloidal gold particles are concentrated at apposed epithelial cell boundaries.

vascular endothelium, where they are released from lipoproteins by the action of LPL, into the extravascular space has been recognized as essential to their ready access to the parenchymal cells of most organs [12,47]. We [12] and others [170] have previously identified mAspAT in endothelial cell plasma membranes by Western blotting.

2. *Do two sub-cellular locations mean two genes or two messages?*

mAspAT destined for mitochondria is translated on free ribosomes as pre-mAspAT with an N-terminal leader that is absent from the mature protein in both mitochondria and plasma membranes. Since this leader is encoded by exon 1 of the mAspAT gene, and the splice site interrupts the first codon of the mature protein [65,166], we initially hypothesized that alternative splicing of a different first exon might account for sorting of some of the protein to the cell surface. The alternative mRNA, if it existed, should be especially prominent in 3T3-L1 adipocytes, in which mAspAT on the plasma membrane is markedly increased during differentiation to an adipocyte phenotype. The obvious approach to detecting such alternative splicing is some form of RNase protection assay. Use of this approach first required cloning of the enzyme's uncommon *a* allele, for which 3T3-L1 cells are homozygous. It differs from the *b* allele at only two base pairs and one amino acid. Using probes derived from the *a* allele, RNase protection analyses indicated that only a single message for mAspAT was present in 3T3-L1 fibroblasts and adipocytes, despite differences in sub-cellular protein distribution [22].

3. *How does mAspAT reach the plasma membrane?*

In considering how pre-mAspAT (pmAspAT) might reach the plasma membrane (PM) as mAspAT, three pathways seemed possible: (1) direct import of pmAspAT into the secretory pathway, (2) direct transport across the PM of proteins maintained in a translocation-competent unfolded state bound to specific chaperonins [121], and (3) mitochondrial import of pmAspAT, excision of the pre-sequence, and subsequent export of mAspAT from mitochondria. The last has been described [45,96,171], but considered part of a degradative pathway. That pathway seemed circuitous, so our hypothesis was that pmAspAT destined for the plasma membrane was imported directly into the secretory pathway. This would require an endopeptidase within that pathway capable of excising the pre-sequence. That the final steps involved vesicular trafficking was strongly suggested by the IEM data described above.

5.3. *Pulse/chase/immunoprecipitation studies*

To test our hypothesis, HuH7 cells were incubated with [^{35}S]cysteine/methionine labeling medium. To permit separate analysis of plasma membrane and intracellular proteins, cell surface proteins were biotinylated prior to cell lysis, and bound to avidin-conjugated beads. mAspAT in *both* cellular fractions and the medium was immunoprecipitated with monospecific rabbit anti-mAspAT coupled to protein-A/G agarose beads. Immunoprecipitates were analyzed by SDS-PAGE/ fluorography.

5.3.1. *Transit of mAspAT to the plasma membrane is very rapid*

In the first experiments, cells and medium were harvested at chase times of 0–240 min after a 10 min pulse. Both mAspAT on the PM and intracellular mAspAT were already highly labeled at 0 chase time, indicating rapid transfer of mAspAT to the PM. Highly labeled mAspAT was detected in the medium by 5 min of chase, just slightly later than albumin, the major export protein of these cells. The demonstration of rapid and substantial export of newly synthesized mAspAT into the medium supported earlier findings in EtOH-treated HepG2 cells [181] that mAspAT is selectively exported from cells in quantities that far exceed those of other intracellular proteins.

5.3.2. *If mitochondrial import of proteins is inhibited, pre-mAspAT is sorted to the plasma membrane*

For these studies, HuH7 cells were incubated in labeling medium for 150 min in the presence and absence of nonactin, a K^+ ionophore that selectively blocks mitochondrial import of proteins [126]. In nonactin treated cells, the protein immunoprecipitated with anti-mAspAT migrated in SDS-PAGE with a MW of 46 kDa, compared with a 43 kDa protein seen in control cells. As with Hsp60 [124], the size difference suggests that the 46 kDa band represents pmAspAT that has not lost its $\sim$3 kDa pre-sequence. pmAspAT is also found on the PM of nonactin treated cells. These results suggest that there is no endopeptidase capable of cleaving the pre-sequence that converts pmAspAT to mAspAT other than that in mitochondria, so that passage through mitochondria must be a part of the route by which mAspAT reaches the PM.

5.3.3. *Brefeldin A blocks the movement of both mAspAT and pre-mAspAT to the plasma membrane*

In the last of the pulse/chase experiments completed thus far, HuH7 cells were incubated with control medium, or with brefeldin A, which collapses the Golgi/ER components of the secretory pathway. In the presence of brefeldin A, labeled mAspAT no longer appeared on the PM or in the medium.

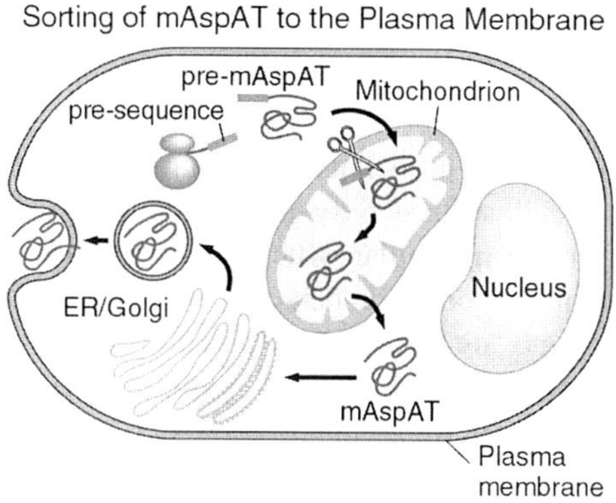

Fig. 9. Diagram of possible route for mAspAT to reach the cell surface. After translation on free polysomes, the protein is imported into mitochondria, where the leader sequence is removed by a specific endopeptidase. Some of the protein is exported, and enters the ER/Golgi secretory pathway. Vesicles bearing mAspAT fuse with the plasma membrane, where it can participate in LCFA uptake.

The conclusions of these studies are summarized in Fig. 9. The data indicate that, after synthesis on free ribosomes, pmAspAT partially enters mitochondria, where its N-terminal pre-sequence is excised. It then exits the mitochondria, reaching the plasma membrane by a brefeldin A-inhibitable vesicular pathway. Details about this vesicular pathway, and in particular, how mAspAT enters it, remain to be determined.

5.4. Tracking the movement of mAspAT with green fluorescent protein (GFP)

A useful method for visualizing protein trafficking is to tag the protein with GFP [86] and examine the cells with fluorescent microscopy. A vector was constructed with the rat mAspAT cDNA cloned in frame with an enhanced GFP cDNA, coupled to a metallothionein promoter. This plasmid, named pMT-AAT/GFP, was transfected into HuH7 cells. Initial examination showed that the main pattern of GFP fluorescence was punctate as expected for mitochondria. Cells were stained with a mitochondrion-specific red fluorescent dye (MitoTracker CMXRos) for confirmation. With superposition of images, virtually all of the punctate intracellular staining became yellow, indicating that most pMT-AAT/GFP co-localized with MitoTracker in mitochondria. Especially bright fluorescence was seen in pMT-AAT/GFP transfected cells in medium containing 100 μM $ZnCl_2$, to activate the MT promoter. In such cells, in addition to intense fluorescence in mitochondria, we also identified non-mitochondrial fluorescence on free PM surfaces and in vesicle-like structures beneath the PM at cell boundaries. The latter appear to correspond to mAspAT collections seen by IEM at similar cell boundary locations in other epithelia (See Fig. 8). By Western blotting, homogenates of pMT-AAT/GFP transfectants show two protein bands, of 43 and 75 kDa, reactive with anti-mAspAT. The 75 kDa band also reacts with anti-GFP, and represents the fusion protein. Using real time fluorescence microscopy with pMT-AAT/GFP and related fluorescent trafficking mutants may permit direct visualization of the intracellular trafficking of mAspAT, establishing more concretely how mAspAT reaches the PM.

4. How does $FABP_{pm}$/$mAspAT_{pm}$ bind fatty acids?

As a relatively hydrophilic mitochondrial matrix enzyme, mAspAT seems an unlikely candidate for a cell surface fatty acid binding protein. Therefore, we have examined the literature on the molecular structures of mAspAT and other proteins that bind LCFA and to attempt to better understand our observations. Other hydrophilic proteins that bind LCFA and are crucial to fatty acid utilization have been known to bind LCFA for much longer and their LCFA binding properties characterized in much greater detail than mAspAT. The most intensively studied of these are serum albumin and the cytosolic fatty acids binding proteins (cFABPs), for which the combined use of X-ray crystallography and binding site mutagenesis has greatly clarified both the important structural features and the electrostatic interactions with specific amino acids that are necessary for LCFA binding. The primary binding of LCFA by human serum albumin occurs at five sites consisting of long hydrophobic pockets capped by basic side chains [19,34,35,98]. The Arg and Lys residues in these chains are implicated in binding by the crystal structures and have been identified previously as crucial to LCFA binding by bovine serum albumin [53,104]. The cFABP family is composed of small (~ 14 kDa) proteins which bind either 2 (the liver cFABP) [158–160] or 1 (all others studied) [7,56,105,113] LCFA in a large central pocket within a structure designated as a β-barrel, formed by orthogonal β-sheets and capped by a helical lid. Basic amino acids (Arg and Lys) are again important for efficient binding of the carboxyl head group of the fatty acid, and the affinity of the protein for the LCFA is dependent on both the entropy and enthalpy of binding, which can cause point mutations to have apparently anomalous effects on binding efficiency [108,110,112]. A hydrophobic pocket or cavity capped by Arg or Lys residues appears to be a characteristic property of proteins that bind LCFA, as it is also shared by β-lactoglobulin, another protein that is known to bind LCFA and is readily co-crystallized with palmitic acid bound in a pocket virtually identical to that of cFABP [179]. The crystal structures of these three classes of proteins reveal numerous hydrophobic residues in structured arrays (either parallel α-helices or antiparallel β-sheets) in close proximity to the hydrophobic chain of LCFA, while positively-charged residues of the protein are found very near to the LCFA carboxyl group. The bond to these residues appears to be crucial to high affinity binding since the LCFA curls back upon itself with increasing chain length while maintaining a fixed position for the carboxylate head group. The affinity of albumin and the cFABPs for LCFA decreases by a factor of 10 as the pH decreases from 7 to 5 [31,60,136,138]. This pH change converts the charged carboxylate group to an uncharged, protonated form. Furthermore, cFABPs bind to anionic membranes by an electrostatic mechanism that appears to be crucial to their release of LCFA [37,123,178]. We have shown that mAspAT has very similar properties, including a hydrophobic pocket, formed by parallel α-helices and bounded by basic residues, that is virtually identical to one of albumin's high affinity LCFA binding sites. It is also a basic protein that exhibits strong binding to anionic membranes [43] at low salt concentrations, and that loses its affinity for LCFA at pH 5.5.

5.5. mAspAT contains structural features important for LCFA binding

If mAspAT is an LCFA transporter, we predict that it would contain a specific LCFA binding site, and are employing two approaches to identify that site. The first is development of a molecular model of its tertiary structure, from which predictions about the location and properties of an LCFA binding site can be derived. Second, experimental approaches to identification of such a site are simultaneously carried out. The experimental strategy is strongly guided by the predictions of the model.

5.5.1. Computer molecular modeling identifies a previously unrecognized 500 Å^3 hydrophobic cleft in mAspAT

A molecular model of the tertiary structure of mAspAT, initially developed using computer graphical techniques and version 4.5 of MacroModel [87], has been refined repeatedly using newer analytical approaches. The crystal structure of chicken heart mAspAT complexed with its cofactor/ligand 2-α-methyl-aspartate-pyridoxal-5′-phosphate at 2.3 Å resolution (Protein Data Bank [*pdb*] designation 1AMA) [84] was selected for initial evaluation. A cleft 500 Å^3 in volume was identified in mAspAT's tertiary conformation between residues 166 and 270, suitable in size to accommodate a LCFA molecule. It is similar in estimated volume to the LCFA binding sites on albumin (350–500 Å^3) [104], and to the volume (450–670 Å^3) within the larger β-barrel/β-clam LCFA binding sites of various lipid binding proteins that are available to hydrophobic ligands after allowing for the presence of ordered, hydrogen-bonded water molecules [8,169]. The cleft was identified as a putative LCFA binding site by computer/graphical algorithms that are designed to identify hydrophobic surfaces between helix–helix interfaces (F. Guarnieri, unpublished). Viewed *en face*, the cleft resembles a deep oval depression between α-helices composed of residues 201–215 and 233–246. A positively charged Arg^{201} is situated at the terminus of the cleft, while Ala^{219} is buried in a strand connecting the two helices. Viewed tangentially (Fig. 10), the site appears as a deep fissure between the two helices, accessed from the exterior through a narrow slit-like entrance between the solvent-accessible amino acids of these two α-helices. Other high resolution crystal structures of chicken heart mAspAT [84] show many conformational changes, but the hydrophobic cleft was preserved in all cases.

5.5.2. Comparison with cytosolic AspAT (cAspAT)

cAspAT, a highly conserved enzyme with ~50% homology to mAspAT in any given species [177], catalyzes the same enzymatic reaction. Modeling of its crystal structure (*pdb* 1AAT) indicates that its tertiary configuration is very similar to mAspAT, including a virtually identical cleft (Fig. 10). However, cAspAT does not bind LCFA [154,155], as assessed either by co-chromatography or in the Lipidex assay [46]. A cAspAT/mAspAT protein sequence alignment indicated a degree of homology among the 105 amino acids of the binding site region (residues 166–270) similar to that of the proteins as a whole. However, 23 residues are invariant in all known mammalian mAspAT's, and different, but equally invariant, in the corresponding cAspAT's. Eight changes represent the replacement in cAspAT of residues that are hydrophobic in mAspAT either by

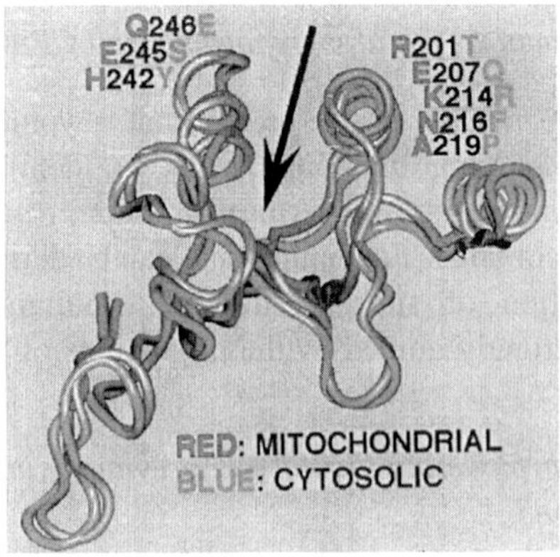

Fig. 10. Molecular model of the proposed LCFA binding site of mAspAT. The α-carbon backbone of mAspAT is shown in red. Two α-helixes flank the cleft. The arrow demonstrates the direction of entry of a fatty acid molecule into the cleft. The similar cAspAT molecule is shown in blue. Several key amino acid substitutions are shown along the models, color-coded to indicate which amino acid is found at that position in each isoform. Although similar, the cleft in cAspAT is much less hydrophobic, narrower, and lacks a key basic residue, all features thought to be important for LCFA binding.*(For a colored version of this figure, see plate section, page 583.)*

non-charged polar ($n = 5$) or charged ($n = 1$) amino acids, or the replacement of non-charged polar amino acids by charged, acidic residues ($n = 2$). Twelve substitutions, including 6 which alter either hydrophobicity or charge, occur in amino acids 200–216 or 240–250, which define the right and left sides, respectively, of the entrance to the hydrophobic cleft. Four of these changes (two on each side) are in highly exposed residues, including an R201T substitution that eliminates the critical positive charge at the terminus of the cleft. Not all of the significant changes involve hydrophobicity or charge. ALA^{219} is deeply buried within a β-strand linking the two sides of the hydrophobic cleft. An A219P substitution at this site in cAspAT is predicted to cause a significant local deformation of the tertiary configuration, with narrowing of the cleft (Fig. 10). Thus, the model identifies changes in cAspAT that lead not only to a loss of hydrophobicity at critical residues or decrease ionic stabilization of the LCFA terminal carboxyl group which has proven to be essential to the binding of LCFA to albumin [19,34,35], but also those which create steric hindrances to LCFA binding. These data potentially explain the difference in LCFA binding by these otherwise similar protein structures and offer a guide for subsequent mapping of the binding site through mutagenesis studies.

5.6. Confirming the location and significance of the LCFA binding site in mAspAT

5.6.1. Enzymatically active mAspAT and its mutants can be produced from pmAspAT expressed in E. coli

Multiple reports indicate that expression of recombinant mAspAT is very difficult (e.g. Refs. [41,75,83]). One of the few laboratories that *has* succeeded is that of Dr Joseph Mattingly, Jr., who provided advice, vectors, and protocols that allowed us led to successfully express recombinant, 46 kDa pmAspAT in *E. coli* at high yields, with an enzyme specific activity ~125 IU mg^{-1}. This is ~75% of the specific activity of enzyme purified from rat liver [155,156]. We have also employed site-directed mutagenesis to produce protein with mutations in the putative LCFA binding site having similar enzymatic activity. As catalytic activity requires proper folding, these data suggest that these recombinant proteins are at least partially folded.

5.6.2. Binding of LCFA by the recombinant R201T mutant and the R201T/A219P double mutant is less than that by wild-type mAspAT

Since, as discussed above, R201T and A219P alterations in mAspAT would have important effects on the properties of the hydrophobic cleft, these were the first 2 mutants constructed, expressed, and tested. The initial estimates of LCFA binding were by co-chromatography of [^{3}H]-OA with the proteins on gel permeation HPLC columns. The amount of [^{3}H]-OA that co-migrated with the R201T single mutant and the R201T/A219P double mutant was approximately 1/2 to 1/3 of that migrating with wild type (WT) recombinant mAspAT, and only slightly greater than non-specific LCFA binding to cAspAT.

5.6.3. Transfection of mAspAT and its binding site mutants into 3T3 cells allows assessment of the functional significance of LCFA binding for cellular LCFA uptake

If LCFA binding to plasma membrane transporters is important for cellular uptake, over-expression of various mAspAT binding site mutants should have different effects on uptake. We have transfected 3T3 fibroblasts with vectors encoding WT-pmAspAT and its R201T mutant and established stably transfected cell lines, designated ZSVAAT and ZSVR201T. Plasma membrane expression of the transfected proteins, assessed by immunofluorescence, and the kinetics of [^{3}H]-OA uptake in these cell lines have been compared with non-transfected control 3T3 cells. mAspAT was readily detected by immunofluorescence in mitochondria of permeabilized 3T3 cells and both the ZSVAAT and ZSVR201T transfectants. In non-permeabilized cells, mAspAT was not detected by routine fluorescence microscopy on the plasma membranes of control 3T3 fibroblasts. Roughly comparable levels of plasma membrane immunofluorescence were seen in the ZSVAAT and ZSVR201T transfectants, indicating similar degrees of cell surface expression of both WT-mAspAT and its R201T mutant. As in earlier studies [62], over-expression of WT-mAspAT led to a significant increase in [^{3}H]-OA uptake compared with controls. By contrast, over-expression of the R201T mutant led to a small, non-significant increase in uptake. The calculated V_{max} values (pmol/s/50,000 cells) were: 3T3, 0.26 ± 0.01; ZSVAAT, 1.52 ± 0.38 ($p < 0.025$ vs. 3T3); and ZSVR201T, 0.49 ± 0.22 ($p > 0.2$ vs. 3T3). Thus, these studies: (1) support the existence of an LCFA binding site in the region originally predicted by our model, (2) support the model's prediction of a role for the positively arginine at position 201 in LCFA binding, and (3) imply that the affinity of the mAspAT binding site for LCFA may be a significant factor in the cellular uptake of LCFA. These data represent a promising step toward determining the LCFA binding site on mAspAT.

6. Significance

The ultimate significance of putative LCFA transport systems lies in the roles they play in human physiology, pathophysiology, and disease. Thus, the work summarized above in adipocytes and HepG2 cells suggests important roles for regulated plasma membrane expression of mAspAT in LCFA transport systems that contribute to the pathogenesis of obesity, type 2 diabetes, and EtOH-associated fatty liver, whereas identification of mAspAT, FAT, and an FATP in particular human placental plasma membranes [23] suggests that they are important for transplacental LCFA transport. Facilitated transport mechanisms increasingly appear to be critical providers of energy substrates to skeletal and cardiac muscle [21,78,157]. Training-induced increases in muscle exercise tolerance correlate with increased expression of $mAspAT_{pm}$ [168]. The existence of protein mediated transport systems opens the possibility of mutations that may lead to disease. It also implies the existence of regulatory processes, dysregulation of which may also cause disease. FAT/CD36 deficiency, for example, is a putative basis of hereditary hypertrophic cardiomyopathy [157]. Finally, a selective LCFA uptake defect is reportedly the basis of a syndrome of recurrent bouts of nonketotic, hypoglycemic acidosis and liver failure requiring transplantation in children [95]. Independent confirmation of this defect has not yet been reported, but would provide indisputable proof of a specific LCFA transport mechanism. Current work in the field of fatty acids transport provides convincing evidence that cellular LCFA is a regulatable, protein mediated process of considerable, potential clinical importance.

References

[1] Abumrad, N., Harmon, C., Ibrahimi, A., 1998. Membrane transport of long-chain fatty acids: evidence for a facilitated process. J. Lipid Res. 39, 2309–2318.

[2] Abumrad, N.A., el Maghrabi, M.R., Amri, E.Z., Lopez, E., Grimaldi, P.A., 1993. Cloning of a rat adipocyte membrane protein implicated in binding or transport of long-chain fatty acids that is induced during preadipocyte differentiation. Homology with human CD36. J. Biol. Chem. 268, 17665–17668.

[3] Abumrad, N.A., Park, J.H., Park, C.R., 1984. Permeation of long-chain fatty acid into adipocytes. Kinetics, specificity, and evidence for involvement of a membrane protein. J. Biol. Chem. 259, 8945–8953.

[4] Abumrad, N.A., Perkins, R.C., Park, J.H., Park, C.R., 1981. Mechanism of long chain fatty acid permeation in the isolated adipocyte. J. Biol. Chem. 256, 9183–9191.

[5] Amri, E.Z., Bonino, F., Ailhaud, G., Abumrad, N.A., Grimaldi, P.A., 1995. Cloning of a protein that mediates transcriptional effects of fatty acids in preadipocytes. Homology to peroxisome proliferator-activated receptors. J. Biol. Chem. 270, 2367–2371.

[6] Baillie, A.G., Coburn, C.T., Abumrad, N.A., 1996. Reversible binding of long-chain fatty acids to purified FAT, the adipose CD36 homolog. J. Membr. Biol. 153, 75–81.

[7] Balendiran, G.K., Schnutgen, F., Scapin, G., Borchers, T., Xhong, N., Lim, K., Godbout, R., Spener, F., Sacchettini, J.C., 2000. Crystal structure and thermodynamic analysis of human brain fatty acid-binding protein. J. Biol. Chem. 275, 27045–27054.

[8] Banaszak, L., Winter, N., Xu, Z., Bernlohr, D.A., Cowan, S., Jones, T.A., 1994. Lipid-binding proteins: a family of fatty acid and retinoid transport proteins. Adv. Protein Chem. 45, 89–151.

[9] Berk, P.D., 1996. How do long-chain free fatty acids cross cell membranes? Proc. Soc. Exp. Biol. Med. 212, 1–4.

[10] Berk, P.D., Potter, B.J., Sorrentino, D., Stump, D., Kiang, C.L., Zhou, S.L., Horio, Y., Wada, H., 1990. Hepatocellular fatty acid uptake is mediated by a plasma membrane fatty acid binding protein closely related to mitochondrial glutamic oxaloacetic transaminase. Ann. NY Acad. Sci. 585, 379–385.

[11] Berk, P.D., Stremmel, W., 1986. Hepatocellular uptake of organic anions. Prog. Liver Dis. 8, 125–144.

[12] Berk, P.D., Stump, D.D., 1999. Mechanisms of cellular uptake of long chain free fatty acids. Mol. Cell. Biochem. 192, 17–31.

[13] Berk, P.D., Wada, H., Horio, Y., Potter, B.J., Sorrentino, D., Zhou, S.L., Isola, L.M., Stump, D., Kiang, C.L., Thung, S., 1990. Plasma membrane fatty acid-binding protein and mitochondrial glutamic-oxaloacetic transaminase of rat liver are related. Proc. Natl Acad. Sci. USA 87, 3484–3488.

[14] Berk, P.D., Zhou, S., Kiang, C., Stump, D.D., Fan, X., Bradbury, M.W., 1999. Selective up-regulation of fatty acid uptake by adipocytes characterizes both genetic and diet-induced obesity in rodents. J. Biol. Chem. 274, 28626–28631.

[15] Berk, P.D., Zhou, S.L., Bradbury, M., Stump, D., Kiang, C.L., Isola, L.M., 1996. Regulated membrane transport of free fatty acids in adipocytes: role in obesity and non-insulin dependent diabetes mellitus. Trans. Am. Clin. Climatol. Assoc. 108, 26–40.

[16] Berk, P.D., Zhou, S.L., Kiang, C.L., Stump, D., Bradbury, M., Isola, L.M., 1997. Uptake of long chain free fatty acids is selectively up-regulated in adipocytes of Zucker rats with genetic obesity and non-insulin-dependent diabetes mellitus. J. Biol. Chem. 272, 8830–8835.

[17] Berk, P.D., Zhou, S.L., Stump, D., Kiang, C.L., Isola, L.M., 1993. Recent studies of the cellular uptake of long chain free fatty acids. Trans. Am. Clin. Climatol. Assoc. 105, 179–189.

[18] Bernlohr, D.A., Coe, N.R., Simpson, M.A., Hertzel, A.V., 1997. Regulation of gene expression in adipose cells by polyunsaturated fatty acids. Adv. Exp. Med. Biol. 422, 145–156.

[19] Bhattacharya, A.A., Grune, T., Curry, S., 2000. Crystallographic analysis reveals common modes of binding of medium and long-chain fatty acids to human serum albumin. J. Mol. Biol. 303, 721–732.

[20] Bojesen, I.N., Bojesen, E., 1994. Binding of arachidonate and oleate to bovine serum albumin. J. Lipid Res. 35, 770–778.

[21] Bonen, A., Luiken, J.J., Liu, S., Dyck, D.J., Kiens, B., Kristiansen, S., Turcotte, L.P., van der Vusse, G.J., Glatz, J.F., 1998. Palmitate transport and fatty acid transporters in red and white muscles. Am. J. Physiol. 275, E471–E478.

[22] Bradbury, M.W., Berk, P.D., 2000. Mitochondrial aspartate aminotransferase: direction of a single protein with two distinct functions to two subcellular sites does not require alternative splicing of the mRNA. Biochem. J. 345(Pt 3), 423–427.

[23] Campbell, F.M., Bush, P.G., Veerkamp, J.H., Dutta-Roy, A.K., 1998. Detection and cellular localization of plasma membrane-associated and cytoplasmic fatty acid-binding proteins in human placenta. Placenta 19, 409–415.

[24] Campbell, F.M., Clohessy, A.M., Gordon, M.J., Page, K.R., Dutta-Roy, A.K., 1997. Uptake of long chain fatty acids by human placental choriocarcinoma (BeWo) cells: role of plasma membrane fatty acid-binding protein. J. Lipid Res. 38, 2558–2568.

[25] Caserta, F., Tchkonia, T., Civelek, V.N., Prentki, M., Brown, N.F., McGarry, J.D., Forse, R.A., Corkey, B.E., Hamilton, J.A., Kirkland, J.L., 2001. Fat depot origin affects fatty acid handling in cultured rat and human preadipocytes. Am. J. Physiol. Endocrinol. Metab. 280, E238–E247.

[26] Cechetto, J.D., Gupta, S., 2000. Immunoelectron microscopy provides evidence that tumor necrosis factor receptor-associated protein 1 (TRAP-1) is a mitochondrial protein which also localizes at specific extramitochondrial sites. Exp. Cell Res. 260, 30–39.

[27] Cechetto, J.D., Sadacharan, S.K., Berk, P.D., Gupta, R.S., 2002. Immunogold localization of mitochondrial aspartate aminotransferase in mitochondria and on the cell surface in normal rat tissues. Histol. Histopathol. 17, 353–364.

[28] Cechetto, J.D., Soltys, B.J., Gupta, R.S., 2000. Localization of mitochondrial 60-kD heat shock chaperonin protein (Hsp60) in pituitary growth hormone secretory granules and pancreatic zymogen granules. J. Histochem. Cytochem. 48, 45–56.

[29] Chen, H., Charlat, O., Tartaglia, L.A., Woolf, E.A., Weng, X., Ellis, S.J., Lakey, N.D., Culpepper, J., Moore, K.J., Breitbart, R.E., Duyk, G.M., Tepper, R.I., Morgenstern, J.P., 1996. Evidence that the diabetes

gene encodes the leptin receptor: identification of a mutation in the leptin receptor gene in db/db mice. Cell 84, 491–495.

[30] Chua, S.C. Jr., Chung, W.K., Wu-Peng, X.S., Zhang, Y., Liu, S.M., Tartaglia, L., Leibel, R.L., 1996. Phenotypes of mouse diabetes and rat fatty due to mutations in the OB (leptin) receptor. Science 271, 994–996.

[31] Cistola, D.P., Sacchettini, J.C., Banaszak, L.J., Walsh, M.T., Gordon, J.I., 1989. Fatty acid interactions with rat intestinal and liver fatty acid-binding proteins expressed in *Escherichia coli.* A comparative 13C NMR study. J. Biol. Chem. 264, 2700–2710.

[32] Civelek, V.N., Hamilton, J.A., Tornheim, K., Kelly, K.L., Corkey, B.E., 1996. Intracellular pH in adipocytes: effects of free fatty acid diffusion across the plasma membrane, lipolytic agonists, and insulin. Proc. Natl Acad. Sci. USA 93, 10139–10144.

[33] Coe, N.R., Smith, A.J., Frohnert, B.I., Watkins, P.A., Bernlohr, D.A., 1999. The fatty acid transport protein (FATP1) is a very long chain acyl-CoA synthetase. J. Biol. Chem. 274, 36300–36304.

[34] Curry, S., Brick, P., Franks, N.P., 1999. Fatty acid binding to human serum albumin: new insights from crystallographic studies. Biochim. Biophys. Acta 1441, 131–140.

[35] Curry, S., Mandelkow, H., Brick, P., Franks, N., 1998. Crystal structure of human serum albumin complexed with fatty acid reveals an asymmetric distribution of binding sites. Nat. Struct. Biol. 5, 827–835.

[36] Daniels, C., Noy, N., Zakim, D., 1985. Rates of hydration of fatty acids bound to unilamellar vesicles of phosphatidylcholine or to albumin. Biochemistry 24, 3286–3292.

[37] Davies, J.K., Thumser, A.E., Wilton, D.C., 1999. Binding of recombinant rat liver fatty acid-binding protein to small anionic phospholipid vesicles results in ligand release: a model for interfacial binding and fatty acid targeting. Biochemistry 38, 16932–16940.

[38] Di Girolamo, M., Mendlinger, S., Fertig, J.W., 1971. A simple method to determine fat cell size and number in four mammalian species. Am. J. Physiol. 221, 850–858.

[39] Distel, R.J., Robinson, G.S., Spiegelman, B.M., 1992. Fatty acid regulation of gene expression. Transcriptional and post-transcriptional mechanisms. J. Biol. Chem. 267, 5937–5941.

[40] Faergeman, N.J., DiRusso, C.C., Elberger, A., Knudsen, J., Black, P.N., 1997. Disruption of the *Saccharomyces cerevisiae* homologue to the murine fatty acid transport protein impairs uptake and growth on long-chain fatty acids. J. Biol. Chem. 272, 8531–8538.

[41] Fluckiger, J., Christen, P., 1988. Degradation of the precursor of mitochondrial aspartate aminotransferase in chicken embryo fibroblasts. J. Biol. Chem. 263, 4131–4138.

[42] Fujii, S., Kawaguchi, H., Yasuda, H., 1987. Isolation and partial characterization of an amphiphilic 56-kDa fatty acid binding protein from rat renal basolateral membrane. J. Biochem. (Tokyo) 101, 679–684.

[43] Furuya, E., Yoshida, Y., Tagawa, K., 1979. Interaction of mitochondrial aspartate aminotransferase with negatively charged lecithin liposomes. J. Biochem. (Tokyo) 85, 1157–1163.

[44] Garlid, K.D., Orosz, D.E., Modriansky, M., Vassanelli, S., Jezek, P., 1996. On the mechanism of fatty acid-induced proton transport by mitochondrial uncoupling protein. J. Biol. Chem. 271, 2615–2620.

[45] Giannattasio, S., Marra, E., Vacca, R.A., Iannace, G., Quagliariello, E., 1992. Import of mutant forms of mitochondrial aspartate aminotransferase into isolated mitochondria. Arch. Biochem. Biophys. 298, 532–537.

[46] Glatz, J.F., Veerkamp, H., 1983. A radiochemical procedure for the assay of fatty acid binding by proteins. Anal. Biochem. 132, 89–95.

[47] Goresky, C.A., Stremmel, W., Rose, C.P., Guirguis, S., Schwab, A.J., Diede, H.E., Ibrahim, E., 1994. The capillary transport system for free fatty acids in the heart. Circ. Res. 74, 1015–1026.

[48] Greenwalt, D.E., Scheck, S.H., Rhinehart-Jones, T., 1995. Heart CD36 expression is increased in murine models of diabetes and in mice fed a high fat diet. J. Clin. Invest. 96, 1382–1388.

[49] Gutknecht, J., 1988. Proton conductance caused by long-chain fatty acids in phospholipid bilayer membranes. J. Membr. Biol. 106, 83–93.

[50] Halaas, J.L., Gajiwala, K.S., Maffei, M., Cohen, S.L., Chait, B.T., Rabinowitz, D., Lallone, R.L., Burley, S.K., Friedman, J.M., 1995. Weight-reducing effects of the plasma protein encoded by the obese gene. Science 269, 543–546.

[51] Hamilton, J.A., 1998. Fatty acid transport: difficult or easy? J. Lipid Res. 39, 467–481.

[52] Hamilton, J.A., Civelek, V.N., Kamp, F., Tornheim, K., Corkey, B.E., 1994. Changes in internal pH caused by movement of fatty acids into and out of clonal pancreatic beta-cells (HIT). J. Biol. Chem. 269, 20852–20856.
[53] Hamilton, J.A., Era, S., Bhamidipati, S.P., Reed, R.G., 1991. Locations of the three primary binding sites for long-chain fatty acids on bovine serum albumin. Proc. Natl Acad. Sci. USA 88, 2051–2054.
[54] Hamilton, J.A., Kamp, F., 1999. How are free fatty acids transported in membranes? Is it by proteins or by free diffusion through the lipids? Diabetes 48, 2255–2269.
[55] Harmon, C.M., Luce, P., Beth, A.H., Abumrad, N.A., 1991. Labeling of adipocyte membranes by sulfo-*N*-succinimidyl derivatives of long-chain fatty acids: inhibition of fatty acid transport. J. Membr. Biol. 121, 261–268.
[56] Hertzel, A.V., Bernlohr, A., 2000. The mammalian fatty acid-binding protein multigene family: molecular and genetic insights into function. Trends Endocrinol. Metab. 11, 175–180.
[57] Hirsch, D., Stahl, A., Lodish, H.F., 1998. A family of fatty acid transporters conserved from mycobacterium to man. Proc. Natl Acad. Sci. USA 95, 8625–8629.
[58] Horio, Y., Sakakibara, R., Tanaka, T., Taketoshi, M., Obaru, K., Shimada, K., Morino, Y., Wada, H., 1986. Molecular cloning of rat mitochondrial glutamic oxaloacetic transaminase mRNA and regulation of its expression in regenerating liver. Biochem. Biophys. Res. Commun. 134, 803–811.
[59] Hotamisligil, G.S., Shargill, N.S., Spiegelman, B.M., 1993. Adipose expression of tumor necrosis factor-alpha: direct role in obesity-linked insulin resistance. Science 259, 87–91.
[60] Hsu, K.T., Storch, J., 1996. Fatty acid transfer from liver and intestinal fatty acid-binding proteins to membranes occurs by different mechanisms. J. Biol. Chem. 271, 13317–13323.
[61] Hui, T.Y., Frohnert, B.I., Smith, A.J., Schaffer, J.E., Bernlohr, D.A., 1998. Characterization of the murine fatty acid transport protein gene and its insulin response sequence. J. Biol. Chem. 273, 27420–27429.
[62] Isola, L.M., Zhou, S.L., Kiang, C.L., Stump, D.D., Bradbury, M.W., Berk, P.D., 1995. 3T3 fibroblasts transfected with a cDNA for mitochondrial aspartate aminotransferase express plasma membrane fatty acid-binding protein and saturable fatty acid uptake. Proc. Natl Acad. Sci. USA 92, 9866–9870.
[63] Jeffery, C.J., 1999. Moonlighting proteins. Trends Biochem. Sci. 24, 8–11.
[64] Jezek, P., Hanus, J., Semrad, C., Garlid, K.D., 1996. Photoactivated azido fatty acid irreversibly inhibits anion and proton transport through the mitochondrial uncoupling protein. J. Biol. Chem. 271, 6199–6205.
[65] Joh, T., Nomiyama, H., Maeda, S., Shimada, K., Morino, Y., 1985. Cloning and sequence analysis of a cDNA encoding porcine mitochondrial aspartate aminotransferase precursor. Proc. Natl Acad. Sci. USA 82, 6065–6069.
[66] Kahn, B.B., 1996. Lilly lecture 1995. Glucose transport: pivotal step in insulin action. Diabetes 45, 1644–1654.
[67] Kamp, F., Hamilton, A., 1992. pH gradients across phospholipid membranes caused by fast flip-flop of un-ionized fatty acids. Proc. Natl Acad. Sci. USA 89, 11367–11370.
[68] Kamp, F., Hamilton, J.A., Kamp, F., Westerhoff, H.V., Hamilton, J.A., 1993. Movement of fatty acids, fatty acid analogues, and bile acids across phospholipid bilayers. Biochemistry 32, 11074–11086.
[69] Kamp, F., Zakim, D., Zhang, F., Noy, N., Hamilton, J.A., 1995. Fatty acid flip-flop in phospholipid bilayers is extremely fast. Biochemistry 34, 11928–11937.
[70] Kava, R., Greenwood, M.R., Johnson, P.R., 1990. Zucker (*fa/fa*) rat. ALAR News 32, 4–8.
[71] Kleinfeld, A.M., 1995. Fatty acid transport across membranes. In: DiSalvo, E.A., Simon, S. (Eds.), Stability and Permeability of Lipid Bilayers. CRC Press, Boca Raton, FL.
[72] Kleinfeld, A.M., Chu, P., Romero, C., 1997. Transport of long-chain native fatty acids across lipid bilayer membranes indicates that transbilayer flip-flop is rate limiting. Biochemistry 36, 14146–14158.
[73] Kleinfeld, A.M., Chu, P., Storch, J., 1997. Flip-flop is slow and rate limiting for the movement of long chain anthroyloxy fatty acids across lipid vesicles. Biochemistry 36, 5702–5711.
[74] Kleinfeld, A.M., Storms, S., Watts, M., 1998. Transport of long-chain native fatty acids across human erythrocyte ghost membranes. Biochemistry 37, 8011–8019.
[75] Lain, B., Iriarte, A., Martinez-Carrion, M., 1994. Dependence of the folding and import of the precursor to mitochondrial aspartate aminotransferase on the nature of the cell-free translation system. J. Biol. Chem. 269, 15588–15596.

[76] Lee, G.H., Proenca, R., Montez, J.M., Carroll, K.M., Darvishzadeh, J.G., Lee, J.I., Friedman, J.M., 1996. Abnormal splicing of the leptin receptor in diabetic mice. Nature 379, 632–635.

[77] Lewin, M.J., Bado, A., 2001. Gastric leptin. Microsc. Res. Tech. 53, 372–376.

[78] Luiken, J.J., Schaap, F.G., Van Nieuwenhoven, F.A., van der Vusse, G.J., Bonen, A., Glatz, J.F., 1999. Cellular fatty acid transport in heart and skeletal muscle as facilitated by proteins. Lipids 34(Suppl.), S169–S175.

[79] Man, M.Z., Hui, T.Y., Schaffer, J.E., Lodish, H.F., Bernlohr, D.A., 1996. Regulation of the murine adipocyte fatty acid transporter gene by insulin. Mol. Endocrinol. 10, 1021–1028.

[80] Maniscalco, W.M., Stremmel, W., Heeney-Campbell, M., 1990. Uptake of palmitic acid by rabbit alveolar type II cells. Am. J. Physiol. 259, L206–L212.

[81] Martin, G., Schoonjans, K., Staels, B., Auwerx, J., 1998. PPAR gamma activators improve glucose homeostasis by stimulating fatty acid uptake in the adipocytes. Atherosclerosis 137(Suppl.), S75–S80.

[82] Mattingly, J.R. Jr., Rodriguez-Berrocal, F.J., Gordon, J., Iriarte, A., Martinez-Carrion, M., 1987. Molecular cloning and in vivo expression of a precursor to rat mitochondrial aspartate aminotransferase. Biochem. Biophys. Res. Commun. 149, 859–865.

[83] Mattingly, J.R. Jr., Youssef, J., Iriarte, A., Martinez-Carrion, M., 1993. Protein folding in a cell-free translation system. The fate of the precursor to mitochondrial aspartate aminotransferase. J. Biol. Chem. 268, 3925–3937.

[84] McPhalen, C.A., Vincent, M.G., Picot, D., Jansonius, J.N., Lesk, A.M., Chothia, C., 1992. Domain closure in mitochondrial aspartate aminotransferase. J. Mol. Biol. 227, 197–213.

[85] Memon, R.A., Feingold, K.R., Moser, A.H., Fuller, J., Grunfeld, C., 1998. Regulation of fatty acid transport protein and fatty acid translocase mRNA levels by endotoxin and cytokines. Am. J. Physiol. 274, E210–E217.

[86] Misteli, T., Spector, L., 1997. Applications of the green fluorescent protein in cell biology and biotechnology. Nat. Biotechnol. 15, 961–964.

[87] Mohamadi, F., Richards, N.G.J., Guida, W.C., 1990. MacroModel – an integrated software system for modeling organic and bioorganic molecules using molecular mechanics. J. Comp. Chem. 11, 440–467.

[88] Motojima, K., Passilly, P., Peters, J.M., Gonzalez, F.J., Latruffe, N., 1998. Expression of putative fatty acid transporter genes are regulated by peroxisome proliferator-activated receptor alpha and gamma activators in a tissue- and inducer-specific manner. J. Biol. Chem. 273, 16710–16714.

[89] Naggert, J.K., Fricker, L.D., Varlamov, O., Nishina, P.M., Rouille, Y., Steiner, D.F., Carroll, R.J., Paigen, B.J., Leiter, E.H., 1995. Hyperproinsulinaemia in obese fat/fat mice associated with a carboxypeptidase E mutation which reduces enzyme activity. Nat. Genet. 10, 135–142.

[90] Nalpas, B., Vassault, A., Le Guillou, A., Lesgourgues, B., Ferry, N., Lacour, B., Berthelot, P., 1984. Serum activity of mitochondrial aspartate aminotransferase: a sensitive marker of alcoholism with or without alcoholic hepatitis. Hepatology 4, 893–896.

[91] Noben-Trauth, K., Naggert, J.K., North, M.A., Nishina, P.M., 1996. A candidate gene for the mouse mutation tubby. Nature 380, 534–538.

[92] Noy, N., Zakim, D., 1993. Physical chemical basis for the uptake of organic compounds by cells. In: Berk, P.D., Tavoloni, N. (Eds.), Hepatic Transport and Bile Secretion. Raven Press, New York.

[93] Nunes, R., Kiang, C.L., Sorrentino, D., Berk, P.D., 1988. 'Albumin-receptor' uptake kinetics do not require an intact lobular architecture and are not specific for albumin. J. Hepatol. 7, 293–304.

[94] Ockner, R.K., Weisiger, R.A., Gollan, J.L., 1983. Hepatic uptake of albumin-bound substances: albumin receptor concept. Am. J. Physiol. 245, G13–G18.

[95] Odaib, A.A., Shneider, B.L., Bennett, M.J., Pober, B.R., Reyes-Mugica, M., Friedman, A.L., Suchy, F.J., Rinaldo, P., 1998. A defect in the transport of long-chain fatty acids associated with acute liver failure. N. Engl. J. Med. 339, 1752–1757.

[96] Passarella, S., Marra, E., Atlante, A., Barile, M., Doonan, S., Quagliariello, E., 1990. Uptake of aspartate aminotransferase into mitochondria in vitro causes efflux of malate dehydrogenase and vice versa. Biochim. Biophys. Acta 1022, 273–282.

[97] Peterson, R.G., Shaw, W.N., Neel, M.-A., Little, L.A., Eichberg, J., 1990. Zucker diabetic fatty rat as a model for non-insulin-dependent diabetes mellitus. ALAR News 32, 16–19.

[98] Petitpas, I., Grune, T., Bhattacharya, A.A., Curry, S., 2001. Crystal structures of human serum albumin complexed with monounsaturated and polyunsaturated fatty acids. J. Mol. Biol. 314, 955–960.
[99] Pohl, J., Ring, A., Stremmel, W., 2002. Uptake of long-chain fatty acids in HepG2 cells involves caveolae: analysis of a novel pathway. J. Lipid Res. 43, 1390–1399.
[100] Poirier, H., Degrace, P., Niot, I., Bernard, A., Besnard, P., 1996. Localization and regulation of the putative membrane fatty-acid transporter (FAT) in the small intestine. Comparison with fatty acid-binding proteins (FABP). Eur. J. Biochem. 238, 368–373.
[101] Potter, B.J., Berk, P.D., 1993. The liver plasma membrane fatty acid binding protein. In: Tavoloni, N., Berk, P.D. (Eds.), Hepatic Transport and Bile Secretion. Raven Press, New York.
[102] Potter, B.J., Stump, D., Schwieterman, W., Sorrentino, D., Jacobs, L.N., Kiang, C.L., Rand, J.H., Berk, P.D., 1987. Isolation and partial characterization of plasma membrane fatty acid binding proteins from myocardium and adipose tissue and their relationship to analogous proteins in liver and gut. Biochem. Biophys. Res. Commun. 148, 1370–1376.
[103] Prieto, R.M., Stremmel, W., Sales, C., Tur, J.A., 1996. Oleic acid uptake by jejunal and ileal rat brush border membrane vesicles. Eur. J. Med. Res. 1, 199–203.
[104] Reed, R.G., 1986. Location of long chain fatty acid-binding sites of bovine serum albumin by affinity labeling. J. Biol. Chem. 261, 15619–15624.
[105] Reese-Wagoner, A., Thompson, J., Banaszak, L., 1999. Structural properties of the adipocyte lipid binding protein. Biochim. Biophys. Acta 1441, 106–116.
[106] Richieri, G.V., Anel, A., Kleinfeld, A.M., 1993. Interactions of long-chain fatty acids and albumin: determination of free fatty acid levels using the fluorescent probe ADIFAB. Biochemistry 32, 7574–7580.
[107] Richieri, G.V., Kleinfeld, M., 1989. Free fatty acid perturbation of transmembrane signaling in cytotoxic T lymphocytes. J. Immunol. 143, 2302–2310.
[108] Richieri, G.V., Low, P.J., Ogata, R.T., Kleinfeld, A.M., 1998. Thermodynamics of fatty acid binding to engineered mutants of the adipocyte and intestinal fatty acid-binding proteins. J. Biol. Chem. 273, 7397–7405.
[109] Richieri, G.V., Ogata, R.T., Kleinfeld, A.M., 1992. A fluorescently labeled intestinal fatty acid binding protein. Interactions with fatty acids and its use in monitoring free fatty acids. J. Biol. Chem. 267, 23495–23501.
[110] Richieri, G.V., Ogata, R.T., Kleinfeld, A.M., 1995. Thermodynamics of fatty acid binding to fatty acid-binding proteins and fatty acid partition between water and membranes measured using the fluorescent probe ADIFAB. J. Biol. Chem. 270, 15076–15084.
[111] Richieri, G.V., Ogata, R.T., Kleinfeld, A.M., 1996. Kinetics of fatty acid interactions with fatty acid binding proteins from adipocyte, heart, and intestine. J. Biol. Chem. 271, 11291–11300.
[112] Richieri, G.V., Ogata, R.T., Kleinfeld, A.M., 1999. Fatty acid interactions with native and mutant fatty acid binding proteins. Mol. Cell. Biochem. 192, 77–85.
[113] Richieri, G.V., Ogata, R.T., Zimmerman, A.W., Veerkamp, J.H., Kleinfeld, A.M., 2000. Fatty acid binding proteins from different tissues show distinct patterns of fatty acid interactions. Biochemistry 39, 7197–7204.
[114] Rose, H., Conventz, M., Fischer, Y., Jungling, E., Hennecke, T., Kammermeier, H., 1994. Long-chain fatty acid-binding to albumin: re-evaluation with directly measured concentrations. Biochim. Biophys. Acta 1215, 321–326.
[115] Schaffer, J.E., Lodish, H.F., 1994. Expression cloning and characterization of a novel adipocyte long chain fatty acid transport protein. Cell 79, 427–436.
[116] Schemmel, R., Mickelsen, O., Gill, J.L., 1970. Dietary obesity in rats: body weight and body fat accretion in seven strains of rats. J. Nutr. 100, 1041–1048.
[117] Schmider, W., Fahr, A., Voges, R., Gerok, W., Kurz, G., 1996. Irreversible inhibition of hepatic fatty acid salt uptake by photoaffinity labeling with 11,11-azistearate. J. Lipid Res. 37, 739–753.
[118] Schwieterman, W., Sorrentino, D., Potter, B.J., Rand, J., Kiang, C.L., Stump, D., Berk, P.D., 1988. Uptake of oleate by isolated rat adipocytes is mediated by a 40-kDa plasma membrane fatty acid binding protein closely related to that in liver and gut. Proc. Natl Acad. Sci. USA 85, 359–363.

[119] Sfeir, Z., Ibrahimi, A., Amri, E., Grimaldi, P., Abumrad, N., 1997. Regulation of FAT/CD36 gene expression: further evidence in support of a role of the protein in fatty acid binding/transport. Prostaglandins Leukot. Essent. Fatty Acids 57, 17–21.

[120] Skulachev, V.P., 1991. Fatty acid circuit as a physiological mechanism of uncoupling of oxidative phosphorylation. FEBS Lett. 294, 158–162.

[121] Sloan, I.S., Horowitz, P.M., Chirgwin, J.M., 1994. Rapid secretion by a nonclassical pathway of overexpressed mammalian mitochondrial rhodanese. J. Biol. Chem. 269, 27625–27630.

[122] Smalheiser, N.R., 1996. Proteins in unexpected locations. Mol. Biol. Cell 7, 1003–1014.

[123] Smith, E.R., Storch, J., 1999. The adipocyte fatty acid-binding protein binds to membranes by electrostatic interactions. J. Biol. Chem. 274, 35325–35330.

[124] Soltys, B.J., Gupta, S., 1996. Immunoelectron microscopic localization of the 60-kDa heat shock chaperonin protein (Hsp60) in mammalian cells. Exp. Cell Res. 222, 16–27.

[125] Soltys, B.J., Gupta, S., 1999. Mitochondrial-matrix proteins at unexpected locations: are they exported? Trends Biochem. Sci. 24, 174–177.

[126] Soltys, B.J., Gupta, S., 2000. Mitochondrial proteins at unexpected cellular locations: export of proteins from mitochondria from an evolutionary perspective. Int. Rev. Cytol. 194, 133–196.

[127] Soltys, B.J., Kang, D., Gupta, R.S., 2000. Localization of P32 protein (gC1q-R) in mitochondria and at specific extramitochondrial locations in normal tissues. Histochem. Cell Biol. 114, 245–255.

[128] Sorrentino, D., Berk, P.D., 1993. Free fatty acids, albumin and the sinusoidal plasma membrane: concepts, trends and controversies. In: Tavoloni, N., Berk, P.D. (Eds.), Hepatic Transport and Bile Secretion. Raven Press, New York.

[129] Sorrentino, D., Robinson, R.B., Kiang, C.L., Berk, P.D., 1989. At physiologic albumin/oleate concentrations oleate uptake by isolated hepatocytes, cardiac myocytes, and adipocytes is a saturable function of the unbound oleate concentration. Uptake kinetics are consistent with the conventional theory. J. Clin. Invest. 84, 1325–1333.

[130] Sorrentino, D., Stump, D., Potter, B.J., Robinson, R.B., White, R., Kiang, C.L., Berk, P.D., 1988. Oleate uptake by cardiac myocytes is carrier mediated and involves a 40-kD plasma membrane fatty acid binding protein similar to that in liver, adipose tissue, and gut. J. Clin. Invest. 82, 928–935.

[131] Sorrentino, D., Stump, D.D., Van Ness, K., Simard, A., Schwab, A.J., Zhou, S.L., Goresky, C.A., Berk, P.D., 1996. Oleate uptake by isolated hepatocytes and the perfused rat liver is competitively inhibited by palmitate. Am. J. Physiol. 270, G385–G392.

[132] Sorrentino, D., Van Ness, K., Stump, D., Berk, P.D., 1994. Oleate uptake kinetics in the perfused rat liver are consistent with pseudofacilitation by albumin. J. Hepatol. 21, 551–559.

[133] Sorrentino, D., Zhou, S.L., Kokkotou, E., Berk, P.D., 1992. Sex differences in hepatic fatty acid uptake reflect a greater affinity of the transport system in females. Am. J. Physiol. 263, G380–G385.

[134] Sorrentino, D., Zifroni, A., Van Ness, K., Berk, P.D., 1994. Unbound ligand drives hepatocyte taurocholate and BSP uptake at physiological albumin concentration. Am. J. Physiol. 266, G425–G432.

[135] Spector, A.A., Fletcher, J.E., Ashbrook, J.D., 1971. Analysis of long-chain free fatty acid binding to bovine serum albumin by determination of stepwise equilibrium constants. Biochemistry 10, 3229–3232.

[136] Spector, A.A., John, K., Fletcher, J.E., 1969. Binding of long-chain fatty acids to bovine serum albumin. J. Lipid Res. 10, 56–67.

[137] Stahl, A., Hirsch, D.J., Gimeno, R.E., Punreddy, S., Ge, P., Watson, N., Patel, S., Kotler, M., Raimondi, A., Tartaglia, L.A., Lodish, H.F., 1999. Identification of the major intestinal fatty acid transport protein. Mol. Cell 4, 299–308.

[138] Stewart, J.M., Dewling, V.F., Wright, T.G., 1998. Fatty acid binding to rat liver fatty acid-binding protein is modulated by early glycolytic intermediates. Biochim. Biophys. Acta 1391, 1–6.

[139] Stoll, G.H., Voges, R., Gerok, W., Kurz, G., 1991. Synthesis of a metabolically stable modified long-chain fatty acid salt and its photolabile derivative. J. Lipid Res. 32, 843–857.

[140] Stremmel, W., 1988. Fatty acid uptake by isolated rat heart myocytes represents a carrier-mediated transport process. J. Clin. Invest. 81, 844–852.

[141] Stremmel, W., 1988. Uptake of fatty acids by jejunal mucosal cells is mediated by a fatty acid binding membrane protein. J. Clin. Invest. 82, 2001–2010.

[142] Stremmel, W., Berk, P.D., 1986. Hepatocellular influx of [14C]oleate reflects membrane transport rather than intracellular metabolism or binding. Proc. Natl Acad. Sci. USA 83, 3086–3090.
[143] Stremmel, W., Diede, H.E., Rodilla-Sala, E., Vyska, K., Schrader, M., Fitscher, B., Passarella, S., 1990. The membrane fatty acid-binding protein is not identical to mitochondrial glutamic oxaloacetic transaminase (mGOT). Mol. Cell. Biochem. 98, 191–199.
[144] Stremmel, W., Kochwa, S., Berk, P.D., 1983. Studies of oleate binding to rat liver plasma membranes. Biochem. Biophys. Res. Commun. 112, 88–95.
[145] Stremmel, W., Lotz, G., Strohmeyer, G., Berk, P.D., 1985. Identification, isolation, and partial characterization of a fatty acid binding protein from rat jejunal microvillous membranes. J. Clin. Invest. 75, 1068–1076.
[146] Stremmel, W., Pohl, L., Ring, A., Herrmann, T., 2001. A new concept of cellular uptake and intracellular trafficking of long-chain fatty acids. Lipids 36, 981–989.
[147] Stremmel, W., Potter, B.J., Berk, P.D., 1983. Studies of albumin binding to rat liver plasma membranes. Implications for the albumin receptor hypothesis. Biochim. Biophys. Acta 756, 20–27.
[148] Stremmel, W., Strohmeyer, G., Berk, P.D., 1986. Hepatocellular uptake of oleate is energy dependent, sodium linked, and inhibited by an antibody to a hepatocyte plasma membrane fatty acid binding protein. Proc. Natl Acad. Sci. USA 83, 3584–3588.
[149] Stremmel, W., Strohmeyer, G., Borchard, F., Kochwa, S., Berk, P.D., 1983. Isolation and partial characterization of fatty acid binding protein in rat liver plasma membranes. Hepatology 3, 824.
[150] Stremmel, W., Strohmeyer, G., Borchard, F., Kochwa, S., Berk, P.D., 1985. Isolation and partial characterization of a fatty acid binding protein in rat liver plasma membranes. Proc. Natl Acad. Sci. USA 82, 4–8.
[151] Stremmel, W., Theilmann, L., 1986. Selective inhibition of long-chain fatty acid uptake in short-term cultured rat hepatocytes by an antibody to the rat liver plasma membrane fatty acid-binding protein. Biochim. Biophys. Acta 877, 191–197.
[152] Stuhlsatz-Krouper, S.M., Bennett, N.E., Schaffer, J.E., 1998. Substitution of alanine for serine 250 in the murine fatty acid transport protein inhibits long chain fatty acid transport. J. Biol. Chem. 273, 28642–28650.
[153] Stump, D.D., Fan, X., Berk, P.D., 2001. Oleic acid uptake and binding by rat adipocytes define dual pathways for cellular fatty acid uptake. J. Lipid Res. 42, 509–520.
[154] Stump, D.D., Nunes, R.M., Sorrentino, D., Isola, L.M., Berk, P.D., 1992. Characteristics of oleate binding to liver plasma membranes and its uptake by isolated hepatocytes. J. Hepatol. 16, 304–315.
[155] Stump, D.D., Zhou, S.L., Berk, P.D., 1993. Comparison of plasma membrane FABP and mitochondrial isoform of aspartate aminotransferase from rat liver. Am. J. Physiol. 265, G894–G902.
[156] Stump, D.D., Zhou, S.L., Potter, B.J., Berk, P.D., 1990. Purification of rat liver mitochondrial aspartate aminotransferase and separation of its isoforms utilizing high-performance liquid chromatography. Protein Expr. Purif. 1, 49–53.
[157] Tanaka, T., Sohmiya, K., Kawamura, K., 1997. Is CD36 deficiency an etiology of hereditary hypertrophic cardiomyopathy? J. Mol. Cell Cardiol. 29, 121–127.
[158] Thompson, J., Ory, J., Reese-Wagoner, A., Banaszak, L., 1999. The liver fatty acid binding protein – comparison of cavity properties of intracellular lipid-binding proteins. Mol. Cell. Biochem. 192, 9–16.
[159] Thompson, J., Reese-Wagoner, A., Banaszak, L., 1999. Liver fatty acid binding protein: species variation and the accommodation of different ligands. Biochim. Biophys. Acta 1441, 117–130.
[160] Thompson, J., Winter, N., Terwey, D., Bratt, J., Banaszak, L., 1997. The crystal structure of the liver fatty acid-binding protein. A complex with two bound oleates. J. Biol. Chem. 272, 7140–7150.
[161] Trigatti, B.L., Anderson, R.G., Gerber, G.E., 1999. Identification of caveolin-1 as a fatty acid binding protein. Biochem. Biophys. Res. Commun. 255, 34–39.
[162] Trigatti, B.L., Mangroo, D., Gerber, G.E., 1991. Photoaffinity labeling and fatty acid permeation in 3T3-L1 adipocytes. J. Biol. Chem. 266, 22621–22625.
[163] Trimble, M.E., 1989. Mediated transport of long-chain fatty acids by rat renal basolateral membranes. Am. J. Physiol. 257, F539–F546.
[164] Trotter, P.J., Ho, S.Y., Storch, J., 1996. Fatty acid uptake by Caco-2 human intestinal cells. J. Lipid Res. 37, 336–346.

[165] Truett, G.E., Bahary, N., Friedman, J.M., Leibel, R.L., 1991. Rat obesity gene fatty (fa) maps to chromosome 5: evidence for homology with the mouse gene diabetes (db). Proc. Natl Acad. Sci. USA 88, 7806–7809.

[166] Tsuzuki, T., Obaru, K., Setoyama, C., Shimada, K., 1987. Structural organization of the mouse mitochondrial aspartate aminotransferase gene. J. Mol. Biol. 198, 21–31.

[167] Turcotte, L.P., Kiens, B., Richter, E.A., 1991. Saturation kinetics of palmitate uptake in perfused skeletal muscle. FEBS Lett. 279, 327–329.

[168] Turcotte, L.P., Swenberger, J.R., Tucker, M.Z., Yee, A.J., 1999. Training-induced elevation in FABP(PM) is associated with increased palmitate use in contracting muscle. J. Appl. Physiol. 87, 285–293.

[169] Veerkamp, J.H., Peeters, R.A., Maatman, R.G., 1991. Structural and functional features of different types of cytoplasmic fatty acid-binding proteins. Biochim. Biophys. Acta 1081, 1–24.

[170] Vyska, K., Meyer, W., Stremmel, W., Notohamiprodjo, G., Minami, K., Machulla, H.J., Gleichmann, U., Meyer, H., Korfer, R., 1991. Fatty acid uptake in normal human myocardium. Circ. Res. 69, 857–870.

[171] Waksman, A., Rendon, A., Cremel, G., Pellicone, C., Goubault de Brugiere, J.F., 1977. Intramitochondrial intermembranal reversible translocation of aspartate aminotransferase and malate dehydrogenase through the inner mitochondrial membrane. Biochemistry 16, 4703–4707.

[172] Watkins, P.A., Lu, J.F., Steinberg, S.J., Gould, S.J., Smith, K.D., Braiterman, L.T., 1998. Disruption of the *Saccharomyces cerevisiae* FAT1 gene decreases very long-chain fatty acyl-CoA synthetase activity and elevates intracellular very long-chain fatty acid concentrations. J. Biol. Chem. 273, 18210–18219.

[173] Weisiger, R., Gollan, J., Ockner, R., 1980. An albumin receptor on the liver cell may mediate uptake of sulfobromophthalein and bilirubin: bound ligand, not free, is the major uptake determinant. Gastroenterology 79, 1065.

[174] Weisiger, R.A., 1993. The role of albumin binding in hepatic organic anion transport. In: Berk, P.D., Tavoloni, N. (Eds.), Hepatic Transport and Bile Secretion. Raven Press, New York.

[175] Weisiger, R., Gollan, J., Ockner, R., 1981. Receptor for albumin on the liver cell surface may mediate uptake of fatty acids and other albumin-bound substances. Science 211, 1048–1051.

[176] Weisiger, R.A., Gollan, J.L., Ockner, R.K., 1982. The role of albumin in hepatic uptake processes. Prog. Liver Dis. 7, 71–85.

[177] Winefield, C.S., Farnden, K.J., Reynolds, P.H., Marshall, C.J., 1995. Evolutionary analysis of aspartate aminotransferases. J. Mol. Evol. 40, 455–463.

[178] Wu, F., Corsico, B., Flach, C.R., Cistola, D.P., Storch, J., Mendelsohn, R., 2001. Deletion of the helical motif in the intestinal fatty acid-binding protein reduces its interactions with membrane monolayers: Brewster angle microscopy, IR reflection–absorption spectroscopy, and surface pressure studies. Biochemistry 40, 1976–1983.

[179] Wu, S.Y., Perez, M.D., Puyol, P., Sawyer, L., 1999. Beta-lactoglobulin binds palmitate within its central cavity. J. Biol. Chem. 274, 170–174.

[180] Zhang, Y., Proenca, R., Maffei, M., Barone, M., Leopold, L., Friedman, J.M., 1994. Positional cloning of the mouse obese gene and its human homologue. Nature 372, 425–432.

[181] Zhou, S.L., Gordon, R.E., Bradbury, M., Stump, D., Kiang, C.L., Berk, P.D., 1998. Ethanol up-regulates fatty acid uptake and plasma membrane expression and export of mitochondrial aspartate aminotransferase in HepG2 cells. Hepatology 27, 1064–1074.

[182] Zhou, S.L., Stump, D., Isola, L., Berk, P.D., 1994. Constitutive expression of a saturable transport system for non-esterified fatty acids in *Xenopus laevis* oocytes. Biochem. J. 297(Pt 2), 315–319.

[183] Zhou, S.L., Stump, D., Kiang, C.L., Isola, L.M., Berk, P.D., 1995. Mitochondrial aspartate aminotransferase expressed on the surface of 3T3-L1 adipocytes mediates saturable fatty acid uptake. Proc. Soc. Exp. Biol. Med. 208, 263–270.

[184] Zhou, S.L., Stump, D., Sorrentino, D., Potter, B.J., Berk, P.D., 1992. Adipocyte differentiation of 3T3-L1 cells involves augmented expression of a 43-kDa plasma membrane fatty acid-binding protein. J. Biol. Chem. 267, 14456–14461.

Role of FATP in parenchymal cell fatty acid uptake

Wolfgang Stremmel,* Robert Ehehalt, Thomas Herrmann, Jürgen Pohl and Axel Ring

Department of Gastroenterology, University of Heidelberg, Bergheimer Str. 58, 69115 Heidelberg, Germany
**Correspondence address: Department of Internal Medicine IV, University of Heidelberg, Bergheimer Str. 58, 69115 Heidelberg, Germany. Tel.: +49-62-21-56-87-00; fax: +49-62-21-56-41-16*
E-mail: wolfgang_stremmel@med.uni-heidelberg.de(W.S.)

Abbreviations

DRM: detergent resistant membrane; FAT: fatty acid translocase; FATP: fatty acid transport protein; FABP: fatty acid binding protein; VLCFA: very long chain fatty acids; FACS: fatty acid acyl-CoA synthetase.

In 1994, Schaffer and Lodish [1] used an expression cloning strategy to identify a specific cDNA in 3T3-L1 adipocytes that, when expressed in cultured COS7 cells, augmented uptake of long chain fatty acids. Enrichment of the fluorescent fatty acid analogue BODIPY 3823 was observed with two cDNAs: one was identified as long chain fatty acid acyl-CoA synthetase (FACS), the other was a novel cDNA, encoding the 646 amino acid fatty acid transport protein (FATP) with a predicted molecular weight of 71 kDa. Hydropathy analysis of the deduced amino acid sequence of FATP followed by the Kyte-Doolittle algorithm suggested six potential membrane-spanning regions. The protein was localized to the plasma membrane by subcellular fractionation. Immunofluorescence studies revealed staining at or near the periphery of the cells.

For functional analysis, cell lines expressing FATP at high levels were incubated with radiolabeled oleate. Over a time course of 10 min, uptake values were 3–4 fold increased compared to controls. With increasing levels of unbound oleate, uptake kinetics attributable to the expression of FATP showed a K_m of 0.2 μM and a V_{max} of 0.6 fmol per cell per min [1].

In 2001, the group of Schaffer revised the originally suggested structure prediction. They proposed only one transmembrane domain and multiple membrane-associated domains peripherally associated with the inner leaflet of the plasma membrane [2]. There are two large stretches of non-membrane-associated amino acid residues orientated

Advances in Molecular and Cell Biology, Vol. 33, pages 81–87

ISSN: 1569-2558/DOI: 10.1016/S1569-2558(03)33005-X

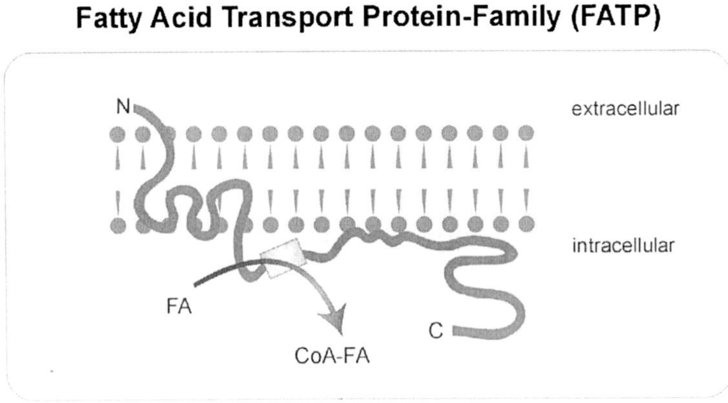

Fig. 1. Schematic membrane topology of FATP. (*For a colored version of this figure, see plate section, page 584.*)

toward the cytosol, one of them carrying a highly conserved AMP binding motive (Fig. 1). Only a very short segment of the amino terminus faces the extracellular side of the membrane bilayer. However, specific binding sites for fatty acids within the FATP structure have not yet been identified. Despite this analysis, there still seems to be uncertainty about the definite membrane topology of the protein, because predictions based on different computer modelings revealed different results [3]. Nevertheless, a classical transmembrane orientation as is expected for membrane transporters cannot be recognized.

Meanwhile a family of FATPs with six members (FATP1–6) with highly conserved structures has been identified [4,5]. The originally described FATP was renamed FATP1. The N-terminal 51 amino acids of the non-glycosylated protein family are variable [4] while the remaining part of the proteins are highly comparable. Thus, the N-termini may represent specific targeting signals for different subcellular organelles or may determine the tissue specificity of individual FATPs. Whether the different FATP family members convey different functions in cellular lipid metabolism or complement each other in individual cell types remains to be elucidated. It is remarkable that specific fatty acid metabolizing tissues contain an array of various FATPs in certain concentrations (Table 1) [6]. A differential regulation by insulin and TNFα of FATP1 and 4 may indicate different responses to metabolic stimuli for tuning cells according to the metabolic demand. In this context, the reported differences in the cellular localization of FATPs are of significance. FATP1, 4 and 6 were reported to reside predominantly at or near the plasma membrane [7]. For FATP1 in adipose tissue it was noted that its subcellular localization can change upon insulin stimulation from an intracellular perinuclear compartment to the plasma membrane [8]. In other studies, FATP4 was proposed to reside in endomembranes in the vicinity of the plasma membrane [3,9]. Further characterization of these endomembranes close to the plasma membrane has not been performed. FATP2 was found in peroxisomes [10] and also at the ER [11,12].

The catalytic region of this membrane-associated protein family is the highly conserved AMP binding region [4]. Carefully conducted studies by a number of investigators demonstrated, in mammalian cells, acyl-CoA synthetase activity of FATP1, 2, 4 and 5

Table 1
Expression patterns of FATP family members

Gene	Intestine	Colon	Heart	Brain	Lung	Liver	Skeletal muscle	Kidney	Adipose
mmFATP1	− − −	− − −	+++	++	+	+	+++	++	+++
mmFATP2	+	− − −	− − −	− − −	− − −	+++	− − −	+++	
mmFATP3	− − −	− − −	− − −	− − −	++	+++	− − −	− − −	
mmFATP4	+++	− − −	+	++	++	+++	− − −	++	
mmFATP5	− − −	− − −	− − −	− − −	− − −	+++	− − −	− − −	
mmFATP6			+++						
hsFATP1	− − −		+++						
hsFATP2	+							+++	
hsFATP3	− − −				+++				
hsFATP4	+++	+++	++	+++	+	++	+	+++	+
hsFATP5	− − −					+++			
hsFATP6	− − −								

mRNA expression levels are denoted by (+) or (− − −), blank areas indicate a lack of data. (Data were collected from Refs. [1,4,5,9].)

toward various lipid compounds, particularly for fatty acids [10–15]. A preference for very long chain fatty acids (VLCFA) has been shown for FATP1, 2 and 4. It is challenging to speculate that facilitation of cellular fatty acid uptake may be due to this preferential activation of VLCFA. Indeed, the original data about the functional characterization of FATP1 and 4 expressing cells verified a moderate but clear increase in cellular uptake of the long chain fatty acid [3H] oleate [1,9]. Thus, although FATP1 and 4 reveal high acyl-CoA synthetase activity for VLCFA, uptake of long chain fatty acids in their unesterified form was observed. How can this be explained?

VLCFA are thought to be essential components of lipid rafts, which represent dynamic assemblies of proteins and lipids within cellular membranes and play an important role in membrane compartmentation, membrane sorting and trafficking. Rafts are small platforms (30–50 nm diameter), composed of sphingolipids and cholesterol in the exoplasmic leaflet, connected to glycerophospholipids and cholesterol in the cytoplasmic leaflet of the lipid bilayer (Fig. 3C) [16]. It is not clear how the inner leaflet is coupled to the outer leaflet, but one likely possibility is that VLCFA substitutes of sphingolipids extend from the outer to the inner leaflet [17]. In fact, sphingolipids with fatty acid side chains of 24:0 and 24:1 are found preferentially in these domains. By interdigitation with the saturated side chains of the glycerophospholipids present in the inner leaflet, a tight communication is established [17]. As FATP1, 2 and 4 preferentially activate VLCFA to acyl-CoA compounds we therefore assume that they must play an important role in the sphingolipid-dependent membrane bilayer interdigitation process within raft microdomains. Caveolae represent specialized microdomains formed through coalescence of raft domains [16,18] and recruitment of caveolin, their major structural protein. In previous studies, we were able to show that caveolae are indeed connected to the overall cellular uptake process of long chain fatty acids [19,20]. When human microvascular endothelial cells were exposed to the fluorescent long chain fatty acid derivative NBD stearate and subsequently processed for caveolin-1 immunostaining, co-localization of 12-NBD stearate with caveolin-1 was observed (Fig. 2). Dissociation of rafts by cholesterol depletion [20] or transfection of the dominant-negative caveolin mutant Cav^{DGV} (J.P. and W.S., personal communication) significantly reduced internalization of long chain fatty acids. There are only a definite number of proteins described, which have a high affinity to lipid rafts. One of them is FAT/CD36, which is known as a membrane fatty acid facilitating transporter [21]. It was shown that FAT/CD36 can translocate on demand from an intracellular site to the plasma membrane [22], where it is preferentially located in lipid rafts [23,24]. We hypothesize that an increase in cellular fatty acid uptake observed in FATP-over-expressing cells represents an indirect event involving recruitment of membrane transporters (e.g. FAT/CD36) to the plasma membrane (Fig. 3). Thus, VLCFA taken up into cells via the known membrane transporters (FAT/CD36, $FABP_{PM}$) might be activated by FATPs to very long chain acyl-CoA and could incorporate into sphingolipids of raft

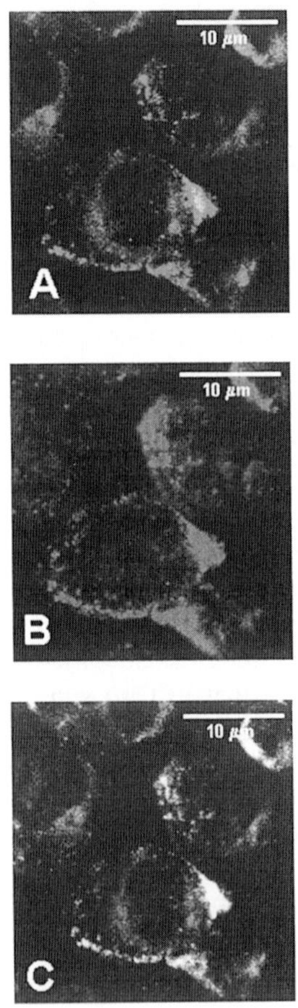

Fig. 2. Immunocytochemistry staining of caveolae after incubation with 12-NBD stearate. Panel A: staining of the plasma membrane and intracellular vesicles by 12-NBD stearate. Panel B: staining of caveolin-1 on the plasma membrane. Panel C: co-localization of caveolin-1 and 12-NBD stearate at the plasma membrane. (*For a colored version of this figure, see plate section, page 585.*)

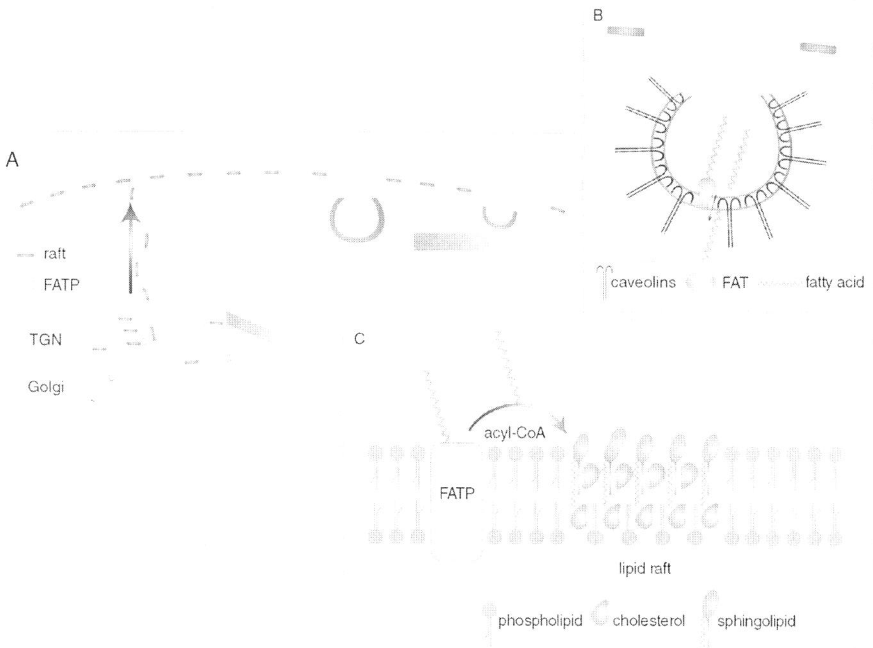

Fig. 3. Proposed model for cellular fatty acid uptake. FATP, stabilization of lipid rafts by acylated VLCFA and FAT/CD36 incorporation into lipid rafts could cooperate to facilitate efficient LCFA uptake. Intracellular VLCFA are esterified to acyl-CoA and thereafter bound to sphingolipids in the outer leaflet of the biomembrane of lipid rafts. This might indirectly facilitate on-demand recruitment of FAT/CD36 from an intracellular compartment to lipid rafts located on the plasma membrane and result in augmented uptake of long chain fatty acids. (*For a colored version of this figure, see plate section, page 586.*)

microdomains. Exposure of cells to insulin and cholesterol is known to result in FAT/CD36 translocation from an intracellular compartment to the plasma membrane [22,23]. This might indicate recruitment of FAT/CD36 to lipid rafts, which finally results in augmented uptake of long chain fatty acids. This is in accordance with the observed moderate increase of oleate uptake in FATP1-overexpressing cells [1]. While the increase in V_{max} represents the recruitment of additional transport facilitating proteins, the K_m of 0.2 μM is compatible to the K_m for oleic acid uptake by 3T3-L1 adipocytes ($K_m = 0.2$ μM), which is mainly mediated by FAT/CD36.

Since we hypothesize that FATPs with VLACS activity are essential for the constitution of raft lipids, a localization from ER to the post-Golgi compartment, the site of lipid raft constitution, can be postulated. Whether these FATPs move from there (via caveolae) to the plasma membrane or, more likely, remain as non-raft proteins at the post-Golgi compartment is one of the open questions.

In case FATPs with VLACS activity can indeed stimulate lipid raft constitution within cells, an alternative explanation of the induced accelerated uptake could be discussed. With more caveolae available, the caveolae-dependent uptake route of certain molecules, in particular long chain fatty acids, could be augmented even independently of FAT/CD36. According to this hypothesis, uptake of long chain fatty acids from the circulation might include the following subsequent steps: a) dissociation of fatty acids bound to albumin in the vicinity of the plasma membrane through interaction with membrane-associated fatty acid binding proteins ($FABP_{PM}$, FAT/CD36), b) protonization due to the acid microclimate of the cell surface, c) partition in the outer layer of the plasma membrane bilayer, d) flip–flop across the membrane bilayer according to their concentration gradient, e) binding to caveolin-1 at the inner surface and f) budding-off with the caveolar membrane for intracellular transport.

For functional analysis of FATP4 we generated an FATP4 knockout mouse [25]. FATP4 null mice died in the early neonatal period, exhibiting features of a lethal restrictive dermopathy. Lipid analysis demonstrated a disturbed fatty acid composition of epidermal ceramides, in particular a decrease in the C26:0 and C26:0-OH fatty acid substitutes. Ceramides represent a subclass of sphingolipids, which require VLCFA for proper function in particular to maintain a normal epidermal barrier. Even if the raft assembly would still be operative due to compensation by other acyl-CoA synthetases, it may not be sufficient to incorporate certain very long chain (e.g. C26:0 or C26:0-OH) fatty acids into the ceramide backbone in the required quantity. A detailed molecular analysis is required to address these open issues.

So far, the biological role of FATP in the cellular uptake process of long chain fatty acids is changing from the traditional view as a direct membrane fatty acid carrier protein [1] to an enzyme that incorporates VLCFA into the endomembrane vesicular trafficking compartment, which is directly or indirectly involved in cellular uptake of regular native long chain fatty acids. Thereby it helps to constitute fatty acid transporting caveolar vehicles or the integration of relevant fatty acid transporters into the plasma membrane. Further studies to test this hypothesis are on their way.

Acknowledgements

This study was supported by grant STR 216/11-1 of the Deutsche Forschungsgemeinschaft and generous gifts of the Dietmar Hopp foundation.

References

[1] Schaffer, J.E., Lodish, H.F., 1994. Expression cloning and characterization of a novel adipocyte long chain fatty acid transport protein. Cell 79, 427–436.

[2] Lewis, S.E., Listenberger, L.L., Ory, D.S., Schaffer, J.E., 2001. Membrane topology of the murine fatty acid transport protein 1. J. Biol. Chem. 276, 37042–37050.

[3] Herrmann, T., van der Hoeven, F., Gröne, H.-J., Stewart, A.F., Langbein, L., Kaiser, I., Liebisch, G., Gosch, I., Buchkremer, F., Drobnik, W., Schmitz, G., Stremmel, W., 2003. Mice with targeted disruption of the fatty acid transport protein 4 (Fatp 4, Slc27a4) gene show features of lethal restrictive dermopathy. J. Cell. Biol. 161, 1105–1115.

[4] Hirsch, D., Stahl, A., Lodish, H.F., 1998. A family of fatty acid transporters conserved from mycobacterium to man. Proc. Natl Acad. Sci. USA 95, 8625–8629.

[5] Gimeno, R.E., Ortegon, A.M., Patel, S., Punreddy, S., Ge, P., Sun, Y., Lodish, H.F., Stahl, A., 2003. Characterization of a heart-specific fatty acid transport protein. J. Biol. Chem. 278, 16039–16044.

[6] Frohnert, B.I., Bernlohr, D.A., 2000. Regulation of fatty acid transporters in mammalian cells. Prog. Lipid Res. 39, 83–107.
[7] Stahl, A., Gimeno, R.E., Tartaglia, L.A., Lodish, H.F., 2001. Fatty acid transport proteins: a current view of a growing family. Trends Endocrinol. Metab. 12, 266–273.
[8] Stahl, A., Evans, J.G., Pattel, S., Hirsch, D., Lodish, H.F., 2002. Insulin causes fatty acid transport protein translocation and enhanced fatty acid uptake in adipocytes. Dev. Cell 2, 477–488.
[9] Stahl, A., Hirsch, D.J., Gimeno, R.E., Punreddy, S., Ge, P., Watson, N., Patel, S., Kotler, M., Raimondi, A., Tartaglia, L.A., Lodish, H.F., 1999. Identification of the major intestinal fatty acid transport protein. Mol. Cell 3, 299–308.
[10] Uchiyama, A., Aoyama, T., Kamijo, K., Uchida, Y., Kondo, N., Orii, T., Hashimoto, T., 1996. Molecular cloning of cDNA encoding rat very long-chain acyl-CoA synthetase. J. Biol. Chem. 271, 30360–30365.
[11] Steinberg, S.J., Wang, S.J., McGuinness, M.C., Watkins, P.A., 1999. Human liver-specific very-long-chain acyl-coenzyme A synthetase: cDNA cloning and characterization of a second enzymatically active protein. Mol. Genet. Metab. 68, 32–42.
[12] Steinberg, S.J., Wang, S.J., Kim, D.G., Mihalik, S.J., Watkins, P.A., 1999. Human very-long-chain acyl-CoA synthetase. Cloning, topography, and relevance to branched-chain fatty acid metabolism. Biochem. Biophys. Res. Commun. 257, 615–621.
[13] Coe, N.R., Smith, A.J., Frohnert, B.I., Watkins, P.A., Bernlohr, D.A., 1999. The fatty acid transport protein (FATP1) is a very long chain acyl-CoA synthetase. J. Biol. Chem. 274, 36300–36304.
[14] Berger, J., Truppe, C., Neumann, H., Forss-Petter, S., 1998. A novel relative of the very-long-chain acyl-CoA synthetase and fatty acid transporter protein genes with a distinct expression pattern. Biochem. Biophys. Res. Commun. 247, 255–260.
[15] Herrmann, T., Buchkremer, F., Gosch, I., Hall, A.M., Bernlohr, D.A., Stremmel, W., 2001. Mouse fatty acid transport protein 4 (FATP4): characterization of the gene and functional assessment as a very long chain acyl-CoA synthetase. Gene 270, 31–40.
[16] Simons, K., Ehehalt, R., 2002. Cholesterol, lipid rafts, and disease. J. Clin. Invest. 110, 597–603.
[17] Rietveld, A., Simons, K., 1998. The differential miscibility of lipids as the basis for the formation of functional membrane rafts. Biochim. Biophys. Acta 1376, 467–479.
[18] Simons, K., Ikonen, E., 1997. Functional rafts in cell membranes. Nature 387, 569–572.
[19] Ring, A., Pohl, J., Völkl, A., Stremmel, W., 2002. Evidence for vesicles that mediate long-chain fatty acid uptake by human microvascular endothelial cells. J. Lipid Res. 43, 2095–2104.
[20] Pohl, J., Ring, A., Stremmel, W., 2002. Uptake of long-chain fatty acids in HepG2 cells involves caveolae: analysis of a novel pathway. J. Lipid Res. 43, 1390–1399.
[21] Abumrad, N.A., El-Maghrabi, M.R., Amri, E.Z., Lopez, E., Grimaldi, P.A., 1993. Cloning of rat adipocyte membrane-protein implicated in binding or transport of long chain fatty acids that is induced during preadipocyte differentiation: homology with human CD 36. J. Biol. Chem. 268, 17665–17668.
[22] Luiken, J.J., Koonen, D.P., Willems, J., Zorzano, A., Becker, C., Fischer, Y., Tandon, N.N., van der Vusse, G.J., Bonen, A., Glatz, J.F., 2002. Insulin stimulates long-chain fatty acid utilization by rat cardiac myocytes through cellular redistribution of FAT/CD36. Diabetes 51, 3113–3119.
[23] Kolleck, I., Guthmann, F., Ladhoff, A.M., Tandon, N.N., Schlame, M., Rüstow, B., 2002. Cellular cholesterol stimulates uptake of palmitate by redistribution of fatty acid translocase in type II pneumocytes. Biochemistry 41, 6369–6375.
[24] Frank, P.G., Marcel, Y.L., Connelly, M.A., Lublin, D.M., Franklin, V., Williams, D.L., Lisanti, M.P., 2002. Stabilization of caveolin-1 by cellular cholesterol and scavenger receptor class B type I. Biochemistry 41, 11931–11940.

Uptake of fatty acids by parenchymal cells: role of FAT/CD36

Jan F.C. Glatz,[a,*] Joep F.F. Brinkmann,[b] Arend Bonen,[c]
Ger J. van der Vusse[b] and Joost J.F.P. Luiken[a]

[a]*Department of Molecular Genetics, Cardiovascular Research Institute Maastricht (CARIM), Maastricht University, P.O. Box 616, NL-6200 MD Maastricht, The Netherlands*
[b]*Department of Physiology, Cardiovascular Research Institute Maastricht (CARIM), Maastricht University, 6200 MD Maastricht, The Netherlands*
[c]*Department of Human Biology and Nutritional Sciences, University of Guelph, Guelph, Ont., Canada N1G 2W1*
[*]*Correspondence address: Tel.: +31-43-388-1998; fax: +31-43-388-4574*
E-mail: glatz@gen.unimaas.nl(J.F.C.G.)

Abbreviations

LCFA: long-chain fatty acids; FABP: fatty acid-binding protein; $FABP_c$: cytoplasmic FABP; $FABP_{pm}$: plasma membrane FABP; FAT: fatty acid translocase (CD36); FATP: fatty acid-transport protein.

1. Introduction

Long-chain fatty acids (LCFA) are important to the cell as a source of energy and as building blocks for membrane phospholipids, and are involved in post-translational modification of proteins. In addition, LCFA, their coenzyme A derivatives and metabolites – e.g. prostaglandines, leukotrienes, and thromboxanes – are able to regulate gene expression, as well as enzyme, ion channel and membrane receptor activity. Inside cells, LCFA can be stored in triacylglycerols (TAGs), which serve mainly as storage of metabolic energy. The tissues most actively involved in LCFA metabolism are liver, adipose tissue, and cardiac and skeletal muscle. In blood plasma LCFA are present either complexed to albumin or esterified in the TAG core of circulating lipoproteins. Following their dissociation from albumin or hydrolysis of the TAGs by action of lipoprotein lipase, LCFA are available for uptake by tissue cells. In heart and muscle LCFA first have to pass the capillary endothelium before reaching the interstitial space, but in liver and adipose tissue interendothelial clefts allow the penetration of plasma albumin and, therefore, the direct delivery of LCFA to parenchymal cells.

Advances in Molecular and Cell Biology, Vol. 33, pages 89–98

ISSN: 1569-2558/DOI: 10.1016/S1569-2558(03)33006-1

In recent years much progress has been made with respect to our understanding of the first step of tissue LCFA utilization, i.e. LCFA uptake. Compelling evidence is now available that LCFA can enter cells either by passive diffusion through the lipid bilayer or by protein-mediated transmembrane transport [1–5] and that the latter represents a site of control, allowing changes in the presence and/or activity of these membrane proteins to regulate LCFA uptake. In this chapter we briefly review the three membrane-associated fatty acid-binding proteins (FABPs) identified to date and which facilitate transmembrane transport of LCFA, and then discuss novel data indicating that one of these, namely fatty acid translocase (FAT)/CD36, is also involved in the acute regulation of LCFA uptake. Moreover, at least in heart and skeletal muscle, alterations in LCFA utilization as observed in disease appear to be related to abnormalities in the presence or functioning of this latter protein.

2. Membrane-associated fatty acid-binding proteins

Although LCFA are capable of diffusing through cellular membranes by a spontaneous diffusional flip-flop process, there is ample evidence for an additional protein-facilitated component in the transport of LCFA across the cellular membrane in adipose tissue [6], liver [7], and cardiac [8,9] and skeletal muscle [10,11]. This led to the identification of three membrane proteins with the ability to non-covalently bind LCFA. These proteins are 43-kD plasma membrane LCFA binding protein ($FABP_{pm}$), a family of some five 60-kD fatty acid-transport proteins (FATP1-5), and 88-kD fatty acid translocase (FAT)/CD36. Both $FABP_{pm}$ and FATP have been detected in virtually all tissues examined, but FAT/CD36 shows a more restricted expression, as in most species it is absent in liver and brain [12]. Interestingly, the three proteins are simultaneously expressed in heart and skeletal muscles [13], making striated muscle types as useful tissues for the study of their putative involvement in LCFA uptake. Because in this chapter we will focus on the significance of FAT/CD36 for cellular LCFA uptake and metabolism, $FABP_{pm}$ and the FATPs are discussed only when relevant for the discussion.

FAT/CD36 was identified as a candidate LCFA transporter based on its binding of both DIDS (4,4′-di-isothiocyanostilbene-2,2′-sulphonate) and reactive sulfo-*N*-succinimidyl LCFA esters [14,15]. These compounds were inhibitors of LCFA uptake in isolated adipocytes. The corresponding cDNA was isolated from a rat adipose tissue library and called "fatty acid translocase" [16], the rat homolog of the ubiquitously expressed human platelet CD36 (also known as PASIV, GPIV, and GPIIIb in the earlier literature [17]). FAT/CD36 is a heavily glycosylated integral membrane protein, with an apparent molecular mass of 88 kDa. The protein was initially isolated from platelet membranes as a thrombospondin receptor [18]. FAT/CD36 is expressed abundantly in tissues active in LCFA metabolism, such as heart, (red) skeletal muscle, adipose tissue, and intestine [13,16, 19]. Its expression profile corresponds with a role in LCFA uptake, as is illustrated in Fig. 1 for rat heart and red and white skeletal muscle. Moreover, FAT/CD36 expression is modulated by conditions that alter lipid metabolism such as high-fat feeding [19–21]. Remarkably, FAT/CD36 expression is absent or low in the liver [19]. Expression of FAT/CD36 in fibroblasts that normally lack the protein resulted in an increase in LCFA

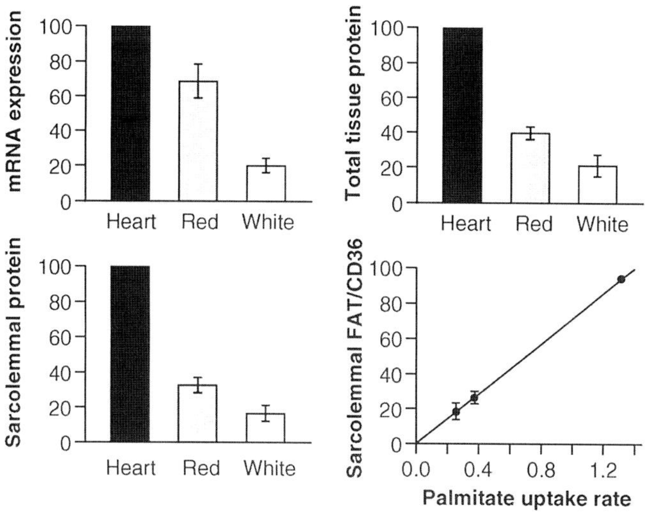

Fig. 1. Relative expression of FAT/CD36 in adult rat heart, red skeletal muscle and white skeletal muscle and its relation to the rate of cellular fatty acid uptake. FAT/CD36 expression was measured on the mRNA level by Northern blot analysis and on the protein level by quantitative Western blotting. Palmitate uptake rates were measured in giant vesicles obtained from the three muscle types. In each bar graph the highest value was arbitrarily set at 100%. Data are obtained from Refs. [12,13,31].

uptake [22]. It has to be mentioned, however, that a similar approach using a rat heart muscle-derived cell line did not yield a direct correlation between the FAT/CD36 protein and LCFA uptake [23]. The possible absence of obligatory partner proteins for LCFA transport in the cells used, with cytoplasmic FABP as a likely candidate, might explain this deviant observation (see below).

3. Molecular mechanism of cellular fatty acid uptake

For both $FABP_{pm}$, FATP and FAT/CD36 evidence is available, mostly from transfection studies with cells in culture, that in their presence the rate of cellular LCFA uptake is markedly increased [22,24,25]. However, the exact nature by which each of these three proteins facilitates transmembrane translocation of LCFA is not yet known, partly because their membrane topology is still poorly understood. Therefore, the precise molecular mechanism by which LCFA are taken up by cells is not yet clear. Based on presently available knowledge, the following mechanism has been proposed (schematically depicted in Fig. 2). $FABP_{pm}$ is a peripheral membrane protein [26] and is believed to function in the trapping of LCFA, whereafter LCFA may cross the membrane either by passive diffusion or facilitated by either FAT/CD36 or (one of) the FATPs. Because FAT/CD36 is a highly glycosylated, integral membrane protein that is believed to have only two transmembrane spanning regions [29], it is unlikely that it facilitates LCFA transport by channeling LCFAs through a pore in the membrane. In view of this notion it has been suggested that FAT/CD36 also operates mainly as a plasmalemmal receptor for LCFA [3,27]. As a consequence, the LCFA concentration in close vicinity of the cellular

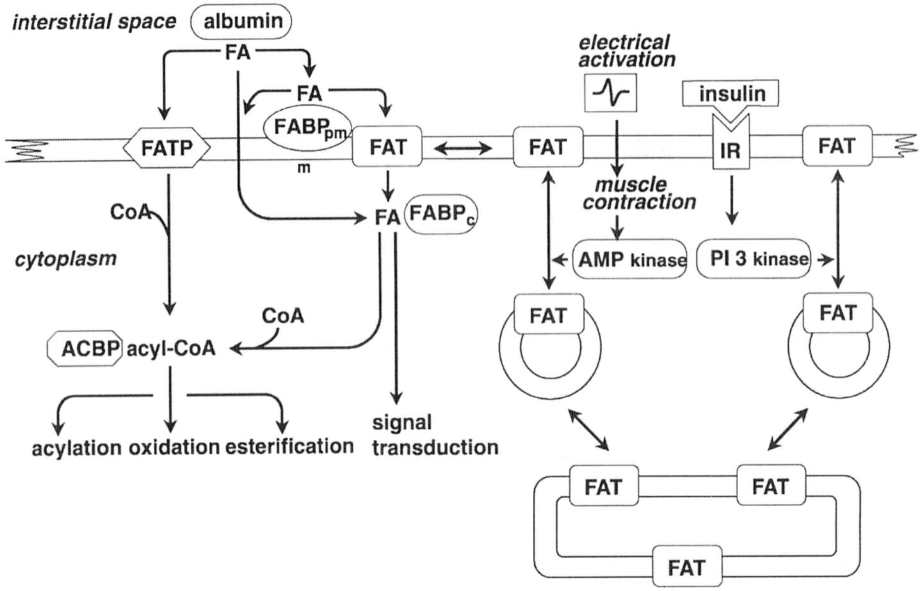

Fig. 2. Schematic presentation of the cellular uptake and utilization of long-chain fatty acids (LCFA or FA) illustrating the presumable roles of various lipid binding proteins in this process. Following their dissociation from plasma albumin, the transmembrane translocation of LCFA most likely takes place either by passive diffusion through the lipid bilayer, or facilitated by membrane-associated proteins, or by a combination of both. This includes $FABP_{pm}$ acting as scavenger and FAT/CD36 acting as scavenger and/or transporter of LCFA. FATP most likely is involved in fatty acyl-CoA synthesis. Intracellularly, LCFA will be bound by cytoplasmic FABP and, after activation to fatty acyl-CoA, by acyl-CoA binding protein (ACBP) [53]. LCFA uptake can be modulated by recycling of FAT/CD36 between the plasma membrane and an endosomal compartment. Alterations in redistribution of FAT/CD36 can be mediated by insulin, following the binding of this hormone to its receptor and involving PI3 kinase, or by muscle contraction, which activates AMP-activated protein kinase. $FABP_{pm}$, plasma membrane fatty acid-binding protein; FATP, fatty acid-transport protein; FAT, fatty acid translocase (CD36); $FABP_c$, cytoplasmic fatty acid-binding protein; ACBP, acyl-CoA binding protein; IR, insulin receptor.

membrane will increase, which will promote LCFA diffusion through the membrane. On the other hand, FAT/CD36 has also been postulated to facilitate the transfer of LCFA from the outer to the inner leaflet of the phospholipid bilayer ("flip-flop") by decreasing the activation energy for movement of the polar LCFA carboxy group through the membrane bilayer [27]. Inside cells, LCFA are bound to cytoplasmic FABP ($FABP_c$), which acts as an acceptor protein and is viewed as an intracellular counterpart of plasma albumin [3,28, 29]. Interestingly, there is evidence for an interaction between FAT/CD36 and $FABP_c$ [30], and it is conceivable that such interaction would accelerate the transfer of LCFA between these two proteins. The role of $FABP_c$ as intracellular LCFA acceptor and cytoplasmic carrier protein has now clearly been established [28,31]. With respect to the role of FATPs in the cellular LCFA uptake process, it was recently found that FATP1-5 also possess acyl-coenzyme A synthetase activity, particularly for very long-chain fatty acids [32–34], so that metabolic trapping might explain their facilitation of LCFA uptake.

Taken together, the current concept is that cellular LCFA uptake occurs by both passive diffusion and protein-mediated LCFA uptake. The membrane-associated FABPs function primarily in the trapping of LCFA from extracellular donors and their release to intracellular targets, whereby the actual transmembrane translocation step occurs by passive diffusion of LCFA through the lipid bilayer. The driving force and direction of LCFA migration most likely are governed by the transmembrane gradient of LCFA concentration.

4. Acute regulation of functional FAT/CD36 expression by protein translocation

In studies with cardiac and skeletal muscle it has been demonstrated that FAT/CD36 not only facilitates but also acutely regulates cellular LCFA uptake. Heart and skeletal muscle, being key LCFA oxidizers and principal sites for the removal of LCFA from the circulation, have a variable metabolic rate and should be able to adapt LCFA transport quickly so as to respond appropriately to changes in energy demand. By comparing resting and in vivo electrically stimulated skeletal muscles, or electrically stimulated isolated cardiac myocytes, it was shown that contraction is able to induce the translocation of FAT/CD36 from an intracellular storage pool to the plasma membrane [35,36]. This translocation was accompanied by a marked increase in the rate of LCFA uptake. This mechanism of acute – within minutes – regulation of LCFA uptake is analogous to the manner in which muscular glucose uptake is regulated by the transporter GLUT4 [35].

The translocational mechanism is substantiated by several observations. Most importantly, LCFA uptake by muscle giant sarcolemmal vesicles is increased after just 5 min of in vivo electrical stimulation. The giant sarcolemmal vesicle preparation is a model representing the plasma membrane, and is suitable for measuring LCFA uptake in the absence of metabolism. Giant-vesicle palmitate uptake increased further over a 30 min contraction period, and decreased time-dependently after termination of the electrical stimulus. Moreover, the increase in vesicular LCFA uptake was shown to be proportional to the rate of muscle stimulation, and could be inhibited by sulpho-*N*-succinimidyl oleate, an inhibitor specific for FAT/CD36. By comparison with GLUT4, gradient centrifugation showed that FAT/CD36 in skeletal muscle is present in both an intracellular pool and a membrane-associated pool (Fig. 3). Almost equal quantities of FAT/CD36 were found in the two pools under basal conditions; electrical stimulation increased the amount of FAT/CD36 in the sarcolemmal fraction by 1.8-fold and decreased that in the intracellular pool by 55% [35].

More recently it was discovered that also insulin is able to induce the recruitment of FAT/CD36 to the plasma membrane to increase LCFA uptake in both cardiac and skeletal muscle [37,38]. The insulin-inducible FAT/CD36 translocation was blocked in the presence of wortmannin, indicating that the translocation is mediated by the signaling protein phosphatidylinositol 3-kinase (PI3-kinase). Strikingly, the effects of muscle contraction and of insulin on both LCFA uptake and FAT/CD36 translocation to the sarcolemma appear additive. This latter finding relates to the fact that the contraction-induced FAT/CD36 translocation is mediated through AMP-activated protein kinase signaling [39]. Together, these observations suggest that there are (at least) two separate

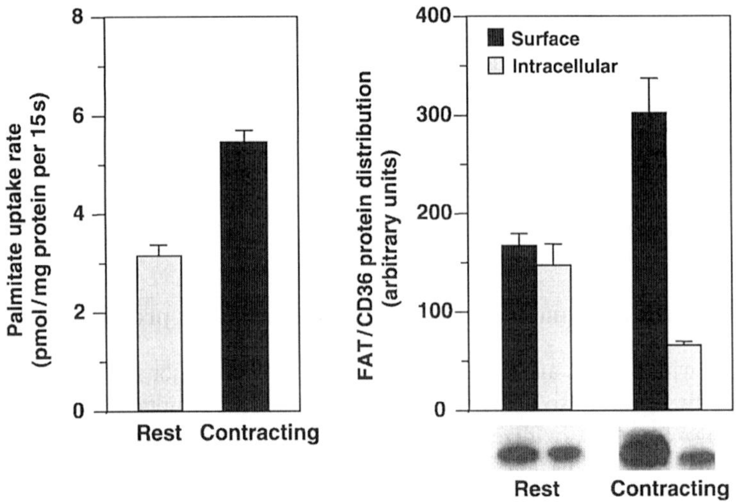

Fig. 3. Translocation of FAT/CD36 from intracellular pools to the sarcolemma upon short-term electrical stimulation of adult rat hindlimb muscles. Muscles were stimulated via the sciatic nerve to contract for 30 min at 40 tetani/min, while the contralateral muscles from the same animal served as non-contracting controls. Left panel. Palmitate uptake rate, measured in giant sarcolemmal vesicles prepared from these muscles, was 65% higher in the stimulated compared to control muscles. Right panel. Distribution of FAT/CD36 in resting and contracting muscles. After density gradient fractionation of the muscles into separate surface and intracellular compartments, FAT/CD36 was visualized and quantified by Western blot analysis. Adapted from Ref. [35]. (*For a colored version of this figure, see plate section, page 587.*)

intracellular pools from which FAT/CD36 can be recruited, one being sensitive to contraction and the other to insulin, or, alternatively, that there is a single depot from which FAT/CD36 can be mobilized following two independent signal transduction cascades (Fig. 3). This mechanism, again, resembles the well-documented manner by which muscle contraction and insulin can independently translocate the glucose transporter GLUT4 from intracellular stores to the sarcolemma [40]. Interestingly, selective agents are able to mobilize FAT/CD36 to the sarcolemma in the absence of GLUT4 translocation, and vice versa, indicating that downstream from the main signaling enzymes, i.e. PI3-kinase and AMP-activated protein kinase, each of the signal transduction pathways branches off into two independent pathways, one leading to mobilization of intracellularly stored GLUT4 and the other to mobilization of intracellularly stored FAT/CD36 (Luiken, unpublished observations).

5. Chronic changes in fatty acid metabolism and in FAT/CD36

The direction and rate of LCFA movement across the plasma membrane are determined by the transmembrane gradient of LCFA [28,41] and, therefore, depend on the plasma supply of LCFA and the metabolic state of the cell. However, given the fact that both membrane-associated and cytoplasmic FABPs increase the rate of LCFA transport across the membrane, long-term alterations in the presence and/or activity of these FABPs would also have an impact on the actual (mean) rate of transport. In line with this notion are the observations that chronic changes in tissue LCFA utilization, such as induced by exercise training, nutrition, and pharmacological manipulations, are paralleled by concomitant changes in the tissue content of membrane as well as cytoplasmic FABPs (reviewed in Refs. [3,28,29]). For instance, FAT/CD36 expression is up-regulated with chronic (i.e. 7 days) stimulation of skeletal muscle concomitant with an increase in the maximal rate of LCFA transport [42]. Conversely, changes in cellular content of FAT/CD36, such as experimentally induced by genetic manipulations (transgenic animals), lead to parallel limitations of the rate of LCFA uptake and utilization. For instance, mice with a disrupted gene encoding FAT/CD36 showed a 50–80% decreased in vivo uptake rate of iodonated LCFA analogs [43].

The factors that regulate the chronic adjustments in the transcription of FAT/CD36 are not yet fully elucidated. Interestingly, LCFA themselves are capable of controlling transcriptional activity of the gene encoding FAT/CD36 [44]. Studies with cultured rat neonatal cardiomyocytes showed that exposure of the cells to physiological levels of LCFA in the surrounding medium induced approximately threefold increases in the mRNA levels of FAT/CD36, as well as that of $FABP_c$ and fatty acyl-CoA synthetase, together with a similar increase in the rate of cellular LCFA uptake and oxidation [45]. Thus, LCFA, or their derivatives, can control their own rate of cellular utilization through fine-tuning of gene expression, including that for FAT/CD36. Most likely, LCFA are natural ligands for members of the family of peroxisome-proliferator activated receptors (PPARs), in particular PPARα, resulting in the up-regulation of transcription of many genes involved in lipid utilization [44].

6. Alterations in fatty acid uptake and FAT/CD36 in disease

Various metabolic abnormalities and diseases characterized by alterations in the rate of LCFA utilization, mostly those affecting the functioning of cardiac and skeletal muscles, have been linked to FAT/CD36, either through the deficiency or mutated state of FAT/CD36 or to malfunctioning of the intracellular translocation of the protein. For instance, in humans, the shift in myocardial substrate utilization seen during the development of hypertrophy, notably a decrease in the use of LCFA together with an increase in that of glucose [46], has been associated with a mutation or deficiency in FAT/CD36 [47,48]. The fact that FAT/CD36 null mice show impaired myocardial LCFA uptake [43] and develop cardiac hypertrophy [49] underscores the notion that FAT/CD36-mediated LCFA uptake may represent an important control site for myocardial LCFA utilization, and that FAT/CD36 deficiency and defective myocardial LCFA uptake are causally linked.

In diabetes mellitus myocardial as well as skeletal muscle glucose utilization are markedly reduced and compensated for by an increase in LCFA utilization. In streptozotocin-induced diabetic rats, a widely used model for type-1 diabetes, this is accompanied by a slight increase in the expression of the genes encoding for $FABP_{pm}$, FATP, FAT/CD36, and $FABP_c$ [12], while at the protein level there is a 2–3-fold increase in muscle content of FAT/CD36, and a 1.3–1.8-fold increase in that of $FABP_c$ [21]. In the obese Zucker rat, a well-established model for obesity and type-2 diabetes [50], the increased rate of heart and skeletal muscle LCFA uptake could not be associated with changes in FAT/CD36 mRNA nor tissue protein contents, but in both tissues there was an increased abundance of FAT/CD36 at the sarcolemma (1.6-fold in heart, 1.8-fold in muscle) [51]. Thus, it appears that in heart and muscle of obese Zucker rats the total cellular pool of FAT/CD36 is similar to that found in lean animals, but a larger proportion of the protein is permanently relocated to the cell surface at the expense of the intracellular storage compartment, resulting in higher LCFA uptake rates. This altered relocation could be the result of either an increased mobilization of FAT/CD36 or an impairment in the rate of endocytosis. In any case, the machinery regulating the subcellular distribution of FAT/CD36 might play a pivotal role in the etiology of obesity and type-2 diabetes.

7. Concluding remarks

Ten years after its identification as membrane-associated LCFA binding protein [14], FAT/CD36 is now generally accepted to play a central role in the cellular uptake of LCFA. Although the mechanism by which FAT/CD36 exerts this action is not yet unraveled, the recent observation that LCFA uptake by heart and muscle is subject to short-term regulation involving the translocation of FAT/CD36 from an intracellular depot to the plasma membrane establishes that FAT/CD36 not only facilitates the uptake process but also is a site of regulation of cellular LCFA metabolism. The physiological significance of FAT/CD36 could very recently been confirmed by the finding that the myocardial recovery from ischemia is impaired in FAT/CD36 null mice and could be restored by myocyte FAT/CD36 re-expression [52]. The significance of FAT/CD36 is also clear from studies of several diseases, as illustrated above for cardiac hypertrophy and for obesity and type-2 diabetes. Hence, malfunctioning of the FAT/CD36-mediated LCFA uptake process may be a critical factor in the pathogenesis of these and perhaps of other metabolic diseases in which lipid metabolism is altered. Future studies should be directed towards further unraveling the mechanism and regulation of cellular LCFA uptake, especially the signaling cascade(s) involved in the cellular redistribution of FAT/CD36. The emerging evidence for a concerted action of the various proteins involved and for a role of protein–protein interaction therein should also receive attention.

Acknowledgements

The authors thank Drs N.A. Abumrad and A. Ibrahimi, State University of New York at Stony Brook, NY, USA, for stimulating discussions, and J. Willems for his help in preparing the illustrations. Work in the author's laboratories was supported by the Netherlands Heart Foundation (grants 97.149 and D98.012), and the Ontario Heart and Stroke Foundation. Joost J.F.P. Luiken is the recipient of a VIDI-Innovational Research grant from the Netherlands Organisation for Scientific Research (NWO-ZonMw grant nr. 016.036.305).

References

[1] Hamilton, J.A., Kamp, F., 1999. Diabetes 48, 2255–2269.
[2] Berk, P.D., Stump, D.D., 1999. Mol. Cell. Biochem. 192, 17–31.
[3] Glatz, J.F.C., Storch, J., 2001. Curr. Opin. Lipidol. 12, 267–274.
[4] Stremmel, W., Pohl, L., Ring, A., Herrmann, T., 2001. Lipids 36, 981–989.
[5] Dutta-Roy, A.K., 2000. Cell. Mol. Life Sci. 57, 1360–1372.
[6] Abumrad, N.A., Perkins, R.C., Park, J.H., Park, C.R., 1981. J. Biol. Chem. 256, 9183–9191.
[7] Stremmel, W., Berk, P.D., 1986. Proc. Natl Acad. Sci. USA 83, 3086–3090.
[8] Stremmel, W., 1988. J. Clin. Invest. 81, 844–852.
[9] Luiken, J.J.F.P., van Nieuwenhoven, F.A., America, G., van der Vusse, G.J., Glatz, J.F.C., 1997. J. Lipid Res. 38, 745–758.
[10] Turcotte, L.P., Kiens, B., Richter, E.A., 1991. FEBS Lett. 279, 327–329.
[11] Bonen, A., Luiken, J.J.F.P., Liu, S., Dyck, D.J., Kiens, B., Kristiansen, S., Turcotte, L.P., van der Vusse, G.J., Glatz, J.F.C., 1998. Am. J. Physiol. 275, E471–E478.
[12] Van Nieuwenhoven, F.A., Willemsen, P.H.M., Van der Vusse, G.J., Glatz, J.F.C., 1999. Int. J. Biochem. Cell Biol. 31, 489–498.
[13] Luiken, J.J.F.P., Turcotte, L.P., Bonen, A., 1999. J. Lipid Res. 40, 1007–1016.
[14] Harmon, C.M., Abumrad, N.A., 1993. J. Membr. Biol. 133, 434–439.
[15] Abumrad, N.A., Harmon, C., Ibrahimi, A., 1998. J. Lipid Res. 39, 2309–2318.
[16] Abumrad, N.A., el-Maghrabi, M.R., Amri, E.Z., Lopez, E., Grimaldi, P.A., 1993. J. Biol. Chem. 268, 17665–17668.
[17] Greenwalt, D.E., Lipsky, R.H., Ockenhouse, C.F., Ikeda, H., Tandon, N.N., Jamieson, G.A., 1992. Blood 80, 1105–1115.
[18] Asch, A.S., Barnwell, J., Silverstein, R.L., Nachman, R.L., 1987. J. Clin. Invest. 79, 1054–1061.
[19] Van Nieuwenhoven, F.A., Verstijnen, C.P., Abumrad, N.A., Willemsen, P.H.M., Van Eys, G.J., Van der Vusse, G.J., Glatz, J.F.C., 1995. Biochem. Biophys. Res. Commun. 207, 747–752.
[20] Greenwalt, D.E., Scheck, S.H., Rhinehart-Jones, T., 1995. J. Clin. Invest. 96, 1382–1388.
[21] Pelsers, M.M.A.L., Lutgerink, J.T., Van Nieuwenhoven, F.A., Tandon, N.N., van der Vusse, G.J., Arends, J.W., Hoogenboom, H.R., Glatz, J.F.C., 1999. Biochem. J. 337, 407–414.
[22] Ibrahimi, A., Sfeir, Z., Magharaie, H., Amri, E.Z., Grimaldi, P., Abumrad, N.A., 1996. Proc. Natl Acad. Sci. USA 93, 2646–2651.
[23] Van Nieuwenhoven, F.A., Luiken, J.J.F.P., De Jong, Y.F., Grimaldi, P.A., Van der Vusse, G.J., Glatz, J.F.C., 1998. J. Lipid Res. 39, 2039–2047.
[24] Isola, L.M., Zhou, S.L., Kiang, C.L., Stump, D.D., Bradbury, M.W., Berk, P.D., 1995. Proc. Natl Acad. Sci. USA 92, 9866–9870.
[25] Schaffer, J.E., Lodish, H.F., 1994. Cell 79, 427–436.
[26] Stump, D.D., Zhou, S.L., Berk, P.D., 1993. Am. J. Physiol. 265, G894–G902.
[27] Abumrad, N.A., Coburn, C.T., Ibrahimi, A., 1999. Biochim. Biophys. Acta 1441, 4–13.
[28] Glatz, J.F.C., Van der Vusse, G.J., 1996. Prog. Lipid Res. 35, 243–282.
[29] Storch, J., Thumser, E.A., 2000. Biochim. Biophys. Acta 1486, 28–44.
[30] Spitsberg, V.L., Matitashvili, E., Gorewit, R.C., 1995. Eur. J. Biochem. 230, 872–878.
[31] Glatz, J.F.C., Van Breda, E., Van der Vusse, G.J., 1998. Adv. Exp. Med. Biol. 441, 207–218.
[32] Watkins, P.A., Pevsner, J., Steinberg, S.J., 1999. Prostaglandins Leukot. Essent. Fatty Acids 60, 323–328.
[33] Coe, N.R., Smith, A.J., Frohnert, B.I., Watkins, P.A., Bernlohr, D.A., 1999. J. Biol. Chem. 274, 36300–36304.
[34] Herrmann, T., Buchkremer, F., Gosch, I., Hall, A.M., Bernlohr, D.A., Stremmel, W., 2001. Gene 270, 31–40.
[35] Bonen, A., Luiken, J.J.F.P., Arumugam, Y., Glatz, J.F.C., Tandon, N.N., 2000. J. Biol. Chem. 275, 14501–14508.
[36] Luiken, J.J.F.P., Willems, J., van der Vusse, G.J., Glatz, J.F.C., 2001. Am. J. Physiol. Endocrinol. Metab. 281, E704–E712.
[37] Luiken, J.J.F.P., Dyck, D.J., Han, X.X., Tandon, N.N., Arumugam, Y., Glatz, J.F.C., Bonen, A., 2002. Am. J. Physiol. Endocrinol. Metab. 282, E491–E495.

[38] Luiken, J.J.F.P., Koonen, D.P.Y., Willems, J., Zorzano, A., Becker, C., Fischer, Y., Tandon, N.N., van der Vusse, G.J., Bonen, A., Glatz, J.F.C., 2002. Diabetes 51, 3113–3119.
[39] Luiken, J.J.F.P., Coort, S.L.M., Willems, J., Coumans, W.A., Bonen, A., van der Vusse, G.J., Glatz, J.F.C., 2003. Diabetes 52, 1627–1634.
[40] Goodyear, L.J., Kahn, B.B., 1998. Annu. Rev. Med. 49, 235–261.
[41] Van der Vusse, G.J., Roemen, T.H.M., 1995. J. Appl. Physiol. 78, 1839–1843.
[42] Bonen, A., Dyck, D.J., Ibrahimi, A., Abumrad, N.A., 1999. Am. J. Physiol. 276, E642–E649.
[43] Coburn, C.T., Knapp, F.F., Febbraio, M., Beets, A.L., Silverstein, R.L., Abumrad, N.A., 2000. J. Biol. Chem. 275, 32523–32529.
[44] Van Bilsen, M., Van der Vusse, G.J., Reneman, R.S., 1998. Eur. J. Physiol. 437, 2–14.
[45] Van der Lee, K.A., Vork, M.M., De Vries, J.E., Willemsen, P.H.M., Glatz, J.F.C., Reneman, R.S., Van der Vusse, G.J., Van Bilsen, M., 2000. J. Lipid Res. 41, 41–47.
[46] Van der Vusse, G.J., Van Bilsen, M., Glatz, J.F.C., 2000. Cardiovasc. Res. 45, 279–293.
[47] Tanaka, T., Sohmiya, K., Kawamura, K., 1997. J. Mol. Cell. Cardiol. 29, 121–127.
[48] Okamoto, F., Tanaka, T., Sohmiya, K., Kawamura, K., 1998. Jpn. Circ. J. 62, 499–504.
[49] Hahn, R.T., Febbraio, M., Bella, J.N., Silverstein, R.L., 1998. Circulation 98(Suppl. 1), 1236 (abs).
[50] Halaas, J.L., Gajiwala, K.S., Maffei, M., Cohen, S.L., Chait, B.T., Rabinowitz, D., Lallone, R.L., Burley, S.K., Friedman, J.M., 1995. Science 269, 543–546.
[51] Luiken, J.J.F.P., Arumugam, Y., Dyck, D.J., Bell, R.C., Pelsers, M.M.A.L., Turcotte, L.P., Tandon, N.N., Glatz, J.F.C., Bonen, A., 2001. J. Biol. Chem. 276, 40567–40573.
[52] Irie, H., Krukenkamp, I.B., Brinkmann, J.F.F., Saltman, A.E., Jou, W., Glatz, J.F.C., Abumrad, N.A., Ibrahimi, A., 2003. Proc. Natl Acad. Sci. USA 100, 6819–6824.
[53] Knudsen, J., Jensen, M.V., Hansen, J.K., Faergeman, N.J., Neergaard, T.B., Gaigg, B., 1999. Mol. Cell. Biochem. 192, 95–103.

Properties and physiological significance of fatty acid binding proteins

Norbert H. Haunerland[a] and Friedrich Spener[b,*]

[a]*Department of Biological Sciences, Simon Fraser University, Burnaby, BC, Canada V5A 1S6*
[b]*Department of Biochemistry, University of Münster, Wilhelm-Klemm-Str. 2, 48149 Münster, Germany*
[*]*Correspondence address: Tel.: +49-251-83-33100; Fax: +49-251-83-32132*
E-mail: spener@uni-muenster.de(F.S.)

Abbreviations

CRABP: cellular retinoic acid binding protein; CRBP: cellular retinol binding protein; FABP: fatty acid binding protein; DHA: docosahexaenoic acid; I-BABP: intestinal bile acid binding protein; iLBP: intracellular lipid binding protein; ITC: isothermal titration calorimetry; PPAR: peroxisome proliferator activated receptor; PPRE: peroxisome proliferator responsive element; RAR: retinoic acid receptor; RARE: retinoic acid responsive element; RXR: retinoid X receptor.

1. Introduction

In 1978, Ockner et al. [1] discovered a small protein in the cytosol of certain rat tissues that bound fatty acids and consequently named it "fatty acid binding protein" (FABP). Since then, such FABPs have been found in many tissues of many different organisms which include mammals, fish, birds, and insects. Some of these proteins were originally characterized in a different context (organic anion binding protein, Z-protein) and only later were found to be FABPs. All FABPs are members of a large multigene family now called "intracellular lipid binding proteins" (iLBPs) with various functions in the transport and metabolism of their ligand fatty acids and other lipophilic ligands. Many excellent reviews have been published on different aspects of these proteins (for a recent review see Ref. [2]), which are remarkably conserved throughout the animal kingdom. While their roles in different cells, tissues, and organisms may vary, common features become apparent in the context of metabolic tasks and conditions. The purpose of this review is to summarize current knowledge about these proteins, and to provide insight into their roles in different organisms.

Advances in Molecular and Cell Biology, Vol. 33, pages 99–122

ISSN: 1569-2558/DOI: 10.1016/S1569-2558(03)33007-3

2. FABPs as members of the iLBP family

FABPs as members of the iLBP family have traditionally been named after the tissue from which they were first isolated. Liver-type, heart-type, and intestinal-type FABP (L-FABP, H-FABP, I-FABP) have been the first to be discovered [1], and later the aP2 protein was recognized as adipocyte-type (A-) FABP [3]. With the increasing availability of ESTs and gene array data, it has become clear that most iLBPs are not confined to a single tissue. This, however, does not necessarily mean that they are un-specifically expressed, as tissues always contain different cell types. For example, heart tissue contains not only cardiomyocytes, but also significant amounts of epithelial and smooth muscle cells as well as some adipocytes. Moreover, even defined cells such as adipocytes express more than one FABP-type [4]. This is even more apparent when FABPs expressed in non-mammalian animals are considered: for example, the most prominent FABP-type expressed in shark liver [5] clearly belongs to the same subfamily (see below) as H-FABP, while the FABPs found in the livers of other fish species and chicken are basic proteins, yet distantly related to the mammalian L-FABP [6].

In this review, the widely accepted nomenclature for FABP that is based on the tissue occurrence will be used. The numerical classification used by Genbank may be more accurate, but less intuitive. In Table 1 the classical names, alternative designations found in the literature and the GenBank designations are summarized, as is the occurrence of the proteins in tissues of mature animals.

FABPs are expressed in vertebrate (mainly mammals, fish, birds) and invertebrate species. Pertaining to the latter, two FABPs are expressed in the midgut of the tobacco hornworm (*Manduca sexta*) [7] and believed to be involved in lipid digestion. The FABP from the flight muscle of locusts has been especially well characterized [8,9]. It is present in high concentration and shares many characteristics with its mammalian H-FABP counterparts. They have a high sequence homology to other insect proteins that have been identified only at cDNA levels, namely from the fruit fly (*Drosophila melanogaster*) [10] and the mosquito *Anopheles gambiae* [11]. A protein found in the brain of the tobacco hornworm, initially identified as a cellular retinoic acid binding protein (CRABP) [12], belongs to the same subfamily as H-FABP as well (see below). Surprisingly, FABPs have also been found to be prominent arthropod allergens, e.g. in the dust mites *Blomia tropicalis* [13] and *Acarus siro* [14]. In the fluke *Schistosoma mansoni* [15] and various other parasitic worms [16], FABPs are considered essential for lipid absorption, since these animals are unable to synthesize complex lipids de novo [17].

Given the wide distribution of iLBPs throughout the animal kingdom, it is apparent that they belong to an ancient gene family. Major gene duplications gave rise to the separate subfamilies. Multiple alignments of iLBP sequences and construction of phylogenetic trees by the Clustal W algorithm illustrate this relationship as shown in Fig. 1. Four major subfamilies for the mammalian proteins have been categorized based on this sequence homology and, in addition, on ligand binding characteristics [18] (see Table 1 and Fig. 1):

(I) The intracellular retinoid binding proteins [19] can be further subdivided into the cellular retinoic acid binding proteins (CRABP I and II) and the cellular retinol binding proteins (CRBP I and II).

Table 1
Nomenclature and expression pattern for intracellular FABPs

iLBP-type	Alternatives names	Gene name (human)	Mammalian expression	Non-mammalian expression
L-FABP (liver)		FABP1	Liver, intestine, kidney, lung, pancreas	
I-FABP (intestinal)		FABP2	Intestine	
H-FABP (heart)	M-FABP (muscle) MDGI	FABP3	Heart, skeletal muscle, kidney, lung, mammary, placenta, testis, stomach, ovary	Fish muscle, bird muscle, insect muscle, fish ovary
A-FABP (adipocyte)	ALBP aP2	FABP4	Adipose tissue	Fish muscle (?)
E-FABP (epidermal)	E-FABP KLBP mal1	FABP5	Skin, adipose tissue, lung, brain, heart, skeletal muscle, testis, retina, kidney	
I-BABP (intestinal)	ILBP Gastrotropin	FABP6	Ileum	
Brain FABP	B-FABP R-FABP	FABP7	Brain, neurons	Bird brain, retina
M-FABP (myelin)	mP2 Myelin P2	FABP8	Schwann cells	
T-FABP (testis)	T-FABP	FABP9	Testis	
Lb-FABP (liver basic)	L-FABP	FABP10		Fish, chicken, iguana liver
Midgut FABP				Insect midgut

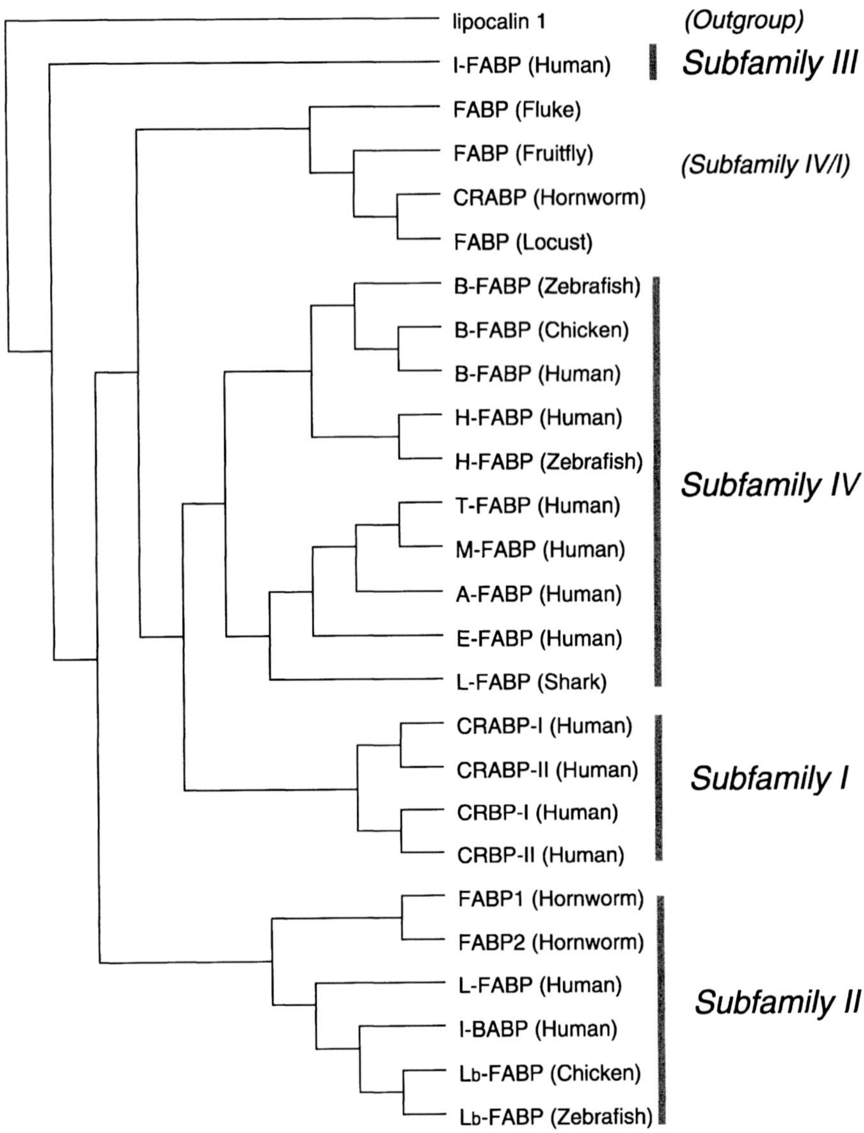

Fig. 1. Phylogenetic tree for the iLBP family. Sequences for the vertebrate and invertebrate members of the iLBP gene family were aligned with Clustal W. The tree was constructed with the neighbor joining method, using lens lipocalin as an outgroup. For mammalian iLBPs only the human paralogs are shown. For the subfamily concept see Sections 2 and 3 in the text.

(II) L-FABP and I-BABP (intestinal bile acid binding protein) are closely related based on sequence homology and both stand out because of their unusual ligand binding specificities. L-FABP, which binds a broad range of ligand molecules (acyl-CoAs, heme, squalene, bilirubin and certain eicosanoids), is the only FABP that forms a complex with two fatty acid molecules at the same time [20–22].

(III) I-FABP is rather singular in sequence characteristics and binds one fatty acid molecule.
(IV) This iLBP subfamily comprises the largest number of different types of FABPs, i.e. H-, A-, E- (epidermal-type), M- (myelin-type), T- (testis-type), and B- (brain-type) FABP. They all bind only a single fatty acid molecule.

Generally, the non-mammalian FABPs fall into one of the subfamilies as defined above and shown in Table 1 and Fig. 1, attesting to the considerable evolutionary conservation of this protein family. Various papers have discussed the phylogenetic relationship between the different members of the FABP family [3,23,24]. From phylogenetic analysis it is likely that a common ancestor gene branched out into two major families more than 900 million years ago, long before the vertebrate–invertebrate divergence. Thus, subfamily II includes not only L-FABP and I-BABP, but also the insect midgut FABPs. The FABP from insect muscle is assembled not only with the H-FABP expressed in mammalian heart and skeletal muscle cells, but also with the cellular retinoid binding proteins, since subfamilies I and IV are believed to have split after the vertebrate–invertebrate divergence [25].

3. Structure and conformation of FABPs and their ligands

The iLBPs are small proteins of 127–134 amino acids, whose expression in *E. coli* made available substantial quantities of recombinant protein for biophysicists and structural biologists to gain deeper insights into structure and binding properties of these proteins. Thus, three-dimensional structures have been determined by X-ray crystallography [22,26–30] and/or NMR [31–35] for all types of the mammalian iLBPs, with the exception of T-FABP. In addition, the crystal [36] and solution structure [37] of the chicken basic liver-type (Lb-) FABP are known. Of the invertebrate FABPs, the three-dimensional structures of a midgut FABP from tobacco hornworm [38] and of the H-FABP from desert locust [9] have been solved. From this wealth of data it has become clear that the tertiary structure of all iLBPs is highly conserved, despite the considerable differences in their primary structure. Sequence identities in this protein family range from 25% for some paralogous members to over 90% for some orthologs. The common structural feature is a 10-stranded β-barrel, made of two orthogonal antiparallel 5-stranded sheets that form the "clam"-shaped binding cavity [39]. The opening of this clam, considered the portal domain, is framed on one side with the N-terminal helix-turn-helix domain, a further common structural motif of all iLBPs (Fig. 2). The 10 antiparallel strands that form the barrel is the salient feature of iLBPs within the "calycin" superfamily of lipid binding proteins, whose other families, the avidins and lipocalins, are characterized by 8-stranded antiparallel barrels forming the binding cavity [40].

In the binding pocket of iLBPs the deprotonated carboxyl group of the bound ligand is generally buried inside the cavity for electrostatic interaction with one or two arginine residues, in addition to be hydrogen bonded by a tyrosine- or serine-OH and an ordered water molecule [27]. Nonetheless, important differences between individual iLBP-types exist, which influence binding kinetics and affinity as well as the mechanism of ligand

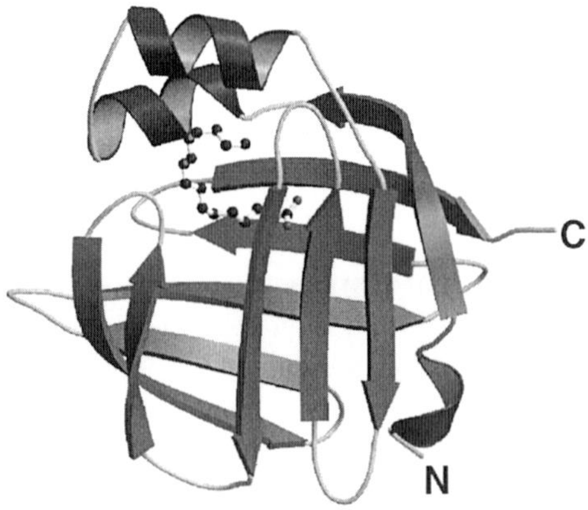

Fig. 2. Three-dimensional structure of *holo* E-FABP (with palmitic acid) [29]. All iLBPs have the characteristic β-barrel structure, in which 10 antiparallel β-strands form the "clam"-shaped ligand binding site, framed by the helix-turn-helix domain as part of the portal. In E-FABP, fatty acid is bound in a U-shaped conformation, characteristic for subfamily IV iLBPs.

transfer [18,41]. FABP-type specific affinities for fatty acids are due to different volumes of the binding cavities and to the amino acid side chains facing one side of the fatty acid's hydrocarbon chain directly, and indirectly the other side via ordered water molecules. This view is not uncontested, however (see Section 4).

A close-up inspection of protein structure and ligand conformation by crystallographic techniques fosters the above-mentioned subfamily concept for iLBPs:

(I) The conformation of the characteristic isoprenoid tail of the retinoid ligands is extended and the α-ionone ring located close to the helix-turn-helix domain, whereas the functional group is always deeply immersed into the binding cavity. Here Arg111 and132 and Tyr134 directly bind all-*trans* retinoic acid in the case of CRABP I and II (cellular retinoic acid binding proteins) [42] which is a scenario similar to that of straight-chain fatty acid binding in proteins of subfamily IV. In CRBP I and II (cellular retinol binding proteins), which bind either all-*trans* retinol or retinal, Gln108 interacts with the functional group of the ligand [43,44] and in CRBP III and IV, variants binding only retinol, Gln108 is replaced by His [45,46].

(II) Of the two fatty acids bound by L-FABP, one is coordinated in a bent conformation electrostatically via Arg122 and an extensive hydrogen-bonding network involving Ser124 and 39 located at the bottom of the protein cavity, which again is reminiscent of fatty acid binding in subfamily IV. The second fatty acid in L-FABP adopts a rather linear shape, with the acyl chain in the cavity extending down towards the center of the other fatty acid molecule and the carboxylate sticking out of the fatty acid portal, thus being solvent exposed and pH sensitive [22]. Interestingly, although I-BABP contains the respective residues (Arg121, Ser123 and 38), it binds fatty acid only weakly, but a bile acid molecule with high affinity. Again, the bulk steroid molecule is inside the cavity and the carboxylate group at the protein–solvent interface [47].

(III) The fatty acid bound by I-FABP adopts a slightly bent conformation, reverse in direction to the second fatty acid in L-FABP, thus the carboxylate group is located deep inside the protein cavity directly coordinated to the side-chain of Arg106 similar to the ligands' carboxylate bound by proteins belonging to subfamilies I and IV [26].

(IV) The FABP-types of this subfamily all bind only a single fatty acid molecule in a U-shaped conformation. While the carboxylate group is bound electrostatically and hydrogen bonded via Arg106 and 126 as well as Tyr128 (H-FABP numbering), the hydrocarbon chain is located close to Phe57 (Leu60 in E-FABP) at the fatty acid portal [27]. Several unique features in this iLBP subfamily have been reported only recently. First, human E-FABP contains six cysteine residues, of which C120 and C127 form a disulfide bridge inside the protein cavity [29]. Secondly, human B-FABP binds oleic acid in the common U-form conformation, but very long-chain docosahexaenoic acid (DHA) in a helical conformation [30]. It remains to be seen whether the latter is a consequence of chain-length, or a specific feature for binding $n-3$ fatty acids. The three-dimensional structure of insect muscle FABP has been solved for the *apo*-protein only [9]. It is remarkably similar to mammalian H-FABP, although steric limitations seem to predict a somewhat different shape of the ligand in the binding pocket.

4. The binding and transfer of fatty acids by FABPs

As far as we know, the obvious task of FABPs is to bind fatty acids. A total of eight FABP-types are expressed in various mammalian tissues each carrying out distinct metabolic tasks. Is fatty acid binding to these FABPs a mere variation of a common structural "leitmotiv", with little consequence for binding affinities? Or do the small structural differences in the binding sites lead to binding selectivities for distinct fatty acid structures? It is not easy to decide which view is correct, and literature data on this aspect are somewhat controversial.

The ADIFAB reagent is a covalently modified I-FABP, with a fluorescent label that changes its emission maximum upon the binding of fatty acids [48,49]. On the one hand, data elaborated with this ADIFAB assay have been interpreted in terms of the "solubility hypothesis", which states in a first approximation that the solubility of a given fatty acid in the bulk aqueous phase drives its affinity for any FABP. The binding site of I-FABP is considered to act similar to a non-polar solvent, and hence its affinity for different fatty acids is mainly determined by the entropic contribution of the hydrophobic effect. Recently, however, thermodynamic parameters for ligand double bonds were incorporated into the calculation of dissociation constants to reflect physico-chemical properties of a given FABP binding site, in fact, the enthalpic contribution to binding. For all FABP-types and their ligand fatty acids tested so far, the values for K_ds found with the ADIFAB method are between 2 and 200 nM.

On the other hand, far greater variations in binding constants were found with other methods. The earliest assays used charcoal to remove unbound fatty acid from the solution and calculated binding constants from the ratio of charcoal- and protein-bound radioactivity [50]. Soon charcoal was replaced by a lipophilic dextrane derivative,

Lipidex 1000 [51]. This material has strong affinity to fatty acids at 37 °C, and can be used to delipidate FABP. At 0 °C, however, protein-bound fatty acids were shown to remain bound to FABP, while unbound fatty acids were adsorbed to Lipidex. Determination by this method afforded dissociation constants between 0.2 and 0.4 μM which are now considered too high, because of the low temperature and the time required to separate Lipidex from FABP [18]. More reliable values can be obtained by measuring dissociation constants without physically separating free from bound ligands, such as fluorescence-based methods like the ADIFAB assay. Another popular approach is isothermal titration calorimetry (ITC), which measures the heat absorbed or released upon binding of the ligand to the protein [18]. For mono- and polyunsaturated fatty acids, dissociation constants in the 10–300 nM range have been determined, whereas remarkably larger values were found for saturated fatty acids, for which the ADIFAB method suggests very strong affinity. The reasons for these discrepancies are not clear, but could be related to solubility problems. A comparison is shown in Table 2, taking the example of B-FABP.

It follows from this short discussion (for more details, see Ref. [18]) that absolute values of dissociation constants depend on the method used for their determination. Their relative values, however, are comparable from method to method, in particular for Lipidex and ITC data. Some of the latter can be explained on the basis of crystallographic studies [52]. Moreover, further insights into binding can be gained by inspecting the dynamic properties of FABPs through various NMR techniques, Fourier transform infrared spectroscopy and recent molecular dynamics calculations [18]. These studies lead to the following conclusions: (i) Differences in the backbone dynamics of various FABPs can be correlated to preferences for specific fatty acids and their relative binding affinities.

Table 2
Dissociation constants for human B-FABP/ligand complexes determined by the ADIFAB and ITC method

Ligand fatty acid class	K_d (nM)	
	ADIFAB, 37°C[a]	ITC, 30°C[b]
Saturated		
Palmitic acid	7	7100[c]
Stearic acid	2.3	13,500[c]
Monounsaturated		
Oleic acid	7	46.7 ± 1.4
Polyunsaturated $n-6$		
Linoleic acid	11	115 ± 19
Arachidonic acid	18	207 ± 19
Polyunsaturated $n-3$		
Docosahexaenoic acid	13	53.4 ± 4.1
α-Linolenic acid	21	27.5 ± 1.3

[a]Ref. [49].
[b]Ref. [30].
[c]By Lipidex assay and referenced to $K_d = 47\ nM$ for oleic acid as obtained by ITC.

(ii) The *apo*-conformation of the protein can adapt to a particular ligand fatty acid and is thus stabilized by reduced backbone flexibility in some *holo*-FABPs [53], even "structured" water molecules as part of the tertiary structure may add to this stability. (iii) In the portal region, the backbone structures generally display an increased conformational variability.

Finding the correct answer to the questions raised at the start of this section is not easy. Certainly, preferences for interactions of certain FABP-types with structurally defined fatty acid classes can be recognized, such as E-FABP with saturated fatty acids, I-FABP with saturated and monounsaturated fatty acids, H-FABP with $n-6$ polyunsaturated fatty acids, L-FABP with mono- and $n-3$ polyunsaturated fatty acids, and B-FABP with $n-3$ polyunsaturated fatty acids. This would have functional implications. A tenet to this statement is that all binding data published originate from in vitro assays that may not reflect the complexity seen within a cell in vivo. According to Weisiger [52], "free" unbound fatty acids in the aqueous cellular compartments originate from their spontaneous membrane-to-membrane transfer that is very slow and depends on the mean diffusional excursion (d_m) of a fatty acid from the membrane. The bulk of the "free" fatty acid molecules in the cell, however, is bound to membranes and to intracellular binding proteins, particularly FABPs. When intracellular transfer of fatty acids beyond d_m is needed, certain FABPs act as "membrane-inactive" binding proteins, and catalyze the diffusional transfer step by increasing fatty acid concentration in the soluble (= diffusible) pool; others act as "membrane-active" binding proteins that catalyze fatty acid dissociation from donor membranes and rebinding to acceptor membranes through FABP-membrane collisions. This intriguing concept received convincing support by elegant studies at the molecular level, which demonstrated that L-FABP and CRABP II belong to the membrane-inactive, non-collisional group, while all other FAPB-types investigated are membrane active and catalyze collisional transfer [54].

This collisional transfer of fatty acids from the FABP to zwitterionic and anionic membranes relies on interactions with positively charged amino acid residues in the helix-turn-helix motif and in turns belonging to the portal domain of respective FABPs [55–57]. Thus, modulation of fatty acid transfer rates in either direction depends on electrostatic interactions of the protein with membrane lipid or protein; additional hydrophobic interactions appear to be at work as well. If this concept is true, why does a cell need membrane-inactive FABP, such as L-FABP at all? It has been proposed that membrane-active FABPs would lose diffusional mobility and thus ability to catalyze efficient fatty acid transfer in cells densely packed with membranes that require efficient fatty acid transfer between membranes over some distance. Hepatocytes and enterocytes are such cell types, and both express L-FABP [58].

5. Metabolic actions of FABPs

In contrast to the very detailed knowledge of the structure and binding characteristics of FABPs, much less is known about their biological functions. The fact that they bind fatty acids suggests that these proteins participate in various aspects of lipid transport and metabolism. Many studies have demonstrated that FABPs modulate metabolic reactions in vitro, but this does not imply that similar effects occur in living cells. Given the poor solubility of fatty acids in water, one can expect, for example, that the presence of FABP in a buffer increases the availability of fatty acid to enzymes, thus leading to increased metabolic rates in vitro. FABPs are believed to serve the following cellular tasks:

- uptake of fatty acids into the cell;
- formation of cytosolic pool for fatty acids to be rapidly utilized and, concomitantly, to avoid detergent effects on cellular proteins and structures;
- targeting of fatty acids to specific metabolic pathways and modulation of enzymatic activities;
- involvement in fatty acid signaling and gene regulation;
- affecting cellular growth and differentiation;

For the first three tasks indirect evidences are available and will be generally addressed first in this section, followed by a detailed account of the specific FABP-types. The other two tasks will be dealt with in Sections 7 and 8.

Uptake of fatty acids into the cell. The various mechanisms and accompanying phenomena of fatty acid uptake are being dealt with in more detail in Chapters 2, 4, 5, and 6 of this book. In these processes FABPs would be at the receiving end in the cytosol. But the need for such cellular proteins in mediating fatty acid uptake, however, remains controversial [59]. General experimental approaches have been transfection of immortalized cultured cells with a certain FABP and determination of fatty acid uptake either by radioactivity or fluorescence. Thus, L-FABP enhanced initial uptake of oleic acid into L-cell fibroblasts [60] as did A-FABP in transfected CHO-cells, but not a non-binding mutant [61]. When endogenous L-FABP concentrations were decreased by transfecting HepG2 cells with antisense L-FABP cDNA, fatty acid uptake decreased accordingly [62]. On the other hand, expression of L-FABP mRNA in oocytes of *Xenopus laevis* had no effect on fatty acid uptake [63] as had the transfection of L6 myoblast with A- and H-FABP [64]. By the same token transfection with I-FABP cDNA of rat hBRIE 380 cells, murine L-cell fibroblasts, and human Caco-2 cells did not change the uptake kinetics of fatty acids [65–68]. The effect of FABP on fatty acid uptake obviously differs with respect to FABP-type and/or cell-type. Reasons can be the unknown coupling of the uptake process to cellular utilization of the fatty acid incorporated and, of course, the unknown proportions of the mechanisms contributing to the translocation of the fatty acid through the membrane.

Cytosolic pool for fatty acids. Due to the amphipathic nature of fatty acids, their accumulation in large quantities would result in the formation of micelles in the cytosol and damage to cellular membrane structures. FABP may protect against such damage, especially in cells that encounter large fatty acid fluxes. The protein may also modulate the regulatory effects of fatty acids on enzymes or on nuclear transcription factors.

Cytosolic fatty acid transport and targeting. Given the poor solubility of fatty acids in aqueous media, protein-mediated transport of fatty acids may be necessary to achieve high fluxes of fatty acids within cells. Indeed, tissues that metabolize large amounts of fatty acids, such as muscle of adipose tissue, have a high FABP content. FABP increases the total concentration of fatty acids in the cytosol, and it may transport fatty acids more rapidly through the aqueous phase (see Section 4). The proteins may also deliver fatty acids to specific intracellular compartments or enzymes, for example, to mitochondria for β-oxidation, or to acyl-CoA synthetases for esterification and subsequent storage as triglycerides.

It is difficult to conclusively determine how a particular FABP functions in a living cell, especially since many cells express more than one member of the FABP gene family. However, functional conclusions can be drawn from metabolic differences in cells, tissues, and animals with different FABP content. At the cellular level, such differences can be induced through the transfection of cell lines with various FABPs. FABP levels can also be modified through experimental conditions, such as diet, hormones, or exercise. More recently, dramatic progress with respect to functional aspects has come from gene disruption studies. Knock-out mice for L-FABP, H-FABP, I-FABP, A-FABP, and E-FABP have shed light at the different functions of these proteins, but also revealed that other members of the gene family may compensate at least partly for the loss of one particular FABP. Other cues were obtained from comparing FABP orthologs in different animals. This approach is especially useful for animals that have adapted to extreme rates of lipid metabolism. In assessing the potential functions of FABPs, it is important to distinguish between the individual members of this gene family, and to consider the metabolic functions of the tissues in which they are expressed. Depending on the tissue, fatty acids need to be directed to different compartments, or to different pathways. Data from experimentally modified animals or different, specially adapted species support functions of FABP in intracellular fatty acid trafficking, but the details of underlying mechanisms have yet to be determined.

L-FABP: Liver is a major place of biosynthesis and detoxification, and L-FABP has long been speculated to function in directing fatty acids or related metabolites to the appropriate sub-cellular compartments. It may increase fatty acid acylation rates by making fatty acid more accessible to acyl-CoA synthetase [69]. Circumstantial evidence for a transport function was obtained from comparative studies between hepatocytes from male and female rats. In female cells, where FABP expression is 20% higher than in males, the fatty acid diffusion rate was markedly increased [70]. Other studies have also demonstrated that L-FABP modulates the uptake of fatty acids. In L-FABP knock-out mice, hepatic uptake of fatty acids from the blood was reduced by 50%. This is most likely a direct consequence of the markedly reduced fatty acid binding capacity (−80%) in the cytosol of liver cells, which do not express any other FABP. The cells, however, maintained normal levels of non-esterified fatty acids, triglycerides, and total lipids [71]. Due to its wide range of ligands that includes xenobiotics, it has been suggested that L-FABP may also play a role in mitogenesis [72] (see Section 8).

I-FABP: Three different members of the FABP gene family are strongly expressed in the small intestine, albeit in different regions: cells of the proximal area of the small intestine express mostly L-FABP, while I-FABP is found in the medial region. The distal region expresses the intestinal bile acid binding protein (I-BABP). Since the small intestine is involved in dietary lipid absorption, it is plausible that these proteins mediate the uptake of lipids and their subsequent release into the bloodstream. The link between fatty acid uptake and I-FABP content is supported by various observations in cultured cells: Fatty acid uptake into undifferentiated stem cells was increased 1.7-fold following transfection with I-FABP, while the reduction of I-FABP levels in cultured enterocytes by epidermal growth factor treatment resulted in reduced fatty acid uptake [73,74]. Other evidence supports a pivotal role of I-FABP in lipid absorption in vivo: A common mutation in this FABP gene doubles the affinity of I-FABP for fatty acids and results in increased fatty acid uptake, a finding that may explain why Pima Indians, a high incidence population group, are predisposed to type 2 diabetes [75,76]. However, targeted gene disruption of the I-FABP gene in knock-out mice did not impair their intestinal lipid absorption [77]. This, however, may be due to the overexpression of L-FABP in the intestine of these animals [78]. Like in other FABP knock-out models, an alternative FABP seems to compensate for the loss of I-FABP in the intestine of I-FABP null mice.

A-FABP: In adipocytes, free fatty acids are mostly incorporated into triacylglycerol for subsequent storage. A-FABP is therefore thought to direct fatty acids towards esterification at intracellular membranes where the long-chain acyl-CoA synthetases are located. Supporting data have been produced in experiments with primary and cultured adipocytes (reviewed in Ref. [79]). Alternatively, a role for A-FABP may arise during lipolysis, when free fatty acids are released from lipid droplets catalyzed by hormone sensitive lipase. As this enzyme is subject to feedback inhibition by fatty acid, it seems logical that rapid removal of fatty acids is required for efficient lipid mobilization. Indeed, A-FABP interacts directly with hormone sensitive lipase, making it possible to sustain rapid transport of fatty acids to the plasma membrane for export, or towards re-esterification at other organelles [80].

In order to study A-FABP function in vivo, a targeted disruption of its gene was generated in mice [81]. The mice appeared to be of normal phenotype, developed normally and were fertile. The morphology of adipocytes, and their fatty acid composition and uptake rates were unaltered. These findings, however, cannot be taken as indication that this FABP is not essential, as its loss greatly increased the expression of E-FABP in adipocytes, which normally makes up only 1% of total FABP in these cells [82]. While no changes in lipid metabolism were apparent in these animals when reared normally, differences were seen after diet-induced obesity. In contrast to wild-type mice, A-FABP null mice showed no increase in serum triglyceride levels, and remained sensitive to insulin. The concentrations of free fatty acid in the adipocytes were elevated, while lipolysis was reduced by 40% [83].

A-FABP is also expressed in macrophages which take up oxidized LDL and contribute to the development of atherosclerosis. Atherosclerotic lesions from hypercholesterolemic, ApoE-deficient mice contained high levels of A-FABP, and it has been demonstrated that oxidized LDL induces A-FABP expression. Double knock-out mice lacking both the ApoE and the A-FABP gene developed smaller lesions with fewer macrophages, indicating that macrophage A-FABP plays an important role in the formation of atherosclerotic lesions [84–86].

E-FABP: Epidermal FABP is the most universally expressed member of this gene family. It is the most abundant FABP in the skin. It may play a role in the maintenance of the water-permeability barrier of the epidermis, as suggested by recent studies with knock-out mice [87]. E-FABP null mice were of normal phenotype, and no differences were visible in histological examinations. No differences were seen in the epidermal fatty acid composition, but the basal *trans*-epidermal water loss was lower that that in wild-type animals. When the lipid barrier was damaged by acetone treatment, the recovery period required to reach the basal level was much longer than in wild-type animals [88]. A significant increase in H-FABP expression was observed in the liver of neonatal mice, where E-FABP is normally strongly expressed [87]. Adipocytes of E-FABP knock-out mice showed a higher capacity for insulin-stimulated glucose transport; higher systemic insulin sensitivity was also observed [89]. In contrast, transgenic mice overexpressing E-FABP were less sensitive to insulin. The expression of E-FABP and A-FABP in adipocytes is interdependent: When E-FABP is overexpressed, the levels of A-FABP are reduced [90], while A-FABP knock-out mice reveal highly elevated levels of E-FABP expression [82].

B-FABP: This protein is found at its highest levels in developing brain [91]. The protein is expressed in glia cells, and its expression is regulated in response to interactions with neurons [92,93]. Unlike most other FABPs, B-FABP does not bind palmitic acid, but requires a longer hydrocarbon chain and a higher degrees of desaturation [94]. Its natural ligand appears to be DHA, the very long-chain fatty acid that is essential for the development of the nervous system. The expression of B-FABP in the brain coincides with its requirement for DHA, and therefore B-FABP is believed to be involved in the signaling pathways between developing neurons and glia cells [95]. B-FABP is also prominent in neural development of avian species, for example, in the neurogenesis of glial cells in chicken retina [96]. In contrast to the mammalian central nervous system, which is fully developed at maturity, the brain of birds shows significant levels of neurogenesis in the adult stage. The presence of B-FABP in adult bird brain, and its anatomical distribution lends credence to its role in neural migration and synaptic reorganization [97].

H-FABP: Perhaps, the clearest link between FABP and fatty acid metabolism is seen for H-FABP. This protein is the only FABP expressed in various muscle tissues, in both vertebrates and invertebrate species [98,99]. The protein is highly conserved, even between insects and mammals, and is found in all muscles that metabolize fatty acids. A strong correlation exists between the fatty acid oxidation capacity of a muscle and its H-FABP content, as illustrated in Fig. 3. Smooth muscle that depends largely on carbohydrates possesses very low levels of this FABP, while the content in red muscles is increased. With higher β-oxidation rates typical for various red muscles, equally increased levels of H-FABP can be found [100]. Cardiac tissue, which depends mostly on lipid for energy supply and encounters the highest β-oxidation rates of all mammalian muscles, also has the highest FABP content (up to 5% of all cytosolic proteins). This observation applies also to non-mammalian muscles, which need to sustain high metabolic rates for long periods: Approximately 9% of all cytosolic proteins are H-FABP in flight muscles of the Western sandpiper, a migratory shorebird found along the Pacific coast of North and South America; this high FABP content again reflects the fatty acid oxidation rates sustained in these muscles [101]. Higher metabolic demands exist for migratory insects as well, which retrieve energy during endurance flights exclusively through β-oxidation [8]. A classical example is the flight muscle of desert locust, which oxidizes almost 1 μM of fatty acid per minute and gram tissue, as H-FABP makes up almost one-fifth of all soluble proteins.

In all these muscles, elevated levels of H-FABP expression have been observed as a consequence of endurance training or otherwise increased fatty acid utilization. For

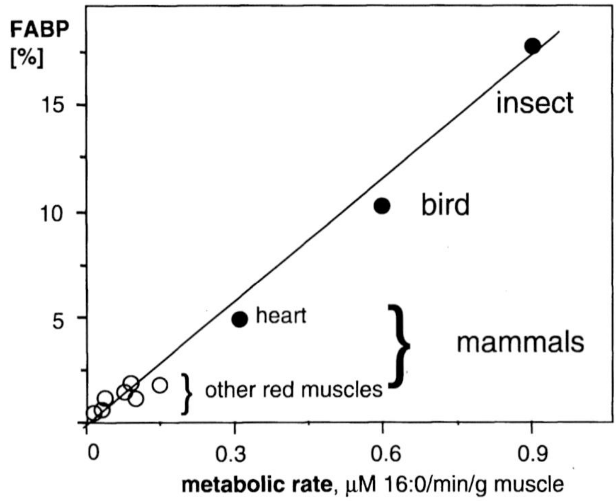

Fig. 3. Correlation between fatty acid oxidation capacity and H-FABP content [% of cytosolic protein] in different muscles. Metabolic rates, expressed as the oxidation of μM of palmitate per minute and gram tissue, for mammalian muscles were taken from Ref. [100], for other muscles from Ref. [159]. FABP values for mammalian muscles were obtained from Refs. [100,161], for locust flight muscle from Ref. [8] and for sandpiper flight muscle from Ref. [162].

example, chronic electrical stimulation in rat soleus muscle led to a 30% increase in H-FABP expression [100], and in vivo experiments confirmed this finding: after 8 weeks of swimming, the concentration of H-FABP in rat skeletal muscle increased by 30%, though not in the heart [102]. Diets enriched with polyunsaturated fatty acids led to similar effects in skeletal muscle. In spite of the already extreme H-FABP content of locust flight muscles, its further expression still can be induced, both in response to exercise and to increased fatty acid supply alone [103]. As discussed in more detail below, H-FABP may act as a fatty acid sensor and modulate the expression of its own gene. This would assure that H-FABP levels are appropriate for the fatty acid transfer rates required to fuel muscle activity.

Studies in H-FABP knock-out mice confirm the importance of H-FABP for fatty acid transport and metabolism. The absence of H-FABP did not result in phenotypical differences, and the histology of skeletal and cardiac muscle appeared normal [104]. However, fatty acid uptake was reduced markedly in cardiac tissue (−80%) and isolated cardiomyocytes (−45%). Because of the impaired fatty acid uptake, cardiac muscle contraction in these animals relied on glucose oxidation, which can provide sufficient energy to resting animals [105]. Higher metabolic rates, however, could not be sustained. When exercised, H-FABP null mice fatigued quickly, a finding that lends support to the essential role of H-FABP in cardiac metabolism. Since no other FABPs are expressed in cardiac cells, a compensation mechanism as observed in other knock-out models may not be possible.

In contrast to vertebrates, fish appear to express both H-FABP and a protein more similar to A-FABP in their heart and skeletal muscle [106]. This is noteworthy because fish muscles also serve as the major lipid storage organ. The presence of A-FABP and H-FABP would be consistent with distinct roles of these proteins in lipid metabolism: A-FABP could direct fatty acids towards storage, for example, during the early stages of migration when food intake exceeds the energy demand. H-FABP should be more prominent during spawning when vast quantities of energy are needed.

6. Regulation of FABP gene expression

From the functional data discussed above, it is not surprising that cells in tissues with prominent roles in fatty acid metabolism are especially rich in FABP. Moreover, FABP levels often increase as a consequence of increased fatty acid exposure. How is this achieved at the molecular level?

All FABPs share an identical gene structure of four conserved exons and three introns of variable size [4,107]. This overall gene structure is of ancient origin, as it is even found in non-mammalian species. The exon/intron boundaries are in identical positions in all FABPs, with the only exception that the second intron has been lost in several, but not all insect FABPs [108]. All FABP promoters contain a classical TATA box. The elements that control the tissue-specific expression of FABP are currently only poorly understood, but potential enhancer sequences have been characterized for several genes. These include two HNF1α regulatory elements in the L-FABP promoter [109], a fat-specific enhancer required for A-FABP expression in adipocytes [110], and several binding sites for members of the POU transcription factor family that control B-FABP expression [111]. A concise promoter region that contained an atypical MEF2 binding site was shown to be responsible for the muscle-specific expression of H-FABP [112].

Better understood is the up-regulation of various FABP genes by fatty acids. It has long been known that the induction of FABP expression in response to lipid-rich diet [113] or endurance training [114] is the result of increased intracellular concentrations of fatty acids, which in turn activate nuclear transcription factors [115,116]. The best known of such transcription factors are the subtypes of the peroxisome proliferators activated receptor (PPAR α, β, γ), so called because of their activation by xenobiotic peroxisome proliferators in rodents [117]. Long-chain fatty acids and certain eicosanoids are considered as their natural ligands. PPARs bind as heterodimers with the subtypes α, β, γ of the retinoid receptor RXR to direct-repeat elements (peroxisome proliferators response elements, PPREs) in the promoter region of the genes that they regulate.

While circumstantial evidence suggests that PPARs are involved in the regulation of various FABP genes, proof has been provided for A-FABP [118] and L-FABP [119] only. In reporter gene and transactivation assays Tontonoz et al. [118] have shown that the murine A-FABP gene is regulated by the binding of PPARγ2 and RXRα to a direct-repeat element 5.2 kb upstream of the FABP gene. The expression of the rodent L-FABP gene in the liver is under the control of PPARα bound to a PPRE around 110 bp upstream of the transcriptional start site; interestingly, its expression in intestinal cells is controlled by PPARβ, which binds to the same response element as PPARα in the liver [120].

Several studies have demonstrated that treatment of muscle cells with the PPARα agonist Wy14,643 resulted in elevated FABP mRNA levels, and concluded that the H-

FABP gene is also under the control of PPARα [121]. Although a direct-repeat sequence reminiscent of a PPRE can be found in the distal promoter of rodent H-FABP genes, the involvement of this element could not be demonstrated. The absence of a functional PPRE in the human H-FABP promoter raises the possibility that PPARs may act indirectly through cross-talk with other nuclear receptors. Alternatively, the observed induction of gene expression by PPAR agonists could instead be a consequence of increased fatty acid uptake into the myocyte, caused by the induction of the membrane fatty acid transporter FAT/CD36 that is known to be controlled by PPARα [121]. While it has been proposed that transcription factors other than PPARs may be involved in fatty acid mediated gene control [122], such factors have not been extensively studied. To this end, insights can be obtained from invertebrates, which do not express PPARs [123], but the ortholog of H-FABP, which can be induced by fatty acids [103]. It is interesting to note that a different fatty acid response element (FARE) has been identified in the promoter of the H-FABP gene from locust muscle [108,124]. Unlike PPRE, the locust FARE is an IR-3 element, a palindromic sequence containing two hexanucleotide half-sites (AGTGGT, ATGGGA) separated by three nucleotides reminiscent of a steroid hormone response element. Reporter gene constructs containing the locust FABP promoter were expressed in rat myoblasts cells, and treatment with fatty acids resulted in a twofold increase in expression. Deletion of the element did not affect the basal expression rate, but completely eliminated induction by fatty acid. Nuclear proteins from rat myoblasts bound to the element in gel-shift experiments, but additional fatty acid was required to achieve the same effect with nuclear proteins from locust muscle [124]. Perhaps, higher concentrations of fatty acids are required in the latter tissue, because its large FABP content may prevent full access of a signaling fatty acid to the nuclear receptor.

The locust FARE appears to be conserved in evolution: similar elements can be found not only within the proximity of putative FABP genes from other insects (*D. melanogaster* and *A. gambiae*), but also in the promoters of all mammalian H-FABP genes. In the latter case, however, the hexanucleotide half-sites (consensus sequence AGAAGA and AGGTGA) are pointing outwards, forming an everted repeat sequence [125]. It remains to be seen whether these elements alone are responsible for the regulation of the H-FABP gene by a fatty acid, and which transcription factors are involved. In any case, it appears that indeed there is more than one way by which fatty acids can control gene expression.

7. The role of FABPs in fatty acid signaling and gene transcription

The induction of A- and L-FABP mRNA expression by fatty acids and retinoids, involving heterodimers of PPAR and RXR subtypes, is a paradigm for all genes having a PPRE. It follows the general scheme for gene activation by lipophilic ligands that bind to nuclear receptors of the steroid hormone receptor superfamily [126]. In A- and L-FABP expressing cells, fatty acids thus induce their own intracellular binding proteins, a finding that insinuates that these proteins may be the vehicles for targeted transfer of the hydrophobic activators into the nucleus, where they become agonists of transcription factors [126,127]. Other examples from the iLBP family include CRABP (subfamily I) and I-BABP (subfamily II). CRABPs transport retinoic acid to the nucleus, and their genes are under the control of retinoic acid response elements (RARE), which in turn are activated by the complex of retinoic acid with RAR and RXR [128]. I-BABP is up-regulated by its ligand as well, via the farnesoid X receptor FXR, a nuclear receptor that is activated by bile acid [129].

The members of the iLBP family are well suited to deliver ligands into the nucleus: as small cytosolic proteins of ~15 kDa, FABPs may pass nuclear pores readily or controlled by a specific recognition signal. Indeed, immunolabeling techniques allowed to detect nuclear localization of L-FABP in hepatocytes already in 1989 [130], of B-FABP in astrocytes [131], of A-FABP in 3T3-L1 adipocytes [132], and of H-FABP in mammalian [133] and insect myocytes [8]. In locust muscle, the cytosolic levels of FABP increase rapidly after adult ecdysis, and the nuclear levels were shown to increase proportionally. Thus, it is conceivable that FABPs transfer fatty acids to PPARs or other nuclear receptors, which in turn are activated to enhance transcription. While the ligand exchange could be simply a matter of fatty acid affinities between binding protein and nuclear receptor, recent studies point towards direct interactions between FABP and PPARs [134]. L-FABP and PPARα co-localize in the nucleus of mouse hepatocytes and, as shown in vitro, the binding protein interacts via protein–protein contacts with PPARα and γ. These contacts are required for the activation of gene expression in response to treatment of HepG2 cells with PPAR ligands, including long-chain fatty acids. Tan et al. [135] obtained similar results using the COS cell model: A-FABP and E-FABP interact directly with PPARγ and β, respectively, and co-expression of the binding protein and respective PPAR subtypes enhance gene activation. Moreover, it appears translocation of the FABP into the nucleus itself is a regulated process, with a massive import in response to ligand binding. The primary structures of FABPs do not carry nuclear import signals; therefore, other mechanisms must be operative. In the case of L-FABP, the negatively charged carboxylate group of the second fatty acid molecule at the surface of the *holo*-protein has been considered such a recognition signal [136,137].

While complete mechanistic details are not yet understood, it seems that FABPs act as fatty acid sensors and mediators in the regulation of gene expression, as illustrated in Fig. 4. This does not mean that the mechanism by protein–protein contacts is exclusive for the ligand to become agonist. Moreover, for reasons not yet known, conflicting data have been reported for the ligand dependence of these protein–protein contacts. On the one hand, the interaction of L-FABP with PPARα or γ has been shown to be independent of the presence of ligand [134]; on the other hand, A-FABP interacted with PPARγ and E-FABP with PPARβ only in the presence of ligand [135,138]. It is interesting to note the parallels between these FABPs and other iLBPs. It was found that CRABP II, but not CRABP I interacts with the retinoic acid receptor (RARα); this collisional contact leads to the transfer of all-*trans* retinoic acid from the binding protein to the nuclear receptor [139]. Although the affinity of 9-*cis* retinoic acid to CRABP II is much lower than that of the *trans* isomer, it can be transferred by the same collisional mechanism to RXRα [140]. Therefore, L-, A-, E-FABP, and CRABP II appear to play complementary roles in gene regulation; protein–protein contacts are necessary between nuclear receptors and these binding proteins and thus can be addressed as co-activators of nuclear receptors [140].

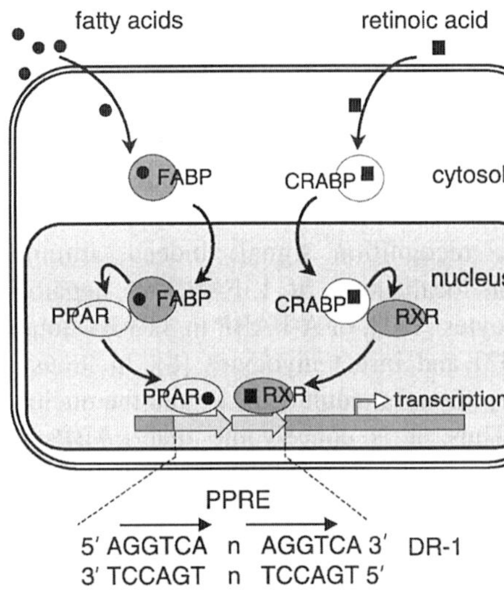

Fig. 4. The path of signaling fatty acids to the nucleus (bold arrows). Protein–protein contacts between iLBP (L-, A-, E-FABP, CRBP II) and the nuclear receptors are shown. The binding proteins deliver fatty acids and retinoic acid to the nucleus, where they are transferred by collision to their respective transcription factors (specific subtypes of PPAR and RXR). Nuclear receptor heterodimers then bind to PPRE for gene transcription.

8. Role of FABPs in cell growth and differentiation

Siding with the notion that FABPs target their lipophilic ligands, e.g. fatty acids or xenobiotics, to the nucleus to affect the cell cycle, we would expect either mitogenesis or growth arrest, the latter with or without differentiation. This modulation brought upon by the binding protein can be seen in the light of its cytosolic sensor function in signaling (Section 7), which may be operative only at low concentrations of the ligand [135]. However, if directed nuclear transport does not take place, the effect will be adverse in either direction, as FABP in a concentration-dependent manner would buffer the lipophilic ligands and prevent them from interacting with their nuclear targets.

L-FABP of subfamily II increased proliferation affected by mitogens and carcinogens in transfected liver and hepatoma cells [72,141]. Carcinogenic peroxisome proliferators became more potent in cells co-transfected by L-FABP, leading to higher cell proliferation rates due to targeting [142].

In contrast, FABPs of subfamily IV reveal growth inhibitory action, for which only a few other peptides are known such as interferons and transforming growth factor β. Thus, loss of A-FABP was correlated with progression of human bladder transitional cell carcinoma [143] and E-FABP, upon application to skin, reduced proliferation of melanoma cells, while normal skin fibroblasts were unaffected [144]. The gene product of a "mammary derived growth inhibitor-related inhibitor gene" (MRG), later identified as B-FABP, suppressed tumor growth in a nude mouse model and breast cancer cell proliferation after transfection with MRG [145,146]. Finally, transfection of MCF-7 cells, a human breast cancer line, with cDNA encoding bovine H-FABP reduced cell growth; in addition, the H-FABP producing transfectants reduced in vivo tumorigenicity [147]. At present it is not clear whether or not growth inhibition is due to the FABP itself or to its putative ligand. But it is also tempting to speculate in the case of B-FABP that the high affinity-ligand DHA (Table 2) would exert the inhibitory effect.

The background of these observations during the last 15 years was the discovery of bovine "mammary derived growth inhibitor" (MDGI) in 1987 [148]. It was soon recognized as a variant of H-FABP [149] and finally identified as a preparation of H-FABP contaminated with small amounts of closely related A-FABP [150]. MDGI was a potent inhibitor of epithelial proliferation in various mammalian organ and cell cultures [151]. MDGI, and H-FABP alone also showed anti-proliferation activity in breast cancer cells and H-FABP expression seemed to be reduced in malignant breast tumors [152]. When administered extracellularly, however, the anti-tumor activity of H-FABP was not due to a bound ligand, but could be mapped to a C-terminal fragment of the protein [153]. More details on MDGI-activities of FABPs can be found in a review published in 1998 [154].

In mammary gland organ culture, growth inhibition was associated with functional differentiation in the presence of MDGI or H-FABP; in fact, this differentiation is preceded by heavy expression of H-FABP in the mammary epithelial cells, which then promotes milk protein synthesis in the differentiated cells [155]. Based on this observation, it was argued that H-FABP acts as a differentiation factor. A-FABP as well was assumed originally to be such a factor as it was expressed in the course of differentiation from preadipocytes to adipocytes of both primary cells and the 3T3-L1 cell model. Yet it was soon recognized that the fatty acids themselves (transported by E-FABP in the preadipocyte?) are the trigger of differentiation and, as a result A-FABP and PPARγ among others are expressed. In fully differentiated adipocyte culture, removal of fatty acids from the medium and re-supplementation decreased and replenished A-FABP mRNA levels, respectively [156]. From today's perspective we can ascribe to A-FABP a carrier function in fatty acid signaling to the nucleus to interact with PPARγ and a transport function needed during the time of heavy triacylglycerol accumulation. Indeed, tissue-specific enhancer and proximal promoter regions of the A-FABP gene interact with adipogenic transcription factors in a time-dependent manner [157]. In line with this, H-FABP in C2C12 cells was induced upon differentiation from the myoblast to the myotube stage [158]. A careful follow-up study demonstrated later that E-FABP in myoblasts is down-regulated during differentiation, while H-FABP was induced at later stages of differentiation when energy retrieval in the cells shifts from glycolysis to β-oxidation, indicative of a metabolic transport function of the binding protein [159].

9. Outlook

Much progress has been made in the last decade in the study of the structure and binding behavior of the FABPs. Much of the current research activity is directed to understand the control of their gene expression, and the interactions of FABPs with other proteins in the cell. Undoubtedly, these studies will help to more fully understand the pleiotropic roles of these intracellular transport proteins, especially with respect to signal transduction, both at the molecular and the cellular level. It is the belief of the authors that analysis of this conserved gene family in various organisms will continue to provide new insights into their regulatory functions.

Acknowledgements

The authors gratefully acknowledge support of the work carried out in their laboratories and reviewed here by grants from the Heart & Stroke Foundation and the Natural Research and Engineering Council of Canada (given to N.H.H.), Deutsche Forschungsgemeinschaft (SFB 310, SP 135/10-3, SP 135/10-2) and Stiftunggsfonds Unilever (given to F.S.).

References

[1] Ockner, R.K., Manning, J.A., Poppenhausen, R.B., Ho, W.K., 1972. Science 177, 56–58.
[2] Stewart, J.M., 2000. Cell. Mol. Life Sci. 57, 1345–1359.
[3] Matarese, V., Stone, R.L., Waggoner, D.W., Bernlohr, D.A., 1989. Prog. Lipid Res. 28, 245–272.
[4] Bernlohr, D.A., Simpson, M.A., Hertzel, A.V., Banaszak, L.J., 1997. Annu. Rev. Nutr. 17, 277–303.
[5] Medzihradszky, K.F., Gibson, B.W., Kaur, S., Yu, Z.H., Medzihradszky, D., Burlingame, A.L., Bass, N.M., 1992. Eur. J. Biochem. 203, 327–339.
[6] Di Pietro, S.M., Veerkamp, J.H., Santomé, J.A., 1999. Eur. J. Biochem. 259, 127–134.
[7] Smith, A.F., Tsuchida, K., Hanneman, E., Suzuki, T.C., Wells, M.A., 1992. J. Biol. Chem. 267, 380–384.
[8] Haunerland, N.H., Andolfatto, P., Chisholm, J.M., Wang, Z.X., Chen, X.M., 1992. Eur. J. Biochem. 210, 1045–1051.
[9] Haunerland, N.H., Jacobson, B.L., Wesenberg, G., Rayment, I., Holden, H.M., 1994. Biochemistry 33, 12378–12385.
[10] Adams, M.D., Celniker, S.E., Holt, R.A., Evans, C.A., Gocayne, J.D., Amanatides, P.G., Scherer, S.E., Li, P.W., Hoskins, R.A., Galle, R.F., et al., 2000. Science 287, 2185–2195.
[11] dellaTorre, A., Favia, G., Mariotti, G., Coluzzi, M., Mathiopoulos, K.D., 1996. Genetics 143, 1307–1311.
[12] Mansfield, S.G., Cammer, S., Alexander, S.C., Muehleisen, D.P., Gray, R.S., Tropsha, A., Bollenbacher, W.E., 1998. Proc. Natl Acad. Sci. USA 95, 6825–6830.
[13] Puerta, L., Kennedy, M.W., Jimenez, S., Caraballo, L., 1998. J. Allergy Clin. Immunol. 101, S170–S170.
[14] Eriksson, T.L.J., Whitley, P., Johansson, E., van Hage-Hamsten, M., Gafvelin, G., 1999. Int. Arch. Allergy Immunol. 119, 275–281.
[15] Tendler, M., Brito, C.A., Vilar, M.M., SerraFreire, N., Diogo, C.M., Almeida, M.S., Delbem, A.C.B., daSilva, J.F., Savino, W., Garratt, R.C., Katz, N., Simpson, A.J.G., 1996. Proc. Natl Acad Sci. USA 93, 269–273.
[16] Esteves, A., Joseph, L., Paulino, M., Ehrlich, R., 1997. Int. J. Parasit. 27, 1013–1023.
[17] McDermott, L., Kennedy, M.W., McManus, D.P., Bradley, J.E., Cooper, A., Storch, J., 2002. Biochemistry 41, 6706–6713.
[18] Hanhoff, T., Lücke, C., Spener, F., 2002. Mol. Cell. Biochem. 239, 45–54.
[19] Ross, A.C., 1993. FASEB J. 7, 317–327.
[20] Haunerland, N., Jagschies, G., Schulenberg, H., Spener, F., 1984. Fatty-acid-binding proteins. Hoppe-Seyler's, Z. Physiol. Chem. 365, 365–376.
[21] Keuper, H.J.K., Klein, R.A., Spener, F., 1985. Chem. Phys. Lipids 38, 159–173.
[22] Thompson, J., Winter, N., Terwey, D., Bratt, J., Banaszak, L., 1997. J. Biol. Chem. 272, 7140–7150.
[23] Schleicher, C.H., Cordoba, O.L., Santomé, J.A., DellAngelica, E.C., 1995. Biochem. Mol. Biol. Int. 36, 1117–1125.
[24] Schaap, F.G., van der Vusse, G.J., Glatz, J.F.C., 2002. Mol. Cell. Biochem. 239, 69–77.
[25] Price, H.M., Ryan, R.O., Haunerland, N.H., 1992. Arch. Biochem. Biophys. 297, 285–290.
[26] Sacchettini, J.C., Gordon, J.I., Banaszak, L.J., 1989. J. Mol. Biol. 208, 327–339.

[27] Zanotti, G., Scapin, G., Spadon, P., Veerkamp, J.H., Sacchettini, J.C., 1992. 3-Dimensional structure of recombinant human muscle fatty acid-binding protein. J. Biol. Chem. 267, 18541–18550.
[28] Xu, Z.H., Bernlohr, D.A., Banaszak, L.J., 1992. Biochemistry 31, 3484–3492.
[29] Hohoff, C., Börchers, T., Rüstow, B., Spener, F., van Tilbeurgh, H., 1999. Biochemistry 38, 12229–12239.
[30] Balendiran, G.K., Schnütgen, F., Scapin, G., Börchers, T., Xhong, N., Lim, K., Godbout, R., Spener, F., Sacchettini, J.C., 2000. J. Biol. Chem. 275, 27045–27054.
[31] Hodsdon, M.E., Ponder, J.W., Cistola, D.P., 1996. J. Mol. Biol. 264, 585–602.
[32] Lassen, D., Lücke, C., Kromminga, A., Lezius, A., Spener, F., Rüterjans, H., 1993. Mol. Cell. Biochem. 123, 15–22.
[33] Gutierrez-Gonzalez, L.H., Ludwig, C., Hohoff, C., Rademacher, M., Hanhoff, T., Rüterjans, H., Spener, F., Lücke, C., 2002. Biochem. J. 364, 725–737.
[34] Lücke, C., Zhang, F.L., Hamilton, J.A., Sacchettini, J.C., Rüterjans, H., 2000. Eur. J. Biochem. 267, 2929–2938.
[35] Rademacher, M., Zimmerman, A.W., Rüterjans, H., Veerkamp, J.H., Lücke, C., 2002. Mol. Cell. Biochem. 239, 61–68.
[36] Schievano, E., Mammi, S., Peggion, E., 1999. Biopolymers 50, 1–11.
[37] Vasile, F., Ragona, L., Catalano, M., Zetta, L., Perduca, M., Monaco, H., Molinari, H., 2003. J. Biomol. NMR 25, 157–160.
[38] Benning, M.M., Smith, A.F., Wells, M.A., Holden, H.M., 1992. J. Mol. Biol. 228, 208–219.
[39] Banaszak, L., Winter, N., Xu, Z.H., Bernlohr, D.A., Cowan, S., Jones, T.A., 1994. Adv. Prot. Chem. 45, 89–151.
[40] Flower, D.R., 1993. FEBS Lett. 333, 99–102.
[41] Zimmerman, A.W., Veerkamp, J.H., 2002. Cell. Mol. Life Sci. 59, 1096–1116.
[42] Kleywegt, G.J., Bergfors, T., Senn, H., Lemotte, P., Gsell, B., Shudo, K., Jones, T.A., 1994. Structure 2, 1241–1258.
[43] Cowan, S.W., Newcomer, M.E., Jones, T.A., 1993. J. Mol. Biol. 230, 1225–1246.
[44] Winter, N.S., Bratt, J.M., Banaszak, L.J., 1993. J. Mol. Biol. 230, 1247–1259.
[45] Folli, C., Calderone, V., Ramazzina, I., Zanotti, G., Berni, R., 2002. J. Biol. Chem. 277, 41970–41977.
[46] Folli, C., Calderone, V., Ottonello, S., Bolchi, A., Zanotti, G., Stoppini, M., Berni, R., 2001. Proc. Natl Acad. Sci. USA 98, 3710–3715.
[47] Lücke, C., Zhang, F.L., Hamilton, J.A., Sacchettini, J.C., Rüterjans, H., 2000. Eur. J. Biochem. 267, 2929–2938.
[48] Richieri, G.V., Ogata, R.T., Kleinfeld, A.M., 1992. J. Biol. Chem. 267, 23495–23501.
[49] Richieri, G.V., Ogata, R.T., Kleinfeld, A.M., 1994. J. Biol. Chem. 269, 23918–23930.
[50] Warner, M., Neims, A.H., 1975. Can. J. Physiol. Pharmacol. 53, 493–500.
[51] Glatz, J.F., Veerkamp, J.H., 1983. Anal. Biochem. 132, 89–95.
[52] Weisiger, R.A., 1996. Comp. Biochem. Phys. B 115, 319–331.
[53] Lücke, C., Rademacher, M., Zimmerman, A.W., van Moerkerk, H.T.B., Veerkamp, J.H., Rüterjans, H., 2001. Biochem. J. 354, 259–266.
[54] Storch, J., Veerkamp, J.H., Hsu, K.T., 2002. Mol. Cell. Biochem. 239, 25–33.
[55] Herr, F.M., Aronson, J., Storch, J., 1996. Biochemistry 35, 1296–1303.
[56] Corsico, B., Cistola, D.P., Frieden, C., Storch, J., 1998. Proc. Natl Acad. Sci. USA 95, 12174–12178.
[57] Herr, F.M., Matarese, V., Bernlohr, D.A., Storch, J., 1995. Biochemistry 34, 11840–11845.
[58] Weisiger, R.A., 2002. Mol. Cell. Biochem. 239, 35–43.
[59] Hamilton, J.A., Guo, W., Kamp, F., 2002. Mol. Cell. Biochem. 239, 17–23.
[60] Murphy, E.J., Prows, D.R., Jefferson, J.R., Schroeder, F., 1996. Biochim. Biophys. Acta 1301, 191–198.
[61] Sha, R.S., Kane, C.D., Xu, Z.H., Banaszak, L.J., Bernlohr, D.A., 1993. J. Biol. Chem. 268, 7885–7892.
[62] Wolfrum, C., Buhlmann, C., Rolf, B., Börchers, T., Spener, F., 1999. Biochim. Biophys. Acta 1437, 194–201.
[63] Fitscher, B.A., Klaaßen-Schlüter, C.M., Stremmel, W., 1995. Biochim. Biophys. Acta 1256, 47–51.
[64] Prinsen, C.F.M., Veerkamp, J.H., 1998. Biochem. J. 329, 265–273.
[65] Holehouse, E.L., Liu, M.L., Aponte, G.W., 1998. Biochim. Biophys. Acta 1390, 52–64.
[66] Prows, D.R., Murphy, E.J., Moncecchi, D., Schroeder, F., 1996. Chem. Phys. Lipids 84, 47–56.

[67] Baier, L.J., Bogardus, C., Sacchettini, J.C., 1996. J. Biol. Chem. 271, 10892–10896.
[68] Prows, D.R., Murphy, E.J., Schroeder, F., 1995. Lipids 30, 907–910.
[69] McCormack, M., Troxler, R.F., Brecher, P., 1986. Fed. Proc. 45, 1593–1593.
[70] Luxon, B.A., Weisiger, R.A., 1993. Am. J. Physiol. 265, G831–G841.
[71] Martin, G.G., Danneberg, H., Kumar, L.S., Atshaves, B.P., Erol, E., Bader, M., Schroeder, F., Binas, B., 2003. J. Biol. Chem. 278, 21429–21438.
[72] Sorof, S., 1994. Cancer Metast. Rev. 13, 317–336.
[73] Atshaves, B.P., Foxworth, W.B., Frolov, A., Roths, J.B., Kier, A.B., Oetama, B.K., Piedrahita, J.A., Schroeder, F., 1998. Am. J. Physiol. 43, C633–C644.
[74] Darimont, C., Gradoux, N., De Pover, A., 1999. Am. J. Physiol. 276, G606–G612.
[75] Baier, L.J., Sacchettini, J.C., Knowler, W.C., Eads, J., Paolisso, G., Tataranni, P.A., Mochizuki, H., Bennett, P.H., Bogardus, C., Prochazka, M., 1995. J. Clin. Invest. 95, 1281–1287.
[76] Tataranni, P.A., Baier, L.J., Paolisso, G., Howard, B.V., Ravussin, E., 1996. Lipids 31, S267–S270.
[77] Vassileva, G., Huwyler, L., Poirier, K., Agellon, L.B., Toth, M.J., 2000. FASEB J. 14, 2040–2046.
[78] Agellon, L.B., Toth, M.J., Thomson, A.B.R., 2002. Mol. Cell. Biochem. 239, 79–82.
[79] Coe, N.R., Bernlohr, D.A., 1998. Biochim. Biophys. Acta 1391, 287–306.
[80] Shen, W.J., Sridhar, K., Bernlohr, D.A., Kraemer, F.B., 1999. FASEB J. 13, A1381–A1381.
[81] Hotamisligil, G.S., Johnson, R.S., Distel, R.J., Ellis, R., Papaioannou, V.E., Spiegelman, B.M., 1996. Science 274, 1377–1379.
[82] Shaughnessy, S., Smith, E.R., Kodukula, S., Storch, J., Fried, S.K., 2000. Diabetes 49, 904–911.
[83] Coe, N.R., Simpson, M.A., Bernlohr, D.A., 1999. J. Lipid Res. 40, 967–972.
[84] Layne, M.D., Patel, A., Chen, Y.H., Rebel, V.I., Carvajal, I.M., Pellacani, A., Ith, B., Zhao, D.Z., Schreiber, B.M., Yet, S.F., Lee, M.E., Storch, J., Perrella, M.A., 2001. FASEB J. 15, U236–U254.
[85] Perrella, M.A., Pellacani, A., Layne, M.D., Patel, A., Zhao, D.Z., Schreiber, B.M., Storch, J., Feinberg, M.W., Hsieh, C.M., Haber, E., Lee, M.E., 2001. FASEB J. 15, U450–U465.
[86] Makowski, L., Boord, J.B., Maeda, K., Babaev, V.R., Uysal, K.T., Morgan, M.A., Parker, R.A., Suttles, J., Fazio, S., Hotamisligil, G.S., Linton, M.F., 2001. Nat. Med. 7, 699–705.
[87] Owada, Y., Suzuki, I., Noda, T., Kondo, H., 2002. Mol. Cell. Biochem. 239, 83–86.
[88] Owada, Y., Takano, H., Yamanaka, H., Kobayashi, H., Sugitani, Y., Tomioka, Y., Suzuki, I., Suzuki, R., Terui, T., Mizugaki, M., Tagami, H., Noda, T., Kondo, H., 2002. J. Invest. Dermatol. 118, 430–435.
[89] Maeda, K., Uysal, K.T., Makowski, L., Gorgun, C.Z., Atsumi, G., Parker, R.A., Bruning, J., Hertzel, A.V., Bernlohr, D.A., Hotamisligil, G.S., 2003. Diabetes 52, 300–307.
[90] Hertzel, A.V., Bennaars-Eiden, A., Bernlohr, D.A., 2002. J. Lipid Res. 43, 2105–2111.
[91] Owada, Y., Kondo, H., 2003. In: Duttaroy, A.K., Spener, F. (Eds.), Cellular Proteins and Their Fatty Acids in Health and Disease. Wiley-VCH, Weinheim, Germany, pp. 253–266.
[92] Veerkamp, J.H., Zimmerman, A.W., 2001. J. Mol. Neurosci. 16, 133–142.
[93] Kurtz, A., Zimmer, A., Schnütgen, F., Bruning, G., Spener, F., Müller, T., 1994. Development 120, 2637–2649.
[94] Xu, L.Z., Sanchez, R., Sali, A., Heintz, N., 1996. J. Biol. Chem. 271, 24711–24719.
[95] Pu, L.X., Igbavboa, U., Wood, W.G., Roths, J.B., Kier, A.B., Spener, F., Schroeder, F., 1999. Mol. Cell. Biochem. 198, 69–78.
[96] Bisgrove, D.A., Monckton, E.A., Godbout, R., 1997. Mol. Cell. Biol. 17, 5935–5945.
[97] Rousselot, P., Heintz, N., Nottebohm, F., 1997. J. Comp. Neurol. 385, 415–426.
[98] Zanotti, G., 1999. Biochim. Biophys. Acta 1441, 94–105.
[99] Haunerland, N.H., 1994. Comp. Biochem. Phys. B 109, 199–208.
[100] Veerkamp, J.H., van Moerkerk, H.T.B., 1993. Mol. Cell. Biochem. 123, 101–106.
[101] Guglielmo, C.G., Haunerland, N.H., Williams, T.D., 1998. Comp. Biochem. Phys. B 119, 549–555.
[102] Clavel, S., Farout, L., Briand, M., Briand, Y., Jouanel, P., 2002. Eur. J. Appl. Physiol. 87, 193–201.
[103] Chen, X.M., Haunerland, N.H., 1994. Insect Biochem. Mol. Biol. 24, 573–579.
[104] Binas, B., Danneberg, H., McWhir, J., Mullins, L., Clark, A.J., 1999. FASEB J. 13, 805–812.
[105] Schaap, F.G., Binas, B., Danneberg, H., van der Vusse, G.J., Glatz, J.F.C., 1999. Circ. Res. 85, 329–337.
[106] Vayda, M.E., Londraville, R.L., Cashon, R.E., Costello, L., Sidell, B.D., 1998. Biochem. J. 330, 375–382.
[107] Veerkamp, J.H., Maatman, R., 1995. Prog. Lipid. Res. 34, 17–52.

[108] Wu, Q.W., Andolfatto, P., Haunerland, N.H., 2001. Insect Biochem. Mol. Biol. 31, 553–562.
[109] Akiyama, T.E., Ward, J.M., Gonzalez, F.J., 2000. J. Biol. Chem. 275, 27117–27122.
[110] Ross, S.R., Graves, R.A., Greenstein, A., Platt, K.A., Shyu, H.L., Mellovitz, B., Spiegelman, B.M., 1990. Proc. Natl Acad. Sci. USA 87, 9590–9594.
[111] Josephson, R., Müller, T., Pickel, J., Okabe, S., Reynolds, K., Turner, P.A., Zimmer, A., McKay, R.D.G., 1998. Development 125, 3087–3100.
[112] Qian, Q.H., Kuo, L., Yu, Y.T., Rottman, J.N., 1999. Circ. Res. 84, 276–289.
[113] Malewiak, M.I., Bass, N.M., Griglio, S., Ockner, R.K., 1988. Int. J. Obesity 12, 543–546.
[114] Vanbreda, E., Keizer, H.A., Vork, M.M., Surtel, D.A.M., Dejong, Y.F., van der Vusse, G.J., Glatz, J.F.C., 1992. Eur. J. Physiol. 421, 274–279.
[115] Duplus, E., Glorian, M., Forest, C., 2000. J. Biol. Chem. 275, 30749–30752.
[116] Wolfrum, C., Spener, F., 2000. Eur. J. Lipid Sci. Technol. 102, 746–762.
[117] Desvergne, B., Wahli, W., 1999. Endocr. Rev. 20, 649–688.
[118] Tontonoz, P., Hu, E., Graves, R.A., Budavari, A.I., Spiegelman, B.M., 1994. Genes Dev. 8, 1224–1234.
[119] Issemann, I., Prince, R., Tugwood, J., Green, S., 1992. Biochem. Soc. Trans. 20, 824–827.
[120] Poisirer, H., Niot, I., Monnot, M.C., Braissant, O., Meunier-Durmort, C., Costet, P., Pineau, T., Wahi, W., Willson, T.M., Besnard, P., 2001. Biochem. J. 355, 481–488.
[121] van der Lee, K.A.J.M., Vork, M.M., De Vries, J.E., Willemsen, P.H.M., Glatz, J.F.C., Reneman, R.S., van der Vusse, G.J., Van Bilsen, M., 2000. J. Lipid Res. 41, 41–47.
[122] Duplus, E., Forest, C., 2002. Biochem. Pharmacol. 64, 893–901.
[123] Detera-Wadleigh, F.T.G., Sevilla, D., 1994. Mol. Phylog. Evol. 3, 192–205.
[124] Wu, Q.W., Haunerland, N.H., 2001. Eur. J. Biochem. 268, 5894–5900.
[125] Wu, Q.W., Chang, W.H., Rickers-Haunerland, J., Higo, T., Haunerland, N.H., 2002. Mol. Cell. Biochem. 239, 173–180.
[126] Berger, J., Moller, D.E., 2002. Annu. Rev. Med. 53, 409–435.
[127] Wolfrum, C., Spener, F., 2003. In: Duttaroy, A.K., Spener, F. (Eds.), Cellular Proteins and their Fatty Acids in Health and Disease. Wiley-VCH, Weinheim, Germany, pp. 267–284.
[128] Noy, N., 2000. Biochem. J. 348, 481–495.
[129] Hwang, S.T., Urizar, N.L., Moore, D.D., Henning, S.J., 2002. Gastroenterology 122, 1483–1492.
[130] Bordewick, U., Heese, M., Börchers, T., Robenek, H., Spener, F., 1989. Biol. Chem. Hoppe-Seyler 370, 229–238.
[131] Young, J.K., Baker, J.H., Müller, T., 1996. Glia 16, 218–226.
[132] Helledie, T., Antonius, M., Sorensen, R.V., Hertzel, A.V., Bernlohr, D.A., Kolvraa, S., Kristiansen, K., Mandrup, S., 2000. J. Lipid Res. 41, 1740–1751.
[133] Börchers, T., Unterberg, C., Rüdel, H., Robenek, H., Spener, F., 1989. Biochim. Biophys. Acta 1002, 54–61.
[134] Wolfrum, C., Borrmann, C.M., Börchers, T., Spener, F., 2001. Proc. Natl Acad. Sci. USA 98, 2323–2328.
[135] Tan, N.S., Shaw, N.S., Vinckenbosch, N., Liu, P., Yasmin, R., Desvergne, B., Wahli, W., Noy, N., 2002. Mol. Cell. Biol. 22, 5114–5127.
[136] Lawrence, J.W., Kroll, D.J., Eacho, P.I., 2000. J. Lipid Res. 41, 1390–1401.
[137] Huang, H., Starodub, O., McIntosh, A., Kier, A.B., Schroeder, F., 2002. J. Biol. Chem. 277, 29139–29151.
[138] Helledie, T., Jørgensen, C., Antonius, M., Krogsdam, A.M., Kratchmarova, I., Kristiansen, K., Mandrup, S., 2002. Mol. Cell. Biochem. 239, 157–164.
[139] Dong, D., Ruuska, S.E., Levinthal, D.J., Noy, N., 1999. J. Biol. Chem. 274, 23695–23698.
[140] Delva, L., Bastie, J.N., Rochette-Egly, C., Kraiba, R., Balitrand, N., Despouy, G., Chambon, P., Chomienne, C., 1999. Mol. Cell. Biol. 19, 7158–7167.
[141] Keler, T., Sorof, S., 1993. J. Cell. Physiol. 157, 33–40.
[142] Khan, S.H., Sorof, S., 1994. Proc. Natl Acad. Sci. USA 91, 848–852.
[143] Celis, J.E., Ostergaard, M., Basse, B., Celis, A., Lauridsen, J.B., Ratz, G.P., Andersen, I., Hein, B., Wolf, H., Orntoft, T.F., Rasmussen, H.H., 1996. Cancer Res. 56, 4782–4790.
[144] Farooqui, J.Z., Robb, E., Boyce, S.T., Warden, G.D., Nordlund, J.J., 1995. J. Invest. Dermatol. 104, 739–743.

[145] Shi, Y.E., Ni, J., Xiao, G.W., Liu, Y.E., Fuchs, A., Yu, G.L., Su, J., Cosgrove, J.M., Xing, L., Zhang, M., Li, J.Y., Aggarwal, B.B., Meager, A., Gentz, R., 1997. Cancer Res. 57, 3084–3091.
[146] Hohoff, C., Spener, F., 1998. Cancer Res. 58, 4015–4016.
[147] Buhlmann, C., Börchers, T., Pollak, M., Spener, F., 1999. Mol. Cell. Biochem. 199, 41–48.
[148] Böhmer, F.D., Kreft, R., Otto, A., Wernstedt, C., Hellmann, U., Kurtz, A., Müller, T., Rohde, K., Elzold, G., Lehmann, W., Lengen, P., Heldin, C.-H., Grosse, R., 1987. J. Biol. Chem. 262, 15137–15143.
[149] Böhmer, F.D., Sun, Q., Pepperle, M., Müller, T., Eriksson, U., Wang, J.L., Grosse, R., 1987. Biochem. Biophys. Res. Commun. 148, 1425–1431.
[150] Specht, B., Bartetzko, N., Hohoff, C., Kuhl, H., Franke, R., Börchers, T., Spener, F., 1996. J. Biol. Chem. 271, 19943–19949.
[151] Böhmer, F.D., Sun, Q., Pepperle, M., Müller, T., Eriksson, U., Wang, J.L., Grosse, R., 1987. Biochem. Biophys. Res. Commun. 148, 1425–1431.
[152] Huynh, H.T., Larsson, C., Narod, S., Pollak, M., 1995. Cancer Res. 55, 2225–2231.
[153] Wang, H.L., Kurtz, A., 2000. Oncogene 19, 2455–2460.
[154] Hohoff, C., Spener, F., 1998. Fett-Lipid 100, 252–263.
[155] Yang, Y.M., Spitzer, E., Kenney, N., Zschiesche, W., Li, M.L., Kromminga, A., Müller, T., Spener, F., Lezius, A., Veerkamp, J.H., Smith, G.H., Salomon, D.S., Grosse, R., 1994. J. Cell Biol. 127, 1097–1109.
[156] Amri, E.Z., Bertrand, B., Ailhaud, G., Grimaldi, P., 1991. J. Lipid Res. 32, 1449–1456.
[157] Gerhold, D.L., Liu, F., Jiang, G., Li, Z., Xu, J., Lu, M., Sachs, J.R., Bagchi, A., Fridman, A., Holder, D.J., Doebber, T.W., Berger, J., Elbrecht, A., Moller, D.E., Zhang, B.B., 2002. Endocrinology 143, 2106–2118.
[158] Rump, R., Buhlmann, C., Börchers, T., Spener, F., 1996. Eur. J. Cell Biol. 69, 135–142.
[159] Bleck, B., Buhlmann, C., Hohoff, C., Müller, M., Börchers, T., Spener, F., 2002. Eur. J. Lipid Sci. Technol. 104, 88–97.
[160] Crabtree, B., Newsholme, E.A., 1975. In: Usherwood, P.N.R. (Ed.), Insect Muscle. Academic Press, London, pp. 405–491.
[161] Crisman, T.S., Claffey, K.P., Saouaf, R., Hanspal, J., Brecher, P., 1987. J. Mol. Cell Cardiol. 19, 423–431.
[162] Guglielmo, C.G., Haunerland, N.H., Hochachka, P.W., Williams, T.D., 2002. Am. J. Physiol. 282, R1405–R1413.

Long chain acyl-CoA esters and acyl-CoA binding protein (ACBP) in cell function

Jens Knudsen,* Mark Burton and Nils Færgeman

Department of Biochemistry and Molecular Biology, University of Southern Denmark, Campusvej 55 DK-5230 Odense M, Denmark
**Correspondence address:E-mail: jkk@bmb.sdu.dk(J.K.)*

1. Introduction

The physiological role of long-chain fatty acyl-CoA (LCACoA) was originally thought to be an intermediate in lipid metabolism only. However, LCACoA esters are increasingly being recognised as modulators of a wide range of cellular functions [1]. This requires a tight regulation of the intracellular LCACoA concentration in a way that simultaneously allows large variation in the rate of lipogenesis and β-oxidation and sufficient supply of LCACoA for special purposes like protein acylation. The major players in the control of the cellular LCACoA concentration are fatty acid supply, the delicate balance between the activity of acyl-CoA synthetases (ACS) and acyl-CoA thioesterase (TE) and finally the concentration of cellular LCACoA binding proteins.

A number of proteins have been reported to bind LCACoA including liver fatty acid binding protein (L-FABP), sterol carrier protein 2 (SCP-2) and acyl-coenzyme A binding protein (ACBP) [2]. In contrast to FABPs and SCP-2, ACBP binds only LCACoA.

ACBP was discovered independently by four different groups as (1) a brain peptide, diazepam binding inhibitor (DBI), found to be able to inhibit diazepam binding to the GABA receptor [3], an effect which could not be reproduced by other groups [4]; (2) a bovine adrenal peptide stimulating cholesterol transport into Leydig cell mitochondria [5], a function which is compatible with its ability to bind LCACoA; (3) a pig intestinal peptide regulating insulin release [6] and finally; (4) a liver LCACoA binding peptide, ACBP [7]. When we in the present review use the name ACBP it will be synonymous with DBI.

ACBP is a ~10 kDa cytosolic protein, which binds LCACoA esters with high specificity and affinity (K_D, 1–10 nM). The protein, or the gene encoding the protein, has been detected in all eukaryotic species tested so far, but it appears to be absent from prokaryotes with the exception of a few bacteria. The extreme high degree of conservation of ACBP among all eukaryotic species and the fact that ACBP is expressed in all cells and

Advances in Molecular and Cell Biology, Vol. 33, pages 123–152

ISSN: 1569-2558/DOI: 10.1016/S1569-2558(03)33008-5

tissues suggest that its function is associated with one or more basal cellular function(s) common to all cells. However, the precise biological function of ACBP is not known at present. A large body of in vitro experimental evidence indicates that ACBP is able to act as an intracellular acyl-CoA transporter and pool former. Recent work indicates that ACBP is required for fatty acid chain elongation, sphingolipid synthesis, protein sorting and vesicular trafficking in yeast (See Refs. [8,9] for recent review). A database search shows that ACBP in addition to being a functional protein on its own also occurs as a domain in a large number of proteins including enzymes and potential regulatory proteins. In this review we will shortly discuss important aspects of ACBP phylogeny, structure and ligand binding. The major emphasis will focus on reviewing and discussing the literature on the role of LCACoA esters and ACBP in metabolism and cell function.

2. ACBP is a highly conserved protein expressed in all eukaryotic species

ACBP consists on an average of 82–92 amino acid residues depending on the species. ACBP has a highly conserved amino acid sequence and is found in all eukaryotic species examined, ranging from yeast and plants to reptiles, birds and mammals. Mouse and rat ACBP share 97% identity at the amino acid level, while diverse species as man and *Saccharomyces cerevisiae* exhibit 48% identity. Using the ACBP sequence as query a database search (March 2003) using BlastP, T-BlastN and BlastN revealed 148 basal ACBP sequences (74–105 residues) from 101 different species, published either as proteins or as gene sequences. In addition to the basal ACBP sequences, the database search also identified 77 sequences in which ACBP occurs as a domain in a larger protein. The retrieved sequences were aligned with ClustalW and a phylogenetic analysis was carried out using the Mega 2.1 software package [10]. This shows that ACBP has evolved in a large group of eukaryote organisms from the unicellular algae *Chlamydonas reinhardtii* to man (Fig. 1). The presence of the ACBP gene in algae indicates that ACBP is required for very basal biological functions common to all eukaryotes. Highly conserved ACBP genes, encoding proteins very similar to the basal vertebrate ACBP, occur in eight prokaryotic species (our unpublished results). However, five of these strains are pathogenic bacteria or belonging to groups which include pathogenic bacteria. They could therefore have obtained the gene by horizontal gene transfer. These sequences are not included in the present phylogenetic analysis. The sequencing of the full genome of many organisms including the nematode *C. elegans*, fruit fly (*Drosophila melanogaster*), *Arabidopsis thaliana*, humans, mouse and pufferfish allows a more detailed analysis of the number of different ACBP isoforms expressed in diverse organisms. Three isoforms, basal ACBP, brain specific ACBP and testis specific ACBP have been detected in vertebrates (Table 1). Both mouse, rat and bovine have functional genes for all three isoforms [11], however, the testis ACBP gene in humans and other primates has undergone a frame shift and become non-functional [12]. Only the liver and the testis and the liver and brain isoform have been detected in rat and dog [13,14] and chicken and duck [15], respectively. The presence of only three genes in the human and mouse genome suggests that only these three isoform exist in vertebrates. The arthropods seem to differ in this aspect. Fruit fly expresses one basal isoform grouping closely together with the basal vertebrate and testis

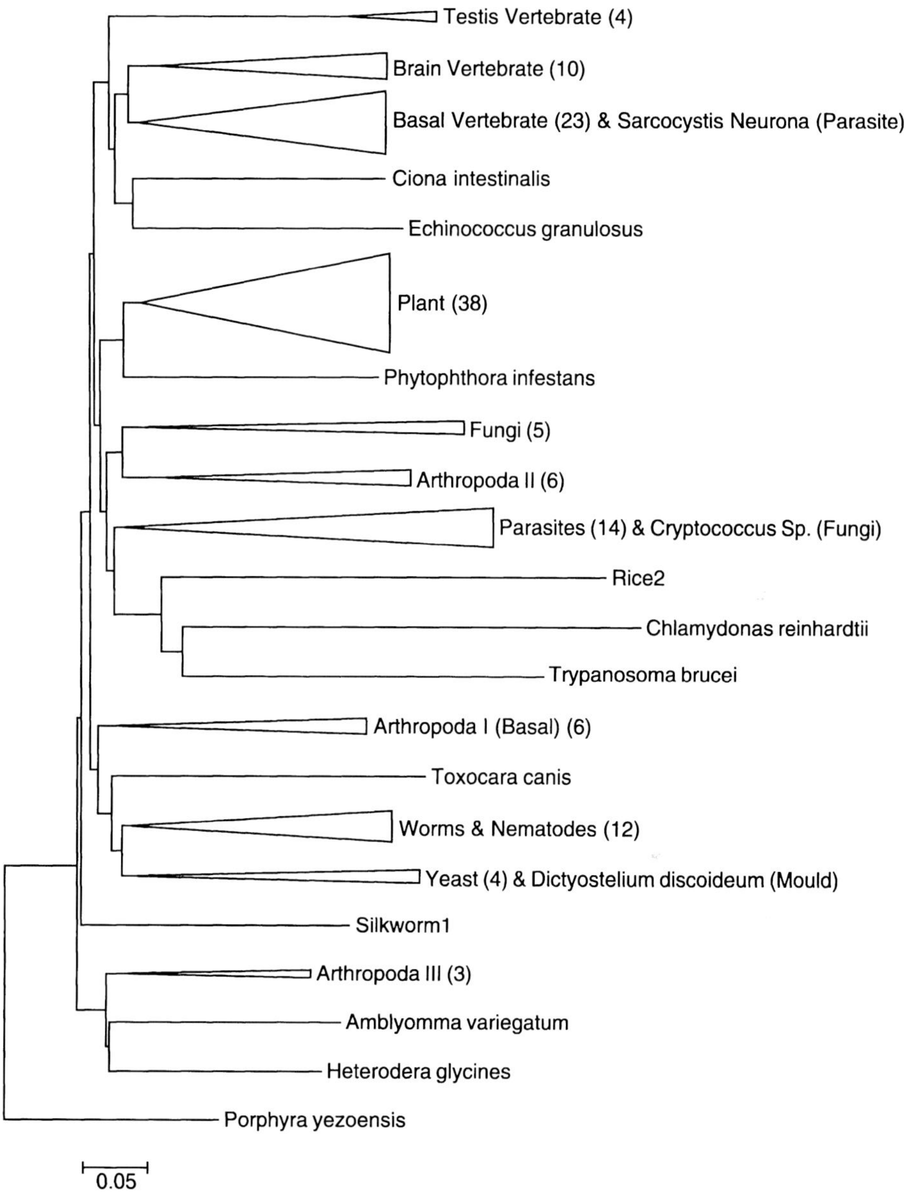

Fig. 1. The ACBP-phylogenetic tree. Rooted neighbour-joining phylogenetic tree showing the evolutionary relationship between the members of the basic ACBP family. The various ACBP-clusters are indicated by a triangle followed by description and number of sequences contained in the respective cluster. The bar indicates the evolutionary distance expressed in number of substitutions per site (modified from Burton et al., 2003).

Table 1
Vertebrate ACBP isoforms

Species	Liver ACBP	Brain ACBP	Testis ACBP
Human	X	X	(1 & 2) not functional, frame shift
Chimpanzee	X	n.d.	Not functional, frame shift
Mouse	X	X	X
Rat	X	X	X
Dog	X	n.d.	X
Bovine	X	X	X
Chicken	X	X	n.d.
Duck	X	X	n.d.
Frog	n.d.	X	n.d.
Pufferfish	X	X	n.d.
Rainbow Trout	X (1 & 2)	X	n.d.
Salmon	X	X	n.d.
Zebrafish	X	X	n.d.

The accession numbers used refers to GenBank accession numbers, with the exception of dog liver-ACBP that is located at the institute for genomic research (TIGR) (http://tigrblast.tigr.org/tgi/). Vertebrate Brain-ACBP: *Anas platyrhynchos* (Duck) (S73733), *Bos taurus* (Bovine) (AV591060), *Gallus gallus* (Chicken) (BU352179), *Danio rerio* (Zebrafish) (AW826792), *Homo sapiens* (Human) (BC029526), *Mus musculus* (Mouse) (AK020348), *Oncorhynchus mykiss* (Rainbow Trout) (CA350729), *Pan troglodytes* (Chimpanzee) (CB295016), *Rana ridibunda* (Frog) (U09205), *Takifugu rubripes* (Pufferfish) (CA333520), *Salmo salar* (Salmon) (CA047983). Vertebrate Liver ACBP: *B. taurus* (Bovine) (M15886), *Canis familiaris* (Dog) (TIGR: TC_85), *D. rerio* (Zebrafish) (BC045916), *G. gallus* (Chicken) (BX276485), *H. sapiens* (Human) (M14200), *M. musculus* (Mouse) (NM_007830), *O. mykiss* 1 (Rainbow Trout 1) (CA379529), *O. mykiss* 2 (Rainbow Trout 2) (CA376648), *Rattus norvegicus* (Rat) (NM_031853), *S. salar* (Salmon) (CA043566), *T. rubripes* (Pufferfish) (AL836649), Vertebrate Testis ACBP: *B. taurus* (Bovine) (AF229798), *C. familiaris* (Dog) (BM537873), *H. sapiens* 1 (Human 1) (AF229803), *H. sapiens* 2 (Human 2) (AF229804), *M. musculus* (Mouse) (NM_021294), *P. troglodytes* (Chimpanzee) (AF229799), *R. norvegicus* (Rat) (NM_021596).

ACBP isoforms. In addition to the general basal ACBP isoform, five additional ACBP isoforms have evolved in fruit fly. Silkworm (*Bombyx mori*) and the malaria mosquito have also genes encoding for two different ACBP forms. Analysis of the yeasts *S. cerevisiae* and *S. pompe* genomes shows that they only contain one ACBP gene.

The plant ACBPs have evolved independently of the vertebrate and mammalian forms (Fig. 1) and in this group three species rice, *Digitalis lanata* [10,16] and *A. thaliana* have been shown to contain two ACBP genes or express two ACBP isoforms. *C. elegans* expresses only one 86 residue basal isoform, however, this organism also expresses ACBP domain proteins with 114 and 125 residues with a highly conserved ACBP domain. Whether these proteins should be regarded as basal ACBP isoforms or ACBP domain proteins with a different function than basal ACBP is not known at present. It will be an interesting challenge for future research to elucidate the biological significance of having two or more very similar basal isoforms. Preliminary experiments show that fruit fly ACBP 1–6 have a different ability to complement growth of ACBP depleted *S. cerevisiae* (our unpublished results). This suggests that different ACBP isoforms have distinct properties and different ability to support specific functions.

In addition to exist as an individual protein, the ACBP domain is also found as a conserved domain in larger multidomain proteins. The ACBP domain containing sequences include three major groups. The first group (14 sequences) contains both an ACBP domain and a complete (8) or partial (6) domain belonging to the enoyl-CoA hydratase/isomerase superfamily. These proteins are all peroxisomal Δ^3-Δ^2-enoyl-CoA isomerases (PECI) required for peroxisomal β-oxidation of unsaturated fatty acids [17]. The function of the ACBP domain in these enzymes is unknown; it may present the substrate for the isomerase or participate in the catalytical process. Preliminary data show that human PECI is active without the ACBP domain (our unpublished results), indicating that the ACBP domain is not part of the active site. The second group of sequences (17) contains both an ACBP domain and one or more ankyrin binding repeats. Ankyrin binding repeats are known to be involved in specific protein–protein interaction, and could thereby potentially target ACBP–ankyrin repeat proteins to specific cellular sites. The function of these proteins is presently unknown. The *C. elegans* 385 residue ACBP–ankyrin repeat protein also contain a highly conserved BolA DNA binding domain in addition to the two other domains. This protein may represent an acyl-CoA regulated transcription factor, like the *Escherichia coli* transcription factor FadR [18,19]. The third group contains genes encoding potential ACBP domain proteins with 145–504 residues. The *A. thaliana* and sunflower genome contain a gene encoding a hypothetical 668-residue protein containing three kelch domains.

A structure similarity search using the program "Finding 3-D Similarities in Protein Structures" (http://cl.sdsc.edu/) shows that a perfect matching ACBP domain also appears as a sub-domain in the FERM domain present in radixin, which plays a role in formation of membrane-associated cytoskeleton by linking actin filaments and adhesion proteins [20].

3. ACBP structure and ligand binding specificity

The three-dimensional structures of bovine and *Plasmodium falciparum* ACBP (PfACBP) have been solved by both NMR and X-ray crystallography and by X-ray crystallography, respectively [21–23]. The bovine ACBP crystal and NMR structures are almost identical and overall very similar to the PfACBP crystal structure. The fold of the peptide backbone of bovine ACBP shows the protein as an up–down–down–up four-α-helix bundle with an overhand loop connecting helix A2 and A3 (Fig. 2). The bundle arrangement of ACBP is unique amongst known four-helix folds. The bundle arrangement is skew, since helix A3 is disjoint to helix A1 and A4, resulting in just four helix–helix interfaces instead of the usual six seen in the well-known super-coiled four-helix bundles. The structure of bovine ACBP is shaped as a relatively flat disc.

The structural basis for the high-affinity binding of LCACoA to ACBP was studied by solving the structure of palmitoyl-CoA in complex with bovine ACBP using NMR spectroscopy. The structures of apo- and holo-forms of ACBP are practically identical [24]. No distances in holo-ACBP are significantly longer than the ones in apo-ACBP, but the C-terminal of A4 is significantly closer to the C-termini of A2 and A3 and the area around Phe49. This suggests that the binding of ligand induces a tightening of the

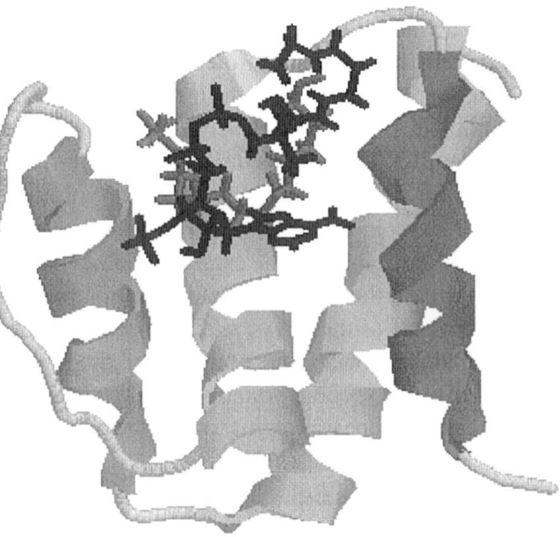

Fig. 2. The three-dimensional structure of ACBP. Three-dimensional image of the structure of recombinant bovine acyl-CoA binding protein in complex with palmitoyl-CoA. The protein is shown in a ribbon presentation, α-helix 1 (3–15) is shown blue, α-helix 2 (21–36) in turquoise, α-helix 3 (51–62) in pink and α-helix 4 (65–84) in green. The palmitoyl-CoA is shown in a ball-and-stick presentation. The hexadecanoyl-chain is shown in red and the CoA group in blue. (*For a colored version of this figure, see plate section, page 588.*)

structure of bovine ACBP. The ligand-binding site of ACBP is divided distinctly into three subsites: one for the adenine ring, one for the 3′-phosphate and one for the palmitoyl part of the ligand. The first two sites are similar in the bovine and PfACBP. The 3′-phosphate group, which contributes with 40% of the total binding energy [25], interacts strongly with ACBP through a massive network of hydrogen bonds and salt bridges to Tyr28, Lys32 and Lys54. The aromatic ring of Tyr31, which stacks with the adenine ring and which is structurally supported by the aromatic rings of Tyr73 and Phe5, forms the hydrophobic pocket [24].

The ω-end ($C_{12}-C_{16}$) of palmitic acid acyl-chain makes several non-polar interactions to residues in the cleft between helix A2 and A3, especially to the side chains of Met24, Leu25 and Ala53. Comparison of the high-resolution X-ray structures of the PfACBP and bovine ACBP crystal structures reveals a number of minor differences between the two molecules [23]. The insertion of two additional residues in the loop between α-helix 1 and α-helix 2 in PfACBP together with the alterations Ala53Lys, Lys50Ile and Asp21Asn change the binding pocket and close the tunnel at the end of the acyl-chain, with the result that PfACBP exhibits preference for shorter ligands (C14). This modification in PfACBP chain length specificity may have occurred in order to ensure the synthesis of massive amounts of the di-C14:0-GPI-anchored protein coat lining and protecting the parasite. This indicates that minor changes in ligand preference may have major biological significance. In this connection it is interesting that the sequence differences Asn19Ser and Glu23Ala in the two digitalis (*D. lanata*) ACBP isoforms change the preference for unsaturated fatty acids (our unpublished results). Thus, although the general overall pattern of chain length specificity looks similar for all ACBPs, the biological importance of small changes in ligand preference should not be overlooked.

4. Genomic organisation in mammals

So far the genomic structure of the rat [26], the human [27] and the mouse (Personal communication Dr Susanne Mandrup, University of Southern Denmark) ACBP genes is known. In addition, a number of processed pseudogenes have been characterised [26,28]. Cumulatively these data show that the mammalian ACBP gene has all the hallmarks of a typical housekeeping gene, i.e. a 5′ CpG island, several transcriptional initiation sites and processed pseudogenes [26]. The gene covers approximately 8 kb and is composed of four exons, which give rise to a transcript of 0.45 kb.

A functional sterol regulatory element (SRE) has been identified in the proximal promoter of both the human [29] and rat ACBP gene (Personal communication Dr Susanne Mandrup, SDU, DK).

The sterol regulatory element binding protein (SREBP-1)/(ADD1), which has been shown to be involved in the coordinated induction of fatty acid synthesis and glycolysis by insulin in the liver [30] and in the androgen-induced lipogenic gene expression in LNCaP cells [31], is likely to play a key role in regulation of ACBP expression in liver. The activation of ACBP expression during adipocyte differentiation is likely to be mediated primarily by the peroxisome proliferator-activated receptor γ (PPARγ) with SREBP-1 playing a more modulatory role. A peroxisome proliferator response element (PPRE) has been identified in intron 1 of the rat ACBP gene. The element is functionally conserved in the human gene, it binds PPARγ/RXR in the chromatin context and mediates inducibility by PPARγ specific ligands in adipocytes [32].

5. Tissue expression pattern

A large number of studies using in situ hybridisation or immune localisation methods show that ACBP is heterogeneously distributed in rat brain, particularly in the area of postrema, the cerebellar cortex and epidyma of the third ventricle. Intermediate levels were found in the olfactory bulb, pontine nuclei, inferior colliculi, arcuate nucleus and pineal gland. Relatively low but significant levels were observed in mesencephalic and telencephalic areas [33–43]. Both duck and frog express two different ACBPs of which one is only expressed in brain [15,44]. It should be noticed that the authors of the papers reporting immune localisation of brain ACBP do not specify which ACBP isoform, which is recognised by the antibodies used. It is therefore unclear which ACBP isoform they have detected.

A number of external factors increase ACBP in selected areas of the rat brain. These factors include acute stress [45,46], psychological but not physical stress [47], alcohol intake [48–51], nicotine [52–54], diazepam [55,56] and continuous treatment with morphine [57] and butanol [58]. Finally increased ACBP levels have been detected in liver and brain tumours [59,60].

A very large number of papers report the finding of ACBP in almost all peripheral tissues studied. ACBP appears to be ubiquitously expressed from early stages of mammalian embryogenesis [61]. However, the level of ACBP differs markedly among different cell types. High concentrations are found in steroid producing cells (glomerulosa and fasciculate cells of adrenal cortex, Leydig cells of testis) [33,34,62–64]; keratinocytes and cells from sweat and sebaceous glands [65]. Intermediate concentrations are found in epithelial cells involved in water and electrolyte transport (intestinal mucosa, distal convoluted tubules of kidney) [33,34,64].

Hepatocytes contain moderate amounts of ACBP, however, the total amount of ACBP in liver is relatively high due to the diffuse presence of ACBP in all hepatocytes [64]. ACBP was detected throughout the intestinal tract with highest concentration in the duodenum and antrum. The protein was detected in goblet cells and enterocytes of the epithelial layer of the intestine. In the stomach ACBP was found to be restricted to the deep layer of epithelial cells [66]. Recent work shows that ACBP is coexpressed closely with fatty acid binding proteins (FABPs) throughout the intestine and in glial cells [67,68]. We have recently shown that expression of ACBP is about threefold higher in slow twist, oxidative fibre than in fast twist glycolytic white gastrocnemius skeletal muscle fibre [69] and that the ACBP level is about 30% higher in all muscles from obese Zucker rats [70].

Immunohistochemical studies in fruit fly, using antibodies against rat ACBP, detected high levels of immune reactivity in specific tissues including potassium transporting cells in the urine secreting Malpighian tubules [71]. However, fruit fly expresses six different basal ACBP isoforms and two acyl-CoA domain proteins (our unpublished results). The rat antibodies used could in principle have recognised one or more of these eight different proteins. These results are therefore not very informative and illustrate the difficulties with a large proportion of the published results in the literature concerning cell and tissue distribution of ACBP/DBI. ACBP expression in tobacco hornworm (*Manduca sexta*) increased about 15-fold in ecdysteroidogenic cells from day 1 to day 7 of the fifth instar. In the larval midgut ACBP expression was highest during times of active feeding [72,73]. Two ACBP isoforms are identified in silkworm (*B. mori*), a midgut ACBP (mgACBP) and pheromone gland ACBP (pgACBP). In adult mgACBP is expressed at high levels both in the midgut and in the pheromone gland in contrast to pgACBP, which is only expressed at high levels in the pheromone gland. All larval tissues express very low levels of pgACBP and only the midgut express mgACBP at significant levels. Developmental studies showed that expression of both pgACBP and mgACBP in the pheromone gland is initiated one day before adult eclosion [74]. The silkworm pheromone gland synthesise the pheromone bombykol ((*E*,*Z*)-10,12-hecadecadiene-1-ol) from palmitoyl-CoA as a starting point, indicating a role for ACBP in pheromone synthesis.

The expression level of the mammalian ACBP is slightly affected by feeding status. Fasting rats for 24 h resulted in a 33% decrease in liver ACBP levels [75] and in reduced ACBP mRNA level as well [76], whereas the level in heart and kidney was unaffected. High-fat diet for 48 h increased liver ACBP levels by 36%. Hepatic levels of ACBP continued to increase and remain elevated with prolonged exposure to high fat (28 days).

Heart ACBP does not respond to short-term fat feeding but was increased after prolonged exposure [75]. Androgens, which stimulate growth of the human prostate cancer cell line LNCaP [77], also stimulate de novo fatty acid synthesis, cholesterol synthesis and lipid accumulation and induce ACBP expression in this cell line [78]. Similarly, androgens induce ACBP expression in several accessory sex organs in the male rat [79]. ACBP expression is also significantly induced during in vitro differentiation of 3T3-L1 preadipocytes [80], a process which is accompanied by a marked triglyceride accumulation and de novo fatty acid synthesis. Peroxisome proliferators, perfluordecanoic acid and 3-thia fatty acids have been shown to induce liver ACBP [32,76,81,82]. This could indicate that ACBP expression is linked to lipid metabolism. However, growing LNCaP cells and differentiating 3T3-L1 cells undergo dramatic structural and functional changes, and it is therefore possible that other functions besides general lipid synthesis require increased expression of ACBP in these cells.

Testis specific ACBP (ELP) is highly expressed in the late haploid stage of male germ cell development only, with the first immune histochemical staining being present in elongated spermatids [11,83]. This expression pattern was confirmed in rat male germ cells using a human ACBP antibody [83]. During the elongation process and spermatozoa formation, the spermatid undergoes dramatic morphological changes. This could indicate a role for ELP in membrane remodelling. Expression of ACBP in rat testis Leydig cells has been shown to be highly dependent on pituitary hormones suggesting a role of ACBP in the steroidogenic process [84]. In primary cerebella, astroglial and C-6 cells the highest expression of ACBP was observed in actively dividing cells [85].

In plant *A. thaliana* ACBP expression is highest in seed and silique and low expression in leaf and root [86]. In *Brassica napus* ACBP is expressed at a very similar level in a wide range of tissues [87].

6. Intracellular localisation of ACBP

Subcellular fractionation studies show that ACBP is predominately localised to the cytoplasm of liver cells [7]. Fractionation studies using Percoll gradients show that ACBP is highly enriched synaptosomes. In synaptosomal lysate ACBP was shown to be enriched with synaptic vesicles partly purified on a sucrose gradient [42]. Early immune histochemical studies using an antibody raised against an ACBP octadecapeptide (33–50) showed that immunoreactivity occurs in many different parts of the rat brain. At the electron microscopy level the immune reactivity was identified in the neuronal perikarya, processes and terminals. In the axon terminals the immunoreactivity seems to be associated with synaptic vesicles. Antibodies towards full length ACBP identify immunoreactivity in the same cells [36]. Ultrastructural localisation of ACBP using an identical octadecapeptide antibody and immunogold labelling showed that ACBP is found exclusively in the cytosol of Leydig cells [62]. Electron microscopic immune histochemical studies localised ACBP to the cytoplasmic fraction in sweat and sebaceous gland cells with no staining of the nucleolus [65]. ACBP shows an even distribution throughout the cytoplasma without relegation to specific organelles in primary astrocytes and C-6 cells [85]. Predominant cytosolic localisation was also confirmed in rat ovary cells

[63]. In rat testis ACBP is found localised throughout the cytoplasma of Leydig cells, Sertoli cells and in seminiferous tubules cells. ACBP was also found in the smooth endoplasmic reticulum, Golgi apparatus and the outer membrane of the mitochondria [84]. ACBP immune reactivity was found to be homogeneously distributed over the cytoplasm in frog adrenal chromaffin cells and concentrated in the periphery of large cytoplasmatic vesicles in Stilling cells [88]. Two interesting electron microscopic immune gold labelling studies on tobacco hornworm *M. secta* intestinal columnar cells and prothoracic gland ecdysteroidogenic cells confirmed the cytosolic localisation of ACBP but also showed that immune gold labelling over nucleolus indicating that ACBP is able to diffuse in and out of the nucleus [72]. On day 2 of the fifth instar ACBP was found to be evenly distributed over the entire cytosol of ecdysteroidogenic cells. This pattern changed to be more concentrated around mitochondria organisation centres on day 7. The nuclear localisation has later been confirmed in rat liver cells and hepatoma cells. ACBP was found to be mainly nuclear localised during early stages of NIH-3T3-L1 adipocyte differentiations and became evenly distributed between cytoplasm and nucleus in fully differentiated cells [89]. A predominant nuclear localisation in non-differentiated cells would require specific targeting of ACBP to the nucleus and there is no evidence for this to occur. A possible explanation for the predominant nuclear detection could therefore be related to technical problems in fixing and detecting low levels of ACBP in the cytosol in contrast to the ACBP retained in the nucleus. Undifferentiated 3T3-L1 cells have a very low expression of ACBP. Immunogold labelling of yeast cell showed an even distribution of ACBP in the cytosol and over the nucleus (our unpublished results). The yeast studies also showed that the immunogold particles were absent over mitochondria and peroxisomes. In L-cell fibroblast and McA-RH777 hepatoma cells ACBP was found to co-localise with acyl-CoA cholesteryl acyltransferase and endoplasmic reticulum, respectively [90]. This study also showed the presence of ACBP in the nucleus with a non-random distribution. In this study, the fixation of the cells was performed with methanol–acetone, which is unable to fix ACBP in a quantitative manner (personal communication, Dr Susanne Mandrup, SDU, DK). The result could therefore be showing solely the ACBP retained in specific positions or compartments without the general cytosol ACBP background.

In conclusion, ACBP is a cytosolic protein with access to the nuclear compartment. The protein may be specifically associated with specific compartments/proteins, cytoplasmatic vesicles, synaptic vesicles or mitochondria. However, the protein is absent in mitochondria and peroxisomes in yeast (our unpublished results). There are three serious problems related to evaluation of the existing immunolocalisation data: (1) all vertebrates express more than one basal ACBP and ACBP domain proteins, which could potentially cross-react with the used antibody (see above), (2) it has at least in our hand been impossible to make polyclonal rabbit antibodies against higher vertebrate ACBPs, which do not recognise a minimum of two additional bands around 50 kDa when tested in a western blot against liver extract for the species in question (our unpublished result) and (3) the antibodies used in the respective studies are very seldom fully characterised. All together these factors signal that published data should be evaluated very carefully and be repeated with monoclonal antibodies specific for one ACBP isoform only.

7. Functions of ACBP

7.1. ACBP is an essential protein in human cell lines and T. brucei

The high evolutionary conservation of ACBP as well as the ubiquitous expression in mammals suggests that ACBP is involved in basal cellular functions in eukaryotic cells and therefore an essential protein. Indeed this turns out to be the case. We have recently shown by using siRNA, that ACBP is essential for growth of HepG2, HeLa and Chang cells [91]. ACBP was also shown to be essential for *T. brucei* viability, using targeted homologous recombination [92]. *T. brucei* expresses as mentioned above, a slightly different ACBP with increased affinity for C10–C14 acyl-CoA esters. *T. brucei* are shielded from their host's defence by a coat of variant surface glycoproteins molecules, each attached to the plasma membrane by a glycosylphosphatidylinositol (GPI) anchor. The GPI anchor undergoes a remodelling step from a di-palmitoyl to a di-myristoyl derivative during synthesis [93]. It has been demonstrated that *T. brucei* ACBP enhances the fatty acid remodelling of the GPI anchor in vitro [93].

7.2. Regulatory functions

ACBP has been shown to modulate the activity of a number of acyl-CoA regulated functions. It has a strong attenuating effect on the inhibition by LCACoA esters of acetyl-CoA carboxylase [94] and the mitochondrial adenine nucleotide translocase [94,95]. ACBP attenuates/modulates the acyl-CoA mediated inhibition of glucose-6-phosphate metabolism and glucose-6-phosphate transporter in oilseed rape (*B. napus*) plastids [95–97]. LCACoA esters have been found to stimulate Ca^{2+} release by the ryanodine-sensitive Ca^{2+} release channel in longitudinal tubules and terminal cisternae of sarcoplasmic reticulum from rabbit skeletal muscle [98] and duckling sarcoplasmic reticulum [99,100]. Both palmitoyl-CoA and the corresponding non-hydrolysable analogue induced release of Ca^{2+} with an EC_{50} of 6 μM, suggesting that the acyl-CoA interact directly with the ryanodine-sensitive Ca^{2+}-channels. Addition of equimolar ACBP relative to acyl-CoA completely prevented acyl-CoA mediated Ca^{2+} release. However, the ACBP/acyl-CoA complex itself was shown to increase the sensitivity of the rabbit muscle terminal cisternae ryanodine receptor Ca^{2+} release channel to caffeine 20-fold. This effect was proportional to the concentration of the complex and independent of the calculated concentration of unbound palmitoyl-CoA [101], strongly indicating that the acyl-CoA/ACBP complex can either donate acyl-CoA directly to the ryanodine receptor or act as a regulator of the receptor itself. It has been reported that rat ACBP is a potent activator of m-calpain in vitro, by lowering the Ca^{2+} requirement of calpain about 50-fold in an acyl-CoA independent manner [102].

7.3. Regulation of gene expression

Expression of the yeast Δ-9-desaturase OLE1 is strongly suppressed by addition of unsaturated fatty acids to the growth medium [103]. Depletion of the yeast ACBP homologue Acb1p which, shows 48% identity to human ACBP, results in a fivefold increase in the OLE1 mRNA level [8,103]. This could indicate that the repressing fatty acid needs to be activated to its CoA derivative and afterwards transported from the site of synthesis to the nucleus. This hypothesis is compatible with our finding that Acb1p is also found in the nucleus of yeast (our unpublished results). Acyl-CoA esters may be natural antagonists of the PPARs. In a recent study, several acyl-CoA esters were reported to bind directly to PPARα and PPARγ and to inhibit the recruitment of co-activators in vitro [104]. Interestingly, a non-hydrolysable thioether palmitoyl-CoA analogue is able to antagonise ligand-induced DNA binding of a PPARα–RXRα heterodimer [105]. Furthermore, the LCACoA ester analogue abolishes ligand-induced interaction with SRC-1, a steroid receptor co-activator, and equally increases recruitment of a co-repressor, NcoR [105]. Consistently, low μM concentrations of LCACoA esters were found to displace the dual agonist KRP-297 from binding to both PPARα and PPARγ, and to inhibit KRP-297-induced SRC-1 recruitment to both PPARα and PPARγ [104]. The possible role of acyl-CoA esters in gene regulation combined with the fact that ACBP is also found in the nucleus (see above) raises the question if ACBP is mediating/modifying the regulatory function of acyl-CoA in the nucleus. Preliminary experiments indicate that this could be the case. Transient overexpression of ACBP in CV-1 cells inhibited the ability of 3-thia fatty acids (TTA) to activate all PPAR subtypes [89]. The mechanism by which ACBP causes this effect is not clear. One possibility is that increased levels of ACBP increases the acyl-CoA buffering capacity which allows an increased flux of fatty acids through the acyl-CoA pool and thereby increases the rate of TTA esterification and decreases the available concentration of TTA for PPAR activation. This notion is supported by the observation that overexpression of either bovine or yeast ACBP in *S. cerevisiae* led to an increased intracellular acyl-CoA level, suggesting that ACBP in vivo can act as an acyl-CoA pool former [106,107].

7.4. Beta oxidation

Carnitine palmitoyl-transferase 1 (CPT1) is able to efficiently use the acyl-CoA/ACBP complex as substrate [94,108,109]. Biochemical and kinetic studies also indicate that CPT1 prefers the acyl-CoA/ACBP complex rather than the free acyl-CoA as substrate [94, 109]. The observation that decreasing acyl-CoA/ACBP ratios did not affect mitochondrial β-oxidation strongly suggest, that CPT1 recognises and interacts with the acyl-CoA/ACBP complex [94]. Finally, ACBP can desorb acyl-CoA from an immobilised phospholipid bilayer and subsequently transport and donate the bound acyl-CoA for mitochondrial β-oxidation, when present in 15-fold excess over total acyl-CoA in the incubation [94]. This further argues for direct interaction between CPT1 and ACBP.

7.5. Glycerolipid synthesis and cholesterol ester synthesis

ACBP was originally identified on its ability to induce synthesis of medium-chain acyl-CoA esters by goat mammary gland fatty acid synthase in vitro [7]. ACBP can also alleviate product inhibition of the mitochondrial long-chain ACS, and simultaneously desorb and solubilise the acyl-CoA esters formed [110]. These properties combined with the ability to efficiently protect acyl-CoA from hydrolysis by cellular acyl-CoA hydrolases [110–112] and increase cellular acyl-CoA concentrations when overexpressed in yeast [106], suggest a role of ACBP as an intracellular acyl-CoA pool former. The role of ACBP in glycerolipid and cholesterol ester synthesis in vitro is less clear.

A number of in vitro studies show that ACBP either stimulates or inhibits the activity of red cell acyl-CoA-lysophospholipid acyltransferase [113], glycerol-3-phosphate acyltransferase [111,112,114] and acyl-CoA/cholesterol acyltransferase (ACAT) [90] depending on substrate concentrations and incubation conditions. However, these effects could, when tested, be mimicked by FABP, bovine serum albumin or both, although these proteins bind acyl-CoA with a 2–3 orders of magnitude of lower affinity than ACBP. It is therefore not possible to differentiate between specific and non-specific effects of ACBP on phosphatidic acid and cholesterol ester synthesis. Whether or not ACBP can donate acyl-CoA directly to the above mentioned transacylases is not clear from these studies. To answer this question it will require measurement of transacylase activity at an increasing concentration of binding protein at constant acyl-CoA concentrations. Such studies have been performed with plant and liver microsomal glycerol-3-phosphate acyltransferase [87, 94,110]. At low substrate concentrations, increasing amount of ACBP inhibits phosphatidic acid and triglyceride synthesis. With increasing substrate concentrations the effect becomes biphasic with an initial stimulation followed by an inhibition of phosphatidic acid synthesis with increasing ACBP concentration [87,94]. At equimolar ACBP and acyl-CoA concentrations the phosphatidic acid synthesis by rat liver microsomal membranes plateaus out and remains almost constant at increasing ACBP concentrations up to a molar ratio of 2, suggesting that the acyl-CoA/ACBP complex can donate acyl-CoA to glycerol-3-phosphate acyltransferase [110]. This view is further supported by the fact that ACBP in 15-fold excess of acyl-CoA in the assay can desorb acyl-CoA from an immobilised phospholipid bilayer and subsequently transport and donate the bound acyl-CoA to glycerol-3-phosphate acyltransferase in microsomes immobilised on a different membrane [94,110].

Although in vitro studies clearly define multiple metabolic pathways in which ACBP could take part, it is still unknown to what extent ACBP actually plays a role in these pathways in vivo. ACBP expression is significantly upregulated during adipocyte differentiation, and expression of high levels of ACBP antisense RNA in the 3T3-L1 preadipocyte cell line decreased endogenous ACBP levels, expression of the adipogenic transcription factors C/EBPα and PPARγ, and accumulation of triglycerides [115]. This would suggest a role of ACBP in triglyceride synthesis. However, clones of the McA-RH 7777 rat hepatoma cell line overexpressing ACBP showed only a marginal increase in esterification compared to control cells [116]. ACBP from *S. cerevisiae*, Acb1p, has been shown to facilitate removal of newly synthesised acyl-CoA esters from the yeast fatty acid synthase in vitro [117], which is consistent with the observation that depletion of Acb1p in *S. cerevisiae* results in accumulation of $C_{18:0}$-CoA and diminished levels of $C_{14:0}$-CoA. Moreover, disruption of Acb1p results in reduced levels of unsaturated acyl-CoA esters like $C_{16:1}$-CoA and $C_{18:1}$-CoA, implying that Acb1p is involved in intracellular acyl-CoA pool formation. Despite these changes in the acyl-CoA composition, depletion of Acb1p in *S. cerevisiae* does not affect general glycerolipid synthesis and glycerolipid turnover [8]. The in vivo data, therefore, do not support a role for ACBP in general glycerolipid synthesis.

7.6. *Fatty acid elongation and sphingolipid synthesis*

The overall fatty acid composition is only slightly affected by depletion of Acb1p in yeast. The only major difference is a dramatic (~50%) reduction in total lipid content of $C_{26:0}$ fatty acid indicating that fatty acid elongation is reduced in the Acb1p depleted strain [8]. Although depletion of Acb1p did not effect incorporation of (^3H]myo-inositol into phosphatidylinositol and phosphatidylinositol turnover, the depleted strain shows a dramatic reduction in incorporation of (^3H]myo-inositol into sphingolipids. The major fatty acid component of yeast sphingolipids is a C26 α-hydroxylated fatty acid [118]. A possible explanation for the reduced sphingolipid synthesis could therefore be due to decreased fatty acid elongation and reduced supply of C_{26} fatty acids for ceramide synthesis. However, mass spectrometric analysis of plasma membrane lipids revealed that the relative levels of the sphingolipids IPC and MIPC are 25–40% increased in Acb1p depleted cells. It is therefore not clear if the decreased sphingolipid synthesis is caused by reduced fatty acid elongation or decreased sphingolipid turnover.

Furthermore, the relative levels of lysophosphatidic acid (LPA), lysophosphatidylserine and lysophosphatidylinositol were found to be 1.7–2.2-fold increased in plasma membranes from cells exhausted of Acb1p [8]. The increased level of LPA was caused by an increase in the unsaturated LPA species only and was accompanied by a large decrease in the content of $C_{16:1}/C_{18:1}$-PA. This could indicate that Acb1p is required for remodelling of plasma membrane phospholipids. A role in phospholipid remodelling is further supported by comparative electrospray mass spectrometry analysis of total lipid extracts from wild type and Acb1p depleted yeast cells. In the ACBP depleted strain we have observed a shift from phospholipid species containing one saturated and one unsaturated long acyl-chain to shorter double unsaturated species (our unpublished results).

7.7. *Protein acylation*

Both enzymatic and autocatalytic mechanisms have been proposed to account for protein thiolacylation (palmitoylation). ACBP strongly suppresses non-enzymatic thiolation of cysteinyl-containing peptides. At physiological ACBP, acyl-CoA esters and membrane lipids concentrations the rate of non-enzymatic thiolation is shown to be very low [119]. In contrast, the protein *S*-acyltransferase remains active in the presence of physiological concentrations of ACBP and acyl-CoA, suggesting that ACBP can donate acyl-CoA for protein thioacylation.

7.8. *ACBP in membrane trafficking and interaction with membranes*

The requirement of acyl-CoA esters for membrane fusion was initially demonstrated by Glick and Rothman [120] and later confirmed by the observations that both fusion [121] and budding [122,123] of transport vesicles were stimulated by long-chain acyl-CoA esters. Haas and Wickner [124] reported that myristoyl- and palmitoyl-CoA stimulated homotypic vacuole fusion in vitro. Analogously, Deeney et al. [125] reported that low micromolar concentrations (1 – 10 μM) of LCACoA esters enhance insulin release and exocytosis in permeabilised mouse β-cells in a protein kinase- and ATP-independent manner [125]. The fact that a non-hydrolysable long-chain acyl-CoA ester analogue inhibited homotypic vacuole fusion [124], budding as well as fusion of transport vesicles [121 – 123] and homotypic vacuole fusion [126], lends credence to the suggestion that fatty acylation of at least one protein and/or lipid remodelling must take place. The requirement of acyl-CoA in membrane fission and fusion raises the question if ACBP plays a role as acyl-CoA transporter to these processes. Acb1p-depleted yeast exhibits strongly perturbed plasma membrane structures, accumulation of 50–60 nm vesicles and autophagocytotic-like bodies [8]. Furthermore, the strain exhibits multilobed vacuoles, which are unable to undergo homotypic vacuole fusion in vitro (our unpublished results). These results strongly imply that the acyl-CoA/ACBP complex exerts a function in vesicular trafficking, most likely by donating acyl-CoA to an acyl-CoA requiring step in vesicle fusion.

Chao et al. reported that membrane charge and curvature determine ACBP interaction with membranes and acyl-CoA/ACBP targeting to membranes [127]. The authors only performed their membrane interaction experiments at low unphysiological salt concentrations (10 mM Tris or phosphate buffer). Despite that, they concluded that the ACBP membrane interaction was electrostatic. We have in the past examined the interaction between rat ACBP and phospholipid vesicles using isotermal titration calorimetry and found that ACBP interact with acidic lipids at low concentrations. However, addition of 100 mM NaCl completely abolished the interaction (our unpublished results), suggesting that ACBP does not interact with this kind of membranes at physiological salt concentrations.

7.9. Steroidogenesis

The highest cellular ACBP concentrations are found in steroid producing cells (glomerulosa and fasciculate cells of adrenal cortex, Leydig cells of testis) [33,34,62,63]. In a search for cytoplasmatic steroidogenic factors in bovine adrenals an 8.2 kDa protein was shown to stimulate cholesterol transport into mitochondria (see Ref. [128] for review). This protein was found to be identical to ACBP except for the loss of the two C-terminal amino acids. Dose response curves with full length ACBP indicated that a threefold stimulation of cholesterol transport into mitochondria from adrenocortical and Leydig cells could be obtained with low concentration of ACBP (0.1 – 1 μM). It was shown that ACBP could displace benzodiazepines from the mitochondrial peripheral benzodiazepine receptor (PBR). Protein crosslinking experiment suggested that ACBP interacted directly with PBR. Dose dependent treatment of steroidogenic MA-10 and R2C cells with ACBP antisense suggested that ACBP is required for both acute stimulation of steroidogenesis and in constitutive steroid synthesis. The exact mechanism by which ACBP stimulate cholesterol transport and pregnenolone synthesis is not known. Several reports have shown that arachidonic acid and its metabolites play an essential role in the regulation of steroidogenesis [129–131]. It has recently been shown that arachidonic acid regulation of steroid synthesis requires a concerted action of a cytosolic arachidonyl-CoA synthetase and the mitochondrial inner membrane thioesterase (MCT1) [132]. These authors suggest that regulatory arachidonic acid species are recruited from arachidonyl-CoA by CTE1 in an unknown mitochondrial compartment. The authors speculate that the arachidonyl-CoA synthesised in the cytosol by mitochondrial associated arachidonyl-CoA synthetases and becomes trapped by cytosolic ACBP. ACBP is then suggested to bind to PBR and thereby facilitate transport of arachidonyl-CoA to the mitochondrial compartment. An equally likely explanation could be that the high levels ACBP in steroidogenic cells is required for efficient reesterification of the massive amount of both fatty acid and cholesterol released as a result of the ACTH stimulation. Finally, high ACBP concentrations are also found in other tissues with many mitochondria and high energy metabolism like epithelial cells involved in water and electrolyte transport (intestinal mucosa, distal convoluted tubules of kidney [33,34,64]). The purpose of high levels of ACBP in these tissues could simply be to support β-oxidation and energy metabolism.

7.10. Are ACBP biological active outside the cell?

As mentioned in Section 1, ACBP was originally identified as a brain peptide suggested to affect GABA receptor function. Injection of ACBP and partial peptides of ACBP in to rat brain has been reported to cause a number of neurophysiological effects (see Ref. [133] for review). This includes anti-nociptive effects [134], changes in circulating *levels* of testosterone and estradiol [135] and changes in fluid intake and taste [136]. It has been reported that ACBP in the intestinal lumen mediates feedback regulation of pancreatic secretion and postprandial release of cholecystokinin [137–139]. Low nM concentrations of ACBP and a synthetic peptide covering residue 17–50 of ACBP both inhibit glucose-induced insulin secretion in isolated pancreatic islands and in rat after intravenous injection [140–142]. The physiological significance of these observations and the mechanisms by which ACBP causes these diverse effects are not known. It has not been established that ACBP is secreted from cells. It is therefore an open question if ACBP has any physiological effects as signal molecule outside the cell.

8. Other regulatory functions of long-chain fatty acyl-CoA esters

8.1. Regulation of protein kinase activity

LCACoA esters have been shown to regulate the activity of a number of signal transduction cascades such as ion channels, protein kinases, vesicular trafficking and transcription factors (see Ref. [1] for review). Here we attempt to review the most recent literature regarding the current view on the role of LCACoA esters with special focus on their role in the regulation of insulin secretion, induction of insulin resistance and transcriptional regulation.

The individual PKC species are affected differentially by LCACoA esters. Increased fatty acid levels in the blood have been suggested to lead to insulin resistance and non-insulin dependent diabetes. In rat soleus muscles the levels of $C_{18:2}$-CoA and diacylglycerol (DAG) are increased 3–5 h after infusion of a lipid emulsion, primarily containing $C_{18:2}$ fatty acid. These increases were associated with activation of membrane-associated PKC-θ and a significant impairment of insulin-stimulated IRS-1 tyrosine phosphorylation and IRS-1 associated PI3-kinase activity and glucose uptake. These changes correlated with an increase in phosphorylation of a specific serine residue on IRS-1 [143].

Earlier, both palmitoyl-CoA and oleoyl-CoA (27 μM) have been found to enhance particulate PKC-activity in human skin fibroblasts in the presence of Ca^{2+}, phosphatidylserine and DAG [144], while partially purified cytosolic PKC activity was enhanced 60–70% by 13.5 μM palmitoyl- or oleoyl-CoA in the absence of DAG. LCACoA esters (20–30 μM) stimulate PKC activity purified from rat brain, with or without added DAG but in the presence of phosphatidylserine and Ca^{2+}, the effect was lower in the absence of DAG [145]. In permeabilised pancreatic β-cells $C_{14:0}$-, $C_{16:0}$- and $C_{18:1}$-CoA (3–30 μM) potentiates aPKC activity [146]. Since activation of PKC stimulates glucose-induced insulin release [147], it has been suggested that LCACoA esters by stimulating PKC activity play a role in the fatty acid mediated stimulation of insulin secretion [146]. In contrast to aPKC $C_{14:0}$- and $C_{18:1}$-CoA (10 μM) were found to inhibit nPKC activity in permeabilised pancreatic β-cells [146]. Consistingly, nPKC was also found to be inhibited by LCACoA (1–10 μM) in neutrophils [148].

Whether the effect of acyl-CoA on PKC is caused by binding of acyl-CoA to the enzyme through protein acylation or indirect through a stimulation of the synthesis of DAG is unknown. It has been shown that palmitoylation of the regulatory subunit of rabbit brain PKC associates the enzyme with cellular membranes in a manner, which is distinct from the non-palmitoylated form of the enzyme [149]. Membrane partitioning may therefore represent a one mechanism by which PKC regulates its membrane-associated targets.

8.2. Modulation of ion channel activity

The elevated concentration of long-chain acyl-CoA esters observed after long-term exposure of β-cells to non-esterified fatty acids has been proposed to activate K_{ATP} channel activity, leading to hyperpolarisation of the β-cells and inhibition of insulin secretion [150]. LCACoA esters (1 μM) activate pancreatic K_{ATP} channels, Kir6.2/SUR1, by interacting with the Kir6.2 subunit, in both native β cells [151] and in a heterologous expression system [152,153]. The effects of acyl-CoA esters on the ATP sensitivity of cardiac K_{ATP} channels are by an order of magnitude stronger than the effects observed in pancreatic K_{ATP} channels [154], even though cardiac and pancreatic K_{ATP} channels share the Kir6.2 subunit, implying that acyl-CoA esters also act through the β subunit of the K_{ATP} channel, SUR2A, in the heart. The total cellular concentration of acyl-CoA esters has been reported to rise in cardiac muscle during the initial phase of low-flow ischemia. An increase in the cardiac muscle acyl-CoA levels may contribute to the opening of K_{ATP} channels during the initial phase of low-flow ischemia [155]. The resulting shortening of the action potential would lead to a decrease in contractility and energy expenditure of cardiac muscle cells during the initial phase of ischemia. Analogously, nanomolar concentrations of oleoyl-CoA ($K_{1/2}$ value 80 nM) and palmitoyl-CoA ($K_{1/2}$ value 260 nM) have been found to inhibit potassium fluxes through reconstituted mitochondrial potassium ATP-sensitive channels (mitoK_{ATP}) in a Mg^{2+}-dependent manner [156], which could be important in regulation of energy metabolism in the mitochondria. It is interesting that the determined $K_{1/2}$ values for the regulation of the mitochondrial ion channel lies within the concentration range of the predicted cellular concentration of unbound acyl-CoA esters.

8.3. Modulation of gene expression

FadR is an acyl-CoA responsive transcription factor, orchestrating the coordinated regulation of fatty acid biosynthesis and degradation in *E. coli*. In the absence of long-chain acyl-CoAs FadR binds to specific DNA sequences as a homodimer [157], and represses the transcription of genes involved in fatty acid degradation (*fadE*, *fadF*, *fadG* and *fadBA*) and import (*fadD* and *fadL*), and activate transcription of biosynthetic genes (*fabA* and *fabB*). FadR binds LCACoAs with a K_D of 50–400 nM [158]. By X-ray crystallography it has elegantly been shown that binding of myristoyl-CoA leads to profound conformational changes throughout the protein resulting in large backbone shifts and rearrangement and separation of the DNA binding domains in the homodimer and subsequently in loss of DNA binding [18,159]. Overall, these conformational changes result in derepression of the catabolic genes, while transcription of the anabolic genes is deactivated.

Hertz et al. [160] have recently reported that saturated and unsaturated LCFA, bind to HNF4α and modulate its transcriptional activity in COS-7 cells. Palmitoyl-CoA, which bind to the ligand-binding domain of HNF4α with a K_D of 2.6 μM, was found to enhance the binding of HNF4α to its DNA-binding site in vitro in band shift assays. Polyunsaturated fatty acids (PUFAs) such as C18:3($n-3$) or C20:5($n-3$), decreased the binding of HNF4α to its site and inhibited the transcriptional activity of HNF4α. The inhibition correlated with the ability of PUFA-CoA esters to displace labelled palmitoyl-CoA from the ligand-binding site of HNF4α (EC_{50} values from 0.6 to 3.9 μM). Polyunsaturated fatty acids and polyunsaturated fatty acyl-CoA inhibited binding of HNF4α to and activation of the glucose-6-phosphatase promoter activity in HepG2 hepatoma cells, while palmitic acid and palmitoyl-CoA were ineffective [161]. This would lead to decreased glucose-6-phosphatase activity and thus decreased hepatic glucose production, and may therefore explain some of the observed beneficial effects of PUFAs on insulin resistance. However, molecular modelling of the HNF4α structure suggests that the van der Waals volume of LCACoA esters is almost three times the van der Waals volume of the ligand-binding site of HNF4α, suggesting that the ligand-binding site is too small to accommodate the acyl-CoA ligand [162]. It is therefore not clear if LCACoA esters are bonafide ligands for HNF4α. The possible roles of LCACoA esters in regulating the function of PPARs have already been discussed.

8.4. Regulation of in vivo long-chain acyl-CoA concentrations

The total cellular concentrations of fatty acyl-CoA esters have been reported to be in the range of 5–160 μM depending on the metabolic state (See Ref. [1] for a review). The actual free concentration of fatty acyl-CoA available for metabolic utilisation and regulatory purposes is only poorly understood. The Dr Jekyl and Mr Hyde character of long-chain acyl-CoA esters suggests that the concentration of LCACoA esters must be tightly controlled. In this regulatory scheme, intracellular acyl-CoA binding proteins and acyl-CoA hydrolases are assumed to play important roles.

ACBP binds medium- and LCACoA esters with very high affinity, with a preference for C_{14}–C_{22} acyl-CoA esters [25,163,164] (see above). Using titration microcalorimetry we have determined the dissociation constants between bovine L-FABP and oleoyl-CoA to be 1 and 7.4 μM [165]. Employing fluorescence spectroscopy the affinity constants of rat L-FABP have recently been determined to be 8 and 97 nM and 10 and 180 nM for *cis*- and *trans*-parinaroyl-CoA, respectively [166]. Hence, the dissociation constants between FABP and these unphysiological CoA-esters are approximately two orders of magnitude lower than dissociation constants we, and others, have determined previously [165,167]. SCP-2, primarily localised in mitochondria and peroxisomes, binds LCACoA with K_D values in the 2–4 nM range [168]. The physiological significance of the acyl-CoA binding ability of SCP-2 is unknown, but its cellular localisation implies that SCP-2 is able to function in peroxisomal handling of VLCACoA. Glutathione-*S*-transferase (GST) has also been reported to bind LCACoA with dissociation constants in the micromolar range [169].

Albeit the only true recognised function of acyl-CoA thioesterase is in termination of fatty acid synthesis [170], acyl-CoA thioesterases are also believed to contribute to the regulation of the size of the different acyl-CoA pools [171]. Acyl-CoA hydrolases are found in most subcellular compartments and include short-, medium-, long- and very-long chain acyl-CoA hydrolases ([172] and references herein). Acyl-CoA hydrolases typically display K_m values in the low micromolar range for LCACoA esters ([172] and references herein). Some acyl-CoA hydrolases have been shown to respond to metabolic stresses, i.e. increasing by ingestion of hypolipidemic drugs ([172] and references herein). Interestingly, these treatments also result in an increase in the total acyl-CoA level in rat liver ([1] and references herein).

LCACoA esters readily partition into membranes with a K_D of $1.5–5 \times 10^5\ M^{-1}$ [173]. The presence of acyl-CoA binding proteins, with K_Ds in the nM-range, and LCACoA hydrolases, therefore, makes it very unlikely that LCACoA esters partition into cellular membranes under normal physiological conditions. Ultimately, large fluctuations in the concentration of LCACoA esters in the cell will be prevented by the fact that the ACS is product inhibited by LCACoA esters, i.e. palmitoyl-CoA inhibits with a K_i of 4 μM [174].

Acyl-CoA hydrolases are very likely to function as "scavengers" to ensure low levels of LCACoA esters and free CoA for metabolic utilisation. This is substantiated by the observation that the psychrophilic antarctic yeast *Rhodoturula aurantiaca* is inviable at non-permissive temperatures due to inactivation of a temperature-sensitive long-chain acyl-CoA thioesterase and concomitant accumulation of myristoyl-CoA [175]. In yeast, a peroxisomal acyl-CoA hydrolase (Tes1p) has been shown to be required for oxidation of fatty acids since it prevents depletion of the peroxisomal CoA pool when acyl-CoA esters accumulate [176]. The majority of the LCACoA esters is presumed to be localised in the mitochondria ([1] and references herein). Thus, a possible role of acyl-CoA hydrolases in mitochondria may be to prevent accumulation of LCACoA esters and depletion of the CoA pool. Recently, the mitochondrial acyl-CoA thioesterase (MTE-I) and the acyl-CoA synthetase 4 were identified as important players in the concerted arachidonic acid-mediated regulation of steroidogenesis [132] by hydrolysing mitochondrial arachidonoyl-CoA derived from arachidonoyl-CoA, synthesised by the acyl-CoA synthetase 4 associated with the mitochondria-associated membrane [177].

In conclusion, the combined presence of the above-mentioned binding proteins and acyl-CoA hydrolases is likely to keep the intracellular concentration of unbound LCACoA esters at constant low levels. Based on the reported in vitro binding affinity and the cellular level of ACBP it can be predicted that the unbound concentration of LCACoA would be in the low nM range when the acyl-CoA/ACBP ratio is below one. As this ratio exceeds one, FABP may start buffering and keep the level of unbound acyl-CoA to values less than 0.2 μM [1]. The fact that fatty acid synthesis occurs despite that the K_i for inhibition of acetyl-CoA carboxylase is 5.5 nM [178] strongly supports the above predictions. In these predictions we have not considered the presence of the predicted ACBP domain proteins, which may contribute to the buffering of the intracellular concentration of LCACoA esters. It is likely that these domain-proteins bind LCACoA esters with high affinity, and thus by competing with the cytosolic ACBP is able to create local pools of acyl-CoA esters. In fact, one could consider the regulation of free acyl-CoA similar to regulation of cellular Ca^{2+} concentrations, where thioesterases replace Ca^{2+} pumps and acyl-CoA binding proteins replace Ca^{2+} binding proteins.

If the total free acyl-CoA concentration under normal physiological conditions is well below 200 nM and most likely below 10 nM, the role of acyl-CoA as a physiological regulator will be limited to the regulation of a few cellular processes unless acyl-CoA can be donated directly from a binding protein, ACBP and perhaps FABP, to the proteins in question. That this occurs is supported by the observation that the acyl-CoA/ACBP complex at molar ratios below one can donate acyl-CoA for β-oxidation [94], to an acyl-CoA:lysophospholipid acyltransferase on immobilised microsomes [94] and regulate sarcoplasmic ryanodine Ca^{2+} release channels [101]. This situation is not unique for ACBP bound ligand. It has recently been demonstrated that the retinal/cellular retinol binding protein (CRBP) complex rather than free retinal is the preferred substrate for lecithin-retinol acyltransferase [179] and for the microsomal retinol dehydrogenase [180].

9. Conclusion and future directions

The experimental evidence strongly indicates that ACBP is able to create an intracellular protected pool of acyl-CoA esters. Together with acyl-CoA thioesterases ACBP is likely to play a key role in regulating the free cytosolic acyl-CoA concentration and thereby also influence acyl-CoA regulated functions. Very little is known about to what extent this pool is available for, or required for, general lipid synthesis. In yeast ACBP is not required for general glycerolipid synthesis. Whether the same is the case in higher eukaryotes is not known, and in this context it should be kept in mind that cytosolic thioesterases have not been identified in yeast. However, it would not make sense that an acyl-CoA synthesised by an ER-associated ACS should pass over cytosolic ACBP in order to be donated to the first enzyme in glycerolipid synthesis, α-glycerol-P-acyltransferase in the same membrane. Kinetic studies with CPT1 indicate that the acyl-CoA/ACBP is the preferred substrate for CPT1. This observation indirectly indicates a role of ACBP in β-oxidation. This suggestion correlates with the observation that high ACBP expression is found in water and electrolyte transporting cells, in intestinal mucosa and distal convoluted tubules of kidney. The large differences in ACBP expression level between different cell types in multicellular organisms indicate that ACBP in addition to its general housekeeping functions have acquired specialised functions in some cell types, i.e. in steroid producing glomerulosa and fasciculate cells of adrenal cortex, in Leydig cells of testis, in keratinocytes, in cells from sweat and sebaceous glands, in the late haploid stage of male germ cell and in silkworm (*B. mori*) pgACBP. The expression of six different 82–92 residue ACBP isoforms in fruit flies suggests that this organism has developed specialised ACBP isoforms for unique functions through gene duplication. In the future it will be an important challenge to delineate how ACBP can exert its specialised functions.

The nature of the unique functions is not known at present. ACBP is required in fatty acid elongation, ceramide synthesis, membrane remodelling, protein acylation and vacuole fusion in yeast, which are processes that all require LCACoA esters. ACBP may therefore not be required for general glycerolipid synthesis but required for β-oxidation, membrane remodelling, fatty acid elongation and specialised functions like pheromone synthesis (Bombykol ((*E*,*Z*)-10,12-hecadecadiene-1-ol) from palmitoyl-CoA in silkworm pheromone gland. Structural comparison and functional tests of different ACBP isoforms and mutants in in vivo model systems will be a valuable tool. Preliminary experiments show that human ACBP in contrast to some of the fruit fly isoforms can complement growth of Acb1p depleted yeast, suggest that yeast could be a valuable model system. Another important and unanswered question is whether or not ACBP specifically interact with target proteins.

Finally, the identification of multidomain proteins containing an ACBP domain fused to other domains such as ankyrin binding repeats and DNA binding domains open up a very interesting field of research. These ACBP domain proteins could be components in hitherto unknown acyl-CoA regulated signalling pathways.

References

[1] Faergeman, N.J., Knudsen, J., 1997. Role of long-chain fatty acyl-CoA esters in the regulation of metabolism and in cell signalling. Biochem. J. 323, 1–12.

[2] Gossett, R.E., Frolov, A.A., Roths, J.B., Behnke, W.D., Kier, A.B., Schroeder, F., 1996. Acyl-CoA binding proteins: multiplicity and function. Lipids 31, 895–918.

[3] Guidotti, A., Forchetti, C.M., Corda, M.G., Konkel, D., Bennett, C.D., Costa, E., 1983. Isolation, characterization, and purification to homogeneity of an endogenous polypeptide with agonistic action on benzodiazepine receptors. Proc. Natl Acad. Sci. USA 80, 3531–3535.

[4] Knudsen, J., Nielsen, M., 1990. Diazepam-binding inhibitor: a neuropeptide and/or an acyl-CoA ester binding protein? Biochem. J. 265, 927–929.

[5] Yanagibashi, K., Ohno, Y., Kawamura, M., Hall, P.F., 1988. The regulation of intracellular transport of cholesterol in bovine adrenal cells: purification of a novel protein. Endocrinology 123, 2075–2082.

[6] Chen, Z.W., Agerberth, B., Gell, K., Andersson, M., Mutt, V., Ostenson, C.G., Efendic, S., Barros Soderling, J., Persson, B., Jornvall, H., 1988. Isolation and characterization of porcine diazepam-binding inhibitor, a polypeptide not only of cerebral occurrence but also common in intestinal tissues and with effects on regulation of insulin release. Eur. J. Biochem. 174, 239–245.

[7] Mogensen, I.B., Schulenberg, H., Hansen, H.O., Spener, F., Knudsen, J., 1987. A novel acyl-CoA-binding protein from bovine liver. Effect on fatty acid synthesis. Biochem. J. 241, 189–192.

[8] Gaigg, B., Neergaard, T.B., Schneiter, R., Hansen, J.K., Faergeman, N.J., Jensen, N.A., Andersen, J.R., Friis, J., Sandhoff, R., Schroder, H.D., Knudsen, J., 2001. Depletion of acyl-coenzyme A-binding protein affects sphingolipid synthesis and causes vesicle accumulation and membrane defects in *Saccharomyces cerevisiae*. Mol. Biol. Cell 12, 1147–1160.

[9] Mandrup, S., Faergeman, N.J., Knudsen, J., 2003. Structure, function and phylogeny of Acyl-CoA binding protein. In: Duttaroy, A.K., Spener, F. (Eds.), Cellular Proteins and Their Fatty Acids in Health and Disease. Wiley-VCH GmbH & Co. KGaA, pp. 151–172.

[10] Burton, M., Rose, T.M., Faergeman, N.J., Knudsen, J., 2003. Evolution of Acyl-CoA binding protein. Mol. Biol. Evol. (submitted).

[11] Pusch, W., Balvers, M., Hunt, N., Ivell, R., 1996. A novel endozepine-like peptide (ELP) is exclusively expressed in male germ cells. Mol. Cell. Endocrinol. 122, 69–80.

[12] Ivell, R., Pusch, W., Balvers, M., Valentin, M., Walther, N., Weinbauer, G., 2000. Progressive inactivation of the haploid expressed gene for the sperm-specific endozepine-like peptide (ELP) through primate evolution. Gene 255, 335–345.

[13] Knudsen, J., Hojrup, P., Hansen, H.O., Hansen, H.F., Roepstorff, P., 1989. Acyl-CoA-binding protein in the rat. Purification, binding characteristics, tissue concentrations and amino acid sequence. Biochem. J. 262, 513–519.

[14] Pusch, W., Balvers, M., Weinbauer, G.F., Ivell, R., 2000. The rat endozepine-like peptide gene is highly expressed in late haploid stages of male germ cell development. Biol. Reprod. 63, 763–768.

[15] Rose, T.M., Schultz, E.R., Sasaki, G.C., Kolattukudy, P.E., Shoyab, M., 1994. Nucleotide sequence and genomic structure of duck acyl-CoA binding protein/diazepam-binding inhibitor: co-localization with S-acyl fatty acid synthase thioesterase. DNA Cell Biol. 13, 669–678.

[16] Metzner, M., Ruecknagel, K.P., Knudsen, J., Kuellertz, G., Mueller-Uri, F., Diettrich, B., 2000. Isolation and characterization of two acyl-CoA-binding proteins from proembryogenic masses of *Digitalis lanata* Ehrh. Planta 210, 683–685.

[17] Geisbrecht, B.V., Zhang, D., Schulz, H., Gould, S.J., 1999. Characterization of PECI, a novel monofunctional Delta(3), Delta(2)-enoyl-CoA isomerase of mammalian peroxisomes. J. Biol. Chem. 274, 21797–21803.

[18] van Aalten, D.M., DiRusso, C.C., Knudsen, J., 2001. The structural basis of acyl coenzyme A-dependent regulation of the transcription factor FadR. EMBO J. 20, 2041–2050.

[19] van Aalten, D.M., DiRusso, C.C., Knudsen, J., Wierenga, R.K., 2000. Crystal structure of FadR, a fatty acid-responsive transcription factor with a novel acyl coenzyme A-binding fold. EMBO J. 19, 5167–5177.

[20] Hamada, K., Shimizu, T., Matsui, T., Tsukita, S., Hakoshima, T., 2000. Structural basis of the membrane-targeting and unmasking mechanisms of the radixin FERM domain. EMBO J. 19, 4449–4462.

[21] Andersen, K.V., Mandrup, S., Knudsen, J., Poulsen, F.M., 1991. The secondary structure in solution of acyl-coenzyme A binding protein from bovine liver using 1H nuclear magnetic resonance spectroscopy. Biochemistry 30, 10654–10663.

[22] Andersen, K.V., Poulsen, F.M., 1992. Three-dimensional structure in solution of acyl-coenzyme A binding protein from bovine liver. J. Mol. Biol. 226, 1131–1141.

[23] van Aalten, D.M., Milne, K.G., Zou, J.Y., Kleywegt, G.J., Bergfors, T., Ferguson, M.A., Knudsen, J., Jones, T.A., 2001. Binding site differences revealed by crystal structures of *Plasmodium falciparum* and bovine acyl-CoA binding protein. J. Mol. Biol. 309, 181–192.

[24] Kragelund, B.B., Andersen, K.V., Madsen, J.C., Knudsen, J., Poulsen, F.M., 1993. Three-dimensional structure of the complex between acyl-coenzyme A binding protein and palmitoyl-coenzyme A. J. Mol. Biol. 230, 1260–1277.

[25] Faergeman, N.J., Sigurskjold, B.W., Kragelund, B.B., Andersen, K.V., Knudsen, J., 1996. Thermodynamics of ligand binding to acyl-coenzyme A binding protein studied by titration calorimetry. Biochemistry 35, 14118–14126.

[26] Mandrup, S., Hummel, R., Ravn, S., Jensen, G., Andreasen, P.H., Gregersen, N., Knudsen, J., Kristiansen, K., 1992. Acyl-CoA-binding protein/diazepam-binding inhibitor gene and pseudogenes. A typical housekeeping gene family. J. Mol. Biol. 228, 1011–1022.

[27] Swinnen, J.V., Esquenet, M., Rosseels, J., Claessens, F., Rombauts, W., Heyns, W., Verhoeven, G., 1996. A human gene encoding diazepam-binding inhibitor/acyl-CoA-binding protein: transcription and hormonal regulation in the androgen-sensitive human prostatic adenocarcinoma cell line LNCaP. DNA Cell Biol. 15, 197–208.

[28] Mandrup, S., Andreasen, P.H., Knudsen, J., Kristiansen, K., 1993. Genome organization and expression of the rat ACBP gene family. Mol. Cell. Biochem. 123, 55–61.

[29] Swinnen, J.V., Alen, P., Heyns, W., Verhoeven, G., 1998. Identification of diazepam-binding inhibitor/Acyl-CoA-binding protein as a sterol regulatory element-binding protein-responsive gene. J. Biol. Chem. 273, 19938–19944.

[30] Foretz, M., Pacot, C., Dugail, I., Lemarchand, P., Guichard, C., Le Liepvre, X., Berthelier-Lubrano, C., Spiegelman, B., Kim, J.B., Ferre, P., Foufelle, F., 1999. ADD1/SREBP-1c is required in the activation of hepatic lipogenic gene expression by glucose. Mol. Cell. Biol. 19, 3760–3768.

[31] Swinnen, J.V., Ulrix, W., Heyns, W., Verhoeven, G., 1997. Coordinate regulation of lipogenic gene expression by androgens: evidence for a cascade mechanism involving sterol regulatory element binding proteins. Proc. Natl Acad. Sci. USA 94, 12975–12980.

[32] Helledie, T., Grontved, L., Jensen, S.S., Kiilerich, P., Rietveld, L., Albrektsen, T., Boysen, M.S., Nohr, J., Larsen, L.K., Fleckner, J., Stunnenberg, H.G., Kristiansen, K., Mandrup, S., 2002. The gene encoding the Acyl-CoA-binding protein is activated by peroxisome proliferator-activated receptor gamma through an intronic response element functionally conserved between humans and rodents. J. Biol. Chem. 277, 26821–26830.

[33] Alho, H., Fremeau, R.T. Jr., Tiedge, H., Wilcox, J., Bovolin, P., Brosius, J., Roberts, J.L., Costa, E., 1988. Diazepam binding inhibitor gene expression: location in brain and peripheral tissues of rat. Proc. Natl Acad. Sci. USA 85, 7018–7022.

[34] Alho, H., Harjuntausta, T., Schultz, R., Pelto Huikko, M., Bovolin, P., 1991. Immunohistochemistry of diazepam binding inhibitor (DBI) in the central nervous system and peripheral organs: its possible role as an endogenous regulator of different types of benzodiazepine receptors. Neuropharmacology 30, 1381–1386.

[35] Ball, J.A., Ghatei, M.A., Sekiya, K., Krausz, T., Bloom, S.R., 1989. Diazepam binding inhibitor-like immunoreactivity (51–70): distribution in human brain, spinal cord and peripheral tissues. Brain Res. 479, 300–305.

[36] Alho, H., Bovolin, P., Jenkins, D., Guidotti, A., Costa, E., 1989. Cellular and subcellular localization of an octadecaneuropeptide derived from diazepam binding inhibitor: immunohistochemical studies in the rat brain. J. Chem. Neuroanat. 2, 301–318.

[37] Alho, H., Bovolin, P., Slobodyansky, E., 1990. Diazepam binding inhibitor (DBI) processing: immunohistochemical studies in the rat brain. Neurochem. Res. 15, 209–216.

[38] Alho, H., Costa, E., Ferrero, P., Fujimoto, M., Cosenza Murphy, D., Guidotti, A., 1985. Diazepam-binding inhibitor: a neuropeptide located in selected neuronal populations of rat brain. Science 229, 179–182.

[39] Ferrarese, C., Alho, H., Guidotti, A., Costa, E., 1987. Co-localization and co-release of GABA and putative allosteric modulators of GABA receptor. Neuropharmacology 26, 1011–1018.

[40] Tonon, M.C., Desy, L., Nicolas, P., Vaudry, H., Pelletier, G., 1990. Immunocytochemical localization of the endogenous benzodiazepine ligand octadecaneuropeptide (ODN) in the rat brain. Neuropeptides 15, 17–24.

[41] Tong, Y., Toranzo, D., Pelletier, G., 1991. Localization of diazepam-binding inhibitor (DBI) mRNA in the rat brain by high resolution in situ hybridization. Neuropeptides 20, 33–40.

[42] Ferrarese, C., Vaccarino, F., Alho, H., Mellstrom, B., Costa, E., Guidotti, A., 1987. Subcellular location and neuronal release of diazepam binding inhibitor. J. Neurochem. 48, 1093–1102.

[43] Malagon, M., Vallarino, M., Tonon, M.C., Vaudry, H., 1992. Localization and characterization of diazepam-binding inhibitor (DBI)-like peptides in the brain and pituitary of the trout (*Salmo gairdneri*). Brain Res. 576, 208–214.

[44] Lihrmann, I., Plaquevent, J.C., Tostivint, H., Raijmakers, R., Tonon, M.C., Conlon, J.M., Vaudry, H., 1994. Frog diazepam-binding inhibitor: peptide sequence, cDNA cloning, and expression in the brain. Proc. Natl Acad. Sci. USA 91, 6899–6903.

[45] Ferrarese, C., Mennini, T., Pecora, N., Gobbi, M., Appollonio, I., Bernasconi, P., Frigo, M., Regondi, C., Pierpaoli, C., Frattola, L., 1991. Acute noise stress in rats increases the levels of diazepam binding inhibitor (DBI) in hippocampus and adrenal gland. Psychopharmacology (Berl.) 103, 339–342.

[46] Ferrarese, C., Mennini, T., Pecora, N., Pierpaoli, C., Frigo, M., Marzorati, C., Gobbi, M., Bizzi, A., Codegoni, A., Garattini, S., 1991. Diazepam binding inhibitor (DBI) increases after acute stress in rat. Neuropharmacology 30, 1445–1452.

[47] Katsura, M., Mohri, Y., Shuto, K., Tsujimura, A., Ukai, M., Ohkuma, S., 2002. Psychological stress, but not physical stress, causes increase in diazepam binding inhibitor (DBI) mRNA expression in mouse brains. Brain Res. Mol. Brain Res. 104, 103–109.

[48] Katsura, M., Ohkuma, S., Tsujimura, A., Kuriyama, K., 1995. Increase of diazepam binding inhibitor mRNA levels in the brains of chronically ethanol-treated and -withdrawn mice. J. Pharmacol. Exp. Ther. 273, 1529–1533.

[49] Katsura, M., Ohkuma, S., Jun, X., Tsujimura, A., Kuriyama, K., 1995. Ethanol stimulates diazepam binding inhibitor (DBI) mRNA expression in primary cultured neurons. Brain Res. Mol. Brain Res. 34, 355–359.

[50] Katsura, M., Ohkuma, S., Tsujimura, A., Xu, J., Hibino, Y., Ishikawa, E., Kuriyama, K., 1998. Functional involvement of benzodiazepine receptors in ethanol-induced increases of diazepam binding inhibitor (DBI) and its mRNA in the mouse brain. Brain Res. Mol. Brain Res. 54, 124–132.

[51] Alho, H., Miyata, M., Korpi, E., Kiianmaa, K., Guidotti, A., 1987. Studies of a brain polypeptide functioning as a putative endogenous ligand to benzodiazepine recognition sites in rats selectively bred for alcohol related behavior. Alcohol Alcohol.(Suppl. 1), 637–641.

[52] Katsura, M., Ohkuma, S., Chen, D.T., Tsujimura, A., Kuriyama, K., 1994. Nicotine increases diazepam binding inhibitor (DBI) mRNA in primary cultured neurons. Neurosci. Lett. 168, 1–4.

[53] Katsura, M., Shuto, K., Mohri, Y., Tsujimura, A., Ohkuma, S., 2001. Withdrawal from nicotine facilitates diazepam binding inhibitor mRNA expression in mouse cerebral cortex. Brain Res. Mol. Brain Res. 97, 194–198.

[54] Katsura, M., Ohkuma, S., Xu, J., Hibino, Y., Tsujimura, A., Kuriyama, K., 1998. Continuous treatment with nicotine increases diazepam binding inhibitor (DBI) and its mRNA in the mouse brain. Brain Res. Mol. Brain Res. 55, 345–349.

[55] Miyata, M., Mocchetti, I., Ferrarese, C., Guidotti, A., Costa, E., 1987. Protracted treatment with diazepam increases the turnover of putative endogenous ligands for the benzodiazepine/beta-carboline recognition site. Proc. Natl Acad. Sci. USA 84, 1444–1448.

[56] Burgi, B., Lichtensteiger, W., Schlumpf, M., 2000. Diazepam-binding inhibitor/acyl-CoA-binding protein mRNA and peripheral benzodiazepine receptor mRNA in endocrine and immune tissues after prenatal diazepam exposure of male and female rats. J. Endocrinol. 166, 163–171.

[57] Katsura, M., Hara, A., Higo, A., Tarumi, C., Hibino, Y., Ohkuma, S., 1998. Continuous treatment with morphine increases diazepam binding inhibitor mRNA in mouse brain. J. Neurochem. 71, 2638–2641.

[58] Gossett, R.E., Schroeder, F., Gunn, J.M., Kier, A.B., 1997. Expression of fatty acyl-CoA binding proteins in colon cells: response to butyrate and transformation. Lipids 32, 577–585.

[59] Spener, F., Mangold, H.K., 1971. The alkyl moieties in wax esters and alkyl diacyl glycerols of sharks. J. Lipid Res. 12, 12–16.

[60] Venturini, I., Alho, H., Podkletnova, I., Corsi, L., Rybnikova, E., Pellicci, R., Baraldi, M., Pelto-Huikko, M., Helen, P., Zeneroli, M.L., 1999. Increased expression of peripheral benzodiazepine receptors and diazepam binding inhibitor in human tumors sited in the liver. Life Sci. 65, 2223–2231.

[61] Burgi, B., Lichtensteiger, W., Lauber, M.E., Schlumpf, M., 1999. Ontogeny of diazepam binding inhibitor/acyl-CoA binding protein mRNA and peripheral benzodiazepine receptor mRNA expression in the rat. J. Neuroendocrinol. 11, 85–100.

[62] Rheaume, E., Tonon, M.C., Smih, F., Simard, J., Desy, L., Vaudry, H., Pelletier, G., 1990. Localization of the endogenous benzodiazepine ligand octadecaneuropeptide in the rat testis. Endocrinology 127, 1986–1994.
[63] Toranzo, D., Tong, Y., Tonon, M.C., Vaudry, H., Pelletier, G., 1994. Localization of diazepam-binding inhibitor and peripheral type benzodiazepine binding sites in the rat ovary. Anat. Embryol. (Berl.) 190, 383–388.
[64] Bovolin, P., Schlichting, J., Miyata, M., Ferrarese, C., Guidotti, A., Alho, H., 1990. Distribution and characterization of diazepam binding inhibitor (DBI) in peripheral tissues of rat. Regul. Pept. 29, 267–281.
[65] Alho, H., Vaalasti, A., Podkletnova, I., Rechardt, L., 1993. Expression of diazepam-binding inhibitor peptide in human skin: an immunohistochemical and ultrastructural study. J. Invest. Dermatol. 101, 800–803.
[66] Steyaert, H., Tonon, M.C., Tong, Y., Smihrouet, F., Testart, J., Pelletier, G., Vaudry, H., 1991. Distribution and characterization of endogenous benzodiazepine receptor ligand (endozepine)-like peptides in the rat gastrointestinal tract. Endocrinology 129, 2101–2109.
[67] Yanase, H., Shimizu, H., Kanda, T., Fujii, H., Iwanaga, T., 2001. Cellular localization of the diazepam binding inhibitor (DBI) in the gastrointestinal tract of mice and its coexistence with the fatty acid binding protein (FABP). Arch. Histol. Cytol. 64, 449–460.
[68] Yanase, H., Shimizu, H., Yamada, K., Iwanaga, T., 2002. Cellular localization of the diazepam binding inhibitor in glial cells with special reference to its coexistence with brain-type fatty acid binding protein. Arch. Histol. Cytol. 65, 27–36.
[69] Franch, J., Andersen, J.L., Jensen, J., Pedersen, P.K., Knudsen, J., 2002. Acyl-coenzyme A binding protein expression is fibre-type specific in rat skeletal muscle but not affected by moderate endurance training. Pflugers Arch. – Eur. J. Physiol. 443, 387–393.
[70] Franch, J., Knudsen, J., Ellis, B.A., Pedersen, P.K., Cooney, G.J., Jensen, J., 2002. Acyl-CoA binding protein expression is fiber type-specific and elevated in muscles from the obese insulin-resistant Zucker rat. Diabetes 51, 449–454.
[71] Kolmer, M., Roos, C., Tirronen, M., Myohanen, S., Alho, H., 1994. Tissue-specific expression of the diazepam-binding inhibitor in *Drosophila melanogaster*: cloning, structure, and localization of the gene. Mol. Cell. Biol. 14, 6983–6995.
[72] Snyder, M.J., Van Antwerpen, R., 1998. Evidence for a diazepam-binding inhibitor (DBI) benzodiazepine receptor-like mechanism in ecdysteroidogenesis by the insect prothoracic gland. Cell Tissue Res. 294, 161–168.
[73] Snyder, M.J., Antwerpen, R.V., 1997. Cellular distribution, levels, and function of the diazepam-binding inhibitor/acyl-CoA-binding protein in last instar *Manduca sexta* midgut. Cell Tissue Res. 288, 177–184.
[74] Matsumoto, S., Yoshiga, T., Yokoyama, N., Iwanaga, M., Koshiba, S., Kigawa, T., Hirota, H., Yokoyama, S., Okano, K., Mita, K., Shimada, T., Tatsuki, S., 2001. Characterization of acyl-CoA-binding protein (ACBP) in the pheromone gland of the silkworm, *Bombyx mori*. Insect Biochem. Mol. Biol. 31, 603–609.
[75] Bhuiyan, J., Pritchard, P.H., Pande, S.V., Seccombe, D.W., 1995. Effects of high-fat diet and fasting on levels of acyl-coenzyme A binding protein in liver, kidney, and heart of rat. Metabolism 44, 1185–1189.
[76] Sterchele, P.F., Vanden Heuvel, J.P., Davis, J.W., Shrago, E., Knudsen, J., Peterson, R.E., 1994. Induction of hepatic acyl-CoA-binding protein and liver fatty acid-binding protein by perfluorodecanoic acid in rats. Lack of correlation with hepatic long-chain acyl-CoA levels. Biochem. Pharmacol. 48, 955–966.
[77] Esquenet, M., Swinnen, J.V., Heyns, W., Verhoeven, G., 1997. LNCaP prostatic adenocarcinoma cells derived from low and high passage numbers display divergent responses not only to androgens but also to retinoids. J. Steroid Biochem. Mol. Biol. 62, 391–399.
[78] Swinnen, J.V., Esquenet, M., Heyns, W., Rombauts, W., Verhoeven, G., 1994. Androgen regulation of the messenger RNA encoding diazepam-binding inhibitor/acyl-CoA-binding protein in the human prostatic adenocarcinoma cell line LNCaP. Mol. Cell. Endocrinol. 104, 153–162.
[79] Swinnen, J.V., Vercaeren, I., Esquenet, M., Heyns, W., Verhoeven, G., 1996. Androgen regulation of the messenger RNA encoding diazepam-binding inhibitor/acyl-CoA-binding protein in the rat. Mol. Cell. Endocrinol. 118, 65–70.

[80] Hansen, H.O., Andreasen, P.H., Mandrup, S., Kristiansen, K., Knudsen, J., 1991. Induction of acyl-CoA-binding protein and its mRNA in 3T3-L1 cells by insulin during preadipocyte-to-adipocyte differentiation. Biochem. J. 277, 341–344.
[81] Vanden Heuvel, J.P., Kuslikis, B.I., Shrago, E., Peterson, R.E., 1991. Inhibition of long-chain acyl-CoA synthetase by the peroxisome proliferator perfluorodecanoic acid in rat hepatocytes. Biochem. Pharmacol. 42, 295–302.
[82] Skorve, J., Rosendal, J., Vaagenes, H., Knudsen, J., Lillehaug, J.R., Berge, R.K., 1995. Fatty acyl-CoA oxidase activity is induced before long-chain acyl-CoA hydrolase activity and acyl-CoA binding protein in liver of rat treated with peroxisome proliferating 3-thia fatty acids. Xenobiotica 25, 1181–1194.
[83] Kolmer, M., Pelto-Huikko, M., Parvinen, M., Hoog, C., Alho, H., 1997. The transcriptional and translational control of diazepam binding inhibitor expression in rat male germ-line cells. DNA Cell Biol. 16, 59–72.
[84] Schultz, R., Pelto Huikko, M., Alho, H., 1992. Expression of diazepam binding inhibitor-like immunoreactivity in rat testis is dependent on pituitary hormones. Endocrinology 130, 3200–3206.
[85] Alho, H., Varga, V., Krueger, K.E., 1994. Expression of mitochondrial benzodiazepine receptor and its putative endogenous ligand diazepam binding inhibitor in cultured primary astrocytes and C-6 cells: relation to cell growth. Cell Growth Differ. 5, 1005–1014.
[86] Engeseth, N.J., Pacovsky, R.S., Newman, T., Ohlrogge, J.B., 1996. Characterization of an acyl-CoA-binding protein from *Arabidopsis thaliana*. Arch. Biochem. Biophys. 331, 55–62.
[87] Brown, A.P., Johnson, P., Rawsthorne, S., Hills, M.J., 1998. Expression and properties of acyl-CoA binding protein from *Brassica napus*. Plant Physiol. Biochem. 36, 629–635.
[88] Lesouhaitier, O., Feuilloley, M., Lihrmann, I., Ugo, I., Fasolo, A., Tonon, M.C., Vaudry, H., 1996. Localization of diazepam-binding inhibitor-related peptides and peripheral type benzodiazepine receptors in the frog adrenal gland. Cell Tissue Res. 283, 403–412.
[89] Helledie, T., Antonius, M., Sorensen, R.V., Hertzel, A.V., Bernlohr, D.A., Kolvraa, S., Kristiansen, K., Mandrup, S., 2000. Lipid-binding proteins modulate ligand-dependent trans-activation by peroxisome proliferator-activated receptors and localize to the nucleus as well as the cytoplasm. J. Lipid Res. 41, 1740–1751.
[90] Chao, H., Zhou, H., McIntosh, H., Schroeder, F., Kier, A.B., 2003. ACBP and cholesterol differentially alter fatty acyl CoA utilization by microsomal ACAT. J. Lipid Res. 44, 72–83.
[91] Faergeman, N.J., Knudsen, J., 2002. Acyl-CoA binding protein is an essential protein in mammalian cell lines. Biochem. J. 368, 679–682.
[92] Milne, K.G., Guther, M.L., Ferguson, M.A., 2001. Acyl-CoA binding protein is essential in bloodstream form *Trypanosoma brucei*. Mol. Biochem. Parasitol. 112, 301–304.
[93] Milne, K.G., Ferguson, M.A., 2000. Cloning, expression, and characterization of the acyl-CoA-binding protein in African trypanosomes. J. Biol. Chem. 275, 12503–12508.
[94] Rasmussen, J.T., Faergeman, N.J., Kristiansen, K., Knudsen, J., 1994. Acyl-CoA-binding protein (ACBP) can mediate intermembrane acyl-CoA transport and donate acyl-CoA for beta-oxidation and glycerolipid synthesis. Biochem. J. 299, 165–170.
[95] Fox, S.R., Rawsthorne, S., Hills, M.J., 2000. Role of acyl-CoAs and acyl-CoA-binding protein in regulation of carbon supply for fatty acid biosynthesis. Biochem. Soc. Trans. 28, 672–674.
[96] Fox, S.R., Hill, L.M., Rawsthorne, S., Hills, M.J., 2000. Inhibition of the glucose-6-phosphate transporter in oilseed rape (*Brassica napus* L.) plastids by acyl-CoA thioesters reduces fatty acid synthesis. Biochem. J. 352(Pt 2), 525–532.
[97] Johnson, P.E., Fox, S.R., Hills, M.J., Rawsthorne, S., 2000. Inhibition by long-chain acyl-CoAs of glucose 6-phosphate metabolism in plastids isolated from developing embryos of oilseed rape (*Brassica napus* L.). Biochem. J. 348(Pt 1), 145–150.
[98] Fulceri, R., Nori, A., Gamberucci, A., Volpe, P., Giunti, R., Benedetti, A., 1994. Fatty acyl-CoA esters induce calcium release from terminal cisternae of skeletal muscle. Cell Calcium 15, 109–116.
[99] Dumonteil, E., Barre, H., Meissner, G., 1994. Effects of palmitoyl carnitine and related metabolites on the avian Ca(2+)-ATPase and Ca2+ release channel. J. Physiol. 479(Pt 1), 29–39.
[100] Dumonteil, E., Barre, H., Meissner, G., 1993. Sarcoplasmic reticulum Ca(2 +)-ATPase and ryanodine receptor in cold-acclimated ducklings and thermogenesis. Am. J. Physiol. 265, C507–C513.

[101] Fulceri, R., Knudsen, J., Giunti, R., Volpe, P., Nori, A., Benedetti, A., 1997. Fatty acyl-CoA–acyl-CoA-binding protein complexes activate the Ca2 + release channel of skeletal muscle sarcoplasmic reticulum. Biochem. J. 325, 423–428.
[102] Melloni, E., Averna, M., Salamino, F., Sparatore, B., Minafra, R., Pontremoli, S., 2000. Acyl-CoA-binding protein is a potent m-calpain activator. J. Biol. Chem. 275, 82–86.
[103] Choi, J.Y., Stukey, J., Hwang, S.Y., Martin, C.E., 1996. Regulatory elements that control transcription activation and unsaturated fatty acid-mediated repression of the *Saccharomyces cerevisiae* OLE1 gene. J. Biol. Chem. 271, 3581–3589.
[104] Murakami, K., Ide, T., Nakazawa, T., Okazaki, T., Mochizuki, T., Kadowaki, T., 2001. Fatty-acyl-CoA thioesters inhibit recruitment of steroid receptor co-activator 1 to alpha and gamma isoforms of peroxisome-proliferator-activated receptors by competing with agonists. Biochem. J. 353, 231–238.
[105] Elholm, M., Dam, I., Jorgensen, C., Krogsdam, A.M., Holst, D., Kratchmarova, I., Gottlicher, M., Gustafsson, J.A., Berge, R., Flatmark, T., Knudsen, J., Mandrup, S., Kristiansen, K., 2001. Acyl-CoA esters antagonize the effects of ligands on peroxisome proliferator-activated receptor alpha conformation, DNA binding, and interaction with co-factors. J. Biol. Chem. 276, 21410–21416.
[106] Knudsen, J., Faergeman, N.J., Skott, H., Hummel, R., Borsting, C., Rose, T.M., Andersen, J.S., Hojrup, P., Roepstorff, P., Kristiansen, K., 1994. Yeast acyl-CoA-binding protein: acyl-CoA-binding affinity and effect on intracellular acyl-CoA pool size. Biochem. J. 302, 479–485.
[107] Mandrup, S., Jepsen, R., Skott, H., Rosendal, J., Hojrup, P., Kristiansen, K., Knudsen, J., 1993. Effect of heterologous expression of acyl-CoA-binding protein on acyl-CoA level and composition in yeast. Biochem. J. 290, 369–374.
[108] Bhuiyan, A.K., Pande, S.V., 1994. Carnitine palmitoyltransferase activities: effects of serum albumin, acyl-CoA binding protein and fatty acid binding protein. Mol. Cell. Biochem. 139, 109–116.
[109] Abo-Hashema, K.A., Cake, M.H., Lukas, M.A., Knudsen, J., 2001. The interaction of acyl-CoA with acyl-CoA binding protein and carnitine palmitoyltransferase I. Int. J. Biochem. Cell Biol. 33, 807–815.
[110] Rasmussen, J.T., Rosendal, J., Knudsen, J., 1993. Interaction of acyl-CoA binding protein (ACBP) on processes for which acyl-CoA is a substrate, product or inhibitor. Biochem. J. 292, 907–913.
[111] Gossett, R.E., Edmondson, R.D., Jolly, C.A., Cho, T.H., Russell, D.H., Knudsen, J., Kier, A.B., Schroeder, F., 1998. Structure and function of normal and transformed murine acyl-CoA binding proteins. Arch. Biochem. Biophys. 350, 201–213.
[112] Jolly, C.A., Wilton, D.C., Schroeder, F., 2000. Microsomal fatty acyl-CoA transacylation and hydrolysis: fatty acyl-CoA species dependent modulation by liver fatty acyl-CoA binding proteins. Biochim. Biophys. Acta 1483, 185–197.
[113] Fyrst, H., Knudsen, J., Schott, M.A., Lubin, B.H., Kuypers, F.A., 1995. Detection of acyl-CoA-binding protein in human red blood cells and investigation of its role in membrane phospholipid renewal. Biochem. J. 306, 793–799.
[114] Kannan, L., Knudsen, J., Jolly, C.A., 2003. Aging and acyl-CoA binding protein alter mitochondrial glycerol-3-phosphate acyltransferase activity. Biochim. Biophys. Acta 1631, 12–16.
[115] Mandrup, S., Sorensen, R.V., Helledie, T., Nohr, J., Baldursson, T., Gram, C., Knudsen, J., Kristiansen, K., 1998. Inhibition of 3T3-L1 adipocyte differentiation by expression of acyl-CoA-binding protein antisense RNA. J. Biol. Chem. 273, 23897–23903.
[116] Yang, Y., Pritchard, P.H., Bhuiyan, J., Seccombe, D.W., Moghadasian, M.H., 2001. Overexpression of acyl-coA binding protein and its effects on the flux of free fatty acids in McA-RH 7777 cells. Lipids 36, 595–600.
[117] Schjerling, C.K., Hummel, R., Hansen, J.K., Borsting, C., Mikkelsen, J.M., Kristiansen, K., Knudsen, J., 1996. Disruption of the gene encoding the acyl-CoA-binding protein (ACB1) perturbs acyl-CoA metabolism in *Saccharomyces cerevisiae*. J. Biol. Chem. 271, 22514–22521.
[118] Dickson, R.C., Lester, R.L., 1999. Yeast sphingolipids. Biochim. Biophys. Acta 1426, 347–357.
[119] Dunphy, J.T., Greentree, W.K., Manahan, C.L., Linder, M.E., 1996. G-protein palmitoyltransferase activity is enriched in plasma membranes. J. Biol. Chem. 271, 7154–7159.
[120] Glick, B.S., Rothman, J.E., 1987. Possible role for fatty acyl-coenzyme A in intracellular protein transport. Nature 326, 309–312.

[121] Pfanner, N., Glick, B.S., Arden, S.R., Rothman, J.E., 1990. Fatty acylation promotes fusion of transport vesicles with Golgi cisternae. J. Cell Biol. 110, 955–961.
[122] Pfanner, N., Orci, L., Glick, B.S., Amherdt, M., Arden, S.R., Malhotra, V., Rothman, J.E., 1989. Fatty acyl-coenzyme A is required for budding of transport vesicles from Golgi cisternae. Cell 59, 95–102.
[123] Ostermann, J., Orci, L., Tani, K., Amherdt, M., Ravazzola, M., Elazar, Z., Rothman, J.E., 1993. Stepwise assembly of functionally active transport vesicles. Cell 75, 1015–1025.
[124] Haas, A., Wickner, W., 1996. Homotypic vacuole fusion requires Sec17p (yeast alpha-SNAP) and Sec18p (yeast NSF). EMBO J. 15, 3296–3305.
[125] Deeney, J.T., Gromada, J., Hoy, M., Olsen, H.L., Rhodes, C.J., Prentki, M., Berggren, P.O., Corkey, B.E., 2000. Acute stimulation with long chain acyl-CoA enhances exocytosis in insulin-secreting cells (HIT T-15 and NMRI beta-cells). J. Biol. Chem. 275, 9363–9368.
[126] Veit, M., Laage, R., Dietrich, L., Wang, L., Ungermann, C., 2001. Vac8p release from the SNARE complex and its palmitoylation are coupled and essential for vacuole fusion. EMBO J. 20, 3145–3155.
[127] Chao, H., Martin, G.G., Russell, W.K., Waghela, S.D., Russell, D.H., Schroeder, F., Kier, A.B., 2002. Membrane charge and curvature determine interaction with acyl-CoA binding protein (ACBP) and fatty acyl-CoA targeting. Biochemistry 41, 10540–10553.
[128] Papadopoulos, V., Amri, H., Boujrad, N., Cascio, C., Culty, M., Garnier, M., Hardwick, M., Li, H., Vidic, B., Brown, A.S., Reversa, J.L., Bernassau, J.M., Drieu, K., 1997. Peripheral benzodiazepine receptor in cholesterol transport and steroidogenesis. Steroids 62, 21–28.
[129] Wang, X., Stocco, D.M., 1999. Cyclic AMP and arachidonic acid: a tale of two pathways. Mol. Cell. Endocrinol. 158, 7–12.
[130] Wang, X., Walsh, L.P., Reinhart, A.J., Stocco, D.M., 2000. The role of arachidonic acid in steroidogenesis and steroidogenic acute regulatory (StAR) gene and protein expression. J. Biol. Chem. 275, 20204–20209.
[131] Wang, X.J., Dyson, M.T., Mondillo, C., Patrignani, Z., Pignataro, O., Stocco, D.M., 2002. Interaction between arachidonic acid and cAMP signaling pathways enhances steroidogenesis and StAR gene expression in MA-10 Leydig tumor cells. Mol. Cell. Endocrinol. 188, 55–63.
[132] Maloberti, P., Lozano, R.C., Mele, P.G., Cano, F., Colonna, C., Mendez, C.F., Paz, C., Podesta, E.J., 2002. Concerted regulation of free arachidonic acid and hormone-induced steroid synthesis by acyl-CoA thioesterases and acyl-CoA synthetases in adrenal cells. Eur. J. Biochem. 269, 5599–5607.
[133] Aarsland, A., Berge, R.K., Bremer, J., Aarsaether, N., 1990. The hypolipidemic peroxisome-proliferating drug, bis(carboxymethylthio)-1.10 decane, a dicarboxylic metabolite of tiadenol, is activated to an acylcoenzyme A thioester. Biochim. Biophys. Acta 1033, 176–183.
[134] Wang, W., Wu, D.C., Chen, Y.H., He, W., Yu, L.C., 2002. Anti-nociceptive effects of diazepam binding inhibitor in the central nervous system of rats. Brain Res. 956, 393–397.
[135] Dong, E., Matsumoto, K., Watanabe, H., 2002. Diazepam binding inhibitor (DBI) reduces testosterone and estradiol levels in vivo. Life Sci. 70, 1317–1323.
[136] Manabe, Y., Toyoda, T., Kuroda, K., Imaizumi, M., Yamamoto, T., Fushiki, T., 2001. Effect of diazepam binding inhibitor (DBI) on the fluid intake, preference and the taste reactivity in mice. Behav. Brain Res. 126, 197–204.
[137] Li, Y., Hao, Y., Owyang, C., 2000. Diazepam-binding inhibitor mediates feedback regulation of pancreatic secretion and postprandial release of cholecystokinin. J. Clin. Invest. 105, 351–359.
[138] Herzig, K.H., Schon, I., Tatemoto, K., Ohe, Y., Li, Y., Folsch, U.R., Owyang, C., 1996. Diazepam binding inhibitor is a potent cholecystokinin-releasing peptide in the intestine [published erratum appears in Proc. Natl Acad. Sci. USA Nov 26;93(24):14214]. Proc. Natl Acad. Sci. USA 93, 7927–7932.
[139] Herzig, K.H., 1998. Cholecystokinin- and secretin-releasing peptides in the intestine – a new regulatory interendocrine mechanism in the gastrointestinal tract. Regul. Pept. 73, 89–94.
[140] Borboni, P., Condorelli, L., De Stefanis, P., Sesti, G., Lauro, R., 1991. Modulation of insulin secretion by diazepam binding inhibitor and its processing products. Neuropharmacology 30, 1399–1403.
[141] Ostenson, C.G., Ahren, B., Karlsson, S., Knudsen, J., Efendic, S., 1994. Inhibition by rat diazepam-binding inhibitor/acyl-CoA-binding protein of glucose-induced insulin secretion in the rat. Eur. J. Endocrinol. 131, 201–204.

[142] Sjoholm, A., Funakoshi, A., Efendic, S., Ostenson, C.G., Hellerstrom, C., 1991. Long term inhibitory effects of pancreastatin and diazepam binding inhibitor on pancreatic beta-cell deoxyribonucleic acid replication, polyamine content, and insulin secretion. Endocrinology 128, 3277–3282.

[143] Yu, C., Chen, Y., Cline, G.W., Zhang, D., Zong, H., Wang, Y., Bergeron, R., Kim, J.K., Cushman, S.W., Cooney, G.J., Atcheson, B., White, M.F., Kraegen, E.W., Shulman, G.I., 2002. Mechanism by which fatty acids inhibit insulin activation of insulin receptor substrate-1 (IRS-1)-associated phosphatidylinositol 3-kinase activity in muscle. J. Biol. Chem. 277, 50230–50236.

[144] Nesher, M., Boneh, A., 1994. Effect of fatty acids and their acyl-CoA esters on protein kinase C activity in fibroblasts: possible implications in fatty acid oxidation defects. Biochim. Biophys. Acta 1221, 66–72.

[145] Bronfman, M., Morales, M.N., Orellana, A., 1988. Diacylglycerol activation of protein kinase C is modulated by long-chain acyl-CoA. Biochem. Biophys. Res. Commun. 152, 987–992.

[146] Yaney, G.C., Korchak, H.M., Corkey, B.E., 2000. Long-chain acyl CoA regulation of protein kinase C and fatty acid potentiation of glucose-stimulated insulin secretion in clonal beta-cells. Endocrinology 141, 1989–1998.

[147] Deeney, J.T., Cunningham, B.A., Chheda, S., Bokvist, K., Juntti Berggren, L., Lam, K., Korchak, H.M., Corkey, B.E., Berggren, P.O., 1996. Reversible Ca2 + -dependent translocation of protein kinase C and glucose-induced insulin release. J. Biol. Chem. 271, 18154–18160.

[148] Majumdar, S., Rossi, M.W., Fujiki, T., Phillips, W.A., Disa, S., Queen, C.F., Johnston, R.B. Jr., Rosen, O.M., Corkey, B.E., Korchak, H.M., 1991. Protein kinase C isotypes and signaling in neutrophils. Differential substrate specificities of a translocatable calcium- and phospholipid-dependent beta-protein kinase C and a phospholipid-dependent protein kinase which is inhibited by long chain fatty acyl coenzyme A. J. Biol. Chem. 266, 9285–9294.

[149] Ford, D.A., Horner, C.C., Gross, R.W., 1998. Protein kinase C acylation by palmitoyl coenzyme A facilitates its translocation to membranes. Biochemistry 37, 11953–11961.

[150] Larsson, O., Deeney, J.T., Branstrom, R., Berggren, P.O., Corkey, B.E., 1996. Activation of the ATP-sensitive K+ channel by long chain acyl-CoA. A role in modulation of pancreatic beta-cell glucose sensitivity. J. Biol. Chem. 271, 10623–10626.

[151] Branstrom, R., Corkey, B.E., Berggren, P.O., Larsson, O., 1997. Evidence for a unique long chain acyl-CoA ester binding site on the ATP-regulated potassium channel in mouse pancreatic beta cells. J. Biol. Chem. 272, 17390–17394.

[152] Branstrom, R., Leibiger, I.B., Leibiger, B., Corkey, B.E., Berggren, P.O., Larsson, O., 1998. Long chain coenzyme A esters activate the pore-forming subunit (Kir6. 2) of the ATP-regulated potassium channel. J. Biol. Chem. 273, 31395–31400.

[153] Gribble, F.M., Proks, P., Corkey, B.E., Ashcroft, F.M., 1998. Mechanism of cloned ATP-sensitive potassium channel activation by oleoyl-CoA. J. Biol. Chem. 273, 26383–26387.

[154] Liu, G.X., Hanley, P.J., Ray, J., Daut, J., 2001. Long-chain acyl-coenzyme A esters and fatty acids directly link metabolism to K(ATP) channels in the heart. Circ. Res. 88, 918–924.

[155] van der Vusse, G.J., Prinzen, F.W., van Bilsen, M., Engels, W., Reneman, R.S., 1987. Accumulation of lipids and lipid-intermediates in the heart during ischaemia. Basic Res. Cardiol. 82(Suppl. 1), 157–167.

[156] Paucek, P., Yarov Yarovoy, V., Sun, X., Garlid, K.D., 1996. Inhibition of the mitochondrial KATP channel by long-chain acyl-CoA esters and activation by guanine nucleotides. J. Biol. Chem. 271, 32084–32088.

[157] DiRusso, C.C., Black, P.N., Weimar, J.D., 1999. Molecular inroads into the regulation and metabolism of fatty acids, lessons from bacteria. Prog. Lipid Res. 38, 129–197.

[158] Raman, N., DiRusso, C.C., 1995. Analysis of acyl coenzyme A binding to the transcription factor FadR and identification of amino acid residues in the carboxyl terminus required for ligand binding. J. Biol. Chem. 270, 1092–1097.

[159] Xu, Y., Heath, R.J., Li, Z., Rock, C.O., White, S.W., 2001. The FadR·DNA complex. Transcriptional control of fatty acid metabolism in *Escherichia coli*. J. Biol. Chem. 276, 17373–17379.

[160] Hertz, R., Magenheim, J., Berman, I., Bar-Tana, J., 1998. Fatty acyl-CoA thioesters are ligands of hepatic nuclear factor-4alpha. Nature 392, 512–516.

[161] Rajas, F., Gautier, A., Bady, I., Montano, S., Mithieux, G., 2002. Polyunsaturated fatty acyl coenzyme A suppress the glucose-6-phosphatase promoter activity by modulating the DNA binding of hepatocyte nuclear factor 4 alpha. J. Biol. Chem. 277, 15736–15744.

[162] Bogan, A.A., Dallas-Yang, Q., Ruse, M.D.J., Maeda, Y., Jiang, G., Nepomuceno, L., Scanlan, T.S., Cohen, F.E., Sladek, F.M., 2000. Analysis of protein dimerization and ligand binding of orphan receptor HNF4alpha. J. Mol. Biol. 302, 831–851.

[163] Rosendal, J., Ertbjerg, P., Knudsen, J., 1993. Characterization of ligand binding to acyl-CoA-binding protein. Biochem. J. 290, 321–326.

[164] Wadum, M.C., Villadsen, J.K., Feddersen, S., Moller, R.S., Neergaard, T.B., Kragelund, B.B., Hojrup, P., Faergeman, N.J., Knudsen, J., 2002. Fluorescently labelled bovine acyl-CoA-binding protein acting as an acyl-CoA sensor: interaction with CoA and acyl-CoA esters and its use in measuring free acyl-CoA esters and non-esterified fatty acids. Biochem. J. 365, 165–172.

[165] Rolf, B., Oudenampsen Kruger, E., Borchers, T., Faergeman, N.J., Knudsen, J., Lezius, A., Spener, F., 1995. Analysis of the ligand binding properties of recombinant bovine liver-type fatty acid binding protein. Biochim. Biophys. Acta 1259, 245–253.

[166] Frolov, A., Cho, T.H., Murphy, E.J., Schroeder, F., 1997. Isoforms of rat liver fatty acid binding protein differ in structure and affinity for fatty acids and fatty acyl CoAs. Biochemistry 36, 6545–6555.

[167] Paulussen, R.J., Veerkamp, J.H., 1990. Intracellular fatty-acid-binding proteins. Characteristics and function. Subcell. Biochem. 16, 175–226.

[168] Frolow, A., Cho, T., Billheimer, J.T., Schroeder, F., 1996. Sterol carrier protein-2 a new fatty acyl Coenzyme A-binding protein. J. Biol. Chem. 271, 31878–31884.

[169] Silva, C., Loyola, G., Valenzuela, R., Garcia-Huidobro, T., Monasterio, O., Bronfman, M., 1999. High-affinity binding of fatty acyl-CoAs and peroxisome proliferator-CoA esters to glutathione *S*-transferases effect on enzymatic activity. Eur. J. Biochem. 266, 143–150.

[170] Knudsen, J., Clark, S., Dils, R., 1975. Acyl-CoA hydrolase(s) in rabbit mammary gland which control the chain length of fatty acids synthesised. Biochem. Biophys. Res. Commun. 65, 921–926.

[171] Berge, R.K., Aarsland, A., 1985. Correlation between the cellular level of long-chain acyl-CoA, peroxisomal beta-oxidation, and palmitoyl-CoA hydrolase activity in rat liver. Are the two enzyme systems regulated by a substrate-induced mechanism? Biochim. Biophys. Acta 837, 141–151.

[172] Hunt, M.C., Alexson, S.E., 2002. The role Acyl-CoA thioesterases play in mediating intracellular lipid metabolism. Prog. Lipid Res. 41, 99–130.

[173] Peitzsch, R.M., McLaughlin, S., 1993. Binding of acylated peptides and fatty acids to phospholipid vesicles: pertinence to myristoylated proteins. Biochemistry 32, 10436–10443.

[174] Pande, S.V., 1973. Reversal by CoA of palmityl-CoA inhibition of long chain acyl-CoA synthetase activity. Biochim. Biophys. Acta 306, 15–20.

[175] Sabri, A., Bare, G., Jacques, P., Jabrane, A., Ongena, M., Van Heugen, J.C., Devreese, B., Thonart, P., 2001. Influence of moderate temperatures on myristoyl-CoA metabolism and acyl-CoA thioesterase activity in the psychrophilic antarctic yeast *Rhodotorula aurantiaca*. J. Biol. Chem. 276, 12691–12696.

[176] Jones, J.M., Nau, K., Geraghty, M.T., Erdmann, R., Gould, S.J., 1999. Identification of peroxisomal acyl-CoA thioesterases in yeast and humans. J. Biol. Chem. 274, 9216–9223.

[177] Lewin, T.M., Van Horn, C.G., Krisans, S.K., Coleman, R.A., 2002. Rat liver acyl-CoA synthetase 4 is a peripheral-membrane protein located in two distinct subcellular organelles, peroxisomes, and mitochondrial-associated membrane. Arch. Biochem. Biophys. 404, 263–270.

[178] Ogiwara, H., Tanabe, T., Nikawa, J., Numa, S., 1978. Inhibition of rat-liver acetyl-coenzyme-A carboxylase by palmitoyl-coenzyme A. Formation of equimolar enzyme-inhibitor complex. Eur. J. Biochem. 89, 33–41.

[179] Ong, D.E., 1994. Cellular transport and metabolism of vitamin A: roles of the cellular retinoid-binding proteins. Nutr. Rev. 52, S24–S31.

[180] Boerman, M.H., Napoli, J.L., 1996. Cellular retinol-binding protein-supported retinoic acid synthesis. Relative roles of microsomes and cytosol. J. Biol. Chem. 271, 5610–5616.

Physical aspects of fatty acid transport between and through biological membranes

Henry J. Pownall[a,1] and James A. Hamilton[b,*]

[a]*Department of Medicine, Baylor College of Medicine, One Baylor Plaza, MS A601, Houston, TX 77030, USA*
[b]*Department of Biophysics and Physiology, Boston University School of Medicine, 715 Albany Street, W302 Boston, MA 02118-2526, USA*
[*]*Correspondence address: Tel.: +1-617-638-5048; fax: +1-617-638-4041*
E-mail: jhamilt@bu.edu(J.A.H.)

1. Introduction

Transport of molecules that provide energy for living cells is an important component of metabolism in all species extending throughout the animal and plant kingdoms. In mammalian systems energy is transported mainly as glucose and fatty acid. How do glucose and free fatty acids in plasma reach their sites of storage and utilization, and what controls their rates of uptake at these sites? This question is important both in normal physiology and in pathophysiology associated with obesity, type 2 diabetes mellitus, hypertriglyceridemia, and atherosclerosis. Major sites of glucose and fatty acid uptake include liver, adipose tissue, skeletal and cardiac muscle, and intestine. In the fasting state, the brain is the main depot for glucose uptake and oxidation. Glucose is stored in liver and skeletal muscle as the polymer glycogen formed by glycogen synthase. Whereas skeletal muscle is the major site of conversion of glucose to energy, liver is an important site for the conversion of glucose to fatty acid through lipogenesis as well as the principal locus of gluconeogenesis. Glucose and fatty acids are taken up by adipose tissue and converted to triglyceride (TG), a non-polymeric form of storage but the most concentrated cellular form of stored energy. Both glucose and fatty acids are essential to TG storage, fatty acids providing the acyl chains of TG and glucose the glycerol backbone of TG.

2. Solubility of glucose and fatty acids in water and membranes

The uptake of water-soluble glucose into cells involves complex mechanisms and requires a transporter within the plasma membrane. Several transporter isoforms, namely

[1]Tel.: +1-713-798-4160; fax: +1-713-798-5134.

Advances in Molecular and Cell Biology, Vol. 33, pages 153–172

ISSN: 1569-2558 / DOI: 10.1016/S1569-2558(03)33009-7

GLUTs 1–3, are plasma membrane proteins that allow constitutive permeation of some glucose into various tissues [1]. Regulation of glucose transport and influx of high concentrations into skeletal muscle, heart, and adipose tissue is initiated by binding of insulin to its receptor. This triggers a sequence of reactions that ultimately lead to the translocation of the glucose transporter GLUT4 from intracellular storage vesicles to the plasma membrane [2]. GLUT4 mediates the transport of extracellular glucose into the cytoplasm, where it is phosphorylated. In the absence of insulin, GLUT4 returns to its intracellular vesicular storage compartment.

Glucose cannot passively diffuse through the plasma membrane. Thus, it requires a transmembrane transporter, and its uptake can be regulated, as established for the GLUT4 transport system. The major physicochemical property of glucose prohibiting its free diffusion across cell membranes is the presence of several hydroxyl groups that impart a high polarity to this small molecule. The hydrocarbon-like nature of membrane bilayers is a solubility barrier to the diffusional translocation of polar and charged species. Simply stated, glucose is very soluble in water (>4000 mM) but is insoluble in hydrocarbons. Studies in model membranes and in living cells have shown that in the absence of a specific protein transporter, glucose does not cross membrane bilayers.

Fatty acid uptake is also stimulated by insulin and it is tempting to postulate that its transport across cell membranes might be controlled in the same way as for glucose, i.e. by a specific transporter. However, the physical properties of fatty acids are very different from those of glucose. Fatty acids are amphiphilic, and are soluble in both water and hydrocarbons. In water, the pK_a of a medium or long chain fatty acid is about 5.2. However, when present in a lipid bilayer the pK_a carboxyl, which is located at the lipid–water interface, is closer to 7.4 [3]. Therefore, under physiological conditions, fatty acids in a membrane are distributed almost equally into their un-ionized (protonated) and ionized forms. Furthermore, these forms are in rapid equilibrium with each other [4] so that the depletion of one form can rapidly be replenished. Therefore, the pool size of one form can be treated as the total fatty acid pool size if the reactions that remove that form are slower than the rates of equilibration of the two pools.

Another significant difference between glucose and fatty acids is that the latter encompasses a large variety of molecular species, the structure of which greatly modifies their physical properties. For example, the solubility of a fatty acid in an aqueous environment compared to a hydrocarbon environment is determined by the charge on the fatty acid carboxyl group and by the structure of its hydrocarbon chain. The un-ionized form is very soluble in hydrocarbon and only sparingly soluble in water; in contrast, ionized fatty acids are soluble in water and less soluble in hydrocarbons.

The structure of the acyl chain can alter the balance of polar and non-polar character of fatty acids and affect their partitioning between aqueous and hydrocarbon phases and other physicochemical behaviors that are a function of solubility. It is well known that the solubility of alkanes decreases as a function of the number of carbons in the chain. A quantitative analysis shows that the free energy of transfer of an alkane from a pure hydrocarbon to water increases by about 880 cal/mol for each additional methylene unit (reviewed by Tanford [5]). Similarly, as the acyl chain length of a fatty acid is increased, its solubility in hydrocarbons increases.

Under conditions where aqueous and hydrocarbon environments coexist, an increasing fraction of fatty acid will partition from the aqueous to the hydrocarbon environment as the acyl chain length increases.

In the absence of a hydrocarbon-like environment, fully ionized fatty acids can form micelles, which are aggregates of 50–100 fatty acids that are essentially spheres with a center having hydrocarbon-like character. In the presence of lipid–water interfaces such as micelles or phospholipid membranes, fatty acids partition between the lipid–water interfaces and the surrounding aqueous phase (Fig. 1) according to a partition coefficient (K_p) and an attendant-free energy of partitioning (ΔG_p) defined by Eqs. (1) and (2)

$$K_p = X_I/X_w \tag{1}$$

$$\Delta G_p = -RT \ln K_p \tag{2}$$

where X_I and X_w, respectively, are the more fractions of the amphiphile in the lipid–water interface and in water. For fatty acids and numerous other amphiphilic and hydrophobic molecules, ΔG_p varies with hydrophobicity as determined by the number of methylene units and double bonds according to two empirically determined rules: (1) each methylene unit increases ΔG_p by approximately 600–700 cal/mol and (2) each double bond decreases ΔG_p by about 800 kcal/mol [5]. Thus, the net effect of increasing acyl chain length or decreasing unsaturation is to increase its affinity for lipid or hydrocarbon-like surfaces and decrease the aqueous concentration of amphiphile. In the presence of micelles of a pure amphiphile, the concentration in the aqueous phase is the critical micelle concentration (cmc). Thus, according to these two empirical rules, the cmc of a homologous series of amphiphiles will decrease with increasing chain length and increase with increasing unsaturation. It is also important to point out the complexity that the pH at which micelles can form increases with increasing chain length, and that long chain fatty acids can form micelles only at pH > 10, a non-physiological condition, whereas medium chain fatty acids can form micelles at pH 7.4 [6]. The formation of micelles of pure fatty

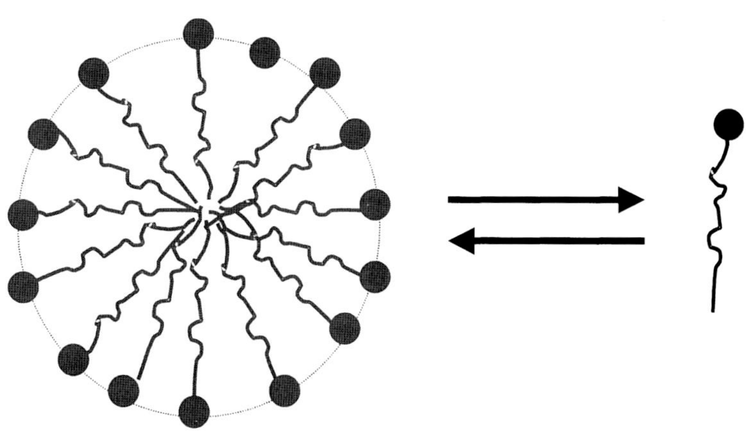

Fig. 1. Schematic representation of desorption of a single-chain lipid from a spherical micelle into the surrounding aqueous phase.

acids is not relevant to their transport in the blood, where they are bound to albumin as individual molecules [7]. However, fatty acids form mixed micelles with other lipids, such as bile acids, in the gut, and this facilitates their transfer to cells.

From the preceding discussion, it is evident that the structure of a fatty acid acyl chain is an important determinant of its transfer between biological surfaces such as membranes and proteins, translocation across biological membranes, and metabolism within cells (Fig. 2). The biophysics of each of these three phenomena is discussed in greater detail below.

3. Transfer of fatty acids between lipid and protein surfaces

Most naturally occurring and common dietary fatty acids have 16–22 carbons and up to six double bonds. The aqueous solubility of these species is sufficiently low (nM–μM at pH 7.4) and their hydrophobicity high enough to support binding to the hydrophobic regions of cell membranes and to some proteins. In the plasma and interstitial compartments, albumin is the major carrier of fatty acids. In cells that metabolize lipids, fatty acid binding proteins (FABPs), which belong to a gene family encoding a group of related proteins, comprise a major fraction of the soluble protein in the cytosol [8].

The rates of transfer of fatty acids from albumin to cells must be very fast because the turnover time of fatty acids in the plasma is very fast, and at the same time fraction that occurs as free monomer is very small. How can this be accomplished? One possible mechanism is that albumin associates directly with cell membranes and transfers one or more (albumin binds up to five fatty acids with medium to high affinity) fatty acid molecule to the cell membrane. Alternatively, fatty acids bound to albumin desorb into the surrounding aqueous phase, after which they associate with another protein or lipid surface. Albumin can and may still approach the cell surface closely, but a direct binding interaction is not required. According to this model, albumin is a carrier that buffers the

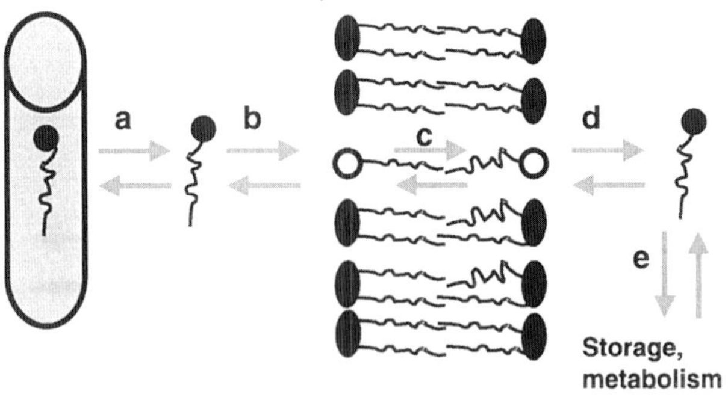

Fig. 2. Schematic representation of the movement of a fatty acid from the plasma compartment where it is bound to albumin into a cell. The steps are: (a) reversible desorption from the protein into the surrounding aqueous phase, (b) reversible association with the outer leaflet of the plasma membrane, (c) reversible translocation across the plasma membrane, (d) reversible desorption from the inner leaflet of the plasma membrane into the cytoplasm, (e) irreversible metabolism, which usually begins with conversion to an acyl Coenzyme A analog and is followed by oxidation or esterification to glycerol or cholesterol.

plasma concentration of unbound fatty acid, an important role given that high fatty acid levels can be cytotoxic. Most evidence favors the latter model of fatty acid transfer: first order, rate limiting desorption from albumin followed by diffusion-controlled association with an acceptor. In a broader sense, the donors and acceptors can be any lipid or protein surface that binds to fatty acids. Thus, the determinants that regulate fatty acid transfer are essentially those that regulate its rate of desorption into water.

Fatty acid desorption and transfer is strongly dependent on the structure of the transferring fatty acid and, to a lesser extent, on the structure of the surfaces (proteins and membranes) with which it is associated. The rates of fatty acid transfer from phospholipid bilayers, lipoprotein or albumin decrease with increasing acyl chain length and decrease with increasing unsaturation [9,10]. According to an analysis of the data shown in Fig. 3, the rate constant for the transfer of a straight-chain fatty acid with n carbons and m double bonds at 37 °C is given by Eq. (3a). The measured (and predicted according to Eq. (3b)) halftimes for the transfer from phospholipid bilayers of the three major fatty acids in plasma (palmitic, oleic, and linoleic acid) [11] are: 3 (8), 17 (37), and 1 (10) ms, respectively. Another study of saturated fatty acids with 14–26 carbons revealed a log-linear relationship between the rate constant and fatty acid chain length, with slightly slower rates of desorption at 24 °C [9]. The desorption of common dietary fatty acids is fast by both measurements but the rates for very long chain saturated fatty acid (>20 carbons) may be slower than some other physiological processes that liberate or utilize fatty acids.

$$\log k_i = 0.63m - 0.67n + 13.19 \tag{3a}$$

$$\log t_{1/2} = 0.62n - 0.59m - 12.0 \tag{3b}$$

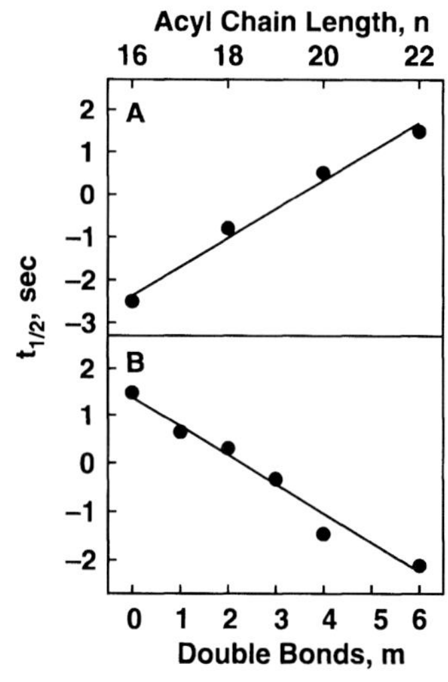

Fig. 3. Effects of acyl chain structure of transfer halftimes of fatty acids at 37 °C. (A) Effect of acyl chain length on the free energy of activation for the transfer of saturated amphiphiles from small SBV to anthraniloyl HSA. (B) Effect of acyl chain unsaturation on the free energy of activation for the transfer of fatty acids from small SBV to anthraniloyl HSA; fatty acid chains contained 22 carbons.

Absolute rate theory gives the free energies of activation for fatty acid transfer [Eq. (4)]; each methylene unit adds a similar increment of 640 cal/mol to the free energy of activation whereas each double bond adds a decrement of 770 cal/mol. Reducing the pH from 7.4 to 2.8 converts free fatty acids entirely to their un-ionized forms, which are less soluble in water. The rates of transfer of fatty acids in their ionized form are 10–50 times faster than their un-ionized form [12,13]. This corresponds to differences in the free energy of activation for transfer of 1.2–2.5 kcal/mol.

$$\Delta G^{\ddagger} = 0.64n - 0.77m + 5.95 \tag{4}$$

These data strongly support a model in which the rate-limiting step is the transfer of fatty acid from the donor to an environment similar or identical to the aqueous phase surrounding the donor. According to a hypothetical reaction coordinate for the transfer of fatty acids from a lipid monolayer, the activated state corresponds to the point along the transfer coordinate at which the fatty acid is first totally surrounded by water. Thus, the lower the solubility of a fatty acid in the aqueous phase separating donor and acceptor species, the slower the rate of transfer. The desorption step is so much slower than the second step (association with an acceptor) that other properties of the system, which usually affect bimolecular reactions (e.g. viscosity), do not affect the transfer rate.

3.1. *Transfer of other lipids*

The relationship between aqueous solubility phase as determined by hydrophobicity and the rates of desorption and transfer has been validated for other more complex lipids and for lipid-associating proteins [14,15]. The half times and free energies of activation for transfer of phosphatidylcholines (PCs) with n methylene units and m double bonds, are described by Eqs. (5a) and (5b).

$$\log T_{1/2} = 0.234n - 0.189m - 5.76 \tag{5a}$$

$$\Delta G^{\ddagger} = 0.44m - 0.54n - 5.64 \tag{5b}$$

Although this equation is similar in form to those describing the transfer of fatty acids, there are some important differences. First, the contribution of each methylene unit to the free energy of activation for the transfer of PCs is much less (320 cal/methylene unit) than that observed for fatty acids (640 cal/methylene unit). Similarly, Smith and Tanford [16] reported that the contributions of each methylene unit to the free energy of transfer of a two-chain amphiphile from a lipid environment to water were less than that of a single-chain amphiphile with the same number of methylene units. This effect was assigned to the formation of hydrophobic bonds that keep the acyl chains at the sn-1 and sn-2 positions of PC associated with each other even when the lipid is an aqueous monomer in water. As a consequence, the exposed hydrophobic area, which is a more reliable predictor of behavior than chain length, is smaller than expected from the consideration of two separate chains, and the contributions of each methylene unit to the free energy of activation for transfer and the free energy of transfer are both about half of what is observed for fatty acids.

The rate constants for the transfer of PC, phosphatidylethanolamine, phosphatidylserine, phosphatidylglycerol, and phosphatidic acid are similar when the acyl chain compositions of these lipids are the same [17]. This is not surprising given that the polar structures are in contact with water when the lipid is associated with a lipid surface and when it is a monomer in water. Thus, there is little change in the environment and energy of the head group during transfer. Moreover, the energetics of transfer of the polar part of lipid from the hydrated interface of a bilayer to the surrounding aqueous phase evidently are very small and do not contribute to the free energy of transfer or the free energy of activation.

Cholesteryl esters and triacylglycerols are two other important classes of lipids. For all practical purposes, naturally occurring cholesteryl esters and triacylglycerols, which typically contain acyl groups of 16 or more carbons, are insoluble in water. By the same token, they are very soluble in hydrocarbons such as hexane. Neither cholesteryl esters nor triacylglycerols undergo spontaneous transfer through water. Thus, their lack of spontaneous transfer supports the hypothesis that the major thermodynamic barrier to the transfer of lipids is their solubility in the aqueous phase separating the donor and acceptor. Nevertheless, the triacylglycerols of very low density lipoproteins are exchanged for the cholesteryl esters of high density lipoproteins, a process that is mediated by cholesteryl ester transfer protein, which also transfers phospholipids among plasma lipoproteins [18,19].

There have been a limited number of studies of spontaneous protein transfer between membrane surfaces. McKeone et al. [20] reported that human apolipoprotein C proteins, which are water soluble, are rapidly transferred between phospholipid single bilayer vesicles (SBV). In contrast, acylated apolipoprotein analogs are transferred at a rate that decreases with increasing length of the acyl chain [21]. Thus as with lipids, the increase in protein hydrophobicity, which is the major determinant of decreasing solubility in water, correlates with a slower rate of spontaneous transfer.

In spite of major differences in the structures of phospholipids, triacylglycerols, cholesteryl esters, long chain alcohols, fatty acids, and apolipoproteins, their rates of spontaneous transfer between lipid and/or protein surfaces are predictable functions of their aqueous solubilities, which are determined by their respective hydrophobicities. Hydrophobicity is a function of acyl chain length and to a lesser extent, the nature of various polar groups that are associated with these molecules. Other factors, such as the qualities of the donor surface and the viscosity of the medium are less important; the identity and structure of the acceptor surface do not play a direct role.

4. Translocation of glucose and lipids across cell membranes

4.1. Glucose transport

Glucose and fatty acids are the major sources of energy that is either stored or utilized by living cells. The highly polar structure of glucose and its high solubility in water inhibit entry into the hydrocarbon-like region between the inner and outer leaflets of the plasma membrane, and glucose would not be expected to enter the cell by simple diffusion. In support of this view, the activation energy for glucose permeation through phospholipid SBV is 25 kcal/mol [22]. Cellular glucose requirements, which can vary greatly according to diurnal effects, nutritional status, and overall health status are addressed by a family of glucose transporters (GLUT) [1]. The activities of GLUT1, GLUT2, and GLUT3, respectively, in erythrocytes and at the blood–brain barrier, in the liver, and in neuronal tissue, are independent of insulin. In contrast, muscle and adipose tissue contain GLUT4, which in response to insulin signaling is transported from the cytosol to the plasma membrane in vesicles [2]. Thus, the barrier to glucose entry into a variety of cells is removed by the occurrence of GLUT in the plasma membrane either constitutively or in response to glucose-activated insulin secretion by the pancreatic T-cells. Insulin also activates enzymes that catalyze glucose metabolism so that the pathways from glucose uptake and metabolism are highly coordinated in insulin-sensitive tissues.

4.2. Lipid transport

Transport of fatty acids in their unesterified form from the extracellular compartment to the cytoplasm is composed of several steps (Fig. 2). Most investigators agree that the first step, diffusion-controlled transfer of fatty acid from the surrounding aqueous phase to the outer leaflet of the cell membrane, is not rate limiting, although this may not be true in all cases. The physiological mechanism for the subsequent translocation of fatty acids from the outer to inner membrane leaflet is the most controversial step of membrane transport. Two mechanisms for the process of translocation, sometimes called flip-flop, have been hypothesized. According to one hypothesis, the thermodynamic barrier to diffusive translocation of fatty acids from the outer to inner leaflet of the cell membrane is so high that plasma membrane fatty acid translocases (FATs) are required (Fig. 2, Step c). As discussed below, this hypothesis is supported by reports from several laboratories showing that fatty acid transport into cells is enhanced by the occurrence or overexpression of one or more protein translocases. The alternative hypothesis is that the thermodynamic barrier to translocation is too low to be rate limiting in the uptake and that metabolism of free fatty acids and that other intracellular processes (Fig. 2, Step d or e) are rate limiting. In many cell types, the overall process of fatty acid uptake is regulated, so that identification of the rate-limiting step for the transfer of fatty acids from outside the cell to their sites of metabolism would localize if not identify the regulated step.

Translocation of a variety of lipids across cell membranes has been studied, using two experimental approaches to identify the rate-limiting step. One is a biophysical approach employing model systems to measure lipid translocation by direct and indirect physicochemical methods. The other is a more physiological/biological approach in which rates of lipid (usually fatty acid) uptake and metabolism are determined in cells or genetically modified animals in which putative lipid transporters have been deleted or overexpressed. Each approach has its advantages and shortcomings, and a thorough understanding of lipid translocation would probably require careful deployment of these two complementary approaches.

5. Biophysical methods

Model membranes and fluorescence probes have been useful in studies of the mechanism by which molecules are transported in living cells. The most utilized model membranes are SBVs, a spherical bilayer of phospholipids surrounding a contained aqueous volume. These have the advantage of structural simplicity and can be prepared by ultrasonic irradiation to yield small SBV, by extrusion methods to yield large SBV, and by various detergent removal methods to yield very large SBV. The compositions and sizes of SBV can be controlled and physical transport of molecules through these simple membranes can be evaluated without the confounding effects of metabolism that occur in living cells.

Studies of lipid transfer have often utilized fluorescent lipid analogs to determine transfer kinetics in real time without separation. However, great care must be exercised in the choice of fluorescent analogs, and the criteria for making these choices differ according to the experimental design, the most demanding criteria being for studies in living cells. First, the fluorescent analog should emulate a natural fatty acid with respect to its physicochemical properties. Solubilities of fluorescent lipids in water or the hydrocarbon environment of membranes should not differ greatly from those of their natural analogs, especially for lipid transport studies. Another criterion is that the metabolism of fluorescent lipids be similar to that of naturally occurring analogs. As with natural fatty acids, various fluorescent fatty acid analogs can form different products [23]. Some analogs may not be suitable for cell studies because they do not form the same spectrum of products that is observed with natural fatty acids. On the other hand, non-metabolizable fluorescence probes can be valuable for distinguishing transport events from metabolic events. The use of fluorescent probes with model membranes has provided new insights into lipid dynamics that have been frequently confirmed by alternative techniques that are not a direct function of probe behavior. Although many of the principles of lipid transfer between lipid surfaces, which were developed using fluorescent lipid analogs [14,17,24–26], have been confirmed with naturally occurring fatty acids [9,10,15], there have been important exceptions that are noted below.

5.1. Glycerolipid translocation

Early studies of membranes showed that phospholipids are asymmetrically distributed in membranes; their very slow rate of translocation clearly helps to maintain the membrane asymmetry. It is well established that permeation of artificial phospholipid bilayers by solutes is a function of the lipid solubility, size, charge, and hydrogen-bonding potential of the solute [27]. Studies of a homologous series of phospholipids have revealed that flip-flop is not nearly as sensitive to acyl chain length as is intermembrane transfer [28]. Addition of four methylene units decreases the intermembrane and transmembrane transfer rates by factors of about 60 and 4, respectively. However, there is a huge barrier to the transmembrane movement of phospholipids, and the rates are very slow. The free energies of activation for flip-flop are much higher than those for intermembrane transfer, reflecting the poor solubility of zwitterionic (phosphorylcholine and phosphorylethanolamine) and charged moieties (phosphate and phosphorylglycerol) in a hydrophobic environment.

In contrast to phospholipids, the neutral, less polar diacylglycerols exhibit very fast flip-flop in a phospholipid membrane. NMR chemical shift measurements of 1,2-dilauroylglycerol in PC SBV give a rate that is faster than $60\ s^{-1}$ [29]. Other studies are consistent with the importance of charge as a barrier to the lipid translocation [30,31]. In their uncharged states, stearylamine and sphingosine rapidly diffuse into the inner monolayer of large unilamellar vesicles. If the pH of the aqueous phase within the vesicles is strongly acidic, this process is essentially irreversible, presumably due to the charge placed on the transferring species by protonation. Similarly, oleic acid that is transported into vesicles in its protonated form is trapped on the opposing leaflet when the aqueous phase in contact with the opposite leaflet is strongly basic. These results are additional evidence that the insolubility of charged species in the hydrophobic interior of a bilayer membrane is an important barrier to translocation.

5.2. Translocation of carboxylates (fatty acids and bile acids)

Bile acids and fatty acids are especially interesting because they can be uncharged or charged, and their ionization behavior in a membrane is markedly different from their ionization in water. Studies of fatty acids have evolved from those utilizing fluorescent analogs to those using a variety of natural fatty acids in conjunction with several kinds of fluorescence reporter probes. Table 1 summarizes the reported translocation rate constants for some carboxylate-containing lipids, many of which are natural fatty acids. Translocation of a medium chain fatty acid attached to a weakly polar hydrocarbon (9[1-pyrenyl]nonanoic acid) from the inner to outer leaflet of small SBV was much faster ($k \geq 5\ s^{-1}$) than intermembrane transfer [12,32]. In contrast, the rate constant for the translocation of a long chain fatty acid with an attached polar group (12-[9-anthroyloxy] stearic acid (12AS)) in large SBV of egg PC is about $0.0005\ s^{-1}$, which corresponds to a translocation halftime on the order of 30 min [26]. The rate constants for 12AS translocation in small SBV were an order of magnitude faster. These relatively small rate constants are reflected in the high free energy of activation for translocation of 22 kcal/mol, a value that is considerably higher than that observed for intermembrane fatty acid transfer. The large difference in transfer rates must be due in part to differences in the structures and properties of the probes. The pyrenyl moiety is very hydrophobic and is very soluble within the hydrocarbon region of membranes. The ester linkage in the anthroyloxy moiety may make these derivatives sufficiently polar so that the free energy of passage into the hydrocarbon part of the bilayer is increased by as much as 2 kcal, which would decrease the rate constant for transfer by a factor of 28 ($e^{-2\ \mathrm{kcal}/RT}$). The different results for the two types of labels appear to support the concept of rate-limiting solubility of the polar part of the transferring species within the hydrocarbon regions of the bilayer. The disparate results also raise concerns about the validity of studies conducted with fluorescent analogs that are poor emulators of natural fatty acids. The studies of Jezek et al. [37] support the hypothesis that small differences in fatty acid polarity due to the addition of a hydroxyl moiety can greatly affect the rate of translocation. However, the effect is a function of the hydroxylation site. ω-Hydroxylated fatty acids appeared to translocate slowly. In contrast, translocation of 2- and 3-hydroxy fatty acids or fatty acids with hydroxyl groups in the middle of the acyl chain is nearly as fast as their underivatized analogs, perhaps due to the formation of hydrogen bonds between the carboxyl and

Table 1
Selected summary of translocation rate constants

	Lipophile	k (s^{-1})		$\Delta G^{\ddagger}_{12}$ (kcal/mol)	Reference
		25 °C	37 °C		
EPC-SUV	$C_{8:0}$; $C_{12:0}$; $C_{16:0}$; $C_{18:0}$; $C_{18:1}$; $C_{18:0}$; $C_{18:2}$	>1			[32]
EPC-SUV	$C_{16:0}$; $C_{18:1}$	>200			[33]
Cholesterol EPC-GUV	$C_{16:0}$	0.25	0.5	18.4	[34]
	$C_{18:1}$	0.14	0.3	18.7	
	$C_{18:2}$	0.43	1.0	18.1	
Cholesterol EPC-LUV	$C_{18:1}$	4.5	6		
	$C_{18:2}$	10	30		
Erythrocyte ghosts	$C_{16:0}$	0.40	0.6	18.1	[35]
	$C_{18:1}$	0.39	0.7	18.1	
	$C_{18:2}$	1.7^{b}	3.0	17.3	
EPC-LUV	$C_{12:0}$; $C_{16:0}$; $C_{18:0}$	~15			[33]
EPC-SUV	Pyrene nonanoic acid	>1			[12,32]
EPC-SUV	5-doxyl stearic acid	>1			[32]
EPC-SUV	5-deoxy cholic acid	>1			
EPC-SUV	Retinoic acid	>1			
EPC-SUV	Chenodeoxy cholic acid	>1			
EPC-SUV	Cholic acid	6×10^{-2} $(3 \times 10^{-2})^{b}$			
EPC-SUV	Taurochenodeoxycholic, taurodeoxycholic, and taurocholic acids	$\leq 3 \times 10^{-4}$			
EPC-LUV	12-AS	5×10^{-4}	2×10^{-3}	22	[26]
EPC-SUV	12-AS	5×10^{-3}	2×10^{-2}	21	
EPC-SUV	12-AS	5×10^{-3}			[32]
EPC-SUV	2-AP	2×10^{-3}			[36]
EPC-LUV	2-AP	3×10^{-3}			[34]
EPC-SUV	11-AUD	1×10^{0}			

[a]*n*:*m* where *n* is the number of carbons and *m* is the number of double bonds in the fatty acid.
[b]faster rate constants have been measured by NMR spectroscopy [64].

hydroxyl moieties in the former case and shielding by adjacent methylene groups in the latter.

Several fluorescence-based methods have been used to measure the transfer of natural fatty acids across membranes. One of these is based on the change of fluorescence of acrylodan-labeled fatty acid binding protein (ADIFAB), which exhibits increased fluorescence intensity upon binding fatty acids. The other is pyranine, a dye whose fluorescence quantum yield increases linearly with increasing pH between 7 and 8. After adsorption to the outer leaflet of phospholipid vesicles, fatty acids translocate to the opposing leaflet in their un-ionized. To reach ionization equilibrium in the interface, a large fraction (~50%) of the translocated fatty acids rapidly releases protons into the contained volume. Thus, flip-flop across the bilayer can be measured by following the change in the internal pH with pyranine that occurs with an increase in the contained proton concentration. In addition, when the fatty acids arrive in the inner leaflet, the fatty acids also equilibrate between the membrane and the aqueous phase. This second step requires desorption of the fatty acid into the contained volume. The fraction of fatty acids in the water phase is very small but can be measured by the increase in ADIFAB fluorescence that occurs upon its binding to fatty acids. The rate of proton transfer from the fatty acid on the inner leaflet to pyranine can be assumed to be very fast, and the change in pyranine fluorescence reflects the translocation of fatty acids from the outer to the inner leaflet. In contrast, changes in ADIFAB fluorescence require both translocation across the bilayer and desorption of fatty acids from the inner leaflet into the contained volume. Even though the association of fatty acid with ADIFAB is close to diffusion controlled, the rate of change of ADIFAB fluorescence could reflect either translocation or desorption, whichever is slowest. The dependence of desorption rate on the chain length and unsaturation of fatty acids is known [Eqs. 3(a) and (3b)], and could be on the order of minutes for some very long chain fatty acids.

With improved methodologies, it became clear that the measured rates of translocation of natural fatty acids are much faster than those of some fluorescent fatty acid analogs. The first results from the pyranine assay [32] measured an upper limit of 1 s (the time resolution of the fluorescence measurement) for the translocation of several fatty acids in small SBV at 25 °C. Subsequent measurements using stopped-flow methods and pyranine yielded rate constants of $>200\ s^{-1}$ [33,34]. Using entrapped ADIFAB, Kleinfeld et al. [35] calculated rate constants of 6 and 30 s^{-1} for flip-flop of the 18-carbon fatty acids oleate and linoleate in cholesterol–PC large unilamellar vesicles at 37 °C but shorter rate constants for flip-flop in "giant" unilamellar cholesterol–PC vesicles (Table 1). Although it is difficult to compare rate constants measured in different laboratories using different probes, the differences in rate constants for translocation in small [33], large, and giant SBVs [34] are sufficiently large to question whether the radius of curvature of the phospholipid bilayer affects fatty acid flip-flop rates.

Measurements of translocation rate constants for long chain fatty acids in planar bilayers have given rate constants between 0.02 and 0.03 s^{-1} [38], but these experiments have low-time resolution and they reveal only a lower limit. In addition, measurements of translocation using ADIFAB must be viewed with the limitation that they are model dependent on a fluorescence ratio and the on and off rates, and are not as direct as those employing pyranine. Although the issue of curvature effects on flip-flop of fatty acids across model membranes remains unresolved, recent studies have shown very fast flip-flop of fatty acids in natural membranes and in cells (see below).

Transport of lipids into a cell by diffusion can be described by a model of the transport of two different monoacyl lipids, palmitic acid and lyso decanoyl PC, from the extracellular compartment into the cell cytoplasm (Fig. 4a) and the associated free energy changes (Fig. 4b). The first step, association with and insertion into the extracellular leaflet of the plasma membrane, is very fast and nearly diffusion controlled for both the fatty acid and the lyso PC. However, the rates and attendant-free energies of activation ($\Delta G^{\ddagger}_{\mathrm{t}1}$) for the second step, translocation from the outer to inner leaflet, are different. The slow rate of translocation of lyso PC has been attributed to the insolubility of its highly polar zwitterionic head group in hydrocarbon-like environments. In contrast, the fatty acid, which in its protonated form carries no charge, is soluble in hydrocarbon-like environments and moves rapidly to the opposite leaflet with little free energy cost. At

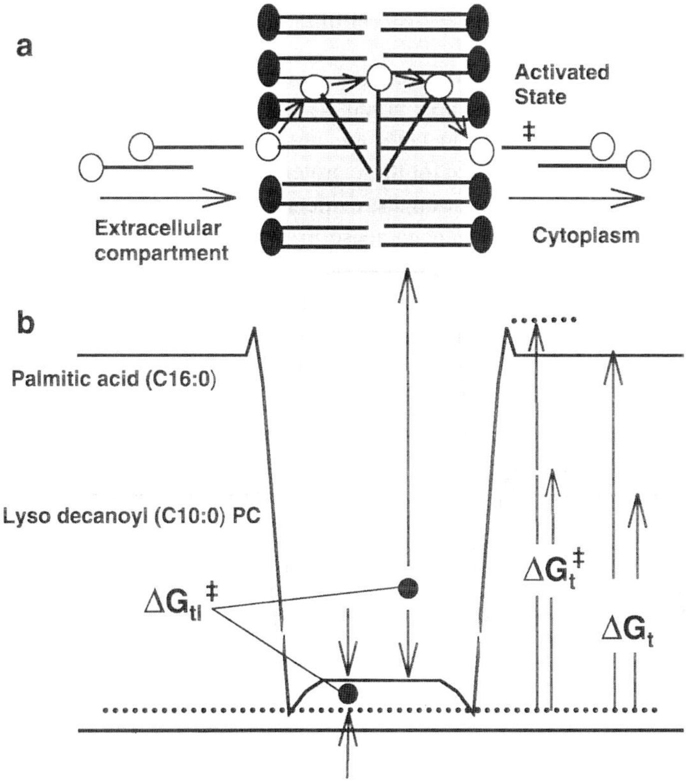

Fig. 4. Mechanism and energetics of lipid transport into a cell. (a) Schematic representation of the transfer of a monoacyl amphiphile from the aqueous phase of the extracellular space to that of the cytoplasm. (b) The reaction coordinates with attendant changes in free energy for transfer of palmitic acid (black line) and lyso PC (gray line) according to the scheme shown above in (a). $\Delta G_{tl}^{\ddagger}$ and $\Delta G_{t}^{\ddagger}$ are the respective free energies of activation for translocation across the hydrocarbon region of the membrane bilayer and for the desorption of the fatty acid from the inner leaflet into the cytoplasm. ΔG_{t} is the free energy of transfer from the inner leaflet to the cytoplasm based on equilibrium distribution measurements.

the intracellular leaflet, the fatty acid and lyso PC desorb into the aqueous phase of the cytoplasm. The free energies of activation for this process ($\Delta G_{t}^{\ddagger}$) are different for palmitic acid and lyso decanoyl PC (Fig. 4b) only because they have different acyl chain lengths, 16 and 10 carbons, respectively. A considerable body of data has shown that the major determinant of the desorption step is acyl chain length; e.g. lyso PCs and fatty acids with the same chain length exhibit similar desorption rates. The model in Fig. 4 helps us to identify the properties of lipids that determine the rate-limiting step for the transport of lipids from the aqueous phase of the extracellular compartment to that of the cytoplasm. For highly polar species, translocation from the outer to inner leaflet of the bilayer is rate limiting. However, for non-polar species this step is likely to be fast, and the final step, desorption from the inner leaflet to the cytoplasm can be rate limiting, particularly for very long chain fatty acids.

6. Translocation of fatty acids across cellular membranes

While some investigators have provided evidence for carrier-free diffusion of fatty acids across membranes, others have proposed that specific fatty acid transporters in the plasma membrane of mammalian cells are required to move fatty acids through the membrane into cells. The evidence for these proteins acting as translocases or permeases is indirect and circumstantial. The strongest evidence for a protein-mediated process would be observation of a dose-dependent increase in fatty acid translocation in phospholipid vesicles reconstituted with a protein that is hypothesized to transport fatty acids across the lipid bilayer. However, even this would not necessarily ensure specificity of fatty acid transport, as proteins could promote diffusion of various molecules by their effects on cell membrane porosity. Although there is dispute about the mechanism by which putative protein transporters of fatty acids affect fatty acid translocation, under some experimental conditions they have an effect on fatty acid influx into cells. That effect is generally seen as an appearance or increase in a component of fatty acid uptake exhibiting saturable kinetics, or an increase in cellular fatty acid when the protein is overexpressed. Three of the prominent putative transporters are discussed below.

6.1. Plasma membrane fatty acid binding protein ($FABP_{pm}$)

First proposed as a transporter of fatty acid anions in the plasma membrane, $FABP_{pm}$, is now known to share identity with the previously reported protein, aspartate aminotransferase; antibodies against aspartate aminotransferase partially inhibited cellular fatty acid uptake [39]. Differentiation of 3T3-L1 cells into adipocytes was associated with increased $FABP_{pm}$ expression [40], and overexpression of $FABP_{pm}$ in 3T3 fibroblasts was associated with increased fatty acid uptake [41]. The same results would be replicated by a variety of proteins, including intracellular FABP and enzymes for adipogenesis.

6.2. Fatty acid transporter protein (FATP)

The most recently proposed protein for the translocation of fatty acids across the plasma membrane, FATP was found by expression cloning that selected cells with enhanced uptake of a fluorescent fatty acid [42]. Moreover, cDNAs for two proteins were found, one a previously identified fatty acyl-CoA synthase. The other cDNA encoded a novel 646 protein with six predicted membrane-spanning regions. Mature adipocytes have 5–7 times more FATP mRNA than fibroblastic precursors, and in fully differentiated 3T3-L1 adipocytes, insulin alone down-regulates FATP mRNA levels [43]. FATP is a highly conserved protein that occurs in many species in a tissue specific way [44]. However, several studies have shed doubt on the proposed activity of this protein. Coe et al. [45] noted that the primary sequence of the murine fatty acid transport protein (FATP1) is very similar to the multigene family of very long chain (C20–C26) acyl-CoA synthetases. This group expressed a FATP1-Myc/His fusion protein in COS1 cells and measured synthase activity. The FATP1-Myc/His forms were distributed between the plasma membrane and intracellular membranes, and cells expressing FATP1-Myc/His exhibited a threefold increase in the very long chain acyl-CoA synthase activity. These results suggest that FATP1 is a very long chain acyl-CoA synthetase that facilitates fatty acid uptake via esterification, and that the mechanism for enhanced uptake would be the support of a concentration gradient between intracellular and extracellular fatty acid.

6.3. Fatty acid translocase (FAT)/CD36

CD36 was first identified on platelets as a receptor for collagen and thrombospondin, and later associated with fatty acid transport by a variety of experimental approaches. For example, differences in FAT/CD36 expression are associated with changes in cellular fatty acid uptake, and CD36 null mice exhibit increased plasma-free fatty acid levels [46]. Although uptake of [^{3}H] oleate was reduced in adipocytes from null mice, the decrease was limited to a low ratio of fatty acid complexed with bovine serum albumin (0.5); no differences were seen at a ratio of 1.5. This result was interpreted to indicate that CD36 was necessary for a high affinity component of the uptake process under conditions where diffusion is assumed to be ineffective, as has been proposed for $FATP_{pm}$. In contrast, CD36 overexpression targeted to skeletal muscle in mice was associated with decreased adiposity and with lower levels of fatty acid and TG in plasma, the latter a likely response to lower levels of plasma fatty acid for hepatic extraction [47]. As with many other proteins for adipogenesis, FAT/CD36 is increased during differentiation of preadipocytes into fat cells [48–50]. In addition, insulin stimulates the movement of CD36 from intracellular pools to the plasma membrane of cardiac myocytes [51].

6.4. Mechanisms for fatty acid translocation

As is apparent from the above discussions, there is currently no universally accepted model for translocation of fatty acids across cellular membranes. Although membrane proteins have been implicated in the transport of fatty acids into cells, except for FATP the mechanisms by which they do this remains undefined. According to newer biophysical studies, fatty acids diffuse by the flip-flop of the un-ionized fatty acid across models of cellular membranes with rate constants that could be as low as $0.3\ s^{-1}$ for giant PC vesicles with cholesterol and as high as $200\ s^{-1}$ for small PC vesicles without cholesterol. These rates of movement through the membrane are fast enough to support intracellular metabolism, which occurs on a time frame of minutes. Although one could argue that a cell membrane is more like the former, the radius of curvature and the cholesterol composition of the cellular translocation sites are not known. These could be cholesterol-rich and as highly curved as the caveolae, which have radius of curvature similar to those of small vesicles [52], or they could be relatively flat surfaces similar to those modeled by giant vesicles. Irrespective of the membrane model, all biophysical studies have shown only a single component for the actual translocation step, even though the succeeding desorption could be faster or slower than translocation.

The transmembrane movement of fatty acids across the plasma membrane of living cells has recently been shown to be fast. The combined steps of adsorption and transmembrane movement are complete within 5 s in adipocytes [53], HepG2 cells, and preadipocytes [54], as detected by acidification of the cytosol. The partition coefficient between the aqueous phase and the plasma membrane of isolated rat adipocytes is the same as that for oleic acid and protein-free PC SBV [53]. At low concentrations of oleic acid, presented to cells without albumin, there is no saturation of binding. Putative inhibitors of transport of fatty acids in membranes such as phloretin have been shown to affect metabolism of fatty acids in HepG2 cells but not membrane transport [54]. Taken together, the new results are strong evidence for fast diffusion of fatty acids into cells. Evidence for diffusion of fatty acids through the plasma membrane of cells from earlier studies has been discussed in detail in previous reviews [55–59].

The proposed mechanism for the translocation of fatty acids is the flip-flop of the un-ionized fatty acid molecule, for which there is a low activation energy for passage through the hydrocarbon interior. In protein-free phospholipid bilayers, this probably occurs in regions of free volume/defects that exist even in a pure lipid phase. In a cell membrane these defects are expected to increase rather than decrease; thus, fatty acids could diffuse rapidly through lipid bilayer even in the presence of proteins, and without catalysis of specific transporters. It is still valid to consider the heterogeneity of the plasma membrane, and to consider whether there are regions where proteins might perform some direct or indirect function to enhance fatty movement in these regions.

Cell studies of fatty acid uptake often reveal a saturable component, which appears to be consistent with a distinct cellular site or protein, and a non-saturable component, which appears to be consistent with carrier-free diffusion. The identity of the site or protein responsible for the saturable component could be FAT/CD36, FATP, or $FABP_{pm}$. However, neither the form of the uptake curve nor the enhancement of fatty uptake by overexpression establishes a mechanism. Moreover, these proteins have not been shown to be fatty acid transporters in an isolated system such as SBVs, and their direct involvement in transport remains unproven. It is possible that they elicit their effects indirectly by changing the quality of the bilayer to enhance diffusion or by increasing the rate of activation of fatty acids by acting as a plasma membrane acyl-CoA synthase. Kinetic studies at very high pressures are consistent PC translocation through transiently formed pores [60]. However, the solubility properties of PCs differ greatly from those of fatty acids and the small number of pores present in artificial membranes would not perceptibly affect fatty acid translocation. On the other hand, proteins could have a general effect on fatty acid translocation through the formation of small pores at the interface between all proteins and phospholipids [61].

6.5. Cellular fatty acid metabolism

An important determinant of fatty acid transport is intracellular metabolism, which can support a gradient in fatty acid concentration across the plasma membrane. This gradient is supported in two ways. First, one of several acyl-CoA synthases could convert fatty acid to its acyl-CoA analog at the plasma membrane. Alternatively, fatty acids could desorb into the cytoplasm and associate with FABP or intracellular membranes. Acyl-CoA synthase activity could modify bound or unbound fatty acid at any of these sites. However, accumulation of fatty acid as acyl-CoA is limited by the sites of storage and potential detergent effects of long chain acyl-CoA. Thus, for continued fatty acid uptake, the acyl-CoA must be converted rapidly to their storage form, triacylglycerol, or channeled to mitochondria for fatty acid oxidation. If translocation across the plasma membrane is rate limiting for fatty acid uptake, increases in intracellular metabolism would not increase uptake, and overexpression of enzymes that metabolize fatty acids should not increase fatty acid uptake.

There are several published examples of how enhanced intracellular metabolism increases cellular fatty acid uptake. One example is the overexpression of acyl-CoA synthases, as reported by Schaffer and Lodish [42]; one of the two identified proteins was prematurely described as a transport protein, FATP [45]. A second example is that overexpression of enzymes in the Kennedy pathway for glycerolipid synthesis can increase cellular uptake of fatty acids and their conversion to TG. Cellular overexpression of mitochondrial glycerol-3-phosphate acyltransferase (mGPAT), the first step in glycerolipid synthesis, and acylglycerol-3-phosphate acyltransferase (AGAT) directed exogenous oleate primarily toward triacylglycerol synthesis [62,63]. These results imply that mGPAT and lysophosphatidic acid acyltransferase produce a separate pool of lysophosphatidic acid and phosphatidic acid that is transported to the endoplasmic reticulum where the remaining enzymes of triacylglycerol synthesis are located. Moreover, this pool is isolated from the pool of lysophosphatidic acid and phosphatidic acid that are precursors to phospholipid synthesis. Thus, mGPAT and AGAT overexpression "pulls" more fatty acid into the cell, while "pushing" more fatty acid toward TG synthesis. These observations support the hypothesis that fatty acid metabolism, not translocation across cell membranes, is rate limiting for uptake of fatty acids into cells.

7. Final perspectives

The fast diffusion of fatty acids across the plasma membrane of several types of living cells has been established by new fluorescence methods. Our current understanding of the basic physical chemistry of fatty acids supports these conclusions drawn from studies of cells. Numerous studies support some role for the putative fatty acid translocation proteins in energy uptake by cells. However, identification of the mechanism(s) by which these proteins promote fatty acid uptake awaits additional study. Unambiguous assignment of fatty acid transport activity to a specific protein will require reconstitution of transport activity in a system containing the pure protein in a simple assay system such as a SBV. This experiment will remove the confounding activities of cells and allow assignment of what step of membrane transport (Fig. 4) is catalyzed by the protein. In the near future the two approaches – biophysical and physiological/biological – should yield in a common model in which the results of all studies can be rationalized.

Acknowledgements

This work was supported by grants from the National Institutes of Health (HL-30914, HL-56865, HL-67188, and HL-26335).

The authors thank Paul Pilch for his helpful comments.

References

[1] Olson, A.L., Pessin, J.E., 1996. Structure, function, and regulation of the mammalian facilitative glucose transporter gene family. Annu. Rev. Nutr. 16, 235–256.

[2] Bryant, N.J., Govers, R., James, D.E., 2002. Regulated transport of the glucose transporter GLUT4. Nat. Rev. Mol. Cell. Biol. 4, 267–277.

[3] Hamilton, J.A., 1995. Studies of minor lipid constituents of membranes by high resolution and solid state 13C NMR spectroscopy: interactions with phospholipid bilayers and transport proteins. In: Beckmann, N. (Ed.), Carbon-13 Magnetic Resonance Spectroscopy of Biological Systems. Academic Press, San Diego, CA, pp. 117–151.

[4] Eigen, M., Kruse, W., Maass, G., De Maeyer, L., 1964. Rate constants of proteolytic reactions in aqueous solution. Prog. React. Kinet. 2, 285–289.

[5] Tanford, C., 1980. The Hydrophobic Effect: Formation of Micelles and Biological Membranes, 2nd ed. Wiley-Interscience, New York.

[6] Cistola, D.P., Atkinson, D., Hamilton, J.A., Small, D.M., 1986. Phase behavior and bilayer properties of fatty acids: hydrated 1:1 acid-soaps. Biochemistry 25, 2804–2812.

[7] Petitpas, I., Grune, T., Bhattacharya, A.A., Curry, S., 2001. Crystal structures of human serum albumin complexed with monounsaturated and polyunsaturated fatty acids. J. Mol. Biol. 314, 955–960.

[8] Lucke, C., Gutierrez-Gonzalez, L.H., Hamilton, J.A., 2003. Intracellular lipid binding proteins: evolution, structure, and ligand binding. In: Dutta-Roy, A., Spener, F. (Eds.), Cellular Proteins and Their Ligand Fatty Acids – Emerging Roles in Gene Expression, Health and Disease: Binding Fatty Acids. Wiley-VCH, Weinheim, Germany, pp. 95–114.

[9] Zhang, F., Kamp, F., Hamilton, J.A., 1996. Dissociation of long and very long chain fatty acids from phospholipid bilayers. Biochemistry 35, 16055–16060.

[10] Massey, J.B., Bick, D.H., Pownall, H.J., 1997. Spontaneous transfer of monoacyl amphiphiles between lipid and protein surfaces. Biophys. J. 72, 1732–1743.

[11] Pownall, H.J., Raynaud, A.S., Harper, E., Choi, S., Rohrbach, K., Pao, Q., Reeves, R.S., Gotto, A.M. Jr., 1995. Effects of twelve weeks of dietary fish oil, polyunsaturated fat, monounsaturated fat, and saturated fat on human plasma lipoprotein structure and composition. In: Pownall, H.J., Spector, A.S. (Eds.), Omega-3 Fatty Acids in Nutrition, Vascular Biology and Medicine. American Heart Association, Inc., Dallas, TX, pp. 64–78.

[12] Doody, M.C., Pownall, H.J., Kao, Y.J., Smith, L.C., 1980. Mechanism and kinetics of transfer of a fluorescent fatty acid between single-walled phosphatidylcholine vesicles. Biochemistry 19, 108–116.

[13] Pownall, H.J., Hickson, D.L., Smith, L.C., 1983. Transport of biological lipophiles: effect of lipophile structure. J. Am. Chem. Soc. 105, 2440–2445.

[14] Massey, J.B., Gotto, A.M. Jr., Pownall, H.J., 1982. Kinetics and mechanism of the spontaneous transfer of fluorescent phosphatidylcholines between apolipoprotein–phospholipid recombinants. Biochemistry 21, 3630–3636.

[15] Pownall, H.J., Bick, D.L., Massey, J.B., 1991. Spontaneous phospholipid transfer: development of a quantitative model. Biochemistry 30, 5696–5700.

[16] Smith, R., Tanford, C., 1972. The critical micelle concentration of L-α-dipalmitoylphosphatidylcholine in water and water–methanol solutions. J. Mol. Biol. 67, 75–83.

[17] Massey, J.B., Gotto, A.M. Jr., Pownall, H.J., 1982. Kinetics and mechanism of the spontaneous transfer of fluorescent phospholipids between apolipoprotein–phospholipid recombinants. Effect of the polar headgroup. J. Biol. Chem. 257, 5444–5448.

[18] Deckelbaum, R.J., Granot, E., Oschry, Y., Rose, L., Eisenberg, S., 1984. Plasma triglyceride determines structure–composition in low and high density lipoproteins. Arteriosclerosis 4, 225–231.

[19] Pownall, H.J., Brauchi, D., Kilinc, C., Osmundsen, K., Pao, Q., Payton-Ross, C., Gotto, A.M. Jr., Ballantyne, C.M., 1999. Correlation of serum triglyceride and its reduction by omega-3 fatty acids with lipid transfer activity and the neutral lipid compositions of high-density and low-density lipoproteins. Atherosclerosis 143, 285–297.

[20] McKeone, B.J., Massey, J.B., Knapp, R.D., Pownall, H.J., 1988. Apolipoproteins C-I, C-II, and C-III: kinetics of association with model membranes and intermembrane transfer. Biochemistry 27, 4500–4505.

[21] Hickson-Bick, D., Knapp, R.D., Sparrow, J.T., Sparrow, D.A., Gotto, A.M. Jr., Massey, J.B., Pownall, H.J., 1988. Kinetics and mechanism of transfer of synthetic model apolipoproteins. Biochemistry 27, 7881–7886.
[22] Bresseleers, G.J., Goderis, H.L., Tobback, P.P., 1984. Measurement of the glucose permeation rate across phospholipid bilayers using small unilamellar vesicles. Effect of membrane composition and temperature. Biochim. Biophys. Acta 772, 374–382.
[23] Naylor, B.L., Picardo, M., Homan, R., Pownall, H.J., 1991. Effects of fluorophore structure and hydrophobicity on the uptake and metabolism of fluorescent lipid analogs. Chem. Phys. Lipids 58, 111–119.
[24] Nichols, J.W., Pagano, R.E., 1981. Kinetics of soluble lipid monomer diffusion between vesicles. Biochemistry 20, 2783–2789.
[25] Pownall, H.J., Hickson, D.L., Smith, L.C., 1983. Transport of biological lipophiles: effect of lipophile structure. J. Am. Chem. Soc. 105, 2440–2445.
[26] Kleinfeld, A.M., Storch, J., 1993. Transfer of long-chain fluorescent fatty acids between small and large unilamellar vesicles. Biochemistry 32, 2053–2061.
[27] Diamond, J.M., Wright, E.M., 1969. Biological membranes: the physical basis of ion and nonelectrolyte selectivity. Annu. Rev. Physiol. 31, 581–646.
[28] Homan, R., Pownall, H.J., 1988. Transbilayer diffusion of phospholipids: dependence on headgroup structure and acyl chain length. Biochim. Biophys. Acta 938, 155–166.
[29] Hamilton, J.A., Bhamidipati, S.P., Kodali, D.R., Small, D.M., 1991. The interfacial conformation and transbilayer movement of diacylglycerols in phospholipid bilayers. J. Biol. Chem. 266, 1177–1186.
[30] Hope, M.J., Cullis, P.R., 1987. Lipid asymmetry induced by transmembrane pH gradients in large unilamellar vesicles. J. Biol. Chem. 262, 4360–4366.
[31] Bai, J., Pagano, R.E., 1997. Measurement of spontaneous transfer and transbilayer movement of BODIPY-labeled lipids in lipid vesicles. Biochemistry 36, 8840–8848.
[32] Kamp, F., Westerhoff, H.V., Hamilton, J.A., 1993. Movement of fatty acids, fatty acid analogues, and bile acids across phospholipid bilayers. Biochemistry 32, 11074–11086.
[33] Kamp, F., Zakim, D., Zhang, F., Noy, N., Hamilton, J.A., 1995. Fatty acid flip-flop in phospholipid bilayers is extremely fast. Biochemistry 34, 11928–11937.
[34] Kleinfeld, A.M., Chu, P., Storch, J., 1997a. Flip-flop is slow and rate limiting for the movement of long chain anthroyloxy fatty acids across lipid vesicles. Biochemistry 36, 5702–5711.
[35] Kleinfeld, A.M., Chu, P., Romero, C., 1997b. Transport of long-chain native fatty acids across lipid bilayer membranes indicates that transbilayer flip-flop is rate limiting. Biochemistry 36, 14146–14158.
[36] Storch, J., Kleinfeld, A.M., 1986. Transfer of long-chain fluorescent free fatty acids between unilamellar vesicles. Biochemistry 25, 1717–1726.
[37] Jezek, P., Modriansky, M., Garlid, K.D., 1997. Inactive fatty acids are unable to flip-flop across the lipid bilayer. FEBS Lett. 408, 161–165.
[38] Pohl, E.E., Peterson, U., Sun, J., Pohl, P., 2000. Changes of intrinsic membrane potentials induced by flip-flop of long-chain fatty acids. Biochemistry 39, 1834–1839.
[39] Isola, L.M., Zhou, S.L., Kiang, C.L., Stump, D.D., Bradbury, M.W., Berk, P.D., 1995. 3T3 fibroblasts transfected with a cDNA for mitochondrial aspartate aminotransferase express plasma membrane fatty acid-binding protein and saturable fatty acid uptake. Proc. Natl Acad. Sci. USA 92, 9866–9870.
[40] Zhou, S.L., Stump, D., Sorrentino, D., Potter, B.J., Berk, P.D., 1992. Adipocyte differentiation of 3T3-L1 cells involves augmented expression of a 43-kDa plasma membrane fatty acid-binding protein. J. Biol. Chem. 267, 14456–14461.
[41] Zhou, S.L., Stump, D., Kiang, C.L., Isola, L.M., Berk, P.D., 1995. Mitochondrial aspartate aminotransferase expressed on the surface of 3T3-L1 adipocytes mediates saturable fatty acid uptake. Proc. Soc. Exp. Biol. Med. 208, 263–270.
[42] Schaffer, J.E., Lodish, H.F., 1994. Expression cloning and characterization of a novel adipocyte long chain fatty acid transport protein. Cell 79, 427–436.
[43] Man, M.Z., Hui, T.Y., Schaffer, J.E., Lodish, H.F., Bernlohr, D.A., 1996. Regulation of the murine adipocyte fatty acid transporter gene by insulin. Mol. Endocrinol. 10, 1021–1028.
[44] Hirsch, D., Stahl, A., Lodish, H.F., 1998. A family of fatty acid transporters conserved from mycobacterium to man. Proc. Natl Acad. Sci. USA 95, 8625–8629.

[45] Coe, N.R., Smith, A.J., Frohnert, B.I., Watkins, P.A., Bernlohr, D.A., 1999. The fatty acid transport protein (FATP1) is a very long chain acyl-CoA synthetase. J. Biol. Chem. 274, 36300–36304.

[46] Febbraio, M., Abumrad, N.A., Hajjar, D.P., Sharma, K., Cheng, W., Pearce, S.F., Silverstein, R.L., 1999. A null mutation in murine CD36 reveals an important role in fatty acid and lipoprotein metabolism. J. Biol. Chem. 274, 19055–19062.

[47] Ibrahimi, A., Bonen, A., Blinn, W.D., Hajri, T., Li, X., Zhong, K., Cameron, R., Abumrad, N.A., 1999. Muscle-specific overexpression of FAT/CD36 enhances fatty acid oxidation by contracting muscle, reduces plasma triglycerides and fatty acids, and increases plasma glucose and insulin. J. Biol. Chem. 274, 26761–26766.

[48] Abumrad, N.A., el-Maghrabi, M.R., Amri, E.Z., Lopez, E., Grimaldi, P.A., 1993. Cloning of a rat adipocyte membrane protein implicated in binding or transport of long-chain fatty acids that is induced during preadipocyte differentiation. Homology with human CD36. J. Biol. Chem. 268, 17665–17668.

[49] Amri, E.Z., Ailhaud, G., Grimaldi, P.A., 1994. Fatty acids as signal transducing molecules: involvement in the differentiation of preadipose to adipose cells. J. Lipid Res. 35, 930–937.

[50] Sfeir, Z., Ibrahimi, A., Amri, E., Grimaldi, P., Abumrad, N., 1997. Regulation of FAT/CD36 gene expression: further evidence in support of a role of the protein in fatty acid binding/transport. Prostaglandins Leukot. Essent. Fatty Acids 57, 17–21.

[51] Luiken, J.J., Koonen, D.P., Willems, J., Zorzano, A., Becker, C., Fischer, Y., Tandon, N.N., Van Der Vusse, G.J., Bonen, A., Glatz, J.F., 2002. Insulin stimulates long-chain fatty acid utilization by rat cardiac myocytes through cellular redistribution of FAT/CD36. Diabetes 51, 3113–3119.

[52] Meyer, H.W., Westermann, M., Stumpf, M., Richter, W., Ulrich, A.S., Hoischen, C., 1998. Minimal radius of curvature of lipid bilayers in the gel phase state corresponds to the dimension of biomembrane structures "caveolae". J. Struct. Biol. 124, 77–87.

[53] Kamp, F., Guo, W., Souto, R., Pilch, P.F., Corkey, B.E., Hamilton, J.A., 2003. Rapid flip-flop of oleic acid across the plasma membrane of adipocytes. J. Biol. Chem. 278, 7988–7995.

[54] Hamilton, J.A., Guo, W., 2001. Fatty acid uptake and metabolism in HepG2 cells. Biophys. J. (Annual Meeting Abstracts) 80, 365–366.

[55] Zakim, D., 2000. Thermodynamics of fatty acid transfer. J. Membr. Biol. 176, 101–109.

[56] Hamilton, J.A., 1998. Fatty acid transport: difficult or easy? J. Lipid Res. 39, 467–481.

[57] Hamilton, J.A., Kamp, F., 1999. How are free fatty acids transported in membranes? Is it by proteins or by free diffusion through the lipids? Diabetes 48, 2255–2269.

[58] Hamilton, J.A., Johnson, R.A., Corkey, B.E., Kamp, F., 2001. Fatty acid transport: the diffusion mechanism in model and biological membranes. J. Mol. Neurosci. 16, 99–108.

[59] Pownall, H.J., Hamilton, J.A., 2003. Fatty acid transport. Acta Physiol. Scand. 178(4), 357–365.

[60] Homan, R., Pownall, H.J., 1987. Effect of pressure on phospholipid translocation in lipid bilayers. J. Am. Chem. Soc. 109, 4759–4760.

[61] Bojesen, I.N., Bojesen, E., 1998. Nature of the elements transporting long-chain fatty acids through the red cell membrane. J. Membr. Biol. 163, 169–181.

[62] Igal, R.A., Wang, S., Gonzalez-Baró, M., Coleman, R.A., 2001. Mitochondrial glycerol phosphate acyltransferase directs incorporation of exogenous fatty acids into triacylglycerol. J. Biol. Chem. 276, 42205–42212.

[63] Ruan, H., Pownall, H.J., 2001. Overexpression of 1-acyl-glycerol-3-phosphate acyltransferase-alpha enhances lipid storage in cellular models of adipose tissue and skeletal muscle. Diabetes 50, 233–240.

[64] Cabral, D.J., Small, D.M., Lilly, H.S., Hamilton, J.A., 1987. Transbilayer movement of bile acids in model membranes. Biochemistry 26(7), 1801–1804.

Computational modeling of cardiac fatty acid uptake and utilization

Mark W.J.M. Musters,[a] Jim B. Bassingthwaighte,[b]
Virjanand Panday,[a] Natal A.W. van Riel[a] and Ger J. van der Vusse[c,*]

[a]*Department of Biomedical Engineering, Eindhoven University of Technology, Eindhoven, The Netherlands*
[b]*Department of Bioengineering, University of Washington, Seattle, WA, USA*
[c]*Department of Physiology, Cardiovascular Research Institute Maastricht (CARIM), Maastricht University, P.O. Box 616, 6200 MD Maastricht, The Netherlands*
[*]*Correspondence address: Tel.: +31-43-3881086; fax: +31-43-3884166*
E-mail: vandervusse@fys.unimaas.nl(G.J. van der Vusse)

1. Introduction

This chapter shows how computer modeling may help in understanding the processes of uptake of fatty acids (FAs), the main source of energy under normal conditions [9], in heart. FA transport from the blood into the heart muscle cells (cardiomyocytes) is diagrammed in a general way in Fig. 1. There is controversy over the mechanistic details, e.g.: is a special protein required to facilitate the FA transport across a membrane or does transmembrane diffusion of FAs suffice? Quantitative computer models incorporating the various possible processes might help to determine which ones best fit with experimental observations. These models are necessarily simplified representations of the physiology in terms of mathematical equations and define a scientific hypothesis explicitly. Model solutions may lead to inspection of the hypothesis if they do not fit the latter. Besides defining a hypothesis, computer models are used also to optimize experimental design. Improving such models to approximate the physiology better is useful if precise experimental data can be obtained. Technology today makes it possible to perform better experiments, deducing better models and improving understanding.

1.1. Benefits and drawbacks of modeling

A computer model has the advantage that it takes relatively little time to get results [32]. In general, physiological experiments can consume a lot of time due to preparation

Advances in Molecular and Cell Biology, Vol. 33, pages 173–221

ISSN: 1569-2558 / DOI: 10.1016/S1569-2558(03)33010-3

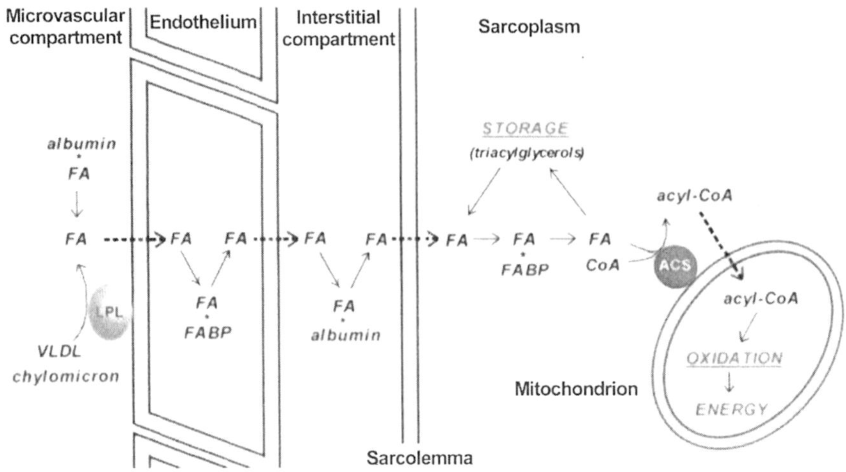

Fig. 1. A simplified representation of the cardiac transport route of FAs. FAs are present in the capillary (microvascular compartment), mostly complexed to albumin or in esterified form as triacylglycerols in VLDLs and chylomicrons. FAs are transported across the endothelial cytoplasm and interstitial compartment (interstitium), where FAs bind to FABP and albumin, respectively. From the interstitium, FAs move across the sarcolemma into the sarcoplasm of the cardiomyocyte. Subsequently, FAs are either stored in the cardiomyocyte (triacylglycerols) or metabolized in the mitochondria, which results in energy production for the heart. (*For a colored version of this figure, see plate section, page 589.*)

and conduction. As a consequence, computer modeling is in general cheaper. Besides a computer and a mathematical package, other expensive experimental materials are not needed to perform computer simulations. Simulations can also be repeated with different parameter values very often, while the number of experiments that can be performed in real life is limited. A fourth advantage of modeling is the integration of heterologous information. In most cases, a physiological model contains data of different sources. Therefore, a computer model provides a quantitatively consistent description and reveals if the physiological theory or data of this system is ill defined. A fifth advantage of a computer model is that a number of laboratory animals that have to be included into the study can be reduced, because the results of the model lead to better experimental design. Unfortunately, the in silico approach has one major drawback: a model is a simplified representation of the reality. Every theory and related model contain a number of assumptions that are supposed to have little influence on the overall results. Sometimes, the implications of these assumptions are underestimated, which could substantially influence the end results, leading to erroneous conclusions.

2. Physiological background

The heart requires energy, which is obtained from oxidizable substrates such as glucose, FAs, lactate, pyruvate, etc. Under normal conditions, about 70% is derived from mitochondrial oxidation of FAs [9]. The transport route of FAs into the cardiomyocyte starts in the blood plasma of the capillaries (Fig. 1). In plasma, more than 99.9% of the unesterified FAs are bound to the protein albumin [55]. Another source of fatty acyl chains are triacylglycerol-containing lipoprotein particles, from which the enzyme lipoprotein lipase releases FAs. The next step is FA transport across the luminal endothelial membrane into the cytoplasm of the endothelial cells. After that, FAs diffuse to the other side of the endothelial membrane: the abluminal membrane. The protein fatty acid-binding protein (FABP) could also be involved in the FA transport across the endothelial cytoplasm. The abluminal membrane is the barrier that FAs have to overcome before entering the interstitium. In this compartment, albumin is also present, which binds to the FAs. Again, diffusion processes of FAs are the main transport mechanism to reach the third membrane of the transport route: the sarcolemma. FAs are moved across this membrane to enter the cytoplasm of the cardiomyocyte (sarcoplasm). FABP in the sarcoplasm binds to FAs and facilitates transport before FAs are converted to acyl-CoA. The acyl chain of acyl-CoA can be stored, e.g. in triacylglycerol pools, or is oxidized inside the mitochondria. Under normal physiological conditions, oxidation prevails over storage. Acyl-CoA is converted stepwise into acetyl-CoA and thereafter oxidized to CO_2 and H_2O to release chemical-bound energy in the form of ATP.

2.1. Blood-borne fatty acids

FAs are present in the blood plasma of the capillaries in two different chemical forms [70]:

- Unesterified FAs. These are present as acids and salts, and mainly bound to albumin. The concentration range of these FAs lies between 0.2 and 1.0 mM under normal physiological conditions [70]. The long non-polar tail makes FA less soluble in an aqueous environment. Albumin is also present in blood plasma with a concentration of ~0.6 mM [70]. Albumin has a high affinity for FA and can bind a number of FAs per albumin molecule. Although 8 and 12 binding sites per albumin molecules have been reported [4,62], up to 3 binding sites for FAs per albumin molecule are physiologically relevant. The affinity constants for the first 3 binding sites of FA to human serum albumin (HSA), which represent the ratio FA bound to HSA and unbound FA (uFA), are 1.45×10^8, 1.30×10^8 and $0.71 \times 10^8\ M^{-1}$ for the 1st (complex of albumin with 1 FA: AF_1), 2nd (2 FAs bound to albumin: AF_2) and 3rd binding site (3 FAs bound to albumin: AF_3), respectively [55]. Moreover, it was shown [55] that due to the strong binding of FAs to albumin, only small concentrations of uFAs are present, ranging from 1.4 to 10.5 nM uFA [55,70], i.e. about 0.001% of the total amount of unesterified FAs present in plasma.
- Esterified FAs, like mono-, di- and triacylglycerols, phospholipids and cholesteryl esters. Very low-density lipoproteins (VLDLs) and chylomicrons, produced in the parenchymal liver cells and in the epithelial cells of the small intestine, respectively, are droplets of triacylglycerols, surrounded by a hydrophilic layer of phospholipids, cholesterol and apoproteins. Lipoprotein lipase (LPL), localized in the glycocalyx of the cellular membrane, catalyses the release of FAs from

chylomicrons and VLDLs by breaching the phospholipid coating and facilitating the hydrolysis of the triacylglycerol core. Although VLDLs and chylomicrons contain sufficient amounts of FA to theoretically fully cover the energy needs of the heart, its contribution under normal situations will not exceed 20–25% at maximum of the total amount of FAs utilized [76].

The corollary is that the normal heart relies heavily on blood-borne unesterified FAs as energy source. The utilization of these substrates is therefore the subject of the present chapter.

2.2. *Transport routes of unesterified fatty acids: qualitative and quantitative considerations*

Blood-borne FAs, either free or bound to albumin, have to cross a distinct number of barriers (Fig. 1), like membranes and aqueous compartments, before they reach the mitochondria inside the cardiomyocytes (Fig. 2). Under physiological conditions, FAs are taken up at a constant rate. Between 40 and 60% of the FAs present in the capillaries [38, 51,78,79] are absorbed by the endothelium during a capillary transit time for blood plasma, which lasts about ~0.8 s [70]. The amount of FAs (in moles) that is transported across a certain cross-section is represented by the so-called FA flux. An average concentration of FAs under normal conditions is ~0.5 mM at an average plasma flow in the capillaries of $0.4 \times 10^{-3}\,\mathrm{l\,min^{-1}\,(g\,ww)^{-1}}$ [1][75]. This leads to a flux of about $200\,\mathrm{nmol\,min^{-1}\,(g\,ww)^{-1}}$ into the microvascular compartment. Since about 40–60% of this flux is transported across the endothelium of an intact, beating heart, the FA flux into

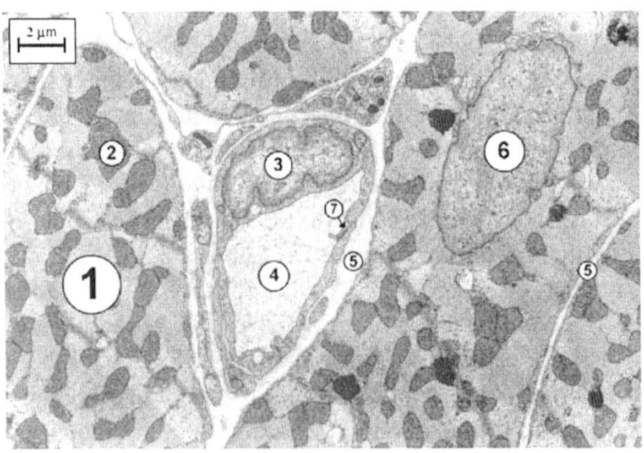

Fig. 2. Electron micrograph of the orientation of a capillary, which is surrounded by cardiomyocytes. The photograph shows (1) a cardiomyocyte, (2) mitochondria inside the cardiomyocyte, (3) the nucleus of an endothelial cell, (4) the microvascular compartment, (5) the interstitial compartment, (6) the nucleus of a cardiomyocyte and (7) a cleft between two adjacent endothelial cells. (Picture taken from Ref. [10].) (*For a colored version of this figure, see plate section, page 590.*)

[1] $(\mathrm{g\,ww})^{-1}$ stands for gram wet weight *cardiac tissue*, unless indicated otherwise.

the cardiomyocyte becomes on the order of 80–120 nmol min^{-1} $(g\ ww)^{-1}$ under normal conditions, i.e. ~100 nmol min^{-1} $(g\ ww)^{-1}$ on average.

2.2.1. *The microvascular compartment*

The capillaries, with an average diameter of 5 μm, are filled with blood plasma, containing oxidizable substrates and binding proteins like FAs and albumin, respectively. In general, the diffusion flux, J, of a substance with concentration C is calculated as follows:

$$J = -DS\frac{\partial C}{\partial x} \tag{1}$$

where D is the diffusion coefficient, S is the surface area of the crossing substrate and $\partial C/\partial x$ is the first position-dependent derivative of substance concentration C (see also Section 3.1.2), reflecting the difference in concentration C (∂C) across a given distance (∂x). Before passing the endothelial barrier, FAs diffuse from the lumen of the capillary to the luminal membrane of the endothelial cells, lining the capillary wall. In the capillary lumen, about 5 nM of uFA and 0.5 mM of the total concentration of FAs in the blood plasma are present under normal conditions [55, 75]. FAs bound to albumin (AF_1, AF_2 and AF_3, when a maximum of three binding sites is used) diffuse at the same speed as free albumin [6]. The maximal diffusional flux of uFAs (assuming an infinite sink for uFAs at the luminal side of the endothelium) is obtained by assuming a difference in concentration of 5 nM across a distance of 0.4 μm (length of endothelial glycocalyx on the luminal side [74]), $D = 3.0 \times 10^{-6}$ $cm^2\ s^{-1}$ and $S = 500$ cm^2 $(g\ ww)^{-1}$ (Table 1) in Eq. (1). This results in a flux J_{uFA} of ~11 nmol min^{-1} $(g\ ww)^{-1}$. This flux of uFAs alone is about 1/10 of the required flux (~100 nmol min^{-1} $(g\ ww)^{-1}$, see Section 2.2). The following calculations show the important influence of albumin-bound FAs on the total flux in case FAs bound to albumin are allowed to diffuse freely from the beginning of the endothelial glycocalyx towards the endothelial cell membrane, again assumed to be an infinite sink for albumin-bound FAs in this situation. This would theoretically increase the maximum FA flux J_{FA} to $\sim 2 \times 10^5$ nmol min^{-1} $(g\ ww)^{-1}$ (Eq. (1) and diffusion coefficient of albumin in plasma used from Table 1, calculation not shown). From a physiological point of view, this high contribution of albumin-bound FAs to the total FA flux cannot be achieved, because the luminal side of the endothelial membrane is not an infinite sink for albumin-bound FAs. Furthermore, the presence of a glycocalyx most likely delays the diffusion rate of FAs bound to albumin [74]. The glycocalyx is part of the extracellular matrix (ECM) that surrounds eukaryotic cells. This glycocalyx is a negatively charged, carbohydrate-rich coating of glycoproteins and glycolipids, which prevents the close approach of cells and macromolecules and forms a stagnant layer [6,63]. It takes albumin, a spherical, slightly negatively charged protein with a radius of ~3.5 nm [40], about 40 min to penetrate the glycocalyx with a thickness of 0.4–0.5 μm [74]. This is too slow to account for a relevant contribution of albumin-bound FAs to the total FA flux to the luminal side of the endothelial cells, since the capillary transit time lasts only 0.8 s. In other words, about 0.8 s instead of 40 min are

Table 1
Model parameters

Parameter	Specification	Value[a]	Description	Reference
$K_{a,1}$	HSA	1.45×10^8 M^{-1}	Affinity constant: uFA with BSA/HSA	[55]
	BSA	1.22×10^8 M^{-1}		
$K_{a,2}$	HSA	1.30×10^8 M^{-1}	Affinity constant: uFA with AF_1	[55]
	BSA	1.53×10^8 M^{-1}		
$K_{a,3}$	HSA	0.71×10^8 M^{-1}	Affinity constant: uFA with AF_2	[55]
	BSA	0.75×10^8 M^{-1}		
k_{-1}, k_{-2}, k_{-3}		6.9 ± 0.6 s^{-1}	Dissociation rate of AF_1, AF_2 and AF_3	[16]
k_1[b]	HSA	1.00×10^9 M^{-1} s^{-1}	Association rate: uFA with BSA/HSA	[16,55]
	BSA	8.42×10^8 M^{-1} s^{-1}		
k_2[b]	HSA	8.97×10^8 M^{-1} s^{-1}	Association rate: uFA with AF_1	[16,55]
	BSA	1.06×10^9 M^{-1} s^{-1}		
k_3[b]	HSA	4.90×10^8 M^{-1} s^{-1}	Association rate: uFA with AF_2	[16,55]
	BSA	5.18×10^8 M^{-1} s^{-1}		
K_p	Low (normal)	6.0×10^5 M^{-1} [L]	Partition coefficient of uFA in lipid/water phase	[27,31]
	High (enhancer)	1.8×10^6 M^{-1} [L]		
k_{off}		0.8 ± 0.1 s^{-1}	Desorption rate of uFA from membrane	[31]
		6.0 s^{-1}		[23]
k_{on}[b]	Low	4.8×10^5 s^{-1}	Absorption rate of uFA in membrane	[31]
		3.6×10^6 s^{-1}		[23]
$k_{on}^{enhancer}$[b]	High	1.4×10^6 s^{-1}		[31]
		1.1×10^7 s^{-1}		[23]
k_{flip}	GUV	0.5 ± 0.075 s^{-1}	Flip rate of uFA from outer to inner leaflet of membrane	[31]
		10 s^{-1}		[31] + adapted
	LUV	20 s^{-1}		[30]
	SUV	70 s^{-1}		[30]

K_a^{FABP}		$9.8 \times 10^7\ M^{-1}$	Affinity constant: uFA with FABP	[57]
k_{-1}^{FABP}		$4.5\ s^{-1}$	Dissociation rate of FABP-bound FA	[57]
$k_{+1}^{FABP\ b}$		$4.4 \times 10^8\ M^{-1}\ s^{-1}$	Association rate FA with FABP	[57]
D_{uFA}	Water (37 °C)	$5.0 \times 10^{-6}\ cm^2\ s^{-1}$	Diffusion coefficient of unbound FA in different media	[40,64,77]
	Plasma	$3.0 \times 10^{-6}\ cm^2\ s^{-1}$		
	Cytoplasm	$1.9 \times 10^{-6}\ cm^2\ s^{-1}$		
$D_{HSA} \approx D_{AF_1} \approx D_{AF_2} \approx D_{AF_3}$	Water (37 °C)	$8.8 \times 10^{-7}\ cm^2\ s^{-1}$	Diffusion coefficient of albumin, AF_1, AF_2 and AF_3	[18,40,74]
	Plasma	$5.2 \times 10^{-7}\ cm^2\ s^{-1}$		
	Glycocalyx	$1.7 \times 10^{-13}\ cm^2\ s^{-1}$		
$D_{FABP} \approx D_{FA}$·FABP	Water (37 °C)	$1.7 \times 10^{-6}\ cm^2\ s^{-1}$	Diffusion coefficient of FABP and FABP-bound FA	[40,68]
	Cytoplasm	$1.7 \times 10^{-7}\ cm^2\ s^{-1}$		
S_{cardio}		$2000\ cm^2\ (g\ ww)^{-1}$	Area of cardiomyocytes	[7]
S_{endo}		$500\ cm^2\ (g\ ww)^{-1}$	Area of endothelium	[7]
x_{mem}		5 nm	Membrane width	[2]
x_{GLYC}	Endothelial luminal side	0.4 μm	Glycocalyx width	[70,74]
	Endothelial abluminal side	0.3 μm		
	Myocyte	0.3 μm		
x_{ENDO}		0.5 μm	Endothelium width	[6]
x_{ISF}		0.5–1.0 μm	Interstitial width	[70]
[L]		0.4 M	Phospholipid concentration in membrane with adsorption layers	[2]
[P]		0.4 mM	Membrane protein concentration in membrane with adsorption layers	[2]
K_a^{CP}		$1 \times 10^6 – 1 \times 10^8\ M^{-1}$	Affinity constant of uFA for carrier protein	Physiological range
$k_{bind}^{CP\ b}$		$1 \times 10^5 – 1 \times 10^9\ M^{-1}\ s^{-1}$	Association rate: uFA with carrier	Physiological range
$k_{release}^{CP}$		$0.1 – 10\ s^{-1}$	Dissociation rate: uFA from carrier-bound FA	Physiological range
k_{flip}^{CP}		$50\ s^{-1}$	Flip rate of carrier from outer to inner leaflet	Physiological range

(continued on next page)

Table 1 *(continued)*

Parameter	Specification	Value[a]	Description	Reference
$K_A^{receptor}$		$4 \times 10^{10}\ M^{-1}$	Affinity constant: albumin-bound FA for receptor	Adapted
$k_1^{receptor}$		$4 \times 10^{16}\ M^{-1}\ s^{-1}$	Association rate: albumin-bound FA with receptor	Adapted
$k_{-1}^{receptor}$		$1 \times 10^{6}\ s^{-1}$	Dissociation rate: albumin-bound FA with receptor	Adapted
$k_2^{receptor}$		$50\ s^{-1}$	Release rate of FA from albumin-bound FA complexed to the receptor	Adapted
$[Alb]_{total}$				
$[FA]_{total}$	Blood plasma	0.5 mM	Total concentration FA	[26,70,75]
	Interstitial fluid	0.09 mM		
	Sarcoplasm	~0 mM		
	Blood plasma	0.5 mM	Total albumin concentration	[70,75]
	Interstitial fluid	0.3 mM		
$[FABP]_{total}$	Endothelial cytoplasm	0.6 μM	Total FABP concentration	[39,70]
	Sarcoplasm	0.3 mM		
[uFA]	Blood plasma	5.1 nM	Concentration unbound FA	[55]
	Interstitium	1.9 nM		

[a]All values have a standard deviation of 10% of their mean value, unless indicated otherwise.
[b]Related to other variables.

available to move sufficient FAs (flux of ~ 100 nmol min^{-1} (g ww)$^{-1}$) from the beginning of the glycocalyx to the surface of the endothelial membrane. Therefore, if the findings of the in vitro experiments are correct [74], one must conclude that albumin-bound FAs diffuse too slowly through the glycocalyx to account for a physiological relevant FA flux.

2.2.2. *Endothelium*

The capillary wall consists of a single layer of endothelial cells, which make up the endothelium, the second compartment that FAs must cross. Four mechanisms of crossing the endothelial cells lining the microvascular compartment have been proposed [7,71]

(1) *Transcytosis*: FAs, bound to albumin, are transported across the endothelium by means of vesicles [25,49,60]. It was shown that the FA–albumin complex binds to uncoated proteins on the endothelial membrane. These parts of the membrane form vesicles, which transport FA–albumin complexes through the endothelium to its abluminal side. Over there, the complex is released into the interstitial compartment. The time for these vesicles to cross the endothelium is 3–5 min [60] and by using Eq. (2) [74], one can deduce the diffusion coefficient D for these vesicles:

$$D = \frac{d^2}{4t} \tag{2}$$

where d is the diffusion distance, which a substrate has to cross within time t. As on the average 4 min is available to cross an endothelial cell with a width of ~ 0.5 μm (Table 1), D will be on the order of $\sim 2.6 \times 10^{-12}$ cm^2 s^{-1}. This value is about six orders in magnitude lower than other diffusion coefficients in Table 1. The calculated D corresponds with a maximum flux of ~ 0.8 nmol FAs min^{-1} (g ww)$^{-1}$ (Eq. (1)). Therefore, transcytosis is quantitatively of no importance for FA transport.

(2) *Clefts*: Intercellular junctions, called clefts, separate the individual endothelial cells. The cleft size is ~ 6–7 nm [48], about the same as the diameter of an albumin molecule [40], and constitutes a total area available for diffusion of ~ 5 cm^2 (g ww)$^{-1}$ [48]. Through these clefts, albumin-bound FAs are allowed to diffuse towards the interstitium. However, the presence of a glycocalyx tethers albumin movement [7], reducing the FA flux of albumin-bound FAs to almost zero. A flux solely established by uFAs is also negligible, as was calculated for uFAs across the glycocalyx in Section 2.2.1. FA transport through the clefts is therefore not sufficient to explain a physiologically relevant flux.

(3) *Diffusion through the endothelial cell*: uFAs are absorbed by the phospholipid bilayer of the endothelial membrane (luminal side). Subsequently, these FAs are transported by three possible mechanisms to the abluminal side of the endothelium:

(a) *Peri-endothelial transfer* [59]: FAs remain inside the membrane and diffuse via the membrane to the abluminal part of the membrane. This process will only occur if FAs are present in huge amounts in the endothelial membrane, i.e. ~7 orders higher than the phospholipid concentration in the membrane, as was shown by calculation [7]. This is obviously not the case in physiological relevant situations. Therefore, peri-endothelial transfer of FAs is inconsequential.

(b) *Transmembrane diffusion* [22,23,28]: (This process is described in more detail in Section 2.3.1.) FAs are absorbed in the membrane and move subsequently to the inner side of the membrane, where they are released into the cytoplasm. After the release from the membrane, FAs diffuse through the endothelial cytoplasm to the abluminal membrane. There, the same process of absorption, movement across the membrane and release take place to transport FAs from the endothelial cytoplasm into the interstitial compartment. The transport through the endothelial cytoplasm could be enhanced by FABP [52, 53], which binds tightly to FAs ($K_a^{FABP} \sim 1 \times 10^8\ M^{-1}$, [56,57]). However, the concentrations of FABP in endothelial cells is most likely too small, about 0.5 μM [39], to contribute to cytoplasmic FA transport [72].

(c) *Protein-mediated transport*: It cannot be excluded that special proteins are present that enhance the FA transport across the endothelial membranes [1,8, 12,19,65,66]. Experimental data showed saturation effects that could be ascribed to the involvement of membrane-associated proteins [75]. Also the use of agents, inhibiting protein-mediated FA transport strengthens this notion [11,37,41,44,45]. A few candidates for these proteins have been proposed: fatty acid translocase/CD36 (FAT/CD36 [1,24]), plasma membrane fatty acid-binding protein ($FABP_{pm}$ [65]), fatty acid transport protein (FATP [58]) and albumin-binding protein (ABP [54,69]). The transport mechanisms of FAs using these proteins remain unclear, but possible mechanisms will be discussed below (Section 2.3).

Summarizing, from the potential FA pathways across the myocardial endothelium, only diffusion of FAs through the endothelial membrane, either by transmembrane diffusion or protein-mediated, seems to be physiologically relevant.

2.2.3. *Interstitial compartment*

The interstitium is the compartment that divides the endothelial cells and cardiomyocytes. The interstitial width is about 0.5 μm [70]. It consists of an aqueous solution, the interstitial fluid, containing among others uFAs and albumin-bound FAs (AF_1, AF_2 and AF_3). The albumin concentration in the interstitium is about half the concentration in blood plasma [26]. Most of the interstitial fluid is "trapped" in a matrix of glycoproteins and proteoglycans, being constituents of the glycocalyx of both endothelial cells and cardiomyocytes. The crossing of the interstitial compartment is most likely due to simple diffusion [7] of both uFAs and albumin-bound FAs. Diffusion of AF_1, AF_2 and AF_3 may, however, be substantially hampered since macromolecules, such as albumin, diffuse slowly through the glycocalyx [74].

2.2.4. Sarcolemma

The final barrier, which FAs must cross before entering the myocardial cytoplasmic compartment, is the sarcolemma. The mechanism of FA transport across this cell membrane might be different from the FA transport across the endothelial membranes.

2.2.5. Metabolism of fatty acids in the cardiomyocyte

The sarcoplasm could be considered as a sink for FAs. The concentration of FAs in the sarcoplasm is very low compared with the extracellular concentrations [70], creating a concentration gradient from the microvascular compartment to the sarcoplasm. In the sarcoplasm, FAs bind to FABP and this complex diffuses to intracellular sites of metabolic conversion. FAs can be stored in the myocardial cytoplasm as triacylglycerols or metabolized in the mitochondria or peroxisomes (Fig. 2). Only the pathway of FA metabolism inside the mitochondria (Fig. 3) will be described. First, FAs react with coenzyme A (CoA), catalyzed by the enzyme acyl-CoA synthetase (ACS) to form reactive acyl-CoA esters, which subsequently reacts with carnitine to yield acylcarnitine. This reaction is catalyzed by carnitine acyltransferase, located in the outer mitochondrial membrane (CAT I). Acylcarnitine is transported across the inner mitochondrial membrane into the mitochondrial matrix by means of a carnitine-dependent shuttle protein, i.e. carnitine-acyl carnitine translocase (CAcT). Carnitine acyltransferase (CAT II) located at the inner side of the inner mitochondrial membrane converts acylcarnitine back to acyl-CoA and carnitine. The next step is the β-oxidation of acyl-CoA, in which acyl-CoA is stepwise degraded to acetyl-CoA. Acetyl-CoA is degraded in the citric acid cycle and finally electron transfer chain activity leads to the formation of ATP. Although the mitochondrial conversion of FAs into acyl-CoA is important for the driving force of FA uptake [33,34,35], in this chapter the sarcoplasm is simplified as a single sink compartment.

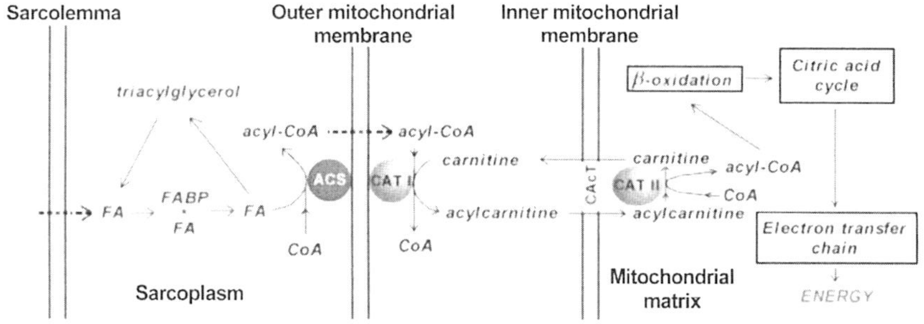

Fig. 3. A more detailed description of the fate of FAs in the cardiomyocyte. In the sarcoplasm, FAs are transported by FABP to intracellular sites of metabolic conversion followed by storage (triacylglycerol) or oxidation. Near the outer mitochondrial membrane, FAs are converted to acyl-CoA, catalyzed by acyl-CoA synthetase (ACS). Acyl-CoA moves across the outer mitochondrial membrane, where carnitine acyltransferase I (CAT I) converts acyl-CoA to acylcarnitine, which is transported through the inner mitochondrial membrane by action of carnitine/acylcarnitine translocase (CAcT). In the mitochondrial matrix, carnitine acyltransferase II (CAT II) catalyzes the reaction of acylcarnitine back to acyl-CoA. This substrate is broken down successively by β-oxidation and the citric acid cycle activity. (*For a colored version of this figure, see plate section, page 591.*)

2.3. *Fatty acid transport across membranes*

Different mechanisms have been proposed to explain FAs crossing a biological membrane. The main controversy is about the necessity of a protein: is transmembrane diffusion of FAs across the myocardial or endothelial membrane high enough to explain a physiological relevant flux of ~ 100 nmol min^{-1} (g $ww)^{-1}$, or is a membrane-associated protein needed to facilitate this transport process?

A biological membrane, like the myocardial or endothelial membrane, is composed of two leaflets: an outer (the layer of the membrane facing the exterior of the cell) and an inner leaflet (the layer of the membrane bordering the cell interior). In a standard membrane, about 50% of the mass of the membrane consist of phospholipids, which are composed of a polar head and two hydrophobic tails [2]. The hydrophobic tails of the two leaflets point to each other, resulting in a phospholipid bilayer, composed of different phospholipids, e.g. phosphatidyl-choline and sphingomyelin. The outer and inner leaflets differ a lot in phospholipid composition (phospholipid asymmetry), but the concentration of phospholipids on both sides of the membrane is globally equal [2], which will be used in our computer model. Proteins constitute the remainder of the mass in the membrane, although the number of proteins in the membrane is much lower than the amount of phospholipid molecules, i.e. about 1 protein per 100 phospholipids [2].

2.3.1. *Transmembrane diffusion of fatty acids*

The first mechanism proposed for FA transport across a membrane is transmembrane diffusion (Fig. 4). It can be divided into three different phases [22,23,28,80]:

1. *Absorption*: First, FAs are absorbed from an aqueous phase, e.g. the blood plasma or interstitial fluid, into the phospholipid part of the membrane.
2. *Flipping*: FAs, present in one of the leaflets, flips to the other leaflet.
3. *Desorption*: Finally, FAs move from the latter leaflet to the aqueous phase.

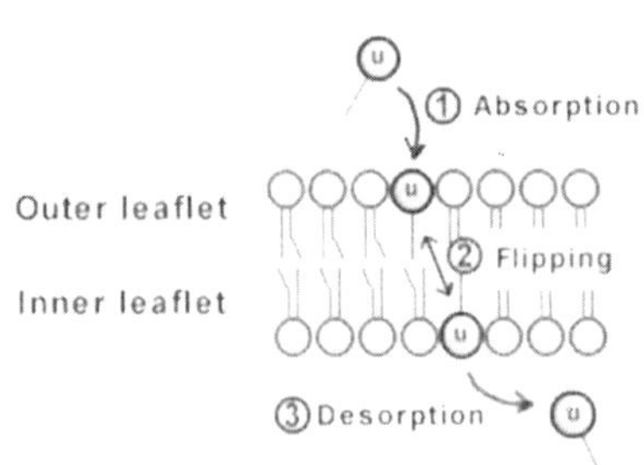

Fig. 4. Transmembrane diffusion of non-protein bound FAs (uFAs, indicated with u in the figure). uFAs present in the aqueous compartment are (1) absorbed into the outer leaflet of the membrane, (2) they flip from the outer to the inner leaflet and (3) are released (desorbed) in the aqueous compartment on the other side of the membrane. *(For a colored version of this figure, see plate section, page 592.)*

The three different stages can be reversed, because the membrane is assumed to have same amounts of phospholipids in both leaflets. Moreover, the flux direction is only determined by the FA gradient across the membrane: from the highest concentration to the lowest.

2.3.2. Facilitated transport of fatty acids by a carrier protein

It has been hypothesized that transmembrane diffusion of FAs across biological membranes is limited by the flipping rate of FAs [1,31]. A so-called carrier protein could enhance the rate of movement of FAs from one leaflet of the membrane to the other. This protein binds to uFAs in the aqueous compartment, transports the uFAs to the other side of the membrane and finally releases it over there. The principle is the same as in Section 2.3.1, because three phases can be identified: absorption, flipping (solely uFAs or complete carrier, depending on which mechanism is proposed) and desorption. However, FA transport by means of a carrier protein does not require the phospholipids inside the membrane as transport medium, unlike transmembrane diffusion of FAs (Fig. 5).

2.3.3. Receptor

FAs are tightly bound to albumin [55]. Release of these albumin–FA complexes could be facilitated by a so-called receptor (Fig. 6). This putative protein, attached to a cell membrane, is supposed to strongly bind to complexes of FA and albumin, creating a change in albumin conformation and thereby enhancing the release of FAs. This leads locally to a higher concentration of FAs in the stagnant water layer near the membrane and, hence, an increase in FA transport across the membrane, due to the ensuing steeper diffusion gradient.

2.3.4. Enhancer: improved uptake of fatty acids in the membrane

An enhancer could also increase the transmembrane FA transport (Fig. 7). This putative protein increases the rate of absorption of uFAs from the aqueous phase directly into the phospholipid layer of the membrane. The remaining pathway of membrane FA transport

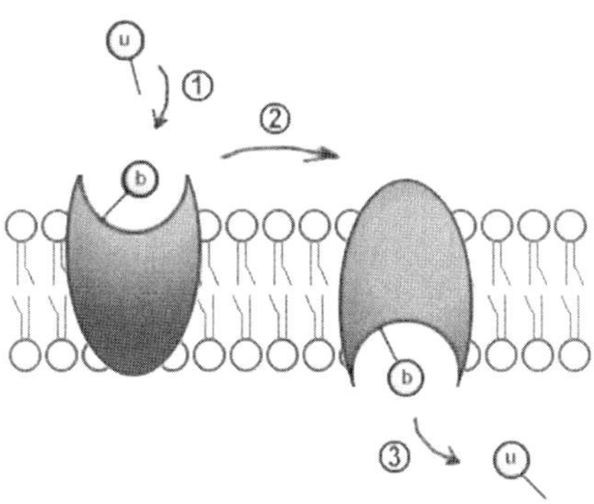

Fig. 5. Carrier protein. A possible representation of a carrier protein is shown here: (1) binding of non-protein bound FAs (uFAs, u) to the carrier (bound FAs are represented with b), (2) transport of uFAs to the other side of the membrane (a confirmation change in carrier protein is shown here as the transport mechanism) and (3) uFAs are released to the other side of the membrane. (*For a colored version of this figure, see plate section, page 593.*)

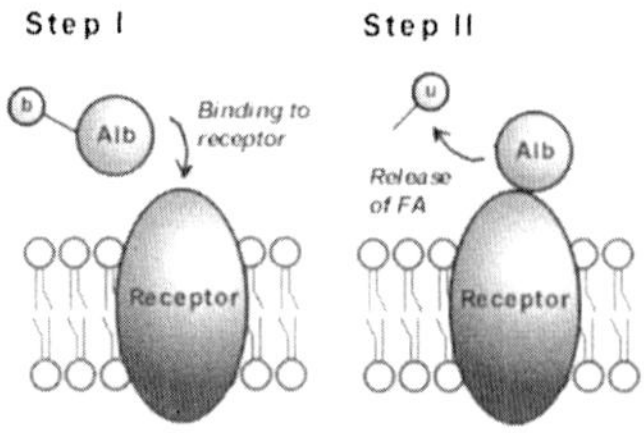

Fig. 6. Receptor. The receptor binds to complexes of FAs (b) and albumin (Alb, Step I). The normally tight binding of albumin to FAs is weakened as a result of binding of the albumin–FA complex and the receptor. This leads to an increased release of FAs from the albumin-FA complex (Step II), which causes a raised concentration of uFAs (u) in the stagnant water layer near the membrane. Eventually, this creates a steeper gradient across the membrane, which increases the FA flux due to transmembrane diffusion. (*For a colored version of this figure, see plate section, page 594.*)

(flipping and desorption) is the same as was already described in Section 2.3.1 about transmembrane diffusion of FAs. The overall effect of the enhancer is an increase in FA transport across the membrane.

2.4. *Differences between the sarcolemmal and endothelial barrier*

The global structure of FAs crossing the microvascular compartment, endothelium, interstitial compartment and sarcolemma has been discussed above. Two main barriers can be identified:

- *The endothelial barrier.* It consists of the microvascular compartment, i.e. the endothelial glycocalyx on the luminal side, the endothelium and that part of the interstitial compartment that is composed of the endothelial glycocalyx on the abluminal side.
- *The sarcolemmal barrier.* This barrier consists of the cardiomyocyte-associated glycocalyx, the sarcolemma and the sarcoplasm.

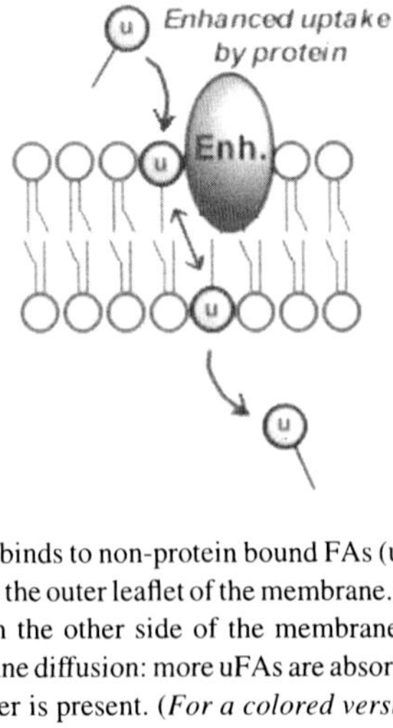

Fig. 7. Enhancer. The enhancer (Enh.) binds to non-protein bound FAs (uFAs, u), present in the aqueous solution, which increases the FA concentration in the outer leaflet of the membrane. Subsequently, the FAs flip from the outer to the inner leaflet and are released on the other side of the membrane. Therefore, the enhancer improves the absorption step of uFAs in transmembrane diffusion: more uFAs are absorbed from the aqueous compartment in the biological membrane, when an enhancer is present. (*For a colored version of this figure, see plate section, page 595.*)

Two differences exist between these two barriers, which might have important physiological implications. The first one is the area available for transmembrane transport. The area of the endothelial barrier is about 500 cm^2 $(g\ ww)^{-1}$ [7,48], while the sarcolemmal barrier covers ~2000 cm^2 $(g\ ww)^{-1}$ [7]. The second difference is the width of both barriers. The sarcolemmal barrier, composed of the glycocalyx of the cardiomyocyte and sarcolemma (sarcoplasm is a sink), is ~0.3 μm and the endothelial barrier, including the microvascular compartment, endothelium and the abluminal side of the endothelium, has a length of ~1.2 μm. These quantitative differences appear to have a significant impact of FA flux rates, as will be discussed in detail in Section 5.3.2.

3. Modeling approach

3.1. Describing physiological conditions in mathematical equations

The mathematical model, required for FA uptake processes, is a dynamic system for which the variables, concentrations of biological substrates in this case, are time dependent [14]. These dependent variables are described by differential equations, which are mathematical representations of the changes in a variable as function of time. Examples of biological processes that induce a change in concentration in time are chemical reactions, diffusion processes and transport mechanisms by means of transport-facilitating proteins. A short guide to formulate mathematical equations of these processes is given in this section.

3.1.1. Ordinary differential equations

Differential equations that contain one or more derivatives of a dependent variable with respect to a single independent variable, time in this chapter, are called ordinary differential equations (ODEs). Each ODE describes the dynamic behavior of one variable by means of a mathematical function of variables, parameters and inputs. A state equation is a special differential equation that cannot be deduced from other differential equations. A model consisting of n state variables must be defined by a set of n equations, ODEs in this case. A set of n equations is described in mathematical terms as follows:

$$\dot{\mathbf{q}} = f(t, \mathbf{q}, \mathbf{u}) \tag{3}$$

with

$$\mathbf{q} = \begin{bmatrix} q_1 \\ q_2 \\ \vdots \\ q_n \end{bmatrix}, \quad \dot{\mathbf{q}} = \begin{bmatrix} \dot{q}_1 \\ \dot{q}_2 \\ \vdots \\ \dot{q}_n \end{bmatrix} \text{ and } \mathbf{u} = \begin{bmatrix} u_1 \\ u_2 \\ \vdots \\ u_n \end{bmatrix}$$

where $\mathbf{q}$ is an n-element column, containing all the variables of the system, $\dot{\mathbf{q}}$ is vector whose elements are the time derivatives of the corresponding elements in $\mathbf{q}$, $\mathbf{u}$ is the input vector of the system, and $f(t, \mathbf{q}, \mathbf{u})$ is a mathematical function that depends on time (t), the variables ($\mathbf{q}$) and inputs ($\mathbf{u}$). The parameters are assumed to have constant values in this equation and will be discussed in Section 3.2. An example to clarify ODEs: the reaction equation of a simple biochemical reaction can be represented as:

$$C_1 + C_2 \underset{k_{-1}}{\overset{k_1}{\rightleftarrows}} C_3$$

It shows the reaction of two substrates with concentrations C_1 and C_2 which form one product with concentration C_3. The three different concentrations, the variables, can be rewritten as differential equations:

$$\dot{C}_1 = -k_1 C_1 C_2 + k_{-1} C_3 \tag{4}$$

$$\dot{C}_2 = -k_1 C_1 C_2 + k_{-1} C_3 \tag{5}$$

$$\dot{C}_3 = k_1 C_1 C_2 - k_{-1} C_3 \tag{6}$$

where C_1, C_2, and C_3 are the two substrate and product concentrations, respectively, k_1 is the rate parameter of C_1 reacting with C_2 to form C_3, and k_{-1} is the rate parameter of the reverse reaction back to C_1 and C_2. $\dot{C}_1$, $\dot{C}_2$ and $\dot{C}_3$ on the left-hand side (LHS) are the time derivatives of substrate and product concentrations C_1, C_2 and C_3, respectively. Accordingly, Eq. (4) is defined as the change of C_1 in time and depends on two reaction rates:

(a) The reaction of C_1 with C_2 to C_3 (the first part of Eq. (4): $k_1 C_1 C_2$).
(b) The reverse reaction of C_3 back to C_1 and C_2 (the second part of Eq. (4): $k_{-1} C_3$).

A minus sign of the first part in Eq. (4) represents a decrease in C_1 as function of time, when C_1 reacts with C_2. The second part is preceded by a plus, because an increase in the conversion from C_3 to C_1 (and C_2) leads to an increase in C_1. In the same way, the plus and minus signs are placed in Eqs. (5) and (6) and notice the similarity between Eqs. (4) and (5) and the opposite signs in Eq. (6). This is necessary, because the law of mass conservation must be obeyed: there is no mass loss or gain in the complete system. In summary, $\dot{C}_1 = \dot{C}_2 = -\dot{C}_3$ leads to the conclusion that three equations have to be solved, including only one state equation, to come to a solution for all three differential equations. Solutions are $C_1(t)$, $C_2(t)$ and $C_3(t)$, which are concentrations as function of time (state trajectories). They are determined by certain initial concentrations and the inputs of the system (Eq. (3)). The three concentrations converge after a certain time span to a final solution (meaning that the system is stable): the so-called steady-state value. In steady state, there is no change in concentration ($\dot{C} = 0$) provided that the input is kept constant.

3.1.1.1. Analytical vs. numerical solution. Possible solutions of differential equations, like ODEs, can be obtained analytically or by a numerical approximation. An analytical solution of a differential equation is an exact representation of the system in the form of a mathematical equation. However, this is only possible in a limited number of cases.

Solutions of differential equations can also be approached by numerical methods, which are often used in more complicated situations, as is the case with most physiological processes. The analytical solution is very cumbersome to acquire or even impossible. Therefore, the equations are solved numerically with mathematical computer programs, e.g. MATLAB (`ode45`) and *Mathematica* (`DSolve`).

3.1.2. Partial differential equations

Besides ODEs that only depend on changes in time, there are differential equations that depend on changes in time *and* other independent variables. These are called partial differential equations (PDEs) and this chapter will only discuss PDEs that depend on changes in time and in distance. An example of a PDE is the diffusion of a substance with concentration C_1 across a given length L (Fig. 8). The diffusion in one dimension is described with the diffusion equation [67]:

$$\dot{C}_1 = D\frac{\partial^2 C_1}{\partial x^2} \tag{7}$$

where C_1 is the substrate concentration, D is the diffusion coefficient and x is the location. $\dot{C}_1$ is the time derivative of concentration, $\partial^2 C_1/\partial x^2$ represents the second position-dependent derivative of C_1. The solution of Eq. (7) is $C_1(t, x)$, a concentration that not only varies with time, but also with the distance. Again, it is possible to obtain an analytical solution, but in complex biological situations a numerical approach is used to come to a solution. For example, the substrate with concentration C_1 diffuses only in the x-direction as shown in Fig. 8. The diffusion space, the rectangular box, is divided in a number of equally spaced smaller rectangles: the segments (Fig. 9). Each segment represents a certain average concentration C_1, a value at a point, in a certain range with a length Δx, which could differ per segment. This method of solving a PDE by creating a volume around one point is called the finite volume method (FVM). The FVM is a way to rewrite the PDE as a coupled set of ODEs, based on a spatial grid (Fig. 9), which has been deduced

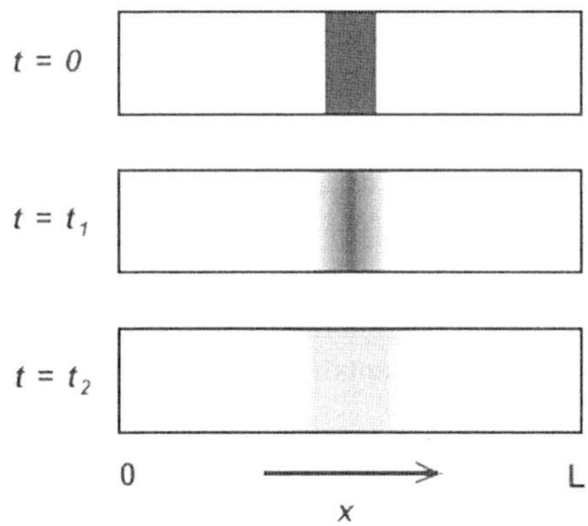

Fig. 8. Diffusion of a substrate in the x-direction of a system with length L. At $t = 0$, a substrate is present in the middle only and at time t_1 and t_2 ($t_2 > t_1$), the substrate diffuses to both sides. (*For a colored version of this figure, see plate section, page 596.*)

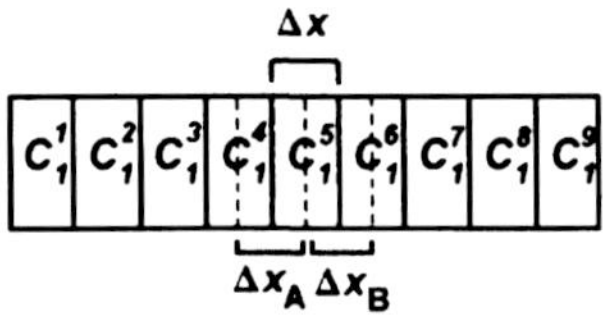

Fig. 9. A system with substrate concentration C_1 is divided in nine segments ($N = 9$, segment number is shown as superscript). The 5th segment has a length of Δx and its distance to the middle of the left and right segment is Δx_A and Δx_B, respectively. In this figure, $\Delta x_A = \Delta x_B$.

for the diffusion equation (Eq. (7)). First, one has to know that the flux J per unit area S is defined in diffusion processes as (Eq. (8), also shown in Eq. (1), now slightly adapted):

$$J = -DS\frac{\partial C_1}{\partial x} \tag{8}$$

where $\partial C_1/\partial x$ is the first position-dependent derivative of substrate concentration C_1, the other parameters have been described previously (Eq. (1)). For segment 5, the concentration C_1^5 depends on the influx of substance from C_1^4 to C_1^5, $J_{4\to5}$, and on the efflux from C_1^5 to C_1^6, $J_{5\to6}$ (Fig. 9). The fluxes $J_{4\to5}$ and $J_{5\to6}$ are determined by discretizing Eq. (8) (dividing in a finite number of segments) and using Fig. 9:

$$J_{4\to5} = DS\frac{C_1^4 - C_1^5}{\Delta x_A} \tag{9}$$

$$J_{5\to6} = DS\frac{C_1^5 - C_1^6}{\Delta x_B} \tag{10}$$

where Δx_A is the distance from the segment of interest, segment 5 in this example, to the neighboring segment on the left, i.e. segment 4, Δx_B is the distance from the segment of interest to the right segment, i.e. segment 6 (Fig. 9).

The net flux, the amount of substrate per unit time moving into minus the amount of substrate per unit time moving out of the compartment, divided by the total volume of segment 5 (V_{seg}^5) in this case results in $\dot{C}_1^5$:

$$\begin{aligned}\frac{J_{4\to5} - J_{5\to6}}{V_{seg}^5} &= \frac{DS}{V_{seg}^5}\left(\frac{C_1^4 - C_1^5}{\Delta x_A} - \frac{C_1^5 - C_1^6}{\Delta x_B}\right)\\ &= \frac{D}{\Delta x}\left(\frac{C_1^4 - C_1^5}{\Delta x_A} - \frac{C_1^5 - C_1^6}{\Delta x_B}\right) = \dot{C}_1^5\end{aligned} \tag{11}$$

Eq. (11) is rewritten in a more general form as Eq. (12), which is a discretized representation of Eq. (7) for the segments $i = 2$ to $N - 1$, with N is the total number of segments:

$$\dot{C}_1^i = \frac{D}{\Delta x}\left(\frac{C_1^{i-1} - C_1^i}{\Delta x_A} - \frac{C_1^i - C_1^{i+1}}{\Delta x_B}\right) \tag{12}$$

where i is an index that indicates the segment number of the system (it varies between 1 and 9 in Fig. 9). However, at $i = 1$ and 9, one cannot use Eq. (12), because the concentration in segment 0 and 10 does not exist, respectively. Still, it is possible to solve this numerical problem by *reflection of diffusion*:

$$\dot{C}_1^1 = \frac{J_{2\to1} - J_{1\to2}}{V_{\text{seg}}^1} = \frac{D}{\Delta x}\left(\frac{C_1^2 - C_1^1}{\Delta x_{\text{A}}} - \frac{C_1^1 - C_1^2}{\Delta x_{\text{B}}}\right) \quad \text{for } i = 1 \tag{13}$$

$$\dot{C}_1^N = \frac{J_{N-1\to N} - J_{N\to N-1}}{V_{\text{seg}}^N} = \frac{D}{\Delta x}\left(\frac{C_1^{N-1} - C_1^N}{\Delta x_{\text{A}}} - \frac{C_1^N - C_1^{N-1}}{\Delta x_{\text{B}}}\right) \quad \text{for } i = N \tag{14}$$

where N is the total number of segments, of which the system consists. Fig. 9 shows an example of nine differential equations ($N = 9$) that needs to be solved to know the complete profile of C_1. But the more segments one uses to approximate Eq. (7), the better it approximates the analytical solution. Needless to say is that a higher number of segments leads to more equations, resulting in a longer calculation time.

3.1.3. *Combination of ODEs and PDEs*

Based on the law of mass balance, the concentrations of the physiological important molecules can be described in terms of differential equations (e.g. Eqs. (4)–(6) and Eqs. (12)–(14)), which are composed of different components, each representing a single process, which involves a change in concentration. For example, the right-hand side (RHS) of Eqs. (4)–(6) only consists of two components. For Eq. (4) this is the binding component, $-k_1 C_1 C_2$, and the release component, $+k_{-1} C_3$. Both are lumped in one single rate component that contains the contribution of both components to the concentration change, dependent on the three substrate concentrations: $-v_{\text{reaction}}(C_1, C_2, C_3)$. The same rate component is valid for Eq. (5) $(-v_{\text{reaction}}(C_1, C_2, C_3))$ and also for Eq. (6) $(v_{\text{reaction}}(C_1, C_2, C_3))$, except for the absence of a minus sign, due to mass balance. Likewise, PDEs can be described in terms of rate components. As mentioned earlier, diffusion processes are characterized by PDEs, which can be rewritten in terms of ODEs for the different segments. These equations are represented in the discretized form as

$$\frac{D}{\Delta x}\left(\frac{C_1^{i-1} - C_1^i}{\Delta x_{\text{A}}} - \frac{C_1^i - C_1^{i+1}}{\Delta x_{\text{B}}}\right)$$

(see also Eqs. (12)–(14)) and can also be lumped in one rate component for all i: $v_{\text{diffusion}}(C_1)$. Similarly, diffusion of another substrate with concentration C_2, in discretized form written as

$$\frac{D}{\Delta x}\left(\frac{C_2^{i-1} - C_2^i}{\Delta x_{\text{A}}} - \frac{C_2^i - C_2^{i+1}}{\Delta x_{\text{B}}}\right)$$

is included in rate component $v_{\text{diffusion}}(C_2)$. Combinations of multiple rate components are also possible: a simple reaction of a substance with concentration C_1 and diffusion of this concentration at the same time, so a combination of Eqs. (4)–(6) and Eqs. (12)–(14), is rewritten in rate components as Eqs. (15)–(17):

$$\dot{C}_1 = -v_{\text{reaction}}(C_1, C_2, C_3) + v_{\text{diffusion}}(C_1) \tag{15}$$

$$\dot{C}_2 = -v_{\text{reaction}}(C_1, C_2, C_3) + v_{\text{diffusion}}(C_2) \tag{16}$$

$$\dot{C}_3 = v_{\text{reaction}}(C_1, C_2, C_3) + v_{\text{diffusion}}(C_3) \tag{17}$$

These differential equations are state equations, which cannot be rewritten in terms of each other, and show very clearly which components influence the dynamic behavior of C_1, C_2 and C_3: the state variables. The rate components have to be determined: which processes are involved in changing the concentration of a given substrate in terms of rate components? If these are determined for the substrates and products, the following question has to be answered: what are the underlying equations of the required rate components? The final answer to this question is a system of one or more differential equations, which need to be solved. These equations are the structure of the model.

3.2. *Obtaining parameter values*

State equations are composed of state variables, parameters and inputs, as mentioned in Section 3.1.1. The parameters in the state equations provide a quantitative description of the different parts of a state equation. Examples of parameters are k_1, k_{-1}, (both from Eqs. (4)–(6)) and D (Eq. (6)), which values can be obtained from experimental data. Unfortunately, not all values of the parameters of interest are experimentally determined or are suitable for model use directly. It depends on what data are available and under which conditions the parameters were obtained, e.g. in vivo (measured in an intact cell or organism), or in vitro (measured in an isolated cell or cell-free homogenate [2]). This subsection shows some important aspects of parameters that need attention, because a model greatly depends on the values of the parameters used. In Sections 3.2.1 and 3.2.2, examples of parameters will be given, which are used in the mathematical model. All these parameter values are listed in Table 1.

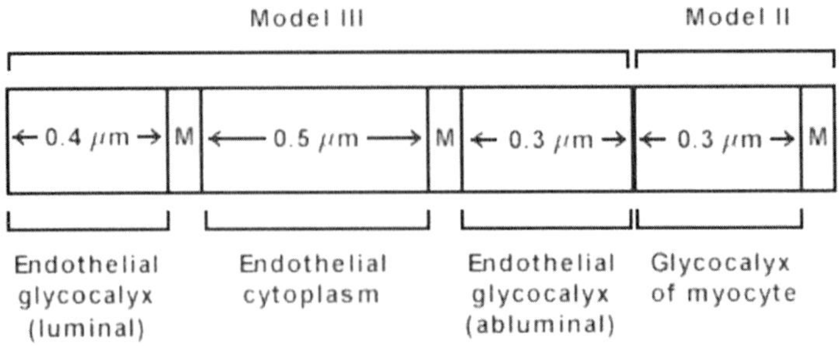

Fig. 10. A schematic representation of the total model. The dimensions of the different parts of the models are also shown here. (*For a colored version of this figure, see plate section, page 597.*)

3.2.1. Calculating parameter values: lipid content of a biological membrane

In some cases, parameter values are needed that cannot be determined directly from experiments, but it requires some post-processing to estimate the value needed. An example is the determination of the concentration of lipids in a biological membrane, which is required to obtain a rate parameter value for the transport model of FAs.

About 5×10^6 phospholipid molecules are located in a $1\ \mu m \times 1\ \mu m$ area of a phospholipid bilayer [2]. The thickness of an average phospholipid bilayer is about 5 nm. Therefore, the volume of a phospholipid bilayer will become $(1 \times 10^{-5}\ dm) \times (1 \times 10^{-5}\ dm) \times (5 \times 10^{-8}\ dm) = 5 \times 10^{-18}\ dm^3$. The quantity of phospholipid molecules in this membrane is 5×10^6 phospholipid molecules, which is equal to 8.3×10^{-18} mol, making use of the number of Avogadro. The average molecular mass of membrane phospholipids is 600 Da, so $(8.3 \times 10^{-18}\ mol) \times (600\ g\ mol^{-1}) = 5 \times 10^{-15}$ g phospholipid is present in $5 \times 10^{-18}\ dm^3$. Dividing this mass by the volume leads to $1\ kg\ dm^{-3}$. This holds for a membrane that only contains phospholipids without proteins. However, about 50 wet weight % of the biological membrane consists of phospholipids [2]. That leads to a phospholipid content in the biological membrane of $0.5\ kg\ dm^{-3}$. Converted back to molar: $(0.5 \times 103)/(600) = 0.8$ M. A double check has been performed to confirm this result by using a typical canine erythrocyte membrane with an area of $195\ \mu m^2$ and a thickness of 7.5 nm, which contains 0.7 pg phospholipids and 0.8 pg membrane proteins [3]. The phospholipid concentration derived from these data (calculation not shown here) is in full agreement with the outcome of the previous calculation. Likewise, one derives a protein concentration in an average biological membrane of ~ 8 mM, because the molar ratio proteins:phospholipids in the membrane is $\sim 1{:}100$ [2].

3.2.1.1. Adsorption layer. In the computer model of FA transport, described in Section 4, the concentration of phospholipids is needed to determine the value for the rate parameter k_{on} of absorption (Eq. (18) [31]):

$$k_{on} = K_p[L]k_{off} \tag{18}$$

where K_p is the so-called partition coefficient of uFA in the phospholipid/water phase, [L] is the concentration of phospholipids present in the membrane with adsorption layers and k_{off} is the rate parameter of desorption. To determine [L], one has to know, besides the phospholipid concentration in the membrane, that the membrane only absorbs uFAs, if these are present in the close vicinity of the membrane. This region is called the adsorption[2] layer with a width of 1 FA molecule (~ 2.5 nm = size of membrane leaflet) and is present on both sides of the membrane. The total volume of the solution of interest consists of the membrane and both adsorption layers. With this knowledge and a phospholipid concentration in the membrane of 0.8 M, one derives a value for [L] of 0.4 M and a membrane-associated protein concentration of 0.4 mM (Table 1).

---

[2] The adsorption layer has been defined as a region *against* the membrane, from which FAs are absorbed *into* the membrane.

3.2.2. *Controversy about parameter values*

Experiments to obtain parameter values have been performed by different research groups. In a number of cases, this has led to differences in values for the same parameters, which eventually resulted in opposing theories. To solve this problem, one has to double check every parameter value used to make sure that the parameter value is unique. If not, experimental conditions of the different experiments have to be scrutinously compared in order to find any crucial differences. Below (Sections 3.2.2.1–3.2.2.3), three examples will illustrate this issue.

3.2.2.1. Fatty acids crossing a membrane. As mentioned above, controversy exists about the mechanism underlying FA transport across membranes to explain a physiological normal flux of about 100 nmol $min^{-1}(g\ ww)^{-1}$, as measured in intact hearts. Experiments were performed on model membranes of small, large and giant unilamellar vesicles, which are abbreviated to SUVs (with a diameter $d \sim 25$ nm), LUVs ($d \sim 100$ nm) and GUVs ($d \sim 200$ nm), respectively [23,31]. In these experiments, different values were found for the rate parameter k_{flip}. This parameter describes the amount of FAs crossing the membrane per unit time and ranges from as low as 0.5 s^{-1} (GUVs [31]) to values higher than 70 s^{-1} (SUVs [30]), depending on the experimental methods used. Unfortunately, it is unclear which experiment approaches the in vivo physiological situation as close as possible. Moreover, the presence of both protein-mediated and transmembrane diffusion of FAs is supported by a substantial number of experiments. In addition, FAs are present in the ionized or un-ionized form in the membrane; ionized FAs "flip" very slowly across a phospholipid bilayer, in contrast to un-ionized FAs [20,28,29]. Which form of FAs (ionized or un-ionized) is predominantly present, depends on the pK_a (acid ionization constant) and pH (measure for the degree in acidity of a solution) of the environment. Its relation is described by the Henderson–Hasselbalch equation:

$$pH - pK_a = \log\left(\frac{[X^-]}{[HX]}\right) \tag{19}$$

where $[X^-]$ and [HX] represent the ionized and un-ionized concentration of substance X, respectively. Eq. (19) shows that if $pH > pK_a$, the ionized form predominates, which is the case when FAs are present in an aqueous environment: the pK_a is $\sim$4.5 and the pH is 7.4 [23]. Therefore, FAs are mainly present as ionized monomeric molecules in aqueous compartments. The pK_a is approximately 7.6 in a phospholipid environment (pH = 7.4 [23]), indicating a higher quantity of un-ionized FAs in the membrane. Consequently, in this chapter, FAs in the phospholipid membrane are assumed to be present in the un-ionized form.

It is of great importance to incorporate the opposing views about transmembrane FA transport in our computer model and compare the in silico findings with experimental data (see Section 5).

3.2.2.2. Diffusion of FABP in the cytosol. Parameters have been obtained by using different methods. Results could, however, be biased by the conditions chosen in the experimental set-up. An illustrative example is the determination of the diffusion coefficient of cytoplasmic FABP present. Experiments were performed with FABP bound to labeled 12-(*N*-methyl)-*N*-[(7-nitrobenz-2-oxa-1,3-diazol-4-yl)amino]-octadecanoic acid (NBD-stearic acid) in order to determine the diffusion coefficient of FABP in liver cells of male rats [5]. The value obtained was about $0.8 \times 10^{-9}\ cm^2\ s^{-1}$, which is 2000 times lower than assessed in another experiment [68]: $1.7 \times 10^{-6}\ cm^2$ (this value is corrected to a temperature of 37 °C; see Section 3.2.2.3). Another remarkable result of the experiments performed with NBD-stearic acid [5] was the differences in diffusion coefficients between male ($0.8 \times 10^{-9}\ cm^2\ s^{-1}$) and female rats ($4.8 \times 10^{-9}\ cm^2\ s^{-1}$). Compared with proteins of the same size as FABP with a molecular weight of 15 kDa [61], the values for FABP in the NBD-stearic acid experiments [5] are surprisingly low. These low values can be ascribed to the NBD-stearic acid used as fluorescent marker. A drawback of this marker is that it does not solely bind to FABP: it was shown that NBD-stearic acid also interacts with cell membranes [46,47], which leads to a decrease in mobility of this compound. This can be interpreted as a decrease in diffusion coefficient and is probably the cause of the low values found [5]. The higher values, obtained in female rats, can be explained by the higher amount of FABP present in female cells, so more NBD-stearic acid binds to FABP, which results in a higher diffusion coefficient compared to experiments performed on male rats [47].

3.2.2.3. Translation of experimental data to physiological relevant conditions. Most experiments performed to obtain parameter values are executed at room temperature, using cellular homogenates, etc. These values must be extrapolated to physiological conditions, e.g. body temperature and quantity per gram tissue. An example is converting the experimental results performed on SUVs [23] to physiological conditions. The value of rate parameter k_{on} was determined as $3.6 \times 10^6\ s^{-1}$ per molar phospholipid in the membrane with adsorption layers ($=$ [L]). The experiments were performed at a phospholipid concentration of 0.2×10^{-3} M [23] leading to a k_{on} of 720 s^{-1}. However, this does not apply on a real physiological situation with a value for [L] of 0.4 M: the latter results in a value for k_{on} of $1.4 \times 10^6\ s^{-1}$. Compared to the k_{on} (720 s^{-1}) in the SUV experiment, this value is considerably higher.

One has to bear in mind that even in experiments performed on biological cells, e.g. on isolated cardiomyocytes [45], a number of assumptions are made that could affect the results. For example, experiments to determine the FA uptake in isolated cardiomyocytes were performed on quiescent cells [45], implicating that these cells did not contract. This leads to a lower energy requirement and therefore a reduced FA uptake. Moreover, the isolated cardiomyocytes were obtained by destroying the glycocalyx, which might have a profound effect on the mobility of albumin [74].

3.2.3. Monte Carlo method

Physiological experiments generate results with a certain probability distribution of the parameters, which can be incorporated in the model. This is implemented by simulating the same model several times (about 500 cycles), but with changed values for the parameters. During each cycle, all the parameters are chosen according to their own probability distribution (e.g. with the command `normrnd` in MATLAB), independent of each other. The results of the model of each cycle is stored and subsequently the simulation is repeated with other values for the parameters, chosen according to their probability distribution and finally the results are stored again. This process is repeated as often as possible and leads to a collection of saved results that has a certain mean and standard deviation, which can be analyzed and compared with experimental data. The name for this method of varying parameters according to their probability distribution is called the *Monte Carlo* method [17] and its advantage is to create a better estimate of the model uncertainty [73].

3.3. *The flux test and unit check*

Till now, only methods have been discussed to create a model, but not to test whether it is correctly implemented. Possible methods are to perform a *flux test* or check the units of the system, i.e. the *unit check*.

3.3.1. *Flux test*

The processes in a stable system reach a steady-state value when a constant input is applied (see Section 3.1.1). This can only be achieved if all derivatives in time of the differential equations become zero (e.g. $\dot{C}_1 = \dot{C}_2 = \dot{C}_3 = 0$ in Eqs. (4)–(6), or $\dot{C}_1^i = 0$ for $i = 1$ to N in Eqs. (12)–(14)). Therefore, the different parts on the RHS of the state equations have to counterbalance each other in steady state. A special case is a system in which diffusion processes are involved. Though the sum of the rate components in these equations has to be equal to zero in steady state, the flux is *not* zero. Section 3.1.2 showed how the influx (Eq. (9)) and efflux (Eq. (10)) have been related to the discretized state equation (Eq. (12)). This state equation has to be equal to zero in steady state (see Section 3.1.1), which results in identical values for both influx and efflux. The flux test checks whether the influx and efflux satisfy these requirements. Therefore, a mathematical model of a discretized diffusion process is only technically correct, if the fluxes across the different segments are equal.

3.3.2. *Unit check*

Differential equations consist of algebraic combinations of parameters, variables and inputs, each having their own units. Therefore, all differential equations must be consistent in units. An example is to check the units of Eq. (4): $\dot{C}_1 = -k_1C_1C_2 + k_{-1}C_3$. The state equation consists of a change in concentration per unit time, $\dot{C}_1$, on the LHS, e.g. the units are M s^{-1} (molar per second). The RHS is composed of two subcomponents, $-k_1C_1C_2$ and $k_{-1}C_3$, which also have the unit molar per second, just like the LHS. Both subcomponents contain the variables C_1, C_2 and C_3, concentrations with unit M (molar). The units for the parameters k_1 and k_{-1} are determined to be (using Eq. (4)): M^{-1} s^{-1} and s^{-1}, respectively, which is in agreement with the literature [31].

4. Computer model of cardiac fatty acid uptake

With the knowledge described in the previous sections of this chapter, one can theoretically create a model of the FA uptake processes across the microvascular compartment, the endothelial cells, the interstitial compartment, the sarcolemma and finally into and through the sarcoplasm of the cardiomyocyte. The source of the complete system is located in the microvascular compartment, at the beginning of the luminal side of the endothelial glycocalyx. Leaving esterified FA present in the core of chylomicrons and VLDLs out of consideration, FAs in this compartment are mainly present as complexed to albumin, only a minor portion is in the unbound form present in the aqueous solution [55]. The sarcoplasm is considered to be a sink, i.e. $[FA] \approx 0$ M.

This section will separate the FA transport pathway from blood plasma to myocardial sarcoplasm in three parts, starting with a model of FAs crossing the sarcolemma. The goal of this model is to elucidate which mechanism is involved in the FA transport across the sarcolemma: is it transmembrane diffusion or protein-mediated? The second model is an extension of the first model by incorporating a glycocalyx to show the effect of a glycocalyx on the FA flux. The third model describes the FA transport from the microvascular compartment to the abluminal glycocalyx of the endothelial compartment. This model is to complete the total uptake process of FA from blood to sarcoplasm and to see how large the flux is across this barrier. When the last two models are linked together, they form the complete cardiac transport pathway of FAs (Fig. 10).

4.1. Model I: The myocardial sarcolemma

The precise transport mechanism of FAs across the sarcolemma is not elucidated yet. Three possible transport mechanisms will be described and implemented in the computer model, presented in Section 5.1.

4.1.1. Transmembrane diffusion

The first mechanism of FAs to overcome the sarcolemma as a constraint is transmembrane diffusion [13,15,21,28] by means of "flip-flop" through the phospholipid bilayer [36]. The three steps are absorption, flipping and desorption, which are described mathematically. The model of FA transport across the sarcolemma consists of four compartments:

1. *The interstitial compartment* in close vicinity of the outer leaflet of the sarcolemma. In this model, the interstitial compartment is only a constant supplier of FAs (only the unbound form will be considered) and no diffusion or reaction processes are involved yet. Because the supply of FAs is constant (no change in FAs per unit time), the ODE describing the change of FAs in this compartment is set to zero.
2. *The outer leaflet* is the phospholipid layer of the sarcolemma, which borders on the interstitial compartment. The FAs, present in this layer, are located in between the phospholipid molecules.
3. *The inner leaflet* is the layer of the sarcolemma that faces the sarcoplasm.

4. *The sarcoplasm* of the cardiomyocyte is the sink compartment for FAs. Because FAs bind directly to FABP and are subsequently converted to acyl-CoA, the uFA concentration is almost negligible [70]. Therefore, the assumption is made that the ODE for this compartment is also equal to zero.

In short, only two ODEs are necessary to describe the FA transport across a membrane (Fig. 11) if one assumes that the source and sink concentrations, the FA concentration in the interstitial compartment and sarcoplasm, respectively, remain constant. The change in concentration of the remaining two compartments is described by the following two state equations:

Outer leaflet:

$$\dot{C}_{\mathrm{FA}}^{\mathrm{outer}} = k_{\mathrm{on}} C_{\mathrm{FA}}^{\mathrm{source}} - k_{\mathrm{off}} C_{\mathrm{FA}}^{\mathrm{outer}} + k_{\mathrm{flip}} (C_{\mathrm{FA}}^{\mathrm{inner}} - C_{\mathrm{FA}}^{\mathrm{outer}}) \tag{20}$$

Inner leaflet:

$$\dot{C}_{\mathrm{FA}}^{\mathrm{inner}} = k_{\mathrm{on}} C_{\mathrm{FA}}^{\mathrm{sink}} - k_{\mathrm{off}} C_{\mathrm{FA}}^{\mathrm{inner}} + k_{\mathrm{flip}} (C_{\mathrm{FA}}^{\mathrm{outer}} - C_{\mathrm{FA}}^{\mathrm{inner}}) \tag{21}$$

where $\dot{C}_{\mathrm{FA}}^{\mathrm{outer}}$ and $\dot{C}_{\mathrm{FA}}^{\mathrm{inner}}$ represent the change of FA concentration per unit time in the outer and inner leaflet, respectively. The state variables are $\dot{C}_{\mathrm{FA}}^{\mathrm{source}}$, $\dot{C}_{\mathrm{FA}}^{\mathrm{outer}}$, $\dot{C}_{\mathrm{FA}}^{\mathrm{inner}}$ and $\dot{C}_{\mathrm{FA}}^{\mathrm{sink}}$, which are the uFA concentration in the interstitial compartment, outer leaflet, inner leaflet and sarcoplasm, respectively. The parameters k_{on}, k_{off} and k_{flip} represent the rate parameters of absorption, desorption and "flip-flop" of FAs in a phospholipid membrane, respectively. The two state equations (Eqs. (20) and (21)) consist only of the three processes involved in transmembrane diffusion of FAs. In Eq. (20), the absorption step is described by $k_{\mathrm{on}} C_{\mathrm{FA}}^{\mathrm{source}}$, which represents the FA uptake from the interstitial compartment into the outer leaflet of the myocardial sarcolemma. Besides absorption, also desorption from the outer leaflet into the interstitial compartment takes place: $-k_{\mathrm{off}} C_{\mathrm{FA}}^{\mathrm{outer}}$ (in Eq. (20)). The remaining part of Eq. (20), $k_{\mathrm{flip}}(C_{\mathrm{FA}}^{\mathrm{inner}} - C_{\mathrm{FA}}^{\mathrm{outer}})$, is the "flip-flop" step, with the "flip" defined as the transition of FAs from the outer to the inner leaflet, $-k_{\mathrm{flip}} C_{\mathrm{FA}}^{\mathrm{outer}}$, and the "flop" as the reverse movement (from inner to outer leaflet), $k_{\mathrm{flip}} C_{\mathrm{FA}}^{\mathrm{inner}}$, assuming that "flipping" and "flopping" of FAs occur at an equal rate ($k_{\mathrm{flip}} = k_{\mathrm{flop}}$). Similarly, one can identify the three processes (absorption, desorption and "flip-flop") in Eq. (21). Absorption of FAs from the sarcoplasm into the inner leaflet is $k_{\mathrm{on}} C_{\mathrm{FA}}^{\mathrm{sink}}$, desorption of FAs is the reverse reaction (from inner leaflet to sarcoplasm): $-k_{\mathrm{off}} C_{\mathrm{FA}}^{\mathrm{inner}}$ and "flip-flop" is represented as $k_{\mathrm{flip}}(C_{\mathrm{FA}}^{\mathrm{outer}} - C_{\mathrm{FA}}^{\mathrm{inner}})$, with $k_{\mathrm{flip}} C_{\mathrm{FA}}^{\mathrm{outer}}$ as the "flop" (from inner to outer leaflet) and $-k_{\mathrm{flip}} C_{\mathrm{FA}}^{\mathrm{inner}}$ as the "flip" (from outer to inner leaflet).

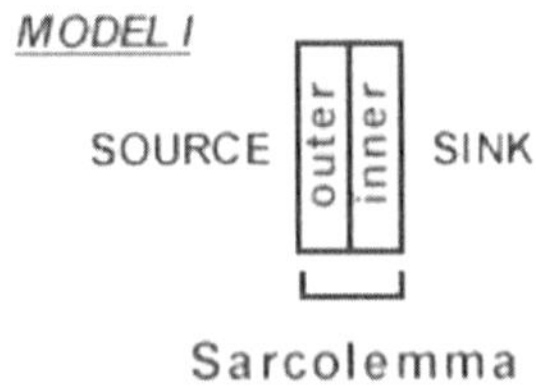

Fig. 11. Model I consists of a sarcolemma only, composed of two leaflets (outer and inner). The source is the interstitial compartment, the sarcoplasm is the sink. (*For a colored version of this figure, see plate section, page 598.*)

Values for the parameters involved in the absorption, flipping and desorption of FAs were derived by conducting experiments on SUVs, LUVs [23,30] and GUVs [31]. The values found are listed in Table 1, which were put in the computer model. Table 1 clearly shows that the parameter values differ substantially among the various experiments: the ones performed on SUVs and LUVs [23,30] resulted in higher parameter values compared to experiments with GUVs [31]. Two possible explanations are the involvement of a fluorescent marker and the kind of vesicles used. In the GUV experiments [31] an anthroyloxy group as marker, bound to FAs, was utilized to obtain a value for k_{flip}. However, this marker could alter the characteristics of FAs bound to this marker, probably leading to artifacts in the measurements [30]. The other explanation are the differences in vesicles used: SUVs and LUVs are black membranes (artificial membranes, only consisting of phospholipids) with a diameter of ~25 and ~100 nm, respectively. This is much smaller than the diameter of normal biological cells (~500 nm on average), which have membranes that also contain other components, e.g. cholesterol. For that reason, GUVs (diameter ~200 nm) with cholesterol [31] match biological cells better than SUVs and LUVs.

It is important that in order to solve the equations, initial conditions are defined. These are values for the state variables ($C_{\text{FA}}^{\text{source}}$, $C_{\text{FA}}^{\text{inner}}$, $C_{\text{FA}}^{\text{outer}}$ and $C_{\text{FA}}^{\text{sink}}$ in this case) at $t = 0$. The initial concentrations for the source and sink remain constant and are shown in Table 1. Now the total model is defined, it has to be solved by an ODE solver. The results obtained are shown in Section 5.

4.1.2. Carrier protein

A second theory about FAs crossing a membrane involves a so-called carrier protein, which transports the FAs through the membrane (Fig. 5). Its transport mechanism shows great similarity with the FA transport by transmembrane diffusion ("flip-flop") due to a three-step transport process. The difference is that with a carrier protein the transport medium is not the phospholipid bilayer, but the carrier protein itself. The three steps to be taken are explained in the section about the physiological background of FA transport (Section 2), from which one could deduce four state variables. These are the concentrations of available sites on the carrier protein at the outer and inner side of the membrane ($C_{\text{CP}}^{\text{outer}}$ and $C_{\text{CP}}^{\text{inner}}$, respectively), and the FA concentration bound to carrier protein sites at the outer and inner membrane ($C_{\text{CP*FA}}^{\text{outer}}$ and $C_{\text{CP*FA}}^{\text{inner}}$, respectively). Rewritten in state equations, FA transport by a carrier protein is mathematically expressed as Eqs. (22)–(25), irrespective of the exact molecular mechanism of transport:

$$\dot{C}_{\text{CP}}^{\text{outer}} = -k_{\text{bind}}^{\text{CP}} C_{\text{FA}}^{\text{source}} C_{\text{CP}}^{\text{outer}} + k_{\text{release}}^{\text{CP}} C_{\text{CP*FA}}^{\text{outer}} + k_{\text{flip}}^{\text{CP}} (C_{\text{CP}}^{\text{inner}} - C_{\text{CP}}^{\text{outer}}) \tag{22}$$

$$\dot{C}_{\text{CP*FA}}^{\text{outer}} = k_{\text{bind}}^{\text{CP}} C_{\text{FA}}^{\text{source}} C_{\text{CP}}^{\text{outer}} - k_{\text{release}}^{\text{CP}} C_{\text{CP*FA}}^{\text{outer}} + k_{\text{flip}}^{\text{CP}} (C_{\text{CP*FA}}^{\text{inner}} - C_{\text{CP*FA}}^{\text{outer}}) \tag{23}$$

$$\dot{C}_{\text{CP}}^{\text{inner}} = -k_{\text{bind}}^{\text{CP}} C_{\text{FA}}^{\text{sink}} C_{\text{CP}}^{\text{inner}} + k_{\text{release}}^{\text{CP}} C_{\text{CP*FA}}^{\text{inner}} + k_{\text{flip}}^{\text{CP}} (C_{\text{CP}}^{\text{outer}} - C_{\text{CP}}^{\text{inner}}) \tag{24}$$

$$\dot{C}_{\text{CP*FA}}^{\text{inner}} = k_{\text{bind}}^{\text{CP}} C_{\text{FA}}^{\text{sink}} C_{\text{CP}}^{\text{inner}} - k_{\text{release}}^{\text{CP}} C_{\text{CP*FA}}^{\text{inner}} + k_{\text{flip}}^{\text{CP}} (C_{\text{CP*FA}}^{\text{outer}} - C_{\text{CP*FA}}^{\text{inner}}) \tag{25}$$

where $k_{\text{bind}}^{\text{CP}}$, $k_{\text{release}}^{\text{CP}}$ and $k_{\text{flip}}^{\text{CP}}$ represent the rate parameters of binding and release of FAs to and the "flip-flop" of the carrier protein, respectively. Moreover, it is assumed that the rate parameters $k_{\text{bind}}^{\text{CP}}$ and $k_{\text{release}}^{\text{CP}}$ are identical on both sides of the membrane and that the "flip" rate equals the "flop" rate. The state variables and source and sink concentrations have already been explained in Section 3. Unfortunately, no experimental data are available about the carrier protein, because its existence is not proven yet. Therefore, values for rate parameters and the initial concentration of carrier protein in the membrane have been based on estimations within physiological range (Table 1). The initial source and sink concentrations of FAs are the same as used in the transmembrane diffusion of FAs, so only concentrations of uFAs are used.

4.1.3. Enhancer

The 3rd transport mechanism is almost similar to transmembrane diffusion, but facilitated by an enhancer protein. The state equations of an enhancer are identical to the state equations for transmembrane diffusion of FAs (Eqs. (20) and (21)). The main difference is the value for the uptake parameter (k_{on}) on the outer side of the membrane, because the enhancer elevates this value, which in turn increases the FA flux across the membrane. The state equation of FAs in the inner leaflet is shown in Eq. (21), the other state equation for FAs in the outer leaflet is virtually identical to Eq. (20):

$$\dot{C}_{\text{FA}}^{\text{outer}} = k_{\text{on}}^{\text{enhancer}} C_{\text{FA}}^{\text{source}} - k_{\text{off}} C_{\text{FA}}^{\text{outer}} + k_{\text{flip}}(C_{\text{FA}}^{\text{inner}} - C_{\text{FA}}^{\text{outer}}) \tag{26}$$

where $k_{\text{on}}^{\text{enhancer}}$ is the rate parameter for absorption of FAs, which is augmented due to the presence of the enhancer.

A remarkable difference between biological and black membranes in the FA uptake is the difference in partition coefficient (K_{p}): the K_{p} for palmitate was shown to be $\sim 6.0 \times 10^5\ \text{M}^{-1}$ [L] in black [57], but $\sim 1.8 \times 10^6\ \text{M}^{-1}$ [L] in biological membranes (adipocytes [27]). The presence of a protein that increases the value of K_{p} could be responsible for the three times higher value in biological membranes than in black membranes. Therefore, a K_{p} of $6.0 \times 10^5\ \text{M}^{-1}$ [L] will be used for all models to come, in which no protein is present and results in a low value for k_{on} (as was shown in Eq. (18): $k_{\text{on}} = k_{\text{off}} K_{\text{p}}[\text{L}]$). The enhancer is modeled with the high value for K_{p} ($1.8 \times 10^6\ \text{M}^{-1}$ [L]), which results in an increased value for k_{on}. The other variables and parameters have been described in Section 4.1.1, just like the initial concentrations.

4.2. Model II: Extension of sarcolemma with a glycocalyx

4.2.1. Diffusion of albumin and its complexes

Model I with FA transport across the sarcolemma by means of transmembrane diffusion has been extended with a glycocalyx. Diffusion processes and binding reactions of uFAs, albumin and albumin-bound FAs have been incorporated in Model IIA. Eqs. (20) and (21)

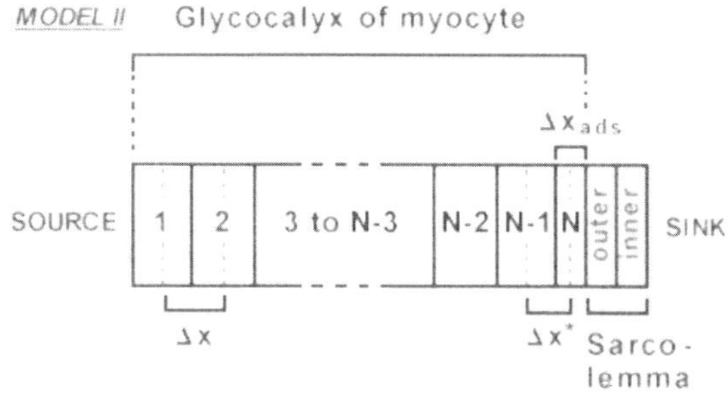

Fig. 12. Model II is an extension of Model I: a glycocalyx has been added. The segments have a thickness Δx and the adsorption layer a thickness Δx_{ads}. The source is located in the middle of the interstitium, the sink remains the same: the sarcoplasm. (*For a colored version of this figure, see plate section, page 599.*)

describe the state equations for the transport across the sarcolemma, but the following equations (Eqs. (27)–(31)) are added to describe the influence of a glycocalyx on the FA transport:

For the segments $i = 2$ to $N - 2$

$$\dot{C}^i_{\mathrm{uFA}} = \frac{D_{\mathrm{uFA}}}{(\Delta x)^2}(C^{i+1}_{\mathrm{uFA}} + C^{i-1}_{\mathrm{uFA}} - 2C^i_{\mathrm{uFA}}) - k_1 C^i_{\mathrm{Alb}} C^i_{\mathrm{uFA}} - k_2 C^i_{\mathrm{AF_1}} C^i_{\mathrm{uFA}} - \cdots$$

$$- k_3 C^i_{\mathrm{AF_2}} C^i_{\mathrm{uFA}} + k_{-1} C^i_{\mathrm{AF_1}} + k_{-2} C^i_{\mathrm{AF_2}} + k_{-3} C^i_{\mathrm{AF_3}} \tag{27}$$

$$\dot{C}^i_{\mathrm{Alb}} = \frac{D_{\mathrm{Alb}}}{(\Delta x)^2}(C^{i+1}_{\mathrm{Alb}} + C^{i-1}_{\mathrm{Alb}} - 2C^i_{\mathrm{Alb}}) - k_1 C^i_{\mathrm{Alb}} C^i_{\mathrm{uFA}} + k_{-1} C^i_{\mathrm{AF_1}} \tag{28}$$

$$\dot{C}^i_{\mathrm{AF_1}} = \frac{D_{\mathrm{AF_1}}}{(\Delta x)^2}(C^{i+1}_{\mathrm{AF_1}} + C^{i-1}_{\mathrm{AF_1}} - 2C^i_{\mathrm{AF_1}}) + k_1 C^i_{\mathrm{Alb}} C^i_{\mathrm{uFA}} - k_2 C^i_{\mathrm{AF_1}} C^i_{\mathrm{uFA}} - k_{-1} C^i_{\mathrm{AF_1}}$$

$$+ k_{-2} C^i_{\mathrm{AF_2}} \tag{29}$$

$$\dot{C}^i_{\mathrm{AF_2}} = \frac{D_{\mathrm{AF_2}}}{(\Delta x)^2}(C^{i+1}_{\mathrm{AF_2}} + C^{i-1}_{\mathrm{AF_2}} - 2C^i_{\mathrm{AF_2}}) + k_2 C^i_{\mathrm{AF_1}} C^i_{\mathrm{uFA}} - k_3 C^i_{\mathrm{AF_2}} C^i_{\mathrm{uFA}} - k_{-2} C^i_{\mathrm{AF_2}}$$

$$+ k_{-3} C^i_{\mathrm{AF_3}} \tag{30}$$

$$\dot{C}^i_{\mathrm{AF_3}} = \frac{D_{\mathrm{AF_3}}}{(\Delta x)^2}(C^{i+1}_{\mathrm{AF_3}} + C^{i-1}_{\mathrm{AF_3}} - 2C^i_{\mathrm{AF_3}}) + k_3 C^i_{\mathrm{AF_2}} C^i_{\mathrm{uFA}} - k_{-3} C^i_{\mathrm{AF_3}} \tag{31}$$

where i is an index, ranging between 1 and N, which represents the segment number of the interstitial compartment (divided in N segments, Fig. 12). The state variables are concentrations of uFA, albumin, AF_1, AF_2 and AF_3 at segment i: C^i_{uFA}, C^i_{Alb}, $C^i_{\mathrm{AF_1}}$, $C^i_{\mathrm{AF_2}}$ and $C^i_{\mathrm{AF_3}}$, respectively. The description of the remaining variables is listed in Table 1. The state equations for segment $i = 1$ are the same as Eqs. (27)–(31), with the difference that the diffusion part on the RHS

$$\left(\frac{D}{(\Delta x)^2}(C^{i+1} + C^{i-1} - 2C^i)\right)$$

is replaced by Eq. (32):

$$\frac{D}{\Delta x}\left(\frac{C^{\text{source}} - C^1}{\Delta x/2} - \frac{C^1 - C^2}{\Delta x}\right) \Leftrightarrow \frac{D}{(\Delta x)^2}(2C^{\text{source}} - 3C^1 + C^2) \tag{32}$$

using the theory outlined in Section 3.1.2. For the state equations at segment $i = N - 1$, the diffusion part on the RHS of Eqs. (27)–(31) is replaced by Eqs. (33) and (34):

$$\frac{D}{\Delta x}\left(\frac{C^{N-2} - C^{N-1}}{\Delta x} - \frac{C^{N-1} - C^N}{\Delta x^*}\right) \tag{33}$$

and

$$\frac{D}{\Delta x_{\text{ads}}}\left(\frac{C^{N-1} - C^N}{\Delta x^*} - \frac{C^N - C^{N-1}}{\Delta x^*}\right) \Leftrightarrow \frac{2D}{\Delta x_{\text{ads}}\Delta x^*}(C^{N-1} - C^N) \tag{34}$$

respectively. The final adaptation of the set of state equations for segment $i = N$ is the absorption of uFAs in the membrane. If one assumes that FAs are absorbed in the membrane within an adsorption layer of length 2.5 nm (~length of an FA molecule), which is half the membrane size, the state equation for uFAs becomes:

$$\begin{aligned}\dot{C}^N_{\text{uFA}} = {} & \frac{2D}{\Delta x \Delta x^*}(C^{N-1}_{\text{uFA}} - C^N_{\text{uFA}}) - k_1 C^N_{\text{Alb}} C^N_{\text{uFA}} - k_2 C^N_{\text{AF}_1} C^N_{\text{uFA}} - \cdots \\ & - k_3 C^N_{\text{AF}_2} C^N_{\text{uFA}} + k_{-1} C^N_{\text{AF}_1} + k_{-2} C^N_{\text{AF}_2} + k_{-3} C^N_{\text{AF}_3} - k_{\text{on}} C^N_{\text{uFA}}\end{aligned} \tag{35}$$

4.2.2. Receptor

The FA supply to the membrane could be enhanced by an albumin–FA complex receptor. This receptor is only useful when albumin and albumin-bound FAs can diffuse freely, like in Model IIA through the glycocalyx. If the assumption is made that the receptor is only present in the adsorption layer of the interstitial compartment ($i = N$), the state equations remain the same as in Model IIA, except for segment $i = N$:

$$\begin{aligned}\dot{C}^N_{\text{uFA}} = {} & \frac{2D_{\text{uFA}}}{\Delta x_{\text{ads}} \Delta x^*}(C^{N-1}_{\text{uFA}} - C^N_{\text{uFA}}) - k_1 C^N_{\text{Alb}} C^N_{\text{uFA}} - k_2 C^N_{\text{AF}_1} C^N_{\text{uFA}} - \cdots \\ & - k_3 C^N_{\text{AF}_2} C^N_{\text{uFA}} + k_{-1} C^N_{\text{AF}_1} + k_{-2} C^N_{\text{AF}_2} + k_{-3} C^N_{\text{AF}_3} + \cdots \\ & + k_2^{\text{receptor}}(C_{\text{R}^*\text{AF}_1} + C_{\text{R}^*\text{AF}_2} + C_{\text{R}^*\text{AF}_3}) - k_{\text{on}} C^N_{\text{uFA}} + k_{\text{off}} C^{\text{outer}}_{\text{FA}}\end{aligned} \tag{36}$$

$$\begin{aligned}\dot{C}^N_{\text{Alb}} = {} & \frac{2D_{\text{Alb}}}{\Delta x_{\text{ads}} \Delta x^*}(C^{N-1}_{\text{Alb}} - C^N_{\text{Alb}}) - k_1 C^N_{\text{Alb}} C^N_{\text{uFA}} + k_{-1} C^N_{\text{AF}_1} + \cdots \\ & + k_2^{\text{receptor}} C_{\text{R}^*\text{AF}_1} - k_1^{\text{receptor}} C^N_{\text{Alb}} C_{\text{receptor}} + k_{-1}^{\text{receptor}} C_{\text{R}^*\text{Alb}}\end{aligned} \tag{37}$$

$$\dot{C}_{\mathrm{AF}_1}^{N} = \frac{2D_{\mathrm{AF}_1}}{\Delta x_{\mathrm{ads}}\Delta x^*}(C_{\mathrm{AF}_1}^{N-1} - C_{\mathrm{AF}_1}^{N}) + k_1 C_{\mathrm{Alb}}^{N} C_{\mathrm{uFA}}^{N} - k_2 C_{\mathrm{AF}_1}^{N} C_{\mathrm{uFA}}^{N} - \cdots$$
$$- k_{-1} C_{\mathrm{AF}_1}^{N} + k_{-2} C_{\mathrm{AF}_2}^{N} + k_2^{\mathrm{receptor}} C_{\mathrm{R^*AF}_2} - k_1^{\mathrm{receptor}} C_{\mathrm{AF}_1}^{N} C_{\mathrm{receptor}}$$
$$+ k_{-1}^{\mathrm{receptor}} C_{\mathrm{R^*AF}_1} \tag{38}$$

$$\dot{C}_{\mathrm{AF}_2}^{N} = \frac{2D_{\mathrm{AF}_2}}{\Delta x_{\mathrm{ads}}\Delta x^*}(C_{\mathrm{AF}_2}^{N-1} - C_{\mathrm{AF}_2}^{N}) + k_2 C_{\mathrm{AF}_1}^{N} C_{\mathrm{uFA}}^{N} - k_3 C_{\mathrm{AF}_2}^{N} C_{\mathrm{uFA}}^{N} - \cdots$$
$$- k_{-2} C_{\mathrm{AF}_2}^{N} + k_{-3} C_{\mathrm{AF}_3}^{N} + k_2^{\mathrm{receptor}} C_{\mathrm{R^*AF}_3} - k_1^{\mathrm{receptor}} C_{\mathrm{AF}_2}^{N} C_{\mathrm{receptor}}$$
$$+ k_{-1}^{\mathrm{receptor}} C_{\mathrm{R^*AF}_2} \tag{39}$$

$$\dot{C}_{\mathrm{AF}_3}^{N} = \frac{2D_{\mathrm{AF}_3}}{\Delta x_{\mathrm{ads}}\Delta x^*}(C_{\mathrm{AF}_3}^{N-1} - C_{\mathrm{AF}_3}^{N}) + k_3 C_{\mathrm{AF}_2}^{N} C_{\mathrm{uFA}}^{N} - k_{-3} C_{\mathrm{AF}_3}^{N} - \cdots$$
$$- k_1^{\mathrm{receptor}} C_{\mathrm{AF}_3}^{N} C_{\mathrm{receptor}} + k_{-1}^{\mathrm{receptor}} C_{\mathrm{R^*AF}_3} \tag{40}$$

where state variables C_{receptor}, $C_{\mathrm{R^*AF}_1}$, $C_{\mathrm{R^*AF}_2}$ and $C_{\mathrm{R^*AF}_3}$ represent the concentrations of the receptor, receptor-AF_1, receptor-AF_2 and receptor-AF_3 complexes, respectively. The state equations for the receptor and its complexes are shown in Appendix A.

4.2.3. No diffusion of albumin and its complexes

Model IIB takes the reduced mobility of albumin and its complexes into account by using exactly the same state equations as described in Model IIA. This is realized by putting the diffusion coefficients for albumin, AF_1, AF_2 and AF_3 to zero. The effect of no albumin diffusion on the total FA flux is compared with Model IIA and shown in the Section 5.2.

4.3. Model III: From blood plasma to interstitium

In the previous sections, models have been described for FA transport across the sarcolemma and glycocalyx. A model of the FA transport from blood plasma to interstitium is composed likewise. One can identify the following five transport barriers for FAs in this pathway: the glycocalyx attached to the luminal side of the endothelial membrane, the luminal membrane, the endothelial cytoplasm, the abluminal membrane and finally a glycocalyx on the abluminal side of the endothelium. The state equations are composed of the same processes as shown in Models I and II: diffusion of substrates (uFA, albumin, AF_1, AF_2, AF_3), binding reactions (binding of uFAs to albumin and complexes) and FA transport across membranes by a "flip-flop" mechanism. Calculations in Section 2 have excluded FA transport by means of transcytosis, peri-endothelial transfer or through the clefts. One process has been added: the binding of uFAs to FABP in the endothelial cytoplasm. The total set of state equations for FA transport from blood plasma to the interstitial compartment (Fig. 13) can be found in Appendix A, just like all the other state equations described above.

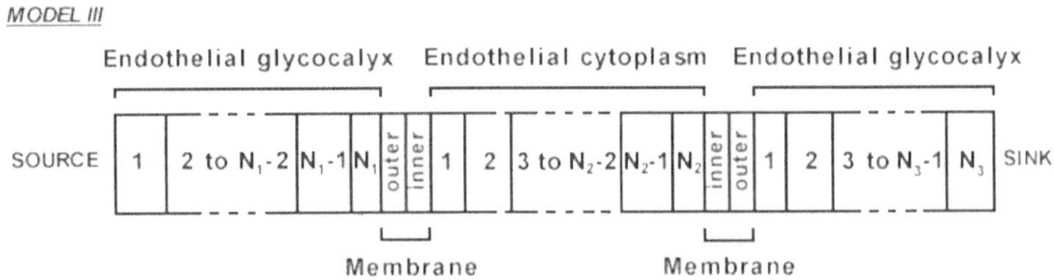

Fig. 13. This figure shows Model III: the endothelial barrier. It consists of two times an endothelial glycocalyx, a luminal and abluminal endothelial membrane and the endothelial cytoplasm. The source is the beginning of the luminal side of the endothelial glycocalyx, the sink is the middle of the interstitium (= source of Model II). (*For a colored version of this figure, see plate section, page 600.*)

5. Results and discussion

The equations of the three models are listed in Appendix A and are solved by the command `ode15s` (see also Appendix A) from the mathematical package MATLAB 6.5 (The MathWorks Inc.). In Models I and II, the *Monte Carlo* method has been incorporated in the model to encompass the probability distribution of the different parameters.

5.1. Results of Model I: The myocardial sarcolemma

Model I is a simplification of the myocardial sarcolemma, consisting of a pure phospholipid bilayer. The source of FAs in this model is the interstitial compartment,

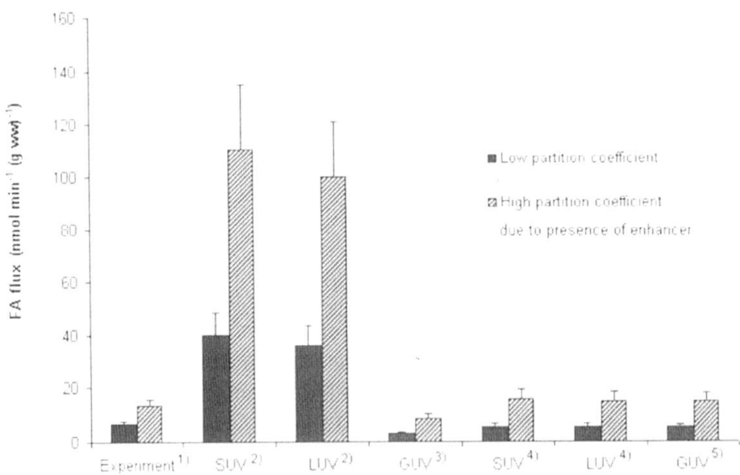

Fig. 14. The results of Model I. This figure shows the FA flux across the sarcolemma. The values vary between ~3 nmol min^{-1} (g ww)$^{-1}$ for GUV [3)] to ~40 nmol min^{-1} (g ww)$^{-1}$ for SUV [2)], using a low partition coefficient. For a high partition coefficient, the flux ranges between ~8 and ~110 nmol min^{-1} (g ww)$^{-1}$. The *Monte Carlo* method was used (500 cycles) to obtain these results. [1)] Experimental results, based on data of Ref. [45]. [2)] Values for k_{flip} and k_{off} from Ref. [30]. [3)] Values for k_{flip} and k_{off} from Ref. [31]. [4)] Values for k_{flip} and k_{off} from Refs. [30] and [31], respectively. [5)] Value for k_{off} from Ref. [31], value for k_{flip} obtained by extrapolation (see Fig. 15). (*For a colored version of this figure, see plate section, page 601.*)

near the sarcolemma, where $[FA]_{total}$ was assumed to be 0.09 mM (the total concentration of FA [45]) and $[Alb]_{total} = 0.3$ mM (total concentration of albumin [45,70]). The sink of this model is the sarcoplasm of the cardiomyocyte, i.e. $[FA]_{sarcoplasm} \approx 0$.

5.1.1. *Experimental data*

The fluxes determined in experiments performed on quiescent, isolated cardiomyocytes [45] will be used as comparison for the model results (Fig. 14). The experiments showed also that sulfo-*N*-succinimidyl-oleate (SSO) influences the FA fluxes. SSO inhibits membrane-associated proteins, putatively facilitating membrane FA transport. The flux in SSO-treated cardiomyocytes (no proteins involved) is used as reference value in the normal situation (low K_p) for the models. SSO-treated cardiomyocytes showed an FA flux of ~11.4 nmol min^{-1} (g ww)$^{-1}$ *cardiomyocyte* [45], which corresponds to ~6.7 nmol min^{-1} (g ww)$^{-1}$ *cardiac tissue*, since 1 g ww cardiac tissue consists of ~0.59 g ww cardiomyocyte (Vinnakota, *personal communication*). The non-SSO treated cardiomyocytes showed a higher FA flux: 13.5 nmol min^{-1} (g ww)$^{-1}$ heart (22.9 nmol min^{-1} (g ww)$^{-1}$ cardiomyocyte [45]), which is taken as reference value for the models with a high K_p (enhancer situation). The flux value found in Ref. [45] in the normal situation is ~7 times lower (13.5 nmol min^{-1} (g ww)$^{-1}$) than under physiological conditions (~100 nmol min^{-1} (g ww)$^{-1}$). A variety of explanations for this difference can be put forward: the absence of insulin [42], no electrical stimulation of the cardiomyocytes [43], no physiologically relevant metabolic requirements, etc.

5.1.2. *Transmembrane diffusion and enhancer*

Models with parameters derived from SUVs and LUVs in Ref. [23] lead to relatively high fluxes of FA of 110 and 100 nmol min^{-1} (g ww)$^{-1}$. These values are higher than those obtained from experiments performed on isolated cells, i.e. 13.5 nmol min^{-1} (g ww)$^{-1}$ [45]. The parameters from GUV experiments [31] show a significantly lower flux rate: 8.4 nmol min^{-1} (g ww)$^{-1}$. The experimental value for k_{flip} is probably underestimated in Ref. [31] as was shown in Section 4.1.1. To account for this underestimation, the value for k_{flip} of the GUV [31] is replaced by the value for k_{flip} of the SUV and LUV experiments. This resulted in a slightly higher flux value: 15.5 and 15.0 nmol min^{-1} (g ww)$^{-1}$, respectively. The final sub-model is still based on the parameters of Ref. [31], but now the value of k_{flip} has been obtained by extrapolation of data from Ref. [23] (see Fig. 15 and Table 1). Compared with the experimental data for both the normal situation and the presence of an enhancer [45], the model of Ref. [31] with the extrapolated k_{flip} gives the best approximation of this situation. The flux in a normal situation is 5.3 nmol min^{-1} (g ww)$^{-1}$, the enhancer leads to a flux of 14.7 nmol min^{-1} (g ww)$^{-1}$. The latter differs only ~8% from the experimental data.

5.1.3. *Carrier protein*

Like for the putative enhancer, the effect of a carrier protein depends on many factors, e.g. the concentration of the carrier protein and the affinity of the carrier for uFAs. These

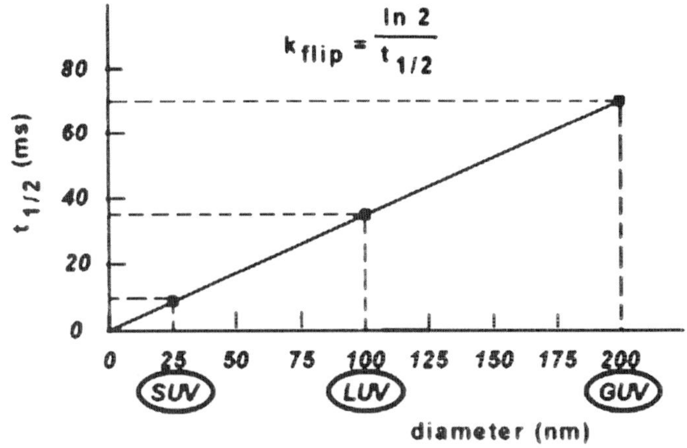

Fig. 15. Derivation of k_{flip} for GUV by extrapolation. Extrapolating $t_{1/2}$ is used to derive the rate parameter k_{flip} for GUV: $k_{flip} = \ln 2/t_{1/2}$. The $t_{1/2}$-values for SUV and LUV are obtained from Ref. [30]. A linear plot results in a $t_{1/2}$ for GUV to be 70 ms, which leads to $k_{flip} \sim 10\ s^{-1}$.

have been varied in the model within physiological range to check its effect (shown in Table 1 and in Fig. 16). Unfortunately, no experimental evidence about carrier proteins that transport FAs across the sarcolemma is available. Linear behavior of the FA flux is shown (Fig. 16), when the affinity constant of the carrier for uFAs and the release of uFAs from the carrier are raised from 1×10^6 to $1 \times 10^8\ M^{-1}$ and from 0.1 to 10 s^{-1}, respectively. The concentration of carrier protein is assumed to be 1 mM. With a high affinity of 1×10^8 M

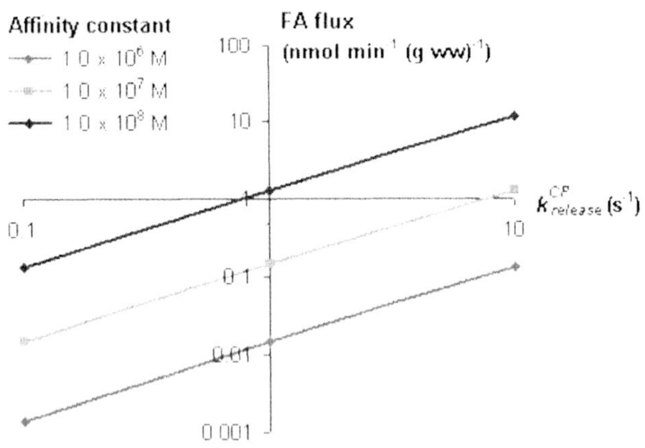

Fig. 16. Influence of carrier protein on the flux. The three plots (logarithmic scale) show that a 10-fold increase in $k^{CP}_{release}$ leads to a 10-fold increase in flux. Likewise, a 10-fold increase in affinity constant of non-protein bound FAs (uFAs) raises the flux also 10-fold. These simulations were performed with a carrier protein concentration of 1 mM. The highest flux in this figure is ~ 12 nmol min^{-1} $(g\ ww)^{-1}$ obtained with an affinity constant of 1.0×10^8 M and a $k^{CP}_{release}$ of 10 s^{-1}. *(For a colored version of this figure, see plate section, page 602.)*

and a fast release ($k_{release}^{CP} = 10\ s^{-1}$), the FA flux is 12.0 nmol min^{-1} (g ww)$^{-1}$, which is about the same flux as obtained with an enhancer (13.5 nmol min^{-1} (g ww)$^{-1}$).

5.1.4. Transport across sarcolemma is most likely by enhanced transmembrane diffusion

In summary, if one considers the outcome of the computational SUV and LUV models (Fig. 14), based on parameters of Ref. [23], pure transmembrane diffusion would be more than sufficient to satisfy a flux of 13.5 nmol min^{-1} (g ww)$^{-1}$, as was obtained from experiments on isolated, quiescent cells [45]. On the other hand, if one compares the adapted results (for k_{flip}) for GUV [31] with the experimental results [45], these correspond much better with the experiments than the computational SUV and LUV models did. This implies that the adapted GUV model [31] gives the best representation of the experimental data of Ref. [45] and the data of this model will be used in Models II and III. The model of the sarcolemma with an enhancer showed a physiologically feasible FA concentration in the sarcolemma ($\sim 6 \times 10^{-4}$ M, more than three orders lower in magnitude compared to the phospholipid concentration in the membrane).

The effect of a carrier protein is only relevant in a relatively high protein concentration of 1 mM, which corresponds with ~25% of all proteins present in the membrane (total protein concentration in membrane with adsorption layers is ~4 mM, see Table 1). Therefore, the involvement of a carrier with properties summarized in Table 1 is not very likely in the FA transport across the sarcolemma.

5.2. Results of Model II: Extension of sarcolemma with a glycocalyx

Model II simulates the effects of diffusion of uFAs and albumin-bound FAs, which are AF_1, AF_2 and AF_3. Model I has been extended with a glycocalyx with a variable thickness of 0.2–0.5 μm. The influence of this glycocalyx on a mobile albumin situation (Model IIA, diffusion of albumin) and the effect of immobile albumin (Model IIB, no diffusion of albumin) has been tested. The effect of a receptor with varying properties (shown in Table 1) on the flux is described for a glycocalyx thickness set at 0.4 μm. The source and sink concentrations of this model are identical to Model I, except that the source is now located in the middle of the interstitial compartment instead of at the adsorption layer of the sarcolemma.

5.2.1. Model IIA and Model IIB

The FA fluxes in steady state for Model IIA (mobile albumin) and Model IIB (immobile albumin) at different glycocalyx thickness are shown in Fig. 17. The FA fluxes for the mobile albumin situation remain the same, varying between ~13.3 and 14.5 nmol min^{-1} (g ww)$^{-1}$, irrespective of the glycocalyx thickness, and are in agreement with the flux obtained in Model I (Fig. 14). The influence of mobile albumin on the total flux (in steady state) is shown in Fig. 18. When albumin diffuses freely, the albumin-bound FAs contribute the most to the FA flux in the interstitial fluid, but near the sarcolemma, primarily uFAs take care of the flux. The summation of the four separate fluxes at a given point in the interstitial compartment gives a required constant flux in steady

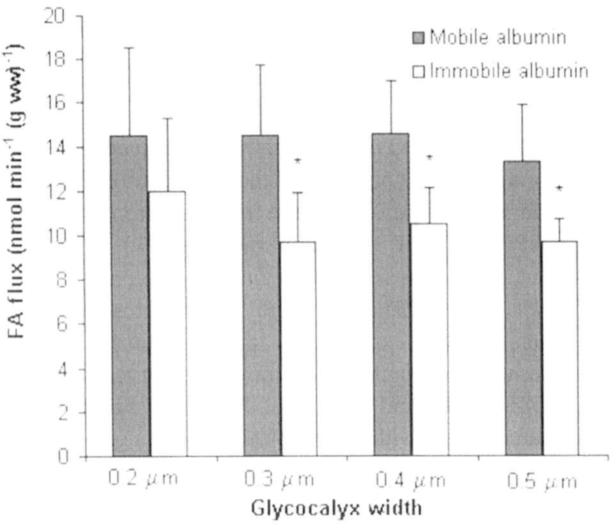

Fig. 17. Effect of glycocalyx width on FA flux. Immobile albumin, due to the glycocalyx, reduces the FA flux to ~70% of the value obtained with mobile albumin. *Monte Carlo* method was used (10 cycles). *Significantly different from value found with mobile albumin, $P < 0.05$, paired Student's *t*-test. (*For a colored version of this figure, see plate section, page 603.*)

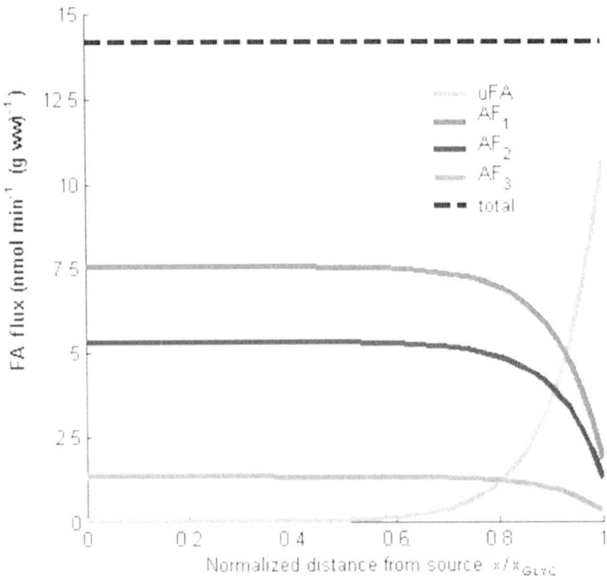

Fig. 18. Contribution of the different components to the total flux. The results of mobile albumin (not influenced by glycocalyx) are shown in this figure. As can be seen, the contributions of albumin-bound FAs (AF_1, AF_2 and AF_3) are higher in the beginning of the glycocalyx, but decreases near the membrane on the right. The uFA flux increases at that point. The black dotted line shows the total flux of all separate components, so the summation of uFA, AF_1, AF_2 and AF_3. (*For a colored version of this figure, see plate section, page 604.*)

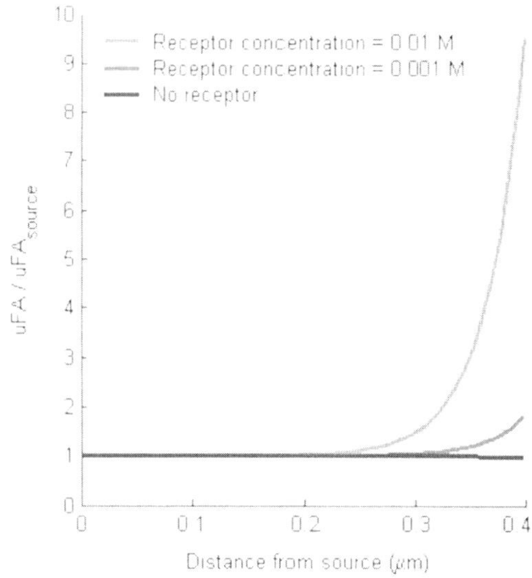

Fig. 19. The normalized concentration profile of non-protein bound FAs (uFAs) at different receptor concentrations of albumin-bound FAs. This profile is located in the glycocalyx of the cardiomyocyte. The source is the middle of the interstitium; the sarcolemma is located at the right part of this figure. For this simulation only, the glycocalyx width of the interstitium is taken to be 0.4 μm, instead of the usual 0.3 μm. Assumption: albumin diffuses freely as in Model IIA. (*For a colored version of this figure, see plate section, page 605.*)

state. However, the presence of a glycocalyx restricts albumin mobility significantly (paired Student's *t*-test, $P > 0.05$). The flux has been reduced to a minimum of $\sim$9.7 nmol min^{-1} (g ww)$^{-1}$. This is about 70% compared to a situation with diffusing albumin, if the glycocalyx width equals 0.3 μm or more (Fig. 17). Unfortunately, only in vitro experiments are performed to investigate the effect of the glycocalyx on the diffusion of macromolecules [74], which might represent an extreme vision (no mobile albumin) of the real physiological situation. To rate these findings on their true physiological significance, in vivo experiments should be performed. Therefore, the only two extremes are presented in this chapter: either the glycocalyx is not a barrier for albumin or albumin is immobilized [74].

5.2.2. Receptor

Model IIA has been extended with a receptor protein. Fig. 19 shows the effect of a receptor on the normalized concentration profile of uFAs in the interstitial compartment at different receptor concentrations. The high uFA concentration near the sarcolemma is explained by the relatively slow diffusion of uFAs originating from albumin–FA complexes. Since the FA release from albumin is facilitated by the receptor, the uFA concentration is further increased. Fig. 20 shows that the higher the concentration of receptor, the greater is the FA flux across the membrane. For a receptor concentration of 0.01 M, which is 2.5 times the total protein concentration in the membrane ($\sim$4 mM, calculated in Section 3), the calculated flux is $\sim$140 nmol min^{-1} (g ww)$^{-1}$, i.e. in the

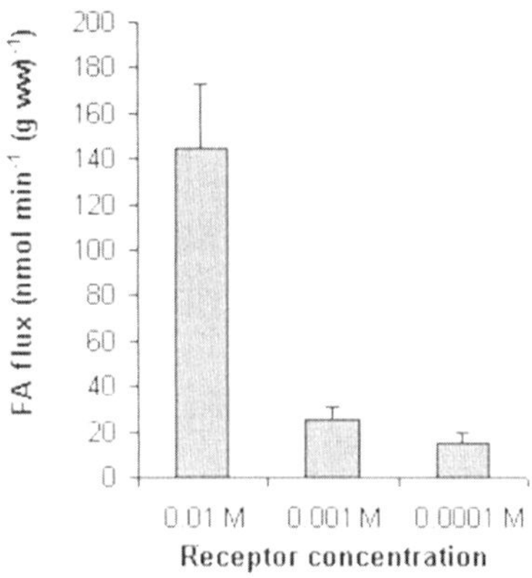

Fig. 20. Effect of receptor protein concentration on the FA flux. The FA flux is high for a receptor concentration of 0.01 M: ~145 nmol min^{-1} $(g\ ww)^{-1}$, but is reduced at lower receptor concentrations. These results were obtained by performing a *Monte Carlo* simulation with 10 cycles. (*For a colored version of this figure, see plate section, page 606.*)

same order of magnitude as the required flux of ~100 nmol min^{-1} $(g\ ww)^{-1}$. Diminishing the receptor concentration in the membrane to 0.001 M (~25% of all membrane proteins) reduces the flux to ~25 nmol min^{-1} $(g\ ww)^{-1}$. The required concentrations of receptor protein present in the membrane are physiologically too high to make it likely that a receptor contributes considerably to the total FA flux. Moreover, in the extreme case of a strong reduction in albumin mobility due to the putative hampering effect of the glycocalyx [74], the significance of a receptor protein is eliminated leading to an even larger reduction in FA flux.

5.2.3. Glycocalyx decreases the flux of fatty acids

In summary, the glycocalyx might reduce the diffusion of albumin considerably (~6 orders in magnitude), which would make the contribution of albumin to the FA transport almost negligible, implicating that only uFAs can diffuse freely. As was shown in Fig. 17, the FA flux in the immobile albumin situation [74] was ~70% of that in the situation with mobile albumin when AF_1, AF_2 and AF_3 contribute substantially to the FA flux (see also Fig. 18). If the experimental findings of Vink and Duling [74] are correct, the contribution of albumin-bound FA diffusion to overall diffusion of FAs to reach the sarcolemma is minor. The influence of a receptor could also be considerable, but only if one assumes receptor concentrations that are physiologically too high. The transport mechanism of a receptor is also based on mobile albumin and its complexes.

5.3. Results of Model III: From blood plasma to the interstitium

The final model represents the endothelial barrier, which is composed of the microvascular compartment, endothelium and the first half of the interstitial compartment

Table 2
Resulting fatty acid flux for different values of key parameters

Parameter	Parameter value	Flux of fatty acids (nmol min^{-1} (g ww)$^{-1}$)
K_p	1.8×10^5 M^{-1} [L]	1.6
	1.8×10^6 M^{-1} [L]	2.8
	1.8×10^7 M^{-1} [L]	3.1
$[FABP]_{total}$	0.6×10^{-5} M	0.9
	0.6×10^{-4} M	0.9
D_{FABP}	1.7×10^{-6} cm^2 s^{-1}	0.9
	1.7×10^{-5} cm^2 s^{-1}	0.9

The flux in the standard situation with parameter values $K_p = 6 \times 10^5$ M^{-1} [L], $[FABP]_{total} = 0.6$ μM and $D_{FABP} = 1.7 \times 10^{-7}$ cm^2 s^{-1} is 0.8 nmol min^{-1} (g ww)$^{-1}$.

(Fig. 13). The source is located in the microvascular compartment, at the beginning of the endothelial glycocalyx on the luminal side, and the sink in the interstitial fluid, at the end of the abluminal glycocalyx of the endothelium. The concentration of $[FA]_{total}$ and $[Alb]_{total}$ present in the source and sink is listed in Table 1, from which one could deduce the concentrations of uFAs, albumin and albumin-bound FAs [55].

5.3.1. Transmembrane diffusion and enhancer

Similar to the sarcolemma, the FA transport across the endothelial membranes has been simulated with transmembrane diffusion and the incorporation of an enhancer. The K_p values for the enhancer, FABP concentration in the endothelial cytoplasm and the diffusion coefficient of FABP have all been varied to test its effect on the overall flux. As can be seen in Table 2, the FA flux is ~0.8 nmol min^{-1} (g ww)$^{-1}$ under standard conditions. Moreover, the FA flux across the endothelium is most sensitive to enhanced K_p: a 30-fold increase in K_p leads to an almost 4-fold increase in FA flux. The increase of FABP concentration and diffusion is of no significance, because even a 100-fold increase in both cases does not influence the FA flux.

5.3.2. Enhancer alone is not sufficient to transport enough fatty acids across the endothelial barrier

In summary, the physiologically required FA flux found in Ref. [75] and also calculated in Section 2, ~100 nmol min^{-1} (g ww)$^{-1}$, is much higher than obtained with endothelial transmembrane diffusion alone or with the aid of an enhancer (Fig. 14). FABP in the endothelial cytoplasm plays no role of importance in the FA transport: elevation of diffusion coefficient and FABP concentration barely raised the FA flux (Table 2).

Explanations for the low FA flux across the endothelial barrier compared to the sarcolemmal barrier are:

- Differences in available area for diffusion. The area of the sarcolemma available for diffusion is 2000 cm^2 (g ww)$^{-1}$ [7], which is four times larger than the area of the endothelium: 500 cm^2 (g ww)$^{-1}$ [7]. A small area of diffusion reduces the FA flux considerably (Eq. (1)).

- The uFA gradient ($\partial C_{uFA}/\partial x$), the driving force for the diffusion flux, is larger across the sarcolemmal barrier than across the endothelial barrier. This is shown by the following calculation: 1.9 nM (source: uFA concentration in interstitial compartment) − 0 nM (sink: uFA concentration in sarcoplasm) = 1.9 nM over ~0.3 μm (see Table 1) results in a gradient of 6.3×10^{-5} M cm^{-1} for the sarcolemmal barrier vs. 5.1 nM (source: uFA concentration in vascular compartment) − 1.9 nM (sink: uFA concentration in interstitial compartment) = 3.2 nM over ~1.2 μm leads to a gradient across the endothelial barrier of 2.7×10^{-5} M cm^{-1}.

These two major differences between the sarcolemmal and endothelial barrier are the main causes for the low uFA fluxes across the endothelial barrier. How an FA flux of ~100 nmol min^{-1} (g ww)$^{-1}$ across the endothelium could be achieved in the intact heart, remains enigmatic and requires further experimentations.

5.4. Conclusion and future perspectives

The goal of this chapter was to develop a computer model that would help to elucidate the controversy of FAs crossing the sarcolemma. The computer model showed clearly that FA transport across the sarcolemma is most likely mediated by transmembrane diffusion with an enhancer. A simplified model of the sarcolemma (transmembrane diffusion of FAs, facilitated by a membrane-associated enhancer) approached the experimental data of FAs crossing the sarcolemma of an isolated cardiac muscle cell very well. The enhancer facilitated the absorption of uFAs in the membrane. Moreover, a carrier protein was not feasible to explain the experimental results, because it had to be present in too large quantities to promote a physiologically relevant FA flux.

The model of the sarcolemma was extended with a glycocalyx, which hampered the diffusion of macromolecules such as albumin [74] and reduced the FA flux with ~30% and has to be taken into account in future experiments. Till now, experiments of the FA transport do not take the glycocalyx into consideration. The possible role of a receptor in the sarcolemma, which facilitates the release of FAs from albumin, is less likely, since very high receptor protein concentrations were required and the mobility of AF_1, AF_2 and AF_3 was impaired due to the presence of the glycocalyx [74].

zThe 3rd model revealed an FA flux across the endothelium, which was too low to account for a physiological relevant flux. Increase in FABP concentration and mobility of albumin in the endothelial glycocalyx could not raise the FA flux to physiological relevant levels: diffusion of uFAs alone through the different compartments is not sufficient either. Therefore, the endothelium barrier is most likely rate governing in overall FA transport in the heart. The present calculations indicate that the FA flux across the endothelium by diffusion is not high enough. The diffusion gradient across the endothelial barrier (Fig. 21) and the area covering this barrier are the most sensitive model parameters. Another transport mechanism has to compensate for the low FA flux

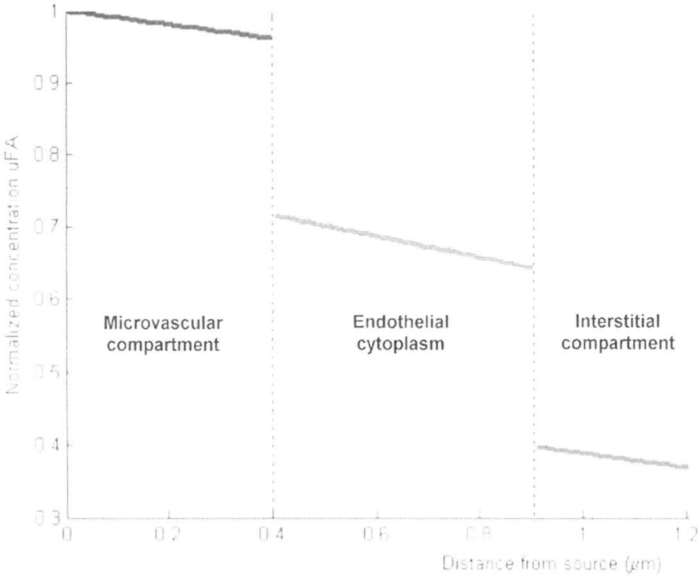

Fig. 21. Normalized concentration profile of uFAs (steady state) across the microvascular, endothelial and interstitial compartment. The concentration profiles are normalized (divided by the source concentration) and linear across the different compartments. (*For a colored version of this figure, see plate section, page 607.*)

across the endothelium, but its nature remains an open question and has to be investigated in future research.

An extension of the models presented in this chapter is the incorporation of the oxidation and storage of FAs in the cardiomyocyte. The models assumed a constant sink, regardless of the metabolic processes in the cardiomyocyte. Small disturbances, however, could exert profound effects on the FA uptake processes, for instance under diabetic conditions. The oxidation of FAs is closely related to the metabolism of carbohydrates, e.g. glucose, or contraction of the heart. An important challenge is to link computer models of all these separate processes to obtain more detailed insight in the interrelationship of metabolic and mechanical processes in the heart.

Appendix A

A.1. In general

Bovine serum albumin (BSA) has a lower affinity constant for FA, compared with human serum albumin (HSA), indicating that less FA bind to BSA than to HSA. This creates significant higher concentrations (paired Student's t-test, $P < 0.05$) of uFA for BSA than for HSA [55]. Because BSA is used in most experimental set-ups, all simulations were also performed with BSA.

A.2. Rate components

Absorption in membrane: $v_{abs}(C_1, C_2) = k_{on}C_1 - k_{off}C_2$

"Flip-flop": $v_{f-f}(C_1, C_2) = k_{flip}(C_1 - C_2)$

Binding to carrier: $v_{cb}(C_1, C_2, C_3) = k_{bind}^{CP}C_1C_2 - k_{release}^{CP}C_3$

Carrier "flip-flop": $v_{cf-f}(C_1, C_2) = k_{flip}^{CP}(C_1 - C_2)$

Diffusion:

$$v_{dif\_1}(C_z^1, C_z^2, C_z^3) = \frac{D_z}{(\Delta x)^2}(2C_z^1 - 3C_z^2 + C_z^3)$$

$$v_{dif\_2}(C_z^1, C_z^2, C_z^3) = \frac{D_z}{(\Delta x)^2}(C_z^1 - 2C_z^2 + C_z^3)$$

$$v_{dif\_3}(C_z^1, C_z^2, C_z^3) = \frac{D_z}{\Delta x}\left(\frac{C_z^1 - C_z^2}{\Delta x} - \frac{C_z^2 - C_z^3}{\Delta x^*}\right)$$

$$v_{dif\_4}(C_z^1, C_z^2) = \frac{2D_z}{\Delta x \Delta x^*}(C_z^1 - C_z^2)$$

Alb + uFA → AF_1: $v_{AF1}(C_1, C_2, C_3) = k_1C_1C_2 - k_{-1}C_3$

AF_1 + uFA → AF_2: $v_{AF2}(C_1, C_2, C_3) = k_2C_1C_2 - k_{-2}C_3$

AF_2 + uFA → AF_3: $v_{AF3}(C_1, C_2, C_3) = k_3C_1C_2 - k_{-3}C_3$

FABP + uFA → FA*FABP: $v_{FA^*FABP}(C_1, C_2, C_3) = k_1^{FABP}C_1C_2 - k_{-1}^{FABP}C_3$

Albumin-complexes binding to receptor: $v_{rb}(C_1, C_2, C_3) = k_1^{receptor}C_1C_2 - k_{-1}^{receptor}C_3$

uFA release from receptor: $v_{rr}(C_1) = k_2^{receptor}C_1$

$$\text{Diffusion}_{\,l}(C^i) = \begin{cases} v_{dif\_1}(C^{source}, C^1, C^2) & \text{for } i = 1 \text{ (Model II: interstitium, Model III: microvascular compartment)} \\ v_{dif\_2}(C^{i+1}, C^i, C^{i-1}) & \text{for } i = 2 \text{ to } N - 2 \text{ (Model II) or to } N_1 - 2 \text{ (Model III)} \\ v_{dif\_3}(C^{N-2}, C^{N-1}, C^N) & \text{for } i = N - 1 \text{ (Model II) or for } i = N_1 - 1 \text{ (Model III)} \\ v_{dif\_4}(C^{N-1}, C^N) & \text{for } i = N \text{ (Model II) or for } i = N_1 \text{ (Model III)} \end{cases}$$

$$\text{Diffusion}_{II}(C^i) = \begin{cases} v_{\text{dif\_4}}(C^{\text{2\_endo}}, C^{\text{1\_endo}}) & \text{for first endothelial segment} \\ v_{\text{dif\_3}}(C^{\text{1\_endo}}, C^{\text{2\_endo}}, C^{\text{3\_endo}}) & \text{for second endothelial segment} \\ v_{\text{dif\_2}}(C^{i+1}, C^i, C^{i-1}) & \text{for } i = 3 \text{ to } N_2 - 2 \\ v_{\text{dif\_3}}(C^{(N_2-2)\text{\_endo}}, C^{(N_2-1)\text{\_endo}}, C^{N_2\text{\_endo}}) & \text{for } i = N_2 - 1 \\ v_{\text{dif\_4}}(C^{N_2-1}, C^{N_2}) & \text{for } i = N_2 \end{cases}$$

$$\text{Diffusion}_{III}(C^i) = \begin{cases} v_{\text{dif\_4}}(C^{\text{source}}, C^1, C^2) & \text{for first segment of interstitium} \\ v_{\text{dif\_3}}(C^{i+1}, C^i, C^{i-1}) & \text{for second segment of intersitium} \\ v_{\text{dif\_2}}(C^{N-2}, C^{N-1}, C^N) & \text{for } i = 2 \text{ to } N_3 - 1 \\ v_{\text{dif\_1}}(C^{N_3-1}, C^{N_3}, C^{\text{sink}}) & \text{for } i = N_3 \end{cases}$$

$$\text{Absorption}_I(C^i_{\text{FA}}) = \begin{cases} -v_{\text{abs}}(C^N_{\text{uFA}}, C^{\text{outer}}_{\text{FA}}) & \text{for } i = N \\ 0 & \text{otherwise} \end{cases}$$

$$\text{Absorption}_{IIA}(C^i_{\text{FA}}) = \begin{cases} -v_{\text{abs}}(C^{N1}_{\text{uFA}}, C^{\text{outer\_1}}_{\text{FA}}) & \text{for } i = N_1 \\ 0 & \text{otherwise} \end{cases}$$

$$\text{Absorption}_{IIB}(C^i_{\text{FA}}) = \begin{cases} -v_{\text{abs}}(C^{\text{endo\_1}}_{\text{uFA}}, C^{\text{outer\_1}}_{\text{FA}}) & \text{for first segment of endothelium} \\ v_{\text{abs}}(C^{N_2}_{\text{uFA}}, C^{\text{inner\_2}}_{\text{FA}}) & \text{for } i = N_2 \\ 0 & \text{otherwise} \end{cases}$$

$$\text{Absorption}_{IIC}(C^i_{\text{FA}}) = \begin{cases} -v_{\text{abs}}(C^{\text{isf\_1}}_{\text{uFA}}, C^{\text{outer\_2}}_{\text{FA}}) & \text{for first segment of interstitium} \\ 0 & \text{otherwise} \end{cases}$$

A.3. Model I: the myocardial sarcolemma

A.3.1. Source

The adsorption layer of the sarcolemma in the interstitial compartment of the cardiomyocyte.

$$[\text{FA}]_{\text{total}} = 0.09 \text{ mM}, \ [\text{Alb}]_{\text{total}} = 0.3 \text{ mM (see Table 1)}$$

This results in $[\text{uFA}] = C^{\text{source}}_{\text{FA}} \sim 1.9$ nM (for BSA) and ~ 1.8 nM (for HSA).

A.3.2. Sink

The adsorption layer of the sarcolemma in the interstitial compartment:

$$[\text{uFA}] = C_{\text{FA}}^{\text{sink}} = 0 \text{ M (see Table 1))}$$

A.3.3. State equations

Transmembrane diffusion

Outer leaflet of sarcolemma

$$\dot{C}_{\text{FA}}^{\text{outer}} = v_{\text{abs}}(C_{\text{FA}}^{\text{source}}, C_{\text{FA}}^{\text{outer}}) + v_{\text{f-f}}(C_{\text{FA}}^{\text{inner}}, C_{\text{FA}}^{\text{outer}})$$

Inner leaflet of sarcolemma

$$\dot{C}_{\text{FA}}^{\text{inner}} = v_{\text{abs}}(C_{\text{FA}}^{\text{sink}}, C_{\text{FA}}^{\text{inner}}) + v_{\text{f-f}}(C_{\text{FA}}^{\text{outer}}, C_{\text{FA}}^{\text{inner}})$$

Enhancer protein

Identical to equations above, except for altered value for k_{on}, namely $k_{\text{on}}^{\text{enhancer}}$ (Table 1).

Carrier protein

Outer leaflet of sarcolemma

$$\dot{C}_{\text{CP}}^{\text{outer}} = -v_{\text{cb}}(C_{\text{FA}}^{\text{source}}, C_{\text{CP}}^{\text{outer}}, C_{\text{CP*FA}}^{\text{outer}}) + v_{\text{cf-f}}(C_{\text{CP}}^{\text{inner}}, C_{\text{CP}}^{\text{outer}})$$

$$\dot{C}_{\text{CP*FA}}^{\text{outer}} = v_{\text{cb}}(C_{\text{FA}}^{\text{source}}, C_{\text{CP}}^{\text{outer}}, C_{\text{CP*FA}}^{\text{outer}}) + v_{\text{cf-f}}(C_{\text{CP*FA}}^{\text{inner}}, C_{\text{CP*FA}}^{\text{outer}})$$

Inner leaflet of sarcolemma

$$\dot{C}_{\text{CP}}^{\text{inner}} = -v_{\text{cb}}(C_{\text{FA}}^{\text{sink}}, C_{\text{CP}}^{\text{inner}}, C_{\text{CP*FA}}^{\text{inner}}) + v_{\text{cf-f}}(C_{\text{CP}}^{\text{outer}}, C_{\text{CP}}^{\text{inner}})$$

$$\dot{C}_{\text{CP*FA}}^{\text{inner}} = v_{\text{cb}}(C_{\text{FA}}^{\text{sink}}, C_{\text{CP}}^{\text{inner}}, C_{\text{CP*FA}}^{\text{inner}}) + v_{\text{cf-f}}(C_{\text{CP*FA}}^{\text{outer}}, C_{\text{CP*FA}}^{\text{inner}})$$

A.4. Model II: extension of sarcolemma with a glycocalyx

A.4.1. Source

The middle of the interstitial compartment, same concentrations used as in Model I.

A.4.2. Sink

Identical to Model I.

A.4.3. State equations

Enhancer incorporated

Interstitial compartment ($i = 1$ to N)

$$\dot{C}_{\text{uFA}}^{i} = \text{Diffusion}_{I}(C_{\text{uFA}}^{i}) + \text{Absorption}_{\text{I}}(C_{\text{uFA}}^{i}) - v_{\text{AF1}}(C_{\text{uFA}}^{i}, C_{\text{Alb}}^{i}, C_{\text{AF1}}^{i}) - v_{\text{AF2}}(C_{\text{uFA}}^{i}, C_{\text{AF1}}^{i}, C_{\text{AF2}}^{i}) - v_{\text{AF3}}(C_{\text{uFA}}^{i}, C_{\text{AF2}}^{i}, C_{\text{AF3}}^{i})$$

$$\dot{C}^{i}_{\mathrm{Alb}} = \mathrm{Diffusion}_{l}(C^{i}_{\mathrm{Alb}}) - v_{\mathrm{AF1}}(C^{i}_{\mathrm{uFA}}, C^{i}_{\mathrm{Alb}}, C^{i}_{\mathrm{AF1}})$$

$$\dot{C}^{i}_{\mathrm{AF1}} = \mathrm{Diffusion}_{l}(C^{i}_{\mathrm{AF1}}) + v_{\mathrm{AF1}}(C^{i}_{\mathrm{uFA}}, C^{i}_{\mathrm{Alb}}, C^{i}_{\mathrm{AF1}}) - v_{\mathrm{AF2}}(C^{i}_{\mathrm{uFA}}, C^{i}_{\mathrm{AF1}}, C^{i}_{\mathrm{AF2}})$$

$$\dot{C}^{i}_{\mathrm{AF2}} = \mathrm{Diffusion}_{l}(C^{i}_{\mathrm{AF2}}) + v_{\mathrm{AF2}}(C^{i}_{\mathrm{uFA}}, C^{i}_{\mathrm{AF1}}, C^{i}_{\mathrm{AF2}}) - v_{\mathrm{AF3}}(C^{i}_{\mathrm{uFA}}, C^{i}_{\mathrm{AF2}}, C^{i}_{\mathrm{AF3}})$$

$$\dot{C}^{i}_{\mathrm{AF3}} = \mathrm{Diffusion}_{l}(C^{i}_{\mathrm{AF3}}) + v_{\mathrm{AF3}}(C^{i}_{\mathrm{uFA}}, C^{i}_{\mathrm{AF2}}, C^{i}_{\mathrm{AF3}})$$

Leaflets of sarcolemma

$$\dot{C}^{\mathrm{outer}}_{\mathrm{FA}} = v_{\mathrm{abs}}(C^{N}_{\mathrm{uFA}}, C^{\mathrm{outer}}_{\mathrm{FA}}) + v_{\mathrm{f-f}}(C^{\mathrm{inner}}_{\mathrm{FA}}, C^{\mathrm{outer}}_{\mathrm{FA}})$$

$$\dot{C}^{\mathrm{inner}}_{\mathrm{FA}} = v_{\mathrm{abs}}(C^{\mathrm{sink}}_{\mathrm{FA}}, C^{\mathrm{inner}}_{\mathrm{FA}}) + v_{\mathrm{f-f}}(C^{\mathrm{outer}}_{\mathrm{FA}}, C^{\mathrm{inner}}_{\mathrm{FA}})$$

Note: For Model IIB, no diffusion of albumin and its complexes is possible, so the values for the diffusion coefficients for albumin and complexes with albumin are set to zero in Model IIB.

Receptor present

Same equations as described in *Enhancer incorporated*, only equations for segment $i = N$ have been adapted and added:

$$\dot{C}^{N}_{\mathrm{uFA}} = v_{\mathrm{dif\_4}}(C^{N-1}_{\mathrm{uFA}}, C^{N}_{\mathrm{uFA}}) - v_{\mathrm{abs}}(C^{N}_{\mathrm{uFA}}, C^{\mathrm{outer}}_{\mathrm{FA}}) - v_{\mathrm{AF1}}(C^{N}_{\mathrm{uFA}}, C^{N}_{\mathrm{Alb}}, C^{N}_{\mathrm{AF1}})$$
$$- v_{\mathrm{AF2}}(C^{N}_{\mathrm{uFA}}, C^{N}_{\mathrm{AF1}}, C^{N}_{\mathrm{AF2}}) - v_{\mathrm{AF3}}(C^{N}_{\mathrm{uFA}}, C^{N}_{\mathrm{AF2}}, C^{N}_{\mathrm{AF3}}) + \cdots + v_{\mathrm{rr}}(C_{\mathrm{R^{*}AF1}})$$
$$+ v_{\mathrm{rr}}(C_{\mathrm{R^{*}AF2}}) + v_{\mathrm{rr}}(C_{\mathrm{R^{*}AF3}})$$

$$C^{N}_{\mathrm{Alb}} = v_{\mathrm{dif\_4}}(C^{N-1}_{\mathrm{Alb}}, C^{N}_{\mathrm{Alb}}) - v_{\mathrm{AF1}}(C^{N}_{\mathrm{uFA}}, C^{N}_{\mathrm{Alb}}, C^{N}_{\mathrm{AF1}}) - v_{\mathrm{rb}}(C^{N}_{\mathrm{Alb}}, C_{\mathrm{receptor}}, C_{\mathrm{R^{*}Alb}})$$
$$+ v_{\mathrm{rr}}(C_{\mathrm{R^{*}AF1}})$$

$$\dot{C}^{N}_{\mathrm{AF1}} = v_{\mathrm{dif\_4}}(C^{N-1}_{\mathrm{AF1}}, C^{N}_{\mathrm{AF1}}) + v_{\mathrm{AF1}}(C^{N}_{\mathrm{uFA}}, C^{N}_{\mathrm{Alb}}, C^{N}_{\mathrm{AF1}}) - v_{\mathrm{AF2}}(C^{N}_{\mathrm{uFA}}, C^{N}_{\mathrm{AF1}}, C^{N}_{\mathrm{AF2}})$$
$$- v_{\mathrm{rb}}(C^{N}_{\mathrm{AF1}}, C_{\mathrm{receptor}}, C_{\mathrm{R^{*}AF1}}) + v_{\mathrm{rr}}(C_{\mathrm{R^{*}AF2}})$$

$$\dot{C}^{N}_{\mathrm{AF2}} = v_{\mathrm{dif\_4}}(C^{N-1}_{\mathrm{AF2}}, C^{N}_{\mathrm{AF2}}) + v_{\mathrm{AF2}}(C^{N}_{\mathrm{uFA}}, C^{N}_{\mathrm{AF1}}, C^{N}_{\mathrm{AF2}}) - v_{\mathrm{AF3}}(C^{N}_{\mathrm{uFA}}, C^{N}_{\mathrm{AF2}}, C^{N}_{\mathrm{AF3}})$$
$$- v_{\mathrm{rb}}(C^{N}_{\mathrm{AF2}}, C_{\mathrm{receptor}}, C_{\mathrm{R^{*}AF2}}) + v_{\mathrm{rr}}(C_{\mathrm{R^{*}AF3}})$$

$$\dot{C}^{N}_{\mathrm{AF3}} = v_{\mathrm{dif\_4}}(C^{N-1}_{\mathrm{AF3}}, C^{N}_{\mathrm{AF3}}) + v_{\mathrm{AF3}}(C^{N}_{\mathrm{uFA}}, C^{N}_{\mathrm{AF2}}, C^{N}_{\mathrm{AF3}}) - v_{\mathrm{rb}}(C^{N}_{\mathrm{AF3}}, C_{\mathrm{receptor}}, C_{\mathrm{R^{*}AF3}})$$

$$\dot{C}_{\mathrm{R^{*}Alb}} = v_{\mathrm{rb}}(C^{N}_{\mathrm{Alb}}, C_{\mathrm{receptor}}, C_{\mathrm{R^{*}Alb}})$$

$$\dot{C}_{\mathrm{R^{*}AF1}} = v_{\mathrm{rb}}(C^{N}_{\mathrm{AF1}}, C_{\mathrm{receptor}}, C_{\mathrm{R^{*}AF1}}) - v_{\mathrm{rr}}(C_{\mathrm{R^{*}AF1}})$$

$$\dot{C}_{\mathrm{R^*AF2}} = v_{\mathrm{rb}}(C^N_{\mathrm{AF2}}, C_{\mathrm{receptor}}, C_{\mathrm{R^*AF2}}) - v_{\mathrm{rr}}(C_{\mathrm{R^*AF2}})$$

$$\dot{C}_{\mathrm{R^*AF3}} = v_{\mathrm{rb}}(C^N_{\mathrm{AF3}}, C_{\mathrm{receptor}}, C_{\mathrm{R^*AF3}}) - v_{\mathrm{rr}}(C_{\mathrm{R^*AF3}})$$

A.5. Model III: from blood plasma to the interstitium

A.5.1. Source

The start of the endothelial glycocalyx on the luminal side
$[\mathrm{FA}]_{\mathrm{total}} = 0.5$ mM, $[\mathrm{Alb}]_{\mathrm{total}} = 0.5$ mM (see Table 1).

A.5.2. Sink

The middle of the interstitium (identical to the source of Model II).

A.5.3. State equations

Microvascular compartment ($i = 1$ to N_1)
See Section A.4.3 and replace Absorption $_I(C^i_{\mathrm{uFA}})$ with Absorption$_{IIA}(C^i_{\mathrm{uFA}})$
Luminal endothelial membrane

$$\dot{C}^{\mathrm{outer\_1}}_{\mathrm{FA}} = v_{\mathrm{abs}}(C^{N1}_{\mathrm{uFA}}, C^{\mathrm{outer\_1}}_{\mathrm{FA}}) + v_{\mathrm{f-f}}(C^{\mathrm{inner\_1}}_{\mathrm{FA}}, C^{\mathrm{outer\_1}}_{\mathrm{FA}})$$

$$\dot{C}^{\mathrm{inner\_1}}_{\mathrm{FA}} = v_{\mathrm{abs}}(C^{\mathrm{1\_endo}}_{\mathrm{uFA}}, C^{\mathrm{inner\_1}}_{\mathrm{FA}}) + v_{\mathrm{f-f}}(C^{\mathrm{outer\_1}}_{\mathrm{FA}}, C^{\mathrm{inner\_1}}_{\mathrm{FA}})$$

Endothelial cytoplasm ($i = 1$ to N_2)

$$\dot{C}^i_{\mathrm{uFA}} = \mathrm{Diffusion}_{II}(C^i_{\mathrm{uFA}}) + \mathrm{Absorption}_{IIB}(C^i_{\mathrm{uFA}}) - v_{\mathrm{FA^*FABP}}(C^i_{\mathrm{uFA}}, C^i_{\mathrm{FABP}}, C^i_{\mathrm{FA^*FABP}})$$

$$\dot{C}^i_{\mathrm{FABP}} = \mathrm{Diffusion}_{II}(C^i_{\mathrm{FABP}}) - v_{\mathrm{FA^*FABP}}(C^i_{\mathrm{uFA}}, C^i_{\mathrm{FABP}}, C^i_{\mathrm{FA^*FABP}})$$

$$\dot{C}^i_{\mathrm{FA^*FABP}} = \mathrm{Diffusion}_{II}(C^i_{\mathrm{FA^*FABP}}) + v_{\mathrm{FA^*FABP}}(C^i_{\mathrm{uFA}}, C^i_{\mathrm{FABP}}, C^i_{\mathrm{FA^*FABP}})$$

Abluminal endothelial membrane

$$\dot{C}^{\mathrm{inner\_2}}_{\mathrm{FA}} = v_{\mathrm{abs}}(C^{N2}_{\mathrm{uFA}}, C^{\mathrm{inner\_2}}_{\mathrm{FA}}) + v_{\mathrm{f-f}}(C^{\mathrm{outer\_2}}_{\mathrm{FA}}, C^{\mathrm{inner\_2}}_{\mathrm{FA}})$$

$$\dot{C}^{\mathrm{outer\_2}}_{\mathrm{FA}} = v_{\mathrm{abs}}(C^{\mathrm{1\_isf}}_{\mathrm{uFA}}, C^{\mathrm{outer\_2}}_{\mathrm{FA}}) + v_{\mathrm{f-f}}(C^{\mathrm{inner\_2}}_{\mathrm{FA}}, C^{\mathrm{outer\_2}}_{\mathrm{FA}})$$

Interstitial compartment ($i = 1$ to N_3)

$$\dot{C}^i_{\mathrm{uFA}} = \mathrm{Diffusion}_{III}(C^i_{\mathrm{uFA}}) + \mathrm{Absorption}_{IIC}(C^i_{\mathrm{uFA}}) - v_{\mathrm{AF1}}(C^i_{\mathrm{uFA}}, C^i_{\mathrm{Alb}}, C^i_{\mathrm{AF1}})$$
$$- v_{\mathrm{AF2}}(C^i_{\mathrm{uFA}}, C^i_{\mathrm{AF1}}, C^i_{\mathrm{AF2}}) - v_{\mathrm{AF3}}(C^i_{\mathrm{uFA}}, C^i_{\mathrm{AF2}}, C^i_{\mathrm{AF3}})$$

$$\dot{C}^i_{\mathrm{Alb}} = \mathrm{Diffusion}_{III}(C^i_{\mathrm{Alb}}) - v_{\mathrm{AF1}}(C^i_{\mathrm{uFA}}, C^i_{\mathrm{Alb}}, C^i_{\mathrm{AF1}})$$

$$\dot{C}^i_{\mathrm{AF1}} = \mathrm{Diffusion}_{III}(C^i_{\mathrm{AF1}}) + v_{\mathrm{AF1}}(C^i_{\mathrm{uFA}}, C^i_{\mathrm{Alb}}, C^i_{\mathrm{AF1}}) - v_{\mathrm{AF2}}(C^i_{\mathrm{uFA}}, C^i_{\mathrm{AF1}}, C^i_{\mathrm{AF2}})$$

$$\dot{C}^i_{\mathrm{AF2}} = \mathrm{Diffusion}_{III}(C^i_{\mathrm{AF2}}) + v_{\mathrm{AF2}}(C^i_{\mathrm{uFA}}, C^i_{\mathrm{AF1}}, C^i_{\mathrm{AF2}}) - v_{\mathrm{AF3}}(C^i_{\mathrm{uFA}}, C^i_{\mathrm{AF2}}, C^i_{\mathrm{AF3}})$$

$$\dot{C}^i_{\mathrm{AF3}} = \mathrm{Diffusion}_{III}(C^i_{\mathrm{AF3}}) + v_{\mathrm{AF3}}(C^i_{\mathrm{uFA}}, C^i_{\mathrm{AF2}}, C^i_{\mathrm{AF3}})$$

A.6. ODE solver configuration

The steady-state solution of Model I was obtained by putting the two time-derivatives, $\dot{C}_{\mathrm{outer}}$ and $\dot{C}_{\mathrm{inner}}$, to zero. The two unknowns, C_{outer} and C_{inner}, could then be deduced from the two equations by substitution in MATLAB.

Models II and III are too complex to solve by substitution, so it has been solved with an ordinary differential equation solver. Both models are composed of "stiff" differential equations. Definition from Mathworks: [50] "An ordinary differential equation problem is stiff if the solution being sought is varying slowly, but there are nearby solutions that vary rapidly, so the numerical method must take small steps to obtain satisfactory results." These "stiff" differential equations are solved in MATLAB with the command ode15s, a "stiff" solver. The time span ranges from 0 to 5×10^6 s, long enough to get a steady-state solution for every compartment. The absolute tolerance, a measure to indicate the accuracy of the solver, is set to 1×10^{-10}, because else the solution converges to an erroneous end solution. Albumin and FABP are equally distributed across the corresponding compartments and taken as initial concentrations.

References

[1] Abumrad, N., Harmon, C., Ibrahimi, A., 1998. J. Lipid Res. 39, 2309.

[2] Alberts, B., Bray, D., Lewis, J., Raff, M., Roberts, K., Watson, J.D., 1994. In: Robertson, M., Adams, R., Goertzen, D., (Eds.), Membrane Structure. Garland Publishing, New York, pp. 478–506.

[3] Anonymous, 2003. Course: Structure of the Plasma Membrane. http://cr.middlebury.edu/biology/scb/pdf/PlasmaMemStructure.pdf.

[4] Ashbrook, J.D., Spector, A.A., Santos, E.C., Fletcher, J.E., 1975. J. Biol. Chem. 250, 2333.

[5] Atshaves, B.P., Foxworth, W.B., Frolov, A., Roths, J.B., Kier, A.B., Oetama, B.K., Piedrahita, J.A., Schroeder, F., 1998. Am. J. Physiol. 274, C633–C644.

[6] Barta, E., Sideman, S., Bassingthwaighte, J.B., 2000. Ann. Biomed. Eng. 28, 331.

[7] Bassingthwaighte, J.B., Noodleman, L., van der Vusse, G.J., Glatz, J.F., 1989. Mol. Cell Biochem. 88, 51.

[8] Berk, P.D., Stump, D., 1999. Mol. Cell Biochem. 192, 17.

[9] Bing, R.J., Siegel, A., Ungar, I., Gilbert, M., 1954. Am. J. Med., 504.

[10] Bloom, W., Fawcett, D.W., 1975. A Textbook of Histology. W.B. Saunders Company, Philadelphia, PA.

[11] Bonen, A., Luiken, J.J.F.P., Arumugam, Y., Glatz, J.F., Tandon, N.N., 2000. J. Biol. Chem. 275, 14501.

[12] Brinkmann, J.F., Abumrad, N.A., Ibrahimi, A., van der Vusse, G.J., Glatz, J.F., 2002. Biochem. J. 367, 561.

[13] Broring, K., Haest, C.W., Deuticke, B., 1989. Biochim. Biophys. Acta 986, 321.

[14] Close, C.M., Frederick, D.K., 1995. Modeling and Analysis of Dynamic Systems. Wiley, New York.

[15] Cooper, R.B., Noy, N., Zakim, D., 1989. J. Lipid Res. 30, 1719.

[16] Demant, E.J., Richieri, G.V., Kleinfeld, A.M., 2002. Biochem. J. 363, 809.
[17] Efron, B., Tibshirani, R., 1993. An Introduction to the Bootstrap. Chapman & Hall, New York.
[18] Gaigalas, A.K., Hubbard, J.B., McCurley, M., Woo, S., 1992. J. Phys. Chem. 96, 2355.
[19] Guthmann, F., Haupt, R., Looman, A.C., Spener, F., Rustow, B., 1999. Am. J. Physiol 277, L191–L196.
[20] Gutknecht, J., 1988. J. Membr. Biol. 106, 83.
[21] Hamilton, J.A., 1998. J. Lipid Res. 39, 467.
[22] Hamilton, J.A., Guo, W., Kamp, F., 2002. Mol. Cell Biochem. 239, 17.
[23] Hamilton, J.A., Kamp, F., 1999. Diabetes 48, 2255.
[24] Harmon, C.M., Abumrad, N.A., 1993. J. Membr. Biol., 43.
[25] Hutter, J.F., Piper, H.M., Spieckermann, P.G., 1984. Basic Res. Cardiol. 79, 274.
[26] Julien, P., Gailis, L., Lepage, M., Roy, P.E., 1979. Artery 5, 37.
[27] Kamp, F., Guo, W., Souto, R., Pilch, P.F., Corkey, B.E., Hamilton, J.A., 2003. J. Biol. Chem. 278, 7988.
[28] Kamp, F., Hamilton, J.A., 1992. Proc. Natl Acad Sci. USA 89, 11367.
[29] Kamp, F., Hamilton, J.A., Kamp, F., Westerhoff, H.V., Hamilton, J.A., 1993. Biochemistry 32, 11074.
[30] Kamp, F., Zakim, D., Zhang, F., Noy, N., Hamilton, J.A., 1995. Biochemistry 34, 11928.
[31] Kleinfeld, A.M., Chu, P., Romero, C., 1997. Biochemistry 36, 14146.
[32] Kohl, P., Noble, D., Winslow, R.L., Hunter, P.J., 2000. Phil. Trans. R. Soc. Lond. A 358, 579.
[33] Kohn, M.C., 1983. Ann. Biomed. Eng. 11, 533.
[34] Kohn, M.C., Garfinkel, D., 1983. Ann. Biomed. Eng. 11, 533.
[35] Kohn, M.C., Garfinkel, D., 1983. Ann. Biomed. Eng. 11, 511.
[36] Kornberg, R.D., McConnell, H.M., 1971. Biochemistry 10, 1111.
[37] Kusaka, Y., Tanaka, T., Okamoto, F., Terasaki, F., Matsunaga, Y., Miyazaki, H., Kawamura, K., 1995. J. Mol. Cell Cardiol. 27, 1605.
[38] Liedtke, A.J., 1981. Prog. Cardiovasc. Dis. 23, 321.
[39] Linssen, M.C., Vork, M.M., De Jong, Y.F., Glatz, J.F., van der Vusse, G.J., 1990. Mol. Cell Biochem. 98, 19.
[40] Luby-Phelps, K., 2000. Int. Rev. Cytol. 192, 189.
[41] Luiken, J.J., Bonen, A., Glatz, J.F., 2002. Prostaglandins Leukot. Essent. Fatty Acids 67, 73.
[42] Luiken, J.J., Koonen, D.P.Y., Willems, J., Zorzano, A., Becker, C., Fischer, Y., Tandon, N.N., van der Vusse, G.J., Bonen, A., Glatz, J.F., 2002. Diabetes 51, 3113.
[43] Luiken, J.J., Willems, J., van der Vusse, G.J., Glatz, J.F., 2001. Am. J. Physiol. Endocrinol. Metab. 281, E704–E712.
[44] Luiken, J.J.F.P., Turcotte, L.P., Bonen, A., 1999. J. Lipid Res. 40, 1007.
[45] Luiken, J.J.F.P., Van Nieuwenhoven, F.A., America, G., van der Vusse, G.J., Glatz, J.F., 1997. J. Lipid Res. 38, 745.
[46] Luxon, B.A., Holly, D.C., Milliano, M.T., Weisiger, R.A., 1998. Am. J. Physiol. 274, G52–G61.
[47] Luxon, B.A., Weisiger, R.A., 1993. Sex differences in intracellular fatty acid transport: role of cytoplasmic binding proteins. Am. J. Physiol. 265(5 (Pt1)), G831–G841.
[48] MacLeod, R.S., 2003. Bioengineering Course About Systems Physiology: Microcirculation (Presentation). http://www.cvrti.utah.edu/~macleod/bioen/be6000/notes/microcirculation-bw.pdf.
[49] Michel, C.C., Curry, F.E., 1999. Physiol. Rev. 79, 703.
[50] Moler, C, 2003. Stiff Differential Equations. http://www.mathworks.com/company/newsletter/clevescorner/may03_cleve.shtml.
[51] Most, A.S., Brachfeld, N., Gorlin, R., Wahren, J., 1969. J. Clin. Invest. 48, 1177.
[52] Ockner, R.K., 1990. Mol. Cell Biochem. 98, 3.
[53] Ockner, R.K., Manning, J.A., Poppenhausen, R.B., Ho, W.K., 1972. Science 177, 56.
[54] Popov, D., Hasu, M., Ghinea, N., Simionescu, N., Simionescu, M., 1992. J. Mol. Cell Cardiol. 24, 989.
[55] Richieri, G.V., Anel, A., Kleinfeld, A.M., 1993. Biochemistry 32, 7574.
[56] Richieri, G.V., Ogata, R.T., Kleinfeld, A.M., 1994. J. Biol. Chem. 269, 23918.
[57] Richieri, G.V., Ogata, R.T., Kleinfeld, A.M., 1996. J. Biol. Chem. 271, 31068.
[58] Schaffer, J.E., Lodish, H.F., 1994. Cell 79, 427.
[59] Scow, R.O., Blanchette-Mackie, E.J., 1985. Prog. Lipid Res. 24, 197.
[60] Simionescu, M., Gafencu, A., Antohe, F., 2002. Microsc. Res. Tech. 57, 269.

[61] Sober, H.A., 1968. Handbook of Biochemistry. Chemical Rubber Co., Cleveland, OH.
[62] Spector, A.A., Fletcher, J.E., Ashbrook, J.D., 1971. Biochemistry 10, 3229.
[63] Stace, T.M., Damiano, E.R., 2001. Biophys. J. 80, 1670.
[64] Stewart, J.M., Driedzic, W.R., Berkelaar, J.A., 1991. Biochem. J. 275(Pt 3), 569.
[65] Stremmel, W., Strohmeyer, G., Berk, P.D., 1986. Proc. Natl Acad. Sci. USA 83, 3584.
[66] Stump, D.D., Fan, X., Berk, P.D., 2001. J. Lipid Res. 42, 509.
[67] Taylor, G., 1953. Proc.R. Soc. Lond. Sect. A 219, 186.
[68] Tipping, E., Ketterer, B., 1981. Biochem. J. 195, 441.
[69] Tiruppathi, C., Finnegan, A., Malik, A.B., 1996. Proc. Natl Acad. Sci. USA 93, 250.
[70] van der Vusse, G.J., Glatz, J.F., Stam, H.C., Reneman, R.S., 1992. Physiol. Rev. 72, 881.
[71] van der Vusse, G.J., van Bilsen, M., Glatz, J.F., 2000. Cardiovasc. Res. 45, 279.
[72] Van Nieuwenhoven, F.A., Verstijnen, C.P., Van Eys, G.J., Van Breda, E., De Jong, Y.F., van der Vusse, G.J., Glatz, J.F., 1994. J. Mol. Cell Cardiol. 26, 1635.
[73] Vicini, P., Cobelli, C., 1997. Ann. Biomed. Eng. 25, 815.
[74] Vink, H., Duling, B.R., 2000. Am. J. Physiol. Heart Circ. Physiol. 278, H285–H289.
[75] Vyska, K., Meyer, W., Stremmel, W., Notohamiprodjo, G., Minami, K., Machulla, H.J., Gleichmann, U., Meyer, H., Korfer, R., 1991. Circ. Res. 69, 857.
[76] Wang, X., Hole, D.G., Da Costa, T.H., Evans, R.D., 1998. Am. J. Physiol. 275, E265–E271.
[77] Weisiger, R.A., Pond, S., Bass, L., 1991. Am. J. Physiol. 261, G872–G884.
[78] Weiss, E.S., Hoffman, E.J., Phelps, M.E., Welch, M.J., Henry, P.D., Ter Pogossian, M.M., Sobel, B.E., 1976. Circ. Res. 39, 24.
[79] Wisneski, J.A., Gertz, E.W., Neese, R.A., Mayr, M., 1987. J. Clin. Invest. 79, 359.
[80] Zhang, F., Kamp, F., Hamilton, J.A., 1996. Biochemistry 35, 16055.

Regulation of fatty acid oxidation by malonyl CoA in cardiac muscle

Gary D. Lopaschuk* and Arzu Onay-Besikci

Pediatrics, Faculty of Medicine and Dentistry, University of Alberta, 423 Heritage Medical Research Building, Edmonton, Alta., Canada T6G 2S2
**Correspondence address: Tel.: +1-780-492-2170; fax: +1-780-492-9753*
E-mail: gary.lopaschuk@ualberta.ca(G.D.L.)

1. Introduction

While carbohydrates (mainly glucose and lactate) are important fuel sources, the adult heart normally obtains the majority of its energy demands from the metabolism of fatty acids [1–3]. The control of cardiac fatty acid metabolism is complex, but normally ensures an adequate energy supply to meet the high energy demands of the heart. There are a number of key sites that control fatty acid metabolism in the heart, including: 1) the supply of fatty acids from either free fatty acids bound to albumin or as fatty acids released from triacylglycerol contained in chylomicrons or VLDL, 2) the uptake of fatty acids across the cardiac sarcolemmal membrane by either carrier-mediated pathways (FAT/CD36, FABPpm, or FATP transporters) or facilitated diffusion, 3) the activation of fatty acids to long chain acyl CoA in the cytoplasm, 4) the transport of fatty acid moieties across the mitochondrial membrane, 5) the β-oxidation of fatty acids, and 6) the competition for acetyl CoA derived from β-oxidation with acetyl CoA derived from carbohydrate oxidation. Other chapters in this volume of *Lipobiology* address the issue of fatty acid supply, cellular uptake, and cytoplasmic activation. Therefore, in this chapter we will focus on the control of mitochondrial fatty acid uptake. The key role of malonyl CoA in this process will also be discussed.

It is increasingly becoming recognized that alterations in fatty acid oxidation contribute to a number of cardiac pathologies, including ischemic heart disease, diabetic cardiomyopathies, cardiac hypertrophy, and heart failure (see Ref. [4] for review). How malonyl CoA control of mitochondrial fatty acid uptake is altered in these pathologies will also be discussed.

Advances in Molecular and Cell Biology, Vol. 33, pages 223–241

ISSN: 1569-2558/DOI: 10.1016/S1569-2558(03)33011-5

2. Energy production in the heart

The heart has a very high energy demand due to the continuous work performed by the cardiac muscle. Under normal physiological conditions, the main fuels involved in maintaining the high energy requirements of the heart are fatty acids, glucose, and lactate (Fig. 1). Glucose metabolism starts with the reactions of glycolysis in the cytoplasm. Under aerobic conditions, the pyruvate generated from glycolysis is then taken up by the mitochondria and is converted to acetyl CoA by the pyruvate dehydrogenase

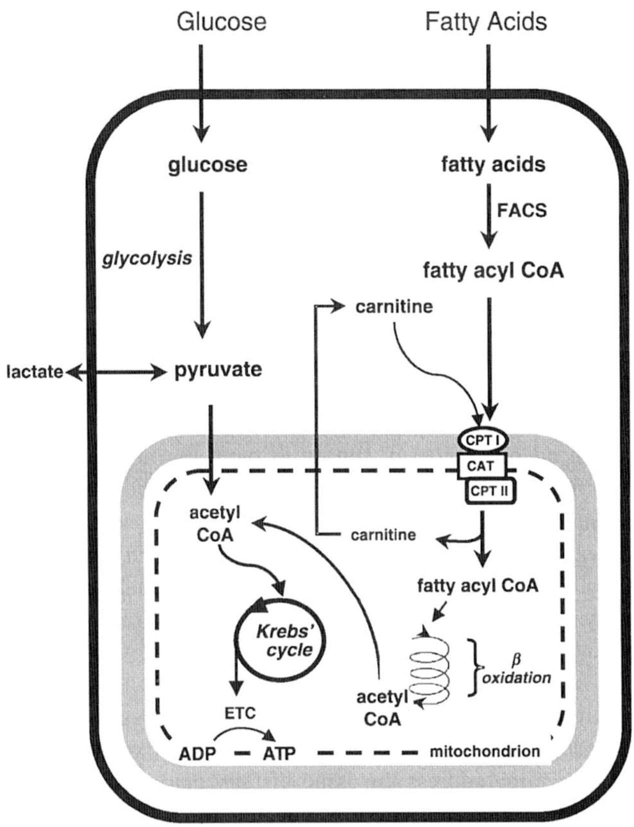

Fig. 1. Overview of myocardial energy metabolism. Glucose and fatty acids are primary sources for energy production in the adult heart. Following uptake, the major metabolic pathway of glucose is that of glycolysis. The pyruvate derived from glycolysis is either converted to lactate or transported into the mitochondria where it is decarboxylated and converted to acetyl CoA. Fatty acids are transported into the cytoplasm and activated to their CoA esters by a FACS. The acyl CoA is converted to acylcarnitine by CPT I translocated across the inner mitochondrial membrane by carnitine acyltranslocase (CAT) and then converted back to acyl CoA by CPT II. The intramitochondrial acyl CoA then enters the fatty acid β-oxidation spiral. Acetyl CoA derived from oxidation of glucose or β-oxidation of fatty acids enters the Krebs cycle. The electron-carrying species $NADH_2$ (produced either in glycolysis or in the Krebs cycle) and FADH (produced in Krebs cycle) are shuttled through a sequence of transformations called the electron transport chain (ETC). The hydrogen on $NADH_2$ and FADH is transferred to water in the presence of oxygen, and ATP is produced.

complex (PDC). Lactate can also be taken up by the heart and converted to pyruvate via lactate dehydrogenase.

Although glucose is an important source for ATP production, under normal physiological conditions the adult heart obtains approximately 60–80% of its overall energy requirements from the oxidation of fatty acids [1,5]. Fatty acids cross plasma membrane by diffusion or by protein-mediated transport (for reviews see Refs. [6,7]). Once in the cytoplasm, fatty acids are converted to their CoA esters, a process known as activation of fatty acids. The acyl moiety is then transferred into the mitochondria by a complex of enzymes involving carnitine palmitoyltransferase I (CPT I), carnitine/acylcarnitine translocase, and carnitine palmitoyltransferase II (CPT II) (Fig. 1). Once inside the mitochondrial matrix, acyl CoA passes through β-oxidation. Each successive cycle of the β-oxidation produces one molecule of acetyl CoA, and one each of NADH and $FADH_2$.

Acetyl CoA derived from glucose, fatty acid, or lactate then enters the Krebs' cycle, resulting in the liberation of CO_2, NADH, and $FADH_2$. The NADH derived from glycolysis, the Krebs' cycle, and β-oxidation, as well as $FADH_2$ from Krebs cycle and β-oxidation then enter the electron transfer chain. The H^+ on NADH and $FADH_2$ is transferred to H_2O in the presence of O_2, and ADP is converted to ATP (Fig. 1).

3. Fatty acid import into the mitochondria

Once fatty acids are taken up by the cardiomyocyte (bound to fatty acid binding protein, FABP), fatty acyl CoA synthase (FACS) activates fatty acids to long-chain acyl CoA. This long-chain acyl CoA can be either used for complex lipid synthesis (triacylglycerol, phospholipids, sphingolipids, diacylglycerol) or used as a substrate for fatty acid β-oxidation. In order to be oxidized, the fatty acid moiety from long-chain acyl CoA is transferred to carnitine and is taken up into the mitochondria by the carnitine shuttle where it is converted back to long-chain acyl CoA. As will be discussed, malonyl CoA has an important role in regulating this carnitine shuttle.

The carnitine palmitoyltransferase system consists of three enzymes that operate sequentially to effect the transport of long-chain fatty acids from the cytosol into the mitochondrial matrix. The first component of this system, CPT I, is an integral protein of the outer mitochondrial membrane (Fig. 2) and catalyzes the transfer of a fatty acyl group from cytosolic acyl CoA to carnitine. The acylcarnitine is then able to enter the mitochondrial matrix by means of a specific carnitine/acylcarnitine carrier protein, carnitine/acylcarnitine translocase. A distinct gene product, CPT II, loosely associated with the mitochondrial inner membrane, regenerates fatty acyl CoA and releases free carnitine [8] (Fig. 2).

4. Malonyl CoA inhibition of CPT I as a regulator of mitochondrial fatty acid uptake

Malonyl CoA is a three-carbon CoA ester that is involved in many physiological and pathophysiological events in the cell. In the heart, malonyl CoA is an important intermediate between two opposing pathways, fatty acid synthesis and fatty acid oxidation [9].

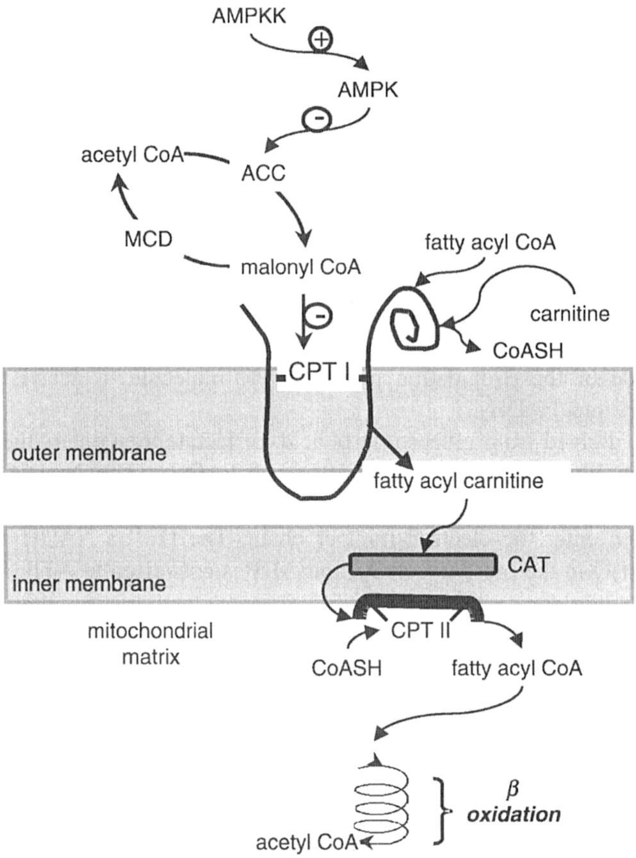

Fig. 2. Regulation of fatty acid oxidation by malonyl CoA. Malonyl CoA is a potent inhibitor of CPT I. ACC in the cytosol synthesizes malonyl CoA from acetyl CoA. 5′-AMPK phosphorylates and inactivates ACC. AMPK activity is increased by AMPKK. Malonyl CoA is degraded by the action of MCD.

Malonyl CoA is also found in non-lipogenic tissues such as the heart and skeletal muscle. In these tissues, malonyl CoA primarily modulates fatty acid oxidation due to its potent inhibitory effect on CPT I. This results in a decrease in the uptake of fatty acids into the mitochondria, thereby reducing mitochondrial fatty acid oxidation.

CPT I is present in the heart as two isoforms, a liver isoform (L-CPT I or CPT Iα) and a muscle isoform (M-CPT I or CPT Iβ). L-CPT I displays a higher affinity for carnitine and is less sensitive to the inhibition by malonyl CoA compared to the M-CPT I (reviewed in Ref. [8]), whereas their affinities for long-chain acyl CoA are similar [10,11]. The isoforms also show different tissue distribution. L-CPT I is widely expressed with the highest level of expression in the liver, kidney, ovary, intestine, pancreatic islet cells, spleen, and to a lesser extent in the heart. M-CPT I is the dominant isoform expressed in the heart and is also expressed in skeletal muscle, brown and white adipose tissue, adult heart, and testis [8,12,13].

CPT I is an integral protein that has two transmembrane domains [14]. Both N-terminus and the catalytic C-terminus are exposed to the cytosol [14–16]. Two transmembrane segments are connected with a short loop of 27 amino acids that is localized in the intermembrane space between the outer and inner mitochondrial membranes (Fig. 2). The extreme N-terminus of the protein is necessary for malonyl CoA sensitivity, since malonyl CoA sensitivity of CPT I is completely lost by the deletion of highly conserved six amino acid residues in this region [17]. In a recent study, Jackson et al. [18] showed that a short segment within the N-terminus has a strong negative effect on malonyl CoA sensitivity. Deletion of this region increased malonyl CoA sensitivity by 50-fold. These authors also shed light on the unexplained feature of L- and M-CPT I isoforms in their inverse relationship between the sensitivity to the inhibition by malonyl CoA and affinity for carnitine [11]. They showed that the extreme N-terminus has reciprocal effects on affinities for carnitine and malonyl CoA: this sequence, although entirely conserved between the two isoforms, controls IC_{50} for malonyl CoA in L-CPT I (but not M-CPT I), whereas it controls the K_m for carnitine in M-CPT I but not in L-CPT I [19]. The interaction between the N- and C-terminus is important for the malonyl CoA sensitivity. This same group constructed chimeras using combinations of three segments (N-terminus plus first transmembrane domain, loop plus second transmembrane segment, and C-terminus) from each L- and M-CPT I [18]. They showed that the precise N-terminus to C-terminus pairings affected the sensitivity to malonyl CoA and K_m for palmitoyl CoA of the chimeric CPTs, whereas the interaction between the transmembrane domains affected the affinity for carnitine [18]. Interestingly, when expressed in *Pichia pastoris*, the pig L-CPT I showed a much higher sensitivity to malonyl CoA inhibition compared to human or rat L-CPT I. In contrast, the affinity of the pig L-CPT I for carnitine and palmitoyl CoA was similar to human and rat L-CPT [13]. The physiological significance of these differences is presently not known.

Although present within the general outer membrane of mitochondria, CPT I has been proposed to be concentrated within the contact sites between the outer and inner membranes [20]. Mitochondrial sub-fractionation studies suggested that the kinetic characteristics are also dramatically different between the CPT I localized to the contact sites and the outer membrane enzyme. In the outer membrane, malonyl CoA inhibition affects the V_{max} and not the K_m for palmitoyl CoA, whereas in the contact sites the V_{max} is unchanged but the K_m for palmitoyl CoA is increased [21]. Consequently, inhibition by malonyl CoA in the contact sites appears to be competitive. Contact sites are defined areas of outer and inner membranes that are in a very stable contact and cannot be separated by mechanical forces [22]. The contact sites of the mitochondria are areas where mitochondrial inner and outer membranes come to within 4 nm of each other, and they make up 5–10% of the outer membrane [23]. Because of the presence of about 40% of both CPT I and CPT II at these sites, it has been suggested that these contact sites play a role in the transport of acylcarnitines to the mitochondria [20,24]. Similarly, Hoppel et al. [25] prepared mitoplasts with an intact inner mitochondrial membrane and CPT I activity by French press treatment. Morphological examination of these mitoplasts showed that the outer membrane retained in these preparations containing contact sites. They observed that long-chain acyl CoA synthase (LCAS) and CPT I are disproportionally retained at these contact sites when compared to other membrane markers. Competitive inhibition of CPT I by malonyl CoA in these contact sites would mean that increases in long chain acyl CoA could overcome malonyl CoA inhibition of CPT I activity. In the outer membrane, non-competitive inhibition would ensure that long-chain acyl CoA that is required for other cytoplasmic functions is not converted to acylcarnitine [26].

The finding that a decrease in pH, which is associated with the protonation of the imidazole group of histidine, indicated the involvement of this amino acid in the interaction with malonyl CoA [27–29]. In a recent study, His483 on CPT I was shown to be responsible for malonyl CoA sensitivity of L-CPT I [30]. Examination of the primary structures of the carnitine acyltransferases revealed another completely conserved histidine residue as the candidate for the catalytic site [31]. According to the recent theory by McGarry's group, the hydroxyl group of the carnitine is deprotonated by this histidine residue. An aspartate residue that is conserved amongst the carnitine acyltransferases acts as a charge relay in this system [31].

5. Regulation of malonyl CoA levels in the heart

Malonyl CoA synthesis. Malonyl CoA in the heart is produced by the carboxylation of acetyl CoA by acetyl CoA carboxylase (ACC). Two major isoforms of ACC have been identified, a 265 kDa (ACC265, ACC1, or ACCα) and a 280 kDa (ACC280, ACC2, or ACCβ) isoform. Both isoforms are expressed in the heart [3,32]. The two isoforms are distinct gene products [32,33]. The major difference between the isoforms is the extended N-terminus of ACC280. In lipogenic tissues, such as white adipose tissue and lactating mammary gland, ACC265 is the dominant form expressed. In oxidative tissues, such as rat heart and skeletal muscle, ACC280 predominates. Having said that, an approximate equal distribution of both isoforms is observed in the rabbit heart [3]. The significance of this differential species distribution between ACC265 and ACC280 is not known. It is also not known if these two isoforms have different roles in the heart. Although the 280 isoform of both the skeletal muscle and the heart ACC have been grouped together and called ACC280, it is not entirely clear that heart and skeletal muscle ACCs are identical.

Our studies in isolated working hearts show that ACC is directly involved in the regulation of fatty acid oxidation [3,34–37]. The role of ACC in the regulation of skeletal muscle fatty acid oxidation has also been shown [38,39]. The importance of the regulation of ACC in muscle was recently shown in a report indicating that knockout mice lacking the muscle isoform of ACC have less fat accumulation in adipose tissue and higher rates of fatty acid oxidation in skeletal muscle compared to normal mice [40]. These high fatty acid oxidation rates correlate with a decrease in malonyl CoA levels, which highlight the role of ACC280-produced malonyl CoA in the regulation of fatty acid oxidation (Table 1).

One of the key mechanisms by which cardiac ACC is regulated is via phosphorylation and inhibition by AMP-activated protein kinase (AMPK) and protein kinase A (PKA). AMPK is a serine/threonine kinase that responds to metabolic stresses that deplete cellular ATP (see Ref. [41] for review). Under conditions of metabolic stress, AMPK is activated and inhibits energy-requiring anabolic processes such as lipid liposynthesis. AMPK is a heterotrimetic enzyme consisting of α-, β-, and γ-subunits [41]. Each subunit has more than one isoform. So far, two isoforms for α (α_1 and α_2), two isoforms for β (β_1 and β_2),

Table 1
Conditions/treatments that modify malonyl CoA levels and fatty acid oxidation rates in the heart

Condition/treatment	Malonyl CoA levels	AMPK activity	ACC	MCD	Fatty acid oxidation
Fetal to newborn transition	↓ [2,3,65]	↑ [37]	↓ [2,3,65]	↑ [65]	↑ [65]
Adiponectin	ns	↑	↓	ns	↑
Cardiac work	↓ [70]	nc	nc [70]	↑ [70]	↑ [70]
Diabetes	↓ [71,108]	nc [71]	nc [71]	↑ [71]	↑ [71]
Ischemia/reperfusion	↓ [34,35,65]	↑ [35]	↓ [35,65]	nc [65]	↑ [34,35,65,78,79]
Hypertrophy	ns	↑	ns	↓ [104]	↓ [104]
PPARα activation	ns	ns	ns	↑ [104]	↑
PPAR KO	↑ [74]	nc [74]	nc [74]	↓ [74]	↓ [74]

nc, no change; ns, not studied.

and three isoforms for γ (γ_1, γ_2, and γ_3) subunits have been identified [42–44]. The heart expressed all of these isoforms except the γ3 isoform, which is localized to skeletal muscle. These isoforms appear to play a role in a tissue-specific manner. The α-subunit is the catalytic subunit containing the kinase domain that transfers a phosphate from ATP to the target protein [41]. The β- and γ-subunits are considered regulatory components of the enzyme [41,44].

We have characterized AMPK activity in the heart tissue and found that AMPK activity is comparable to the activity in the liver [35,36]. AMPK itself is regulated by both allosteric and covalent mechanisms. The increase in AMP results in allosteric activation of AMPK. AMP binding makes AMPK a better substrate for the phosphorylation by an upstream kinase, AMPK-kinase (AMPKK) [45,46]. AMP also directly activates AMPKK [47]. Initially, this activation has been shown to phosphorylate AMPK by Thr172 on the α-subunit [47]. However, recent evidence suggests that AMPKK phosphorylates both the α- and β-subunits of AMPK [48]. In addition to making AMPK a better substrate for AMPKK, AMP binding also makes AMPK a poor substrate for protein phosphatase 2C, the phosphatase that appears to be responsible for dephosphorylation of AMPK and thus keeps AMPK in the active state [49]. Recently, AMPK activation independent of AMP:ATP ratio has been demonstrated [50,51].

Once activated, AMPK phosphorylates a number of targets resulting in increases in glucose transport and fatty acid oxidation [41,52,53]. Of importance to fatty acid oxidation in the heart is that AMPK can phosphorylate and inactivate ACC [54–56]. Both AMPK and PKA can phosphorylate ACC in vitro. In rat ACC265, AMPK phosphorylates serine residues 79, 1200, and 1215 and PKA phosphorylates serine residues 77 and 1200 [57]. Hardie's group has shown that the removal of the N-terminus of ACC265 phophorylated by either kinase results in full activation of ACC [58]. Based on this finding, they suggested that only Ser77 and Ser79 have an inhibitory effect on ACC activity upon phosphorylation. However, Ha et al. have mutated Ser79 and Ser1200 and shown that only the latter mutation abolished the inhibitory effects of PKA [59]. The relative importance of phosphorylation of Ser79 and Ser1200 to the regulation of ACC activity remains an open question.

PKA phosphorylation of ACC280 in vitro has been shown to produce significant inactivation [60] or has no effect on activity [61]. However, it is clear that ACC280 can also be phosphorylated by PKA. Whether these conflicting results between studies are due to two possible distinct isoforms in the skeletal muscle and the heart is not known.

Citrate is an allosteric activator of ACC in vivo [57]. Increases in malonyl CoA concentration are not always accompanied by measurable changes in ACC activity [62,63]. However, there is a strong positive correlation between malonyl CoA and increased concentrations of citrate and malate. Therefore, allosteric activation of ACC280 by citrate rather than a change in its phosphorylation state is proposed to be responsible for the increase in malonyl CoA levels [62,63].

Malonyl CoA degradation. Unlike lipogenic tissues where malonyl CoA can be metabolized by fatty acid synthase or fatty acid elongases, the heart does not contain measurable activities of these enzymes [64]. In the heart, we have proposed that the main pathway for degradation of malonyl CoA is via degradation by malonyl CoA decarboxylase (MCD), which decarboxylates malonyl CoA to acetyl CoA (Fig. 2).

Studies from our laboratory showed that the heart has an active MCD [65]. Although originally reported to be a mitochondrial enzyme [66], recent evidence suggests that MCD has multiple intracellular locations such as peroxisomes, mitochondria, and cytosol [64,67,68]. Determination of the subcellular localization of this enzyme will increase our understanding as to how β-oxidation continues despite a high intracellular malonyl CoA concentration.

We have purified MCD and showed that heart MCD is expressed as a 50 kDa protein, that forms a tetramer in the intact cell [68]. In collaboration with Dr Marc Prentki (University of Montreal), we also successfully cloned and sequenced rat MCD [68]. The human MCD cDNA has also now been sequenced [69]. MCD has two putative $5'$ start sites that code for a 54 and 50 kDa protein. While islet cells express both isoforms of MCD, cardiac cells primarily express the 50 kDa isoform [68]. Sequence analysis of MCD suggests that a mitochondrial targeting sequence exists on the amino terminus. We expressed MCD in cardiac cells and found that both isoforms of MCD are expressed, with the 50 kDa isoform localized to the mitochondria (Sambandam et al., manuscript submitted).

Since our initial identification of MCD in cardiac muscle, a number of studies have implicated MCD as being an important regulator of fatty acid oxidation. In the heart, conditions where fatty acid oxidation increases are associated with high MCD activity, including fasting, diabetes, ischemia, increased work, and newborn heart development [65,68,70,71]. In skeletal muscle, liver, and pancreatic islet cells, high MCD activity is also associated with high fatty acid oxidation rates [39,56,72,73].

MCD is subjected to both acute regulation (i.e. phosphorylation control by as yet unidentified kinases) [65] and chronic transcriptional control [74]. We and others have shown that cardiac MCD activity and expression is increased in diabetes, fasting, high fat feeding, and newborn hearts [65,68,70,71]. Part of this transcription control is regulated by PPARα [74]. In PPARα null mice, MCD expression and activity is decreased [74]. This is accompanied by a decrease in fatty acid oxidation rates.

6. Ischemic-induced changes in malonyl CoA control of fatty acid oxidation

During ischemia, we [35,36] and others [50] have shown that AMPK is rapidly phosphorylated and activated. We have also recently shown that AMPK is activated during both mild and severe ischemia (J.Y. Altarejos and G.D. Lopaschuk, unpublished observations). This is accompanied by an increase in T172 phosphorylation of the α-catalytic subunit of AMPK and an increase in AMPK activity. This activation of AMPK is due in part to an ischemia-induced increase in the ratio of AMP/ATP and Cr/PCr. However, AMPK activation can occur independent of nucleotide changes [50].

Activation of AMPK during ischemia may represent an attempt by the myocyte to increase energy supply. This includes a stimulation of fatty acid oxidation [35,36]. In addition, AMPK also promotes glucose uptake and glycolysis during ischemia [50,75, 76]. AMPK activation promotes the translocation of GLUT4 to the sarcolemma of the myocyte [75,76] and also phosphorylates and activates phosphofructokinase kinase 2 [50,77], which produces fructose 2,6-bisphosphate, a potent stimulator of glycolysis.

As a result, AMPK may be an important mediator of the increased glycolysis that occurs during ischemia. However, one of the consequences of this is an accelerated production of deleterious glycolytic by-products, particularly lactate and protons. Although overall oxidative metabolism during ischemia is primarily dependent on the degree of ischemia (i.e. oxygen supply), activation of AMPK during ischemia could result in fatty acid oxidation dominating, at the expense of glucose oxidation, as the source of residual oxidative metabolism [4].

Upon reperfusion of the reversibly damaged ischemic myocardium, AMPK activation persists [35,36]. The reperfused heart is also exposed to high levels of fatty acids in the plasma. This activation of AMPK and high circulating fatty acid levels have two consequences: 1) fatty acid oxidation recovers and provides the majority of the acetyl CoA for the Krebs cycle at the expense of glucose oxidation [78–81] and 2) glycolytic rates remain elevated [82,83]. As a result, proton production from uncoupled glucose metabolism persists into reperfusion, decreasing the rates of pH recovery [83].

During reperfusion of ischemic hearts, high AMPK activity correlates with a decrease in ACC activity, which is accompanied by a dramatic drop in malonyl CoA levels [34,35]. We believe that this drop in malonyl CoA is primarily responsible for the high rates of fatty acid oxidation that occurs during this period. However, of equal importance is that MCD activity is maintained [84] during early reperfusion. A maintained degradation of malonyl CoA in the presence of decrease in malonyl CoA synthesis by ACC contributes to the dramatic decrease in malonyl CoA levels [84]. This drop in malonyl CoA concentration relieves the inhibition on CPT I and due to the high levels of circulating plasma fatty acid levels, fatty acids are almost unrestricted in their import into the mitochondria. This sets in motion the sequelae of events mentioned above that leads to enhanced ischemic damage.

7. Diabetes-induced changes in malonyl CoA control of fatty acid oxidation

In uncontrolled diabetics, myocardial glucose use is reduced and fatty acid oxidation accounts for most of the myocardial oxygen consumption [80,85–87]. This is due in part to a decrease in the number and transport of GLUT 4 to the sarcolemmal membrane. However, a major reason for the decrease in glucose metabolism is the elevated levels of plasma fatty acids seen in the diabetic [88]. Use of fatty acids for mitochondrial oxidative metabolism decreases both PDC activity and phosphofructose-1 kinase, key enzymes in glucose oxidation and glycolysis, respectively [1,86,89].

While high circulating levels of fatty acids in the diabetic decrease glucose metabolism, it is clear that other metabolic changes in the heart are also responsible for low rates of glucose metabolism. For instance, the decrease in PDC activity seen in diabetic rat hearts exceeds what would be expected if the inhibition occurred solely to a high level of circulating fatty acids [86]. This is supported by the observation that glucose oxidation rates are significantly lower in diabetic rat hearts compared to control hearts, even if the hearts are perfused with similar concentrations of fatty acids [86,87]. As a result, the combination of high levels of circulating fatty acids and direct alterations in insulin control of fatty acid and glucose oxidation can result in the diabetic rat heart becoming almost entirely dependent on fatty acid oxidation for its energy requirements [71,85–87,90].

High circulating fatty acid levels and increased rates of cardiac fatty acid oxidation have been observed in a number of experimental models of diabetes. We have shown that fatty acid oxidation rates are high in streptozotocin-diabetic rats [71], BB Wistar rats [91], and insulin-resistant JCR/LA rats [120]. Recent studies have also shown an increase in fatty acid oxidation contribution to energy production in db/db mice [92]. This increase in fatty acid oxidation inhibits glucose oxidation and glycolysis [92] and contributes to a decrease in cardiac efficiency in the diabetic heart [92].

Changes in malonyl CoA control offatty acid oxidation in diabetes. Under conditions of increased fatty acid oxidation such as diabetes and starvation, L-CPT I of the liver becomes more active and less sensitive to the inhibition by malonyl CoA. The decrease in the sensitivity is accompanied by a decrease in tissue of malonyl CoA that allows maximal fatty acid oxidation [93]. The change in the sensitivity of CPT I to malonyl CoA is unique to the liver, but not to the muscle isoform of the enzyme [94]. The mechanism of this desensitization is not known. However, changes in the lipid environment of the membrane have been implicated by several studies [95]. Desensitization of the enzyme also occurs when mitochondria are allowed to warm from 0 to 22 °C [8,96]. Again, loss of sensitivity on warming has not been observed in adult rat heart ($\approx$98% M-CPT I) [97]. If and how these in vitro changes in the sensitivity of CPT I relate to the decrease in sensitivity in vivo in diabetes and starvation is not known.

We also have shown that MCD expression and activity in hearts from 6-week streptozotocin-induced diabetic rats are significantly higher than control hearts [71]. This suggests that an increased MCD activity contributes to the high fatty acid oxidation rates seen in diabetic hearts. This increase in MCD expression may involve an increase in PPARα activity in the diabetic [98]. PPARα is an important transcriptional regulator of MCD in the heart [74]. In addition, overexpression of PPARα results in an increase in fatty acid oxidation in the mouse heart, and a phenotype similar to that seen in the diabetic mouse heart [99].

Lipotoxicity in the heart. Lipid accumulation within the cell is associated with a condition called "lipotoxicity". Excessive accumulation of lipids (particularly triacylglycerol) within non-adipose tissue increases the intracellular pool of long-chain fatty acyl CoA, thereby providing fatty acid substrate for non-oxidative processes, including ceramide synthesis and diacylglycerol synthesis [100,101]. This can lead to cell dysfunction and potentially cell death through apoptosis. Lipotoxicity has been identified in a number of tissues, with the heart as being an important organ susceptible to lipotoxicity [100]. The potential for lipotoxicity in obesity and insulin-resistance has also recently been described [121]. A marked accumulation of triacylglycerol occurs within the myocardium of obese and insulin-resistant rats [102], which is associated with the development of contractile dysfunction. We have shown that insulin-resistant rat hearts have elevated levels of TG, which is associated with a decrease in glucose uptake and glycolysis [103]. However, there is presently confusion as to the relative importance of an increased fatty acid supply versus a decreased fatty acid oxidative capacity as the major contributor to lipotoxicity. In obesity, Kelly's group has implicated a downregulation of PPARα and an underexpression of fatty acid oxidative enzymes as contributing to lipotoxic heart disease [102]. Young et al. [104] have also suggested that an impairment of myocardial fatty acid oxidation during obesity, insulin-resistance, and diabetes may also accelerate contractile dysfunction, due in part to a decrease in PPARα responsiveness. However, in diabetic and obese/insulin-resistant rat hearts, a decrease in fatty acid oxidation is not observed, but rather an increase in fatty acid oxidation. We therefore propose that the primary determinant of lipotoxicity and triacylglycercol accumulation in obese, insulin-resistant and diabetic hearts is due to an increased fatty acid supply and uptake to the myocardium, as opposed to a decreased oxidation of the fatty acids. Even if fatty acid oxidation is accelerated in the heart, if fatty acid supply and uptake exceeds oxidation, an accumulation of lipids within the myocardium would be expected. This would explain in uncontrolled diabetes, triacylglycerol levels in the heart are markedly elevated, despite the heart obtaining almost all of its energy from fatty acid oxidation.

8. Control of mitochondrial fatty acid uptake in cardiac hypertrophy

Cardiac hypertrophy is another condition where chronic alterations in malonyl CoA control of CPT I could contribute to alterations in energy metabolism. Previous studies have shown that the expression of genes encoding fatty acid oxidation enzymes is downregulated in the hypertrophied heart [105,106]. This is accompanied with a shift of energy metabolism towards the fetal pattern (increase in ATP production from carbohydrates) [105–107]. Studies showed that many enzymes involved in the transport and oxidation of fatty acids such as fatty acid transport protein (FATP), fatty acid translocase (FAT), fatty acyl-CoA synthase (FAS), CPT I, enoyl-CoA hydratase, and 3-hydroxy-acyl-CoA dehydrogenase are regulated by the expression of PPARα. Recently, PPARα regulation of MCD has been demonstrated by Young et al. [104]. In that study, feeding rats with a PPARα ligand increased cardiac MCD expression. Further evidence that PPARα is involved in the regulation of MCD expression is provided by our study of MCD expression and activity in PPARα null mice [74]. Expressions of both mRNA and protein as well as the activity of MCD significantly decrease in hearts from PPARα null mice, which was accompanied by an increase in the tissue levels of malonyl CoA. The decrease in palmitate oxidation rate is compensated for by an increase in the rates of glucose oxidation and glycolysis.

9. Summary

Inhibiting fatty acid oxidation or stimulating glucose oxidation has been proposed to protect the heart from fatty acid-induced ischemic damage. Pharmacological agents designed to directly stimulate myocardial glucose oxidation have been proven extremely efficacious in reducing ischemic injury [79,82,108–119]. The emerging role of malonyl CoA in the regulation of fatty acid oxidation introduces a new area for drug development for the manipulation of cardiac metabolism. Drugs that target ACC, AMPK, or MCD in the heart therefore have future potential in the treatment of the heart disease.

References

[1] Neely, J.R., Morgan, H.E., 1974. Relationship between carbohydrate metabolism and energy balance of heart muscle. Annu. Rev. Physiol. 36, 413–459.

[2] Lopaschuk, G.D., Belke, D.D., Gamble, J., Itoi, T., Schönekess, B.O., 1994a. Regulation of fatty acid oxidation in the mammalian heart in health and disease. Biochim. Biophys. Acta 1213, 263–276.

[3] Lopaschuk, G.D., Witters, L.A., Itoi, T., Barr, R., Barr, A., 1994b. Acetyl-CoA carboxylase involvement in the rapid maturation of fatty acid oxidation in the newborn rabbit heart. J. Biol. Chem. 269, 25871–25878.

[4] Lopaschuk, G.D., 2002. Metabolic abnormalities in the diabetic heart. Heart Fail. Rev. 7, 149–159.

[5] Shipp, J., Opie, L.H., Challoner, D., 1961. Fatty acid and glucose metabolism in the perfused heart. Nature 189, 1018–1019.

[6] Schaffer, J.E., 2002. Fatty acid transport: the roads taken. Am. J. Physiol. Endocrinol. Metab. 282, E239–E246.

[7] Eaton, S., 2002. Control of mitochondrial β-oxidation flux. Prog. Lipid Res. 41, 197–239.

[8] McGarry, J.D., Brown, N.F., 1997. The mitochondrial carnitine palmitoyltransferase system. From concept to molecular analysis. Eur. J. Biochem. 244, 1–14.

[9] McGarry, J.D., Foster, D.W., 1980. Regulation of hepatic fatty acid oxidation and ketone body production. Annu. Rev. Biochem. 49, 395–420.

[10] Saggerson, E.D., Carpenter, C.A., 1982. Regulation of hepatic carnitine palmitoyltransferase activity during the foetal-neonatal transition. FEBS Lett. 150, 177–180.

[11] McGarry, J.D., Mills, S.E., Long, C.S., Foster, D.W., 1983. Observations on the affinity for carnitine, and malonyl-CoA sensitivity, of carnitine palmitoyltransferase I in animal and human tissues. Demonstrations of the presence of malonyl-CoA in nonhepatic tissues. Biochem. J. 214, 21–28.

[12] Kerner, J., Hoppel, C., 1998. Genetic disorders of carnitine metabolism and their nutritional management. Annu. Rev. Nutr. 18, 179–206.

[13] Nicot, C., Hegardt, F.G., Woldegiorgis, G., Haro, D., Marrero, P.F., 2001. Pig liver carnitine palmitoyltransferase I, with low Km for carnitine and high sensitivity to malonyl-CoA inhibition, is a natural chimera of rat liver and muscle enzymes? Biochemistry 40, 2260–2266.

[14] Fraser, F., Corstorphine, C.G., Zammit, V.A., 1997. Topology of carnitine palmitoyltransferase I in the mitochondrial outer membrane. Biochem. J. 323, 711–718.

[15] Fraser, F., Corstorphine, C.G., Zammit, V.A., 1996. Evidence that both the acyl-CoA- and malonyl-CoA binding sites of mitochondrial overt carnitine palmitoyltransferase (CPT I) are exposed on the cytosolic face of the outer membrane. Biochem. Soc. Trans. 24, 184S.

[16] Zhou, Y.T., Grayburn, P., Karim, A., Shimbukuro, M., Higa, M., Baetens, D., Orci, L., Unger, R.H., 2000. Lipotoxic heart disease in obese rats: implication for human obesity. PNAS 97, 1784–1789.

[17] Shi, J., Zhu, H., Arvidson, D.N., Cregg, J.M., Woldegiorgis, G., 1998. Deletion of the conserved first 18 N-terminal amino acid residues in rat liver carnitine palmitoyltransferase I abolishes malonyl-CoA sensitivity and binding. Biochemistry 37, 11033–11038.

[18] Jackson, V.N., Cameron, J.M., Fraser, F., Zammit, V.A., Price, N.T., 2000a. Use of six chimeric proteins to investigate the role of intramolecular interactions in determining the kinetics of carnitine palmitoyltransferase I isoforms. J. Biol. Chem. 275, 19560–19566.

[19] Jackson, V.N., Zammit, V.A., Price, N.T.J., 2000b. Identification of positive and negative determinants of malonyl-CoA sensitivity and carnitine affinity within the amino termini of rat liver- and muscle-type carnitine palmitoyltransferase I. Biol. Chem. 275, 38410–38416.

[20] Fraser, F., Zammit, V.A., 1998. Enrichment of carnitine palmitoyltransferases I and II in the contact sites of rat liver mitochondria. Biochem. J. 329, 225–299.

[21] Fraser, F., Padovese, R., Zammit, V.A., 2001. Distinct kinetics of carnitine palmitoyltransferase I in contact sites and outer membranes of rat liver mitochondria. J. Biol. Chem. 276, 20182–20185.

[22] Brdiczka, D., Knoll, G., Riesinger, I., Weiler, U., Klug, G., Benz, R., Krause, J., 1986. Microcompartmentation at the mitochondrial surface: its function in metabolic regulation. Adv. Exp. Med. Biol. 194, 55–69.

[23] Pfanner, N., Rassow, J., Klei, I.J.V.D., Neupert, W., 1992. A dynamic model of the mitochondrial protein import machinery. Cell 68, 999–1002.

[24] Brdiczka, D., 1991. Contact sites between mitochondrial envelope membranes. Structure and function in energy- and protein-transfer. Biochim. Biophys. Acta 1071, 291–312.

[25] Hoppel, C.L., Kerner, J., Turkaly, P., Turkaly, J., Tandler, B., 1998. The malonyl-CoA-sensitive form of carnitine palmitoyltransferase is not localized exclusively in the outer membrane of rat liver mitochondria. J. Biol. Chem. 273, 23495–23503.

[26] Ramsay, R.R., Gandour, R.D., van der Leij, F.R., 2001. Molecular enzymology of carnitine transfer and transport. Biochim. Biophys. Acta 1546, 21–43.

[27] Stephens, T.W., Cook, G.A., Harris, R.A., 1983. Effect of pH on malonyl-CoA inhibition of carnitine palmitoyltransferase I. Biochem. J. 212, 521–524.

[28] Mills, S.E., Foster, D.W., McGarry, J.D., 1984. Effects of pH on the interaction of substrates and malonyl-CoA with mitochondrial carnitine palmitoyltransferase I. Biochem. J. 219, 601–618.

[29] Morillas, M., Clotet, J., Rubi, B., Serra, D., Arino, J., Hegardt, F.G., Asins, G., 2000. Inhibition by etomoxir of rat liver carnitine octanoyltransferase is produced through the co-ordinate interaction with two histidine residues. Biochem. J. 351, 495–502.

[30] Morillas, M., Gomez-Puertas, P., Rubi, B., Clotet, J., Arino, J., Valencia, A., Hegardt, F.G., Serra, D., Asins, G., 2002. Structural model of a malonyl-CoA-binding site of carnitine octanoyltransferase and carnitine palmitoyltransferase I: mutational analysis of a malonyl-CoA affinity domain. J. Biol. Chem. 277, 11473–11480.

[31] Brown, N.F., Anderson, R.C., Caplan, S.L., Foster, D.W., McGarry, J.D., 1994. Catalytically important domains of rat carnitine palmitoyltransferase II as determined by site-directed mutagenesis and chemical modification. Evidence for a critical histidine residue. J. Biol. Chem. 269, 19157–19162.

[32] Abu-Elheiga, L., Almarza-Ortega, D.B., Baldini, A., Wakil, S.J., 1997. Human acetyl-CoA carboxylase 2. Molecular cloning, characterization, chromosomal mapping, and evidence for two isoforms. J. Biol. Chem. 272, 10669–10677.

[33] Abu-Elheiga, L., Jayakumar, A., Baldini, A., Chirala, S.S., Wakil, S.J., 1995. Human acetyl-CoA carboxylase: characterization, molecular cloning, and evidence for two isoforms. Proc. Natl Acad. Sci. USA 92, 4011–4015.

[34] Saddik, M., Gamble, J., Witters, L.A., Lopaschuk, G.D., 1993. Acetyl-CoA carboxylase regulation of fatty acid oxidation in the heart. J. Biol. Chem. 268(34), 25836–25845.

[35] Kudo, N., Barr, A.J., Barr, R.L., Desai, S., Lopaschuk, G.D., 1995. High rates of fatty acid oxidation during reperfusion of ischemic hearts are associated with a decrease in malonyl-CoA levels due to an increase in 5′-AMP-activated protein kinase inhibition of acetyl-CoA carboxylase. J. Biol. Chem. 270, 17513–17520.

[36] Kudo, N., Gillespie, J.G., Kung, L., Witters, L.A., Schulz, R., Clanachan, A.S., Lopaschuk, G.D., 1996. Characterization of 5′AMP-activated protein kinase activity in the heart and its role in inhibiting acetyl-CoA carboxylase during reperfusion following ischemia. Biochim. Biophys. Acta 1301, 67–75.

[37] Makinde, A.O., Gamble, J., Lopaschuk, G.D., 1997. Upregulation of 5′-AMP-activated protein kinase is responsible for the increase in myocardial fatty acid oxidation rates following birth in the newborn rabbit. Circ. Res. 80, 482–489.

[38] Merrill, G.F., Kurth, E.J., Hardie, D.G., Winder, W.W., 1997. AICA riboside increases AMP-activated protein kinase, fatty acid oxidation, and glucose uptake in rat muscle. Am. J. Physiol. 273, E1107–E1112.

[39] Saha, A.K., Schwarsin, A.J., Roduit, R., Masse, F., Kaushik, V., Tornheim, K., Prentki, M., Ruderman, N.B., 2000. Activation of malonyl-CoA decarboxylase in rat skeletal muscle by contraction and the AMP-activated protein kinase activator 5-aminoimidazole-4-carboxamide-1-beta-D-ribofuranoside. J. Biol. Chem. 275, 24279–24283.

[40] Abu-Elheiga, L., Matzuk, M.M., Abo-Hashema, K.A., Wakil, S.J., 2001. Continuous fatty acid oxidation and reduced fat storage in mice lacking acetyl-CoA carboxylase 2. Science 291, 2613–2616.

[41] Hardie, D.G., Carling, D., 1997. The AMP-activated protein kinase – fuel gauge of the mammalian cell? Eur. J. Biochem. 246, 259–273.

[42] Beri, R.K., Marley, A.E., See, C.G., Sopwith, W.F., Aguan, K., Carling, D., 1994. Molecular cloning, expression and chromosomal localization of human AMP-activated protein kinase. FEBS Lett. 356, 117–121.

[43] Hardie, D.G., Carling, D., Carlson, M., 1998. The AMP-activated/SNF1 protein kinase subfamily: metabolic sensors of the eukaryotic cell? Annu. Rev. Biochem. 67, 821–855.

[44] Cheung, P.C.F., Salt, I.P., Davies, S.P., Hardie, D.G., Carling, D., 2000. Characterization of AMP-activated protein kinase λ subunit isoforms and their role in AMP binding. Biochem. J. 346, 659–669.

[45] Hawley, S.A., Selbert, M.A., Goldstein, E.G., Edelman, A.M., Carling, D., Hardie, D.G., 1995. 5′AMP activates the AMP-activated protein kinase cascade, and Ca^{2+}/calmodulin-dependent protein kinase I cascade, via three independent mechanisms. J. Biol. Chem. 270, 27186–27191.

[46] Hamilton, S.R., O'Donnell, J.B. Jr., Hammet, A., Stapleton, D., Habinowski, S.A., Means, A.R., 2002. AMP-activated protein kinase kinase: detection with recombinant AMPK α1 subunit. Biochem. Biophys. Res. Commun. 293, 892–898.

[47] Hawley, S.A., Davison, M., Woods, A., Davies, S.P., Beri, R.K., Carling, D., Hardie, D.G., 1996. Characterization of the AMP-activated protein kinase kinase from rat liver and identification of threonine 172 as the major site at which it phosphorylates AMP-activated protein kinase. J. Biol. Chem. 271, 27879–27887.

[48] Warden, S.M., Richardson, C., O'Donnell, J.B. Jr., Stapleton, D., Kemp, B.E., Witters, L.A., 2001. Post-translational modifications of the β-1 subunit of AMP-activated protein kinase affect enzyme activity and cellular localization. Biochem. J. 354, 275–283.

[49] Davies, S.P., Helps, N.R., Cohen, P.T.W., Hardie, D.G., 1995. 5′-AMP inhibits dephosphorylation, as well as promoting phosphorylation, of the AMP-activated protein kinase. Studies using bacterially expressed human protein phosphatase-2C alpha and native bovine protein phosphatase-2AC. FEBS Lett. 377, 421–425.

[50] Beauloye, C., Marsin, A.S., Bertrand, L., Krause, U., Hardie, D.G., Vanoverschelde, J.L., Hue, L., 2001. Insulin antagonizes AMP-activated protein kinase activation by ischemia or anoxia in rat heart, without affecting total adenine nucleotides. FEBS Lett. 505, 348–352.

[51] Daniel, T., Carling, D., 2002. Expression and regulation of the AMP-activated protein kinase/SNF1 kinase complexes in yeast and mammalian cells. Studies using chimeric catalytic subunits. Biochem. J. 365, 629–638.

[52] Ruderman, N.B., Saha, A.K., Vavvas, D., Witters, L.A., 1999. Malonyl-CoA, fuel sensing, and insulin resistance. Am. J. Physiol. 276, E1–E18.

[53] Mu, J., Brozinick, J.T. Jr., Valladares, O., Bucan, M., Birnbaum, M.J., 2001. A role for AMP-activated protein kinase in contraction- and hypoxia-regulated glucose transport in skeletal muscle. Mol. Cell 7, 1085–1094.

[54] Witters, L.A., Kemp, B.E., 1992. Insulin activation of acetyl-CoA carboxylase accompanied by inhibition of the 5′-AMP-activated protein kinase. J. Biol. Chem. 267, 2864–2867.

[55] Witters, L.A., Gao, G., Kemp, B.E., Quistorff, B., 1994. Hepatic 5′-AMP-activated protein kinase: zonal distribution and relationship to acetyl-CoA carboxylase activity in varying nutritional states. Arch. Biochem. Biophys. 308, 413–419.

[56] Park, S.H., Gammon, S.R., Knippers, J.D., Paulsen, S.R., Rubink, D.S., Winder, W.W., 2002. Phosphorylation-activity relationships of AMPK and acetyl-CoA carboxylase in muscle. J. Appl. Physiol. 92, 2475–2482.

[57] Munday, M.R., Campbell, D.G., Carling, D., Hardie, D.G., 1988. Identification by amino acid sequencing of three major regulatory phosphorylation sites on rat acetyl-CoA carboxylase. Eur. J. Biochem. 175, 331–338.

[58] Davies, S.P., Sim, A.T., Hardie, D.G., 1990. Location and function of three sites phosphorylated on rat acetyl-CoA carboxylase by the AMP-activated protein kinase. Eur. J. Biochem. 187, 183–190.

[59] Ha, J., Daniel, S., Broyles, S.S., Kim, K.H., 1994. Critical phosphorylation sites for acetyl-CoA carboxylase activity. J. Biol. Chem. 269, 22162–22168.

[60] Dyck, J.R., Kudo, N., Barr, A.J., Davies, S.P., Hardie, D.G., Lopaschuk, G.D., 1999. Phosphorylation control of cardiac acetyl-CoA carboxylase by cAMP-dependent protein kinase and 5′-AMP activated protein kinase. Eur. J. Biochem. 262, 184–190.
[61] Boone, A.N., Rodrigues, B., Brownsey, R.W., 1999. Multiple-site phosphorylation of the 280 kDa isoform of acetyl-CoA carboxylase in rat cardiac myocytes: evidence that cAMP-dependent protein kinase mediates effects of beta-adrenergic stimulation. Biochem. J. 341, 347–354.
[62] Saha, A.K., Laybutt, D.R., Dean, D., Vavvas, D., Sebokova, E., Ellis, B., Klimes, I., Kraegen, E.W., Shafrir, E., Ruderman, N.B., 1999. Cytosolic citrate and malonyl-CoA regulation in rat muscle in vivo. Am. J. Physiol. 276, E1030–E1037.
[63] Bavenholm, P.N., Pigon, J., Saha, A.K., Ruderman, N.B., Efendic, S., 2000. Fatty acid oxidation and the regulation of malonyl-CoA in human muscle. Diabetes 49, 1078–1083.
[64] Alam, N., Saggerson, E.D., 1998. Malonyl-CoA and the regulation of fatty acid oxidation in soleus muscle. Biochem. J. 334, 233–241.
[65] Dyck, J.R., Barr, A.J., Barr, R.L., Kolattukudy, P.E., Lopaschuk, G.D., 1998. Characterization of cardiac malonyl-CoA decarboxylase and its putative role in regulating fatty acid oxidation. Am. J. Physiol. 275, H2122–H2129.
[66] Kolattukudy, P.E., Poulose, A.J., Kim, Y.S., 1981. Malonyl-CoA decarboxylase from avian, mammalian, and microbial sources. Methods Enzymol. 71, 150–163.
[67] Sacksteder, K.A., Morrell, J.C., Wanders, R.J., Matalon, R., Gould, S.J., 1999. MCD encodes peroxisomal and cytoplasmic forms of malonyl-CoA decarboxylase and is mutated in malonyl-CoA decarboxylase deficiency. J. Biol. Chem. 74, 24461–24468.
[68] Dyck, J.R., Berthiaume, L.G., Thomas, P.D., Kantor, P.F., Barr, A.J., Barr, R., Singh, D., Hopkins, T.A., Voilley, N., Prentki, M., Lopaschuk, G.D., 2000. Characterization of rat liver malonyl-CoA decarboxylase and the study of its role in regulating fatty acid metabolism. Biochem. J. 350, 599–608.
[69] Gao, J., Waber, L., Bennett, M.J., Gibson, K.M., Cohen, J.C., 1999. Cloning and mutational analysis of human malonyl-coenzyme A decarboxylase. J. Lipid Res. 40, 178–182.
[70] Goodwin, G.W., Taegtmeyer, H., 1999. Regulation of fatty acid oxidation of the heart by MCD and ACC during contractile stimulation. Am. J. Physiol. 277, E772–E777.
[71] Sakamoto, J., Barr, R.L., Kavanagh, K.M., Lopaschuk, G.D., 2000. Contribution of malonyl-CoA decarboxylase to the high fatty acid oxidation rates seen in diabetic heart. Am. J. Physiol. Heart Circ. Physiol. 278, H1196–H1204.
[72] Antinozzi, S.P., Segall, L., Prentki, M., McGarry, J.D., 1998. Molecular or pharmacologic perturbation of the link between glucose and lipid metabolism is without effect on glucose-stimulated insulin secretion. A re-evaluation of the long-chain acyl-CoA hypothesis. J. Biol. Chem. 273, 16146–16154.
[73] Ruderman, N.B., Dean, D., 1998. Malonyl CoA, long-chain acyl-CoA and insulin resistance in skeletal muscle. J. Basic Clin. Physiol. Pharmacol. 9, 295–308.
[74] Campbell, F.M., Kozak, R., Wagner, A., Altarejos, J.Y., Dyck, J.R., Belke, D.D., Severson, D.L., Kelly, D.P., Lopaschuk, G.D., 2002. A role for peroxisome proliferator-activated receptor alpha (PPARalpha) in the control of cardiac malonyl-CoA levels: reduced fatty acid oxidation rates and increased glucose oxidation rates in the hearts of mice lacking PPARalpha are associated with higher concentrations of malonyl-CoA and reduced expression of malonyl-CoA decarboxylase. J. Biol. Chem. 277, 4098–4103.
[75] Russell, R.R., Renfu, Y., Caplan, M.J., Hu, X., Ren, J., Shulman, G.I., Sinusas, A.J., Young, L.H., 1998. Additive effects of hyperinsulinemia and ischemia on GLUT1 and GLUT4 translocation in vivo. Circulation 98, 2180–2186.
[76] Russell, R.R., Bergeron, R., Shulman, G.I., Young, L.H., 1999. Translocation of myocardial GLUT4 and increased glucose uptake through activation of AMPK by AICAR. Am. J. Physiol. 277, H643–H649.
[77] Marsin, A.S., Bertrand, L., Rider, M.H., Deprez, J., Beauloye, C., Vincent, M.F., Van der Berghe, G., Carling, D., Hue, L., 2000. Phosphorylation and activation of heart PFK-2 by AMPK has a role in the stimulation of glycolysis during ischemia. Curr. Biol. 10, 1247–1255.
[78] Liedtke, A.J., DeMaison, L., Eggleston, A.M., Cohen, L.M., Nellis, S.H., 1988. Changes in substrate metabolism and effects fatty acids in reperfused myocardium. Circ. Res. 62, 535–542.

[79] Lopaschuk, G.D., Spafford, M.A., Davies, N.J., Wall, S.R., 1990. Glucose and palmitate oxidation in isolated working rat hearts reperfused after a period of transient global ischemia. Circ. Res. 66, 546–553.
[80] Saddik, M., Lopaschuk, G.D., 1991. Myocardial triglyceride turnover during reperfusion of isolated rat hearts subjected to a transient period of global ischemia. J. Biol. Chem. 267, 3825–3831.
[81] Benzi, R.H., Lerch, R., 1992. Dissociation between contractile function and oxidative metabolism in postischemic myocardium. Circ. Res. 71, 567–576.
[82] Liu, B., Clanachan, A.S., Schulz, R., Lopaschuk, G.D., 1996. Cardiac efficiency is improved after ischemia by altering both the source and fate of protons. Circ. Res. 79, 940–948.
[83] Liu, Q., Dockerty, J.C., Rendell, J.C., Clanachan, A.S., Lopaschuk, G.D., 2002. High levels of fatty acids delay the recovery of intracellular pH and cardiac efficiency in post-ischemic hearts by inhibiting glucose oxidation. J. Am. Coll. Cardiol. 39, 718–725.
[84] Awan, M.M., Saggerson, E.D., 1993. Malonyl CoA metabolism in cardiac myocytes and its relevance to the control of fatty acid oxidation. Biochem. J. 295, 61–66.
[85] Goodale, W.T., Olson, R.E., Hackel, D.B., 1959. The effects of fasting and diabetes mellitus on myocardial metabolism in man. Am. J. Med. 27, 212–220.
[86] Randle, P.J., Hales, C.N., Garland, P.B., Newsholme, E.A., 1963. The glucose-fatty acid cycle. Its role in insulin sensitivity and the metabolic disturbances of diabetes mellitus. Lancet 1, 785–789.
[87] Wall, S.R., Lopaschuk, G.D., 1989. Glucose oxidation rates in fatty acid-perfused isolated working hearts from diabetic rat. Biochim. Biophys. Acta 1006, 97–103.
[88] Reaven, G.M., Hollenbeck, C., Jeng, C.Y., Wu, M.S., Chen, Y.I., 1988. Measurements of plasma glucose, free fatty acid, lactate, and insulin for 24 h in patients with NIDDM. Diabetes 37, 1020–1024.
[89] French, T.J., Goode, A.W., Holness, M.J., MacLennan, P.A., Sugden, M.C., 1988. The relationship between changes in lipid fuel availability and tissue fructose 2,6-bisphosphate concentrations and pyruvate dehydrogenase complex activities in the fed state. Biochem. J. 256, 935–939.
[90] Garland, P.B., Randle, P.J., 1964. Regulation of glucose uptake by muscle. Effects of alloxan diabetes, starvation, hypophysectomy, and adrenalectomy and of fatty acids, ketone bodies and pyruvate on the glycerol output and concentrations of free fatty acids, long-chain fatty acyl coenzyme A, glycerol phosphate and citrate cycle intermediates in rat hearts and diaphragm muscles. Biochem. J. 93, 678–687.
[91] Broderick, T.L., Barr, R.L., Quinney, H.A., Lopaschuk, G.D., 1992. Acute insulin withdrawal from diabetic BB rats decreases myocardial glycolysis during low-flow ischemia. Metabolism 41, 332–338.
[92] Aasum, E., Hafstad, A.D., Severson, D.L., Larsen, T.S., 2003. Age-dependent changes in metabolism, contractile function, and ischemic sensitivity in hearts from db/db mice. Diabetes 52, 434–441.
[93] Drynan, L., Quant, P.A., Zammit, V.A., 1996. The role of changes in the sensitivity of hepatic mitochondrial overt carnitine palmitoyltransferase in determining the onset of the ketosis of starvation in the rat. Biochem. J. 318, 767–770.
[94] Mynatt, R.L., Lappi, M.D., Cook, G.A., 1992. Myocardial carnitine palmitoyltransferase of the mitochondrial outer membrane is not altered by fasting. Biochim. Biophys. Acta 128, 105–111.
[95] Zammit, V.A., Corstorphine, C.G., Gray, S.R., 1984. Changes in the ability of malonyl-CoA to inhibit carnitine palmitoyltransferase I activity and to bind to rat liver mitochondria during incubation in vitro. Differences in binding at 0 degree C and 37 degrees C with a fixed concentration of malonyl-CoA. Biochem. J. 222, 335–342.
[96] Zammit, V.A., Fraser, F., Corstorphine, C.G., 1997. Regulation of mitochondrial outer-membrane carnitine palmitoyltransferase (CPT I): role of membrane-topology. Adv. Enzyme Regul. 37, 295–317.
[97] Weis, B.C., Cowan, A.T., Brown, N., Foster, D.W., McGarry, J.D., 1994. Use of a selective inhibitor of liver carnitine palmitoyltransferase I (CPT I) allows quantification of its contribution to total CPT I activity in rat heart. Evidence that the dominant cardiac CPT I isoform is identical to the skeletal muscle enzyme. J. Biol. Chem. 269, 26443–26448.

[98] Leone, T.C., Weinheimer, C.J., Kelly, D.P., 1999. A critical role for the peroxysome proliferator-activated receptor α (PPARα) in the cellular fasting response: the PPARα-null mouse as a model of fatty acid oxidation disorders. Proc. Natl Acad. Sci. USA 96, 7473–7478.

[99] Finck, B.N., Lehman, J.J., Leone, T.C., Welch, M.J., Bennett, M.J., Kovacs, A., Han, X., Gross, R.W., Kozak, R., Lopaschuk, G.D., Kelly, D.P., 2002. The cardiac phenotype induced by PPARalpha overexpression mimics that caused by diabetes mellitus. J. Clin. Invest. 109, 121–130.

[100] Unger, R.H., Orci, L., 2000. Lipotoxic diseases of nonadipose tissues in obesity. Int. J. Obes. Relat. Metab. Disord. 24, S28–S32.

[101] Zierz, S., Engel, A.G., 1987. Different sites of inhibition of carnitine palmitoyltransferase by malonyl-CoA, and by acetyl-CoA and CoA, in human skeletal muscle. Biochem. J. 245, 205–209.

[102] Finck, B.N., Han, X., Courtois, M., Aimond, F., Nerbonne, J.M., Kovacs, A., Gross, R.W., Kelly, D., 2003. A critical role for PPARα-mediated lipotoxicity in the pathogenesis of diabetic cardiomyopathy: modulation by dietary fat content. PNAS 100, 1226–1231.

[103] Atkinson, L.L., Kozak, R., Kelly, S.E., Onay-Besikci, A., Russell, J.C., Lopaschuk, G.D., 2003. Potential mechanisms and consequences of cardiac triacylglycerol accumulation in insulin-resistant rats. Am. J. Physiol. Endocrinol. Metab. 284, E923–E930.

[104] Young, M.E., Goodwin, G.W., Ying, J., Guthrie, P., Wilson, C.R., Laws, F.A., Taegtmeyer, H., 2001. Regulation of cardiac and skeletal muscle malonyl-CoA decarboxylase by fatty acids. Am. J. Physiol. Endocrinol. Metab. 280, E471–E479.

[105] Sack, M.N., Rader, T.A., Park, S., Bastin, J., McCune, S.A., Kelly, D.P., 1996. Fatty acid oxidation enzyme gene expression is downregulated in the failing heart. Circulation 94, 2837–2842.

[106] Sack, M.N., Disch, D.L., Rockman, H.A., Kelly, D.P., 1997. A role for Sp and nuclear receptor transcription factors in a cardiac hypertrophic growth program. Proc. Natl Acad. Sci. USA 94, 6438–6443.

[107] Depre, C., Shipley, G.L., Chen, W., Han, Q., Doenst, T., Moore, M.L., Stepkowski, S., Davies, P.J.A., Taegtmeyer, H., 1998. Unloaded heart in vivo replicates fetal gene expression of cardiac hypertrophy. Nat. Med. 4, 1269–1275.

[108] Gamble, J., Lopaschuk, G.D., 1994. Glycolysis and glucose oxidation during reperfusion of ischemic hearts from diabetic rats. Biochim. Biophys. Acta 1225, 191–199.

[109] Barak, C., Reed, M.K., Maniscalco, S.P., Sherry, A.D., Malloy, C.R., Jessen, M.E., 1998. Effects of dichloroacetate on mechanical recovery and oxidation of physiologic substrates after ischemia and reperfusion in the isolated heart. J. Cardiovasc. Pharmacol. 31, 336–344.

[110] Bersin, R.M., Stacpoole, P.W., 1997. Dichloroacetate as metabolic therapy for myocardial ischemia and failure. Am. Heart J. 134, 841–855.

[111] Itoi, T., Huang, L., Lopaschuk, G.D., 1993. Glucose use in neonatal rabbit hearts reperfused after global ischemia. Am. J. Physiol. Heart Circ. Physiol. 265, H427–H433.

[112] Montague, T., DeAlmeida, J., Lopaschuk, G.D., Witkowski, F., Walker, D., Ackman, M., Humen, D., Dzavik, V., Teo, K., 1994. Enhanced glucose oxidation in exercise-induced myocardial ischemia. Can. J. Cardiol. 10, 913–919.

[113] Nicholl, T.A., Lopaschuk, G.D., McNeill, J.H., 1991. Effects of free fatty acids and dichloroacetate on isolated working diabetic rat heart. Am. J. Physiol. Heart Circ. Physiol. 261, H1053–H1059.

[114] Wambolt, R.B., Lopaschuk, G.D., Brownsey, R.W., Allard, M.F., 2000. Dichloroacetate improves postischemic function of hypertrophied rat hearts. J. Am. Coll. Cardiol. 36, 1378–1385.

[115] Schönekess, B.O., Allard, M.F., Lopaschuk, G.D., 1995. L-carnitine improvement of hypertrophied heart function is accompanied by an increase in carbohydrate oxidation. Circ. Res. 77, 726–734.

[116] Stacpoole, P.W., 1989. The pharmacology of dichloroacetate. Metabolism 38, 1124–1144.

[117] Clarke, B., Wyatt, K.M., McCormack, J.G., 1996. Ranolazine increases active pyruvate dehydrogenase in perfused normoxic rat hearts: evidence for an indirect mechanism. J. Mol. Cell. Cardiol. 28, 341–350.

[118] Fantini, E., Demaison, L., Sentex, E., Grynberg, A., Athias, P., 1994. Some biochemical aspects of the protective effect of trimetazidine on rat cardiomyocytes during hypoxia and reoxygenation 26, 949–958.

[119] Belardinelli, R., 2000. Trimetazidine and the contractile response of dysfunctional myocardium in ischemic cardiomyopathy. Rev. Port. Cardiol. 19, V35–V39.
[120] Lopaschuk, G.D., Russell, J.C., 1991. Myocardial function and energy substrate metabolism in the insulin-resistant JCR:LA corpulent rat. J. Appl. Physiol. 71, 1302–1308.
[121] Chen, H.C., Stone, S.J., Zhou, P., Buhman, K.K., Farese, R.V. Jr., 2002. Dissociation of obesity and impaired glucose disposal in mice overexpressing acyl coenzyme a:diacylglycerol acyltransferase 1 in white adipose tissue. Diabetes 51, 3189–3195.

Alterations in muscular fatty acid handling in diabetes

Joost J.F.P. Luiken,[a,b,*] Arend Bonen[a,b] and Jan F.C. Glatz[a,b]

[a]*Department of Molecular Genetics, Cardiovascular Research Institute Maastricht (CARIM), Maastricht University, P.O. Box 616, NL-6200 MD Maastricht, The Netherlands*
[b]*Department of Human Biology and Nutritional Sciences, University of Guelph, Guelph, Ont., Canada N1G 2W1*
[*]*Correspondence address: Tel.: +31-43-388-1998; fax: +31-43-388-4574*
E-mail: j.luiken@gen.unimaas.nl(J.J.F.P.L.)

1. Introduction

Long-chain fatty acids (FA) are important substrates for skeletal muscle metabolic energy production. By virtue of its mass and its highly variable metabolic rate, skeletal muscle is a key player in the regulation of lipid homeostasis. The relative contribution of FA to metabolic energy production differs among the various muscle fiber types. Notably, FA utilization is relatively low in fast-twitch (type II) fibers, whereas slow-twitch oxidative (type I) fibers display considerably greater FA fluxes. Although FA present an important oxidizable fuel, especially in type I fibers, oxidation is not their only destination. The non-oxidized portion of FA is incorporated into distinct cellular lipid fractions. Thus, skeletal muscle possesses neutral lipid stores from which FA can be liberated to fulfill the metabolic demands during prolonged submaximal exercise [1]. However, it has been estimated that under normal physiological conditions, exogenous FA, rather than FA stored in intracellular neutral lipid pools, are preferentially used for energy production [2].

The driving force of FA to be taken up by skeletal muscle from the blood circulation is the concentration gradient between vascular space and the intracellular compartment of the myocyte [3]. Hence, myocytes strive to maintain a low intracellular concentration of non-metabolized (free) FA [4]. This is achieved through rapid metabolic conversion of FA upon their cellular entry. Surprisingly, our understanding of FA uptake by skeletal muscle is modest, especially when compared to glucose, the other main substrate for muscular energy production. This is in large part due to the hydrophobic nature of FA and their ability to readily partition in biological membranes, which confirms the argument that a specific protein-mediated transport system is not required for the transmembrane passage of FA. However about two decades ago, evidence began to accumulate that FA uptake in several other tissues (adipocytes, see Ref. [5]; liver, see Ref. [6]; heart, see Ref. [7]) is saturable, and involves a carrier-mediated system. In addition, in perfused muscle,

Advances in Molecular and Cell Biology, Vol. 33, pages 243–258

ISSN: 1569-2558 / DOI: 10.1016/S1569-2558(03)33012-7

FA uptake has been found to be saturable [8]. Nonetheless, all this kinetic evidence in favor of a protein-mediated transport process, in theory, could be explained by a saturation of metabolism leading to a rise in intracellular FA and a breakdown of the concentration gradient. However, a rise in intracellular FA would be very undesirable because of their toxic action spectrum. These adverse effects include detergent action on biological membranes and the ability to interfere with enzyme activity [9]. On the other hand, on a theoretical basis, the existence of a transport system for FA may be essential to operate in concert with the highly variable metabolic demands. This then would allow the efficient channeling of exogenous FA directly into metabolic pathways. In addition, changes in the FA transport system may be beneficial to cope with altered metabolic fluxes and elevated levels of circulating FA in pathological conditions such as diabetes.

Skeletal muscle plays an important role in the etiology of diabetes, because, by virtue of its mass, it is a principal site of insulin action. This metabolic disease, which is dramatically on the rise in the Western world, is classified into insulin-dependent (type-1) and non-insulin-dependent diabetes (type-2). There is a general consensus that in muscle tissues of diabetic animals, glucose utilization is markedly reduced [10–12]. The decrease in the availability of glucose is associated with an increase in the utilization of FA. The primary defect in diabetes-associated changes in glucose handling has been pinpointed to the level of cellular glucose uptake, which is mainly mediated by the glucose transporter GLUT4 in skeletal muscle. However, the molecular mechanism leading to the decrease in glucose uptake may be different in type-1 diabetes versus type-2 diabetes. For instance, in obese Zucker rats (model for obesity and type-2 diabetes), muscular GLUT4 levels are unaltered, but an impaired translocation of GLUT4 from endosomes to the sarcolemma in response to insulin is likely to cause a decrease in insulin-sensitive glucose uptake [12,13]. In contrast, in streptozotocin (STZ)-injected rats (model for type-1 diabetes), reduced glucose transport is due to reduced sarcolemmal GLUT4 but cannot fully be explained by translocation defects. For instance, other factors such as altered expression levels and alterations in the intrinsic activity of glucose transporters might also play a role [14–17]. Another striking difference between both types of diabetes concerns the metabolic fate of incoming FA. Whereas in STZ rats, there is a greater dependence on FA for mitochondrial β-oxidation [11,18], in skeletal muscle of rodent models of type-2 diabetes, an increase in intracellular triacylglycerol pools has been observed, which are strongly linked to the occurrence of insulin resistance [12,19]. Extrapolating these findings to the FA uptake process, it can be postulated that there are differences in regulation of FA transporters between both types of diabetes.

Using the novel model of giant sarcolemmal vesicles to study skeletal muscle substrate uptake, our research group has provided functional evidence that the bulk of FA uptake is mediated by a transport system consisting of (at least) two sarcolemmal FA-binding proteins. In this chapter, we will outline the experiments that were instrumental in the identification of these sarcolemmal proteins. We will also focus on the role of these proteins in the regulation of FA utilization by major physiological stimuli, and the evidence for their altered functioning in type-1 and type-2 diabetes. A better understanding of the molecular mechanism by which these proteins act could help to design therapeutic strategies to help fight both forms of diabetes.

2. Giant vesicles as model to study FA uptake by skeletal muscle

Data emanating from FA transport assays in intact muscle preparations, such as incubated muscle strips and perfused muscles, are difficult to interpret, not only due to the possible feedback effects of metabolism, but also due to the accumulation of radiolabeled substrate in the interstitial space [20,21].

The development for a suitable model system to measure "true" substrate uptake rates, i.e. the rate of transsarcolemmal transport, was boasted by the coincidental observation that large sarcolemmal structures cleave off from frog cutaneous pectoris muscle in the presence of iso-osmotic KCl in combination with proteolytic treatment [22]. The molecular mechanisms leading to excision of large membrane spheres from the surface of muscle cells is not yet understood [23]. A second breakthrough was made by Danish researchers who devised a preparative procedure to obtain such vesicles from rat hind-limb muscle in sufficient yield to use them for transport studies [24,25]. Subsequently, these vesicles were successfully used to characterize muscle uptake of glucose and lactate [24–27].

We have characterized the giant sarcolemmal vesicles for determination of FA uptake, and found them to have several advantages over other systems: (i) metabolism is absent, as FA esterification- and oxidation intermediates/products could not be detected; (ii) their large diameter (10–15 μm) avoids the problem of backflux of substrates during assay of the initial uptake rate; (iii) the vesicles are fully oriented right-side-out; and (iv) they contain ample quantities of soluble small (15 kDa) cytoplasmic heart-type FA-binding protein (H-FABPc) to act as intravesicular FA sink [28,29]. These combined properties of giant vesicles have made them particularly suitable for identification of an FA transport system that is subject to regulation by physiological stimuli, and that undergoes alterations in diabetes.

3. Identification of the sarcolemmal FA transport system in skeletal muscle

The classical criteria to discriminate between passive diffusion versus protein-mediated transport are to determine whether the uptake process under study is (i) saturable as function of the exogenous substrate concentration, (ii) inhibitable by agents that interfere with protein-mediated processes and (iii) sensitive to competition by other substrates. The absence of metabolism makes giant vesicles excellently suitable to investigate these issues. In order to assess the degree of saturability, we have investigated palmitate uptake as function of the unbound palmitate concentration, since it is the unbound palmitate fraction that traverses the plasma membrane rather than the palmitate/albumin complex. When giant vesicles were prepared from both red and white regions of rat hind-limb muscle, both preparations displayed saturation at increasing exogenous FA concentrations (Fig. 1A, see also Ref. [30]). While both giant vesicle preparations display the same apparent K_m, the V_{max} in red muscle vesicles was 2-fold higher than that in white muscle. These differences in transport capacities of red and white skeletal muscles scale well with the known differences in their capacities for metabolism [2], indicating a fine tuning between both processes. Moreover, the similarity in K_m but the difference in V_{max}

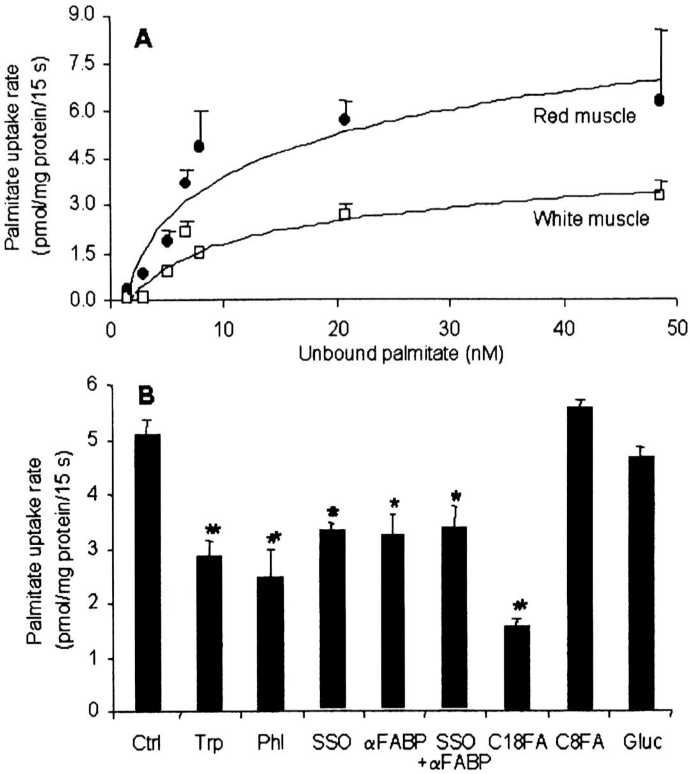

Fig. 1. Characterization of FA uptake by giant vesicles from rat skeletal muscle. (A) Kinetics of palmitate uptake in giant vesicles from white (□) and red (•) gastrocnemius muscle (data redrawn from Ref. [98]). (B) Sensitivity of palmitate uptake by giant vesicles from mixed muscles to inhibition by trypsin (Trp), phloretin (Phl), SSO, a polyclonal antiserum against FABPpm (αFABP), a combination of SSO and anti-FABPpm antibodies, oleate (C18FA), octanoate (C8FA) and glucose (Gluc). *Significantly different from Ctrl ($P < 0.05$). Data are redrawn from Ref. [30].

tentatively suggests that the same transport system is operative in both muscle types, but this system is more abundant in red muscle.

Using the protein modifying agents phloretin and trypsin at optimal concentrations, i.e. at concentrations at which they exert the maximal effect without disrupting vesicular integrity, FA uptake was inhibitable to ~50% [30]. In addition, manipulations to specifically affect FA uptake, such as antisera and inhibitors directed against membrane proteins identified in other tissues (see below), inhibited FA uptake to the same extent. Apparently, none of these agents was able to completely block the FA uptake process. Likely, the inhibitable component of vesicular FA uptake (~50% of the total uptake rate) represents the protein-mediated transport, and the non-inhibitable component (~50% of the total uptake rate) represents the passive diffusion process.

Besides providing evidence for a protein-mediated process, competition experiments can be used to characterize the structural requirements for binding to the transport site. Uptake of palmitate was inhibited by another FA species oleate, but not by the medium-chain fatty acid octanoate nor by glucose, indicating that the transport system responsible for uptake of palmitate is specific for (long-chain) FA [30].

Clearly, the kinetics of FA uptake into giant vesicles from skeletal muscle fulfill, at least in part, the laws of a protein-mediated process. For identification of the membrane-associated proteins responsible for muscle FA uptake, one can use the list of FA-binding proteins that were identified in other tissues and have been demonstrated to increase FA uptake upon overexpression in cell lines. The major membrane proteins are (i) a 40 kDa plasma membrane FA-binding protein (FABPpm) identified in 1985 by affinity chromatography of rat liver plasma membranes [31], (ii) an 88 kDa FA translocase (FAT/CD36) identified in 1991 by radiolabeling of adipocyte plasma membranes with sulfo-*N* succinimidylesters of FA [32] and (iii) a 62 kDa FA transport protein (FATP) identified in 1994 by expression cloning of a cDNA library from adipocytes [33]. FABPpm was the first of these proteins to be detected in skeletal muscle, and was found to be upregulated during physiological conditions known to be associated with an increase in FFA utilization [34,35]. This suggests that FABPpm may be involved in the regulation of FA handling by skeletal muscle. Subsequently, all three proteins have been shown to be present in skeletal muscle, both at the mRNA and the protein levels [29,30,36]. A comparison between FA uptake rates among the different muscle types (Fig. 1A) and the sarcolemmal contents of the three candidate transporters (Fig. 2) reveals that both FABPpm and FAT/CD36 show a positive correlation with FA uptake rates, whereas FATP does not. Specifically, FAT/CD36 displays a very good linear relationship with FA uptake rates, also when the comparison is extended to heart muscle [29]. Hence, FABPpm and

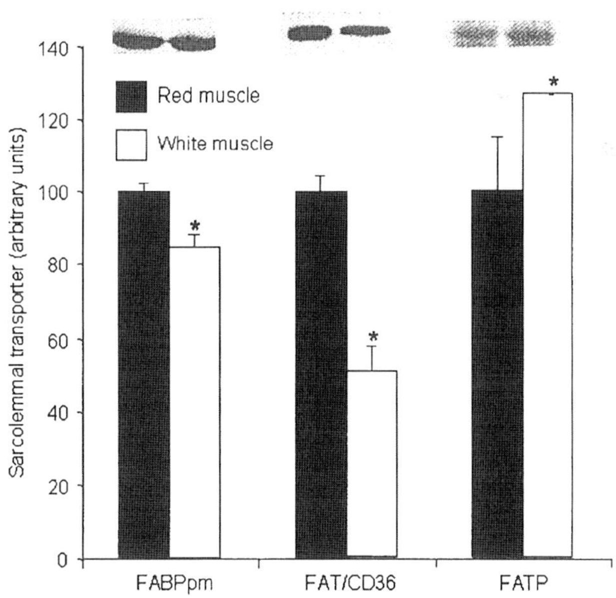

Fig. 2. Presence of FABPpm, FAT/CD36 and FATP in giant vesicles from red and white rat skeletal muscle. The red muscle data for each transporter were set at 100. FABPpm was detected at 43 kDa, FAT/CD36 at 88 kDA and FATP at 63 kDa. *White muscle compared with red muscle ($P < 0.05$). Data are redrawn from Ref. [29].

FAT/CD36, but not FATP, may be functionally important FA transporters in muscle. However, it does not exclude a more specific function for FATP, for instance in (i) preferential uptake of lesser occurring FA or (ii) coupling FA uptake directly to a certain metabolic process. Additional evidence for the importance of FABPpm and FAT/CD36 in muscle FA uptake is provided by the ability of anti-FABPpm antibodies and sulfo-*N*-succinimidyloleate (SSO) to inhibit FA uptake into skeletal muscle, as mentioned above. Interestingly, these manipulations inhibited FA uptake to the same extent as general protein modifying agents, indicating that when the transport function of either FABPpm or FAT/CD36 is prevented, protein-mediated transport is fully annihilated [30]. This tentatively suggests that FABPpm and FAT/CD36 cooperate in governing the transsarcolemmal passage of FA. This notion is bolstered by the observation that a combination of anti-FABPpm antiserum and SSO does not inhibit FA uptake further than the individual manipulations. When integrating the topographical positioning of each of the proteins within the phospholipid bilayer, with FABPpm being a peripheral membrane protein located at the cell surface [37] and FAT/CD36 being an integral membrane protein having one or two membrane spanning regions [38], we have proposed the following model [29]: FABPpm serves as the extracellular receptor for FA, while FAT/CD36 mediates the transmembrane passage of FA. The mechanism of this assisted passage has not yet been elucidated, but in view of the limited number of putative membrane spanning regions, it is unlikely that FAT/CD36 acts as a classical pore or channel. Rather FAT/CD36 could lower the physicochemical barrier for flipflop of FA through the bilayer [39]. Upon entering the cytoplasmic side of the bilayer, FA will bind to H-FABPc through a collisional process [40]. In view of the facts that FAT/CD36 has been detected in FABPc immunoprecipitates from bovine milk fat globule membranes [41] and that gene deletion of H-FABPc in mice results in a marked impairment of muscular FA uptake [42], H-FABPc could be an integral member of the transsarcolemmal transport system.

In the past few years, several FAT/CD36 transgenic mouse models have been generated, which were very helpful in the establishment of a key role of FAT/CD36 in muscle FA uptake. In FAT/CD36 null mice, there is an increase in circulating triacylglycerol and FA [43], whereas in transgenic mice that overexpress FAT/CD36, their circulating levels are reduced [44]. This suggests that skeletal muscle FAT/CD36 is a key protein in regulating the circulating concentrations of FA. Furthermore, in the soleus of transgenic mice that overexpress FAT/CD36, FA oxidation was markedly increased during contractions, compared to the soleus of wild type animals [44]. More direct evidence from these FAT/CD36 transgenics was provided by in vivo measurement of FA using the fluorescent non-metabolizable analog BMIPP, demonstrating a lesser accumulation of this analog into skeletal muscle in FAT/CD36 null mice [45]. In order to shed more light on the role of FABPpm and FATP in muscle FA uptake, transgenic animal for these proteins would be instrumental, but these have, to our knowledge, not yet been generated.

4. Regulation of muscle FA uptake by FA transporters

The involvement of proteins in cellular FA uptake provides the physiologically important advantage of altering the amount and/or activity of these proteins as a result of which the uptake process can be adjusted. Such a regulation at the sarcolemmal level is not feasible when passive diffusion would have been the sole determinant of muscle FA uptake. For the purpose of this chapter, we will focus on acute regulation of FA uptake, i.e. changes occurring within minutes. The long-term regulation via alterations in protein synthesis of FA transporters has been detailed in other reviews [21,46].

4.1. Regulation by contractions

Acute regulation of FA uptake by skeletal muscle was deduced from observations in which the sum of FA oxidation and esterification (an indication of steady state uptake) was increased during muscle contraction [47]. To examine the effect of contractions on initial FA uptake, rat hind-limb muscle was electrically stimulated for up to 30 min via the sciatic nerve. Compared to the control lateral leg, giant vesicles isolated from the contracting leg muscles showed a marked (1.5-fold) increase in FA uptake (Fig. 3, see also Ref. [48]). This increase was sensitive to inhibition by SSO, indicating that contraction-inducible FA uptake is mainly mediated by FAT/CD36. Kinetic studies indicated an increased V_{max} at an unaltered K_m, suggesting that the increase in FA uptake was due to increased amounts of FAT/CD36 rather than an increase in its intrinsic activity. Indeed, Western analysis of

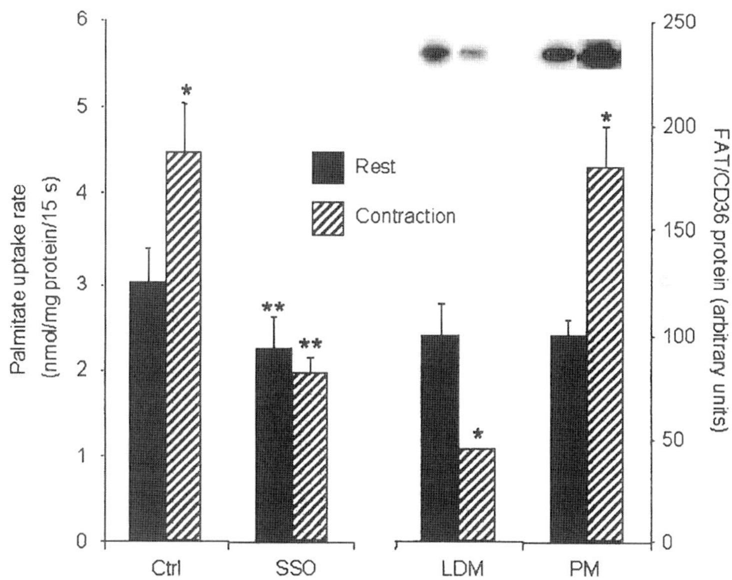

Fig. 3. Cellular contractions stimulate muscle FA uptake in an SSO-sensitive manner and induce translocation of FAT/CD36 from intracellular stores to the sarcolemma. Rat hind-limb muscles were electrically stimulated to contract via the sciatic nerve for 30 min at 40 tetani/min, and subsequently used for giant vesicle isolation followed by treatment with SSO prior to the assay of palmitate uptake (left panel) or for subcellular fractionation followed by Western blotting (right panel). The contralateral muscles from the same animal served as non-contracting control muscles. In the fractionation studies, the light density microsome (LDM) and the plasma membrane (PM) fraction from non-contracting muscles were set at 100. *Contraction compared with rest ($P < 0.05$). **SSO-treated vesicles compared with untreated vesicles ($P < 0.05$). Data are redrawn from Ref. [48].

giant vesicle membranes confirmed a greater abundance of FAT/CD36 within the sarcolemma of contracting muscle [48]. The effects of contractions were rapid and already detectable after 5 min of stimulation, excluding the occurrence of de novo protein synthesis to explain elevated sarcolemmal amounts of FAT/CD36. Subcellular fractionation showed that the underlying mechanism involves a translocation of FAT/CD36 from an intracellular, presumably endosomal, compartment to the sarcolemma. During the postcontraction recovery period, FA uptake returned to basal levels and likely FAT/CD36 was internalized from the sarcolemma [48]. This mechanism resembles the well-characterized mobilization of GLUT4 from an intracellular storage compartment to the plasma membrane.

4.2. Regulation by insulin

Insulin is another major regulator of muscle substrate utilization. With respect to FA handling by muscle, insulin has been found to increase esterification of FA into triacylglycerols [49], as well as to elevate the total sum of FA metabolites in muscle [50]. Using subcellular fractionation, we demonstrated that insulin is another major factor able to recruit FAT/CD36 from intracellular stores to the sarcolemma within minutes (Fig. 4). Insulin-stimulated FAT/CD36 translocation is prevented in the presence of the compound

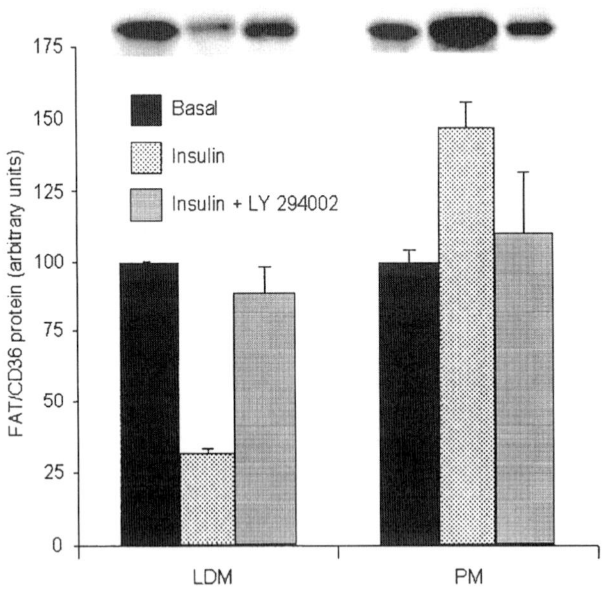

Fig. 4. Insulin induces translocation of FAT/CD36 from intracellular stores to the sarcolemma. Rat hind-limb muscles were perfused in the absence (basal) or presence of insulin, or with insulin in combination with the PI3K inhibitor LY 294002 prior to subcellular fractionation. The light density microsome (LDM) and the plasma membrane (PM) fraction from non-treated muscles were set at 100. *Significantly different from basal ($P < 0.05$). Data are redrawn from Ref. [51].

LY 294002 (Fig. 4), indicating that this translocation is dependent on the activation of phosphatidylinositol-3 kinase [51]. These studies disclosed an entirely novel role of insulin in muscle substrate utilization, which in fact mirrors its effects on glucose handling. Thus, insulin stimulates deposition of FA and glucose into triacylglycerols and glycogen, respectively, through simultaneous substrate transporter translocation.

When insulin was administered to contracting muscles, insulin further stimulated FA uptake, indicating that the effects of insulin and contractions on FA uptake are additive [50]. Given the fact that both stimuli operate via translocation of FAT/CD36, this additivity suggests that either (i) FAT/CD36 is stored in two distinct intracellular compartments or (ii) FAT/CD36 is stored in one intracellular compartment from which it is recruited via two independent signaling pathways. We regard the first mechanism as more likely, since insulin- and contraction-sensitive stores have also been implicated in the regulation of GLUT4. Nonetheless, the additivity of insulin and contractions on both FA and glucose uptakes adds to the striking similarities between regulation of glucose and FA handling by skeletal muscle.

Whether FABPpm is redistributed by insulin or by contractions is not known. In addition, we do not know whether FABPpm's intracellular localization can be extended from mitochondria, where it is operative as aspartate aminotransferase, to the endosomal compartment.

5. FA uptake into skeletal muscle of type-1 and type-2 diabetic rat models

The ability of insulin to regulate muscle FA uptake through transporter recycling allows the suggestion that in syndromes of insulin deficiency or insulin resistance muscle FA uptake will be dysregulated as a result of a change in sarcolemmal transporter contents. Due to the large contribution of muscle to body mass, this could have important implications for whole body lipid homeostasis and could be a critical factor in the etiology of diabetes. These considerations made us investigate the skeletal muscle FA uptake, as well as the expression and distribution of muscle FA transporters in type-1 and type-2 diabetic rats.

5.1. Type-1 diabetes

For induction of insulin deficiency, we have used two doses of STZ. The lowest dose gave rise to relatively moderate symptoms of type-1 diabetes with a partial reduction in circulating insulin in combination with a slightly (but significantly) elevated plasma glucose. The highest dose brought about a severe form of this disease with a greatly reduced insulin level and more than 4-fold rise in plasma glucose [52]. FA uptake was 1.4-fold elevated in skeletal muscle giant vesicles from moderately type-1 diabetic animals (Fig. 5). However, in severely type-1 diabetic rats, muscle FA uptake was not significantly different from the moderately diabetic rats, indicating that this process is not related to the severity of insulin deficiency. The changes in FA uptake were not accompanied by significant changes in the mRNA levels of both FABPpm and FAT/CD36. With respect to the total tissue protein contents of both transporters, these were slightly but significantly increased with FABPpm and more pronouncedly

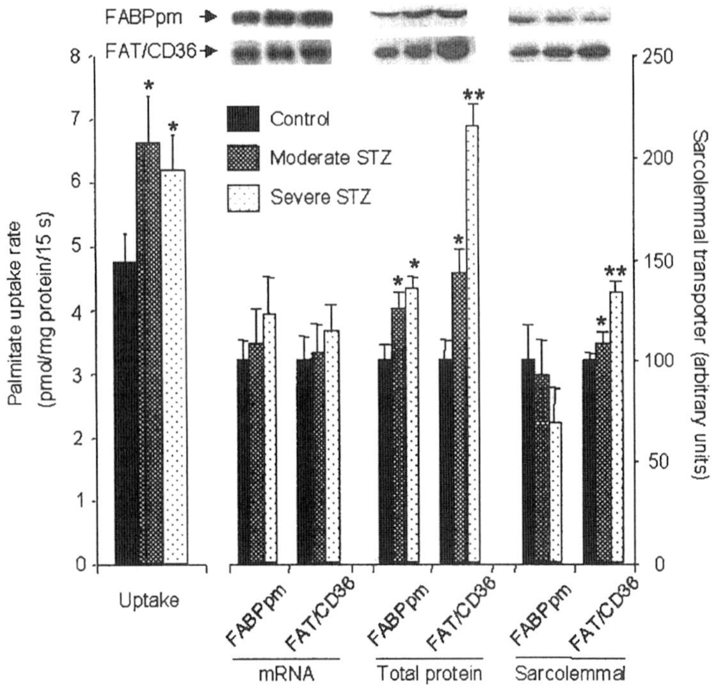

Fig. 5. Influence of moderate and severe STZ-induced diabetes on FA uptake (left panel) and on total tissue- and cell-surface expression of FA transporters (right panel) in rat hind-limb muscles. Moderate and severe diabetes were induced by a single tail vein injection of 35 and 55 mg/kg body mass, respectively. Rats were sacrificed 3 months thereafter. Palmitate uptake and sarcolemmal abundance of FABPpm and FAT/CD36 were determined in giant vesicles. Total tissue expression of these transporters at the mRNA- and the protein levels were determined in total tissue homogenates. FABPpm and FAT/CD36 were detected at 2.4 and 2.9 kB in Northern blots, respectively, and at 43 and 88 kDa in Westerns. *Significantly different from control ($P < 0.05$). **Severe diabetes versus moderate diabetes ($P < 0.05$). Data are redrawn from Ref. [52].

increased with FAT/CD36, the latter also showing a relationship with the severity of the disease. In agreement with these findings, the sarcolemmal levels of FAT/CD36 were also increased proportionally to the severity of diabetes, albeit to a lesser extent in comparison to the total tissue protein contents. However, sarcolemmal levels of FABPpm did not increase in type-1 diabetes, and tended to decrease in the severe form [52]. A number of important conclusions can be drawn: (i) changes in transporter protein levels do not parallel changes in mRNA level, suggesting the importance of posttranscriptional regulation of transporter expression in type-1 diabetes; (ii) the lack of a relationship between total tissue contents and sarcolemmal levels of FABPpm suggests that the subcellular distribution has undergone alterations, with perhaps an internalization in severely type-1 diabetic muscle; (iii) sarcolemmal FAT/CD36 levels scale better with FA uptake rates than sarcolemmal FABPpm levels, suggesting that FABPpm is not the primary transporter regulating FA uptake. Possibly, plasmalemmal FABPpm may already be present in excess, and becomes only a rate-limiting factor at elevated FA fluxes.

Taken together, the increase in muscle FA uptake in moderate type-1 diabetes can best be explained by a posttranscriptional upregulation of only FAT/CD36. In severe type-1 diabetes, a further increase in FAT/CD36-mediated FA uptake is prevented by FABPpm internalization, which could later serve as a compensatory mechanism to protect the type-1 diabetic myocyte against a further harmful influx of FA.

When moderately and severely type-1 diabetic muscles were subjected to electrical stimulation, FA uptake is increased by 1.6–2.0-fold, just as in control muscles (Luiken and Bonen, unpublished observations). We expect that in type-1 diabetes, FAT/CD36 translocation is not malfunctioning, and that contractions maintain the ability to recruit FAT/CD36 to the cell surface.

5.2. Type-2 diabetes

The obese Zucker rat has been proven to be an established rodent model of insulin resistance and type-2 diabetes. In skeletal muscle giant vesicles from obese Zucker rats, FA uptake was markedly (1.8-fold) increased (Fig. 6). However, expression of FABPpm and FAT/CD36 at both the messenger and the protein levels was not altered in obese

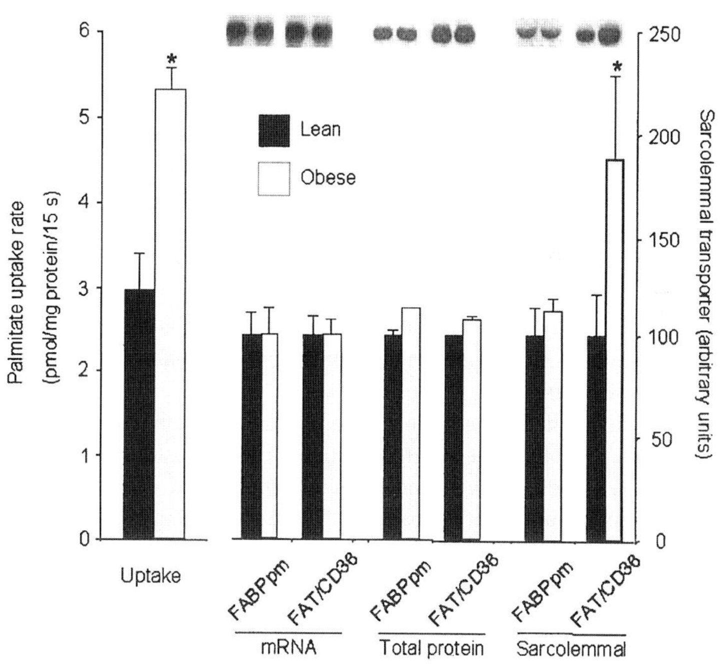

Fig. 6. FA uptake (left panel) and total tissue- and cell-surface expression of FA transporters (right panel) in rat hind-limb muscles of lean (*fa*/+) and obese (*fa*/*fa*) Zucker rats. Palmitate uptake and sarcolemmal abundance of FABPpm and FAT/CD36 were determined in giant vesicles. Total tissue expression of these transporters at the mRNA- and the protein levels were determined in total tissue homogenates. FABPpm and FAT/CD36 were detected at 2.4 and 2.9 kB in Northern blots, respectively, and at 43 and 88 kDa in Westerns. *Obese compared to lean ($P < 0.05$). Data are redrawn from Ref. [53].

Zucker rats (Fig. 6). Thus, neither transcriptional nor posttranscriptional regulation of transporter expression could account for the increase in muscle FA uptake in obesity/type-2 diabetes. In, addition the sarcolemmal content of FABPpm was not changed. Rather, the increased FA uptake was attributable to an elevated sarcolemmal content of FAT/CD36. Because the total cellular pool of FAT/CD36 was not changed, it appears that in obese muscles a larger part of the intracellular pool is permanently relocated at the sarcolemma [53]. Alterations in transporter redistribution associated with obesity have also been observed for GLUT4 in muscle. Notably, in obese skeletal muscle, the insulin-inducible mobilization is inhibited [54], indicating that the GLUT4 translocation machinery is impaired. Combining this with the relocation of FAT/CD36 to the sarcolemma, it appears that the patterns of subcellular distribution of FAT/CD36 and GLUT4 at the sarcolemma and the intracellular compartments are diametrically opposed in obesity.

Recently, FA uptake has been measured in skeletal muscle of type-2 diabetic humans in vivo using continuous infusion with the stable isotope tracer [U-^{13}C]palmitate [55,56]. Using this technique, skeletal muscle FA uptake was ~40% lower in type-2 diabetics than in healthy individuals. Possible explanations for the discrepancy with the Zucker study could include (i) a difference in muscle FA handling between human and rodent type-2 diabetes or (ii) a difference in the methods to measure FA uptake rates. The first explanation is unlikely, given that in both rodent and human type-2 diabetes there is a strong association between increased triacylglycerol content in skeletal muscle and the development of insulin resistance [57,58]. Rather, the complexity of the in vivo situation makes it likely that many other factors might influence FA uptake, as opposed to giant vesicles in which these factors are eliminated. For instance, ketone bodies whose circulating levels are elevated in type-2 diabetics have a mitigating effect on FA utilization [59]. Furthermore, when giant vesicles were prepared from muscle biopsies of obese and type-2 diabetic humans, these vesicles displayed a more than 2-fold greater FA uptake rate and a greater sarcolemmal content of FAT/CD36 compared to skeletal muscle vesicles from healthy individuals (Bonen, MacDonald, Steinberg, Glatz, Luiken and Heigenhauser, unpublished results), adding to the similarity between human and rodent type-2 diabetes in muscle substrate handling and pinpointing to a potentially pivotal role of FAT/CD36 in the etiology of human diabetes.

More recent studies indicate that induction of contractile activity in skeletal muscle of obese Zucker rats, in contrast to their lean littermates, did not result in an increase in giant-vesicular FA uptake (Dyck, Luiken and Bonen, unpublished observations). Instead, a significant decrease in FA uptake was observed. These changes in obese muscle FA uptake during contractions were paralleled by similar changes in sarcolemmal FAT/CD36 content (Dyck, Luiken and Bonen, unpublished results). This suggests that in obese animals, FAT/CD36 may be re-internalized to its intracellular pool. The physiological relevance of this phenomenon is at present incompletely understood, but it might be an attempt of obese muscles to restrict FA uptake during contractions.

6. Comparison of FA transporter functioning in type-1 and type-2 diabetes

In both type-1 and type-2 diabetes muscle FA uptake is increased, and in both cases FAT/CD36 is responsible for this increase. Interestingly, the underlying mechanism is different. Whereas in type-1 diabetes, the increase in sarcolemmal content of FAT/CD36 is due to increased protein expression, the increase in sarcolemmal FAT/CD36 in type-2 diabetes is caused by an altered subcellular distribution at the expense of intracellular stores. We do not know whether this difference in sarcolemmal upregulation of FAT/CD36 is connected to the fact that contraction-inducible FA uptake is maintained in type-1 diabetes but not in type-2 diabetes. The notion arises that the inability of type-2 diabetic muscle to further increase FA uptake upon contraction is caused by the absence of intracellular FAT/CD36 that is available for recruitment to the sarcolemma. Such a depletion of intracellular FAT/CD36 stores is not occurring in type-1 diabetes, and therefore contraction-inducible FA uptake is not hampered.

A related issue is the difference in metabolic fate of FA in type-1 and type-2 diabetes. It has been observed that FA oxidation is increased in cardiac [11,18] and skeletal muscle [60] of STZ-treated rats. In muscle tissues of obese Zucker rats, on the other hand, no change [61,62] or even a decrease [63] in FA oxidation has been reported in addition to the lack of alterations in mitochondrial capacity and oxygen consumption [64]. Furthermore, there are multiple reports on muscle triacylglycerol accumulation in type-2 diabetes in both rodents and humans [12,19,65,66], whereas the picture is less clear in type-1 diabetes [67–69]. Thus, while in type-2 diabetes, FA are preferentially channeled into esterification into triacylglycerols, such a preferential channeling is less evident in type-1 diabetic muscle that is increasingly dependent on FA oxidation. Combined with the differential recruitability of FAT/CD36 in both types of diabetes, one might speculate (i) whether the "apparent" lack in preferential channeling of FA into cellular triacylglycerols in type-1 diabetic muscle is related to its ability to increase FA uptake upon contractions, i.e. a condition that favors oxidative metabolism, and reciprocally (ii) whether the loss of contraction-inducible FA uptake in type-2 diabetic muscle is another factor leading to increased partitioning of FA into storage.

Integrating these findings concerning uptake of FA and their subsequent metabolic fate in both types of diabetes has potential consequences for a possible therapeutic treatment at the level of the FA uptake process. In type-2 diabetes, the increase in sarcolemmal FAT/CD36 is directly responsible for the build-up of intracellular lipid. Our recent findings in obese and type-2 diabetic human muscle biopsies in which sarcolemmal contents of FAT/CD36 correlated well with the size of intramuscular triacylglycerol stores (Bonen, MacDonald, Steinberg, Glatz, Luiken and Heigenhauser, unpublished results) are strongly confirmatory to this notion. In type-1 diabetic muscle, like type-2, upregulation of sarcolemmal FAT/CD36 enhances muscle FA uptake, but is probably more connected to the increased dependence of type-1 diabetic muscle on FA oxidation. Moreover, the mechanism of enhancing sarcolemmal FAT/CD36 through upregulation of expression, then, is linked to maintaining intracellular pools for oxidative purposes. The important implication is that FAT/CD36 is an attractive target for therapeutic intervention in type-2 diabetes, but probably not in type-1 diabetes

7. Concluding remarks

The evidence presented in this chapter favors the notion that uptake of FA into skeletal muscle is governed by a transport system that consists of FABPpm acting as FA receptor and FAT/CD36 acting as flippase. Likely, FABPpm and FAT/CD36 are not constantly in mutual contact, as FABPpm could be constitutively present at the sarcolemma, while in resting muscles the majority of FAT/CD36 is localized in intracellular stores. Upon exposure to insulin and/or electrical stimulation, FAT/CD36 migrates to the sarcolemma to interact with FABPpm in order to facilitate FA uptake. Similar to skeletal muscle, we found an intracellular storage pool of FAT/CD36 in cardiac muscle, from which this protein can be mobilized upon an increase in mechanical activity as well as upon insulin treatment [70–72]. Using cardiac myocytes we have begun unraveling the signaling pathways responsible for FAT/CD36 translocation. The effect of insulin on FAT/CD36 translocation is dependent on the activation of phosphatidylinositol-3 kinase, while cellular contractions induce activation of AMP-dependent protein kinase. We speculate that both the signaling enzymes are also important for FAT/CD36 translocation in skeletal muscle. Further unraveling of the signaling processes that regulate FAT/CD36 translocation may provide novel subcellular targets to restore substrate handling by skeletal muscle in diabetes.

Acknowledgements

The authors thank J. Williems for his help in preparing the illustrations. Work in the author's laboratories was supported by the Netherlands Heart Foundation (grant D98.012) and the Ontario Heart and Stroke Foundation. Joost J.F.P. Luiken is the recipient of a VIDI-Innovational Research grant from the Netherlands Organisation for Scientific Research (NWO-Zon-Mw grant nr. 016.036.305).

References

[1] van der Vusse, G.J., Reneman, R.S., 1996. Lipid metabolism in muscle. In: Rowell, L.B., Shepherd, J.T. (Eds.), Handbook of Physiology. Am. Phys. Soc. Oxford Press, New York, pp. 954–994.
[2] Dyck, D.J., Peters, S.J., Glatz, J., Gorski, J., Keizer, H., Kiens, B., Liu, S., Richter, E.A., Spriet, L.L., van der Vusse, G.J., Bonen, A., 1997. Am. J. Physiol. 272, E340–E351.
[3] Kiens, B., Roemen, T.H.M., van der Vusse, G.J., 1999. Am. J. Physiol. 276, E352–E357.
[4] van der Vusse, G.J., Roemen, T.H.M., 1995. J. Appl. Physiol. 78, 1839–1843.
[5] Abumrad, N.A., Perkins, R.C., Park, J.H., Park, C.R., 1981. J. Biol. Chem. 256, 9183–9191.
[6] Stremmel, W., Berk, P.D., 1986. Proc. Natl Acad. Sci. USA 83, 3086–3090.
[7] Stremmel, W., 1988. J. Clin. Invest. 81, 844–852.
[8] Turcotte, L.P., Kiens, B., Richter, E.A., 1991. FEBS Lett. 279, 327–329.
[9] Glatz, J.F.C., van der Vusse, G.J., 1990. Mol. Cell. Biochem. 98, 237–251.
[10] Pan, D.A., Lillioja, S., Kriketos, A.D., Milner, M.R., Baur, L.A., Bogardus, C., Jenkins, A.B., Storlien, L.H., 1997. Diabetes 46, 983–988.
[11] Rodrigues, B., Cam, M.C., McNeill, J.H., 1998. Mol. Cell. Biochem. 180, 53–57.
[12] Shulman, G.I., 2000. J. Clin. Invest. 106, 171–176.
[13] Kolter, T., Uphues, I., Eckel, J., 1997. Am. J. Physiol. 273, E59–E67.
[14] Klip, A., Marette, A., 1992. J. Cell. Biochem. 48, 51–60.
[15] Kahn, B.B., 1992. J. Clin. Invest. 89, 1367–1374.
[16] Kainulainen, H., Breiner, M., Schurmann, A., Marttinen, A., Virjo, A., Joost, H.G., 1994. Biochim. Biophys. Acta 1225, 275–282.
[17] Kawanaka, K., Higuchi, M., Ohmori, H., Shimegi, S., Ezaki, O., Katsuta, S., 1996. Horm. Metab. Res. 28, 75–80.

[18] Stanley, W.C., Lopaschuk, G.D., McCormack, J.G., 1997. Cardiovasc. Res. 34, 25–33.
[19] Goodpaster, B.H., Kelley, D.E., 2002. Curr. Diab. Rep. 2, 216–222.
[20] Luiken, J.J.F.P., Glatz, J.F.C., Bonen, A., 2000. Can. J. Appl. Physiol. 25, 333–352.
[21] Bonen, A., Luiken, J.J.F.P., Glatz, J.F.C., 2002. Mol. Cell. Biochem. 239, 181–192.
[22] Standen, N.B., Stanfield, P.R., Ward, T.A., 1984. J. Physiol. 346, 81P.
[23] Koonen, D.P.Y., Coumans, W.A., Arumugam, Y., Bonen, A., Glatz, J.F.C., Luiken, J.J.F.P., 2002. Mol. Cell. Biochem. 239, 121–130.
[24] Juel, C., 1991. Biochim. Biophys. Acta 1065, 15–20.
[25] Ploug, T., Wojtaszewski, J., Kristiansen, S., Hespel, P., Galbo, H., Richter, E.A., 1993. Am. J. Physiol. 264, E270–E278.
[26] McDermott, J.C., Bonen, A., 1993. Mol. Cell. Biochem. 122, 113–121.
[27] McCullagh, K.J., Juel, C., O'Brien, M., Bonen, A., 1996. Mol. Cell. Biochem. 156, 51–57.
[28] Luiken, J.J.F.P., van Nieuwenhoven, F.A., America, G., van der Vusse, G.J., Glatz, J.F.C., 1997. J. Lipid Res. 38, 745–758.
[29] Luiken, J.J.F.P., Turcotte, L.P., Bonen, A., 1999. J. Lipid Res. 40, 1007–1016.
[30] Bonen, A., Luiken, J.J.F.P., Liu, S., Dyck, D.J., Kiens, B., Kristiansen, S., Turcotte, L.P., van der Vusse, G.J., Glatz, J.F.C., 1998. Am. J. Physiol. 275, E471–E478.
[31] Stremmel, W., Strohmeyer, G., Borchard, F., Kochwa, S., Berk, P.D., 1985. Proc. Natl Acad. Sci. USA 82, 4–8.
[32] Harmon, C.M., Luce, P., Beth, A.H., Abumrad, N.A., 1991. J. Membr. Biol. 121, 261–268.
[33] Schaffer, J.E., Lodish, H.F., 1994. Cell 79, 427–436.
[34] Turcotte, L.P., Srivastava, A.K., Chiasson, J.L., 1997. Mol. Cell. Biochem. 166, 153–158.
[35] Kiens, B., Kristiansen, S., Jensen, P., Richter, E.A., Turcotte, L.P., 1997. Biochem. Biophys. Res. Commun. 231, 463–465.
[36] van Nieuwenhoven, F.A., Willemsen, P.H., van der Vusse, G.J., Glatz, J.F.C., 1999. Int. J. Biochem. Cell. Biol. 31, 489–498.
[37] Stump, D.D., Zhou, S.L., Berk, P.D., 1993. Am. J. Physiol. 265, G894–G902.
[38] van Nieuwenhoven, F.A., van der Vusse, G.J., Glatz, J.F.C., 1996. Lipids 31(Suppl.), S223–S227.
[39] Glatz, J.F.C., van der Vusse, G.J., 1996. Prog. Lipid Res. 35, 243–282.
[40] Kim, H.K., Storch, J., 1992. J. Biol. Chem. 267, 20051–20056.
[41] Spitsberg, V.L., Matitashvili, E., Gorewit, R.C., 1995. Eur. J. Biochem. 230, 872–878.
[42] Schaap, F.G., Binas, B., Danneberg, H., van der Vusse, G.J., Glatz, J.F.C., 1999. Circ. Res. 85, 329–337.
[43] Febbraio, M., Abumrad, N.A., Hajjar, D.P., Sharma, K., Cheng, W., Pearce, S.F., Silverstein, R.L., 1999. J. Biol. Chem. 274, 19055–19062.
[44] Ibrahimi, A., Bonen, A., Blinn, W.D., Hajri, T., Li, X., Zhong, K., Cameron, R., Abumrad, N.A., 1999. J. Biol. Chem. 274, 26761–26766.
[45] Coburn, C.T., Knapp, F.F. Jr., Febbraio, M., Beets, A.L., Silverstein, R.L., Abumrad, N.A., 2000. J. Biol. Chem. 275, 32523–32529.
[46] Luiken, J.J.F.P., Miskovic, D., Arumugam, Y., Glatz, J.F.C., Bonen, A., 2001. Int. J. Sport Nutr. Exerc. Metab. 11, S92–S96.
[47] Dyck, D.J., Bonen, A., 1998. Am. J. Physiol. 275, E888–E896.
[48] Bonen, A., Luiken, J.J.F.P., Arumugam, Y., Glatz, J.F.C., Tandon, N.N., 2000. J. Biol. Chem. 275, 14501–14508.
[49] Muoio, D.M., Dohm, G.L., Tapscott, E.B., Coleman, R.A., 1999. Am. J. Physiol. 276, E913–E921.
[50] Dyck, D.J., Steinberg, G., Bonen, A., 2001. Am. J. Physiol. Endocrinol. Metab. 281, E600–E607.
[51] Luiken, J.J.F.P., Dyck, D.J., Han, X.X., Tandon, N.N., Arumugam, Y., Glatz, J.F.C., Bonen, A., 2002. Am. J. Physiol. Endocrinol. Metab. 282, E491–E495.
[52] Luiken, J.J.F.P., Arumugam, Y., Bell, R.C., Calles-Escandon, J., Tandon, N.N., Glatz, J.F.C., Bonen, A., 2002. Am. J. Physiol. Endocrinol. Metab. 283, E612–E621.
[53] Luiken, J.J.F.P., Arumugam, Y., Dyck, D.J., Bell, R.C., Pelsers, M.M.A.L., Turcotte, L.P., Tandon, N.N., Glatz, J.F.C., Bonen, A., 2001. J. Biol. Chem. 276, 40567–40573.
[54] Brozinick, J.T. Jr., Etgen, G.J. Jr., Yaspelkis, B.B. III, Kang, H.Y., Ivy, J.L., 1993. Am. J. Physiol. 265, E419–E427.

[55] Blaak, E.E., Wagenmakers, A.J., Glatz, J.F.C., Wolffenbuttel, B.H., Kemerink, G.J., Langenberg, C.J., Heidendal, G.A., Saris, W.H., 2000. Am. J. Physiol. Endocrinol. Metab. 279, E146–E154.
[56] Blaak, E.E., Wagenmakers, A.J., 2002. Diabetes 51, 784–789.
[57] Roden, M., Perseghin, G., Petersen, K.F., Hwang, J.H., Cline, G.W., Gerow, K., Rothman, D.L., Shulman, G.I., 1996. J. Clin. Invest. 97, 642–648.
[58] Perseghin, G., Scifo, P., De Cobelli, F., Pagliato, E., Battezzati, A., Arcelloni, C., Vanzulli, A., Testolin, G., Pozza, G., Del Maschio, A., Luzi, L., 1999. Diabetes 48, 1600–1606.
[59] Hasselbaink, D.M., Glatz, J.F.C., Luiken, J.J.F.P., Roemen, T.H.M., van der Vusse, G.J., 2003. Biochem. J. 371, 753–760.
[60] Bitar, M.S., 2001. Horm. Metab. Res. 33, 577–584.
[61] Turcotte, L.P., Swenberger, J.R., Tucker, M.Z., Yee, A.J., 2001. Diabetes 50, 1389–1396.
[62] Young, M.E., Guthrie, P.H., Razeghi, P., Leighton, B., Abbasi, S., Patil, S., Youker, K.A., Taegtmeyer, H., 2002. Diabetes 51, 2587–2595.
[63] Bessesen, D.H., Rupp, C.L., Eckel, R.H., 1995. Obes. Res. 3, 179–189.
[64] Wardlaw, G.M., Kaplan, M.L., 1984. Am. J. Physiol. 247, R911–R917.
[65] Kraegen, E.W., Cooney, G.J., Ye, J., Thompson, A.L., 2001. Exp. Clin. Endocrinol. Diabetes 109, S516–S526.
[66] Kelley, D.E., Goodpaster, B.H., Storlien, L., 2002. Annu. Rev. Nutr. 22, 325–346.
[67] Murthy, V.K., Shipp, J.C., 1977. Diabetes 26, 222–229.
[68] Ramanadham, S., Doroudian, A., McNeill, J.H., 1990. Can. J. Cardiol. 6, 75–82.
[69] Han, X., Abendschein, D.R., Kelley, J.G., Gross, R.W., 2000. Biochem. J. 352, 79–89.
[70] Luiken, J.J.F.P., Willems, J., van der Vusse, G.J., Glatz, J.F.C., 2001. Am. J. Physiol. Endocrinol. Metab. 281, E704–E712.
[71] Luiken, J.J.F.P., Koonen, D.P.Y., Willems, J., Zorzano, A., Becker, C., Fischer, Y., Tandon, N.N., van der Vusse, G.J., Bonen, A., Glatz, J.F.C., 2002. Diabetes 51, 3113–3119.
[72] Luiken, J.J.F.P., Coort, S.L.M., Willems, J., Coumans, W.A., Bonen, A., van der Vusse, G.J., Glatz, J.F.C., 2003. Contraction-induced FAT/CD36 translocation in rat cardiac myocytes is mediated through AMP-activated protein kinase signaling. Diabetes 52, 1627–1634.

Fatty acid metabolism in cardiac hypertrophy and failure

Heinrich Taegtmeyer* and Leonard Golfman

Division of Cardiology, University of Texas Houston Medical School, 6431 Fannin, MSB 1.246, Houston, TX 77030, USA
**Correspondence address: Tel.: +1-713-500-6569; fax: +1-713-500-6556*
E-mail: Heinrich.Taegtmeyer@uth.tmc.edu(H.T.)

1. Historical perspective

Perhaps the best way to introduce this chapter is a brief historical overview. Studies performed by Bing and coworkers in the 1950s established the predilection of the human heart for fatty acids as fuel for respiration [1]. Shortly thereafter the same group reported that myocardial uptake of fatty acids, glucose, and ketone bodies was not altered in patients with congestive heart failure [2]. However, measurements of atheriovenous concentration differences across the heart under non-steady state conditions are notoriously difficult and can mask intramyocardial damages in substrate metabolism [3]. Earlier studies, still valid today, have proposed that contractile failure may develop as a consequence of structural and biochemical alterations that accompany cardiac hypertrophy [4]. The hypothesis is based on two major premises: (1) the anatomical nature of cardiac hypertrophy and the role of hypoxia and (2) functional defects in mitochondria in failing heart [5]. Confirmation of these early studies had to wait for the development of new and more precise techniques for the assessment of cardiac function and metabolism.

The first evidence for defective fatty acid metabolism of the human heart in dilated cardiomyopathies came from studies using positron emission tomography (PET) of the heart after intravenous injection of [^{11}C] palmitate [6]. In these studies, the mean left ventricular myocardial accumulation of [^{11}C] palmitate was significantly less and also significantly more heterogeneous than that of the control subjects. Soon thereafter, a series of rare, genetic defects in fatty acid oxidation became defined as inherited causes of cardiomyopathy [7]. Here, the time of clearance of palmitate from the myocardium is significantly prolonged, indicative of decreased rates of β-oxidation and the entry of fatty acids into a slow-turnover (i.e. triglyceride) pool [7,8].

More recently, investigators have used fatty acid tracer analogs to assess fatty acid metabolism in dilated, failing hearts of patients before and after treatment with a

Advances in Molecular and Cell Biology, Vol. 33, pages 259–270

ISSN: 1569-2558/DOI: 10.1016/S1569-2558(03)33013-9

β-blocker [9]. The investigators demonstrated enhanced uptake and retention of the fatty acid tracer suggesting an imbalance of fatty acid transport (presumably unchanged) and oxidation (presumably increased). The reason is that, unlike [^{11}C] palmitate, the tracer analog [^{18}F]-fluoro-G-thia-heptadecanoic acid is not oxidized when taken up by the cell, and is, instead, subject to esterification with glycerol to triglycerides. Treatment of heart failure patients with a β-blocker (carvedilol) leads to a readily detectable decrease of fatty acid uptake by the heart and an increase in glucose uptake suggesting increased reliance on glucose as an energy-providing substrate (Fig. 1). The findings are consistent with the concept of a change in the biological properties of the failing human heart with β-blockade that includes substrate preference and substrate-induced improvement in cardiac efficiency [10,11]. Because uptake of the fatty acid analog was higher in fasting than in non-fasting hearts, it can be speculated there is a loss of synchronization between fatty acid transport and β-oxidation. Taken together, all available evidence supports the concept of impaired fatty acid metabolism in the failing human heart that can be readily detected by PET.

The observation of increased lipid deposition in cardiac myocytes is not new. In the classical text on cellular pathology, Virchow describes "lipid atrophy" in the failing human heart [12]. Virchow already made the distinction between fatty infiltration (adipocyte infiltration) and fatty degeneration or "metamorphosis" of the cardiomyocyte. Experimentally, defective lipid metabolism in the failing heart was first observed in the 1960s [13,14]. Following initial observations in a model of diphtheria infection, the severe clinical manifestations of inborn errors in human fatty acid oxidizing enzymes, including dilated cardiomyopathies and sudden death, revealed the importance of *normal* fatty acid metabolism for *normal* function of the *normal* heart. At least a dozen separate inherited disorders of mitochondrial fatty acid β-oxidation have been described in humans [15]. The most commonly recognized defect is medium-chain acyl-CoA dehydrogenase (MCAD) deficiency [15]. In addition, polymorphisms of the PPARα gene have been shown to influence human left ventricular growth in response to exercise and hypertension [16].

Another example for mutational defects in fatty acid metabolism is defects in fatty acid transport. Clinical studies in a Japanese population have shown a high prevalence of point mutations in the FAT/CD36 gene. Patients being homozygous for this mutation showed FAT/CD36 deficiency and developed a hypertrophic cardiomyopathy [17].

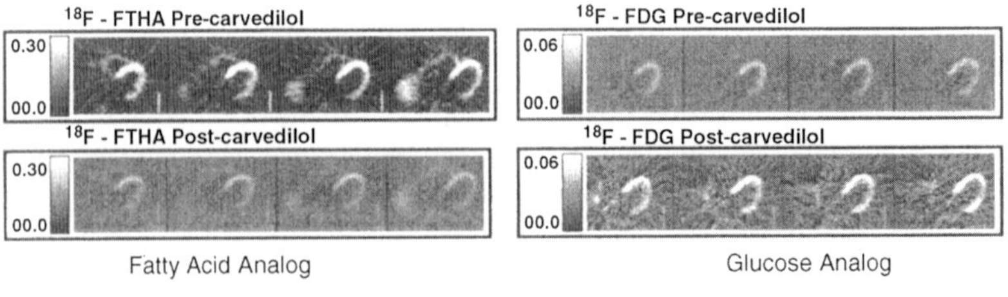

Fig. 1. Positron emission tomography of a fatty acid analog and a glucose analog in dilated cardiomyopathy before and after β-blocker treatment. Representative parameter slope images of [^{18}F] FTHA (left) and [^{18}F] FDG (right) are shown, before (top) and after (bottom) 3 months of carvedilol treatment. The ejection fraction increased from 25 to 37%. (From Wallhaus et al., Circulation, 2001, reproduced with permission.)

2. Metabolic remodeling of the heart

The above examples of defective fatty acid metabolism in the heart, taken from the clinical literature, lead us to a brief consideration of the normal responses of the heart to changes in its physiologic environment. Studies carried out in isolated perfused heart preparations have brought to light that the heart responds to changes in its environment by redirecting substrate fluxes through those pathways that are energetically most efficient in a given metabolic or physiologic environment [18,19]. Other examples include the suppression of glucose oxidation when there is an oversupply of fatty acids [20], and conversely, the suppression of long-chain fatty acid oxidation in the presence of high concentrations of glucose and insulin [21]. An acute increase in the workload of the heart results in instantaneous mobilization and oxidation of glycogen and a shift from fat to carbohydrates (glucose and lactate) as the main fuel for respiration [18]. In the heart in vivo, a shift to greater efficiency of energy conversion has been documented [22].

It is not always appreciated that the heart constantly adapts to alterations in its environment, allowing maintenance of cardiac output in response to physiologic and pathophysiologic stimuli. As outlined above, this adaptation can either be acute (alterations in preexisting proteins) or chronic (alterations in gene expression, remodeling), depending upon the intensity and duration of the stimulus [23,24]. It is reasonable to oppose that metabolic adaptation precedes or parallels functional adaptation of the heart and that the heart's inability to adapt appropriately to a particular stimulus leads to contractile dysfunction and failure (maladaptation).

Fig. 2 delineates the sequence of events leading from altered environment to altered gene and protein expression. One influence on cardiac metabolism, gene expression, and function that has been studied intensely is the workload of the heart. Changes in workload affect the heart both acutely and chronically. When the workload is increased, the heart rapidly responds by increasing the flux of carbon through specific metabolic pathways. In doing so, the heart balances the rates of energy-generating reactions (e.g. oxidative phosphorylation) with that of energy-consuming reactions (e.g. cross-bridge cycling).

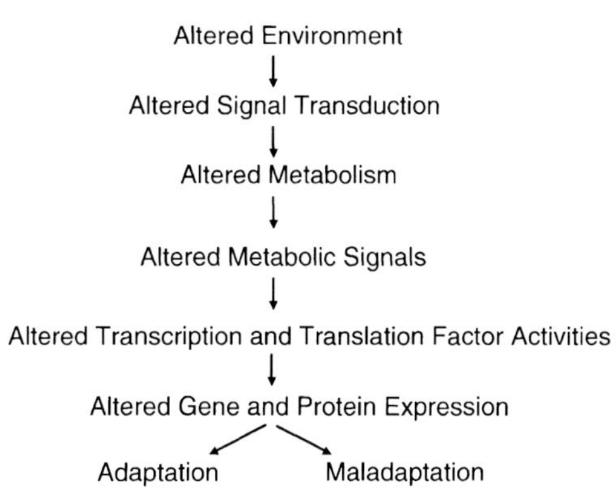

Fig. 2. The proposed sequence of steps leading from altered environment to altered gene expression in fatty acid metabolism of the heart. See text for details.

More directly, increased workload on the heart rapidly results in increased glycogenolysis, glucose uptake, glycolytic flux, and pyruvate oxidation (whether derived from extracellular glucose, extracellular lactate, or intracellular glycogen), with little effect on fatty acid oxidation [18]. Thus, there is acute substrate switching: The normal heart oxidizes preferentially fatty acids while the stressed heart oxidizes carbohydrates to meet the increased energy demand.

If the change in workload of the heart is sustained, the heart responds with characteristic alterations in morphology, increased cell size, and gene expression (e.g. re-expression of fetal genes), paralleled by a process of metabolic remodeling. In general terms, the hypertrophied heart increases its reliance on glucose as fuel, while decreasing fatty acid oxidation [25]. This upregulation of glucose metabolism occurs even before there is an increase in left ventricular mass. Somewhat unexpectedly, with a sustained *decrease* in workload the heart also responds with an up-regulation of glucose metabolism [26,27]. It appears that any sustained changes in workload result in similar patterns of metabolic remodeling and in a switch from an energetically less efficient (fatty acids) to an energetically more efficient (glucose) fuel for respiration.

3. Metabolic flexibility

Two questions arise when considering metabolic adaptation of the hypertrophied heart: (1) What is the mechanism for substrate switching? (2) Why does substrate switching occur? In the past, mechanistic studies of substrate switching have focused on regulation of the pyruvate dehydrogenase complex (PDC). Recent evidence suggests, however, that the nuclear receptor PPARα plays a key role in substrate switching, whereas PDC activity remains relatively unaffected in response to sustained pressure overload [28,29]. PPARα binds to the promoter, and subsequently induces the transcription, of multiple genes encoding for proteins involved in fatty acid metabolism. These include about a dozen enzymes, including fatty acid translocase (FAT/CD36), heart-specific fatty acid binding protein (hFABP), acyl-CoA synthetase I (ACSI), malonyl-CoA decarboxylase (MCD), muscle-specific carnitine palmitoyltransferase I (mCPTI), medium-chain acyl-CoA dehydrogenase (MCAD), long-chain acyl-CoA dehydrogenase (LCAD), and very long-chain acyl-CoA dehydrogenase (VLCAD). PPARα is itself activated by fatty acids, thereby forming a positive feed-forward mechanism for fatty acid induced fatty acid oxidation. Pressure overload represses this mechanism, by decreasing the expression PPARα itself, as well as the expression of its co-activator, PGC1. PPARα-DNA binding activity is also reduced through covalent modification (phosphorylation) in response to pressure overload. The result is decreased expression of fatty acid metabolizing genes in the hypertrophied heart, and therefore decreased fatty acid oxidation capacity. In addition, several other transcription factors (e.g. Sp1, Coup-TF) known to be activated in response to pressure overload have been implicated in the repression of fatty acid oxidation genes [30,31].

The mechanism by which reliance on glucose as a metabolic fuel increases in the hypertrophied heart is less clear than the mechanisms underlying suppressed fatty acid oxidation. One possibility is that the increased glucose oxidation is the result of decreased fatty acid oxidation through a mechanism complementing the one described nearly 40 years ago by Randle and his group [20]. Here, the focus is on both the short- and long-term regulation of the PDC [32]. In the heart, a major regulator is the enzyme pyruvate dehydrogenase kinase 4 (PDK4), which itself is under the transcriptional control of PPARα. In the normal heart, fatty acids suppress glucose oxidation to a greater extent than glycolysis and glycolysis to a greater extent than glucose uptake. Conversely, glucose also suppresses fatty acid oxidation [21] through a mechanism that involves malonyl-CoA-induced inhibition of the fatty acid transporter protein CPTI [33]. Although increased glycolytic rates in the hypertrophied heart are consistent with the Randle hypothesis, rates of pyruvate oxidation are not increased to the same extent, and insulin's effects on glucose oxidation are attenuated. Therefore, a loss of coordination between glucose oxidation and glycolysis exists. Other potential factors involved in increased glucose utilization in response to pressure overload include chronic activation of AMPK and elevated cytosolic levels of Ca^{2+} in the hypertrophied heart.

4. Substrate switching in hypertrophied heart

Why does substrate switching occur in the hypertrophied heart? A classical explanation for this phenomenon is at the energetic level. Glucose is a more efficient energy source compared to fatty acids (i.e. more ATP generated per O_2 consumed) [34]. This is particularly important when oxygen demand outstrips oxygen supply, as is the case in hypoxic heart [35], and as may be the case in the hypertrophied heart. Furthermore, evidence exists suggesting that glycolytically derived ATP is preferentially utilized by ion channels, the activities of which are increased in the hypertrophied heart [36]. However, substrate switching may have roles beyond its stereotypical function as a provider of ATP, potentially creating essential signals, within the cardiomyocyte, required for adaptation and remodeling. For example, as mentioned above, fatty acids affect the expression of several target genes through transcription factor activation. Work performed in the liver as a model system for metabolically controlled gene transcription has shown that glucose metabolites alter the expression of specific genes, such as pyruvate kinase, acetyl-CoA-carboxylase α, and fatty acid synthase [37]. Specific glucose-sensing transcription factors (e.g. SP1, USF) become activated through reversible covalent modification (phosphorylation and/or glycosylation) when glucose metabolites accumulate within the cell. These transcription factors are present in the heart, and both are activated in response to pressure overload. Evidence is accumulating for their involvement in the induction of several fetal genes. Glucose has also been suggested to be involved in the induction of growth factors and the activation of translation factors, and may therefore be involved in the trophic response to pressure overload. Indeed, heart-specific over-expression of GLUT1 or the knock-out of the hFABP in the heart results in substrate switching and cardiac hypertrophy, suggesting increased glucose utilization is sufficient to trigger the hypertrophic response.

As outlined above, it appears that metabolic adaptation may be important for the heart, not only as a means of efficient ATP generation, but also for the generation of intracellular signals involved in the hypertrophic response to pressure overload. If true, then substrate switching would be expected to be essential for the adaptation of the heart and subsequent maintenance of cardiac output in the face of pressure overload. As PPARα seems to play a key role in substrate switching, we postulated that activation of PPARα in the hypertrophied heart would prevent substrate switching. This is indeed the case; pharmacological activation of PPARα induced PPARα-regulated genes, prevented downregulation of fatty acid oxidation, and blocked the increased reliance on glucose in response to pressure overload in rats [38]. Such PPARα activation resulted in contractile dysfunction in the hypertrophied heart, suggesting substrate switching is essential for maintenance of contractile function. Furthermore, the induction of the fetal gene, skeletal α-actin, was abolished by PPARα activation. The activation of this contractile protein in response to pressure overload has previously been shown to be dependent upon the activation of the glucose sensing transcription factor Sp1. Despite the block of metabolic and gene expression adaptation, the trophic response to pressure overload was unaffected by PPARα activation [38].

Metabolic adaptation is therefore essential for the compensation of contractile function. As mentioned above, prevention of metabolic adaptation in response to pressure overload results in contractile dysfunction. Does this metabolic maladaptation occur in the pathogenesis of contractile dysfunction and heart failure? Some evidence suggests that this is indeed the case. Perhaps the best example is the metabolic syndrome, where contractile dysfunction ensues in the presence of hyperglycemia, hyperinsulinemia, and dyslipidemia. The elevation in plasma non-esterified fatty acids results in activation of PPARα in the heart and subsequent attenuation of metabolic adaptation in the face of pressure overload. This syndrome is associated with contractile dysfunction.

5. Metabolic remodeling of the heart in obesity, and insulin resistance: from adaptation to maladaptation

Recently, there has been an increasing awareness of a possible link between obesity and cardiovascular disease. Like heart failure, obesity is a growing public health concern, and each year contributes to the death of 300,000 people in the United States [39]. The prevalence of obesity in the United States has increased rapidly during the past decade [40]. In 2000 the prevalence of obesity (body mass index $> 30\ kg/m^2$) among the American population was nearly 20% [41]. In morbidly obese patients, over-feeding the heart with energy-providing substrates results in the accumulation of triglycerides in cardiac myocytes (Fig. 2) and impaired contractile function. This is not a new concept. Lipotoxicity or "fat in all the wrong places" [42] was already described by Virchow who observed in his *Cellular Pathology* that "genuine fatty degeneration (metamorphosis) of the heart is a real transformation of its substance, going on in the interior of the fibers" (vide supra).

Lipotoxicity of the heart has also been demonstrated in a variety of animal models of obesity as well as in lipoatrophy, a disease characterized by lipid storage in non-adipocyte peripheral tissues that is readily reversible by transplantation of adipose tissue [43]. Neutral lipid accumulates in hypertrophied and in hypoxic cardiac myocytes and in the myocardium of fasted mice overexpressing PPARα. Neutral lipids accumulate in the heart

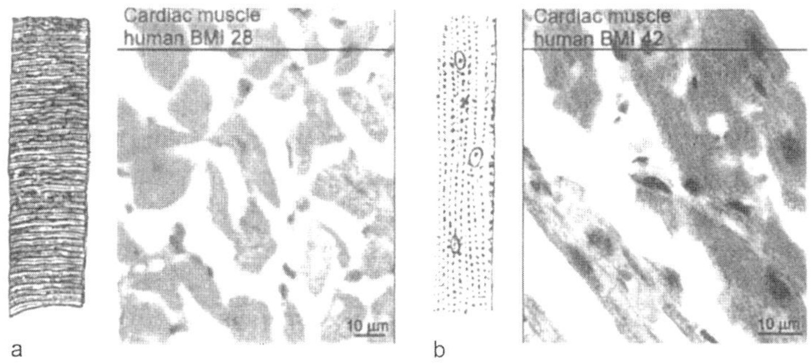

Fig. 3. Accumulation of triglycerides (oil red O stain) in the heart of an overweight (BMI 28, left) and a clinically obese patient (BMI 42, right). Note that the observation had, in principle, already been made by Virchow in 1858, who coined the term "fatty atrophy of the heart" [12]. (a) Normal cardiac myocytes (left); (b) cardiomyocyte showing fatty atrophy (right). (Modified after Unger et al., FASEB J, 2001, reproduced with permission.) *(For a colored version of this figure, see plate section, page 608.)*

when substrate availability exceeds oxidation and fatty acids spill over into β-oxidation-distinct pathways. A case in point is the fasted obese Zucker rat, where there is a loss of synchronization of fatty acid availability, uptake, and β-oxidation (Fig. 3). "Spillover" of fatty acids [44] then results in cell damage, either by the generation of reactive oxygen species and/or by ceramide-induced apoptosis. The latter process appears to be mediated by saturated fatty acids. In spite of this emerging evidence for lipotoxicity of the heart, it is not clear whether there is a causal relation between obesity and heart failure.

6. Adaptation and maladaptation of the heart in diabetes

The question whether there is a diabetes specific cardiomyopathy has also been debated for decades and is not yet answered. Patients with diabetes mellitus have an increased lifetime risk of congestive heart failure and are over-represented in large CHF databases [45]. Circumstantial evidence for a "diabetic cardiomyopathy" in humans rests on a number of morphological, functional, and biochemical observations.

During the last three decades, it has become more and more evident that diabetes is as much a disorder of fatty acid metabolism as it is a disorder of glucose metabolism [46]. In the postprandial, resting state, the normal heart relies approximately 70% on fatty acids as an energy source. In the diabetic milieu, increased fatty acid availability is accompanied by increased fatty acid oxidation[47,48]. Current concepts support the view that the greater reliance on fat oxidation in the diabetic heart is due to PPARα and subsequent induction of PPARα-regulated fatty acid oxidation enzymes [49]. For example, induction of MCD in the heart of streptozotocin-induced diabetic rats likely plays a role in the decreased malonyl-CoA levels (an inhibitor of fatty acyl-CoA entry into the mitochondrion) and increased fatty acid oxidation [50]. In addition to increased reliance on fatty acids as an energy source, with diabetes, there is a severe depression of glucose oxidation in the heart. This is likely due to inhibition of PDC. PDC activity is tightly controlled by a phosphorylation/dephosphorylation cycle, with the extent of phosphorylation of this complex determined by the relative activities of the pyruvate dehydrogenase kinase (PDK) versus the pyruvate dehydrogenase phosphatase (PDP). Induced changes in PDK activity relative to PDP activity during transitions to different nutritional, hormonal and disease states lead to substantial changes in the phosphorylation state and therefore activity state of PDC. It is now recognized that altered expression of PDK4, one of the four isoenzymes of PDK expressed in mammalian tissues, plays an important role in determining the activity state of PDC. Starvation and diabetes cause an increase in PDK4 expression in the heart [51]. Furthermore, activation of PPARα by fatty acids induces PDK4, which phosphorylates and inhibits PDC [38,52].

Two isoforms of PDP (a Ca^{2+}-sensitive isoform, PDP1 and a Ca^{2+}-insensitive isoform, PDP2) are expressed in the rat heart. Starvation and STZ-induced diabetes cause decreases in PDP2 mRNA abundance, PDP2 protein amount and PDP activity in the rat heart, while refeeding or insuling treatment effectively reverse these effects of starvation and diabetes, respectively [53].

These above studies on the regulation of PDC underscore the importance of opposite changes in expression of specific PDK and PDP isoenzymes that contribute to hyperphosphorylation and thus inactivation of PDC in the heart during altered metabolic and nutritional states such as diabetes, starvation and attendant changes in FFA availability. A surprise finding is that the heart in diabetes shows striking similarities at the level of gene expression when compared to the pressure-overloaded heart [49] which is exemplified by the re-expression of various fetal genes by the heart during diabetes. In the previous section, it was proposed that increased glucose metabolites within the cardiomyocyte are involved in the induction of fetal genes [49,54]. At first sight this does not seem to be consistent for the heart during diabetes. However, due to hyperglycemia, the rate of glucose transport is equivocal for the hearts of normal and diabetic animals. Due to the severe attenuation of pyruvate oxidation, glucose oxidation rates are lower than rates of glucose uptake, resulting in the accumulation of glycolytic intermediates in the heart during diabetes. These conditions would be optimal for activation of glucose sensing mechanisms. Thus, metabolic adaptation during diabetes may play a role beyond prevention of lipid accumulation.

The same questions that were asked for metabolic adaptation to pressure overload need to be addressed for the heart in diabetes. Why does this adaptation occur, and is it essential for contractile function in the diabetic milieu? As mentioned above, overfeeding the heart with fat likely plays a key role in increased fatty acid utilization. Like glucose and glucose metabolites, fatty acids and fatty acid metabolites have several functions in the cell, in addition to their role as an energy source (Fig. 4). These functions include mediating signal transduction (e.g. activation of various protein kinase C (PKC) isoforms, initiation of apoptosis), acting as ligands for nuclear transcription factors (e.g. PPARα), and making up essential components of biological membranes. Not surprisingly, the levels of intracellular fatty acids and their derivatives are tightly regulated in the heart at the level of fatty acid uptake, fatty acid metabolism, and triglyceride export from the heart [55–57]. Loss of this regulation and subsequent elevation of intracellular fatty acids and lipids have been associated with various pathologies, including insulin resistance, pancreatic β-cell dysfunction, and cardiomyopathy. For example, when PPARα-knockout mice are fasted, the animals die [58]. Fatal cardiac dysfunction in this model appears to be due to the accumulation of lipids within the cardiomyocytes (so-called lipotoxicity). Thus, a failure

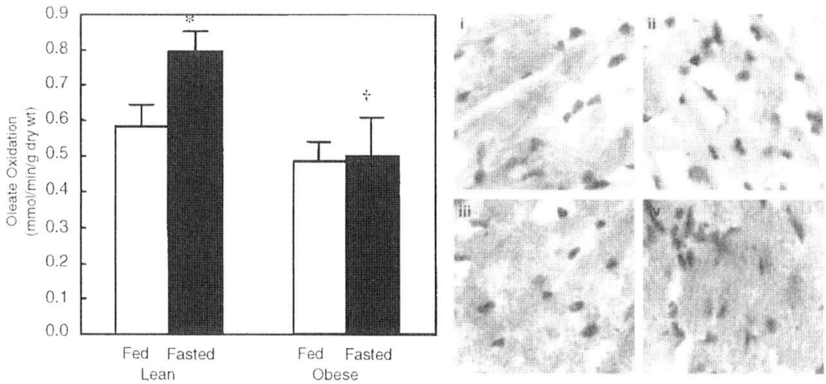

Fig. 4. Impaired fatty acid oxidation and accumulation of triglycerides in an animal model of obesity (obese Zucker rat). Note that impaired oleate oxidation corresponds to increased triglycerides in the heart. (Young et al., Diabetes, 2002, reproduced with permission.) (*For a colored version of this figure, see plate section, page 609.*)

of the myocardium to respond to increased fatty acid availability, due to loss of PPARα in this case, results in heart failure. The phenomenon of lipotoxicity has been observed in both rodent models and humans. As already stated, the exact mechanism of lipotoxicity is not known, but mediators such as chronically activated PKCs, generation of ROS, and/or ceramide-induced apoptosis are early candidates (see below). It is therefore plausible that metabolic adaptation of the heart in diabetes is an attempt to prevent lipotoxicity.

It is appropriate to ask the question: Does metabolic maladaptation of the heart during diabetes cause contractile dysfunction? Diabetes is often associated with a plethora of complications, including endothelial dysfunction and hypertension, observations that lead to the description of the metabolic syndrome. Induction of diabetes in rats with pressure overload-induced hypertrophy results in rapid cardiac failure. This phenomenon is similar to the observation that reactivation of PPARα in the hypertrophied heart results in contractile dysfunction. With pressure overload, substrate switching by the hypertrophied heart is essential for maintenance of function. However, the attenuation of reliance on fatty acids as a substrate due to pressure overload will accelerate lipid deposition within the cardiomyocyte in the diabetic milieu, thereby accelerating lipotoxicity. A compromise may therefore be set, balancing the need for glucose metabolism with that of the utilization of fatty acids, to reduce the rate of lipid deposition, at the expense of contractile function. This may be the case in the obese Zucker diabetic fatty (ZDF) rat, an animal model of the metabolic syndrome, which possesses both a volume and a pressure overload- (hypertension) induced hypertrophied heart. Forcing a hypertrophied heart to utilize fatty acids in the diabetic milieu will result in a maladapted heart exhibiting contractile dysfunction. Consistent with this hypothesis is the observation that treatment of ZDF rats with thiazoladinediones (TZDs; insulin sensitizers and lipid lowering agents) reduces intramyocardial lipid deposition and improves contractile function [59]. Furthermore, we have previously shown that progression of Type I diabetes is associated with a dramatic decrease in the expression of PPARα (and the PPARα-regulated gene, UCP3) within the heart, through a still unknown mechanism. Thus, continued exposure to high fatty acids levels, accompanied by decreased PPARα expression, will accelerate lipid accumulation and lipotoxicity. The treatment of this metabolic maladaptation with agents such as TZDs holds the promise to redistribute lipid from non-adipocytes to adipocytes and thus lower the lipid burden for the heart.

7. Summary and conclusions

It had been known for half a century that fatty acids are the main energy-providing substrates for the normal heart. Factors that control and regulate fatty acid metabolism are well described [60]. Although derangements in fatty acid metabolism of the heart have already been noted in the nineteenth century [12], a link between derangements in fatty acid metabolism, altered gene expression, and impaired contractile function of the heart has been appreciated only very recently. As a result, fatty acid metabolism may become a target for the pharmacological treatment of the failing heart.

Acknowledgements

Work in the author's laboratory is supported by grants from the National Institutes of Health and the American Heart Association, National Center. We thank John Lemm for assistance in editing.

References

[1] Bing, R.J., Siegel, A., Ungar, I., Gilbert, M., 1954. Metabolism of the human heart. 2. Studies on fat, ketone and amino acid metabolism. Am. J. Med. 16, 504–515.
[2] Blain, J.M., Schafer, H., Siegle, A.L., Bing, R.J., 1956. Studies in myocardial metabolism. VI. Myocardial metabolism in congestive heart failure. Am. J. Med. 20, 820.
[3] Bing, R.J., 1965. Cardiac metabolism. Physiol. Rev. 45, 171–213.
[4] Linzbach, A.J., 1960. Heart failure from the point of quantitative anatomy. Am. J. Cardiol. 5, 370.
[5] Schwartz, A., Lee, K.S., 1962. Study of heart mitochondria and glycolytic metabolism in experimentally induced cardiac failure. Circ. Res. 10, 321.
[6] Geltman, E.M., Smith, J.L., Beecher, D., Ludbrook, P.A., Ter-Pogossian, M.M., Sobel, B.E., 1983. Altered regional myocardial metabolism in congestive cardiomyopathy detected by positron tomography. Am. J. Med. 74, 773–785.
[7] Kelly, D.P., Strauss, A.W., 1994. Inherited cardiomyopathies. N. Engl. J. Med. 330, 913–919.
[8] Bergmann, S.R., Herrero, P., Sciacca, R., Hartman, J.J., Rubin, P.J., Hickey, K.T., Epstein, S., Kelly, D.P., 2001. Characterization of altered myocardial fatty acid metabolism in patients with inherited cardiomyopathy. J. Inherit. Metab. Dis. 24, 657–674.
[9] Wallhaus, T.R., Taylor, M., DeGrado, T.R., Russel, T.C., Stanko, P., Nickles, R.J., Stone, C.K., 2001. Myocardial free fatty acid and glucose use after carvedilol treatment in patients with congestive heart failure. Circulation 103, 2441–2446.
[10] Eichhorn, E.J., Bristow, M.R., 1996. Medical therapy can improve the biological properties of the chronically failing heart. A new era in the treatment of heart failure. Circulation 94, 2285–2296.
[11] Beanlands, R.S., Nahmias, C., Gordon, E., Coates, G., deKemp, R., Firnau, G., Fallen, E., 2000. The effects of beta(1)-blockade on oxidative metabolism and the metabolic cost of ventricular work in patients with left ventricular dysfunction: a double-blind placebo-controlled, positron-emission tomography study. Circulation 102, 2070–2075.

[12] Virchow, R., 1858. Die Zellularpathologie und ihre Begründung auf physiologische und pathologische Gewebelehre. A. Hirschwald, Berlin.
[13] Wittels, B., Bressler, R., 1964. Biochemical lesion of diphtheria toxin in the heart. J. Clin. Invest. 43, 630.
[14] Wittels, B., Spann, J.F., 1968. Defective lipid metabolism in the failing heart. J. Clin. Invest. 47, 1787–1794.
[15] Coates, P.M., Tanaka, K., 1992. Molecular basis of mitochondrial fatty acid oxidation defects. J. Lipid Res. 33, 1099–1110.
[16] Jamshidi, Y., Montgomery, H., Hense, H., Myerson, S., Torra, I., Staels, B., World, M., Doering, A., Erdmann, J., Hengstenberg, C., Humphries, S., Schunkert, H., Flavell, D., 2002. Peroxisome proliferator-activated receptor alpha gene regulates left ventricular growth in response to exercise and hypertension. Circulation 105, 950–955.
[17] Okamoto, F., Tanaka, T., Sohmiya, K., Kawamura, K., 1998. CD36 abnormality and impaired myocardial long-chain fatty acid uptake in patients with hypertrophic cardiomyopathy. Jpn. Circ. J. 62, 499–504.
[18] Goodwin, G.W., Taylor, C.S., Taegtmeyer, H., 1998. Regulation of energy metabolism of the heart during acute increase in heart work. J. Biol. Chem. 273, 29530–29539.
[19] Goodwin, G.W., Taegtmeyer, H., 2000. Improved energy homeostasis of the heart in the metabolic state of exercise. Am. J. Physiol. Heart Circ. Physiol. 279, H1490–H1501.
[20] Randle, P.J., Garland, P.B., Hales, C.N., Newsholme, E.A., 1963. The glucose fatty-acid cycle. Its role in insulin sensitivity and the metabolic disturbances of diabetes mellitus. Lancet 1, 785–789.
[21] Taegtmeyer, H., Hems, R., Krebs, H.A., 1980. Utilization of energy providing substrates in the isolated working rat heart. Biochem. J. 186, 701–711.
[22] Korvald, C., Elvenes, O.P., Myrmel, T., 2000. Myocardial substrate metabolism influences left ventricular energetics in vivo. Am. J. Physiol. Heart Circ. Physiol. 278, H1345–H1351.
[23] Meerson, F.Z., 1983. The Failing Heart: Adaptation and Deadaptation. Raven Press, New York.
[24] Taegtmeyer, H., 2000. Genetics of energetics: transcriptional responses in cardiac metabolism. Ann. Biomed. Eng. 28, 871–876.
[25] Allard, M.F., Schonekess, B.O., Henning, S.L., English, D.R., Lopaschuk, G.D., 1994. Contribution of oxidative metabolism and glycolysis to ATP production in hypertrophied hearts. Am. J. Physiol. 267, H742–H750.
[26] Depre, C., Shipley, G., Davies, P., Frazier, O., Taegtmeyer, H., 1998. Glucose transporter isoform expression in the failing human heart. Circulation 98, I-627 (abstract).
[27] Doenst, T., Goodwin, G.W., Cedars, A.M., Wang, M., Stepkowski, S., Taegtmeyer, H., 2001. Load-induced changes in vivo alter substrate fluxes and insulin responsiveness of rat heart in vitro. Metabolism 50, 1083–1090.
[28] Barger, P.M., Brandt, J.M., Leone, T.C., Weinheimer, C.J., Kelly, D.P., 2000. Deactivation of peroxisome proliferator-activated receptor-alpha during cardiac hypertrophic growth. J. Clin. Invest. 105, 1723–1730.
[29] Kelly, D.P., 2002. Peroxisome proliferator-activated receptor alpha as a genetic determinant of cardiac hypertrophic growth: culprit or innocent bystander? Circulation 105, 1025–1027.
[30] van Bilsen, M., Van der Vusse, G.J., Reneman, R.S., 1998. Transcriptional regulation of metabolic processes: implications for cardiac metabolism. Pflugers Arch. 437, 2–14.
[31] Taegtmeyer, H., Razeghi, P., Young, M.E., 2002. Mitochondrial proteins in hypertrophy and atrophy: a transcript analysis in rat heart. Clin. Exp. Pharmacol. Physiol. 29, 346–350.
[32] Sugden, M.C., Langdown, M.L., Harris, R.A., Holness, M.J., 2000. Expression and regulation of pyruvate dehydrogenase kinase isoforms in the developing rat heart and in adulthood: role of thyroid hormone status and lipid supply. Biochem. J. 352, 731–738.
[33] McGarry, J.D., Mills, S.E., Long, C.S., Foster, D.W., 1983. Observations on the affinity for carnitine and malonyl-CoA sensitivity of carnitine palmitoyl transferase I in animal and human tissues. Demonstration of the presence of malonyl-CoA in non-hepatic tissues of the rat. Biochem. J. 214, 21–28.
[34] Opie, L.H., 1998. The Heart: Physiology, from Cell to Circulation, 3rd ed. Lippincott-Raven, New York.
[35] Razeghi, P., Young, M.E., Abbasi, S., Taegtmeyer, H., 2001. Hypoxia in vivo decreases peroxisome proliferator-activated receptor alpha-regulated gene expression in rat heart. Biochem. Biophys. Res. Commun. 287, 5–10.
[36] Xu, K.Y., Zweier, J.L., Becker, L.C., 1995. Functional coupling between glycolysis and sarcoplasmic reticulum Ca^{2+} transport. Circ. Res. 77, 88–97.

[37] Girard, J., Ferre, P., Foufelle, F., 1997. Mechanisms by which carbohydrates regulate expression of genes for glycolytic and lipogenic enzymes. Annu. Rev. Nutr. 17, 325–352.

[38] Young, M.E., Laws, F.A., Goodwin, G.W., Taegtmeyer, H., 2001. Reactivation of peroxisome proliferator-activated receptor alpha is associated with contractile dysfunction in hypertrophied rat heart. J. Biol. Chem. 276, 44390–44395.

[39] McGinnis, J.M., Foege, W.H., 1993. Actual causes of death in the United States. JAMA 270, 2207–2212.

[40] Mokdad, A.H., Serdula, M.K., Dietz, W.H., Bowman, B.A., Marks, J.S., Koplan, J.P., 1999. The spread of the obesity epidemic in the United States, 1991–1998. JAMA 282, 1519–1522.

[41] Mokdad, A.H., Bowman, B.A., Ford, E.S., Vinicor, F., Marks, J.S., Koplan, J.P., 2001. The continuing epidemics of obesity and diabetes in the United States. JAMA 286, 1195–1200.

[42] Friedman, J.J., 2002. Fat in all the wrong places. Nature 415, 268–269.

[43] Gavrilova, O., Marcus-Samuels, B., Graham, D., Kim, J.K., Shulman, G.I., Castle, A.L., Vinson, C., Eckhaus, M., Reitman, M.L., 2000. Surgical implantation of adipose tissue reverses diabetes in lipoatrophic mice. J. Clin. Invest. 105, 271–278.

[44] Young, M.E., Guthrie, P.H., Razeghi, P., Leighton, B., Abbasi, S., Patil, S., Taegtmeyer, H., 2002. Impaired long chain fatty acid oxidation and contractile dysfunction in the obese Zucker rat heart. Diabetes 51, 2587–2595.

[45] Garcia, M.J., McNamara, P.M., Gordon, T., Kannel, W.B., 1974. Morbidity and mortality in diabetics in the Framingham population. Sixteen year follow-up study. Diabetes 23, 105–111.

[46] McGarry, J.D., 1992. What if Minkowski had been ageusic? An alternative angle on diabetes. Science 258, 766–770.

[47] Starling, E.H., Evans, C.L., 1914. The respiratory exchanges of the heart in the diabetic animal. J. Physiol. (London) 49, 67–88.

[48] Ungar, I., Gilbert, M., Siegel, A., Blain, J.M., Bing, R.J., 1955. Studies on myocardial metabolism: IV. Myocardial metabolism in diabetes. Am. J. Med. 18, 385–396.

[49] Depre, C., Young, M.E., Ying, J., Ahuja, H.S., Han, Q., Garza, N., Davies, P.J.A., Taegtmeyer, H., 2000. Streptozotocin-induced changes in cardiac gene expression in the absence of severe contractile dysfunction. J. Mol. Cell. Cardiol. 32, 985–996.

[50] Sakamoto, J., Barr, R.L., Kavanagh, K.M., Lopaschuk, G.D., 2000. Contribution of malonyl-CoA decarboxylase to the high fatty acid oxidation rates seen in the diabetic heart. Am. J. Physiol. Heart Circ. Physiol. 278, H1196–H1204.

[51] Wu, P., Sato, J., Zhao, Y., Jaskiewicz, J., Popov, K.M., Harris, R.A., 1998. Starvation and diabetes increase the amount of pyruvate dehydrogenase kinase isoenzyme 4 in rat heart. Biochem. J. 329, 197–201.

[52] Sugden, M.C., Orfali, K.A., Fryer, L.G., Holness, M.J., Priestman, D.A., 1997. Molecular mechanisms underlying the long-term impact of dietary fat to increase cardiac pyruvate dehydrogenase kinase: regulation by insulin, cyclic AMP and pyruvate. J. Mol. Cell. Cardiol. 29, 1867–1875.

[53] Huang, B., Wu, P., Popov, K.M., Harris, R.A., 2003. Starvation and diabetes reduces the amount of pyruvate dehydrogenase phosphatase in rat heart and kidney. Diabetes 52, 1371–1376.

[54] Dillmann, W.H., 1980. Diabetes mellitus induces changes in cardiac myosin of the rat. Diabetes 29, 579–582.

[55] Schaffer, J.E., 2002. Fatty acid transport: the roads taken. Am. J. Physiol. Endocrinol. Metab. 282, E239–E246.

[56] Goodwin, G.W., Cohen, D.M., Taegtmeyer, H., 2001. [5-3H]glucose overestimates glycolytic flux in isolated working rat heart: role of the pentose phosphate pathway. Am. J. Physiol. Endocrinol. Metab. 280, E502–E508.

[57] Bjorkegren, J., Veniant, M., Kim, S.K., Withycombe, S.K., Wood, P.A., Hellerstein, M.K., Neese, R.A., Young, S.G., 2001. Lipoprotein secretion and triglyceride stores in the heart. J. Biol. Chem. 276, 38511–38517.

[58] Barger, P.M., Kelly, D.P., 2000. PPAR signaling in the control of cardiac energy metabolism. Trends Cardiovasc. Med. 10, 238–245.

[59] Zhou, Y.T., Grayburn, P., Karim, A., Shimabukuro, M., Higa, M., Baetens, D., Orci, L., Unger, R.H., 2000. Lipotoxic heart disease in obese rats: implications for human obesiy. Proc. Natl. Acad. Sci., USA, 97, 1784–1789

[60] van der Vusse, G.J., Glatz, J.F.C., Stam, H.C.G., Reneman, R.S., 1992. Fatty acid homeostasis in the normoxic and ischemic heart. Physiol. Rev. 72, 881–940..

Physiological significance of uncoupling protein-3: a role in fatty acid handling?

Matthijs K.C. Hesselink[a,*] and Patrick Schrauwen[b]

[a]*Department of Movement Sciences, Nutrition and Toxicology Research Institute Maastricht (NUTRIM), Maastricht University, P.O. Box 616, 6200 MD Maastricht, The Netherlands*
[b]*Department of Human Biology, Nutrition and Toxicology Research Institute Maastricht (NUTRIM), Maastricht University, P.O. Box 616, 6200 MD Maastricht, The Netherlands*
[*]*Correspondence address: Tel.: +31-43-3881317; fax: +31-43-3670972*
E-mail: matthijs.hesselink@bw.unimaas.nl(M.K.C.H.)

1. Introduction

In all living systems, combustion of nutrients to carbon dioxide and water is the main pathway to release the energy needed to fuel cellular processes like ion pumping, muscular contraction, protein synthesis and degradation of nutrients in the digestive tract. In all these processes, hydrolysis of adenosine triphosphate (ATP), the universal phosphor donor, liberates the energy needed. Therefore, it is of utmost importance that ATP levels are maintained, even under conditions of severe energy stress. The vast majority of ATP is synthesized in a process referred to as the mitochondrial oxidative phosphorylation. Degradation of nutrients like proteins, carbohydrates and lipids ultimately results in the production of the coenzymes nicotinamide adenine dinucleotide (NAD^+) and flavin adenine dinucleotide (FAD), which can in turn be reduced to NADH and $FADH_2$ in exchange for an electron. This process, referred as the electron transfer or respiratory chain, is located in the inner mitochondrial membrane. According to the chemiosmotic theory defined by Mitchell [1], the electron transfer chain results in a net proton gradient across the inner mitochondrial membrane. If the proton gradient is high enough, the protons may flow back to the mitochondrial matrix via the F_0-F_1-ATPase, releasing the energy needed to phosphorylate adenosine diphosphate (ADP) and generate ATP (oxidative phosphorylation). In tightly coupled mitochondria, there is no proton leak across the inner mitochondrial membrane and all the energy built up in the respiratory chain can be used for (is coupled to) the generation of ATP.

However, it has long been recognized that, even in the absence of ADP, isolated mitochondria show respiration. Obviously, in this condition the potential energy of the proton gradient is not used for phosphorylation of ADP indicating that there is proton transfer across the inner mitochondrial membrane that is not coupled to phosphorylation of

Advances in Molecular and Cell Biology, Vol. 33, pages 271–293

ISSN: 1569-2558 / DOI: 10.1016/S1569-2558(03)33014-0

ADP (referred to as proton leak or mitochondrial uncoupling). This process, called state 4 respiration (see Fig. 1), indicates that uncoupling of mitochondria is an existing phenomenon that occurs in mitochondria derived from numerous tissues [e.g., brown adipose tissue (BAT), skeletal muscle and hepatocytes]. For most tissues, the physiological significance of uncoupling is still unknown, but much is learned from the mitochondrial uncoupling that occurs in BAT. This tissue plays a recognized role in adaptive thermogenesis, when the energy generated in the electron transfer chain is released as heat rather than used for phosphorylation of ADP, thereby allowing adaptation to cold. Reversely, during hibernation, when preservation of energy is essential, BAT mitochondria become more tightly coupled, attenuating energy expenditure and saving energy stores. Moreover, only in BAT the protein responsible for mitochondrial uncoupling has been identified and was named thermogenin (nowadays called uncoupling protein-1, UCP1) [2]. Thus, this protein is responsible for adaptive thermogenesis in rodents.

As the amount of BAT in adult humans is scarce, it could be argued that proton leak, and its concomitant increase in basal metabolic rate, are trivial processes in adult humans. It should however be noted that computations have been made indicating that in vivo proton leak in liver and skeletal muscle mitochondria may account for ~20% of the basal metabolic rate [3]. In extension to this, it was also shown that in contracting skeletal

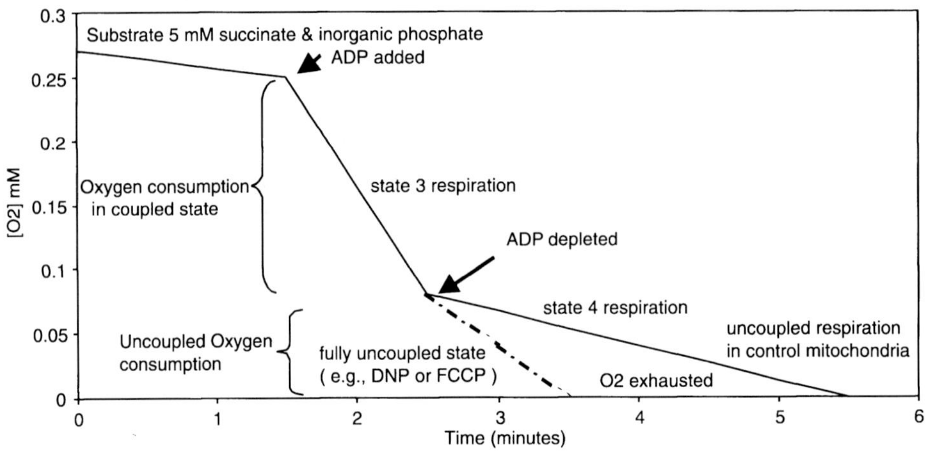

Fig. 1. Typical recording of oxygen uptake of mitochondria in vitro. In the presence of substrate, inorganic phosphate and oxygen and with ATP hydrolysis blocked, mitochondria start respiring at a relatively high rate (state 3 respiration). Under state 3 conditions, oxygen is consumed and coupled to phosphorylation of ADP to ATP, which is driven by the energy liberated in the electron transfer chain (coupled respiration). Upon depletion of ADP (as all ADP has been phosphorylated and ATP hydrolysis is blocked), mitochondria continue respiration at a much lower rate referred to as state 4 respiration. As the energy generated in the electron transfer cannot be coupled to oxidative phosphorylation (as ADP is depleted), this is uncoupled respiration. If uncoupling agents like dinitrophenol (DNP) or FCCP (*p*-trifluoromethoxy carbonyl cyanide phenyl hydrazone) are added to the medium, the rate of uncoupled respiration is increased (fully uncoupled respiration), as indicated by the rapid depletion of oxygen from the medium (dotted line). Note that uncoupled respiration is an intrinsic trait of mitochondria, also of "normal healthy" mitochondria.

muscle, proton leak makes up 15% of the basal metabolic rate [4]. Whilst these studies clearly underscored the impact of proton leak on basal metabolic rate, the protein(s) responsible for the proton leak had not been identified. Not surprisingly, the discovery of the UCP1 homologues, UCP2 [5] and UCP3 [6], in 1997 was warmly welcomed.

The ubiquitous expression of UCP2 and the expression of UCP3 in skeletal muscle and their homology with the "true" [7] uncoupling protein, UCP1, made UCP2 and UCP3 attractive targets for interventions aiming to manipulate energy expenditure. Extensive research towards the regulation and the putative functions of these novel uncoupling proteins has resulted in a vast amount of papers in the last 6 years. Notwithstanding the overwhelming number of studies published, there appears to be no consensus on the primary function of UCP2 and UCP3 yet.

As UCP3 is expressed almost exclusively in skeletal muscle, which makes up ~40% of the body mass in lean individuals and which is responsible for the majority of the basal metabolic rate, the pioneering studies towards UCP3 focused mainly on UCP3 and its putative role in energy expenditure. In the present chapter, we aim to review the literature currently available on UCP3 in relation to its regulation and putative function. In addition, we will present some previously unpublished data indicating that the primary physiological role of UCP3 may not be regulation of energy expenditure.

2. Effects of non-native expression of UCP3 in cell systems

It was mainly based on the similarity in amino acid sequence of UCP3 with UCP1 (57%; [6]) suggestions were made that UCP3 may possess uncoupling activity in vivo. Indeed it was shown that expression of recombinant human UCP3 in yeast lowered the mitochondrial membrane potential [8], increased basal oxygen consumption with 31%, induced a 20% increase in state 4 respiration in isolated mitochondria and retarded growth [9]. It was also shown that overexpressing UCP3 in yeast induced a decline in cellular respiration coupled to oxidative phosphorylation from 57 to 11%; this decline was concerted with a 33% increase in cellular heat production [10]. These findings appear to confirm uncoupling activity of UCP3 in vitro. However, it was also noted that, in a heterologous yeast expression system, the uncoupling activity of UCP3 was hardly affected by GDP or ATP [11], which was hypothesized based on the similarity of UCP3 with UCP1.

While the pioneering studies of (over)expression of UCP3 in cell systems have certainly been valuable, these studies have also been criticized [12]. The main criticism relates to computation of the respiratory control values (RC, the ratio of fully uncoupled respiration over state 4 respiration, see Fig. 1). Adding a chemical uncoupler to the incubation medium and measuring maximal respiration rates is referred to as fully uncoupled respiration. Thus, if UCP3 is considered a true uncoupler, it should increase state 4 respiration but leave the fully uncoupled state unaffected. In the in vitro overexpression studies mentioned, however, the fully uncoupled respiration in UCP3 overexpression conditions was lower compared to control values. These results can be interpreted as a malfunctioning of the electron transfer chain [12]. Interestingly, it was shown that yeast overexpressing UCP1 at modest concentrations possessed the uncoupling behavior anticipated (i.e. GDP inhibitable uncoupling and unaffected rates of the fully uncoupled state), while yeast expressing UCP1 about 10-fold higher were growth retarded, partly GDP insensitive and had decreased chemically uncoupled respiration [13]. These observations are similar to UCP3-expressing yeast and may indicate that expression of UCP3 in yeast induces artifactual uncoupling, possibly induced by improper incorporation of UCP3 in the inner mitochondrial membrane. With respect to this, it is important to note that it was shown that after sonication and differential centrifugation, UCP3 and its degradation products were primarily present in extramitochondrial fractions [14], rather than folded properly into the mitochondria.

In conclusion, yeast systems expressing UCP3 in vitro have taught us that UCP3 expression affects the mitochondrial membrane potential, decreases chemically uncoupled respiration, mildly increases state 4 respiration, whilst the uncoupling observed cannot be inhibited by GDP or ATP, like UCP1. Care should be taken when extrapolating these findings to in vivo (human) conditions.

3. Effects of genetically manipulating UCP3 expression in mice

3.1. UCP3 overexpression

Encouraging data that UCP3 might be a true uncoupler in vivo were reported by Clapham et al. [15], showing that mice overexpressing UCP3 (UCP3-tg) were lean despite the fact that they are hyperphagic compared with their wild type littermates. Additional phenotypical changes, next to the 66-fold upregulation of UCP3 mRNA in skeletal muscle, include 25% increase in resting oxygen consumption, decreased fasting blood glucose and insulin levels, improved glucose tolerance, decreased total cholesterol and a 44–57% reduction in adipose tissue over total animal volume. Shortly after this work, Li et al. overexpressed the BAT-specific UCP1 in skeletal muscle (by coupling UCP1 cDNA to a myosin light chain promoter) [16] and showed a similar phenotype (reduced body mass, decreased fasting blood glucose and triglycerides and prevention of diet-induced obesity and insulin resistance). The straightforward interpretation of these data is that UCP3 overexpression results in increased energy expenditure and the accompanied phenotype. However, along the same lines as the overexpression of UCP3 in yeast was challenged (the lack of inducible uncoupling and the lack of inhibition by GDP), the increased proton leak and related phenotype in the UCP3-tg mice were considered artifactual [17]. Quantifying the absolute concentration of UCP3 in mitochondria isolated from UCP3-tg and wild type mice [15] revealed that UCP3-tg mice (3200 ng UCP3/mg mitochondrial protein) had ~22-fold more UCP3 protein than age-paired wild types (140 ng/mg mitochondrial protein). According to Cadenas et al. [17], the uncoupling observed at this supraphysiological level of expression does not represent native uncoupling, as it can neither be induced by superoxide in the presence of fatty acids, nor inhibited by purine nucleotides as was shown for UCP3 expressed at physiological levels [18]. Next to the lack of inducible uncoupling in UCP3-tg mitochondria, the previously reported decreased proton conductance in UCP3-ko mitochondria [19,20] could not be confirmed [17], justifying the conclusion that UCP3 is not a significant contributor to proton conductance of muscle mitochondria [17] and thus not to energy expenditure. The inability of overexpressed UCP3 to catalyze proton conductance was attributed to poor insertion or folding of UCP3 in the mitochondria, while the authors leave open the option that a not yet identified endogenous cofactor is required to provoke UCP3 to affect proton conductance. Yet, no attempts have been made to examine if in mice expressing supraphysiological levels of UCP3, the protein was indeed poorly folded or misfolded into the mitochondria.

In an attempt to do so, we studied the localization of hUCP3 in the medial gastrocnemius muscle of mice overexpressing UCP3 [15] by immunogold electron microscopy. Similar to previous reports [15,17], the UCP3-tg mice had lower body mass compared to wild type littermates (22.4 ± 0.9 versus 25.9 ± 0.8 g, respectively) and a 5.8-fold higher UCP3 protein level. Using our previously validated antibody specifically detecting hUCP3 (code 1331) [21,22], together with our antibody detecting both hUCP3 and endogenous mice and rat UCP3 (code 1338), we were able to delineate hUCP3 expression form endogenous UCP3 expression in the UCP3-tg mice. It was shown that in wild type mice, only expressing endogenous UCP3, a clear mitochondrial labeling was observed with the mice-specific antibody (1338) while no label was detected using the anti hUCP3 (1331) antibody. In UCP3-tg mice, the vast majority of the label was associated with the inner mitochondrial membrane and the cristae, and did not result in extramitochondrial labeling (see Fig. 2). This may indicate that artifactual uncoupling in UCP3 overexpression systems, as suggested previously [13,17,23], is not due to extramitochondrial expression of UCP3. This is in contrast with observations in yeast where it was shown that expression of UCP3 in yeast resulted in the majority of UCP3 present in extramitochondrial compartments [14]. A major difference between UCP3 expression in yeast and in the UCP3-tg mice is the in vivo condition and therefore the

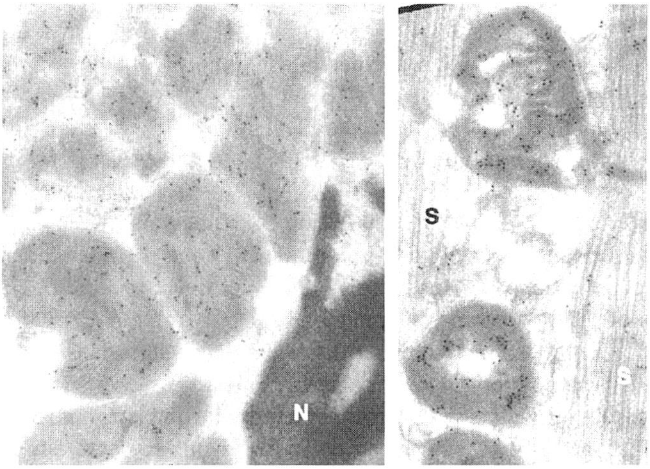

Fig. 2. Immunogold electron microscopy staining of UCP3 in medial gastrocnemius muscle of UCP3-tg mice. The left panel shows gold-labeled subsarcolemmal mitochondria, without staining in other subcellular constituents like the nuclei (N). The right panel shows gold-labeled intramyofibrillar mitochondria with sparse background labeling on sarcomeres (S). The antibody used was tested to specifically recognize human UCP3. Using this antibody it was shown that in UCP3-tg mice the vast majority of the label was restricted to mitochondria.

presence of the complete transcriptional and translational machinery, including all cofactors needed, is in contrast to the yeast system. It should be noted however that our immunoelectron microscopical observations do not exclude that improper folding of UCP3 in the inner mitochondrial membrane resulted in pores in the inner membrane, thus leading to uncoupling in a way similar to protonophores (non-inducible uncoupling). Moreover, the mice in our study showed a 5.8-fold upregulation, which apparently is way lower than other data available on UCP3-tg mice at the protein level [17].

3.2. UCP3 knockout

The generation of mice lacking the UCP3 gene (UCP3-ko) by two independent groups [19,20] has complicated matters. In these mice, decreased state 4 respiration with similar state 3 respiration levels have been reported [20], indicating more tightly coupled mitochondria. Whole body oxygen uptake and fuel partitioning (reflected by respiratory exchange ratio) under resting conditions were not affected in UCP3-ko mice [19,20]. Obviously, the lack of UCP3 was not compensated for by the induction of any other uncoupling proteins known so far. The observation of more tightly coupled mitochondria in UCP3-ko mice did not result in apparent phenotypical changes [19,20]. By creation of a double knockout, lacking UCP1 and UCP3, it was shown that the phenotype observed after deletion of the UCP1 gene was not more pronounced by the additional deletion of UCP3 [19]. Attempts to trigger a phenotype by exposing the UCP3-ko mice to a series of conditions previously related to induction of uncoupling proteins (high-fat feeding, cold exposure, stimulation with thyroid hormone) did not result in differences between the UCP3-ko and their wild type littermates [19,20]. Interestingly however, Vidal-puig et al. reported increased production of the superoxide anion in vitro and augmented mitochondrial aconitase production, indicating in vivo production of reactive oxygen species (ROS) in the UCP3-ko mice [20]. These findings have let the authors suggest that one of the functions of UCP3 could be the prevention of excessive oxidative stress by lowering the mitochondrial membrane potential, thus lowering the probability for electrons to interact with oxygen [20].

The putative role of UCP3 in the prevention of oxidative stress is highlighted by the observation that exogenous superoxide-induced uncoupling in skeletal muscle mitochondria, showing a 2-fold increase in uncoupling if UCP3 levels were doubled by fasting and a lack of effect of superoxide in mitochondria isolated from UCP3-ko mice [18]. The increased uncoupling was inhibited by purine nucleotides and fascinatingly the activation of uncoupling by superoxide was abolished if 0.3% BSA was added to the medium and was restored by adding palmitic acid in the micromolar range (300 μmol), indicating a role for free fatty acids in the activation of UCP3 [18]. Although these studies clearly indicate that UCP3 can be involved in regulation of oxidative stress, one should keep in mind that the findings reported are derived from isolated mitochondria exposed to high levels of exogenous superoxide, which complicates extrapolation of these findings to intact human beings.

In addition to in vitro determination of mitochondrial uncoupling in UCP3-ko mice, an elegant, non-invasive in vivo ^{31}P-NMR approach has been used to assess mitochondrial coupling in UCP3-ko mice. This approach revealed that the unidirectional rates of ATP synthesis were increased in UCP3-ko mice, while TCA cycle flux remained unaltered [24]. These findings were interpreted as evidence for uncoupling activity by UCP3 in vivo [24]. It should be recognized however that increased ATP synthesis rate without alterations in TCA cycle flux is difficult to interpret, but simply interpreting it as increased coupling appears premature.

3.3. Summary

To summarize, detailed examination of UCP3-tg and UCP3-ko mice revealed that the observed phenotype of UCP3-tg mice is most likely due to uncoupling. As the uncoupling in UCP3-tg mice is non-inducible, it is considered artifactual, possibly due to misfolded UCP3 and/or UCP3 loosely imported into the inner mitochondrial membrane. However, immunoelectron microscopy studies showed that in the UCP3-tg mice, almost all the UCP3 expressed was confined to the mitochondria, in contrast to observations in yeast. The UCP3-ko mice lack an apparent phenotype and reports on proton leak are inconsistent. These findings indicate that i) the phenomenon uncoupling per se (inducible or non-inducible) may have significant advantageous phenotypical effects and ii) it is unlikely that UCP3 contributes significantly to the basal proton leak, and thus basal metabolic rate, in skeletal muscle. Rather, a role for UCP3 in regulation of oxidative stress matches the experimental data available.

4. Effects of physiological manipulation of UCP3 expression

4.1. Genetic studies

Shortly after the discovery of the UCP3 gene, the genomic structure and chromosomal localization of hUCP3 were identified [25]. It was shown that UCP3 was mapped within 7 kb of the UCP2 gene on chromosome 11q13, a region that has been linked to obesity and hyperinsulinemia [5]. Several polymorphisms in the UCP3 gene have been identified [26–32] and related to markers of energy metabolism and obesity. We have recently reviewed the effect of UCP3 polymorphisms known so far and concluded that the effect on markers of obesity shows inconsistent results [33].

Nonetheless, one of the early detected and promising observations was an exon 6-splice donor polymorphism, resulting in an apparent null mutation of UCP3, reported in Gullah speaking African Americans with early onset of severe obesity (BMI $= 38\ kg/m^2$) and type 2 diabetes [34]. Interestingly, carriers of the exon 6-splice donor mutation had a significant ($p = 0.0188$, $n = 24$) 50% decrease in fat oxidation [34]. It was this observation tempting the authors to suggest that UCP3 may increase fat oxidation by introducing fatty acids into the mitochondrial matrix [34] and explaining the declined fat oxidation in the carriers of the polymorphism. In later studies in African Americans [35] or in Danish Caucasians [36], carriers of the exon 6-splice donor mutation showed no changes in resting metabolic rate or fuel partitioning.

Overall, the inconsistency in results linking UCP3 polymorphisms to obesity-related phenotypes does not favor a significant role for UCP3 in the regulation of energy expenditure, although a role of UCP3 in substrate metabolism in selected populations cannot be excluded.

4.2. *Effect of thyroid hormone on UCP3 expression*

In Pima Indians, sleeping metabolic rate is a very strong predictor of weight gain [37] and it was in this population that we reported significant correlations of UCP3 mRNA with both BMI (negative) and resting metabolic rate (positive) [38], suggesting that skeletal muscle UCP3 content is related to energy expenditure. In this context, it was already recognized that hyperthyroidism, characterized by increased energy expenditure, was associated with increased proton leak in liver mitochondria while the opposite was observed in hypothyroidism [39]. Interestingly, hypothyroid rats had only 32% of the UCP3 mRNA levels found in controls. Treatment of not only hypothyroid, but also euthyroid rats, with thyroid hormone induced a 4.7- and 6.2-fold increase in skeletal muscle UCP3 mRNA [8]. These findings were extended with assessments of mitochondrial respiration rates in the transition from the hypothyroid to the euthyroid state, showing that state 3 and state 4 respiration rates were ranked hierarchically from hypothyroid < euthyroid < hyperthyroid [40], suggesting that UCP3 is involved in mediating the proton leak observed in hyperthyroidism. More compelling data for a role of UCP3 in thyroid hormone-induced increases in energy expenditure is presented by de Lange et al., showing that after a single dose injection of thyroid hormone, UCP3 mRNA peaked at 24 h post-injection. The increase in mRNA was reflected in increased protein levels 65 h post-injection with a concerted increase in resting metabolic rate assessed in vivo [41].

Administration of thyroid hormone for 10 days in control rats resulted in an 8.1-fold increase in UCP3 mRNA with concerted upregulation of UCP3 at the protein level (2.8-fold) [42]. ^{31}P-NMR spectroscopy revealed that the increase in UCP3 protein was associated with increased TCA cycle flux without an effect on unidirectional ATP synthesis rate, which was interpreted by the authors as a 60% decrease in mitochondrial coupling and has led them to suggest that UCP3 is responsible for the increased energy expenditure and thermogenesis observed after administration of thyroid hormone [42]. In humans treated with thyroid hormone, similar changes in TCA cycle flux and unidirectional ATP synthesis rate were observed [43]. Unfortunately, no UCP3 levels were reported in that study [43]. It should however be noted that, although very elegant and non-invasive, results from ^{31}P-NMR spectroscopy with regard to mitochondrial uncoupling have been non-uniformly interpreted, i.e. increased unidirectional ATP synthesis rate without effect on the TCA cycle flux has been interpreted as decreased uncoupling [24], and at the same time increased TCA cycle flux without an effect on ATP synthesis rate was taken as a marker for increased uncoupling [42,43] while the combination of decreased TCA cycle flux with decreased ATP synthesis rate was presented as a lack of effect on mitochondrial uncoupling [44].

Taken together, administration of thyroid hormone both in hypothyroid and euthyroid states increases energy expenditure in parallel with increases in UCP3, indicating that UCP3 may play a role in thyroid hormone-induced thermogenesis.

4.3. Effect of β-adrenergic stimulation on UCP3 expression

Another condition accompanied by increased thermogenesis is β-adrenergic stimulation. It was shown that ephedrine induced a 60% increase in oxygen consumption and it was estimated that ~50% of this increase was accounted for by muscular thermogenesis [45]. Similar estimates of contribution of skeletal muscle to adrenalin-induced increased energy expenditure were made [46]. Given the skeletal muscle specific expression of UCP3, it is tempting to relate the β-adrenergic increase in thermogenesis to UCP3.

Intraperitoneal injections with the β3-adrenergic agonist CL-316243 in rats induced a prominent increase in UCP3 mRNA in white adipose tissue and, to a lesser extent, in skeletal muscle [8]. Using the same compound, similar observations were made in obese rats treated for 10 days [47]. This finding however, could not be reproduced by others using trecadrine as a β3-agonist [48]. The observation in L6 myotubes that isoproterenol and salbutamol (β2-agonists) increased UCP3 mRNA levels, while propanolol (a β2-antagonist) blunted the observed increase [49], suggests that UCP3 may play a role in β-adrenergic activation-induced thermogenesis. However, observations of increases in UCP3 after β-adrenergic stimulation are biased by increased lipolysis and a concomitant rise in free fatty acid levels, known to induce UCP3 expression (see below). To delineate the direct effect of β-stimulation from the effect of increased lipolysis on skeletal muscle UCP3 expression, we recently studied the effect of salbutamol infusion with and without acipimox (to block lipolysis) in humans. In this study, we showed that salbutamol infusion did not increase UCP3 mRNA levels. In the condition in which lipolysis was blocked by acipimox during salbutamol infusion, free fatty acid levels were significantly decreased. This decrease was paralleled by a decrease in UCP3 mRNA (Hoeks et al., Unpublished observations).

Thus, we conclude that in humans, β-adrenergic-increased energy expenditure was not related to UCP3 expression. Deduced from the other papers available on β-stimulation and UCP3 expression, the more obvious conclusion would be that β-adrenergic induction of the UCP3 gene is due to increased fatty acid levels rather than to a direct effect of β-stimulation.

4.4. The effect of nutritional status and fatty acids on UCP3 expression

Fatty acids have long been recognized as potent regulators of gene transcription for numerous genes. The importance of fatty acids in the induction of UCP3 is illustrated by the observation that in skeletal and heart muscle from fetal mice, UCP3 is not induced until lactation. Lactation with high levels of fat results in upregulation of UCP3 mRNA; whereas a high-carbohydrate diet at weaning induces a decrease in UCP3 mRNA levels [50].

In addition, it was observed that caloric restriction for 5 days resulted in a ~2–3-fold increase in UCP3 mRNA levels in lean and obese humans [51]. In a study in rodents, it was shown that a 48 h fast induced UCP3 mRNA 5.6-fold in rat tibialis anterior muscle and a 24 h fasting induced a 3.5-fold increase in UCP3 mRNA in mice soleus muscle without affecting basal heat production in vitro [52]. These seemingly paradoxical observations, i.e. increased expression of UCP3 under conditions of attenuated energy expenditure, were put into perspective when plasma free fatty acid levels were taken into consideration. In rats fasted between 24 and 72 h, a 10-fold increase in UCP3 was measured [53]. This induction of UCP3 could be mimicked by elevation of fatty acid levels (by infusion of intralipid together with heparin), whereas other physiological responses known to occur during fasting (a fall in leptin and increased corticosterone levels) had no effect on UCP3 mRNA [53]. A more close examination of fatty acids in relation to UCP3 expression again indicated that fasting (30 h) increased UCP3 mRNA significantly while refeeding completely reversed the fasting-induced increase in UCP3 mRNA levels to control values within 2 h of refeeding [54].

The effects of fasting-induced increase in UCP3 on mitochondrial energy metabolism have been examined in control mice and UCP3 ablated mice. Again, fasting induced upregulation of UCP3 mRNA in control mice, but did not affect proton motive force and state 4 respiration [55]. Interestingly, this was the first study to report increased respiratory exchange ratio values if UCP3-ko mice had ad libitum access to food compared to ad libitum fed controls [55]. Fasting attenuated these differences, albeit non-significantly. These observations were interpreted as supportive for the idea that UCP3 plays an important role in fat oxidation [55]. Modifying nutritional status by feeding a high-fat diet has also been reported to profoundly affect UCP3 expression, as indicated amongst others by observations in weaning mice [56]. In 5-week-old rats, UCP3 was induced upon isocaloric high-fat feeding in gastrocnemius muscle and was reported to depend on both chain length of the fatty acids as well as the extent of saturation [57]. Ad libitum feeding rats a high-fat (60% of energy from fat) versus a low-fat diet (12% of energy from fat) resulted in significant elevations in plasma free fatty acid levels and induced a 2-fold increase in UCP3 protein level after 4 weeks of feeding [58]. The increase in UCP3 protein neither affected the mitochondrial ion permeability nor the proton permeability as assessed by mitochondrial swelling, while apparent effects were detected if a chemical uncoupler was added to the incubation medium [58]. The lack of effect of increased UCP3 expression on in vitro proton permeability was extended by the lack of effect of UCP3 on 24 h energy expenditure.

Consumption of a high-fat diet by healthy trained athletes for 4 weeks (41% of energy from fat versus 17% of energy from fat) also resulted in increased UCP3 mRNA expression, with most prominent results in humans with a high percentage type IIa muscle fibers [59]. In a recent study, feeding well-trained subjects high- and low-fat diets (65% of energy from fat versus 70–75% of energy as carbohydrate) for only 5 days in a cross-over design did not affect UCP3 gene expression, in contrast to other genes involved in lipid metabolism [60]. The lack of effect of a high-fat diet on UCP3 may be due to the relatively short period of high-fat consumption, possibly in combination with the selection of subjects recruited (discussed below). In normal healthy subjects, we recently showed upregulation of UCP3 at the protein level after supplying subjects isocaloric high-fat versus low-fat diets (60% of energy as fat versus 30% of energy as fat with identical protein levels). The high-fat diet resulted in a significant increase in UCP3 protein levels within the physiological range within 7 days [61]. Following severe exhaustive ischemic contractile activity, mitochondrial coupling was measured in vivo by measuring creatine phosphate resynthesis rate after blood flow was reinstated. We concluded that, despite a 44% increase in UCP3 protein content, and with all potential physiological modulators of UCP3 activity present, mitochondrial coupling was not affected [61], suggesting that mitochondrial uncoupling may not be the principal function of UCP3 in vivo.

In summary, increasing circulatory free fatty acids, either by fasting or by consumption of a high-fat diet consistently results in induction of the UCP3 gene, with concordant changes in UCP3 protein content. This increase appears to depend on chain length and the extent of saturation of the fatty acids ingested. In none of the cases reporting increased UCP3 mRNA or protein levels, evidence is found linking fatty acid-induced increased UCP3 to increased mitochondrial uncoupling. Thus, it appears that uncoupling per se may not be the primary physiological function of UCP3.

4.5. The effect of exercise on UCP3 expression

With regard to exercise, it is of importance to discern the effects of acute exercise from the effects of exercise training. There appears to be consistency with regard to the upregulating effect of acute exercise on UCP3 levels. 2 h of treadmill running induced a 252% increase in UCP3 mRNA in white gastrocnemius and a 63% increase in red gastrocnemius [62]. It has also been shown that the extent of increase in UCP3 depends on the duration of exercise; after only 30 min of exercise, UCP3 mRNA increased slightly but significantly [63]. When exercise was continued for 200 min the induction in UCP3 was way more prominent, reaching levels of ~700%, with similar induction by either swimming exercise or treadmill running [63]. In this study, increased mRNA matched almost perfectly with increased protein expression [63]. In humans, 4 h of cycling exercise induced an increase in transcriptional activity of UCP3 of almost 600% [64]. The extent of increase was linked to the exercise duration showing the most prominent induction after 4 h of exercise [64].

A straightforward interpretation of this would be that exercise per se induces UCP3 transcription. If this would be the case, one would expect exercise training to result in increased UCP3 protein levels. Strikingly, rats housed in cages equipped with running wheels for 9 weeks, permitting increased spontaneous activity, had UCP3 levels similar to their sedentary counterparts [62]. When subjected to an endurance training program of stepwise incremental load for 4 weeks followed by another 4 weeks at the same level of training volume and intensity, UCP3 was significantly downregulated in soleus and anterior tibialis muscle in rats, with the most prominent decrease in the latter (less oxidative) muscle [65]. In endurance-trained humans, UCP3 mRNA levels also were significantly lower compared to physically fit but untrained controls (maximal power output 5.6 versus 3.9 W/kg bw for untrained subjects) [66]. Later, observations of decreased UCP3 mRNA in subjects following endurance training or in trained subjects were extended to the protein level [67,68].

The complete absence of contractile activity after denervation results in a marked increase in UCP3 mRNA in rat skeletal muscle, within 72 h after cutting the sciatic nerve [62]. Comparable observations were made in skeletal muscles of tetraplegic humans, showing increased UCP3 mRNA levels, which returned to control levels after an 8 week training program of electrically stimulated cycling exercise [69].

The paradoxical increase of UCP3 mRNA following acute exercise and the decrement in UCP3 following exercise training or in endurance-trained athletes indicate that it is not exercise per se that affects UCP3 expression. To distinguish the effects of exercise as such from the exercise-induced increase in fatty acid levels, we monitored UCP3 mRNA expression before and 4 h following a 2 h cycling exercise [70]. Subjects were tested after an overnight fast and were given only plain water during and after exercise. This protocol resulted in substantial increases in plasma free fatty acid levels, peaking to 1000 μmol/L immediately after cessation of exercise, and remained elevated during the 4 h post-exercise period (~800–900 μmol/L) [70]. In the second test, fatty acid levels were blunted by ingestion of glucose (1.4 g/kg body weight, in a 20% solution) before exercise and doses of 0.35 g/kg body weight, in a 10% solution during- and post-exercise. Glucose ingestion successfully suppressed lipolysis as indicated by the lack of effect of exercise on plasma free fatty acids and the decline in fat oxidation observed in the glucose trial [70]. Under both conditions, subjects were able to maintain euglycemia throughout and after exercise. We showed that only in the fasting trial, with very high levels of fatty acids present, UCP3 mRNA was increased 4 h post-exercise while in the glucose condition, with no changes in fatty acid levels but with exercise of the same duration and work load, no effect on UCP3 mRNA was detected [70]. Thus, we concluded that observations of increased UCP3 mRNA following acute exercise are mediated by prolonged increased free fatty acid levels and not by another factor intrinsically related to physical exercise.

Interestingly, Hildebrandt and Neufer [71] reported that 2 h of treadmill running significantly attenuated the fasting-induced increase in transcriptional rate of UCP3, but not other genes related to lipid metabolism like lipoprotein lipase (LPL), long-chain acylCoA dehydrogenase (LCAD) or carnitine palmitoyl transferase 1 (CPT1) [71]. The authors suggested that fasting and exercise might trigger opposing regulatory mechanism(s) [71]. This observation may indicate that it is not just the free fatty acid levels regulating UCP3 expression but the regulation of UCP3 expression that is more delicate.

In summary, there appears to be consensus that acute exercise induces upregulation of UCP3, most likely due to elevated plasma free fatty acid levels. Fascinatingly, the study by Hildebrandt and Neufer indicates that if fatty acid levels are increased already at the onset of exercise, exercise may attenuate the increase in UCP3 mRNA [71]. With respect to this, it is important to note that initiation of exercise with elevated plasma free fatty acid levels readily results in oxidation of fatty acids and consequently lower plasma fatty acid levels. This may indicate that the balance between fatty acid delivery to the cell and mitochondrial oxidation of fatty acids plays a crucial role in the regulation of UCP3 expression. Along these lines, the decline in UCP3 content observed in trained individuals or following training can be explained by the training-induced increase in fat oxidative capacity. Indeed we previously showed that UCP3 mRNA correlated negatively with maximal aerobic capacity (VO_2max) [66]. Although we are not aware of any studies aiming to study the decrease in UCP3 in trained subjects in relation to mitochondrial coupling, indirect negative associations between UCP3 mRNA and mechanical efficiency point towards improved energy efficiency [66]. It should however be noted that patients suffering from chronic respiratory disease (COPD) have decreased mechanical efficiency [72] while having lower UCP3 protein content [73].

5. From regulation to putative physiological function of UCP3

5.1. Hints from studies physiologically manipulating UCP3

As discussed above, several observations have been made suggesting that regulation of UCP3 is closely related to fatty acid metabolism. This has led to the suggestion that UCP3 was involved in the regulation of lipids as fuel substrate rather than as mediators of regulatory thermogenesis [74]. This suggestion was based for the most part on the tissue-dependent differential mRNA expression of the UCP homologues in skeletal muscles of distinct fiber typology, which is consistent with the differential requirement of these tissues for lipids during fasting and their ability to shift from glucose to fat oxidation during refeeding and exercise. Thus, it was shown that increases in UCP3 mRNA, under conditions of elevated free fatty acid levels, were more pronounced in muscles comprised of fibers enzymatically equipped for glycolysis (type IIa and IIx fibers) than in muscles made up of slow (type I) fibers with a high fat oxidative capacity [53,75]. In addition, if examined at the cellular level we showed that in healthy humans as in type 2 diabetics, protein expression of UCP3 was most prominently expressed in glycolytic type IIb (or type IIx) fibers, with somewhat lower expression in type IIa fibers and the lowest expression in the fat oxidative type I fibers [22,67]. Given the low expression of fat oxidative enzymes in type IIx fibers and the high level of UCP3 expression, it is not conceivable that UCP3 serves to facilitate fat oxidation. Also, decreased UCP3 content in skeletal muscle of endurance-trained athletes [67], known for their high fat oxidative capacity, does not also favor the idea that UCP3 plays a major role in modulating fat oxidation. In fact, when lean, previously untrained subjects participated in a 3 month training program their fat oxidative capacity increased significantly [76] while plasma free fatty acid levels remained unaltered. Analyses of UCP3 protein content pre- and post-training revealed that training induced a decrease in UCP3. This decrease in UCP3 negatively correlated with the training-induced increase in fat oxidation, i.e. the subjects with the most prominent increase in fat oxidative capacity showed the most prominent decrease in UCP3 protein [68]. The high capacity to oxidize fats in trained athletes may also be the reason for the observation that consumption of a high-fat diet for 5 days failed to affect UCP3 mRNA levels in athletes [60] while feeding a diet with comparable high levels of fat to normal healthy subjects (with normal fat oxidative capacity) for 7 days resulted in upregulation of UCP3 at the protein level [61].

The reports referred to above, indicate that increased fat oxidative capacity without marked increases in fatty acid supply induce downregulation of UCP3 rather than upregulation, which was anticipated if the role of UCP3 were to modulate fat oxidation. Another model in which fat oxidative capacity is increased while plasma free fatty acid levels are reduced is following weight reduction induced by caloric restriction for

10 weeks. Under these conditions we have reported decreased UCP3 protein levels that correlated well with skeletal muscle FABP content [21].

On the other hand, conditions in which UCP3 is upregulated are characterized by a rapid rise in fatty acid levels without improved fat oxidation capacity, such as fasting, acute exercise, fatty acid infusion (discussed above). In experimentally induced diabetes in rodents by administration of streptozotocin, fatty acid levels rise acutely without instantaneous effects on fat oxidative capacity. Indeed, increased UCP3 levels have been reported at the mRNA level following administration of streptozotocin [77,78].

Altogether, the consistent conclusion of these studies is that UCP3 levels increase if there is an imbalance in fatty acid delivery and fat oxidation, or more precisely if the supply of fatty acids to the mitochondria exceeds the capacity of the mitochondria to oxidize fatty acids.

5.2. Deduction of the principal physiological function of UCP3: a hypothesis

Oxidation of (long-chain, >C12) fatty acids starts with import of fatty acids into the mitochondria. To do so, the fatty acids need to cross the outer and the inner mitochondrial membrane. While non-esterified (free) fatty acids can cross the outer mitochondrial membrane, transport across the inner mitochondrial membrane is more intricate. Transport across the inner mitochondrial membrane is in general facilitated by the carnitine acyltransferase system (CAT1 and CAT 2), which catalyses the transport of fatty acylCoA esters. Outside the mitochondria, fatty acids are esterified by fatty acylCoA synthetase resulting in fatty acylCoA, which in turn is converted into acylcarnitine by CAT1. Acylcarnitine crosses the inner mitochondrial membrane where it is reconverted to fatty acylCoA by CAT2. Only in this form fatty acids can be degraded by β-oxidation to acetylCoA, which in turn may enter the TCA cycle. Any defect in uptake via the carnitine acyltransferase system or downstream in β-oxidation will lead to decreased fat oxidation.

However, it has also been shown that transport of free fatty acids through phospholipid membranes may occur if fatty acids in the unionized form are incorporated into the phospholipid bilayer. Once incorporated, these fatty acids may cross the membrane by flip–flop [79]. In adipocytes, it was shown that the entry of fatty acids occurred at high and low concentrations of fatty acids following kinetics of simple diffusion [79]. This implies that if the load of fatty acids to the mitochondria is very high (e.g. after high-fat feeding, infusion of lipids or following acute exercise in the fasting state) a noteworthy portion of the fatty acids may enter the mitochondria via flip–flop in their non-esterified form. Once in the mitochondrial matrix, the fatty acids will be deprotonated due to the proton gradient, leaving non-esterified fatty acid anions within the mitochondrial matrix. As mitochondria lack long-chain fatty acylCoA synthetase, these fatty acid anions cannot be esterified and therefore not be diverted towards β-oxidation, neither can they flip–flop back due to proton gradient. Thus, prolonged exposure of mitochondria to fatty acids in levels high enough to exceed the oxidative capacity, either by limitations in the carnitine shuttle system or by defects more downstream, may result in redundant accumulation of fatty acid anions within the mitochondrial matrix. Here they may exert deleterious effects on mitochondrial function and may damage mitochondrial DNA. Mitochondrial DNA is more sensitive to damage than nuclear DNA and can be less efficiently repaired, underscoring the need to avoid the deleterious effects of increased levels of non-esterified fatty acid anions within the mitochondrial matrix. With respect to this, it is of interest to note that UCP3 has been associated with a modulating role in oxidative stress by lowering production of ROS [20], thus lowering the risk of mitochondrial DNA damage. As transport of fatty acids appears to be an accepted property of uncoupling proteins, this has led us to hypothesize that UCP3 may function as a fatty acid anion exporter in exchange for a proton. Thus UCP3 may primarily function as part of the mitochondrial defense mechanism against an excess load of fatty acids [68]. Please be aware that by outward transport of fatty acid anions, the mitochondrial proton gradient is lowered and UCP3 might, as a consequence, possess mild uncoupling as a side effect.

Shortly after submitting the paper describing this alternative role of UCP3 [68], a paper with a similar hypothesis was published [80]. In this paper, the authors propose that, once inside the mitochondrial matrix in their esterified form, not all fatty acylCoA esters are diverted towards β-oxidation but that some may be hydrolyzed by a mitochondrial thioesterase (MTE1), resulting in fatty acids and CoA. Again, due to the lack of fatty acylCoA synthetase within the mitochondrial matrix, outward transport of the fatty acid is preferred and UCP3 is the protein hypothesized to be responsible. The authors suggest that the energy needed to esterify the fatty acids before entering the mitochondrial matrix (2 ATP per fatty acid) may contribute to the increased energy expenditure observed in systems expressing UCP3 abundantly [80]. This futile cycle may serve to liberate coenzyme A and regenerate the supply of CoASH, required for other metabolic processes within the mitochondria.

Regardless of the origin of the fatty acid anion within the matrix [either flip–flop across the inner mitochondrial membrane or by hydrolyses of fatty acylCoA by a mitochondrial thioesterase (MTE)], both hypotheses propose that the primary role of UCP3 is the outward translocation of fatty acids away from the mitochondrial matrix. It should be noted that we hypothesize that fatty acid anion export is essential to prevent mitochondrial damage for example by lipid peroxidation or by damage to mitochondrial DNA, while J. Himms-Hagen and M.E. Harper propose that fatty acid anion export serves to liberate CoASH required for other metabolic processes.

6. UCP3 as a mitochondrial fatty acid anion exporter: experimental data

6.1. Concerted induction of UCP3 and MTE1

The observation that mice overexpressing UCP3 possessed increased levels of MTE1 [81] was interpreted by the authors as support for their hypothesis that UCP3 is an outward translocator of fatty acids generated by MTE1. As the levels of UCP3 expression in the UCP3-tg mice exceed physiological levels by far (as discussed above), the authors extended their findings in obese and insulin resistant (*db/db* mice) mice and in their lean non-diabetic controls, with endogenous UCP3 expression. In these mice, the use of selective agonists of PPARα and PPARγ (Wy-14.643 and rosiglitazone, respectively) was used to affect UCP3 gene expression [82]. Interestingly, under basal conditions, *db/db* mice had increased skeletal muscle levels of UCP3 and MTE1 compared to their controls [82]. If treated with rosiglitazone or the Wy compound, UCP3 and MTE1 were concordantly decreased compared to non-diabetic controls. The authors concluded that if changes in UCP3 mRNA occurred, these were complimentary with changes in MTE1 suggesting that they are involved in the same metabolic pathway, either in response to, or as regulators of, fatty acid oxidation [82].

While their statement appears to be correct, it does little to prove that UCP3 serves to export fatty acid anions derived from the thioesterase reaction. It may also be the reflection of an adaptive response of the *db/db* mice to a defective β-oxidation downstream of the carnitine acyltransferase system. Thus, fatty acids physically ready for β-oxidation as fatty acylCoA esters may accumulate within the mitochondria. This is an obnoxious situation, as it may drain the CoASH levels required for proper TCA-cycling. Thus accumulation of acetylCoA levels may further augment β-oxidation with increased fatty acylCoA levels as a result, possibly in harmful concentrations. To avoid this quandary, the adaptive response of skeletal muscle mitochondria from *db/db* mice may be to upregulate MTE1 in order to liberate CoASH and finally maintain a proper β-oxidative flux. The remaining fatty acid anions within the matrix may then be exported in a process facilitated by UCP3.

It should be noted however, that induction of UCP3 in concordance with MTE1 was also observed in rat liver after treatment with fenofibrate, an a-selective activator of peroxisome proliferator-activated receptors (PPARs) [83]. This is remarkable, as under physiological conditions and also in the control rats in the study by Lanni et al. [83], UCP3 expression is silent in liver. Immunohistochemistry revealed that UCP3 was expressed in hepatocytes (the fat oxidizing cells in liver), rather than in Kupfer cells or other cells found in intact liver [83].

The above-mentioned studies clearly indicate that UCP3 and MTE1 are trans-activated under the experimental conditions described. It should be noted however that in both experiments in which endogenous UCP3 was induced [82,83], this was achieved by treatment with agonists of PPARs. Both UCP3 [84,85] and MTE1 [86] have PPAR responsive elements. Thus, the observed concerted upregulation of UCP3 and MTE1 may merely reflect their PPAR responsiveness, rather than a functional coupling between the two genes. Moreover, the observation that under basal conditions expression of UCP3 at the protein level is restricted to skeletal muscle, while MTE1 is more ubiquitously expressed [86] does not designate a straight functional coupling between the two proteins.

Clearly, the concordant response of UCP3 and MTE1 is an intriguing observation, which deserves detailed examination to elucidate if the trans-activation of MTE1 and UCP3 also occurs in humans and follows non-pharmaceutical induction of the UCP3 gene. It is also of relevance to test if UCP3 drives increased expression of MTE1 or if increased expression of MTE1 entails upregulation of UCP3. To us, the latter appears to be more conceivable. We hypothesize that upregulation of MTE1 may play a role under the (patho-physiological) condition of severely hampered β-oxidative capacity and high flux through the carnitine acyltransferase shuttle. The fatty acid moieties remaining after hydrolysis of fatty acylCoA esters by MTE1 will then be exported out of the mitochondrial matrix by UCP3.

6.2. Manipulation of the balance of fat supply and fat oxidation in relation to UCP3 expression

Deduced from observations of previous studies, we recently postulated that UCP3 might serve as a fatty acid anion exporter if there is an imbalance between delivery of fatty acids to mitochondria and the mitochondrial capacity to oxidize fatty acids [68]. In our hypothesis, we propose that following prolonged exposure of mitochondria to fatty acids, fatty acids may enter the mitochondria via flip–flop next to the fatty acyl transferase shuttle system [68]. To examine this hypothesis we have intruded fatty acid transport, uptake and oxidation at distinct levels in a series of experiments, which will be outlined briefly below (Fig. 3).

In the first intervention, we blocked mitochondrial entry of fatty acids via CAT1 by etomoxir. As a consequence, the concentration of sarcoplasmic free fatty acids will rise and the fraction of fatty acids entering the mitochondria in their non-esterified form increases. According to our hypothesis, this will again result in increased UCP3 expression. We administered etomoxir for 36 h to human subjects while they were

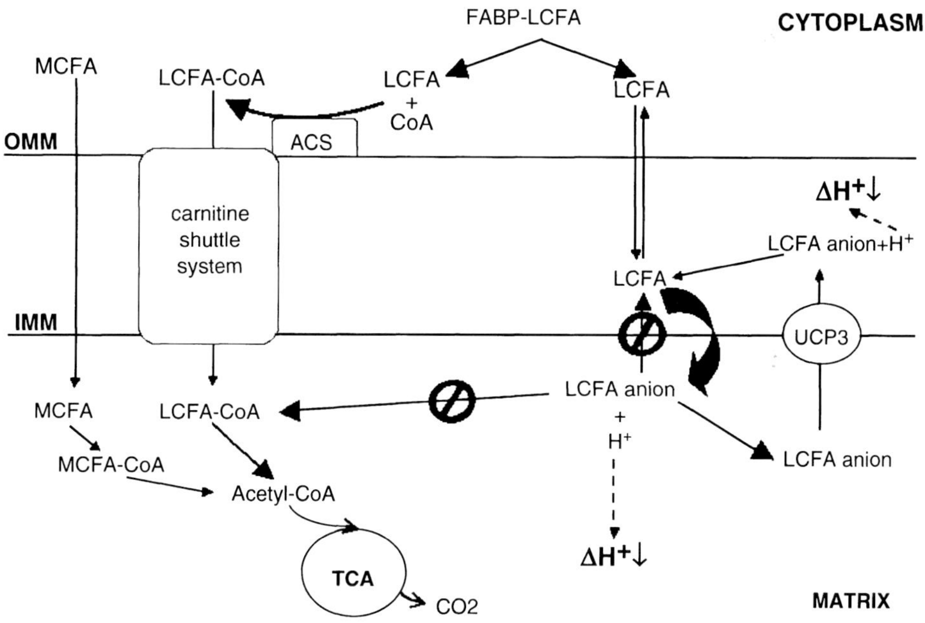

Fig. 3. Schematic model of UCP3's putative function and some of the experimental conditions discussed in Section 6.2. UCP3 protein content increases if the entry of LCFA into the mitochondrial matrix is enhanced: 1) after inhibition of the carnitine acyl transferase shuttle system using etomoxir and 2) on a high-fat diet, associated with increased supply of fatty acids to the mitochondria. Note that a high-fat diet rich in MCFA does not increase UCP3 protein content. MCFA can enter the mitochondrial matrix in their non-esterified form, but can still be oxidized due to the presence of a medium-chain acylCoA synthetase inside the matrix. In the latter situation, UCP3 would not be needed for export of fatty acid anions. OMM, outer mitochondrial membrane; IMM, inner mitochondrial membrane; FA, fatty acid; TCA, tricarboxylic acid cycle, ACS (long-chain fatty) acylCoA synthetase.

consuming high-fat diets (to increase fatty acid supply). Using respiration chambers, we showed that etomoxir effectively interfered with fat oxidation, given the decreased fat and increased carbohydrate oxidation [87]. In all subjects, UCP3 levels were increased after etomoxir treatment compared to controls, resulting in an average increase of 67% at the UCP3 protein level after 36 h! Furthermore, we observed a negative correlation between the decrease in fat oxidation after etomoxir treatment and the increase in UCP3 protein [87]. These data are highly compatible with our hypothesis that UCP3 exports fatty acid anions derived from "flip–flop"-driven entry of non-esterified fatty acids. However, it should be noted that blockade of CAT1 is an experimental condition in which the entry of fatty acylCoA into the mitochondria is reduced, making it unlikely that the induction of UCP3 was related to the action of MTE1.

The commonly observed induction of UCP3 by high-fat feeding also fits with the idea that UCP3 is needed to export the surplus in fatty acids delivered by the diet that cannot be oxidized. However, this is not true for medium chain fatty acids (MCFA), which can be taken up by the mitochondria next to CAT1 and can, in contrast to long-chain fatty acids (LCFA), be esterified within the mitochondrial matrix to their respective CoA esters and can therefore be readily diverted towards β-oxidation. Thus, the majority of MCFA delivered to the mitochondria will be almost instantaneously oxidized, this in contrast to LCFA. Using this approach, we showed that 14 days high-fat feeding with diets comprised entirely of LCFA (HF-LCT, 46 en% as fat of which 79% C16 and 20% C18) resulted in increased UCP3 content in rat gastrocnemius muscle. However, with high-fat diets comprised of MCFA (HF-MCT, 46 en% as fat of which 60% C8 and 40% C10), UCP3 levels were comparable to values in rats fed an isocaloric low-fat diet (Hoeks et al. unpublished observations). It is important to note that this differential response to a diet comprised of LCFA versus MCFA was accompanied by a similar rise in plasma free fatty acid levels in these rats. Again this study is an indication that increased fatty acid availability only affects UCP3 content if the fatty acid load to the mitochondria exceeds the mitochondrial fat oxidative capacity.

Next to manipulation of fatty acid delivery to the mitochondria we also studied the effect of defective fat oxidative capacity.

In rats rendered diabetic by treatment of streptozotocin, decreased fatty acid oxidative capacity was associated with increased UCP3 mRNA levels in the heart [77]. Under healthy conditions, the heart relies almost exclusively on fat oxidation and is well equipped to efficiently handle lipids over a wide range of conditions. When thinking of UCP3 as mitochondrial fatty acid anion exporter, it is not striking that UCP3 has only been reported at very modest protein levels in cardiac muscle. Thus, the observations by Hidaka et al. are fascinating. To examine the role of streptozotocin-induced diabetes on UCP3 expression in the heart more closely, rats were made diabetic with streptozotocin, and the heart was removed for extraction and analyses of non-esterified fatty acids, as well as for UCP3 mRNA and protein levels. This revealed a significant and massive increase in UCP3 mRNA and protein expression, which was closely associated with increased cytosolic fatty acids in the cardiac muscle cells (van der Vusse et al., unpublished findings). This observation directly links the change in UCP3 to the change in sarcoplasmic non-esterified fatty acids.

Patients suffering from riboflavin responsive multiple acylCoA dehydrogenase deficiency (RR-MAD), a rare mitochondrial myopathy, are characterized by severely hampered fat oxidation and excessive muscular storage of lipids, which can be reversed partly by treatment with riboflavin. Untreated RR-MAD patients show decreased oxidative capacity of short (C2), medium (C8) and long (C16) chain fatty acids, excessive intramyocellular storage of lipids and significant increases in UCP3 mRNA and protein [88]. Treatment of riboflavin induced a return of fat oxidative capacity to control values, and induced a significant drop in intramyocellular lipids. These notable changes in fat oxidation were accompanied by a return of UCP3 protein levels to control values [88]. This unique data-set is the first to show in a longitudinal design that restoration of fat oxidative capacity with simultaneous decline in fatty acid supply is followed by a rapid decline in UCP3 protein. Once more, this study supports our hypothesis that UCP3 exports fatty acid anions from the mitochondrial matrix. In this particular myopathy, we anticipate that the increased UCP3 expression observed in the untreated RR-MAD patients may occur in concert with increased MTE1 levels.

In summary, interference in successive steps in fatty acid handling, transport and oxidation persistently showed that UCP3 protein increases if the supply of fatty acids to the mitochondria exceeds fat oxidative capacity. In addition, data are available indicating that the increased UCP3 levels can be normalized if fat oxidative capacity is enhanced. Although we realize that none of the studies presented provide direct and definitive proof for our hypothesis, we interpret these data as compelling circumstantial evidence that UCP3 indeed facilitates outward translocation of fatty acids from the mitochondrial matrix.

7. Summary and conclusion

The cloning of the UCP1 homologues, UCP2 and UCP3, has raised considerable interest in the phenomenon uncoupling. The expression of UCP3 mainly in skeletal muscle mitochondria and the potency of the skeletal muscle as a thermogenic organ made UCP3 an attractive target for studies towards manipulation of energy expenditure to combat pathologies like type 2 diabetes and obesity. Transfecting yeast with UCP3 and overexpressing UCP3 in mice resulted in increased thermogenesis and related desirable phenotypical changes. However, the lack of an apparent phenotype in mice lacking UCP3 triggered the search for alternative functions of UCP3. Data obtained from UCP3-ko mice reveal that UCP3 may be involved in preventing excessive mitochondrial oxidative stress. The observation that fatty acids levels significantly affect UCP3 expression has given UCP3 a position in fatty acid handling and/or oxidation. The original idea that UCP3 may facilitate fat oxidation by introducing fatty acids into the mitochondria appears no longer plausible. Rather, prevailing data indicate that the primary physiological role of UCP3 may be the outward transport of fatty acid anions from the mitochondrial matrix. In doing so, fatty acids are exchanged with protons, explaining the uncoupling activity of UCP3. The fatty acid anions exported may originate from hydrolysis of fatty acid esters by an MTE, or they may have entered the mitochondria as non-esterified fatty acids by incorporating into and flip–flop across the mitochondrial inner membrane. Although the origin of the fatty acid anions may seem trivial to understand the physiological role of

UCP3, it is of importance to note that the concordant regulation of UCP3 and MTE has only been reported after pharmacological interventions, not necessarily reflecting human physiology.

Acknowledgements

The authors would like to thank Prof. Dr G.J. van de Vusse, Dr A. Russel and MSc J. Hoeks for letting them refer to their encouraging data, pending publication when this chapter was prepared for submission. The research of Dr P. Schrauwen has been made possible by fellowships of the Royal Netherlands Academy of Arts and Sciences and the Netherlands Organization for Scientific Research (NWO).

References

[1] Mitchell, P., Moyle, J., 1967. Nature 213, 137–139.
[2] Nicholls, D.G., Bernson, V.S., Heaton, G.M., 1978. Experientia Suppl. 32, 89–93.
[3] Rolfe, D.F., Brand, M.D., 1996. Am. J. Physiol. 271, C1380–C1389.
[4] Rolfe, D.F., Newman, J.M., Buckingham, J.A., Clark, M.G., Brand, M.D., 1999. Am. J. Physiol. 276, C692–C699.
[5] Fleury, C., Neverova, M., Collins, S., Raimbault, S., Champigny, O., Levi-Meyrueis, C., Bouillaud, F., Seldin, M.F., Surwit, R.S., Ricquier, D., Warden, C.H., 1997. Nat. Genet. 15, 269–272.
[6] Boss, O., Samec, S., Paoloni-Giacobino, A., Rossier, C., Dulloo, A., Seydoux, J., Muzzin, P., Giacobino, J.P., 1997. FEBS Lett. 408, 39–42.
[7] Nedergaard, J., Matthias, A., Golozoubova, V., Jacobsson, A., Cannon, B., 1999. J. Bioenerg. Biomembr. 31, 475–491.
[8] Gong, D.W., He, Y., Karas, M., Reitman, M., 1997. J. Biol. Chem. 272, 24129–24132.
[9] Zhang, C.Y., Hagen, T., Mootha, V.K., Slieker, L.J., Lowell, B.B., 1999. FEBS Lett. 449, 129–134.
[10] Hinz, W., Faller, B., Gruninger, S., Gazzotti, P., Chiesi, M., 1999. FEBS Lett. 448, 57–61.
[11] Hinz, W., Gruninger, S., De Pover, A., Chiesi, M., 1999. FEBS Lett. 462, 411–415.
[12] Stuart, J.A., Cadenas, S., Jekabsons, M.B., Roussel, D., Brand, M.D., 2001. Biochim. Biophys. Acta 1504, 144–158.
[13] Stuart, J.A., Harper, J.A., Brindle, K.M., Jekabsons, M.B., Brand, M.D., 2001. Biochem. J. 356, 779–789.
[14] Winkler, E., Heidkaemper, D., Klingenberg, M., Liu, Q., Caskey, T., 2001. Biochem. Biophys. Res. Commun. 282, 334–340.
[15] Clapham, J.C., Arch, J.R., Chapman, H., Haynes, A., Lister, C., Moore, G.B., Piercy, V., Carter, S.A., Lehner, I., Smith, S.A., Beeley, L.J., Godden, R.J., Herrity, N., Skehel, M., Changani, K.K., Hockings, P.D., Reid, D.G., Squires, S.M., Hatcher, J., Trail, B., Latcham, J., Rastan, S., Harper, A.J., Cadenas, S., Buckingham, J.A., Brand, M.D., Abuin, A., 2000. Nature 406, 415–418.
[16] Li, B., Nolte, L.A., Ju, J.S., Ho Han, D., Coleman, T., Holloszy, J.O., Semenkovich, C.F., 2000. Nat. Med. 6, 1115–1120.
[17] Cadenas, S., Echtay, K.S., Harper, J.A., Jekabsons, M.B., Buckingham, J.A., Grau, E., Abuin, A., Chapman, H., Clapham, J.C., Brand, M.D., 2002. J. Biol. Chem. 277, 2773–2778.
[18] Echtay, K.S., Roussel, D., St-Pierre, J., Jekabsons, M.B., Cadenas, S., Stuart, J.A., Harper, J.A., Roebuck, S.J., Morrison, A., Pickering, S., Clapham, J.C., Brand, M.D., 2002. Nature 415, 96–99.
[19] Gong, D.W., Monemdjou, S., Gavrilova, O., Leon, L.R., Marcus-Samuels, B., Chou, C.J., Everett, C., Kozak, L.P., Li, C., Deng, C., Harper, M.E., Reitman, M.L., 2000. J. Biol. Chem. 275, 16251–16257.
[20] Vidal-Puig, A.J., Grujic, D., Zhang, C.Y., Hagen, T., Boss, O., Ido, Y., Szczepanik, A., Wade, J., Mootha, V., Cortright, R., Muoio, D.M., Lowell, B.B., 2000. J. Biol. Chem. 275, 16258–16266.
[21] Schrauwen, P., Schaart, G., Saris, W.H., Slieker, L.J., Glatz, J.F., Vidal, H., Blaak, E.E., 2000. Diabetologia 43, 1408–1416.

[22] Hesselink, M.K., Keizer, H.A., Borghouts, L.B., Schaart, G., Kornips, C.F., Slieker, L.J., Sloop, K.W., Saris, W.H., Schrauwen, P., 2001. FASEB J. 15, 1071–1073.
[23] Harper, J.A., Stuart, J.A., Jekabsons, M.B., Roussel, D., Brindle, K.M., Dickinson, K., Jones, R.B., Brand, M.D., 2002. Biochem. J. 361, 49–56.
[24] Cline, G.W., Vidal-Puig, A.J., Dufour, S., Cadman, K.S., Lowell, B.B., Shulman, G.I., 2001. J. Biol. Chem. 276, 20240–20244.
[25] Solanes, G., Vidal-Puig, A., Grujic, D., Flier, J.S., Lowell, B.B., 1997. J. Biol. Chem. 272, 25433–25436.
[26] Halsall, D.J., Luan, J., Saker, P., Huxtable, S., Farooqi, I.S., Keogh, J., Wareham, N.J., O'Rahilly, S., 2001. Int. J. Obes. Relat. Metab. Disord. 25, 472–477.
[27] Otabe, S., Clement, K., Dina, C., Pelloux, V., Guy-Grand, B., Froguel, P., Vasseur, F., 2000. Diabetologia 43, 245–249.
[28] Cassell, P.G., Saker, P.J., Huxtable, S.J., Kousta, E., Jackson, A.E., Hattersley, A.T., Frayling, T.M., Walker, M., Kopelman, P.G., Ramachandran, A., Snehelatha, C., Hitman, G.A., McCarthy, M.I., 2000. Diabetologia 43, 1558–1564.
[29] Dalgaard, L.T., Sorensen, T.I., Drivsholm, T., Borch-Johnsen, K., Andersen, T., Hansen, T., Pedersen, O., 2001. J. Clin. Endocrinol. Metab. 86, 1398–1402.
[30] Dalgaard, L.T., Hansen, T., Urhammer, S.A., Drivsholm, T., Borch-Johnsen, K., Pedersen, O., 2001. Diabetologia 44, 1065–1067.
[31] Schrauwen, P., Xia, J., Walder, K., Snitker, S., Ravussin, E., 1999. Int. J. Obes. Relat. Metab. Disord. 23, 1242–1245.
[32] Meirhaeghe, A., Amouyel, P., Helbecque, N., Cottel, D., Otabe, S., Froguel, P., Vasseur, F., 2000. Diabetologia 43, 1424–1428.
[33] Schrauwen, P., Hesselink, M., 2002. J. Exp. Biol. 205, 2275–2285.
[34] Argyropoulos, G., Brown, A.M., Willi, S.M., Zhu, J., He, Y., Reitman, M., Gevao, S.M., Spruill, I., Garvey, W.T., 1998. J. Clin. Invest. 102, 1345–1351.
[35] Chung, W.K., Luke, A., Cooper, R.S., Rotini, C., Vidal-Puig, A., Rosenbaum, M., Chua, M., Solanes, G., Zheng, M., Zhao, L., LeDuc, C., Eisberg, A., Chu, F., Murphy, E., Schreier, M., Aronne, L., Caprio, S., Kahle, B., Gordon, D., Leal, S.M., Goldsmith, R., Andreu, A.L., Bruno, C., DiMauro, S., Heo, M., Lowe Jr., W.L., Lowell, B.B., Allison, D.B., Leibel, R.L., 1999. Diabetes 48, 1890–1895.
[36] Urhammer, S.A., Dalgaard, L.T., Sorensen, T.I., Tybjaerg-Hansen, A., Echwald, S.M., Andersen, T., Clausen, J.O., Pedersen, O., 1998. Diabetologia 41, 241–244.
[37] Ravussin, E., Lillioja, S., Knowler, W.C., Christin, L., Freymond, D., Abbott, W.G., Boyce, V., Howard, B.V., Bogardus, C., 1988. N. Engl. J. Med. 318, 467–472.
[38] Schrauwen, P., Xia, J., Bogardus, C., Pratley, R.E., Ravussin, E., 1999. Diabetes 48, 146–149.
[39] Hafner, R.P., Nobes, C.D., McGown, A.D., Brand, M.D., 1988. Eur. J. Biochem. 178, 511–518.
[40] Lanni, A., Beneduce, L., Lombardi, A., Moreno, M., Boss, O., Muzzin, P., Giacobino, J.P., Goglia, F., 1999. FEBS Lett. 444, 250–254.
[41] de Lange, P., Lanni, A., Beneduce, L., Moreno, M., Lombardi, A., Silvestri, E., Goglia, F., 2001. Endocrinology 142, 3414–3420.
[42] Jucker, B.M., Dufour, S., Ren, J., Cao, X., Previs, S.F., Underhill, B., Cadman, K.S., Shulman, G.I., 2000. Proc. Natl Acad. Sci. USA 97, 6880–6884.
[43] Lebon, V., Dufour, S., Petersen, K.F., Ren, J., Jucker, B.M., Slezak, L.A., Cline, G.W., Rothman, D.L., Shulman, G.I., 2001. J. Clin. Invest. 108, 733–737.
[44] Jucker, B.M., Ren, J., Dufour, S., Cao, X., Previs, S.F., Cadman, K.S., Shulman, G.I., 2000. 13C/31P NMR Assessment of Mitochondrial Energy Coupling in Skeletal Muscle of Awake Fed and Fasted Rats: Relationship with Uncoupling Protein 3 Expression. J. Biol. Chem. 275, 39279–39286.
[45] Astrup, A., Bulow, J., Madsen, J., Christensen, N.J., 1985. Am. J. Physiol. 248, E507–E515.
[46] Simonsen, L., Bulow, J., Madsen, J., Christensen, N.J., 1992. Am. J. Physiol. 263, E850–E855.
[47] Nakamura, Y., Nagase, I., Asano, A., Sasaki, N., Yoshida, T., Umekawa, T., Sakane, N., Saito, M., 2001. J. Vet. Med. Sci. 63, 309–314.
[48] Gomez-Ambrosi, J., Fruhbeck, G., Martinez, J.A., 1999. Cell Mol. Life Sci. 55, 992–997.
[49] Nagase, I., Yoshida, T., Saito, M., 2001. FEBS Lett. 494, 175–180.

[50] Brun, S., Carmona, M.C., Mampel, T., Vinas, O., Giralt, M., Iglesias, R., Villarroya, F., 1999. FEBS Lett. 453, 205–209.
[51] Millet, L., Vidal, H., Andreelli, F., Larrouy, D., Riou, J.P., Ricquier, D., Laville, M., Langin, D., 1997. J. Clin. Invest. 100, 2665–2670.
[52] Boss, O., Samec, S., Kuhne, F., Bijlenga, P., Assimacopoulos-Jeannet, F., Seydoux, J., Giacobino, J.P., Muzzin, P., 1998. J. Biol. Chem. 273, 5–8.
[53] Weigle, D.S., Selfridge, L.E., Schwartz, M.W., Seeley, R.J., Cummings, D.E., Havel, P.J., Kuijper, J.L., BeltrandelRio, H., 1998. Diabetes 47, 298–302.
[54] Hwang, C.S., Lane, M.D., 1999. Biochem. Biophys. Res. Commun. 258, 464–469.
[55] Bezaire, V., Hofmann, W., Kramer, J.K., Kozak, L.P., Harper, M.E., 2001. Am. J. Physiol. Endocrinol. Metab. 281, E975–E982.
[56] Brun, S., Carmona, M.C., Mampel, T., Vinas, O., Giralt, M., Iglesias, R., Villarroya, F., 1999. Diabetes 48, 1217–1222.
[57] Samec, S., Seydoux, J., Dulloo, A.G., 1999. Diabetes 48, 436–441.
[58] Chou, C.J., Cha, M.C., Jung, D.W., Boozer, C.N., Hashim, S.A., Pi-Sunyer, F.X., 2001. Obes. Res. 9, 313–319.
[59] Schrauwen, P., Hoppeler, H., Billeter, R., Bakker, A., Pendergast, D., 2001. Int. J. Obes. Relat. Metab. Disord. 25, 449–456.
[60] Cameron-Smith, D., Burke, L.M., Angus, D.J., Tunstall, R.J., Cox, G.R., Bonen, A., Hawley, J.A., Hargreaves, M., 2003. Am. J. Clin. Nutr. 77, 313–318.
[61] Hesselink, M.K., Greenhaff, P.L., Constantin-Teodosiu, D., Hultman, E., Saris, W.H., Nieuwlaat, R., Schaart, G., Kornips, E., Schrauwen, P., 2003. J. Clin. Invest. 111, 479–486.
[62] Cortright, R.N., Zheng, D., Jones, J.P., Fluckey, J.D., DiCarlo, S.E., Grujic, D., Lowell, B.B., Dohm, G.L., 1999. Am. J. Physiol. 276, E217–E221.
[63] Zhou, M., Lin, B.Z., Coughlin, S., Vallega, G., Pilch, P.F., 2000. Am. J. Physiol. Endocrinol. Metab. 279, E622–E629.
[64] Pilegaard, H., Ordway, G.A., Saltin, B., Neufer, P.D., 2000. Am. J. Physiol. Endocrinol. Metab. 279, E806–E814.
[65] Boss, O., Samec, S., Desplanches, D., Mayet, M.H., Seydoux, J., Muzzin, P., Giacobino, J.P., 1998. FASEB J. 12, 335–339.
[66] Schrauwen, P., Troost, F.J., Xia, J., Ravussin, E., Saris, W.H., 1999. Int. J. Obes. Relat. Metab. Disord. 23, 966–972.
[67] Russell, A.P., Wadley, G., Hesselink, M.K.C., Schaart, G., Lo, S., Leger, B., Garnham, A., Kornips, E., Cameron-Smith, D., Giacobino, J.P., Muzzin, P., Snow, R., Schrauwen, P., 2003. Pflugers Arch. 445, 563–569.
[68] Schrauwen, P., Saris, W.H., Hesselink, M.K., 2001. FASEB J. 15, 2497–2502.
[69] Hjeltnes, N., Fernstrom, M., Zierath, J.R., Krook, A., 1999. Diabetologia 42, 826–830.
[70] Schrauwen, P., Hesselink, M.K., Vaartjes, I., Kornips, E., Saris, W.H., Giacobino, J.P., Russell, A., 2002. Am. J. Physiol. Endocrinol. Metab. 282, E11–E17.
[71] Hildebrandt, A.L., Neufer, P.D., 2000. Am. J. Physiol. Endocrinol. Metab. 278, E1078–E1086.
[72] Baarends, E.M., Schols, A.M., Akkermans, M.A., Wouters, E.F., 1997. Thorax 52, 981–986.
[73] Gosker, H.R., Schrauwen, P., Hesselink, M.K.C., Schaart, G., van der Vusse, G.J., Wouters, E.F., Schols, A.M., 2003. Uncoupling protein-3 content is decreased in peripheral skeletal muscle of patients with COPD. Eur. Respir. J. Eur. Respir. J. 22, 89–93.
[74] Samec, S., Seydoux, J., Dulloo, A.G., 1998. FASEB J. 12, 715–724.
[75] Samec, S., Seydoux, J., Dulloo, A.G., 1998. Diabetes 47, 1693–1698.
[76] Schrauwen, P., Wagenmakers, A.J., van Marken Lichtenbelt, W.D., Saris, W.H., Westerterp, K.R., 2000. Diabetes 49, 640–646.
[77] Hidaka, S., Kakuma, T., Yoshimatsu, H., Sakino, H., Fukuchi, S., Sakata, T., 1999. Diabetes 48, 430–435.
[78] Hidaka, S., Yoshimatsu, H., Kakuma, T., Sakino, H., Kondou, S., Hanada, R., Oka, K., Teshima, Y., Kurokawa, M., Sakata, T., 2000. Proc. Soc. Exp. Biol. Med. 224, 172–177.
[79] Hamilton, J.A., Kamp, F., 1999. Diabetes 48, 2255–2269.
[80] Himms-Hagen, J., Harper, M.E., 2001. Exp. Biol. Med. (Maywood) 226, 78–84.

[81] Moore, G.B., Himms-Hagen, J., Harper, M.E., Clapham, J.C., 2001. Biochem. Biophys. Res. Commun. 283, 785–790.
[82] Clapham, J.C., Coulthard, V.H., Moore, G.B., 2001. Biochem. Biophys. Res. Commun. 287, 1058–1062.
[83] Lanni, A., Mancini, F., Sabatino, L., Silvestri, E., Franco, R., De Rosa, G., Goglia, F., Colantuoni, V., 2002. FEBS Lett. 525, 7–12.
[84] Acin, A., Rodriguez, M., Rique, H., Canet, E., Boutin, J.A., Galizzi, J.P., 1999. Biochem. Biophys. Res. Commun. 258, 278–283.
[85] Silva, J.E., Rabelo, R., 1997. Eur. J. Endocrinol. 136, 251–264.
[86] Hunt, M.C., Nousiainen, S.E., Huttunen, M.K., Orii, K.E., Svensson, L.T., Alexson, S.E., 1999. J. Biol. Chem. 274, 34317–34326.
[87] Schrauwen, P., Hinderling, V., Hesselink, M.K., Schaart, G., Kornips, E., Saris, W.H., Westerterp-Plantenga, M., Langhans, W., 2002. Etomoxir-induced increase in UCP3 supports a role of uncoupling protein 3 as a mitochondrial fatty acid anion exporter. FASEB J 16, 1688–1690.
[88] Russell, A.P., Schrauwen, P., Somm, E., Gastaldi, G., Hasselink, M.K.C., Schaart, G., Kornips, E., Lo, S.K., Bufano, D., Giocobino, J.-P., Muzzin, P., Ceccon, M., Angelini, C., Vergani, L., 2003. Decreased fatty acid β-oxidation in riboflavin-responsive multiple acylCoA dehydrogenase deficient patients is associated with an increase in UCP3. J. Clin. Endocrinol. Metab. 88, (in press).

Defects in mitochondrial and peroxisomal fatty acid oxidation

Ronald J.A. Wanders*

Lab Genetic Metabolic Diseases, Departments of Clinical Chemistry and Pediatrics, Academic Medical Centre, Emma Children's Hospital, University of Amsterdam, F0-224 Meibergdreef 9, 1105 AZ Amsterdam, The Netherlands

*Correspondence address: Tel.: +31-20-5665958/5664197; fax: +31-20-6962596
E-mail: r.j.wanders@amc.uva.nl(R.J.A.W.)*

1. Introduction

Fatty acids (FAs) are a major source of energy in human beings and are derived from different sources including diet, de novo synthesis and release from adipose tissue. It is now quite clear that FAs are not just a source of energy but also play a major role in the expression of a variety of genes coding for proteins involved in diverse processes like energy metabolism, cell differentiation, cell growth, among others. FAs exert their effect on gene expression by regulating the activity and abundance of nuclear transcription factors directly. The 1990 discovery of a novel lipid-activated transcription factor, named peroxisome proliferator-activated receptor alpha (PPARα) [1], provided the first evidence that the nucleus did, in fact, contain transcription factors that were dependent on FA ligands, notably polyunsaturated FAs. The importance of PPARα in overall glucose and FA homeostasis became especially clear through studies in PPARα (−/−) mice, which lack the ability to increase rates of FA oxidation (FAO) during periods of food deprivation and develop characteristics of adult-onset diabetes including fatty liver, raised blood triglycerides and hyperglycemia. The PPARα (−/−) mouse model was also instrumental in showing that not all the effects of FAs were PPARα-dependent. Indeed, the effect of dietary polyunsaturated FAs on hepatic lipogenesis by suppressing the expression of genes coding for various enzymes involved in glucose and FA metabolism, including glucokinase, pyruvate kinase, acetyl-CoA carboxylase, FA synthase, stearoyl-CoA desaturase and Δ6- and Δ5-desaturases, was largely independent of PPARα.

Studies in recent years have shown that one of the key players involved in these mechanisms are the SREBPs, which are a family of transcription factors that were first isolated as a result of their property of binding to the sterol regulatory element (SRE) present in many genes. SREBP-2 is a regulator of genes encoding proteins involved in cholesterol metabolism whereas SREBP-1, which exists in two forms, i.e. 1a and 1c of

Advances in Molecular and Cell Biology, Vol. 33, pages 295–317

ISSN: 1569-2558 / DOI: 10.1016/S1569-2558(03)33015-2

which SREBP-1c is the most important, primarily regulates genes involved in lipogenesis. Studies from different groups have shown that FAs have a major effect on SREBP. Indeed, diets rich in 18:2 (n-6) or 20:5 (n-3) and 22:6 (n-3) were found to reduce the protein levels of SREBP-1c in hepatic nuclei drastically. Available evidence indicates that FAs exert their effect on SREBP-1c via a number of different mechanisms. First, unsaturated FAs appear to lower SREBP-1 mRNA by accelerating its degradation [2]. Secondly, unsaturated FAs inhibit the proteolytic processing of SREBP-1a and 1c as shown in HEK-293 cells [3]. A third mechanism was recently discovered by Brown and Goldstein's group [4] who reported that unsaturated FAs inhibit transcription of the SREBP-1c gene by antagonizing ligand-dependent activation of LXR. Transcription of SREBP-1c is under the control of LXR of which there are two forms, α and β, two closely related members of the nuclear hormone receptor superfamily. The LXRs enhance transcription of the SREBP-1c gene by binding to a consensus recognition sequence in the enhancer region of SREBP-1c [5]. LXR activators include the naturally occurring sterols 24(*S*), 25-epoxycholesterol and 22(*R*)-hydroxycholesterol [6]. It appears that unsaturated FAs compete with sterols for the same binding site on LXR thereby blocking activation of LXR and, in turn, of SREBP-1c. The ultimate result is that the expression of genes which are under the control of SREBP-1c is repressed.

The intricate effects of FAs, as described above, make it quite clear that the intracellular levels of FAs must be controlled rigorously. In principle, cells can rapidly dispose off FAs by catalysing their incorporation in phospholipids and especially triacylglycerols but this temporary solution to the problem can only be resolved by the human body via oxidation of FAs to CO_2 and H_2O.

In principle, there are only two ways in which FAs can be degraded to CO_2 and H_2O. The first and quantitatively most important mechanism is FA beta-oxidation, which produces acetyl-CoA in case of unbranched FAs and propionyl-CoA in case of 2-methyl branched-chain FAs after which acetyl-CoA and propionyl-CoA can be degraded to CO_2 and H_2O in the citric acid cycle.

FA alpha-oxidation is quantitatively less important but nevertheless indispensable for human life since some FAs cannot undergo beta-oxidation. This applies especially to 3-methyl branched-chain FAs, which fully rely on alpha-oxidation in order to be oxidized. The importance of the FA alpha-oxidation pathway for human beings is fully clear if it is realized that a defect in the alpha-oxidation machinery is associated with severe neurological abnormalities as observed in Refsum disease patients [7].

In contrast to lower eukaryotes, such as yeasts and plants, in which FA beta-oxidation is confined to peroxisomes, higher eukaryotes including human beings have two beta-oxidation systems, one in mitochondria and one in peroxisomes. Interestingly, the basic mechanism of beta-oxidation in the two organelles is the same with the involvement of four subsequent reactions in which an acyl-CoA ester first undergoes dehydrogenation to an enoyl-CoA ester. Subsequently, H_2O is added across the double bond to produce a 3-hydroxyacyl-CoA ester followed by oxidation to a 3-ketoacyl-CoA ester, which can then undergo thiolytic cleavage to produce acetyl-CoA and an acyl-CoA ester which is now two carbon atoms shorter (Fig. 1).

For a long time, the importance of a second beta-oxidation system, next to the one in mitochondria, remained mysterious but it is now fully clear that the two systems are not

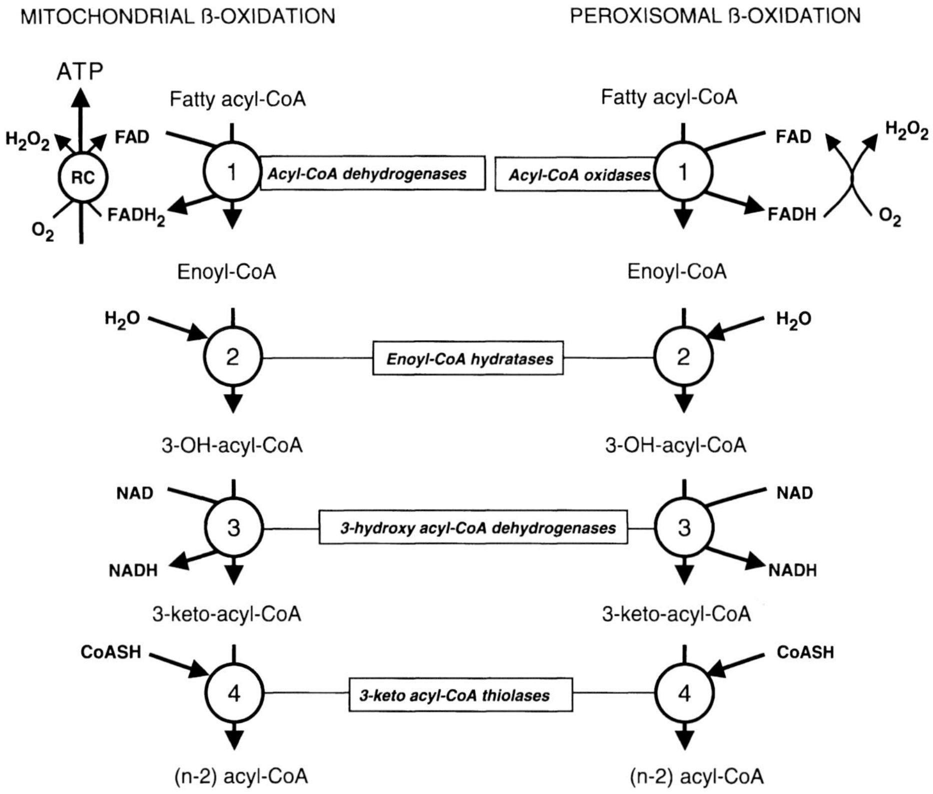

Fig. 1. The mitochondrial and peroxisomal β-oxidation of FAs. The first step in the β-oxidation of FAs in mitochondria is catalysed by acyl-CoA dehydrogenases but in peroxisomes by acyl-CoA oxidases. Both acyl-CoA dehydrogenases and acyl-CoA oxidases are flavoproteins, which use different mechanisms for the reoxidation of enzyme-bound $FADH_2$. In case of mitochondrial acyl-CoA dehydrogenases, enzyme-bound E-$FADH_2$ is reoxidized by electron transfer flavoprotein (ETF) followed by reoxidation of reduced ETF by ETF-dehydrogenase, so that the electrons ultimately enter the respiratory chain at the level of ubiquinone. In case of acyl-CoA oxidases, enzyme-bound $FADH_2$ is directly reoxidized by molecular oxygen to produce H_2O_2. Steps 2, 3 and 4 are catalysed by enoyl-CoA hydratases, 3-hydroxyacyl-CoA dehydrogenases and 3-ketothiolases, which catalyse identical reactions but are different proteins (see text).

only different in terms of the enzymes involved and their regulation, a.o., but also serve different roles in whole cell FAO. Indeed, it is now clearly established that mitochondria are the main site of oxidation of the bulk of FAs including palmitic, oleic, linoleic and linolenic acid with only little contribution of peroxisomes in this respect. However, some FAs cannot be handled by mitochondria and completely rely on the peroxisome for beta-oxidation (Fig. 2). The most important substrates in this respect are the very-long-chain FAs (VLCFAs), such as tetracosanoic (C24:0) and hexacosanoic (C26:0) acids. Other important substrates include 2-methyl FAs like pristanic acid and also di- and trihydroxycholestanoic acids, which are beta-oxidized in peroxisomes to produce the primary bile acids, cholic acid and chenodeoxycholic acid. The different physiological

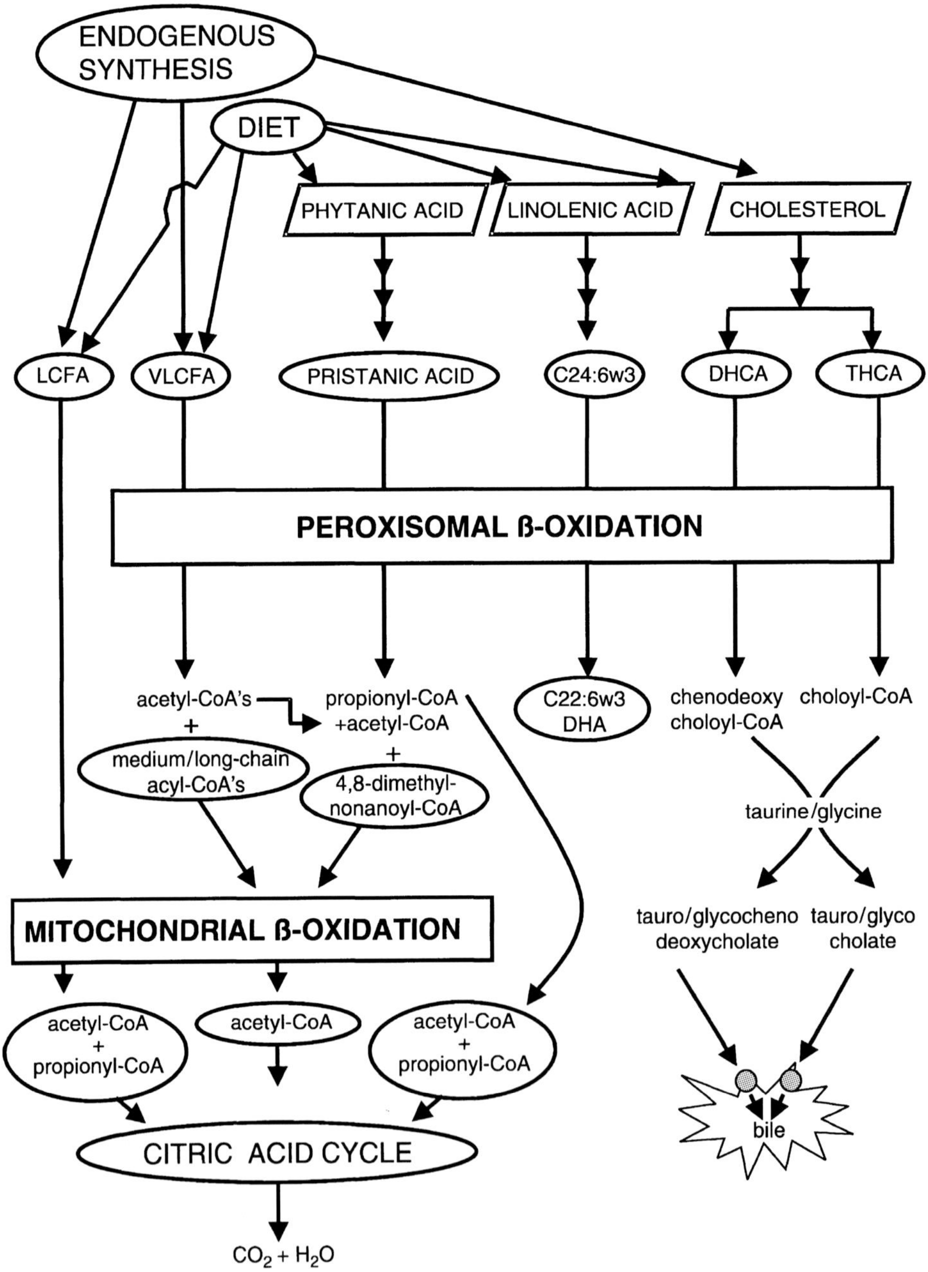

Fig. 2. Simplified scheme depicting the different roles of mitochondria and peroxisomes in cellular FA β-oxidation. Abbreviations: LCFA, long-chain FAs; VLCFA, very-long-chain FAs; DHCA, dihydroxycholestanoic acid; THCA, trihydroxycholestanoic acid.

roles of the mitochondrial and peroxisomal beta-oxidation systems are exemplified by the profound differences in clinical presentation of patients affected by defects in the mitochondrial or peroxisomal system, respectively, as will be discussed later.

In this chapter, the disorders of mitochondrial and peroxisomal FAO will be described. In order to provide the necessary background, a brief update will be given on the enzymology and characteristics of the peroxisomal and mitochondrial FAO systems.

2. Mitochondrial versus peroxisomal FA beta-oxidation

Although mitochondria as well as peroxisomes are capable of FAO there are notable differences as summarized below.

2.1. Incomplete oxidation of FAs in peroxisomes versus complete oxidation of FAs to CO_2 and H_2O in mitochondria

Peroxisomes lack a citric acid cycle, which implies that the acetyl-CoA (and propionyl-CoA) units produced during peroxisomal beta-oxidation cannot be fully oxidized to CO_2 and H_2O in peroxisomes. Nevertheless, the observation is that the peroxisomal acetyl-CoA and propionyl-CoA units can be degraded to CO_2 and H_2O. The mechanism has been resolved and involves the transfer of acetyl and propionyl units from the peroxisomes in the form of carnitine esters. Peroxisomes contain carnitine acetyltransferase (CAT) activity, which allows them to convert acetyl-CoA into acetyl-carnitine that is then transported out of the peroxisomes to mitochondria, which they enter via the carnitine/acylcarnitine translocator (CACT). Once inside mitochondria, the acetylcarnitine is converted back into acetyl-CoA, which is then combusted in the citric acid cycle to CO_2 and H_2O.

2.2. Differential role of carnitine in mitochondrial and peroxisomal FA β-oxidation

The second important difference between peroxisomes and mitochondria is the involvement of carnitine in the two systems. In mitochondria, carnitine plays an essential role in the import of FAs from the cytosolic space into the mitochondrial matrix. In mitochondria, carnitine is involved in the transfer of long-chain FAs across the mitochondrial membrane via the concerted action of carnitine palmitoyl transferase 1 (CPT1), the mitochondrial CACT and carnitine palmitoyl transferase 2 (CPT2). In peroxisomes, however, carnitine plays no role in FA uptake but does play an indispensable role in the transfer of chain-shortened FAs from peroxisomes to mitochondria as described above. With respect to the transfer of FAs across the peroxisomal membrane, there is still a lot of uncertainty. There is increasing evidence, however, which suggests that transport of FAs across the peroxisomal membrane is mediated by one of a variety of different peroxisomal membrane proteins belonging to the family of ATP-binding cassette (ABC) proteins. Peroxisomes contain four of these proteins, which are half ABC-transporters, named ALDP, ALDR, PMP70 and PMP69 [8]. At least for ALDP there are strong indications that this protein either as a homodimer or as a heterodimer transports VLCFAs across the peroxisomal membrane [8].

2.3. Peroxisomal beta-oxidation is energetically less favourable as compared to mitochondrial beta-oxidation

In mitochondria, the first step in the beta-oxidation of FAs is catalysed by a number of acyl-CoA dehydrogenases, all FAD-linked, which donate electrons to the respiratory chain thus generating ATP. In peroxisomes, however, the acyl-CoA oxidases, which are also FAD-linked, donate their electrons directly to molecular oxygen to produce H_2O_2, which is subsequently decomposed to H_2O and O_2. As a consequence, one cycle of beta-oxidation in peroxisomes is at most half as efficient as compared to β-oxidation in mitochondria in terms of ATP production (Fig. 1).

2.4. Regulation of mitochondrial and peroxisomal beta-oxidation

Mitochondrial, but not peroxisomal beta-oxidation is under rapid short-term control via malonyl-CoA, the key regulator of mitochondrial beta-oxidation at the level of CPT1. On the other hand, despite this difference in short-term control, the long-term control of both systems shares some features with PPARα having major effects on both system.

2.5. Involvement of different enzymes produced by distinct genes in peroxisomal and mitochondrial β-oxidation

Although the reactions involved in the peroxisomal and mitochondrial beta-oxidation systems are basically identical, these reactions are catalysed by different enzymes, which are in general encoded by distinct genes with a few exceptions in which a single gene codes for a protein directed to both peroxisomes and mitochondria. Examples are: carnitine-acetyltransferase [9], Δ3,5-, Δ2,4-dienoyl-CoA reductase [10] and 2-methylacyl-CoA racemase (AMACR) [11].

2.6. Mitochondria and peroxisomes have different substrate specificities

From a physiological point of view, the most important difference between the mitochondrial and peroxisomal beta-oxidation systems is that the two systems have different substrates specificities. The realization that the peroxisomal system handles a distinct set of substrates has largely come from studies on a rare genetic disease, called the cerebro-hepato-renal syndrome of Zellweger, in short Zellweger syndrome (ZS). In Zellweger patients, peroxisome biogenesis is defective, due to a genetic defect in one of the many genes involved in peroxisome biogenesis, which leads to the complete absence of peroxisomes. Careful studies in plasma and cells of these patients have revealed a large number of abnormalities which has allowed us to conclude that peroxisomes play an indispensable role in the beta-oxidation of the following FAs (see Fig. 2).

2.6.1. Very-long-chain fatty acids

Peroxisomes play an indispensable role in the oxidation of VLCFAs such as tetracosanoic (C24:0) and hexacosanoic (C26:0) acids, which are derived from dietary sources but can also be synthesized de novo from endogenous shorter-chain FAs.

2.6.2. Pristanic acid (2,6,10,14-tetramethylpentadecanoic acid)

Peroxisomes also catalyse the beta-oxidation of certain 2-methyl branched-chain FAs. The prototype of this group of FAs is pristanic acid, which is derived from dietary sources but is also formed endogenously from phytanic acid via alpha-oxidation (see later). Phytanic acid itself cannot be made endogenously and is of dietary origin only.

2.6.3. Di- and trihydroxycholestanoic acid

These cholestanoic acids are intermediates in the production of the primary bile acids, cholic acid and chenodeoxycholic acid in the liver. After activation to their CoA-esters they undergo beta-oxidation in the peroxisome to produce chenodeoxycholoyl-CoA and choloyl-CoA, respectively, which are then converted into the corresponding tauro- and/or glyco-conjugates with tauro- or glycochenodeoxycholate and tauro- or glycocholate as products (see Fig. 2).

2.6.4. Polyunsaturated FAs

Peroxisomes play a major role in the beta-oxidation of a variety of mono- and polyunsaturated FAs. One of the polyunsaturated FAs oxidized in peroxisomes is C24:6 (n-3), which is derived from exogenous sources but also from linoleic acid and is the direct precursor of docosahexaenoic acid C22:6 (n-3). Formation of docosahexaenoic acid from C24:6 is completely dependent on the proper functioning of the peroxisomal β-oxidation system (Fig. 2).

3. Enzymology of the mitochondrial and peroxisomal beta-oxidation systems

3.1. Mitochondrial β-oxidation enzymes

Studies over the years have clearly shown that the four reactions of the mitochondrial β-oxidation spiral are not catalysed by single enzymes covering the whole spectrum of different substrates. Instead, each reaction is catalysed by multiple enzymes, each having a certain chain-length specificity (see Ref. [12] for detailed information). Full oxidation of a long-chain FA like palmitate to its 8 acetyl-CoA units requires the active participation of:

- Three acyl-CoA dehydrogenases [very-long-chain (VLCAD), medium-chain (MCAD) and short-chain (SCAD) acyl-CoA dehydrogenase];
- Two enoyl-CoA hydratases [short-chain (SCEH) and long-chain (LCEH) enoyl-CoA hydratase]. It should be noted that LCEH is not a mono-functional enzyme but part of the mitochondrial tri-functional protein (MTP), an octamer of 4α- and 4β-subunits with LCEH, long-chain 3-hydroxyacyl-CoA dehydrogenase (LCHAD) and long-chain 3-ketothiolase (LCKT) activity;

- Two 3-hydroxyacyl-CoA dehydrogenases, including short-chain 3-hydroxyacyl-CoA dehydrogenase (SCHAD) and LCHAD, again part of MTP;
- Two 3-kethothiolases, including medium-chain 3-ketothiolase (MCKT) and long-chain thiolase, again part of MTP.

The fact that the mitochondrial enzymes involved in long-chain FA β-oxidation (VLCAD and MTP) are membrane-bound whereas the other enzymes are soluble enzymes localized in the mitochondrial matrix suggests a different spatial organization as shown schematically in Fig. 3.

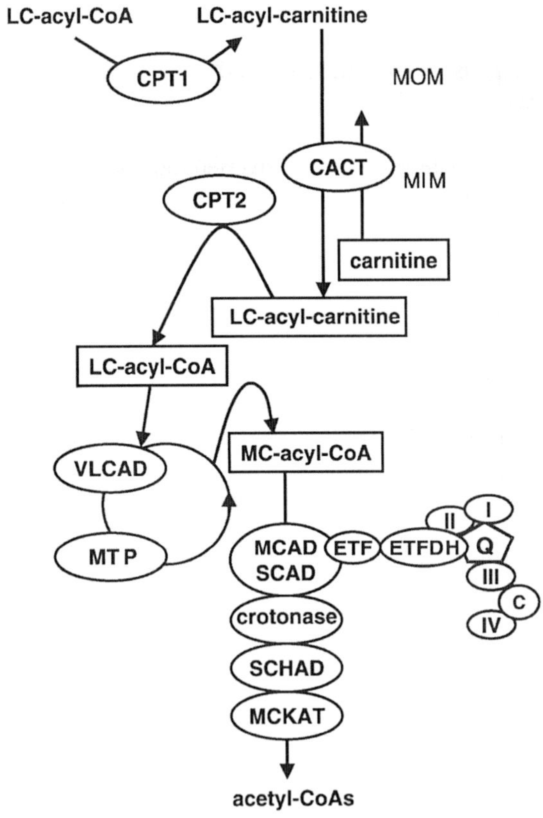

Fig. 3. Schematic representation of the enzymatic organisation of the mitochondrial FAO machinery with VLCAD and MTP as membrane-bound enzymes and MCAD, SCAD, SCEH (= crotonase), SCHAD and MCKT as soluble enzymes. Additional information: ETF, electron transfer flavoprotein; ETFDH, ETF dehydrogenase; Q, coenzyme Q; CI, CII, CIII and CIV, respiratory chain complexes I, II, III and IV; C, cytochrome c.

3.2. Peroxisomal β-oxidation enzymes

Just like mitochondria, peroxisomes contain multiple enzyme proteins to catalyse the β-oxidative chain-shortening of the various acyl-CoA esters including two acyl-CoA oxidases, two so-called bifunctional proteins with enoyl-CoA hydratase and 3-hydroxyacyl-CoA dehydrogenase activities and two 3-ketothiolases (see Ref. [13] for detailed discussion and references).

Acyl-CoA oxidases. Human peroxisomes contain two acyl-CoA oxidases with different substrate-specificities. The first oxidase (SCOX/ACOX1) is the human equivalent of the clofibrate-inducible rat enzyme identified by Hashimoto and co-workers and handles straight-chain acyl-CoAs. The enzyme is not reactive with 2-methyl branched-chain acyl-CoAs like pristanoyl-CoA and trihydroxycholestanoyl-CoA, which are handled by the second oxidase (ACOX2), also called branched-chain acyl-CoA oxidase (BCOX).

Bifunctional proteins. Human peroxisomes contain two bifunctional proteins. Both enzymes catalyse the conversion of enoyl-CoA esters into the respective 3-keto esters but they do this in different ways via an L-3-hydroxy and D-3-hydroxy intermediate, respectively. Different names have been given to these proteins including L-BP/D-BP, L-PBE/D-PBE, MFEI/MFEII and MFP1/MFP2. Studies in humans [14–17] and mutant mice [18] have shown that D-BP is the enzyme involved in the oxidation of C26:0, pristanic acid, as well as di-and trihydroxycholestanoic acid. It remains to be established what the function of L-BP is. Clofibrate feeding to rats leads to a marked induction of L-BP but not D-BP.

3-Ketothiolases. Human peroxisomes contain two thiolases, one of which being the human counterpart of the clofibrate-inducible thiolase identified by Hashimoto and co-workers [19]. The second thiolase is the 58 kDa sterol-carrier-protein/3-ketothiolase identified by Seedorf et al. [20], which goes by the name SCPx, or its alternative pTH2 [8]. SCPx/pTH2 plays an indispensable role in the β-oxidation of 2-methyl branched-chain FAs like pristanic acid and di- and trihydroxycholestanoic acid, as concluded from several studies including studies in SCPx (−/−) mice [21] (see Fig. 4).

4. Mitochondrial FAO disorders

The clinical manifestations of the various mitochondrial FAO disorders are diverse and depend upon the nature of the enzyme block. Patients usually show no external stigmata such as cranial facial dysmorphism. The main organs directly affected by defects in mitochondrial beta-oxidation are the liver (fatty changes, micro-vesicular steatosis); heart (acute heart block, progressive cardiomyopathy, arrhythmias and tachycardia); skeletal muscle (rhabdomyolysis); brain (energy deficit); and the kidneys (renal failure, renal tubular acidosis). Depending upon the primary organ involved, the clinical manifestations of the FAO defects can range from a pure hepatic or cardiac presentation to one dominated by skeletal muscle or kidney involvement. In many cases, these clinical manifestations occur in different combinations and these may vary with the age of the affected individual, making diagnosis of mitochondrial FAO defects difficult in clinical practice. Table 1 lists the mitochondrial FAO disorders known at present.

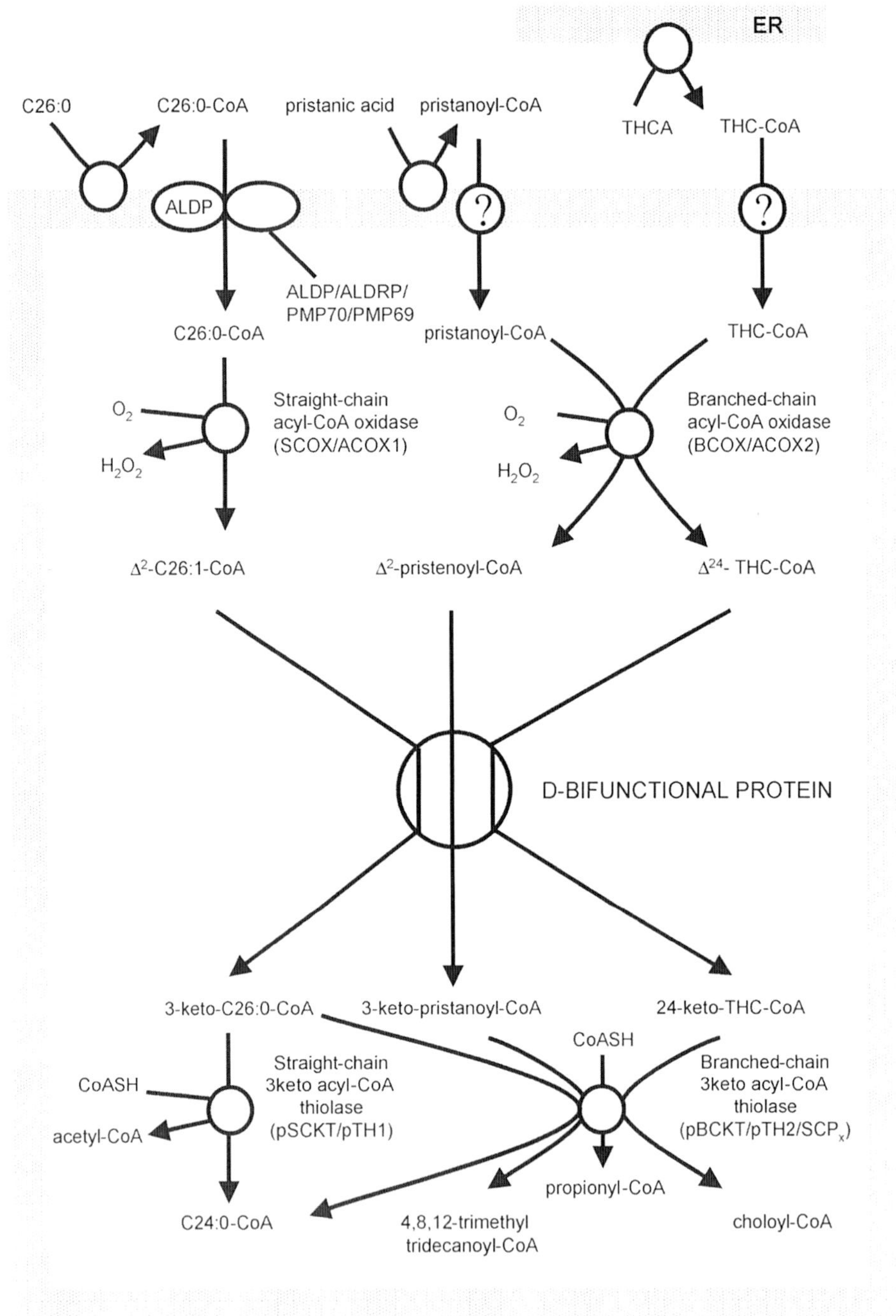
ER
C26:0
C26:0-CoA
pristanic acid
pristanoyl-CoA
THCA
THC-CoA
ALDP
?
?
ALDP/ALDRP/
PMP70/PMP69
C26:0-CoA
pristanoyl-CoA
THC-CoA
O2
H2O2
Straight-chain
acyl-CoA oxidase
(SCOX/ACOX1)
O2
H2O2
Branched-chain
acyl-CoA oxidase
(BCOX/ACOX2)
Δ2-C26:1-CoA
Δ2-pristenoyl-CoA
Δ24- THC-CoA
D-BIFUNCTIONAL PROTEIN
3-keto-C26:0-CoA
3-keto-pristanoyl-CoA
24-keto-THC-CoA
CoASH
acetyl-CoA
Straight-chain
3keto acyl-CoA
thiolase
(pSCKT/pTH1)
CoASH
Branched-chain
3keto acyl-CoA
thiolase
(pBCKT/pTH2/SCPx)
propionyl-CoA
C24:0-CoA
4,8,12-trimethyl
tridecanoyl-CoA
choloyl-CoA

Table 1
Disorders of mitochondrial and peroxisomal β-oxidation

Mitochondrial FAO disorders
Primary carnitine deficiency (Na^+/carnitine transport defect)
Carnitine palmitoyl transferase 1 deficiency
Mitochondrial carnitine/acylcarnitine translocase deficiency
Carnitine palmitoyl transferase 2 deficiency
Very-long-chain acyl-CoA dehydrogenase deficiency
Medium-chain acyl-CoA dehydrogenase deficiency
Short-chain acyl-CoA dehydrogenase deficiency
Long-chain 3-hydroxyacyl-CoA dehydrogenase deficiency
Mitochondrial tri-functional protein deficiency
Medium-chain 3-ketothiolase deficiency
Peroxisomal FAO disorders
X-linked adrenoleukodystrophy
Acyl-CoA oxidase 1 deficiency
D-bifunctional protein deficiency
2-Methylacyl-CoA racemase deficiency

4.1. Primary carnitine (OCTN2) deficiency

Primary carnitine deficiency is a potentially lethal but eminently treatable disorder. Patients show a progressive infantile onset cardiomyopathy and skeletal muscle weakness, recurrent hypoketotic hypoglycaemic encephalopathy and failure to thrive. Plasma and tissue concentrations of carnitine are low (less than 5% of control) with lipid storage in muscle and liver and severe tubular loss of carnitine. The primary defect in this disorder is at the level of the plasma membrane carnitine transporter, which is a sodium/carnitine co-transporter. The co-transport of sodium and carnitine, together with the large sodium-gradient across the plasmalemmal membrane of cells by virtue of the Na^+/K^+-ATPase, allows the uphill

Fig. 4. Enzymology of the peroxisomal FA β-oxidation system. Human peroxisomes contain two acyl-CoA oxidases, *one* specific for straight-chain FAs like C26:0, therefore called straight-chain acyl-CoA oxidase (SCOX or ACOX1) and a *second* one, catalysing the dehydrogenation of 2-methyl branched-chain FAs like pristanoyl-CoA and di- and trihydroxycholestanoyl-CoA. The latter enzyme is called BCOX or ACOX2. The enoyl-CoA esters of C26:0, pristanic acid and DHCA and THCA are all handled by a single bifunctional enzyme harbouring both enoyl-CoA hydratase and 3-hydroxyacyl-CoA dehydrogenase activity. The newly identified bifunctional protein, which forms and dehydrogenates D-3-hydroxyacyl-CoA esters rather than L-3-hydroxyacyl-CoAs, is a single enzyme involved in the oxidation of C26:0, pristanic acid and DHCA and THCA. Finally, human peroxisomes contain two peroxisomal thiolases, one specific for the 3-ketoacyl-CoAs of straight-chain FAs [straight-chain ketothiolase (SCKT)] and a second thiolase, which is able to handle the 3-ketoacyl-CoAs of both straight-chain and 2-methyl branched-chain FAs. The latter enzyme is referred to in literature as SCPx, alternative names are branched-chain 3-ketothiolase (BCKT) on peroxisomal thiolase 2 (pTH2). Current evidence holds that the CoA-esters of C26:0, pristanic acid and DHCA and THCA are transported across the peroxisomal membrane. In case of C26-CoA, there is increasing evidence that this is mediated by ALDP either as a homodimer or as heterodimer with either ALDRP, PMP70 or PMP69 as partner (see Ref. [8] for details).

transport of carnitine from the blood compartment into the interior of the cell. This explains why intracellular carnitine levels are much higher in cells than in blood. Early diagnosis, followed by treatment with high doses of oral carnitine, is critical giving the otherwise progressive and lethal nature of this disorder. This is exemplified by the striking family histories of affected siblings, including sudden infant death syndrome (SIDS). Carnitine supplementation is life saving and reverses myopathy and cardiomyopathy.

4.2. CPT1 deficiency

Although two different isoforms of CPT1 have been described (a liver-type CPT1 and a muscle-type CPT1), only the liver type has been found deficient so far in patients. Since the first report in 1981 [22], some 20 patients with liver-type CPT1 deficiency have been described. The clinical presentation is rather homogeneous with onset in infancy and recurrent episodes of hypoketotic hypoglycaemia and coma, triggered by fasting or intercurrent illness. Neither myopathy nor cardiomyopathy has been observed in any of the patients in line with the notion that CPT1 activity is fully normal in these tissues.

4.3. CACT deficiency

CACT deficiency was first described by Stanley et al. in 1992 [23] in a male infant patient who presented at 36 h of age with seizures, severe apnoea and bradycardia. The patient had recurrent premature ventricular contractions, cardiac arrhythmias and hypertension, but gradually improved on a formula low in long-chain fats. Despite treatment, however, the child developed progressive muscle weakness and died of aspiration, pneumonia and respiratory failure. Approximately 20 cases of CACT deficiency have been described in literature. Although a few patients show a mild phenotype, most (>90%) suffer from the severe form with life-threatening episodes in the newborn period. These are characterised by distress with usually hypoketotic hypoglycaemia, severe cardiac abnormalities including heart block, liver dysfunction and muscle involvement. Most patients die early in life despite treatment.

4.4. CPT2 deficiency

Widely differing clinical presentations of CPT2 deficiency have been described in the literature ranging from a mild adult form to a severe and often fatal neonatal form. The most frequent type of CPT2 deficiency is the muscular form, first described in 1973 [24]. Patients with this defect generally present in adolescence or adulthood with recurrent episodes of muscle pain, rhabdomyolysis and paroxysmal myoglobinuria, triggered by exercise, fasting, infections or cold exposure. It is probably the most common cause of hereditary myoglobinuria. Later, more severe forms of CPT2 deficiency have been described with primarily hepatic and hepatocardiomuscular presentations. The neonatal form is rapidly lethal as exemplified by the patient of Hug et al. [25] who displayed severe hepatopathy, encephalopathy, cardiomegaly and death at 5 days of age, due to arrhythmia and cardiac failure with multi-organ failure. An intermediate type of CPT2 deficiency has been described with late-infantile onset with fasting hypoketotic hypoglycaemia, liver dysfunction, cardiomyopathy and mild signs of muscle involvement.

4.5. VLCAD deficiency

VLCAD deficiency was first described in 1985 by Hale et al. [26] who described three children from unrelated families presenting in early childhood with hypoglycaemia and cardiorespiratory arrest associated with fasting. Since this initial report, many more cases of VLCAD deficiency have been described in literature. These studies have shown that the clinical presentation of VLCAD deficiency includes three main phenotypes: 1) early onset, hypertrophic cardiomyopathy with a high rate of morbidity or mortality; 2) a milder form with episodic hypoketotic hypoglycaemia reminiscent of MCAD deficiency; and 3) a form resembling muscular CPT2 deficiency with stress-induced rhabdomyolysis.

4.6. MCAD deficiency

MCAD deficiency is the most frequent FAO disorder. The typical presentation is between 6 months and 2 years of age and consists of episodes of acute illness, usually after a fasting period of 12 h or more, often associated with inter-current infectious disease. Patients are usually, but not invariably, hypoglycaemic during these episodes. Hypoketonuria is also common, but exceptions have been reported. A large retrospective analysis of 120 MCAD deficient patients by Lafolla et al. [27] showed that the age of onset ranged from 2 days to 6.5 years. In 23 children (19%), the diagnosis of MCAD deficiency was made after death, but no child identified biochemically as MCAD deficient died after the correct diagnosis was made, which stresses the importance of prompt and early diagnosis. A list of the initial signs and symptoms in patients with clinical illness includes lethargy (84%), emesis (66%), encephalopathy (49%), respiratory arrest (48%), hepatomegaly (44%), seizures (43%), apnoea (37%), cardiac arrest (36%) and sudden death (18%). Only 14 children were identified as MCAD deficient at the onset of clinical illness. Of the remaining 106 MCAD patients, 86 (= 72%) had a variety of diagnosis, including Rey syndrome, idiopathic hypoglycaemia and SIDS.

4.7. SCAD deficiency

One of the most puzzling disorders of mitochondrial beta-oxidation is SCAD deficiency. Amendt et al. [28] reported the first case with well-established SCAD deficiency in 1987 after which 8 additional patients have been described (see Ref. [12] for details). Most of these patients presented in the neonatal period with a variable phenotype that included metabolic acidosis, failure to thrive, developmental delay, seizures and myopathy. However, the clinical spectrum is very wide as exemplified by the two cases described by Ribes et al. [29] with mild (case 1) or absent (case 2) clinical signs. Recently, Corydon et al. [30] described 10 additional patients with established SCAD deficiency. In addition to these cases of proven SCAD deficiency, many patients have been described in literature with ethylmalonic-aciduria (EMA) of unknown etiology. In a minority of patients, EMA is the secondary consequence of a defect in the respiratory chain. Gregersen et al. [31] identified two so-called susceptibility alleles (625G → A and 511C → T), which were strongly over-represented (69 versus 14% in the general population) among a clinically heterogeneous group of 133 patients with EMA. Expression studies have shown that the 625G → A and 511C → T variations, leading to amino acid substitutions G185S and R147W, respectively, do not affect the activity of SCAD per se.

Recent studies by Corydon et al. [30], in a group of 10 patients with biochemically and enzymatically proven SCAD deficiency, have shown an even greater complexity. Only a single patient turned out to carry two pathogenic mutations. The remaining 9 patients were doubly heterozygous for the pathogenic mutation in combination with the 625G → A variation, homozygous for one of the two variations 625G → A or 511C → T or doubly heterozygous for the 625G → A and 511C → T variations. These data imply that homozygosity for the 625G → A or 511C → T variations, or both, may be associated with a full deficiency of SCAD activity, as demonstrated by biochemical studies in fibroblasts. This is all the more relevant, because these variations are found in the homozygous or doubly heterozygous form in 14% of the general population, suggesting that the actual incidence of clinically expressed SCAD deficiency could well be much higher than currently recognized.

4.8. LCHAD deficiency and MTP deficiency

In MTP deficiency and LCAD deficiency, the defective enzyme protein is the same. MTP deficiency is characterized by the complete absence of the MTP protein, which explains the combined deficiency of all three enzyme activities catalysed by MTP, i.e. LCEH, LCHAD and long-chain thiolase. In LCHAD deficiency, the protein itself is normally present, but the activity of one of the component enzyme activities, i.e. LCHAD, is defective. Remarkably, LCHAD deficiency seems much more frequent than MTP deficiency. More than 60 patients with LCHAD deficiency have been described versus only 10 patients with MTP deficiency.

LCHAD deficiency was first published in 1989 and since then many patients have been described [32]. The clinical variability among LCHAD deficient patients is remarkable: on the one hand patients present immediately after birth with a rapidly fatal cardiomyopathy whereas on the other hand patients have been described with a much more indolent myopathic presentation. Such patients may present much later as adults with exercise-induced muscle pain and rhabdomyolysis, mimicking adult type CPT2 deficiency as described by Schaefer et al. [33].

After its first description in 1992 by two different groups [34,35] only very few additional patients with MTP deficiency have been described. The clinical presentation of MTP deficiency closely resembles that of LCHAD deficiency although in general the clinical picture is more severe in MTP-deficient patients with early death in most of the cases reported. On the other hand, a variant form of MTP deficiency was described recently in a 23 year old Japanese man with recurrent myoglobinuria in adolescence, showing again the clinical variability also within MTP deficiency [36].

4.9. MCKT deficiency

To date, only a single case of MCKT deficiency has been described, which remains ill-defined [37].

4.10. Short-chain 3-hydroxyacyl-CoA dehydrogenase deficiency

Although several cases of SCHAD deficiency have been described in literature, it remains to be established whether these are true cases of SCHAD deficiency, especially since the defect appeared to be tissue specific in most of these patients, which contrasts with the fact that SCHAD is expressed in all tissues. Recently, the first true case of SCHAD deficiency resulting from inactivating mutations in the SCHAD gene has been described [38]. The patient presented at 4 months with a hypoglycaemic convulsion. Further episodes of hyperketotic hyperglycaemia were associated with inappropriately elevated plasma insulin levels. However, unlike other children with hyperinsulinism, this patient had persistently elevated 3-hydroxybutyrylcarnitine concentrations under fed and fasted conditions that prompted a full study of the SCHAD enzyme, which was found to be deficient. Furthermore, immunoblot analysis revealed the complete absence of the SCHAD protein. Finally, molecular studies revealed a homozygous mutation completely disrupting SCHAD activity. It is very remarkable that SCHAD deficiency appears to be associated with hyperinsulinism. Recently, another patient has been identified with the same combination (Molven et al., in preparation).

5. Laboratory diagnosis of mitochondrial FAO disorders

It is difficult to generate general guidelines for the clinical recognition and laboratory diagnosis of FAO disorders because these conditions are extremely heterogeneous in clinical as well as laboratory terms. This is not only true for the different FAO disorders as a group but also within one particular FAO disorder. In some cases, this is explained by the nature of the mutations found as exemplified by CPT2 deficiency with a rapidly fatal hepatocardiomuscular form at one end of the spectrum and the classical muscular form presenting in adulthood with muscle pain, rhabdomyolysis and paroxysmal myoglobinuria at the other end of the spectrum. In other cases, however, the same underlying molecular defect may give rise to wildly variable clinical presentations as for instance in MCAD deficiency and SCAD deficiency.

An FAO disorder may present with different clinical signs and symptoms depending on the organ primarily expressing the defect. If the heart, for instance, is the principal organ affected, the patient will present with predominantly cardiac symptoms including tachycardia, cardiac arrhythmias, hypertrophic cardiomyopathy and sudden heart block. If skeletal muscle is the primary target organ, patients may show a phenotype dominated by progressive muscle weakness, recurrent myalgia, muscle pain, rhabdomyolysis and paroxysmal myoglobinuria. On the other hand, if the liver is the organ principally affected, then hypoketotic hypoglycaemia will result with all its attending consequences. This implies that the FAO disorder should not only be considered in any patient presenting with a progressive or sudden onset picture of lethargy, drowsiness or coma but also in patients with cardiac and muscle symptomatology.

In all cases in which an FAO disorder is suspected, initial laboratory tests should include analysis of glucose, free FAs and ketone bodies but at the same time acylcarnitine analysis should be done, preferably using tandem-mass spectrometry (MS). The introduction of tandem-MS in the field of inborn errors in general, and FAO disorders in particular, has proven to be of extreme importance in recent years. Acylcarnitines are usually abnormal in *all* types of FAO disorders with a few exceptions even under stable conditions. In some cases, however, the acylcarnitine profile may be normal in patients under stable conditions stressing the point that an FAO disorder is not ruled out definitively by a normal acylcarnitine profile. In such cases, we usually perform a loading test using sunflower oil to stress the system, followed by analysis of glucose, free FAs and ketone bodies (see Ref. [12] for review). If abnormalities are found in the acylcarnitine profile then the enzyme defect should be identified as soon as possible. In the last few years, we have established that virtually all enzyme deficiencies can be identified in lymphocytes making rapid and correct diagnosis of the various FAO disorders feasible now.

6. Enzymatic and molecular basis of mitochondrial FAO disorders

As described above, a particular enzyme deficiency is usually, but not always, associated with a specific set of plasma acylcarnitine abnormalities. If an acylcarnitine profile is indeed suggestive for a particular enzyme defect, the enzyme can be measured in lymphocytes and/or fibroblasts.

Measurement of these enzymes is surely not trivial since they often require the availability of special chemicals including substrates, which are not commercially available. Our laboratory has specialised in the diagnosis of mitochondrial and peroxisomal FAO disorders, which requires the in-house synthesis of these compounds and the generation of antibodies to do immunoblot and immunofluorescence studies.

In some cases, identification of the defect is not so straightforward. In such cases, we perform a full study in fibroblasts, which involves loading of fibroblasts with specific substrates followed by acylcarnitine profiling. Also in cases in which the enzyme has been established in lymphocytes, we advocate a full study in fibroblasts, especially if prenatal diagnosis is requested in the future. Prenatal diagnosis of virtually all mitochondrial FAO disorders can be done by measurement of the enzyme in direct chorionic villous biopsy material.

Furthermore, molecular studies should be performed to identify the nature of the mutations involved. This can be done for all the mitochondrial FAO disorders known today. In most cases, the mutations are private mutations and only in rare cases more frequent mutations are found. In this respect three examples should be mentioned. First, the muscular form of CPT2 deficiency as first described by DiMauro and DiMauro in 1993 is associated with one particular mutation [39]. The mutation involved (S113L) is associated with a high residual activity of CPT2 in line with the observations made in patients with the muscular form of CPT2 deficiency. The second example is MCAD deficiency. Mutation analysis has shown that more than 90% of the mutant alleles in affected patients carry a 985A $\rightarrow$ G mutation.

The carrier frequency of this mutant allele in the north-western European population has been estimated to be approximately 1–in–60. The third example is LCHAD deficiency in which a common mutation has been identified (1528G → C), which changes a glutamate residue into glutamine at position 474 of the α-subunit of MTP. This mutation fully inactivates the 3-hydroxy-acyl-CoA dehydrogenase component of MTP without affecting the hydratase. In our series of more than 80 patients, the frequency of the 1528G → C allele is 87% [40]. Only very few additional mutations have been reported in literature.

7. Peroxisomal FAO disorders

To date, four defined disorders of peroxisomal FAO oxidation have been identified: 1. X-linked adrenoleukodystrophy (XALD); 2. acyl-CoA oxidase 1 (ACOX1) deficiency; 3. D-bifunctional protein (D-BP) deficiency; and 4. 2-methyacyl-CoA racemase (AMACR) deficiency.

1. *X-linked adrenoleukodystrophy*. XALD is a devastating disease showing marked clinical variability even within families. At least six phenotypic variants can be distinguished [41]. The classification of the different phenotypes is somewhat arbitrary and is based on the age of onset and the organs principally involved. The most frequent phenotypes, accounting for approximately 80% of all cases, are childhood cerebral adrenoleukodystrophy (CCALD) and adrenomyeloneuropathy (AMN). CCALD is characterized by rapidly progressive cerebral demyelination. The age of onset ranges from 3 to 10 years. Frequent early neurologic symptoms are behavioural disturbances, a decline in school performance, deterioration of vision and impaired auditory discrimination. The course is relentlessly progressive, and seizures, spastic paraplegia and dementia develop within months. Most patients die within 2–3 years after the onset of neurological symptoms.

AMN presents later in life, with neurologic symptoms usually starting in the third or fourth decade. Neurologic deficits are primarily caused by myelopathy, and to a lesser extent, by neuropathy. Patients gradually develop a spastic paraparesis, often in combination with a disturbed vibration sense in the legs and sphincter dysfunction.

The biochemical hallmark of all forms of XALD is the accumulation of VLCFAs, notably C26:0, in plasma owing to a defect in the peroxisomal β-oxidation of VLCFAs. The defective gene was discovered in 1993 [42] and codes for a peroxisomal membrane protein, which belongs to the ABC family of transporters. This protein, called ALDP, is a half ABC-transporter, which is involved in the transport of VLCFAs across the peroxisomal membrane although the details are not yet resolved [8].

2. *Acyl-CoA oxidase 1 deficiency*. ACOX1 deficiency has been reported in a few patients in literature only. All patients reported showed neurological abnormalities including early-onset seizures, hypotonia, hearing impairment and visual failure resulting from retinopathy. Most patients die early in life (see Ref. [43]).

3. *D-Bifunctional protein deficiency*. D-BP deficiency is the second most frequent disorder of peroxisomal β-oxidation after XALD. The clinical presentation of D-BP patients is usually severe and resembles ZS in many respects, with hypotonia, craniofacial dysmorphia, neonatal seizures, hepatomegaly, developmental delay and

usually early death [43]. A particular feature is that patients with D-BP deficiency often show disordered neuronal migration as observed in ZS. The central role of D-BP in the peroxisomal β-oxidation of both straight-chain and 2-methyl branched-chain FAs explains why there is an accumulation of VLCFAs, pristanic acid and di- and trihydroxycholestanoic acid in most of these patients [14–16,43]. Until recently, peroxisomal thiolase 1 (pTH1) deficiency was supposed to be a separate disorder of peroxisomal FAO. Recently, however, we discovered that the true defect in this patient is at the level of D-BP [44].

4. *2-Methylacyl-CoA racemase deficiency.* AMACR deficiency is a newly identified disorder of peroxisomal β-oxidation in which only the peroxisomal oxidation of the 2-methyl branched-chain FAs, pristanic acid and di- and trihydroxycholestanoic acid, is impaired. Following the description of three patients (see patients in our first report), we have identified a few additional patients, all of whom show a late-onset neuropathy, which resembles classical Refsum disease in some respects.

8. Laboratory diagnosis of the peroxisomal FAO disorders and their enzymatic and molecular basis

As shown in Table 2, plasma VLCFAs are elevated in all peroxisomal FAO disorders except in ACAMR deficiency in which VLCFAs are normal but pristanic acid and di- and trihydroxycholestanoic acid are elevated. VLCFA analysis in plasma is highly reliable with no false negatives although in some cases abnormalities may be very mild. This is especially true for XALD. In case plasma VLCFAs have been found to be abnormal, a full study should be done in fibroblasts to pinpoint the enzymatic defect. If the

Table 2
Biochemical abnormalities in the disorders of peroxisomal β-oxidation

	Peroxisomal disorders			
	DBPD	SCOXD	AMACRD	XALD
Plasma				
VLCFAs	↑	↑	N	↑
Pristanic acid	N-↑ [a]	N	N-↑ [a]	N
Phytanic acid	N-↑ [a]	N	N-↑ [a]	N
Di- and trihydroxycholestanoic acid	↑	N	↑	N
Fibroblasts				
C26:0 β-oxidation	↓	↓	N	↓
Pristanic acid β-oxidation	↓	N	↓	N
Phytanic acid α-oxidation	N	N	N	N

DBPD, D-bifunctional protein deficiency; SCOXD, straight-chain acyl-CoA oxidase deficiency; AMACRD, 2-methyl acyl-CoA racemase deficiency; XALD, X-linked adrenoleukodystrophy.

[a] Since phytanic acid is derived from dietary sources only, the levels of phytanic acid and its α-oxidation product, i.e. pristanic acid, may differ widely and may even be normal in DBP- and AMACR-deficient patients when such patients have not been exposed to phytanic acid containing foodstuffs.

clinical signs and symptoms are suggestive of XALD, a full study in fibroblasts may be omitted and molecular analysis can be initiated right away.

In all other cases, a full study in fibroblasts is warranted, especially since the clinical signs and symptoms of acyl-CoA oxidase deficiency and D-BP deficiency strongly overlap with those of the disorders of peroxisome biogenesis, which include ZS, neonatal adrenoleukodystrophy (NALD) and infantile Refsum disease (IRD). This is immediately clear if is realized that the first cases of acyl-CoA oxidase deficiency were described as pseudo-NALD and the "in retrospect" first case of D-BP deficiency as pseudo-ZS [44].

In our centre, a full study in fibroblasts starts by performing immunofluorescence microscopy analysis using antibodies raised against catalase, a peroxisomal matrix enzyme, to resolve whether we are dealing with a disorder of peroxisome biogenesis or an isolated defect in isolated FAO disorder. In case of a peroxisome biogenesis defect, other parameters of peroxisomal functions are measured followed by complementation analysis to identify the defective gene involved, followed by mutation analysis. On the other hand when peroxisomes have been found to be present, oxidation of different FAs is measured including C26:0 and pristanic acid, followed by measurement of the different individual enzymes including acyl-CoA oxidase and D-BP. Once the enzyme defect has been established, molecular analysis can be performed (see Ref. [43] for review).

9. Peroxisomal FA alpha-oxidation

As described above, the presence of a 3-methyl group in FAs prohibits its β-oxidation. Oxidation of 3-methyl branched-chain FAs requires a process called α-oxidation in which the terminal carboxyl group is removed, thus producing a 2-methyl branched-chain FA plus formyl-CoA, the activated form of formic acid (HCOOH). Phytanic acid (3,7,11,15-tetramethylhexadecaenoic acid) is such a 3-methyl FA which, after α-oxidation, generates pristanic acid (2,6,10,14-tetramethylpentadecanoic acid). Until recently, the mechanism of FA alpha-oxidation was mysterious but studies in recent years have resolved the issue and it is now clear that degradation of phytanic acid involves a fourth step pathway including (1) activation of phytanic acid to phytanoyl-CoA; (2) hydroxylation of phytanoyl-CoA to 2-hydroxyphytanoyl-CoA; (3) cleavage of 2-hydroxyphytanoyl-CoA to pristanal and formyl-CoA; and (4) dehydrogenation of pristanal to pristanic acid (see Fig. 5). Peroxisomes play an indispensable role in FA α-oxidation since at least the first two steps of the pathway are strictly peroxisomal (see Ref. [7] for review).

10. Disorders of peroxisomal FA alpha-oxidation

Refsum disease is the single enzyme deficiency in the phytanic acid alpha-oxidation pathway. Refsum disease was first delineated as a distinct disease entity on a clinical basis by Siegfeld Refsum in the 1940s under the name heredopathia atactica polyneuritiformis. Cardinal manifestations of the disease include retinitis pigmentosa, cerebellar ataxia, chronic polyneuropathy and an elevated protein level in CSF with normal cell count. Less constant features include sensory neural hearing loss, anosmia, ichtyosis, skeletal

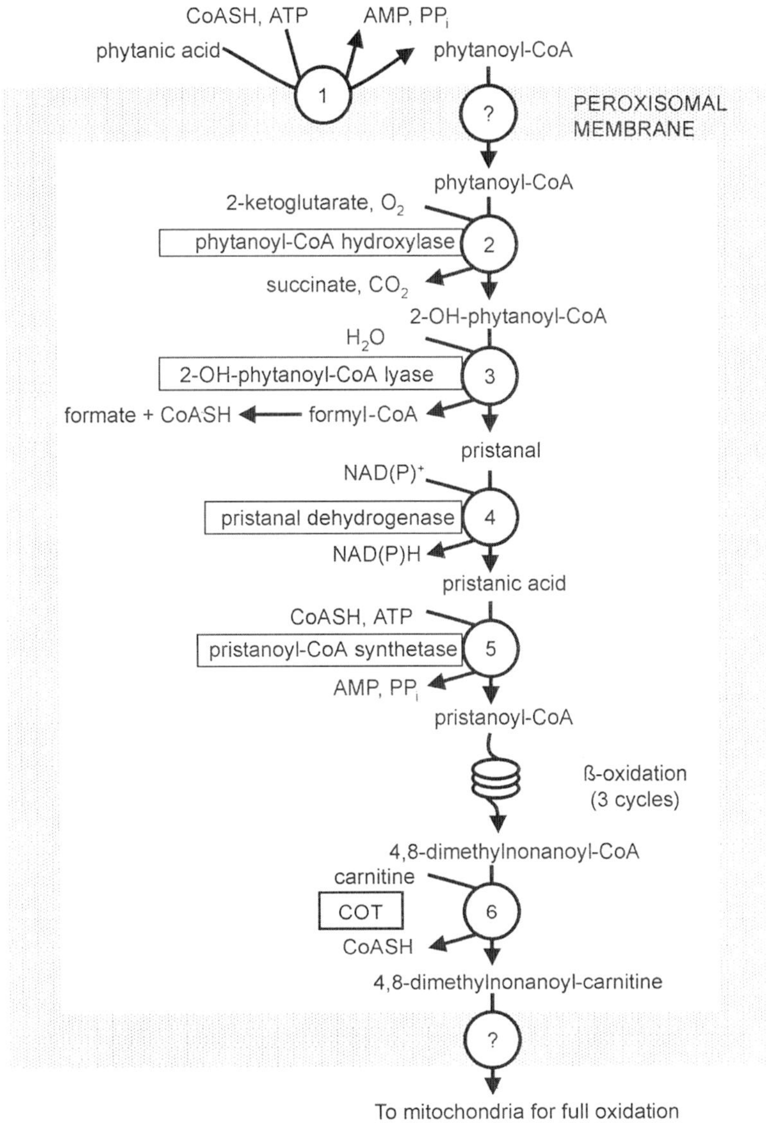

Fig. 5. Structure and enzymology of the peroxisomal FA α-oxidation system. Phytanic acid first undergoes activation to phytanoyl-CoA, probably by the long-chain acyl-CoA synthetase (LACS) present at the outer aspect of the peroxisomal membrane and then enters the peroxisomal matrix via a mechanism not well defined yet. Once inside peroxisomes, phytanoyl-CoA undergoes hydroxylation by the enzyme phytanoyl-CoA hydroxylase, which is deficient in adult Refsum disease (see text). Subsequently, 2-hydroxyphytanoyl-CoA then undergoes cleavage to pristanal and formyl-CoA, followed by oxidation of pristanal to pristanic acid. After activation to its CoA-ester, pristanoyl-CoA undergoes three cycles of β-oxidation in the peroxisome to produce 4,8-dimethylnonanoyl-CoA, which leaves the peroxisomes in the form of a carnitine-ester to undergo full oxidation in mitochondria [46].

malformations and cardiac abnormalities. The clinical picture of Refsum disease is often that of a slowly developing, progressive peripheral neuropathy manifested by severe motor weakness and muscular wasting, especially of the lower extremities. Patients in whom Refsum disease is destined to develop appear to be perfectly normal as infants and do not show any obvious defects in growth and development. Onset has occasionally been detected in early childhood but not until the fifth decade in others. Accumulation of phytanic in plasma and tissues of Refsum patients is the single abnormality in patients identified with Refsum disease. The underlying defect is at the level of the enzyme phytanoyl-CoA hydroxylase, which converts phytanoyl-CoA into 2-hydroxyphytanoyl-CoA. The enzyme defect can be shown in fibroblasts and in tissue samples and molecular analysis has led to the identification of many mutations, often of a private character [45].

In summary, much has been learned in recent years about the peroxisomal and mitochondrial FAO systems and defects therein.

References

[1] Issemann, I., Green, S., 1990. Activation of a member of the steroid hormone receptor superfamily by peroxisome proliferators. Nature 347, 645–650.

[2] Xu, J., Teran-Garcia, M., Park, J.H., Nakamura, M.T., Clarke, S.D., 2001. Polyunsaturated fatty acids suppress hepatic sterol regulatory element-binding protein-1 expression by accelerating transcript decay. J. Biol. Chem. 276, 9800–9807.

[3] Hannah, V.C., Ou, J., Luong, A., Goldstein, J.L., Brown, M.S., 2001. Unsaturated fatty acids down-regulate srebp isoforms 1a and 1c by two mechanisms in HEK-293 cells. J. Biol. Chem. 276, 4365–4372.

[4] Ou, J., Tu, H., Shan, B., Luk, A., DeBose-Boyd, R.A., Bashmakov, Y., Goldstein, J.L., Brown, M.S., 2001. Unsaturated fatty acids inhibit transcription of the sterol regulatory element-binding protein-1c (SREBP-1c) gene by antagonizing ligand-dependent activation of the LXR. Proc. Natl Acad. Sci. USA 98, 6027–6032.

[5] Repa, J.J., Liang, G., Ou, J., Bashmakov, Y., Lobaccaro, J.M., Shimomura, I., Shan, B., Brown, M.S., Goldstein, J.L., Mangelsdorf, D.J., 2000. Regulation of mouse sterol regulatory element-binding protein-1c gene (SREBP-1c) by oxysterol receptors, LXRalpha and LXRbeta. Genes Dev. 14, 2819–2830.

[6] Janowski, B.A., Grogan, M.J., Jones, S.A., Wisely, G.B., Kliewer, S.A., Corey, E.J., Mangelsdorf, D.J., 1999. Structural requirements of ligands for the oxysterol liver X receptors LXRalpha and LXRbeta. Proc. Natl Acad. Sci. USA 96, 266–271.

[7] Wanders, R.J.A., Jakobs, C., Skjeldal, O.H., 2001. Refsum disease. In: Scriver, C.R., Beaudet, A.L., Sly, W.S., Valle, D., (Eds.), The Metabolic and Molecular Bases of Inherited Disease. McGraw-Hill, New York, pp. 3303–3321.

[8] Wanders, R.J.A., Vreken, P., Ferdinandusse, S., Jansen, G.A., Waterham, H.R., van Roermund, C.W.T., van Grunsven, E.G., 2001. Peroxisomal fatty acids alpha- and beta-oxidation in humans: enzymology, peroxisomal metabolite transporters and peroxisomal diseases. Biochem. Soc. Trans. 29, 250–267.

[9] Corti, O., DiDonato, S., Finocchiaro, G., 1994. Divergent sequences in the 5′ region of cDNA suggest alternative splicing as a mechanism for the generation of carnitine acetyltransferases with different subcellular localizations. Biochem. J. 303, 37–41.

[10] Filppula, S.A., Yagi, A.I., Kilpelainen, S.H., Novikov, D., FitzPatrick, D.R., Vihinen, M., Valle, D., Hiltunen, J.K., 1998. Delta3,5-delta2,4-dienoyl-CoA isomerase from rat liver. Molecular characterization. J. Biol. Chem. 273, 349–355.

[11] Amery, L., Fransen, M., De Nys, K., Mannaerts, G.P., Van Veldhoven, P.P., 2000. Mitochondrial and peroxisomal targeting of 2-methylacyl-CoA racemase in humans. J. Lipid Res. 41, 1752–1759.

[12] Wanders, R.J.A., Vreken, P., Den Boer, M.E.J., Wijburg, F.A., Van Gennip, A.H., IJlst, L., 1999. Disorders of mitochondrial fatty acyl-CoA beta-oxidation. J. Inherit. Metab. Dis. 22, 442–487.

[13] Wanders, R.J.A., Tager, J.M., 1998. Lipid metabolism in peroxisomes in relation to human disease. Mol. Aspects Med. 19, 71–154.

[14] van Grunsven, E.G., van Berkel, E., IJlst, L., Vreken, P., de Klerk, J.B., Adamski, J., Lemonde, H., Clayton, P.T., Cuebas, D.A., Wanders, R.J.A., 1998. Peroxisomal D-hydroxyacyl-CoA dehydrogenase deficiency: resolution of the enzyme defect and its molecular basis in bifunctional protein deficiency. Proc. Natl Acad. Sci. USA 95, 2128–2133.

[15] van Grunsven, E.G., van Berkel, E., Mooijer, P.A.W., Watkins, P.A., Moser, H.W., Suzuki, Y., Jiang, L.L., Hashimoto, T., Hoefler, G., Adamski, J., Wanders, R.J.A., 1999. Peroxisomal bifunctional protein deficiency revisited: resolution of its true enzymatic and molecular basis. Am. J. Hum. Genet. 64, 99–107.

[16] van Grunsven, E.G., Mooijer, P.A.W., Aubourg, P., Wanders, R.J.A., 1999. Enoyl-CoA hydratase deficiency: identification of an new type of D-bifunctional protein deficiency. Hum. Mol. Genet. 8, 1509–1516.

[17] Suzuki, Y., Jiang, L.L., Souri, M., Miyazawa, S., Fukuda, S., Zhang, Z., Une, M., Shimozawa, N., Kondo, N., Orii, T., Hashimoto, T., 1997. D-3-hydroxyacyl-CoA dehydratase/D-3-hydroxyacyl-CoA dehydrogenase bifunctional protein deficiency: a newly identified peroxisomal disorder. Am. J. Hum. Genet. 61, 1153–1162.

[18] Baes, M., Huyghe, S., Carmeliet, P., Declercq, P.E., Collen, D., Mannaerts, G.P., Van Veldhoven, P.P., 2000. Inactivation of the peroxisomal multifunctional protein-2 in mice impedes the degradation of not only 2-methyl-branched fatty acids and bile acid intermediates but also of very long chain fatty acids. J. Biol. Chem. 275, 16329–16336.

[19] Hashimoto, T., 1996. Peroxisomal beta-oxidation: enzymology and molecular biology. Ann. N. Y. Acad. Sci. 804, 86–98.

[20] Seedorf, U., Brysch, P., Engel, T., Schrage, K., Assmann, G., 1994. Sterol carrier protein X is peroxisomal 3-oxoacyl coenzyme A thiolase with intrinsic sterol carrier and lipid transfer activity. J. Biol. Chem. 269, 21277–21283.

[21] Seedorf, U., Raabe, M., Ellinghaus, P., Kannenberg, F., Fobker, M., Engel, T., Denis, S., Wouters, F., Wirtz, K.W.A., Wanders, R.J.A., Maeda, N., Assmann, G., 1998. Defective peroxisomal catabolism of branched fatty acyl coenzyme A in mice lacking the sterol carrier protein-2/sterol carrier protein-x gene function. Genes Dev. 12, 1189–1201.

[22] Bougneres, P.F., Saudubray, J.M., Marsac, C., Bernard, O., Odievre, M., Girard, J., 1981. Fasting hypoglycemia resulting from hepatic carnitine palmitoyl transferase deficiency. J. Pediatr. 98, 742–746.

[23] Stanley, C.A., Hale, D.E., Berry, G.T., Deleeuw, S., Boxer, J., Bonnefont, J.P., 1992. Brief report: a deficiency of carnitine-acylcarnitine translocase in the inner mitochondrial membrane. N. Engl. J. Med. 327, 19–23.

[24] DiMauro, S., DiMauro, P.M., 1973. Muscle carnitine palmityltransferase deficiency and myoglobinuria. Science 182, 929–931.

[25] Hug, G., Bove, K.E., Soukup, S., 1991. Lethal neonatal multiorgan deficiency of carnitine palmitoyltransferase II. N. Engl. J. Med. 325, 1862–1864.

[26] Hale, D.E., Batshaw, M.L., Coates, P.M., Frerman, F.E., Goodman, S.I., Singh, I., Stanley, C.A., 1985. Long-chain acyl coenzyme A dehydrogenase deficiency: an inherited cause of nonketotic hypoglycemia. Pediatr. Res. 19, 666–671.

[27] Lafolla, A.K., Thompson, R.J. Jr., Roe, C.R., 1994. Medium-chain acyl-coenzyme A dehydrogenase deficiency: clinical course in 120 affected children. J. Pediatr. 124, 409–415.

[28] Amendt, B.A., Greene, C., Sweetman, L., Cloherty, J., Shih, V., Moon, A., Teel, L., Rhead, W.J., 1987. Short-chain acyl-coenzyme A dehydrogenase deficiency. Clinical and biochemical studies in two patients. J. Clin. Invest. 79, 1303–1309.

[29] Ribes, A., Riudor, E., Garavaglia, B., Martinez, G., Arranz, A., Invernizzi, F., Briones, P., Lamantea, E., Sentis, M., Barcelo, A., Roig, M., 1998. Mild or absent clinical signs in twin sisters with short-chain acyl-CoA dehydrogenase deficiency. Eur. J. Pediatr. 157, 317–320.

[30] Corydon, M.J., Vockley, J., Rinaldo, P., Rhead, W.J., Kjeldsen, M., Winter, V., Riggs, C., Babovic-Vuksanovic, D., Smeitink, J., de Jong, J., Levy, H., Sewell, A.C., Roe, C., Matern, D., Dasouki, M., Gregersen, N., 2001. Role of common gene variations in the molecular pathogenesis of short-chain acyl-CoA dehydrogenase deficiency. Pediatr. Res. 49, 18–23.

[31] Gregersen, N., Winter, V.S., Corydon, M.J., Corydon, T.J., Rinaldo, P., Ribes, A., Martinez, G., Bennett, M.J., Vianey-Saban, C., Bhala, A., Hale, D.E., Lehnert, W., Kmoch, S., Roig, M., Riudor, E., Eiberg, H., Andresen, B.S., Bross, P., Bolund, L.A., Kolvraa, S., 1998. Identification of four new mutations in the SCAD gene in two patients: one of the variant alleles, 511C → T, is present at an unexpectedly high frequency in the general population, as was the case for 625G → A, together conferring susceptibility to ethylmalonic aciduria. Hum. Mol. Genet. 7, 619–627.

[32] den Boer, M.E.J., Wanders, R.J.A., Morris, A.A., IJlst, L., Heymans, H.S.A., Wijburg, F.A., 2002. Long-chain 3-hydroxyacyl-CoA dehydrogenase deficiency: clinical presentation and follow-up of 50 patients. Pediatrics 109, 99–104.

[33] Schaefer, J., Jackson, S., Dick, D.J., Turnbull, D.M., 1996. Trifunctional enzyme deficiency: adult presentation of a usually fatal beta-oxidation defect. Ann. Neurol. 40, 597–602.

[34] Wanders, R.J.A., IJlst, L., Poggi, F., Bonnefont, J.P., Munnich, A., Brivet, M., Rabier, D., Saudubray, J.M., 1992. Human trifunctional protein deficiency: a new disorder of mitochondrial fatty acid beta-oxidation. Biochem. Biophys. Res. Commun. 188, 1139–1145.

[35] Jackson, S., Kler, R.S., Bartlett, K., Briggs, H., Bindoff, L.A., Pourfarzam, M., Gardner-Medwin, D., Turnbull, D.M., 1992. Combined enzyme defect of mitochondrial fatty acid oxidation. J. Clin. Invest. 90, 1219–1225.

[36] Miyajima, H., Orii, K.E., Shindo, Y., Hashimoto, T., Shinka, T., Kuhara, T., Matsumoto, I., Shimizu, H., Kaneko, E., 1997. Mitochondrial trifunctional protein deficiency associated with recurrent myoglobinuria in adolescence. Neurology 49, 833–837.

[37] Kamijo, T., Indo, Y., Souri, M., Aoyama, T., Hara, T., Yamamoto, S., Ushikubo, S., Rinaldo, P., Matsuda, I., Komiyama, A., Hashimoto, T., 1997. Medium chain 3-ketoacyl-coenzyme A thiolase deficiency: a new disorder of mitochondrial fatty acid beta-oxidation. Pediatr. Res. 42, 569–576.

[38] Clayton, P.T., Eaton, S., Aynsley-Green, A., Edginton, M., Hussain, K., Krywawych, S., Datta, V., Malingre, H.E.M., Berger, R., van den Berg, I.E.T., 2001. Hyperinsulinism in short-chain L-3-hydroxyacyl-CoA dehydrogenase deficiency reveals the importance of beta-oxidation in the control of insulin secretion. J. Clin. Invest. 108(3), 457–465.

[39] Taroni, F., Verderio, E., Dworzak, F., Willems, P.J., Cavadini, P., DiDonato, S., 1993. Identification of a common mutation in the carnitine palmitoyltransferase II gene in familial recurrent myoglobinuria patients. Nat. Genet. 4, 314–320.

[40] IJlst, L., Ruiter, J.P.N., Hoovers, J.M., Jakobs, M.E., Wanders, R.J.A., 1996. Common missense mutation G1528C in long-chain 3-hydroxyacyl-CoA dehydrogenase deficiency. Characterization and expression of the mutant protein, mutation analysis on genomic DNA and chromosomal localization of the mitochondrial trifunctional protein alpha subunit gene. J. Clin. Invest. 98, 1028–1033.

[41] Moser, H.W., Smith, K.D., Moser, A.B., 1995. X-linked adrenoleukodystrophy. In: Scriver, C.R., Beaudet, A.L., Sly, W.S., Valle, D., (Eds.), The Metabolic and Molecular Bases of Inherited Disease. McGraw-Hill, New York, pp. 2325–2349.

[42] Mosser, J., Douar, A.M., Sarde, C.O., Kioschis, P., Feil, R., Moser, H., Poustka, A.M., Mandel, J.L., Aubourg, P., 1993. Putative X-linked adrenoleukodystrophy gene shares unexpected homology with ABC transporters. Nature 361, 726–730.

[43] Wanders, R.J.A., Barth, P.G., Heymans, H.S.A., 2001. Single peroxisomal enzyme deficiencies. In: Scriver, C.R., Beaudet, A.L., Sly, W.S., Valle, D., (Eds.), The Metabolic and Molecular Bases of Inherited Disease. McGraw-Hill, New York, pp. 3219–3256.

[44] Ferdinandusse, S., van Grunsven, E.G., Oostheim, W., Denis, S., Hogenhout, E.M., IJlst, L., van Roermund, C.W., Waterham, H.R., Goldfischer, S., Wanders, R.J.A., 2002. Reinvestigation of peroxisomal 3-ketoacyl-CoA thiolase deficiency: identification of the true defect at the level of D-bifunctional protein. Am. J. Hum. Genet. 70, 1589–1593.

[45] Jansen, G.A., Hogenhout, E.M., Ferdinandusse, S., Waterham, H.R., Ofman, R., Jakobs, C., Skjeldal, O.H., Wanders, R.J.A., 2000. Human phytanoyl-CoA hydroxylase: resolution of the gene structure and the molecular basis of Refsum's disease. Hum. Mol. Genet. 9, 1195–1200.

[46] Verhoeven, N.M., Roe, D.S., Kok, R.M., Wanders, R.J.A., Jakobs, C., Roe, C., 1998. Phytanic acid and pristanic acid are oxidized by sequential peroxisomal and mitochondrial reactions in cultured fibroblasts. J. Lipid Res. 39, 66–74.

Transcriptional regulation of cellular fatty acid homeostasis

Marc van Bilsen*

Department of Physiology, Cardiovascular Research Institute Maastricht (CARIM), Maastricht University, P.O. Box 616, 6200 MD Maastricht, The Netherlands
**Correspondence address: Tel.: +31-43-3881204; fax: +31-43-3884166*
E-mail: marc.vanbilsen@fys.unimaas.nl(M. van Bilsen)

1. Introduction

The ability to respond to changes in external nutrient availability by adjusting the expression of proteins involved in the metabolic conversion of these nutrients appears to be a conserved mechanism throughout evolution. Since the pioneering work on the regulation of the lactose operon in *Escherichia coli* in the 1960s, it has become clear that more or less conserved transcriptional mechanisms are operative in organisms ranging from simple unicellular eukaryotes as yeast up to complex multi-cellular organisms as mammals. For instance, in the yeast *Saccharomyces cereviseae* the gene encoding the membrane-bound enzyme Δ-9 desaturase, which converts stearoyl-CoA into its mono-unsaturated equivalent, is transcriptionally controlled based on metabolic needs. The gene gets activated in the presence of excess saturated fatty acids, and becomes inactivated when mono-unsaturated fatty acids are abundantly present [1]. Likewise, in mammalian cells the transcription of its homologue stearoyl-CoA desaturase-1 (SCD-1) is suppressed by poly-unsaturated fatty acids, albeit in a more complex manner [2]. Although higher organisms as mammals have developed a complex neuro-endocrine system, which allows them to respond to changes in food intake and energy need, and that is responsible for the metabolic communication between organs, it is evident that also in these organisms the tissue cells themselves have retained the capacity to respond to changes in nutrient availability. Within these cells molecular pathways exist that sense local changes in glucose, amino acid or fatty acid availability and are able to convey such changes to the transcriptional machinery in the nucleus of the cell. Ultimately, this results in the up-regulation of the expression of specific proteins involved in the transport or metabolic conversion of a particular nutrient, often along with the down-regulation of the expression of genes encoding for proteins involved in the uptake and conversion of alternative

Advances in Molecular and Cell Biology, Vol. 33, pages 319–336

ISSN: 1569-2558/DOI: 10.1016/S1569-2558(03)33016-4

substrates. The present chapter focusses primarily on the mechanisms by which changes in the availability of long-chain fatty acids are translated into a transcriptional response within the cell. Given the fact that important key players in this fatty acid-mediated response, namely certain transcription factors of the nuclear hormone receptor super family, have been identified during the last decade, this chapter will deal with the biology of these transcription factors, referred to as the peroxisome proliferator-activated receptors (PPARs). In addition, attention will be paid to other transcription factors, dubbed sterol regulatory element-binding proteins (SREBPs), that are also involved in certain aspects of fatty acid-mediated gene regulation.

2. Fatty acid-mediated gene expression

Effects of exogenous long-chain fatty acids on gene expression in mammalian cells were first reported in cultured adipocytes [3,4], in which it was shown that the down-regulation of the expression of the adipocyte-specific fatty acid-binding protein aP2 in differentiated adipocytes following impairment of fatty acid synthesis by means of glucose deprivation was prevented by fatty acid supplementation. Next, it was shown that the transcriptional activation of the aP2 gene could not be induced by short-chain fatty acids, but required the addition of long-chain fatty acids [3]. The induction of aP2 expression turned out to be largely independent of the chain length and the degree of saturation of the long-chain fatty acid species and, interestingly, the effect could even be mimicked by the non-metabolisable fatty acid-derivative α-bromopalmitate [4]. Concurrently, Jump and co-workers investigated the effect of fatty acids on liver gene expression [5–7]. Unlike the stimulatory effect of fatty acids on aP2 expression in adipocytes, these in vivo and in vitro studies revealed that fatty acids suppressed the expression fatty acid synthase (FAS) and of Spot 14 gene in liver cells, i.e. genes involved in hepatic lipogenesis. More importantly, this effect was specific for poly-unsaturated fatty acids, with saturated fatty acids being ineffective. The original observations on adipose aP2 and liver FAS gene expression in the beginning of the 1990s have raised lots of interest in the molecular mechanisms responsible for the fatty acid-mediated effects on gene expression in subsequent years.

3. Transcriptional control of lipogenesis

Next to FAS and Spot 14 various other enzymes involved in lipogenesis, including acetyl-CoA carboxylase, ATP citrate lyase, malic enzyme, glucose-6-phosphate dehydrogenase and stearoyl-CoA desaturase-1, were shown to be down-regulated by poly-unsaturated fatty acids. Although during the past couple of years the molecular link between poly-unsaturated fatty acids and the suppression of lipogenic gene expression has been partly elucidated, several key questions still remain to be solved.

It turns out that the SREBPs play an important role in the control of lipogenesis. The SREBPs belong to the so-called helix-loop-helix (HLH) family of transcription factors. SREBP-1 and SREBP-2 are encoded by different genes and serve different functions. SREBP-2 is primarily involved in the transcription of genes involved in cholesterol synthesis [8,9], whereas SREBP-1 plays an important role in the control of the expression of lipogenic enzymes in the liver [10]. SREBPs are synthesised as inactive precursors, which become bound to the endoplasmic reticulum provided that sufficient amounts of sterols are available. When the amount of sterols becomes limiting, however, the precursor protein will be cleaved, thereby releasing the active form. The active SREBP subsequently translocates to the nucleus allowing it to bind to its target genes. In the case of SREBP-1, this results in an activation of transcription of genes encoding for the lipogenic enzymes. What then is the relationship between poly-unsaturated fatty acids, SREBP-1 and the suppression of transcription of lipogenic enzymes? Cumulative evidence suggests that poly-unsaturated fatty acids suppress SREBP-1 activity via two mechanisms. First, by inhibiting the proteolytic cleavage of the inactive precursor protein, thereby preventing its activation. Second, poly-unsaturated fatty acids negatively affect the transcription of the SREBP-1 gene per se. Accordingly, the latter mechanism requires a rephrasing of the earlier question, namely, what is the mechanism by which poly-unsaturated fatty acids inhibit SREBP-1 expression?

Recent findings suggest the involvement of the nuclear hormone receptor, liver X receptor-α (LXRα), in the control of SREBP-1 expression. The LXRs (LXRα and LXRβ) are members of the family of ligand-activated nuclear hormone receptor, and are activated by oxysterols. Following heterodimerisation with the retinoid X receptor-α (RXRα) the LXRs bind to a direct repeat of six nucleotides separated by four spacer nucleotides (DR-4 element) in the promoters of their target genes to activate transcription. In this manner, LXRα also augments the transcription of SREBP-1. An in vitro study showed that poly-unsaturated, but not saturated, fatty acids may compete with oxysterols for binding to the ligand-binding domain of LXRα and, thus, may suppress LXRα activity [11]. Notably, the expression of LXRα in turn has been shown to be stimulated by PPARα, which is known to be activated not only by poly-unsaturated, but also by saturated fatty acids (see below). Collectively, these observations seem to indicate that the expression of lipogenic enzymes in liver is modulated by PPARα, albeit in a very indirect manner involving a cascade of transcription factors from PPARα → LXRα → SREBP-1 → target genes. Nonetheless, in spite of the identification of the key players involved in the transcriptional regulation of lipogenic enzymes, it should be noted that each of them activates, rather than inhibits, its downstream target. Accordingly, although several potential steps that are inhibited by poly-unsaturated fatty acids have been identified [12], it still remains to be established how the suppressive effect of poly-unsaturated fatty acids is actually achieved.

4. Transcriptional control of fatty acid oxidation

As described above, fatty acids were first shown to enhance the expression of genes in adipocytes [3]. More recently, it was shown that fatty acids also modulate gene expression in skeletal muscle cells [13,14] and in cardiac muscle cells [15–18]. Based on the ability of fatty acids to alter gene expression in isolated cell system, it is to be anticipated that also under in vivo conditions fatty acids modulate the expression of proteins involved in fatty acid transport and metabolism. Indeed, a relationship between plasma fatty acid levels and cardiac fatty acid utilisation has long been recognised. An elevation of plasma fatty acid levels, as occurs during peri-natal development, fasting and diabetes, coincides with an increased cardiac fatty acid utilisation. Theoretically, the increased uptake and metabolism of fatty acids could be simply due to the rise in plasma concentration and, hence, an enhanced supply of this substrate to the organs. However, there are clear indications that in addition to mass effects, the cardiac muscle also adapts by adjusting the tissue content of proteins required for fatty acid uptake and the activity of enzymes involved in fatty acid oxidation via transcriptional mechanisms [17,19]. It is tempting to speculate that the increased levels of blood-borne fatty acids are directly responsible for the alterations in cardiac gene expression.

When exposed to physiological concentrations of palmitic and oleic acid, the capacity of the neonatal cardiomyocytes to oxidise fatty acids increased significantly, a response which was associated with a two- to fourfold increase in the mRNA levels of proteins involved in both transsarcolemmal uptake (fatty acid translocase, FAT/CD36), intracellular transport (fatty acid-binding protein, FABP) and metabolic conversion (acyl-CoA synthetase, ACS; muscle-type carnitine palmitoyl-transferase-I, mCPT1; long-chain acyl-CoA dehydrogenase, LCAD) of fatty acids [18]. The fact that the effect of fatty acids could be mimicked by synthetic PPARα ligands pointed to the involvement of the PPARs in this process. Using transfection assays with mCPT1 promoter/reporter constructs, it was subsequently shown that this response required a functional peroxisome proliferator response element (PPRE) site [16,20]. In search for the PPAR species responsible, the cells were treated with PPAR isotype specific synthetic ligands. These studies revealed that PPARα as well as PPARβ/δ are involved in the fatty acid mediated up-regulation of genes in the cardiac muscle cell [21]. PPARγ-specific ligands, on the other hand, were found to be ineffective, thereby arguing against a major role of PPARγ in the regulation of cardiac lipid metabolism. Studies with PPARα-deficient mice provide strong evidence for the involvement of PPARα in cardiac gene expression. First, basic mRNA levels of mCPT1, LCAD, MCAD and acyl-CoA oxidase (AOX) were found to be lower in hearts of PPARα (−/−) mice than in wild-type animals [22]. Furthermore, in contrast to wild-type mice, fasting of PPARα (−/−) mice did not increase the expression of mCPT1, MCAD or AOX in the heart [23]. The latter findings provide compelling evidence for the regulatory role of PPARα in the expression of genes encoding proteins involved in cardiac fatty acid metabolism. On the other hand, despite the abundance of PPARα in both liver and heart, it has been repeatedly observed that the administration of PPARα ligands to adult rodents has marked effects on gene expression in liver, but hardly any in heart (Table 1). Treatment of rodents with various fibrates did not affect cardiac mRNA levels of proteins involved in lipid metabolism, including lipoprotein lipase (LPL), fatty acid transport protein (FATP) and ACS [24–28]. Presently, no satisfying explanation can be offered for these apparently conflicting results. Although it is feasible that aspects of bioavailability of the ligands are involved, it is more likely that the answer resides in the complex manner in which PPARα activity is being regulated in different tissues and species.

Table 1
Tissue-specific changes in gene expression in response to PPAR ligands

	Adipose		Liver		Heart	
	FF	BRL	FF	BRL	FF	BRL
LPL	=	+	++	=	=	=
FATP	=	++	++	=	=	=
ACS	=	++	+	−	=	=

Rats were treated with the PPARα ligand fenofibrate (FF) or the thiazolidinedione PPARγ ligand (BRL 49653). As anticipated, mRNA levels of lipoprotein lipase (LPL), fatty acid transport protein (FATP) and acyl-CoA synthase (ACS) increase in response to BRL, but not in response to FF in the PPARγ containing white adipose tissue. In contrast, PPARα-containing liver tissue is primarily responsive to FF, but not to BRL. Surprisingly, the PPARα-containing cardiac tissue appeared to be unresponsive to FF. Data are derived from Refs. [24,25,107].

5. Peroxisome proliferator-activated receptors

The first PPAR was discovered just over a decade ago [29]. It was identified as a ligand-activated transcription factor that interacted with various drugs that were known for their ability to cause peroxisomal proliferation in liver. Accordingly, the transcription factor was referred to as PPAR. Since the discovery of this PPAR, later renamed PPARα, two other PPAR subtypes were identified, i.e. PPARβ/δ (also called NUC-1 or FAAR) and PPARγ, encoded by different genes.

Like the LXRs the PPARs are members of the family of nuclear hormone receptors. All nuclear hormone receptors share a close structural homology, reflected by the presence of several structural and functional domains (Fig. 1) [30]. The amino-terminal A/B domain is least conserved and includes a ligand-independent transactivation function (AF-1), containing putative phosphorylation sites. The DNA-binding C domain is highly conserved and includes two zinc finger motifs, containing two conserved boxes, the P- and D-box. The P-box is specifically involved in the interaction with specific DNA sequences, whereas the D-box participates in receptor dimerisation. The variable hinge region (D domain) allows conformational changes. The ligand-binding domain (E domain) encompasses a ligand-dependent transactivation function (AF-2). In addition to ligand binding, the E domain is required for nuclear localisation, receptor dimerisation and the interaction with auxiliary proteins. The exact role of the carboxy-terminal end (F domain) remains to be established still.

The PPARs form heterodimers with the RXR [31]. The PPAR/RXR heterodimer is able to bind to a DNA consensus sequence, the PPRE. A PPRE is formed by a direct repeat of six nucleotides, separated by one spacer nucleotide (DR-1) and has the consensus sequence AGGTCA n AGGTCA. It should be noted, however, that the nucleotide sequence of many functional PPREs diverges quite substantially from the consensus sequence. Functional PPREs have been found in a large number of genes encoding proteins involved in lipid metabolism. This includes proteins involved in inter-organ lipid

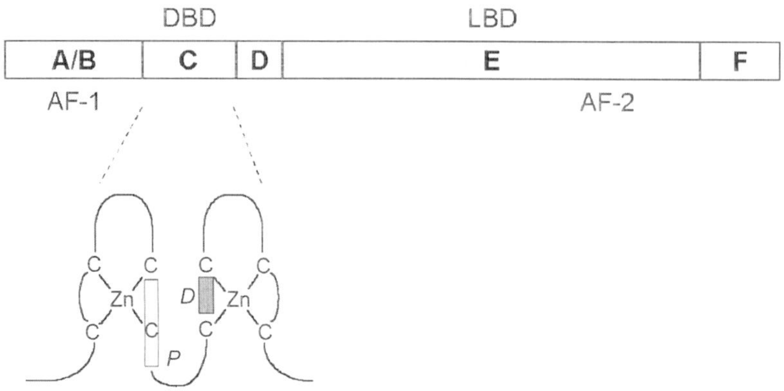

Fig. 1. Schematic representation depicting the conserved domains of the nuclear hormone receptors. For details see text. DBD, DNA-binding domain; LBD, ligand-binding domain; AF-1 and AF-2, transactivating function 1 and 2, respectively.

transport (several apolipoproteins), in lipid processing and uptake (LPL, FAT/CD36, FATP), and metabolism (e.g. AOX, ACS, mCPT1, LCAD). The list of PPAR-responsive genes is still expanding and also includes genes, for which the link with fatty acid metabolism is less well established (e.g. uncoupling protein-2 and -3) [32,33]. Interestingly, one of the pyruvate dehydrogenase kinase isoforms (PDK4) is also PPAR responsive [34]. Through the PDK4-dependent phosphorylation of pyruvate dehydrogenase (PDH), the activity of the PDH complex diminishes, as a result of which the oxidation of glucose is inhibited. Accordingly, the activation of PPAR on the one hand promotes the utilisation fatty acids, while on the other hand suppressing the utilisation of glucose. This suggests that the PPARs have the ability to fine-tune the utilisation of these competitive substrates at the gene level, more or less analogous to the Randle cycle, which is based on enzyme kinetic principles.

Recently, it was shown that the expression of a number of genes involved in hepatic amino acid metabolism is also regulated by PPARα [35]. In addition to the regulation of energy metabolism, PPARs have been described to be involved in processes such as adipogenesis, inflammation, atherosclerosis and carcinogenesis (see Refs. [36–38] for recent reviews). In this context, it is interesting to note that the transcription of the inducible form of cyclooxygenase COX2, which is involved in prostaglandin synthesis, is also enhanced by fatty acids, although in this particular case, the involvement of PPAR in mediating fatty acid-induced transcription is being questioned [39]. Together, these findings suggest that the PPARs are involved in a large variety of cellular processes and, in addition to functioning as "liposensors", serve various other biological functions. In view of their importance in the regulation of many biological processes, it is evident that their activity must be precisely controlled. In Table 2 the various factors that are likely to be involved in the regulation of PPAR function are listed. In the next paragraphs we will discuss the role of each of these factors consecutively.

Table 2
Factors modulating PPAR activity

1.	PPAR expression (tissue specificity, splice variants, protein stability)
2.	Ligand specificity (natural agonists, antagonists)
3.	Ligand availability (FABPs, phospholipases)
4.	Competition for RXR
5.	Competition for PPRE-binding sites
6.	PPRE sequence variation
7.	Co-activators and co-repressors
8.	Phosphorylation

6. Tissue expression of PPAR subtypes

The expression of the three PPAR subtypes is tissue restricted. PPARα is primarily expressed in tissues with a high oxidative capacity such as liver, kidney and heart. The expression of PPARβ/δ appears to be rather ubiquitous. In contrast, the expression of PPARγ is confined to a limited number of tissues. The use of alternative promoters in the PPARγ gene gives rise to two splice variants, designated as PPARγ1 and PPARγ2, that differ at their N-terminus [40]. PPARγ2 is exclusively expressed in adipose tissue. The expression of PPARγ1, however, is more ubiquitous, its expression being highest in adipose tissue, white blood cells and vascular cells. The question whether PPARγ1 is also expressed in tissues as liver, kidney and muscle is still controversial. Depending on the assays used (northern blotting, PCR, western blotting or immunohistochemistry), various groups either proved [41,42] or disproved [43,44] the presence of PPARγ in these tissues. For instance, our own findings on PPARγ in primary cultures of neonatal cardiomyocytes revealed that the γ-isotype was not detectable by northern blotting, but its signal could be picked up using the more sensitive RT-PCR technique and by western blotting [21]. The general consensus is that, if expressed at all, PPARγ is expressed at low levels in tissues as skeletal and cardiac muscle. Nevertheless, some argue that this could still represent PPARγ expression in vascular cells or interspersed fat cells present in these tissues. However, the observations of Takano and co-workers [44] who documented the presence of PPARγ in isolated neonatal cardiomyocytes at the mRNA and protein level do challenge the latter notion. Compelling evidence that PPARγ is expressed in skeletal muscle cells themselves comes from studies (unpublished observations) with gene-targeted mice in which muscle-specific inactivation of the PPARγ gene was established. In these studies the muscle creatine kinase promoter was used to drive expression of the Cre-recombinase, allowing muscle-specific excision of the PPARγ allele in floxed PPARγ mice. These mice have a distinct phenotype compared to wild-type mice thereby demonstrating that PPARγ must be present and functional within the skeletal muscle cell.

RNAse protection has been applied to determine expression levels of the three PPAR isotypes in a large variety of tissues from adult rats in a sensitive and quantitative manner [45]. Using intact tissue (containing both parenchymal and vascular cells) as a source of

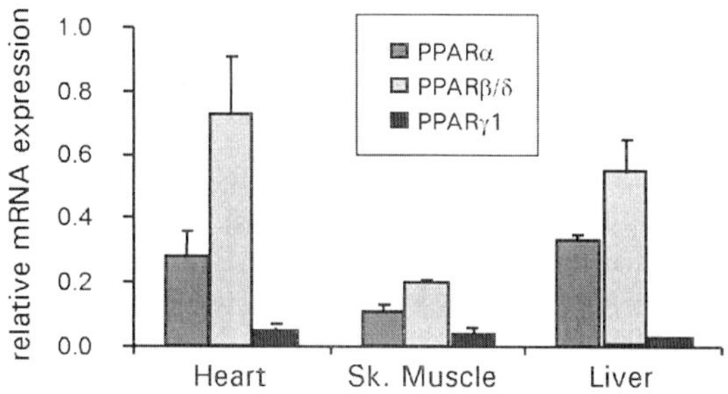

Fig. 2. Comparison of the tissue mRNA content of the different PPAR isotypes in rat heart, skeletal muscle and liver. Data are derived from a study of Escher and co-workers [45]. Despite the fact that these tissue are traditionally considered as tissues rich in PPARα, the PPARβ/δ mRNA content appears to be the highest in each of these tissues. PPARγ1 mRNA is present in all tissues, albeit in small amounts.

RNA, these investigators showed that in both cardiac and skeletal muscle all three isotypes could be detected (Fig. 2). In both striated muscle types the expression levels of the three isotypes were in the following order: PPARβ/δ > PPARα ≫ PPARγ. As mentioned earlier PPARγ2 is considered to be adipocyte specific. Nevertheless, the presence of both the γ1 and γ2 splice variant in muscle has been reported [46] and the apparent correlation between the level of PPARγ expression in skeletal muscle and obesity and type II diabetes [47] could indicate that part of the PPARγ detected is indeed derived from adipocytes within the muscle. Alternatively, expression of PPARγ within skeletal muscle fibres might be enhanced under certain conditions.

Finally, it should be noted that in the case of PPARα, it was shown that both its mRNA and protein content are under dynamic control. The PPARα mRNA level varies diurnally [48], indicating that either the rate of transcription or the stability of the mRNA is subject to circadian rhythms. In addition, it has been shown for PPARα that it is a short-lived protein, the degradation of which is controlled by ubiquination [49]. It remains to be established whether this also applies to the other PPAR isotypes.

7. Subcellular localisation of the PPAR isotypes

It is generally believed that the PPARs are constitutively localised in the nucleus, independent of ligand binding. Nevertheless, a couple of studies have cast doubt on this concept. A PPARα splice variant, present in humans, which gives rise to a truncated protein product was found in the cytoplasm in relatively large amounts [50]. Furthermore, in analogy to the oestrogen receptor, it has been speculated that heat shock proteins (HSPs) keep the PPARs in the cytoplasmic compartment and that ligand binding is required to loosen the HSP/PPAR interaction and allowing the translocation of the PPARs to the nucleus. Accordingly, although direct proof is lacking, the translocation of PPARs from cytoplasm to nucleus potentially represents another level of control of PPAR activity, which cannot be discarded entirely at this moment.

8. PPAR ligands

The resolution of the three-dimensional structures of the ligand-binding domains of the PPAR isotypes by means of X-ray crystallography has been instrumental in the understanding of the ligand–protein interactions. The crystal structures revealed that the hydrophobic ligand-binding pocket of the three PPARs is Y shaped and relatively large as compared to other receptors [51]. This helps us to explain some of the pharmacological features of the PPARs, namely that the PPAR isotypes are rather promiscuous with respect to the molecular structure of their ligands, and that many naturally occurring lipid-like substances bind to the PPARs as low-affinity ligands.

8.1. Natural ligands

The diversity of the compounds that bind to PPARs in combination with modest affinity with which they all bind have kept the discussion alive as to the exact nature of the natural ligands for the PPAR isotypes. Since the discovery of the PPARs long-chain fatty acids have always been considered as primary candidates for the role of natural ligands of the PPARs [52,53]. Next to fatty acids, many fatty acid derivatives have been implicated as natural ligands of PPARα (Table 3). The prostaglandin J2 derivative 15-deoxy-Δ12,14-PGJ2 is considered one of the natural ligands of the PPARγ receptor [54]. In particular, the capacity of individual fatty acid species, differing in chain length and/or degree of saturation, to act as ligands for the three PPAR isotypes has been subject of intense research [52,53,55–57]. Various types of in vitro assays, including classical competition binding assays [51] as well as gel mobility shift assays, making use of the ability of ligands to enhance PPAR–DNA interactions [56], and functional assays using promoter reporter studies in intact cells [58], have been applied to detect potential differences in fatty acid preference of the PPAR isotypes. Although differences exist in the outcome between the various studies, the general picture that emerges is that both saturated, mono- and poly-unsaturated fatty acids interact with PPARα with only modest differences in affinity between the different fatty acid species. The same holds for PPARβ/δ, albeit that the affinity of the various fatty acid species for PPARβ/δ appears to be lower than that for PPARα [51]. PPARγ, on the other hand, displays a clear preference for

Table 3
Putative natural PPARα ligands

Long-chain fatty acids
Long-chain acyl-CoA thioesters
Fatty acid methyl-esters
Branched-chain fatty acids (e.g. phytanic acid)
Hydroxylated fatty acids (HETE)
Leukotrienes
Prostaglandins
Bile acids

Table 4
Binding affinities of the PPAR isotypes for the different fatty acid species

	PPARα	PPARβ/δ	PPARγ
C12:0	>30	>30	>30
C16:0	1.5	7.4	>30
C18:0	1.1	6.0	>30
C18:1	0.6	5.3	4.1
C18:2	1.1	10.0	6.2
C20:4	1.2	3.1	1.6

IC_{50} values (μM) are derived from Xu and co-workers (for details see Ref. [51]) and based on scintillation proximity assays using poly-histidine tagged PPAR ligand-binding domains purified from transfected *E. coli.*

unsaturated fatty acids [51,56,57]. The binding constants of the most common fatty acid species for the three PPARs [51] are depicted in Table 4. Consistent with these data, the three-dimensional structure of the PPARs predicts that the carboxyl group and the first 8 C-atoms of the fatty acid are buried inside the binding pocket and the hydrophobic tail is bent into the upper arm of the Y-shaped pocket. Accordingly, fatty acids with chain lengths less than 14 C-atoms or more than 20 C-atoms are much poorer ligands for the PPARs, or may even have antagonistic properties [59].

Although the far majority of the fatty acids and fatty acid derivatives acts as activators of the PPARs, recent studies indicate that, in addition to C22 fatty acids, long-chain acyl-CoA thioesters also seem to have antagonistic rather than agonistic properties [60,61]. This opens the intriguing possibility that the intracellular molar ratio between fatty acids and fatty acyl-CoA thioesters modulates the transactivating activity of PPARs.

8.2. *Ligand availability*

As shown in Table 4 and as determined in the majority of other studies, the dissociation constants for fatty acids are in the micromolar range [55,62,63]. Obviously, the binding constants of fatty acids for PPARs are relevant to the question whether fatty acids are indeed able to function as bona fide ligands for PPARs. It has been estimated that the total concentration of unesterified fatty acids in the cytoplasmic compartment of cardiomyocytes is not likely to exceed 10 μM [64]. As the majority of the poorly water-soluble fatty acids is intercalated in between membrane phospholipids or bound to cytoplasmic FABPs, the actual free fatty acid concentration is likely to be much lower. Whether the same arguments also apply to the nuclear compartment remains to be established. Anyway, in order to be able to function as bona fide ligands for the PPARs, one would expect binding constants in the nanomolar, rather than micromolar range. However, several recent lines of evidence suggest that the delivery of fatty acids to the nuclear compartment may take place using specific pathways. First, it was postulated that the intracellular fatty acid transporter FABP might facilitate the transfer of fatty acids to PPAR in the nucleus. Indeed H-FABP is present, albeit in low amounts, in nuclei of cardiomyocytes [65] and for liver-type FABP it has been shown that it physically interacts with PPARα [58].

High levels of keratinocyte-type FABP (K-HABP) were shown to be required for PPARβ/δ-mediated responses in keratinocytes [66]. Interestingly, synthetic PPARβ/δ ligands were found to bind to both PPARβ/δ and K-FABP and the binding of these ligands to K-FABP resulted in the translocation of the ligand/K-KABP complex to the nucleus. Alternative routes of fatty acid delivery have also been described. Several studies demonstrated that fatty acids released by the action of cellular phospholipases, the release of arachidonic acid (C20:4) by cytosolic phospholipase A_2 ($cPLA_2$) in particular [67,68], generates ligands for the PPARs. As the activity of $cPLA_2$ is tightly regulated by submicromolar changes in Ca^{2+} concentration and by mitogen-activated protein (MAP) kinase-dependent phosphorylation, this would provide a mechanism by which ligand supply, and thus PPAR activity, is controlled by signal transduction pathways. Finally, there are some indications that also the way in which exogenous fatty acids are delivered to the cell is of influence on PPAR activity. The release of fatty acids by extracellular LPL acting on triglyceride-rich lipoprotein particles appears to be effectively coupled to nuclear PPARα activation in endothelial cells [69]. Altogether, these considerations indicate that ligand availability is not simply a function of exogenous fatty acid concentration, but that more subtle mechanisms are likely to be involved in the delivery of fatty acids to nuclear transcription factors.

8.3. Synthetic ligands

The fibrates, i.e. the classical peroxisome proliferator type of drugs, act as specific ligands for PPARα. The thiazolidinediones (TZDs) are specific activators for PPARγ. Both fibrates and TZDs are already being applied clinically to treat hyperlipidemia and type II diabetes, respectively. More recently, experimental synthetic ligands for PPARβ/δ have also become available [70,71]. Notwithstanding the isotype specificity of the synthetic ligands, it should be noted that at higher ligand concentration, cross-activation of the other PPAR isotypes often occurs. Furthermore, PPAR-independent effects of various ligands have been reported as well. This is particularly true for the TZDs, for which evidence has been provided that they modulate the inflammatory response via PPARγ-dependent as well as PPARγ-independent mechanisms [72]. Also in light of the discussion of the presence or absence of PPARγ in the cardiac muscle cell, it could well be that the reported effects of TZDs on cardiac function and phenotype are primarily accomplished via PPARγ-independent mechanisms. Indeed, the anti-hypertrophic effect of PPARγ ligands in neonatal cardiomyocytes was associated with inhibition of NF-κB activation [73]. In the same model the administration of TZDs, in contrast to PPARα or PPARβ/δ ligands, neither affected the mRNA level or various established PPAR responsive genes, nor affected mCPT1 promoter activity [21]. Furthermore, the effect of TZDs on skeletal muscle metabolism was found to be very fast and was not linked to changes in gene expression, indicating that PPARγ-mediated transcription was not essential [74]. Fibrates can be converted into CoA thioesters. The fibrate–CoA thioesters are able to bind to hepatocyte nuclear factor 4α (HNF-4α) and thereby inhibit the activity of this transcription factor [75], indicating that fibrates may also exert PPARα-independent effects. Furthermore, the administration of specific PPARα ligands to isolated adult myocytes resulted in changes in LPL activity, but not in LPL protein content [76], suggesting that the change in LPL activity was based on post-translational rather than transcriptional events. The possibility of non-specific effects of PPAR ligands should always be kept in mind in evaluating the biological effects of these PPAR ligands.

Differences in the response to treatment with hypolipidaemic fibrates between species were recognised long ago. Whereas the treatment of rats and mice with these PPARα ligands causes severe peroxisome proliferation in liver, ultimately leading to tumorigenesis, no such pathology has been observed in humans. It is generally believed that excessive peroxisome proliferation as a result of PPARα activation finally leads to the generation of reactive oxygen intermediates, which increase the likelihood of developing cancer. The molecular basis for this difference in response between species remains largely enigmatic. It is well known that the expression levels of PPARα in liver vary substantially between species, the level in humans being substantially lower than in rats. It has also been suggested that the subtle differences in amino acid sequence between species are responsible for differences in the transactivating capacity. Furthermore, it is noteworthy that the truncated splice variant of PPARα is found in human, but not in rat tissue. Since the truncated PPARα has been shown to act in a dominant negative fashion [50], it is tempting to speculate that the ratio of splice variants determined the response of PPRE-containing genes and that changes in expression of the two splice variants allow for one level of transcriptional control in responsive species.

Whatever the molecular explanation may be, this difference in response to fibrates indicates that caution should be exerted when extrapolating experimental findings from animals to humans.

9. Auxiliary proteins

Transcriptional activation of PPAR and LXR target genes is highly dependent on other proteins. First of all, as mentioned earlier, these nuclear hormone receptors bind to their cis-regulatory DNA-binding site as heterodimers, using RXR as partner. In addition, the transcriptional activity of these heterodimer complexes is controlled by a variety of co-repressor and co-activator proteins. It is believed that in the unliganded state the PPAR/RXR–DNA complex is transcriptionally inactive by favouring the interaction with co-repressor proteins. The binding of ligands to the AF2 domain results in conformational changes, which diminish the affinity for the co-repressors and/or promote the recruitment of co-activator proteins, which will subsequently activate transcription. Studies with genetically engineered mice suggest that, in its unliganded state, the LXR receptor may actually function as a repressor. This notion is based on studies showing that ablation of the LXR gene led to a higher basal level of expression, whereas at the same time inducibility using the LXR ligand was lost.

9.1. Heterodimerisation with RXR

As both PPARs and LXRs and several other nuclear hormone receptors interact with RXR, it is feasible that the cellular content of this common heterodimerisation partner is limiting. Indeed, PPAR was shown to decrease the activity of thyroid hormone receptor by sequestration of RXR [77,78] and, vice versa, PPAR activity was found to be diminished when cells were co-transfected with the thyroid hormone receptor [79,80]. In addition to sequestration of RXR by other nuclear receptors, evidence was provided that under certain pathological conditions RXRα is rapidly degraded, making the competition between nuclear hormone receptors even more eminent. It has been postulated that the hypoxia-induced degradation of RXRα in cardiomyocytes is responsible for the diminished transcription of PPRE-containing genes as mCPT1 [81]. A reduction of RXRα protein content has also been observed during heart failure [82], a condition, which is also characterised by a diminished cardiac utilisation of fatty acids as energy-providing substrate.

9.2. DNA-binding sites

In addition to PPARs several other nuclear receptors were shown to interact with DR-1 sites, indicating that there may be competition between transcription factors for the same cis-regulatory element. The existence of this regulatory mechanism has been convincingly demonstrated in the case of COUP-TF (chicken ovalbumin upstream promoter transcription factor) and PPAR. For COUP-TF it was shown that its binding to PPRE sequences suppresses transcriptional activity [80,83]. The observation that during normal postnatal cardiac growth, the level of COUP-TF decreases and that of PPARα increases, has been put forward as the molecular mechanism explaining the postnatal shift in cardiac substrate preference from glucose to fatty acids as primary energy source [84].

It was mentioned earlier that there is considerable variation in the nucleotide sequence of functional PPRE-binding sites. It is generally acknowledged that the flanking sequences of the DR-1 site confer additional specificity to the PPRE. Given the fact that multiple PPAR isotypes often coexist in the same cell, it is still unknown whether these isotypes exert redundant or non-redundant functions. In this context it is tempting to speculate that subtle differences in DNA sequence make a gene more responsive to one PPAR isotype than to another. In fact it has been reported that the consensus DNA-binding site for PPARβ/δ clearly differs from that of PPARα and γ [85].

9.3. Co-activators and co-repressors

The PPAR/RXR dimer is known to interact with various other proteins, referred to as co-activators and co-repressors depending on their ability to stimulate or inhibit PPAR activity, respectively (for review see Ref. [86] and references therein). These cofactors are thought to modulate transcriptional activity by interacting both with nuclear receptors and basal transcription factors, thereby affecting the rate of assembly of the basal transcription initiation complex. Furthermore, some cofactors can alter the accessibility of genes through acetylation/deacetylation of histones. The acetylation of histones relaxes chromatin structure, thereby making the genes more accessible for the transcriptional machinery. Co-repressor proteins for PPARs include the receptor-interacting protein-140 (RIP-140), the silencing mediator of retinoid and thyroid receptors (SMRT) and the nuclear receptor co-repressor (NCoR) [87–89]. Co-activators known to interact with PPARs include the steroid receptor co-activator SRC-1, the PPAR-binding protein (PBP), the nuclear receptor-activating protein 250 (RAP250), the PPARγ co-activators PGC-1 and PGC-2 and the integrator protein p300 [90–94]. Many of these cofactors have been shown to be expressed simultaneously in multiple tissues and show limited specificity towards the various nuclear hormone receptors. As the name implies, PGC-1 was originally discovered as a co-activator of PPARγ, but is being studied extensively in relation to PPARα in muscle and liver. Indeed, the expression of PGC-1 itself increases in response to fasting and diabetes, i.e. conditions in which fatty acid oxidation is increased. At the same time, however, PGC-1 also serves as co-activator of HNF-4 [95] and of the nuclear respiratory factors NRF-1 and NRF-2, transcription factors that control mitochondrial biogenesis [96]. In fact the most marked phenotypic feature of mice overexpressing PGC-1 in a cardiac-specific manner is exuberant biogenesis of malformed mitochondria [97].

Conversely, for other cofactors there are strong indications that their interaction with PPARs is isotype specific. For instance, the co-repressor NCoR did not influence PPARγ activity, but repressed PPARα activity [89,98]. As mentioned above, the recruitment of cofactors is likely to be ligand dependent. The inhibitory effect of acyl-CoA esters on PPARα is likely to be based on their ability to enhance recruitment of the co-repressor NCoR at the expense of the co-activator SRC1 [60]. Along the same line, it has been reported that bile acids, like cholic acid and chenodeoxycholic acid, repress PPARα activity by interfering with the recruitment of co-activators [99]. The fact that the genetic inactivation of co-activator proteins as PBP and RAP250 is associated with embryonic lethality [94,100] illustrates their biological importance and furthermore shows that loss of one co-activator is not simply compensated for by another co-activator.

9.4. Post-translational modification of PPARs

In addition to regulatory mechanisms based on protein–proteins interactions, solid evidence has been obtained that both PPARα and PPARγ are subject to phosphorylation on serine residues in the N-terminal AF-1 region. Both PPARγ and PPARα have been shown to be phosphorylated in a MAP kinase-dependent manner [101]. However, the reported effects of the phosphorylation of both PPARα and PPARγ are not consistent. For example, in cardiac muscle cells p38 MAP kinase-dependent phosphorylation of PPARα led to a decline in PPARα activity [102], whereas the phosphorylation (of probably the same serine residues) via the ERK MAP kinase pathway appeared to have opposite effects [103]. Very recently, evidence was provided that PPARβ/δ may also be a target of p38 MAP kinase [104]. Although the contention that phosphorylation is important in the regulation of PPAR activity is well accepted, it poses many new questions as to how phosphorylation events in the AF-1 domain cooperate or interfere with ligand binding and co-factor recruitment in the AF-2 domain [105,106]. Obviously, much more research is warranted into the biological significance of PPAR phosphorylation.

10. Concluding remarks

In the present chapter current knowledge on the role of fatty acids in the regulation of genes involved in fatty acid homeostasis is being reviewed, focussing primarily on the role of LXR and SREBP in lipogenesis and, foremost, on the role of PPAR in the transcriptional regulation of fatty acid oxidation. In the past few years extensive research into PPAR biology has taught us that the regulation of PPAR activity involves the simultaneous action of several levels of control. Currently, we are only beginning to understand how these levels of control converge in the control PPAR activity. The multi-level control of this process probably enables fine-tuning and adjustment of lipid metabolism to variations in energy demand of the different tissues.

Acknowledgements

The work of MvB is supported by grants 1998T015 and 2002B018 from the Netherlands Heart Foundation.

References

[1] Choi, J.Y., Stukey, J., Hwang, S.Y., Martin, C.E., 1996. J. Biol. Chem. 271, 3581–3589.
[2] Ntambi, J.M., 1999. J. Lipid Res. 40, 1549–1558.
[3] Amri, E.Z., Bertrand, B., Ailhaud, G., Grimaldi, P., 1991. J. Lipid Res. 32, 1449–1456.
[4] Grimaldi, P.A., Knobel, S.M., Whitesell, R.R., Abumrad, N.A., 1992. Proc. Natl Acad. Sci. USA 89, 10930–10934.
[5] Clarke, S.D., Armstrong, M.K., Jump, D.B., 1990. J. Nutr. 120, 225–231.
[6] Jump, D.B., Clarke, S.D., MacDougald, O., Thelen, A., 1993. Proc. Natl Acad. Sci. USA 90, 8454–8458.
[7] Jump, D.B., Clarke, S.D., Thelen, A., Liimatta, M., 1994. J. Lipid Res. 35, 1076–1084.
[8] Horton, J.D., Shimomura, I., Brown, M.S., Hammer, R.E., Goldstein, J.L., Shimano, H., 1998. J. Clin. Invest. 101, 2331–2339.
[9] Yahagi, N., Shimano, H., Hasty, A.H., Amemiya-Kudo, M., Okazaki, H., Tamura, Y., Iizuka, Y., Shionoiri, F., Ohashi, K., Osuga, J., Harada, K., Gotoda, T., Nagai, R., Ishibashi, S., Yamada, N., 1999. J. Biol. Chem. 274, 35840–35844.
[10] Shimano, H., Yahagi, N., Amemiya-Kudo, M., Hasty, A.H., Osuga, J., Tamura, Y., Shionoiri, F., Iizuka, Y., Ohashi, K., Harada, K., Gotoda, T., Ishibashi, S., Yamada, N., 1999. J. Biol. Chem. 274, 35832–35839.
[11] Ou, J., Tu, H., Shan, B., Luk, A., DeBose-Boyd, R.A., Bashmakov, Y., Goldstein, J.L., Brown, M.S., 2001. Proc. Natl Acad. Sci. USA 98, 6027–6032.
[12] Jump, D.B., 2002. Curr. Opin. Lipidol. 13, 155–164.
[13] Teboul, L., Gaillard, D., Staccini, L., Inadera, H., Amri, E.-Z., Grimaldi, P.A., 1995. J. Biol. Chem. 270, 28183–28187.
[14] Muoio, D.M., Way, J.M., Tanner, C.J., Winegar, D.A., Kliewer, S.A., Houmard, J.A., Kraus, W.E., Dohm, G.L., 2002. Diabetes 51, 901–909.
[15] Van Bilsen, M., de Vries, J.E., van der Vusse, G.J., 1997. Prostaglandins Leukot. Essent. Fatty Acids 57, 30–45.

[16] Brandt, J.M., Djouadi, F., Kelly, D.P., 1998. J. Biol. Chem. 273, 23786–23792.
[17] van Bilsen, M., van der Vusse, G.J., Reneman, R.S., 1998. Pflugers Arch. 437, 2–14.
[18] Van der Lee, K.A.J.M., Vork, M.M., De Vries, J.E., Willemsen, P.H.M., Glatz, J.F.C., Reneman, R.S., Van der Vusse, G.J., Van Bilsen, M., 2000. J. Lipid Res. 41, 41–47.
[19] Van der Lee, K.A.J.M., Willemsen, P.H.M., Samec, S., Seydoux, J., Dulloo, A.G., Pelsers, M.M.A.L., Glatz, J.F.C., van der Vusse, G.J., van Bilsen, M., 2001. J. Lipid Res. 42, 1752–1758.
[20] Mascaro, C., Acosta, E., Ortiz, J.A., Marrero, P.F., Hegardt, F.G., Haro, D., 1998. J. Biol. Chem. 273, 8560–8563.
[21] Gilde, A.J., van der Lee, K.A., Willemsen, P.H., Chinetti, G., van der Leij, F.R., van der Vusse, G.J., Staels, B., van Bilsen, M., 2003. Circ. Res. 92, 518–524.
[22] Watanabe, K., Fujii, H., Takahashi, T., Kodama, M., Aizawa, Y., Ohta, Y., Ono, T., Hasegawa, G., Naito, M., Nakajima, T., Kamijo, Y., Gonzalez, F.J., Aoyama, T., 2000. J. Biol. Chem. 275, 22293–22299.
[23] Leone, T.C., Weinheimer, C.J., Kelly, D.P., 1999. Proc. Natl Acad. Sci. USA 96, 7473–7478.
[24] Schoonjans, K., Watanabe, M., Suzuki, H., Mahfoudi, A., Krey, G., Wahli, W., Grimaldi, P., Staels, B., Yamamoto, T., Auwerx, J., 1995. J. Biol. Chem. 270, 19269–19276.
[25] Martin, G., Schoonjans, K., Lefebvre, A.-M., Staels, B., Auwerx, J., 1997. J. Biol. Chem. 272, 28210–28217.
[26] Schoonjans, K., Staels, B., Auwerx, J., 1996. J. Lipid Res. 37, 907–925.
[27] Schoonjans, K., Peinado-Onsurbe, J., Lefebvre, A.-M., Heyman, R.A., Briggs, M., Deeb, S., Staels, B., Auwerx, J., 1996. EMBO J. 15, 5336–5348.
[28] Cook, W.S., Yeldandi, A.V., Rao, M.S., Hashimoto, T., Reddy, J.K., 2000. Biochem. Biophys Res. Commun. 278, 250–257.
[29] Issemann, I., Green, S., 1990. Nature 347, 645–650.
[30] Tsai, M.J., O'Malley, B.W., 1994. Annu. Rev. Biochem. 63, 451–486.
[31] Kliewer, S.A., Umesono, K., Noonan, D.J., Heyman, R.A., Evans, R.M., 1992. Nature 358, 771–774.
[32] Van der Lee, K.A.J.M., Willemsen, P.H.M., van der Vusse, G.J., van Bilsen, M., 2000. FASEB J. 14, 495–502.
[33] Schrauwen, P., Saris, W.H., Hesselink, M.K., 2001. FASEB J. 15, 2497–2502.
[34] Wu, P., Peters, J.M., Harris, R.A., 2001. Biochem. Biophys. Res. Commun. 287, 391–396.
[35] Kersten, S., Mandard, S., Escher, P., Gonzalez, F.J., Tafuri, S., Desvergne, B., Wahli, W., 2001. FASEB J. 15, 1971–1978.
[36] Fruchart, J.-C., Duriez, P., Staels, B., 1999. Curr. Opin. Lipidol. 10, 245–257.
[37] Grimaldi, P.A., 1999. Lipids 34, S205–S208.
[38] Vanden Heuvel, J.P., 1999. J. Nutr. 129, 575S–580S.
[39] Lee, J.Y., Sohn, K.H., Rhee, S.H., Hwang, D., 2001. J. Biol. Chem. 276, 16683–16689.
[40] Fajas, L., Auboeuf, D., Raspe, E., Schoonjans, K., Lefebvre, A.M., Saladin, R., Najib, J., Laville, M., Fruchart, J.C., Deeb, S., Vidal Puig, A., Flier, J., Briggs, M.R., Staels, B., Vidal, H., Auwerx, J., 1997. J. Biol. Chem. 272, 18779–18789.
[41] Kliewer, S.A., Forman, B.M., Blumberg, B., Ong, E.S., Borgmeyer, U., Mangelsdorf, D.J., Umesono, K., Evans, R.M., 1994. Proc. Natl Acad. Sci. USA 91, 7355–7359.
[42] Aperlo, C., Pognonec, P., Saladin, R., Auwerx, J., Boulukos, K.E., 1995. Gene 162, 297–302.
[43] Braissant, O., Foufelle, F., Scotto, C., Dauça, M., Wahli, W., 1996. Endocrinology 137, 354–366.
[44] Takano, H., Nagai, T., Asakawa, M., Toyozaki, T., Oka, T., Komuro, I., Saito, T., Masuda, Y., 2000. Circ. Res. 87, 596–602.
[45] Escher, P., Braissant, O., Basu-Modak, S., Michalik, L., Wahli, W., Desvergne, B., 2001. Endocrinology 142, 4195–4202.
[46] Vidal-Puig, A.J., Considine, R.V., Jimenez-Liñan, M., Werman, A., Pories, W.J., Caro, J.F., Flier, J.S., 1997. J. Clin. Invest. 99, 2416–2422.
[47] Park, K.S., Ciaraldi, T.P., Abrams-Carter, L., Mudaliar, S., Nikoulina, S.E., Henry, R.R., 1997. Diabetes 46, 1230–1234.
[48] Young, M.E., Razeghi, P., Taegtmeyer, H., 2001. Circ. Res. 88, 1142–1150.
[49] Blanquart, C., Barbier, O., Fruchart, J.C., Staels, B., Glineur, C., 2002. J. Biol. Chem. 277, 37254–37259.

[50] Gervois, P., Torra, I.P., Chinetti, G., Grötzinger, T., Dubois, G., Fruchart, J.-C., Fruchart-Najib, J., Leitersdorf, E., Staels, B., 1999. Mol. Endocrinol. 13, 1535–1549.
[51] Xu, H.E., Lambert, M.H., Montana, V.G., Parks, D.J., Blanchard, S.G., Brown, P.J., Sternbach, D.D., Lehmann, J.M., Wisely, G.B., Willson, T.M., Kliewer, S.A., Milburn, M.V., 1999. Mol. Cell 3, 397–403.
[52] Gottlicher, M., Widmark, E., Li, Q., Gustafsson, J.A., 1992. Proc. Natl Acad. Sci. USA 89, 4653–4657.
[53] Issemann, I., Prince, R.A., Tugwood, J.D., Green, S., 1993. J. Mol. Endocrinol. 11, 37–47.
[54] Forman, B.M., Tontonoz, P., Chen, J., Brun, R.P., Spiegelman, B.M., Evans, R.M., 1995. Cell 83, 803–812.
[55] Kliewer, S.A., Sundseth, S.S., Jones, S.A., Brown, P.J., Wisely, G.B., Koble, C.S., Devchand, P., Wahli, W., Willson, T.M., Lenhard, J.M., Lehmann, J.M., 1997. Proc. Natl Acad. Sci. USA 94, 4318–4323.
[56] Forman, B.M., Chen, J., Evans, R.M., 1997. Proc. Natl Acad. Sci. USA 94, 4312–4317.
[57] Johnson, T.E., Holloway, M.K., Vogel, R., Rutledge, S.J., Perkins, J.J., Rodan, G.A., Schmidt, A., 1997. J. Steroid Biochem. Mol. Biol. 63, 1–8.
[58] Wolfrum, C., Borrmann, C.M., Borchers, T., Spener, F., 2001. Proc. Natl Acad. Sci. USA 98, 2323–2328.
[59] Lee, J.Y., Hwang, D.H., 2002. Biochem. Biophys. Res. Commun. 298, 667–674.
[60] Elholm, M., Dam, I., Jorgensen, C., Krogsdam, A.M., Holst, D., Kratchmarova, I., Gottlicher, M., Gustafsson, J.A., Berge, R., Flatmark, T., Knudsen, J., Mandrup, S., Kristiansen, K., 2001. J. Biol. Chem. 276, 21410–21416.
[61] Murakami, K., Ide, T., Nakazawa, T., Okazaki, T., Mochizuki, T., Kadowaki, T., 2001. Biochem. J. 353, 231–238.
[62] Krey, G., Braissant, O., L'Horset, F., Kalkhoven, E., Perroud, M., Parker, M.G., Wahli, W., 1997. Mol. Endocrinol. 11, 779–791.
[63] Murakami, K., Ide, T., Suzuki, M., Mochizuki, T., Kadowaki, T., 1999. Biochem. Biophys. Res. Commun. 260, 609–613.
[64] van der Vusse, G.J., Glatz, J.F.C., Stam, H.C.G., Reneman, R.S., 1992. Physiol. Rev. 72, 881–940.
[65] Borchers, T., Unterberg, C., Rudel, H., Robenek, H., Spener, F., 1989. Biochim. Biophys. Acta 1002, 54–61.
[66] Tan, N.S., Shaw, N.S., Vinckenbosch, N., Liu, P., Yasmin, R., Desvergne, B., Wahli, W., Noy, N., 2002. Mol. Cell. Biol. 22, 5114–5127.
[67] Han, C., Demetris, A.J., Michalopoulos, G., Shelhamer, J.H., Wu, T., 2002. Am. J. Physiol. Gastrointest. Liver Physiol. 282, G586–G597.
[68] Pawliczak, R., Han, C., Huang, X.L., Demetris, A.J., Shelhamer, J.H., Wu, T., 2002. J. Biol. Chem. 277, 33153–33163.
[69] Ziouzenkova, O., Perrey, S., Asatryan, L., Hwang, J., MacNaul, K.L., Moller, D.E., Rader, D.J., Sevanian, A., Zechner, R., Hoefler, G., Plutzky, J., 2003. Proc. Natl Acad. Sci. USA 100, 2730–2735.
[70] Berger, J., Leibowitz, M.D., Doebber, T.W., Elbrecht, A., Zhang, B., Zhou, G., Biswas, C., Cullinan, C.A., Hayes, N.S., Li, Y., Tanen, M., Ventre, J., Wu, M.S., Berger, G.D., Mosley, R., Marquis, R., Santini, C., Sahoo, S.P., Tolman, R.L., Smith, R.G., Moller, D.E., 1999. J. Biol. Chem. 274, 6718–6725.
[71] Oliver, W.R. Jr., Shenk, J.L., Snaith, M.R., Russell, C.S., Plunket, K.D., Bodkin, N.L., Lewis, M.C., Winegar, D.A., Sznaidman, M.L., Lambert, M.H., Xu, H.E., Sternbach, D.D., Kliewer, S.A., Hansen, B.C., Willson, T.M., 2001. Proc. Natl Acad. Sci. USA 98, 5306–5311.
[72] Chawla, A., Barak, Y., Nagy, L., Liao, D., Tontonoz, P., Evans, R.M., 2001. Nat. Med. 7, 48–52.
[73] Yamamoto, K., Ohki, R., Lee, R.T., Ikeda, U., Shimada, K., 2001. Circulation 104, 1670–1675.
[74] Brunmair, B., Gras, F., Neschen, S., Roden, M., Wagner, L., Waldhausl, W., Furnsinn, C., 2001. Diabetes 50, 2309–2315.
[75] Hertz, R., Sheena, V., Kalderon, B., Berman, I., Bar-Tana, J., 2001. Biochem. Pharmacol. 61, 1057–1062.
[76] Carroll, R., Severson, D.L., 2001. Am. J. Physiol. Heart Circ. Physiol. 281, H888–H894.
[77] Juge-Aubry, C.E., Gorla-Bajszczak, A., Pernin, A., Lemberger, T., Wahli, W., Burger, A.G., Meier, C.A., 1995. J. Biol. Chem. 270, 18117–18122.
[78] Meier-Heusler, S.C., Zhu, X., Juge-Aubry, C., Pernin, A., Burger, A.G., Cheng, S.-y., Meier, C.A., 1995. Mol. Cell. Endocrinol. 107, 55–66.
[79] Winrow, C.J., Kassam, A., Miyata, K.S., Marcus, S.L., Hunter, J., Capone, J.P., Rachubinski, R.A., 1996. Ann. NY Acad. Sci. 804, 214–230.

[80] Marcus, S.L., Capone, J.P., Rachubinski, R.A., 1996. Mol. Cell. Endocrinol. 120, 31–39.
[81] Huss, J.M., Levy, F.H., Kelly, D.P., 2001. J. Biol. Chem. 276, 27605–27612.
[82] Osorio, J.C., Stanley, W.C., Linke, A., Castellari, M., Diep, Q.N., Panchal, A.R., Hintze, T.H., Lopaschuk, G.D., Recchia, F.A., 2002. Circulation 106, 606–612.
[83] Miyata, K.S., Zhang, B., Marcus, S.L., Capone, J.P., Rachubinski, R.A., 1993. J. Biol. Chem. 268, 19169–19172.
[84] Sack, M.N., Rader, T.A., Park, S., Bastin, J., McCune, S.A., Kelly, D.P., 1996. Circulation 94, 2837–2842.
[85] He, T.C., Chan, T.A., Vogelstein, B., Kinzler, K.W., 1999. Cell 99, 335–345.
[86] McKenna, N.J., Lanz, R.B., O'Malley, B.W., 1999. Endocr. Rev. 20, 321–344.
[87] Miyata, K.S., McCaw, S.E., Meertens, L.M., Patel, H.V., Rachubinski, R.A., Capone, J.P., 1998. Mol. Cell. Endocrinol. 146, 69–76.
[88] Robinson, C.E., Wu, X., Nawaz, Z., Oñate, S.A., Gimble, J.M., 1999. Endocrinology 140, 1586–1593.
[89] Dowell, P., Ishmael, J.E., Avram, D., Peterson, V.J., Nevrivy, D.J., Leid, M., 1999. J. Biol. Chem. 274, 15901–15907.
[90] Zhu, Y., Qi, C., Calandra, C., Rao, M.S., Reddy, J.K., 1996. Gene Expression 6, 185–195.
[91] Zhu, Y., Qi, C., Jain, S., Rao, M.S., Reddy, J.K., 1997. J. Biol. Chem. 272, 25500–25506.
[92] Puigserver, P., Wu, Z., Park, C.W., Graves, R., Wright, M., Spiegelman, B.M., 1998. Cell 92, 829–839.
[93] Gelman, L., Zhou, G., Fajas, L., Raspé, E., Fruchart, J.-C., Auwerx, J., 1999. J. Biol. Chem. 274, 7681–7688.
[94] Antonson, P., Schuster, G.U., Wang, L., Rozell, B., Holter, E., Flodby, P., Treuter, E., Holmgren, L., Gustafsson, J.A., 2003. Mol. Cell. Biol. 23, 1260–1268.
[95] Boustead, J.N., Stadelmaier, B.T., Eeds, A.M., Wiebe, P.O., Svitek, C.A., Oeser, J.K., O'Brien, R.M., 2003. Biochem. J. 369, 17–22.
[96] Wu, Z., Puigserver, P., Andersson, U., Zhang, C., Adelmant, G., Mootha, V., Troy, A., Cinti, S., Lowell, B., Scarpulla, R.C., Spiegelman, B.M., 1999. Cell 98, 115–124.
[97] Lehman, J.J., Barger, P.M., Kovacs, A., Saffitz, J.E., Medeiros, D.M., Kelly, D.P., 2000. J. Clin. Invest. 106, 847–856.
[98] Lavinsky, R.M., Jepsen, K., Heinzel, T., Torchia, J., Mullen, T.-M., Schiff, R., Del-Rio, A.L., Ricote, M., Ngo, S., Gemsch, J., Hisenbeck, S.G., Osborne, C.K., Glass, C.K., Rosenfeld, M.G., Rose, D.W., 1998. Proc. Natl Acad. Sci. USA 95, 2920–2925.
[99] Sinal, C.J., Yoon, M., Gonzalez, F.J., 2001. J. Biol. Chem. 276, 47154–47162.
[100] Zhu, Y., Qi, C., Jia, Y., Nye, J.S., Rao, M.S., Reddy, J.K., 2000. J. Biol. Chem. 275, 14779–14782.
[101] Juge-Aubry, C.E., Hammar, E., Siegrist-Kaiser, C., Pernin, A., Takeshita, A., Chin, W.W., Burger, A.G., Meier, C.A., 1999. J. Biol. Chem. 274, 10505–10510.
[102] Barger, P.M., Browning, A.C., Garner, A.N., Kelly, D.P., 2001. J. Biol. Chem. 276, 44495–44501.
[103] Barger, P.M., Brandt, J.M., Leone, T.C., Weinheimer, C.J., Kelly, D.P., 2000. J. Clin. Invest. 105, 1723–1730.
[104] Hellemans, K., Michalik, L., Dittie, A., Knorr, A., Rombouts, K., De Jong, J., Heirman, C., Quartier, E., Schuit, F., Wahli, W., Geerts, A., 2003. Gastroenterology 124, 184–201.
[105] Camp, H.S., Tafuri, S.R., 1997. J. Biol. Chem. 272, 10811–10816.
[106] Camp, H.S., Tafuri, S.R., Leff, T., 1999. Endocrinology 140, 392–397.
[107] Schoonjans, K., Staels, B., Grimaldi, P., Auwerx, J., 1993. Eur. J. Biochem. 216, 615–622.

Triacylglycerol metabolism in adipose tissue

Keith N. Frayn[a,*] and Dominique Langin[b]

[a]*Oxford Centre for Diabetes, Endocrinology and Metabolism, University of Oxford, Churchill Hospital, Oxford OX3 7JY, UK*
[b]*Unité de Recherches sur les Obésités, Institut National de la Santé et de la Recherche Médicale (INSERM) Unité 586, Institut Louis Bugnard, Centre Hospitalier Universitaire de Toulouse, Université Paul Sabatier, Toulouse, France*
[*]*Correspondence address: Tel.: +44-1865-857220; fax: +44-1865-857217*
E-mail: keith.frayn@oxlip.ox.ac.uk(K.N.F.)

1. Introduction

There are two types of adipose tissue, distinguished histologically and functionally: brown adipose tissue (BAT) and white adipose tissue (WAT). They have in common cells (adipocytes) that store lipid in the form of triacylglycerol (TG). BAT is distinguished from WAT macroscopically by its brown appearance, which in turn reflects the presence of many mitochondria. Under the microscope, brown adipocytes have multilocular fat droplets, whereas white adipocytes typically have one large fat droplet, occupying most of the volume of the cell. The mitochondria in brown adipocytes are densely packed into the cell, and each mitochondrion has complex cristae. In contrast, white adipocytes have relatively few mitochondria and a small volume of cytoplasm. However, significant mitochondrial biogenesis and remodelling appear to be inherent to white adipocyte differentiation. These new data challenge the commonly held view of a cell with little involvement of mitochondria in metabolism [1].

Functionally, BAT is a tissue specialized for production of heat. It is prevalent in small animals, especially when exposed to cold, and in hibernating animals. It is present in human neonates but is not found in significant quantities in adult humans. Its fat stores turn over rapidly, and fatty acids liberated from the stores are oxidized within the adipocytes in a process that generates heat directly. Heat production in BAT is stimulated by the sympathetic innervation of the tissue, which also increases blood flow, so delivering the heat to the rest of the body. We will not consider BAT further in this review and instead refer the reader to recent reviews on this topic [2,3].

WAT, in contrast, is characterized by low O_2 consumption. Typically, WAT accounts for 20–30% of body weight but for less than 2% of whole-body O_2 consumption [4]. This does not mean that it is a metabolically inert tissue, however. Its physiological role is the storage of excess "energy" in the form of TG. The pathways by which WAT accumulates

Advances in Molecular and Cell Biology, Vol. 33, pages 337–356

ISSN: 1569-2558 / DOI: 10.1016/S1569-2558(03)33017-6

and mobilizes fat are dynamic. They are regulated on a minute-to-minute basis, providing an essential function of "buffering" the influx of fat into the circulation that occurs after meals [5], and also delivering fatty acids from the TG stores into the circulation according to the body's need for energy. Indeed, normal health depends upon having WAT in relatively normal amounts. Conditions in which there is either too little WAT (lipodystrophy) or too much (obesity) are associated with insulin resistance, hypertension, dyslipidaemia and an increased risk of developing type 2 diabetes.

WAT is a complex tissue, as befits its important regulatory role. It is far more than just a collection of adipocytes. There are a number of other cell types present in adipose tissue: endothelial cells lining the blood vessels, macrophages, fibroblast-like adipocyte precursor cells and nerve cells, amongst others. They act in an integrated way. For instance, blood flow to WAT is highly regulated; it may increase several-fold on feeding and it increases during exercise. This is not surprising considering that the main metabolic exchanges between adipocytes and the blood are of hydrophobic molecules, TG and fatty acids, which cannot diffuse easily through interstitial spaces; blood flow is necessary to deliver and remove these molecules. Individual adipocytes can expand enormously in conditions of energy excess, but the cellularity of WAT is also plastic, with differentiation of new adipocytes from precursor cells when required for additional fat storage [6,7], and apoptosis of adipocytes when fat stores decrease [8]. Some important aspects of TG metabolism in adipose tissue can, indeed, only be understood by considering adipose tissue as an integrated whole. In particular, the major pathway of fat storage (described in more detail below) involves the action of the enzyme lipoprotein lipase (LPL), attached to the luminal aspect of the capillary endothelium, on circulating lipoprotein-TG. LPL delivers fatty acids that reach the adipocytes for uptake and storage.

WAT is distributed throughout the body in discrete depots, which are homologous amongst all mammals, although different depots assume greater or lesser importance in different species [9,10]. There are broad similarities in the behaviour of the larger WAT depots although it has been argued that some of the smaller depots may be specialized for functions other than "depot" fat storage [11]. Nevertheless, biologically, fat depots in different parts of the body relate to aspects of health or disease in typical ways that will be discussed in more detail below.

The amount of TG stored in WAT is determined by the balance between whole-body energy intake and energy expenditure. If intake consistently exceeds expenditure, then there must be an increase in the body's energy stores. Protein is not an energy store in the usual sense of that term. Although amino acids are indeed oxidized to produce energy during prolonged starvation, there is no specific storage form of protein, and the response to starvation is, in fact, characterized by preservation of body protein to as great an extent as possible. Glycogen is a relatively small energy store in relation to energy balance: the body's glycogen store (about 500 g in liver and skeletal muscle together) is about one day's worth of energy. Indeed, liver glycogen is almost completely depleted after fasting for 24 h [12]. TG in white adipocytes is the body's long-term repository for excess food energy. In turn, it is the most important fuel store for starvation. A typical body fat mass of 15 kg would be sufficient for approximately 50–60 days of total starvation, and this figure accords well with such data as there are on the limits of survival of initially normal-weight humans [13]. However, obese subjects have been totally fasted for much longer periods (over 120 days in some reports; [14]). This shows the evolutionary importance of fat storage in adipose tissue. For much of human history, even until modern times, famine has been an evolutionary driving force [15]. The ability to store excess energy away safely as TG in adipocytes must have given an enormous survival advantage. Nowadays, of course, when dietary energy is not in short supply in many parts of the world, we are facing the consequences of our evolutionary history.

The way in which TG storage in adipocytes has to agree with whole-body energy balance suggests that there must be powerful regulatory mechanisms at work, controlling the processes of fat deposition and fat mobilization. In fact, we now recognize also that adipocytes can feed information back to the rest of the body about the size of the TG store. Some of these complex regulatory mechanisms will be covered in the following sections.

2. Physiology of fat storage in white adipose tissue

TG in the adipocyte may arise basically from one of two routes: de novo lipogenesis from non-lipid precursors, or uptake of fatty acids from the plasma. These plasma fatty acids might in principle derive either from circulating TG, or from the non-esterified fatty acid (NEFA) pool bound to albumin in plasma.

2.1. De novo lipogenesis

If a person, or animal, is over-fed carbohydrate, there is a limit to the amount of glucose or glycogen that can accumulate in the body, and hence there must be a pathway for the synthesis of lipid from carbohydrate. This is the physiological significance of de novo lipogenesis.

The starting point for de novo lipogenesis is acetyl-CoA. If this arises from glucose or amino acid catabolism, it will be mitochondrial. The enzymes for fatty acid synthesis are cytosolic, so mitochondrial acetyl-CoA must first be translated to the cytoplasm. This is done by condensation with oxaloacetate to form citrate (the normal route of entry into the tricarboxylic acid cycle). Citrate can move out of the mitochondria via the mitochondrial tricarboxylate carrier. In the cytosol it is split by ATP:citrate lyase to generate cytosolic acetyl-CoA (the oxaloacetate can be converted to pyruvate and re-enter the mitochondria, where pyruvate can be reconverted to oxaloacetate by pyruvate carboxylase). Cytosolic acetyl-CoA is the precursor for both fatty acid and cholesterol synthesis.

In humans, net de novo lipogenesis may be less important than in rodents. The whole-body respiratory exchange ratio does not exceed 1.00 in most circumstances in normal physiology, implying net fat oxidation rather than fat synthesis. Even following a very large carbohydrate-rich meal, net de novo lipogenesis is not a route for storage of excess energy [16]. Net whole-body lipogenesis is seen in humans only under extreme conditions: for instance, during over-feeding (i.e. energy intake > energy requirements) with a carbohydrate-rich diet [17,18], or during total parenteral nutrition with glucose as the main energy substrate [19].

Two tissues are quantitatively important for whole-body de novo lipogenesis in mammals: liver and WAT. (Mammary gland is also important during lactation.) WAT de novo lipogenesis can readily be demonstrated in small animals, whose diet is usually rich in carbohydrate and low in fat. In humans, it seems to be less important. The activity of ATP:citrate lyase is normally very low in human adipose tissue [20,21]. On the other hand, the expression of ATP:citrate lyase, like that of acetyl-CoA carboxylase and fatty acid synthase, the other enzymes required for synthesis of fatty acids, is up-regulated by insulin [22,23]. It cannot therefore be ruled out that de novo lipogenesis might operate to a significant extent in human adipose tissue when there is a need to dispose of "excess" carbohydrate. In one report of healthy subjects over-fed with a carbohydrate-rich diet, hepatic de novo lipogenesis could account for only a small proportion of whole-body net fat synthesis measured by indirect calorimetry [18]. The authors concluded that adipose tissue was likely to be a major site for de novo lipogenesis. Other studies of lipogenesis in adipose tissue explants taken from patients on intravenous carbohydrate feeding seem to confirm this [24]. However, the opposite conclusion was reached from a recent study of healthy volunteers over-fed carbohydrate for 2 weeks: adipose tissue lipogenesis was very small compared with hepatic [25]. In summary, even in the circumstances in which net whole-body lipogenesis is observed in humans, the liver seems to play the dominant role [26]. In further support of a limited role for WAT de novo lipogenesis under normal, eucaloric conditions, the fatty acid composition of adipose tissue TG in humans reflects dietary fatty acid intake, and indeed has been used as a marker of dietary fat in epidemiological studies [27,28].

There is one situation in which de novo lipogenesis might be important, apart from the high-energy high-carbohydrate diet: deficiency of LPL, the principal enzyme of the alternative route for TG storage in WAT (discussed below). People with complete LPL deficiency (type 1 hyperlipoproteinaemia) still have normal adipose depots and fat-filled white adipocytes [29,30]. In mice lacking LPL specifically in WAT, adipose depots are also normal but the fatty acid composition of adipose tissue suggests that this is achieved by up-regulation of de novo lipogenesis [31]. Limited data on human adipose tissue fatty acid composition in LPL deficiency seem to confirm this [32].

2.2. *Lipoprotein lipase and the uptake of TG-fatty acids*

The major route for fat deposition in human WAT, and probably also in other species, is the uptake of pre-formed fatty acids from circulating TG. In the fasting state circulating TG is mainly present in very low-density lipoprotein (VLDL) particles. After a meal rich in fat, dietary TG enters the circulation in the form of chylomicron particles. Chylomicron particles are larger than VLDL but less numerous, so they are outnumbered by about 20:1 even in the postprandial state [33]. Nevertheless, chylomicron-TG appears to be the favoured substrate for fatty acid uptake into WAT. This provides a simple, energy-efficient pathway for the storage of dietary fatty acids in WAT. It explains why the fatty acid composition of adipose tissue TG in humans reflects dietary fatty acid intake, as discussed above.

LPL is an enzyme with a complex cell biology. It is expressed in a number of extra-hepatic tissues, and is regulated on a tissue-specific basis. WAT LPL is up-regulated by insulin, giving WAT a special role in removal of circulating TG in the postprandial period.

The uptake of labelled fatty acids (fed orally) into adipose tissues in rats subjected to various nutritional interventions is directly proportional to LPL activity measured in the same tissue [34]. LPL is synthesized in the adipocytes themselves (or the parenchymal cells of other tissues) but is then exported to the endothelium, where it is bound to the luminal surface of endothelial cells by interaction with cell-surface glycosaminoglycans, especially heparan sulphate. The TG-rich lipoprotein particles, VLDL and chylomicrons, are too large to penetrate the endothelial lining of capillaries. Therefore, endothelial-bound LPL can act upon these particles in the vascular space. It has been shown by enzyme-kinetic calculations and by binding-site studies that several LPL molecules (perhaps around 40) may act simultaneously upon a TG-rich lipoprotein particle [35]. The action of LPL liberates fatty acids from the particle TG. These fatty acids then move, presumably through a structured pathway (although this is not clearly understood), through the endothelial lining and to the adipocytes, where they are taken up. Details of this process remain elusive. Electron-micrographic studies suggest that the particles become intimately associated with, and partially engulfed by, the endothelial cells [36].

Many studies of LPL action in vitro show that it is specific for the *sn*-1(3) ester bonds in a TG molecule [37]. If albumin (to bind fatty acids) is limiting, LPL action in vitro results in the generation of *sn*-2-monoacylglycerols (2-MAG) [35]. However, it has not been possible to detect MAG generation in studies made by arterio-venous difference measurement in vivo even when TG hydrolysis is proceeding at a high rate [38,39]. This implies either that the 2-MAG are taken up into the tissue, or that non-enzymatic isomerization occurs to produce *sn*-1(3)-MAG which may then be hydrolyzed by LPL. Further studies in this area are needed.

The movement of fatty acids into the adipocyte from the site of LPL action must be governed by concentration gradients (since there is no evidence for active transport of fatty acids). This necessitates the coordination of several metabolic pathways in the adipocyte. When LPL is most active, in the postprandial state with high insulin concentrations, the opposing process of fat mobilization is suppressed (described in detail later). In addition, the pathway of fatty acid esterification in the cell to make new TG is stimulated by insulin. Thus, there is a net "downhill" gradient to move fatty acids into the adipocyte. Regulation of this process appears to be quite different in WAT from the situation in skeletal muscle. In skeletal muscle, there is always a strong gradient favouring fatty acid uptake [40] and there is no evidence, in vivo, for fatty acids generated by LPL "escaping" into the circulation [41]. In WAT, in contrast, this gradient is continuously changing with nutritional or physiological state. In fasting or during exercise, for example, when fat mobilization is active, it is difficult to see how fatty acids would enter the cell when the net flux of fatty acids is strongly "outward" from the adipocytes to the capillaries. Exactly in accord with this view, studies in vivo show that the uptake of LPL-derived fatty acids is regulated and changes with time and nutritional state [41]. If lipid is provided intravenously, without a rise in insulin concentration, then WAT LPL is still active against the infused TG-rich particles, but in that case the fatty acids are released almost quantitatively into the plasma NEFA pool [42].

LPL plays a key role in fat partitioning [43] and, through delivery of fatty acids to skeletal muscle, may play an important role in the genesis of insulin resistance as a result of competition between fatty acid and glucose. LPL-deficient mice are normal at birth, but develop lethal hypertriglyceridaemia within the first day of life [44]. To directly assess the role of LPL in adipose tissue, LPL heterozygous knockout mice have been crossed with transgenic mice expressing human LPL in skeletal muscle and heart [45]. Through backcrossing, mice expressing LPL exclusively in muscle were obtained. Growth and body composition were not altered by the lack of LPL in adipose tissue on a standard genetic background. However, when adipose tissue LPL deficiency was obtained on a leptin-deficient *ob/ob* background, the rate of weight gain was decreased due to an impaired accumulation of lipid in adipose tissue. TG content was increased in skeletal muscle, suggesting partial reallocation of dietary fat storage from adipose tissue to skeletal muscle. As noted earlier, the fatty acid composition of the TG stored in adipose tissue was markedly modified in adipose tissue LPL-deficient mice and suggested that the development of fat stores in adipose tissue LPL-deficient mice relies on de novo lipogenesis. On a genetically obese background, this compensatory mechanism does not keep up with the massive weight gain programmed in *ob/ob* mice due to leptin deficiency.

The other source of fatty acids for WAT in principle is the plasma NEFA pool. However, since plasma NEFA themselves arise from WAT, this would constitute a futile metabolic cycle. It is quite possible to show uptake of NEFA by adipocytes in vitro [46], but this does not seem to occur in vivo under normal conditions. The net exchange of NEFA across WAT in vivo is always a release rather than an uptake. When NEFA uptake has been studied specifically using an infusion of labelled fatty acids, some uptake has been shown only under the non-physiological conditions of high glucose and insulin produced by intravenous glucose infusion [47].

The mechanism by which fatty acids enter the adipocyte is still debated although increasing evidence points to a substantial role for facilitated diffusion, with fatty acid translocase (FAT)/CD36 playing an important role in adipose tissue [48]. The recent demonstration that FAT translocates to the cell membrane when stimulated by insulin in skeletal and cardiac muscle [49,50] suggests a role for this transporter in fatty acid uptake, perhaps more than in fatty acid release. The adipocytes of CD36 null mice lack the high affinity component of long chain fatty acid transport observed in wild type fat cells [51]. Furthermore, there is a defective uptake of fatty acids in vivo, which results in impaired TG synthesis in WAT [52]. The defective fatty acid esterification is most likely due to a limiting supply of acyl-CoA. These aspects are covered more fully in other chapters. Fatty acid binding proteins (FABPs) within the cell sequester the fatty acids, again as discussed in detail in other chapters. However, evidence that the process of membrane transport of fatty acids is intimately associated with their "activation" (esterification with coenzyme A) [53,54] means that it is difficult at present to envisage how fatty acid entry, binding to FABPs and activation all inter-relate. There is also an acyl-CoA binding protein that would sequester the product of activation. Activation is catalyzed by acyl-CoA synthase and uses ATP, which is hydrolyzed to AMP and PPi. The production of PPi, which itself is rapidly hydrolyzed, is considered to render a reaction virtually irreversible. This may be part of the process whereby the concentration gradient for fatty uptake is maintained.

The pathway for fatty acid esterification to form TG in adipocytes involves the sequential addition of acyl-CoA to a glycerol "backbone". There are two major pathways for TG synthesis in mammals: the MAG pathway and the glycerol 3-phosphate (or phosphatidic acid) pathway. The former predominates in enterocytes and is involved in intestinal fat absorption, 2-MAGs having been formed in the intestinal lumen by pancreatic lipase, an enzyme closely related to LPL. This suggests again, by analogy, the possibility of 2-MAG generation by LPL with subsequent MAG uptake by adipocytes. However, the MAG esterification pathway seems to be of minor importance in adipocytes, contributing only around 5–10% of total glycerolipid synthesis when measured in cell homogenates [55,56]. Even that measurement necessitates the use of synthetic (non-hydrolysable) MAG analogues as substrates because there is a highly active MAG lipase in adipocytes [57] that hydrolyses the 2-MAG to glycerol and fatty acid. This makes a major role for the MAG esterification pathway in fat deposition in WAT unlikely. Instead it may be a "salvage" pathway for partially hydrolyzed TG within the adipocyte [55].

The phosphatidic acid pathway starts with glycerol 3-phosphate. The enzymes of this pathway have mostly been cloned within the past few years although several remain only partially characterized, since these are insoluble membrane-bound enzymes that are not easy to work with. The pathway and its regulation have been reviewed recently [58,59] and will not be described in detail here. An important feature is that the esterification pathway is stimulated by insulin. This has been known for many years from studies both in vitro [60] and in vivo [61,62], but the locus of insulin's acute action is still not clear. In the longer term, insulin up-regulates gene expression for several of the enzymes of the esterification pathway [58,59]. In addition, insulin may promote glycerol 3-phosphate availability. Glycerol 3-phosphate is produced from the pathway of glycolysis (from dihydroxyacetone phosphate, by glycerol 3-phosphate dehydrogenase). Since glucose uptake by adipocytes is exquisitely sensitive to stimulation by insulin [63,64], insulin may increase glycerol 3-phosphate supply. However, studies in vivo in humans suggest that control is exerted more on the esterification pathway itself [62].

Data from transgenic animals suggest that conversion of diacylglycerol to TG at the level of diacylglycerolacyltransferase (DGAT) is a critical step in vivo. Mice lacking DGAT are viable and fertile [65]. The animals are capable of synthesizing TG and have normal body weight on a standard chow diet. The fat pad weights are slightly lower than in wild-type control mice. However, DGAT-deficient mice are resistant to diet-induced obesity, which appears to be due to increased energy expenditure. The mechanism underlying the changes in metabolic rate is unclear. It does not result from increased lean body mass or changes in cold-induced thermogenesis. Puzzlingly, the study also shows that TG synthesis can occur without DGAT. This suggests the existence of another enzyme with DGAT activity. Indeed, such an enzyme has been characterized and may partially compensate for the lack of DGAT [66].

The net result is that TG synthesis in adipose tissue is stimulated by insulin at multiple stages: activation of LPL, possibly uptake of fatty acids across the cell membrane, activation of the pathway of de novo fatty acid synthesis and stimulation of fatty acid esterification to form TG. In parallel, insulin inhibits fat mobilization, so the net effect on WAT TG stores is strongly "anabolic".

3. Physiology of fat mobilization

Fat storage in WAT may have several functions. Well-filled fat stores may provide insulation and mechanical cushioning, for instance. But undoubtedly the major physiological role for fat stores is to supply lipid energy when it is needed by other tissues. This is achieved by a highly active, and highly regulated, pathway whereby the TG stored in the adipocyte is hydrolyzed, and fatty acids delivered to the plasma.

Fat mobilization is regulated by multiple mechanisms [67]. In general, fat mobilization and fat deposition are reciprocally regulated, so that fatty acids flow "in" and "out" of the adipocyte according to the nutritional and physiological state of the organism (Fig. 1).

Fat mobilization is stimulated primarily (at least acutely) by catecholamines acting through β-adrenoceptors, 7-transmembrane domain GTP-binding protein-coupled receptors in the cell membrane. These stimulate adenylyl cyclase, producing 3′,5′-cyclic adenosine monophosphate (cAMP) from ATP. Other signals also play a role. A recently described pathway involves atrial natriuretic peptide (ANP, a peptide secreted from the heart in response to volume overload) and brain natriuretic peptide (BNP). These peptides stimulate fat mobilization independently of the cAMP pathway [68]. Human adipocytes express ANP receptors of the A subtype that possess guanylyl cyclase activity. Following an increase in 3′,5′-cyclic guanosine monophosphate (cGMP) levels, hormone-sensitive lipase (HSL), the main enzyme responsible for cellular TG hydrolysis, is phosphorylated by cGMP-dependent protein kinase.

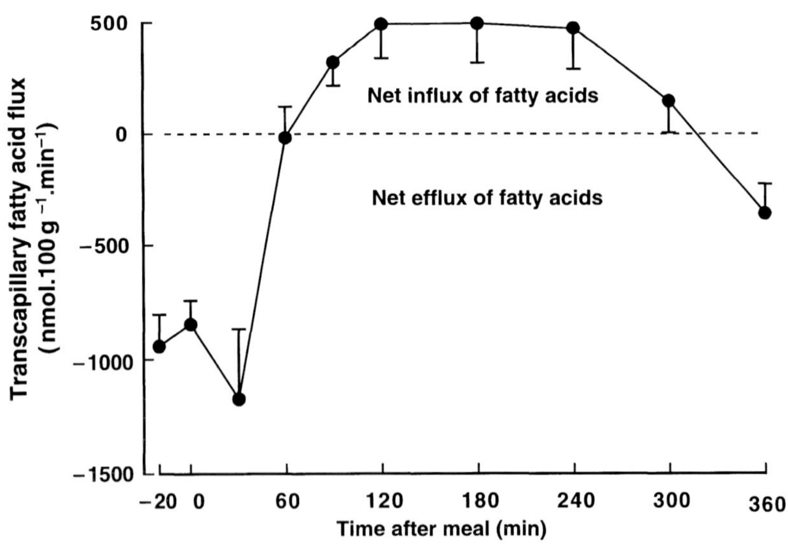

Fig. 1. "Transcapillary flux" of fatty acids in human adipose tissue in vivo, in the fasting and postprandial states, in 35 healthy subjects. The figure shows the changing direction of fatty acid movement, "outwards" from adipocytes to capillaries in the fasting state, "inwards" from capillaries to adipocytes in the fed state. Reproduced from Ref. [5] with permission.

There is also powerful inhibitory control of fat mobilization. Fat mobilization is strongly suppressed by insulin, giving the reciprocal regulation referred to above. Insulin acts through the normal signal chain for acute metabolic events, i.e. via phosphatidylinositol-3′-kinase forming phosphatidylinositol (3′,4′,5′)-trisphosphate (PIP_3), which activates protein kinase B (PKB). PKB then phosphorylates and activates a specific isoform of cAMP-phosphodiesterase (PDE), PDE3B. PDE3B hydrolyses cAMP to AMP, so reducing cAMP concentrations [69]. Therefore, the cellular cAMP concentration is a major integrator for the regulation of fat mobilization. Other antilipolytic pathways involve α_2-adrenergic receptors, A_1-adenosine-receptors, EP_3-prostaglandin E_2 receptors and neuropeptide Y/peptide YY (NPY-1) receptors. The existence of inhibitory nicotinic acid receptors is proposed to explain the well-known antilipolytic action of nicotinic acid. The receptor protein has recently been identified [70,71]. The 7 transmembrane domain receptors are coupled through inhibitory GTP-binding proteins, Gi, to adenylyl cyclase. This multiplicity of controls presumably reflects the importance of precise control of fatty acid delivery to the circulation.

In normal physiological states, the main effectors seem to be catecholamines and insulin. After an overnight fast, the catecholamine effect is actually a tonic inhibition via α-adrenoceptors [72,73]. Insulin's suppressive effect is low in this state. The implication seems to be that the normal state of fat mobilization is "on". The "tone" of fat mobilization after an overnight fast also seems to be set by overnight secretion of growth hormone [74] and by the morning rise in cortisol [75]. During exercise, adrenergic stimulation of lipolysis is undoubtedly important. Exercise in the presence of β-adrenergic blockers, whether introduced locally into the tissue or given systemically, is associated with much lower rates of fat mobilization [76,77]. The main adrenergic stimulus may be circulating adrenaline (from the adrenal medulla) or activation of the sympathetic nerves. Studies of exercise in people who have had spinal cord injuries, so that some of their adipose tissue is innervated whilst some is not, suggest that circulating adrenaline is more important than the sympathetic innervation [78]. However, as the plasma levels of natriuretic peptides are increased during physical exercise, the ANP/BNP lipolytic pathway may contribute to the activation of lipid mobilization.

The key regulatory enzyme in the process of fat mobilization is HSL, a TG-lipase that also has cholesterol esterase activity and is expressed in a number of tissues involved in TG-fatty acid and cholesterol metabolism [79,80]. In white adipocytes, its role is hydrolysis of the TG in the TG droplet. HSL is highly regulated, mainly by reversible phosphorylation of serine residues. It is active when phosphorylated. Ser^{659} and Ser^{660} were shown to be responsible for in vitro activation of HSL by cAMP-dependent protein kinase (protein kinase A, PKA) which, in turn, is activated by binding of cAMP generated as described above. Ser^{565}, which is phosphorylated by AMP-activated protein kinase, may play an antilipolytic role as its phosphorylation prevents HSL activation and impairs lipolysis. Activation of the extracellular signal-regulated kinase pathway is able to activate lipolysis by phosphorylating HSL on Ser^{600}. Whether this pathway is important in vivo remains to be demonstrated. Phosphorylation of HSL by PKA can be shown to increase its activity in vitro, but not to the same extent as cellular lipolysis can be increased by catecholamines. This implies mechanisms other than simply conformational change leading to increased enzyme activity. In vivo, an important step in lipolysis activation seems to be the translocation of HSL from a cytosolic compartment to the surface of the lipid droplet. In unstimulated cells, HSL is diffusively distributed throughout the cytosol. Upon stimulation with a β-adrenergic agonist, the enzyme translocates concomitantly with the onset of lipolysis. Recent data from transgenic models have led to a reassessment of the role of HSL in fat mobilization. In HSL-deficient mice, catecholamine-induced lipolysis is markedly blunted as expected, but basal (or unstimulated) lipolysis is unaltered in isolated adipocytes, suggesting the existence of a lipase different from HSL [81,82]. Although no major change in the weight of fat pads was observed, lipid metabolism was altered in the knockout mice. WAT from HSL-deficient mice accumulated diacylglycerols, demonstrating that the enzyme catalyzed the rate-limiting step in diacylglycerol catabolism [83]. During fasting, i.e. when the enzyme activity is maximal in wild-type animals, HSL −/− mice showed decreased plasma NEFA and TG concentrations [84]. Alteration of TG-rich lipoprotein metabolism was due to a down-regulation of VLDL synthesis in liver and an up-regulation of LPL activity in skeletal muscle and WAT. Recently, mice with a 3-fold increased HSL expression in WAT were produced [85]. Despite an increase in hydrolytic capacities towards TG and diacylglycerols, cAMP-inducible adipocyte lipolysis was lower in isolated adipocytes of transgenic animals. The animals showed a decrease in fat mass that was dependent on the age, the diet and the genetic background revealing complex interactions between these components. The data strengthen the view that hydrolysis of TG stored in the lipid droplet relies on both lipase and non-enzymatic components. Moreover, they suggest that HSL levels may influence adipose tissue re-esterification as the decrease in lipolysis was not observed in vivo.

Access to the lipid droplet constitutes an important mechanism for the control of lipolysis. Perilipins are proteins covering the large lipid droplets in adipocytes [86]. They shield stored TG from cytosolic lipases. It has been hypothesized that, upon phosphorylation, perilipins allow access to the lipid droplet and thereby allow lipases to interact with their substrates. Ablation of perilipin results in mice with decreased fat mass and increased lean body mass [87,88]. The mice are resistant to diet-induced obesity. Moreover, mice lacking perilipin on an *ob/ob* background are protected against the obesity phenotype due to mutation in the leptin receptor. No hepatic steatosis or alteration of the lipid profile was observed, which might be due to the increased metabolic rate of the mutant animals. Basal lipolysis is increased in perilipin-deficient adipocytes, which is in line with a role of perilipin as a suppressor of lipolysis in quiescent cells.

HSL is relatively specific, like LPL and pancreatic lipase, for the *sn*-1(3) ester bonds in TG. The 2-MAGs that result from its action are rapidly hydrolyzed by the constitutively active MAG lipase [57]. There is no evidence that the latter is a regulated step; control seems to reside almost entirely with HSL [57].

Thus, each TG molecule generates three fatty acids, which are released from the cell and enter the plasma where they are bound to albumin and distributed to other tissues. The pathway by which fatty acids leave the adipocyte is not clear. As mentioned earlier, facilitated diffusion by FAT/CD36 seems to be associated with "inward" fatty acid transport. Whether there is a specific fatty acid transporter for fat mobilization is not known as yet.

In the process of lipolysis, glycerol is also produced. Glycerol also moves across the cell membrane by facilitated diffusion, using a member of the aquaporin channel family [89].

Glycerol release from adipocytes or adipose tissue is often taken as a marker of lipolysis, since adipose tissue expresses relatively low levels (according to some sources no) glycerol kinase activity, which would be necessary for reutilization of glycerol released from TG hydrolysis. However, it should be noted en passant that the degree of glycerol utilization in WAT is presently uncertain. Some studies [90,91] but not others [47], show uptake of labelled glycerol by human adipose tissue in vivo. (The uptake is small in relation to release, so that there is still always a net release of glycerol.) Recently, the suggestion has been made that the thiazolidinedione (TZD) anti-diabetic "insulin sensitizing" agents up-regulate glycerol kinase in WAT, thus increasing the capacity for fatty acid re-esterification in WAT and reducing systemic fatty acid delivery [92]. This has not been borne out in the limited data available from humans to date [93].

When insulin is infused, glycerol release from adipose tissue is reduced, but not suppressed so completely as is NEFA release (Fig. 2). The explanation is that insulin also stimulates re-esterification of the fatty acids released, as described above. This means that there is dual control of fat mobilization by insulin. Insulin suppresses HSL activity, but in addition "mops up" any fatty acids produced by stimulating re-esterification. The pathway of fatty acid esterification is, of course, the same as described earlier for the pathway of fat deposition. This shows again how the regulation of fat deposition and fat mobilization is intimately linked (Fig. 3).

4. Site-specific features of adipose tissue TG metabolism

It was noted earlier that the various WAT depots relate in different ways to health or disease. This was first clearly pointed out by Jean Vague, who described the typical masculine upper-body pattern of fat accumulation, which he called android obesity, and distinguished it from typically female, lower-body, fat distribution, which he called gynoid obesity [94]. He pointed out the association between upper-body obesity and a range of chronic diseases including "diabetes, atherosclerosis, gout and uric calculous disease" [95]. It is now recognized that upper-body obesity is associated with insulin resistance and the various adverse features that cluster with insulin resistance, including increased risk of diabetes, cardiovascular disease and hypertension [96,97]. Lower-body fat deposition may actually be protective against insulin resistance, dyslipidaemia and ischaemic heart disease [98–100].

This picture is undoubtedly simplistic: more detailed analysis suggests that, even within the thigh, for instance, there are differences in the relationships between fat accumulation and health risk. The most superficial layers of fat do not relate to insulin resistance, whereas deeper layers and fat within the muscle do correlate with insulin resistance [101]. Subcutaneous abdominal fat can also be subdivided into more superficial and deeper, subfascial layers and it is only the latter that relate strongly (similarly to intra-abdominal depots) to insulin resistance [102].

It should be stressed again that in general there are more similarities between the larger adipose depots than differences, and such differences as do exist are more quantitative than qualitative. For instance, NEFA release from one particular, relatively large depot, the subcutaneous anterior abdominal depot, correlates strongly with systemic plasma NEFA

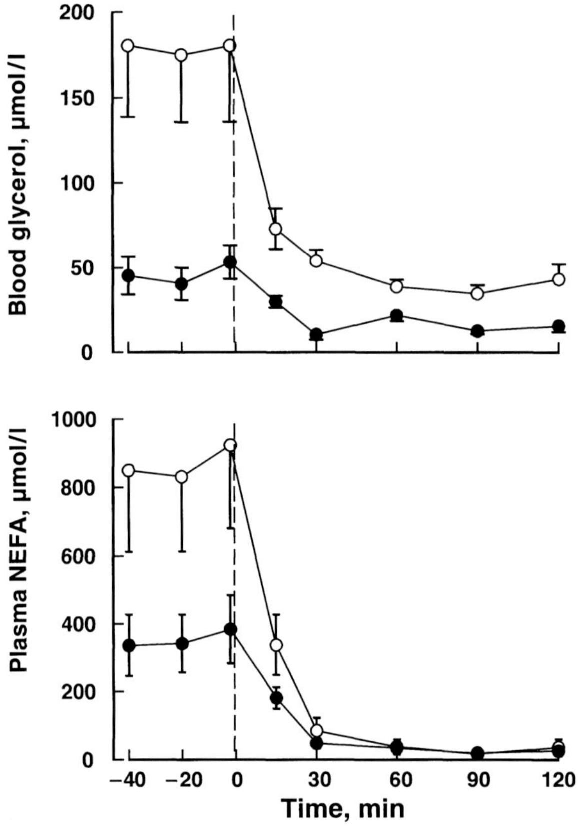

Fig. 2. Effect of insulin on NEFA and glycerol release from subcutaneous adipose tissue in vivo. Results show blood glycerol (top panel) and plasma NEFA (lower panel) concentrations in the fasting state (to time 0 min) and then during infusion of insulin to raise the plasma insulin concentration to a moderately high physiological level. Open circles show concentrations in adipose tissue venous drainage; solid points show arterial concentrations. Net release of both glycerol and NEFA (venous > arterial) is suppressed during insulin infusion, but NEFA release is more completely suppressed than is glycerol release. Replotted from data in Ref. [124].

concentrations [103], arguing for common control of lipolysis in all large depots. Similarly, when blood flow has been studied in various depots, there are qualitative similarities in the responses to nutrient ingestion [104]. Recent evidence for pulsatile behaviour of systemic plasma NEFA concentrations in dogs [105] and humans [106] also shows coordinated regulation of lipolysis throughout all major adipose depots; in this case, the controller seems to be the sympathetic nervous system [105].

It is still not entirely clear whether insulin resistance predisposes to upper-body fat deposition or vice versa: there may well be a vicious cycle at work. But mechanisms whereby upper-body obesity could lead to insulin resistance have at least been proposed. When adipose depots are studied individually – usually ex vivo after adipose tissue biopsy – then metabolic differences appear between them (e.g. Refs. [107–110]).

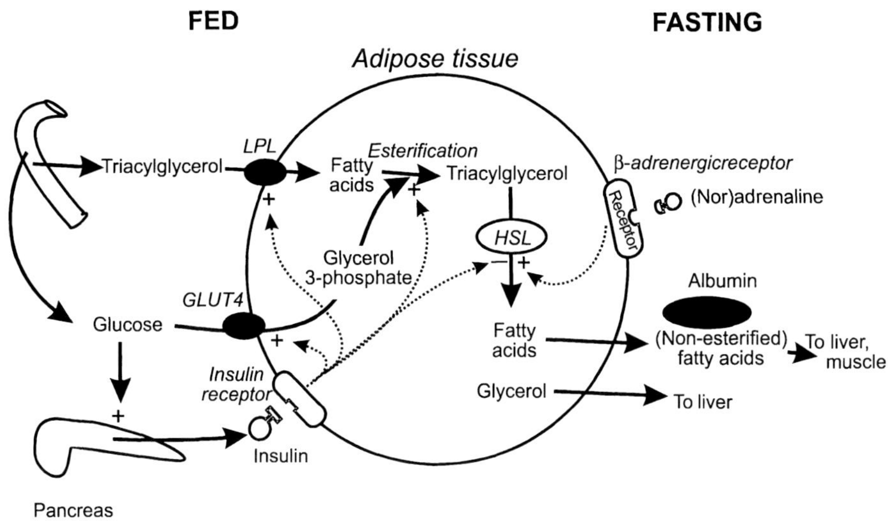

Fig. 3. Coordination of the regulation of fat deposition and fat mobilization in WAT. HSL, hormone-sensitive lipase; LPL, lipoprotein lipase. Reproduced from Ref. [125].

The general finding is that lower-body adipocytes have a relatively "sluggish" metabolism compared with upper-body adipocytes; they are presumably long-term fuel reserves to cover child bearing and child rearing, and these depots are relatively spared during periods of fat loss in women [111]. Upper-body depots, in contrast, have a more active pattern of metabolism: lipolysis occurs at a greater rate when stimulated with catecholamines, and is less readily suppressed by insulin. Intra-abdominal adipose tissue is more active in this respect than subcutaneous abdominal fat. The differences between the depots arise from differences in expression of adrenoceptors and other receptors [112,113]. There are also differences between adipose depots in their LPL activity, that may relate to their different capacities to expand under various physiological and pathological stimuli [113].

A hypothesis has arisen that the different health associations of these various depots reflect their metabolic properties. The larger intra-abdominal depots, the omental and mesenteric fat, drain mainly into the hepatic portal vein. If they are liberating NEFA at a high rate, these fatty acids may have particularly marked effects on liver metabolism. It is well established that elevated NEFA concentrations increase hepatic glucose output and reduce its suppression by insulin [114,115]. Upper-body fat generally might be responsible for undue elevation of the systemic plasma NEFA concentration, which may then have adverse effects on insulin-stimulated glucose utilization [114,116] and on cardiovascular risk factors [117]. Lower-body fat, on the other hand, might be good at removing fat from the circulation.

As mentioned above, these studies have been based primarily upon observations made in isolated adipocytes (or explants of adipose tissue) ex vivo. Because of the highly integrated nature of adipose tissue, discussed earlier, it is important that these findings are verified in vivo. Measurements of regional fatty acid release made by selective venous catheterization certainly bear out the larger contribution of upper-body depots than lower-body depots to total fatty acid delivery [118,119]. Recently, selective catheterization of the venous drainage from the gluteal depot has shown clearly the low rate of NEFA delivery compared with subcutaneous abdominal fat [120]. The details do not completely bear out the hypothesis described above, however. Estimates of the supply of NEFA to the liver suggest that intra-abdominal depots make a small contribution compared with subcutaneous upper-body depots (which are generally considerably larger) [119]. Microdialysis has been used, with measurement of blood flow, to estimate lipolysis rates in different layers of subcutaneous abdominal fat. This shows, contrary to expectations, that the more superficial the fat, the greater the blood flow and lipolysis rate [121]. Also, there are some difficulties with this superficially attractive picture of the different depots supplying fatty acids at different rates that relate to their health associations. The most obvious is that no depot can continue supplying NEFA at a high rate unless it also captures fatty acids at a high rate. But then – considering the intra-abdominal depots, for instance – one could argue that they might protect the liver from an excessive influx of fat in the postprandial period [122]. The associations with health may, of course, relate to entirely different mechanisms such as differential secretion of peptide factors like the cytokine TNF-α, although, again, there is not a lot of evidence to support this idea [123]. All in all, the reasons for the different associations of upper- and lower-body fat with health are not yet entirely clear.

Acknowledgements

The authors were able to collaborate under the European Commission-funded Concerted Action FATLINK: Dietary fat, body weight control and links between obesity and cardiovascular disease (FAIR-CT98-4141).

References

[1] Wilson-Fritch, L., Burkart, A., Bell, G., Mendelson, K., Leszyk, J., Nicoloro, S., Czech, M., Corvera, S., 2003. Mitochondrial biogenesis and remodeling during adipogenesis and in response to the insulin sensitizer rosiglitazone. Mol. Cell. Biol. 23, 1085–1094.
[2] Cannon, B., Houstek, J., Nedergaard, J., 1998. Brown adipose tissue. More than an effector of thermogenesis? Ann. NY Acad. Sci. 856, 171–187.
[3] Cinti, S., 2001. The adipose organ: morphological perspectives of adipose tissues. Proc. Nutr. Soc. 60, 319–328.
[4] Frayn, K.N., 1992. Studies of human adipose tissue in vivo. In: Kinney, J.M., Tucker, H.N. (Eds.), Energy Metabolism: Tissue Determinants and Cellular Corollaries. Raven Press, New York, pp. 267–295.
[5] Frayn, K.N., 2002. Adipose tissue as a buffer for daily lipid flux. Diabetologia 45, 1201–1210.
[6] Gregoire, F.M., Smas, C.M., Sul, H.S., 1998. Understanding adipocyte differentiation. Physiol. Rev. 78, 783–809.
[7] Rosen, E.D., Walkey, C.J., Puigserver, P., Spiegelman, B.M., 2000. Transcriptional regulation of adipogenesis. Genes Dev. 14, 1293–1307.
[8] Sorisky, A., Magun, R., Gagnon, A.M., 2000. Adipose cell apoptosis: death in the energy depot. Int. J. Obes. 24(Suppl. 4), S3–S7.
[9] Pond, C.M., 1986. The natural history of adipocytes. Sci. Prog. Oxf. 70, 45–71.

[10] Pond, C.M., 1992. An evolutionary and functional view of mammalian adipose tissue. Proc. Nutr. Soc. 51, 367–377.
[11] Pond, C.M., 1999. Physiological specialisation of adipose tissue. Prog. Lipid Res. 38, 225–248.
[12] Nilsson, L.H., Hultman, E., 1973. Liver glycogen in man – the effect of total starvation or a carbohydrate-poor diet followed by carbohydrate refeeding. Scand. J. Clin. Lab. Invest. 32, 325–330.
[13] Leiter, L.A., Marliss, E.B., 1982. Survival during fasting may depend on fat as well as protein stores. JAMA 248, 2306–2307.
[14] Forbes, G.B., 1970. Weight loss during fasting: implications for the obese. Am. J. Clin. Nutr. 23, 1212–1219.
[15] Prentice, A.M., 2001. Fires of life: the struggles of an ancient metabolism in a modern world. Br. Nutr. Found. Nutr. Bull. 26, 13–27.
[16] Acheson, K.J., Flatt, J.P., Jéquier, E., 1982. Glycogen synthesis versus lipogenesis after a 500 gram carbohydrate meal in man. Metabolism 31, 1234–1240.
[17] Acheson, K.J., Schutz, Y., Bessard, T., Anantharaman, K., Flatt, J.-P., Jéquier, E., 1988. Glycogen storage capacity and de novo lipogenesis during massive carbohydrate overfeeding in man. Am. J. Clin. Nutr. 48, 240–247.
[18] Aarsland, A., Chinkes, D., Wolfe, R.R., 1997. Hepatic and whole-body fat synthesis in humans during carbohydrate overfeeding. Am. J. Clin. Nutr. 65, 1774–1782.
[19] King, R.F.G.J., Almond, D.J., Oxby, C.B., Holmfield, J.H.M., McMahon, M.J., 1984. Calculation of short-term changes in body fat from measurement of respiratory gas exchange. Metabolism 33, 826–832.
[20] Shrago, E., Spennetta, T., Gordon, E., 1969. Fatty acid synthesis in human adipose tissue. J. Biol. Chem. 244, 2761–2766.
[21] Swierczynski, J., Goyke, E., Wach, L., Pankiewicz, A., Kochan, Z., Adamonis, W., Sledzinski, Z., Aleksandrowicz, Z., 2000. Comparative study of the lipogenic potential of human and rat adipose tissue. Metabolism 49, 594–599.
[22] O'Brien, R.M., Granner, D.K., 1991. Regulation of gene expression by insulin. Biochem. J. 278, 609–619.
[23] O'Brien, R.M., Granner, D.K., 1996. Regulation of gene expression by insulin. Physiol. Rev. 76, 1109–1161.
[24] Chascione, C., Elwyn, D.H., Davila, M., Gil, K.M., Askanazi, J., Kinney, J.M., 1987. Effect of carbohydrate intake on de novo lipogenesis in human adipose tissue. Am. J. Physiol. 253, E664–E669.
[25] Diraison, F., Yankah, V., Letexier, D., Dusserre, E., Jones, P., Beylot, M., 2003. Differences in the regulation of adipose tissue and liver lipogenesis by carbohydrates in humans. J. Lipid Res. 44, 846–853.
[26] Hellerstein, M.K., Schwarz, J.-M., Neese, R.A., 1996. Regulation of hepatic de novo lipogenesis in humans. Annu. Rev. Nutr. 16, 523–557.
[27] Jacobsen, B.K., Trygg, K., Hjermann, I., Thomassen, M.S., Real, C., Norum, K.R., 1983. Acyl patterns of adipose tissue triglycerides, plasma fatty acids, and diet of a group of men participating in a primary coronary prevention program (The Oslo Study). Am. J. Clin. Nutr. 38, 906–913.
[28] van Staveren, W.A., Deurenberg, P., Katan, M.B., Burema, J., de Groot, L.C.P.G.M., Hoffmans, D.A.F., 1986. Validity of the fatty acid composition of subcutaneous fat tissue microbiopsies as an estimate of the long-term average fatty acid composition of the diet of separate individuals. Am. J. Epidemiol. 123, 455–463.
[29] Brun, L.-D., Gagne, C., Julien, P., Termblay, A., Moorjani, S., Bouchard, C., Lupien, P.-J., 1989. Familial lipoprotein lipase-activity deficiency: study of total body fatness and subcutaneous fat tissue distribution. Metabolism 38, 1005–1009.
[30] Peeva, E., Brun, D.L., Ven Murthy, M.R., Després, J.-P., Normand, T., Gagné, C., Lupien, P.-J., Julien, P., 1992. Adipose cell size and distribution in familial lipoprotein lipase deficiency. Int. J. Obes. 16, 737–744.
[31] Weinstock, P.H., Levak Frank, S., Hudgins, L.C., Radner, H., Friedman, J.M., Zechner, R., Breslow, J.L., 1997. Lipoprotein lipase controls fatty acid entry into adipose tissue, but fat mass is preserved by endogenous synthesis in mice deficient in adipose tissue lipoprotein lipase. Proc. Natl Acad. Sci. USA 94, 10261–10266.
[32] Ullrich, N.F., Purnell, J.Q., Brunzell, J.D., 2001. Adipose tissue fatty acid composition in humans with lipoprotein lipase deficiency. J. Invest. Med. 49, 273–275.

[33] Xiang, S.-Q., Cianflone, K., Kalant, D., Sniderman, A.D., 1999. Differential binding of triglyceride-rich lipoproteins to lipoprotein lipase. J. Lipid Res. 40, 1655–1663.
[34] Cryer, A., Riley, S.E., Williams, E.R., Robinson, D.S., 1976. Effect of nutritional status on rat adipose tissue, muscle and post-heparin plasma clearing factor lipase activities: their relationship to triglyceride fatty acid uptake by fat-cells and to plasma insulin concentrations. Clin. Sci. Mol. Med. 50, 213–221.
[35] Scow, R.O., Olivecrona, T., 1977. Effect of albumin on products formed from chylomicron triacylglycerol by lipoprotein lipase in vitro. Biochim. Biophys. Acta 487, 472–486.
[36] Scow, R.O., Blanchette-Mackie, E.J., Smith, L.C., 1976. Role of capillary endothelium in the clearance of chylomicrons. A model for lipid transport from blood by lateral diffusion in cell membranes. Circ. Res. 39, 149–162.
[37] Nilsson-Ehle, P., Egelrud, T., Belfrage, P., Olivecrona, T., Borgström, B., 1973. Positional specificity of purified milk lipoprotein lipase. J. Biol. Chem. 248, 6734–6737.
[38] Fielding, B.A., Humphreys, S.M., Shadid, S., Frayn, K.N., 1995. Arterio-venous differences across human adipose tissue for mono-, di- and tri-acylglycerols before and after a high-fat meal. Endocrinol. Metab. 2, 13–17.
[39] Summers, L.K.M., Fielding, B.A., Herd, S.L., Ilic, V., Clark, M.L., Quinlan, P.T., Frayn, K.N., 1999. Use of structured triacylglycerols containing predominantly stearic and oleic acids to probe early events in metabolic processing of dietary fat. J. Lipid Res. 40, 1890–1898.
[40] van der Vusse, G.J., Reneman, R.S., 1996. Lipid metabolism in muscle. In: Rowell, L.B., Shepherd, J.T., (Eds.), Handbook of Physiology, Section 12. American Physiological Society, New York, pp. 952–994.
[41] Evans, K., Burdge, G.C., Wootton, S.A., Clark, M.L., Frayn, K.N., 2002. Regulation of dietary fatty acid entrapment in subcutaneous adipose tissue and skeletal muscle. Diabetes 51, 2684–2690.
[42] Evans, K., Clark, M.L., Frayn, K.N., 1999. Effects of an oral and intravenous fat load on adipose tissue and forearm lipid metabolism. Am. J. Physiol. 276, E241–E248.
[43] Fielding, B.A., Frayn, K.N., 1998. Lipoprotein lipase and the disposition of dietary fatty acids. Br. J. Nutr. 80, 495–502.
[44] Weinstock, P.H., Bisgaier, C.L., Aalto-Setälä, K., Radner, H., Ramakrishnan, R., Levak-Frank, S., Essenburg, A.D., Zechner, R., Breslow, J.L., 1995. Severe hypertriglyceridemia, reduced high density lipoprotein, and neonatal death in lipoprotein lipase knockout mice. Mild hypertriglyceridemia with impaired very low density lipoprotein clearance in heterozygotes. J. Clin. Invest. 96, 2555–2568.
[45] Weinstock, P.H., Levak-Frank, S., Hudgins, L.C., Radner, H., Friedman, J.M., Zechner, R., Breslow, J.L., 1997. Lipoprotein lipase controls fatty acid entry into adipose tissue, but fat mass is preserved by endogenous synthesis in mice deficient in adipose tissue lipoprotein lipase. Proc. Natl Acad. Sci. USA 94, 10261–10266.
[46] Abumrad, N., Harmon, C., Ibrahimi, A., 1998. Membrane transport of long-chain fatty acids: evidence for a facilitated process. J. Lipid Res. 39, 2309–2318.
[47] Coppack, S.W., Persson, M., Judd, R.L., Miles, J.M., 1999. Glycerol and nonesterified fatty acid metabolism in human muscle and adipose tissue in vivo. Am. J. Physiol. 276, E233–E240.
[48] Abumrad, N., Coburn, C., Ibrahimi, A., 1999. Membrane proteins implicated in long-chain fatty acid uptake by mammalian cells: CD36, FATP and FABPm. Biochim. Biophys. Acta 1441, 4–13.
[49] Luiken, J.J., Dyck, D.J., Han, X.-X., Tandon, N.N., Arumugam, Y., Glatz, J.F., Bonen, A., 2002. Insulin induces the translocation of the fatty acid transporter FAT/CD36 to the plasma membrane. Am. J. Physiol. Endocrinol. Metab. 282, E491–E495.
[50] Luiken, J.J.F.P., Koonen, D.P.Y., Willems, J., Zorzano, A., Becker, C., Fischer, Y., Tandon, N.N., van der Vusse, G.J., Bonen, A., Glatz, J.F.C., 2002. Insulin stimulates long-chain fatty acid utilization by rat cardiac myocytes through cellular redistribution of FAT/CD36. Diabetes 51, 3113–3119.
[51] Febbraio, M., Abumrad, N.A., Hajjar, D.P., Sharma, K., Cheng, W., Pearce, S.F., Silverstein, R.L., 1999. A null mutation in murine CD36 reveals an important role in fatty acid and lipoprotein metabolism. J. Biol. Chem. 274, 19055–19062.
[52] Coburn, C.T., Knapp, F.F. Jr., Febbraio, M., Beets, A.L., Silverstein, R.L., Abumrad, N.A., 2000. Defective uptake and utilization of long chain fatty acids in muscle and adipose tissues of CD36 knockout mice. J. Biol. Chem. 275, 32523–32529.

[53] Gargiulo, C.E., Stuhlsatz-Krouper, S.M., Schaffer, J.E., 1999. Localization of adipocyte long-chain fatty acyl-CoA synthetase at the plasma membrane. J. Lipid Res. 40, 881–892.
[54] Schaffer, J.E., 2002. Fatty acid transport: the roads taken. Am. J. Physiol. Endocrinol. Metab. 282, E239–E246.
[55] Dodds, P.F., Brindley, D.N., Gurr, M.I., 1976. The effects of diet on the esterification of glycerol phosphate, dihydroxyacetone phosphate and 2-hexadecylglycerol by homogenates of rat adipose tissue. Biochem. J. 160, 701–706.
[56] Dodds, P.F., Gurr, M.I., Brindley, D.N., 1976. The glycerol phosphate, dihydroxyacetone phosphate and monoacylglycerol pathways of glycerolipid synthesis in rat adipose-tissue homogenates. Biochem. J. 160, 693–700.
[57] Fredrikson, G., Tornqvist, H., Belfrage, P., 1986. Hormone-sensitive lipase and monoacylglycerol lipase are both required for complete degradation of adipocyte triacylglycerol. Biochim. Biophys. Acta 876, 288–293.
[58] Sul, H.S., Wang, D., 1998. Nutritional and hormonal regulation of enzymes in fat synthesis: studies of fatty acid synthase and mitochondrial glycerol-3-phosphate acyltransferase gene transcription. Annu. Rev. Nutr. 18, 331–351.
[59] Coleman, R.A., Lewin, T.M., Muoio, D., 2000. Physiological and nutritional regulation of enzymes of triacylglycerol synthesis. Annu. Rev. Nutr. 20, 77–103.
[60] Leboeuf, B., 1965. Regulation of fatty acid esterification in adipose tissue incubated in vitro. In: Renold, A.E., Cahill, G.F. (Eds.). Handbook of Physiology Section 5: Adipose Tissue. American Physiological Society, Washington, DC, pp. 385–391.
[61] Campbell, P.J., Carlson, M.G., Hill, J.O., Nurjhan, N., 1992. Regulation of free fatty acid metabolism by insulin in humans: role of lipolysis and reesterification. Am. J. Physiol. 263, E1063–E1069.
[62] Frayn, K.N., Shadid, S., Hamlani, R., Humphreys, S.M., Clark, M.L., Fielding, B.A., Boland, O., Coppack, S.W., 1994. Regulation of fatty acid movement in human adipose tissue in the postabsorptive-to-postprandial transition. Am. J. Physiol. 266, E308–E317.
[63] Taylor, R., Husband, D.J., Marshall, S.M., Tunbridge, W.M.G., Alberti, K.G.M.M., 1984. Adipocyte insulin binding and insulin sensitivity in 'brittle' diabetes. Diabetologia 27, 441–446.
[64] Jensen, M.D., Caruso, M., Heiling, V., Miles, J.M., 1989. Insulin regulation of lipolysis in nondiabetic and IDDM subjects. Diabetes 38, 1595–1601.
[65] Smith, S.J., Cases, S., Jensen, D.R., Chen, H.C., Sande, E., Tow, B., Sanan, D.A., Raber, J., Eckel, R.H., Farese, R.V., 2000. Obesity resistance and multiple mechanisms of triglyceride synthesis in mice lacking Dgat. Nat. Genet. 25, 87–90.
[66] Cases, S., Zhou, P., Yen, E., Tow, B., Lardizabal, K.D., Voelker, T., Farese, R.V.J., 2001. Cloning of DGAT2, a second mammalian diacylglycerol acyltransferase, and related family members. J. Biol. Chem. 276, 38870–38876.
[67] Langin, D., Lafontan, M., 2001. Fat cell lipolysis in the new millenium. Obes. Matters 4, 7–10.
[68] Sengenès, C., Berlan, M., De Glisezinski, I., Lafontan, M., Galitzky, J., 2000. Natriuretic peptides: a new lipolytic pathway in human adipocytes. FASEB J. 14, 1345–1351.
[69] Degerman, E., Belfrage, P., Manganiello, V.C., 1997. Structure, localization, and regulation of cGMP-inhibited phosphodiesterase (PDE3). J. Biol. Chem. 272, 6823–6826.
[70] Wise, A., Foord, S.M., Fraser, N.J., Barnes, A.A., Elshourbagy, N., Eilert, M., Ignar, D.M., Murdock, P.R., Steplewski, K., Green, A., Brown, A.J., Dowell, S.J., Szekeres, P.G., Hassall, D.G., Marshall, F.H., Wilson, S., Pike, N.B., 2003. Molecular identification of high and low affinity receptors for nicotinic acid. J. Biol. Chem. 278, 9869–9874.
[71] Tunaru, S., Kero, J., Schaub, A., Wufka, C., Blaukat, A., Pfeffer, K., Offermanns, S., 2003. PUMA-G and HM74 are receptors for nicotinic acid and mediate its anti-lipolytic effect. Nat. Med. 9, 352–355.
[72] Arner, P., Kriegholm, E., Engfeldt, P., 1990. In situ studies of catecholamine-induced lipolysis in human adipose tissue using microdialysis. J. Pharmacol. Exp. Ther. 254, 284–288.
[73] Galitzky, J., Lafontan, M., Nordenström, J., Arner, P., 1993. Role of vascular alpha-2 adrenoceptors in regulating lipid mobilization from human adipose tissue. J. Clin. Invest. 91, 1997–2003.
[74] Samra, J.S., Clark, M.L., Humphreys, S.M., Macdonald, I.A., Bannister, P.A., Matthews, D.R., Frayn, K.N., 1999. Suppression of the nocturnal rise in growth hormone reduces subsequent lipolysis in subcutaneous adipose tissue. Eur. J. Clin. Invest. 29, 1045–1052.

[75] Samra, J.S., Clark, M.L., Humphreys, S.M., Macdonald, I.A., Matthews, D.R., Frayn, K.N., 1996. Effects of morning rise in cortisol concentration on regulation of lipolysis in subcutaneous adipose tissue. Am. J. Physiol. 271, E996–E1002.

[76] Arner, P., Kriegholm, E., Engfeldt, P., Bolinder, J., 1990. Adrenergic regulation of lipolysis in situ at rest and during exercise. J. Clin. Invest. 85, 893–898.

[77] Van Baak, M.A., Saris, W.H., 2000. The effect of caffeine on endurance performance after nonselective beta-adrenergic blockade. Med. Sci. Sports Exerc. 32, 499–503.

[78] Stallknecht, B., Lorentsen, J., Enevoldsen, L.H., Bülow, J., Biering-Sørensen, F., Galbo, H., Kjaer, M., 2001. Role of the sympathoadrenergic system in adipose tissue metabolism during exercise in humans. J. Physiol. 536, 283–294.

[79] Holm, C., Osterlund, T., Laurell, H., Contreras, J.A., 2000. Molecular mechanisms regulating hormone-sensitive lipase and lipolysis. Annu. Rev. Nutr. 20, 365–393.

[80] Kraemer, F.B., Shen, W.J., 2002. Hormone-sensitive lipase: control of intracellular tri-(di-)acylglycerol and cholesteryl ester hydrolysis. J. Lipid Res. 43, 1585–1594.

[81] Osuga, J., Ishibashi, S., Oka, T., Yagyu, H., Tozawa, R., Fujimoto, A., Shionoiri, F., Yahagi, N., Kraemer, F.B., Tsutsumi, O., Yamada, N., 2000. Targeted disruption of hormone-sensitive lipase results in male sterility and adipocyte hypertrophy, but not in obesity. Proc. Natl Acad. Sci. USA 97, 787–792.

[82] Wang, S.P., Laurin, N., Himms-Hagen, J., Rudnicki, M.A., Levy, E., Robert, M.-F., Pan, L., Oligny, L., Mitchell, G.A., 2001. The adipose tissue phenotype of hormone-sensitive lipase deficiency in mice. Obes. Res. 9, 119–128.

[83] Haemmerle, G., Zimmermann, R., Hayn, M., Theussl, C., Waeg, G., Wagner, E., Sattler, W., Magin, T.M., Wagner, E.F., Zechner, R., 2002. Hormone-sensitive lipase deficiency in mice causes diglyceride accumulation in adipose tissue, muscle, and testis. J. Biol. Chem. 277, 4806–4815.

[84] Haemmerle, G., Zimmermann, R., Strauss, J.G., Kratky, D., Riederer, M., Knipping, G., Zechner, R., 2002. Hormone-sensitive lipase deficiency in mice changes the plasma lipid profile by affecting the tissue-specific expression pattern of lipoprotein lipase in adipose tissue and muscle. J. Biol. Chem. 277, 12946–12952.

[85] Lucas, S., Tavernier, G., Tiraby, C., Mairal, A., Langin, D., 2003. Expression of human hormone-sensitive lipase in white adipose tissue of transgenic mice increases lipase activity but does not enhance in vitro lipolysis. J. Lipid Res. 44, 154–163.

[86] Londos, C., Brasaemle, D.L., Schultz, C.J., Adler-Wailes, D.C., Levin, D.M., Kimmel, A.R., Rondinone, C.M., 1999. On the control of lipolysis in adipocytes. Ann. NY Acad. Sci. 892, 155–168.

[87] Martinez-Botas, J., Anderson, J.B., Tessier, D., Lapillonne, A., Chang, B.H., Quast, M.J., Gorenstein, D., Chen, K.H., Chan, L., 2000. Absence of perilipin results in leanness and reverses obesity in Leprdb/db mice. Nat. Genet. 26, 474–479.

[88] Tansey, J.T., Sztalryd, C., Gruia-Gray, J., Roush, D.L., Zee, J.V., Gavrilova, O., Reitman, M.L., Deng, C.-X., Li, C., Kimmel, A.R., Londos, C., 2001. Perilipin ablation results in a lean mouse with aberrant adipocyte lipolysis, enhanced leptin production, and resistance to diet-induced obesity. PNAS 98, 6494–6499.

[89] Kuriyama, H., Kawamoto, S., Ishida, N., Ohno, I., Mita, S., Matsuzawa, Y., Matsubara, K., Okubo, K., 1997. Molecular cloning and expression of a novel human aquaporin from adipose tissue with glycerol permeability. Biochem. Biophys. Res. Commun. 241, 53–58.

[90] Kurpad, A., Khan, K., Calder, A.G., Coppack, S., Frayn, K., Macdonald, I., Elia, M., 1994. Effect of noradrenaline on glycerol turnover and lipolysis in the whole body and subcutaneous adipose tissue in humans in vivo. Clin. Sci. 86, 177–184.

[91] Samra, J.S., Clark, M.L., Humphreys, S.M., Bannister, P.A., Summers, L.K.M., Macdonald, I.A., Frayn, K.N., 1999. Evidence for glycerol uptake by subcutaneous adipose tissue. Proc. Nutr. Soc. 58, 164A.

[92] Guan, H.P., Li, Y., Jensen, M.V., Newgard, C.B., Steppan, C.M., Lazar, M.A., 2002. A futile metabolic cycle activated in adipocytes by antidiabetic agents. Nat. Med. 8, 1122–1128.

[93] Tan, G.D., Debard, C., Tiraby, C., Humphreys, S.M., Frayn, K.N., Langin, D., Vidal, H., Karpe, F., 2003. A "futile cycle" induced by thiazolidinediones in human adipose tissue? Nat. Med. 9, 811–812.

[94] Vague, J., 1947. La différenciation sexuelle, facteur déterminant des formes de l'obésité. Presse Méd. 30, 339–340.

[95] Vague, J., 1956. The degree of masculine differentiation of obesities: a factor determining predisposition to diabetes, atherosclerosis, gout and uric calculous disease. Am. J. Clin. Nutr. 4, 20–34.
[96] Krotkiewski, M., Björntorp, P., Sjöström, L., Smith, U., 1983. Impact of obesity on metabolism in men and women. Importance of regional adipose tissue distribution. J. Clin. Invest. 72, 1150–1162.
[97] Kissebah, A.H., Krakower, G.R., 1994. Regional adiposity and morbidity. Physiol. Rev. 74, 761–811.
[98] Terry, R.B., Stefanick, M.L., Haskell, W.L., Wood, P.D., 1991. Contributions of regional adipose depots to plasma lipoprotein concentrations in overweight men and women: possible protective effects of thigh fat. Metabolism 40, 733–740.
[99] Kahn, H.S., Austin, H., Williamson, D.F., Arensberg, D., 1996. Simple anthropometric indices associated with ischemic heart disease. J. Clin. Epidemiol. 49, 1017–1024.
[100] Van Pelt, R.E., Evans, E.M., Schechtman, K.B., Ehsani, A.A., Kohrt, W.M., 2002. Contributions of total and regional fat mass to risk for cardiovascular disease in older women. Am. J. Physiol. Endocrinol. Metab. 282, E1023–E1028.
[101] Goodpaster, B.H., Thaete, F.L., Kelley, D.E., 2000. Thigh adipose tissue distribution is associated with insulin resistance in obesity and in type 2 diabetes mellitus. Am. J. Clin. Nutr. 71, 885–892.
[102] Kelley, D.E., Thaete, F.L., Troost, F., Huwe, T., Goodpaster, B.H., 2000. Subdivisions of subcutaneous abdominal adipose tissue and insulin resistance. Am. J. Physiol. Endocrinol. Metab. 278, E941–E948.
[103] Frayn, K.N., Coppack, S.W., Humphreys, S.M., Clark, M.L., Evans, R.D., 1993. Periprandial regulation of lipid metabolism in insulin-treated diabetes mellitus. Metabolism 42, 504–510.
[104] Simonsen, L., Bülow, J., Astrup, A., Madsen, J., Christensen, N.J., 1990. Diet-induced changes in subcutaneous adipose tissue blood flow in man: effect of β-adrenoceptor inhibition. Acta Physiol. Scand. 139, 341–346.
[105] Hücking, K., Hamilton-Wessler, M., Ellmerer, M., Bergman, R.N., 2003. Burst-like control of lipolysis by the sympathetic nervous system in vivo. J. Clin. Invest. 111, 257–264.
[106] Karpe, F., Fielding, B.A., Lawrence, V., Macdonald, I., Coppack, S.W., Frayn, K.N., 2003. Metabolic pulsatility of human adipose tissue. Int. J. Obes. 27(Suppl. 1), S5.
[107] Östman, J., Arner, P., Engfeldt, P., Kager, L., 1979. Regional differences in the control of lipolysis in human adipose tissue. Metabolism 28, 1198–1205.
[108] Mauriège, P., Galitzky, J., Berlan, M., Lafontan, M., 1987. Heterogeneous distribution of beta and alpha-2 adrenoceptor binding sites in human fat cells from various fat deposits: functional consequences. Eur. J. Clin. Invest. 17, 156–165.
[109] Engfeldt, P., Arner, P., 1988. Lipolysis in human adipocytes, effects of cell size, age and of regional differences. Horm. Metab. Res. Suppl. 19, 26–29.
[110] Van Harmelen, V., Lönnqvist, F., Thörne, A., Wennlund, A., Large, V., Reynisdottir, S., Arner, P., 1997. Noradrenaline-induced lipolysis in isolated mesenteric, omental and subcutaneous adipocytes from obese subjects. Int. J. Obes. 21, 972–979.
[111] Jones, P.R., Edwards, D.A., 1999. Areas of fat loss in overweight young females following an 8-week period of energy intake reduction. Ann. Hum. Biol. 26, 151–162.
[112] Arner, P., 1999. Catecholamine-induced lipolysis in obesity. Int. J. Obes. 23(Suppl. 1), 10–13.
[113] Wajchenberg, B.L., 2000. Subcutaneous and visceral adipose tissue: their relation to the metabolic syndrome. Endocr. Rev. 21, 697–738.
[114] Ferrannini, E., Barrett, E.J., Bevilacqua, S., DeFronzo, R.A., 1983. Effect of fatty acids on glucose production and utilization in man. J. Clin. Invest. 72, 1737–1747.
[115] Fanelli, C., Calderone, S., Epifano, L., De Vincenzo, A., Modarelli, F., Pampanelli, S., Perriello, G., De Feo, P., Brunetti, P., Gerich, J.E., Bolli, G.B., 1993. Demonstration of a critical role for free fatty acids in mediating counterregulatory stimulation of gluconeogenesis and suppression of glucose utilization in humans. J. Clin. Invest. 92, 1617–1622.
[116] Boden, G., Chen, X., Ruiz, J., White, J.V., Rossetti, L., 1994. Mechanisms of fatty acid-induced inhibition of glucose uptake. J. Clin. Invest. 93, 2438–2446.
[117] Frayn, K.N., Williams, C.M., Arner, P., 1996. Are increased plasma non-esterified fatty acid concentrations a risk marker for coronary heart disease and other chronic diseases? Clin. Sci. 90, 243–253.
[118] Martin, M.L., Jensen, M.D., 1991. Effects of body fat distribution on regional lipolysis in obesity. J. Clin. Invest. 88, 609–613.

[119] Guo, Z., Hensrud, D.D., Johnson, C.M., Jensen, M.D., 1999. Regional postprandial fatty acid metabolism in different obesity phenotypes. Diabetes 48, 1586–1592.

[120] Goossens, G.H., Tan, G.D., Humphreys, S.M., Vidal, H., Karpe, F., 2003. Upper and lower body adipose tissue function: a direct comparison of fat mobilization. Int. J. Obes. 27(Suppl. 1), S49.

[121] Enevoldsen, L.H., Simonsen, L., Stallknecht, B., Galbo, H., Bülow, J., 2001. In vivo human lipolytic activity in preperitoneal and subdivisions of subcutaneous abdominal adipose tissue. Am. J. Physiol. Endocrinol. Metab. 281, E1110–E1114.

[122] Frayn, K.N., 2000. Visceral fat and insulin resistance – causative or correlative? Br. J. Nutr. 83(Suppl. 1), S71–S77.

[123] Montague, C.T., Prins, J.B., Sanders, L., Zhang, J., Sewter, C.P., Digby, J., Byrne, C.D., O'Rahilly, S., 1998. Depot-related gene expression in human subcutaneous and omental adipocytes. Diabetes 47, 1384–1391.

[124] Coppack, S.W., Frayn, K.N., Humphreys, S.M., Dhar, H., Hockaday, T.D.R., 1989. Effects of insulin on human adipose tissue metabolism in vivo. Clin. Sci. 77, 663–670.

[125] Gurr, M.I., Harwood, J.L., Frayn, K.N., 2002. Lipid Biochemistry: An Introduction, 5th ed. Blackwell Science, Oxford.

Phospholipid biosynthesis

Grant M. Hatch and Patrick C. Choy*

Departments of Biochemistry & Molecular Biology, Pharmacology & Therapeutics, Internal Medicine, Center for Research and Treatment of Atherosclerosis, Faculty of Medicine, The University of Manitoba, 770 Bannatyne Avenue, Winnipeg, Man., Canada R3E 0W3
**Correspondence address: Tel: +1-204-789-3723; fax: +1-204-789-3942*
E-mail: pchoy@ms.umanitoba.ca(P.C.C.)

1. Preamble

1.1. Nomenclature

Phospholipids are the most abundant class of lipids in all mammalian membranes [1]. Glycerol containing phospholipids, the phosphoglycerides, may be divided into three distinct subclasses according to the type of bond at the *sn*-1 position. The 1,2-diacylphosphoglyceride is by far the most abundant type whereas 1-alkyl-2-acyl-phosphoglyceride represents a small fraction (1–2%) of the total phospholipids. The 1-alkenyl-2-acyl-phosphoglyceride is commonly known as plasmalogen and its distribution in mammalian tissues ranges from a low of 2% to a high of 40%. Polyglycerophosphoglycerides, which include cardiolipin, phosphatidylglycerol and bis-monoacylglycerophosphate, are minor components in most cells but are found in significant amounts in some tissues such as the lung and heart [2]. This class of phospholipid contains either two or three glycerol moieties. In this chapter, we will focus on the biosynthesis of diacylphosphoglycerides and polyglycerophosphoglycerides collectively known as the glycerophospholipids.

1.2. Function

As the major component of the membrane, phospholipids provide the cell with a barrier where selective permeability protects cellular metabolism. Within the cell, these phospholipids participate in the membrane network that delineates individual organelles that constitute a uniquely ordered system for intracellular metabolic reactions. In addition to their role in membrane formation, phospholipids are involved in the modulation of membrane bound enzymes [3]. For example, the direct involvement of phosphatidylcholine and phosphatidylethanolamine in the synthesis of lung surfactant, lipoprotein

Advances in Molecular and Cell Biology, Vol. 33, pages 357–385

ISSN: 1569-2558 / DOI: 10.1016/S1569-2558(03)33018-8

secretion and arachidonate production for prostaglandin biosynthesis has been clearly demonstrated [4]. Phosphatidylserine is required for surface recognition of apoptotic cells by phagocytes [5]. Phosphatidic acid and lysophosphatidic acid have been identified as potential mitogenic signals [6]. Cardiolipin plays a critical role in the apoptotic process in many cells [7]. In addition, the role of phosphatidylinositol and its derivatives in signal transmission is well documented [8,9]. It is clear from these studies that the physiological role of phospholipids has evolved well beyond their primary function as building blocks of the biological membrane.

2. General overview of phospholipid biosynthesis

2.1. The *sn*-glycerol-3-phosphate pathway

The *sn*-glycerol-3-phosphate pathway of phospholipid biosynthesis was first elucidated by Kennedy and co-workers and has been extensively reviewed [10–13]. In this pathway, the *sn*-glycerol-3-phosphate is acylated to 1-acyl-*sn*-glycerol-3-phosphate by glycerol-3-phosphate acyltransferase (Fig. 1). The glycerol-3-phosphate acyltransferase is localised to both endoplasmic reticulum and mitochondrial membranes within mammalian cells. The two forms of glycerol-3-phosphate acyltransferase can be distinguished by the sensitivity of the microsomal form to sulfhydryl reagents such as *N*-ethylmaleimide, whereas the mitochondrial form of the enzyme is insensitive. Both microsomal and mitochondrial forms of the rat liver enzyme exhibit a broad pH optimum (between 6.6 and 9.0) and the K_m for *sn*-glycerol-3-phosphate for the mitochondrial enzyme (1 mM) is much greater than the microsomal enzyme (0.1–0.2 mM) [14]. The mitochondrial glycerol-3-phosphate acyltransferase prefers saturated fatty acyl substrates and is

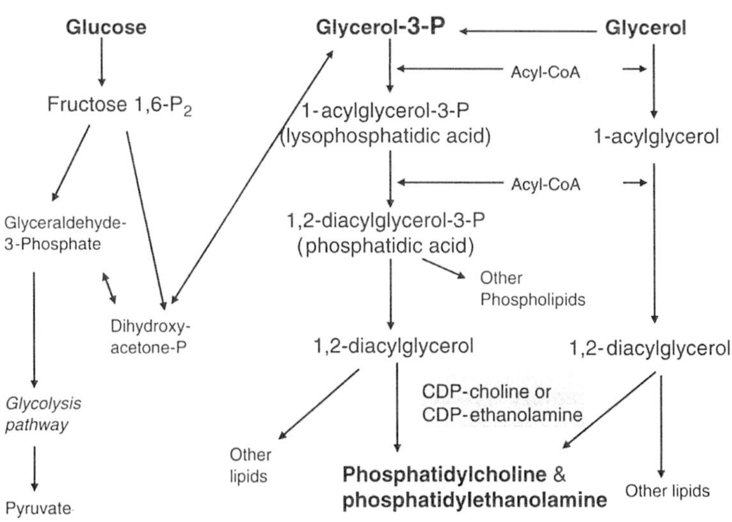

Fig. 1. The *sn*-glycerol-3-phosphate pathway and the pathway for the direct acylation of glycerol.

extensively regulated under physiological and pathophysiological conditions and the nutritional status of the animal [15]. The murine mitochondrial glycerol-3-phosphate acyltransferase cDNA has been cloned [16]. The 6.8 kb mRNA coded for a 90 kDa protein that contains an open reading frame of 827 amino acids and exhibited a 30% homology with the *Escherichia coli* glycerol-3-phosphate acyltransferase [17]. The 90 kDa protein was not detected in 3T3-L1 preadipocytes but was greatly expressed during adipocyte conversion.

The 1-acyl-*sn*-glycerol-3-phosphate product formed from *sn*-glycerol-3-phosphate acyltransferase is then acylated to 1,2-diacyl-*sn*-glycerol-3-phosphate (phosphatidic acid) by 1-acyl-*sn*-glycerol-3-phosphate acyltransferase [10,11]. Overexpression of the 1-acyl-*sn*-glycerol-3-phosphate acyltransferase-alpha in 3T3-L1 adipocytes increases non-esterified fatty acid uptake and metabolic flow into 1,2,3-triacyl-*sn*-glycerol and insulin-mediated glucose utilisation [18]. In contrast, overexpression of the 1-acyl-*sn*-glycerol-3-phosphate acyltransferase-alpha in myotubes increases non-esterified fatty acid uptake but diverts glucose from glycogen synthesis to lipogenesis upon insulin stimulation.

Phosphatidic acid lies at a branch point in phospholipid biosynthesis. Phosphatidic acid may be converted to 1,2-diacyl-*sn*-glycerol by phosphatidic acid phosphohydrolase (PAP) [19]. Several isoforms of mammalian PAP, PAP-1 and PAP-2a, PAP-2b and PAP-2c have been isolated [20,21]. The 1,2-diacyl-*sn*-glycerol produced from the PAP-1 reaction may be utilised for phosphatidylcholine and phosphatidylethanolamine biosynthesis (see below) and 1,2,3-triacyl-*sn*-glycerol biosynthesis. The purified plasma membrane PAP-2 exists as a 51–53 kDa integral membrane glycoprotein and the 1,2-diacyl-*sn*-glycerol produced from this reaction is involved in signal transduction and cell activation [22]. Alternatively, phosphatidic acid may be converted to cytidine-5′-diphosphate-1,2-diacyl-*sn*-glycerol (CDP-DG) for biosynthesis of phosphatidylinositol and the polyglycerophospholipids, phosphatidylglycerol and cardiolipin (see below) in a reaction catalysed by CTP:phosphatidic acid cytidylyltransferase.

2.2. The pathway for the direct acylation of glycerol

The acylation of glycerol for the formation of monoacylglycerol appears to be the most direct pathway in the synthesis of diacylglycerol, and the quest for the existence of this pathway dated back to the early 1960s. Since then, the uptake of glycerol was studied in hepatocytes, liver, kidney, mammary cells, adipose tissue, pneumocytes, aorta, heart, cardiac myocytes and skeletal muscle [23–31]. An example of these studies is illustrated by the work of Kinsella, in which the incorporation of [^{14}C] glycerol into the lipid fraction of bovine mammary cells was examined [26]. After incubation with labelled glycerol, the specific radioactivities of monoacylglycerol, diacylglycerol and triacylglycerol were determined. Although the monoacylglycerol pool had the highest specific radioactivity immediately after pulse-labelling, the dogma precluded the participation of direct acylation of glycerol, and the production of monoacylglycerol was attributed to the catabolism of labelled phosphatidic acid and lysphosphatidic acid.

During the last several years, we have examined the direct acylation of glycerol in several mammalian tissues [32]. Lipid biosynthesis in myoblasts and hepatocytes was re-assessed by conducting pulse-chase experiments with [1,3-^{3}H] glycerol. Several lines of evidence have been developed to supporting the existence of a novel pathway for lipid biosynthesis (Fig. 1). First, pulse-chase studies with both H9c2 myocytes and Chang liver cells demonstrated that glycerol could be directly converted to monoacylglycerol and subsequently other lipids. Second, a novel enzyme for the direct acylation of glycerol in the microsomal fraction of the rat and porcine heart, liver, kidney, skeletal muscle and brain was detected and characterised. The glycerol:acyl-CoA acyltransferase has the ability to transfer various long-chain acyl-CoA species, but displays a high degree of specificity for arachidonoyl-CoA. Although the physiological importance of the direct acylation pathway has not been completely established, there is evidence to suggest that the pathway may serve as a shunt for monoacylglycerol and diacylglycerol biosynthesis. Glycerol is present in the mammalian tissues at a concentration of about 0.1 mM [33]. In human serum, the glycerol level fluctuates from 0.04 to 0.4 mM [34]. Kinetic studies of the enzyme revealed that it has an apparent K_m value of 1.1 mM for glycerol, indicating the rate of acylation is linear with the intracellular glycerol concentrations. As such, the direct acylation of glycerol may occur in limited capacity during normal physiological conditions. Alternatively, the intracellular glycerol concentration is dramatically increased in certain forms of glycerol kinase deficiency, in diabetes, during fasting, extreme cold and ischemia of the heart [35–37]. Since the acylation of glycerol is regulated by glycerol availability, the direct acylation pathway may be greatly potentiated during hyperglycerolemia. Incubation of H9c2 cells in medium containing high glycerol levels caused the direct acylation pathway to become more prominent. In addition, the inhibition of glycerol kinase in Chang liver cells caused glycerol metabolism to be shunted towards the direct acylation pathway, and an elevated level of monoacylglycerol production. Hence, we postulate that the direct acylation pathway is an alternative route for the synthesis of monoacylglycerol and diacylglycerol in mammalian tissues [32].

3. The biosynthesis of phosphatidylcholine

3.1. Introduction

Phosphatidylcholine is the major phospholipid in mammalian tissues [1]. Following the discovery of its biosynthetic pathways in the early 1950s, the metabolism of phosphatidylcholine has been studied extensively. A large number of these studies were done in the liver and lung, but increasingly information is becoming available for the heart. In the early 1980s, the mammalian heart was shown to have the ability to synthesise phosphatidylcholine. It can be argued that the metabolism of phosphatidylcholine in the liver and lung should not be different from other mammalian tissues, and the information obtained from the liver and lung should be applicable to the heart. Substantial portions of the phosphatidylcholine synthesised in the liver and lung, however, are secreted as lipoproteins or as surfactant, respectively [38,39] whereas in the heart, none of the phosphatidylcholine synthesised is known to be exported [40]. The difference in the utilisation of phosphatidylcholine suggests the possibility that its metabolism may be regulated differently in these organs.

3.2. Pathways for the biosynthesis of phosphatidylcholine

In mammalian tissues, several pathways are known for the formation of phosphatidylcholine (Fig. 2). The majority of the phospholipid is formed from choline via the CDP-choline pathway [11]. In this pathway, choline is taken up by the cell through a choline uptake system and subsequently becomes phosphorylated by choline kinase (EC 2.7.1.32). Phosphocholine is then converted to CDP-choline in a reaction catalysed by CTP: phosphocholine cytidylyltransferase (EC 2.7.7.15) [41]. Phosphatidylcholine is formed by the condensation of CDP-choline and 1,2-diacylglycerol in a reaction catalysed by CDP-choline:1,2-diacylglycerol cholinephosphotransferase (EC 2.7.8.2). Phosphatidylcholine can also be formed by the methylation pathway in which phosphatidylethanolamine is converted to phosphatidylcholine by the transfer of methyl groups from *S*-adenosylmethionine in a reaction catalysed by phosphatidylethanolamine-*N*-methyltransferase (EC 2.1.1.17) [42–44]. This pathway contributes significantly to phosphatidylcholine formation in the liver [45] but not in other organs [46]. Another pathway for phosphatidylcholine formation is the Ca^{2+}-mediated base exchange of choline for other phospholipid head groups [47].

Using the isolated hamster heart as a model, we have sought the presence of the three known pathways and their relative contribution to the formation of phosphatidylcholine. Since choline is the common precursor for phosphatidylcholine formation via the CDP-choline pathway and the base-exchange pathway, the contributions of these two pathways were evaluated in the isolated heart by perfusion with labelled choline [40]. The distinct lag in the incorporation of label into phosphatidylcholine suggested that the majority of the labelled phosphatidylcholine was not formed by the base-exchange reaction. A similar approach had been used to identify the base-exchange pathway as a minor route for the

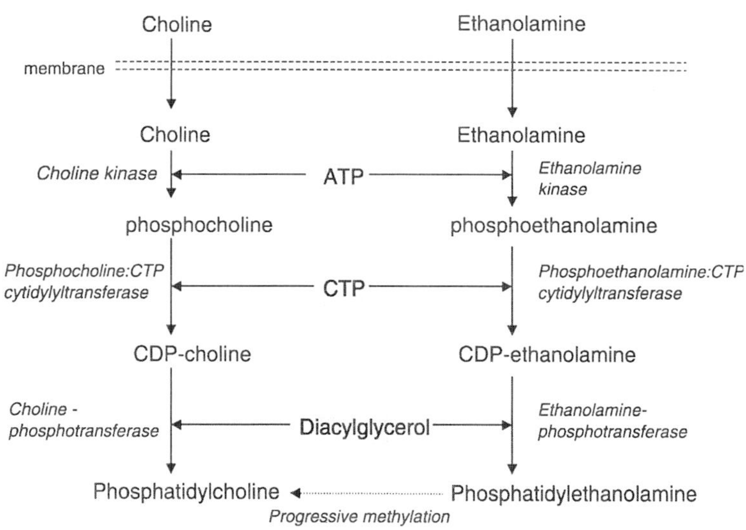

Fig. 2. Biosynthesis of phosphatidylcholine and phosphatidylethanolamine.

formation of phosphatidylcholine in the brain [48]. The rate of phosphatidylcholine formation via the CDP-choline pathway in the hamster heart was estimated to be 39 nmol min^{-1} (g wet $weight)^{-1}$ [40]. In a series of pulse-chase experiments, we have identified the rate-limiting step in the CDP-choline pathway which is the conversion of phosphocholine to CDP-choline, a reaction catalysed by CTP:phosphocholine cytidylyltransferase. The rate-limiting role of cytidylyltransferase in the CDP-choline pathway has also been identified in the liver, lung, brain and other mammalian tissues [49]. When the isolated hamster heart was perfused with labelled ethanolamine, the contribution of phosphatidylethanolamine methylation to phosphatidylcholine formation was found to be relatively minor [40]. The results of this study were in contrast to those obtained for the liver, where the methylation pathway makes a substantial contribution to the overall phosphatidylcholine synthesis [50].

3.3. *Control of phosphatidylcholine biosynthesis*

The biosynthesis of phosphatidylcholine must be tightly controlled in order to maintain its membrane content and composition. To date, at least four points of control have been identified: (i) choline uptake and its subsequent phosphorylation; (ii) the energy status of the organ; (iii) modulation of the rate-limiting enzyme, CTP:phosphocholine cytidylyltransferase and (iv) modulation of the final step in the CDP-choline pathway, catalysed by CDP-choline:1,2-diacylglycerol cholinephosphotransferase.

3.3.1. *Choline uptake*

Choline is an essential nutrient in the diet, and its absence has profound effects on phosphatidylcholine biosynthesis [51,52]. Similar results have been obtained from cell lines with restricted choline supply [53,54]. In the heart, choline is taken up by a saturable mechanism with a K_m of 0.1 mM [40]. It is thus possible that the plasma choline concentration (approximately 0.18 mM) provides a mechanism for the regulation of choline uptake in the heart. In other tissues, two choline receptors have been identified [55]. One receptor, which exhibits a relatively high affinity for choline and requires Na^+ for activity, is associated with the synthesis of acetylcholine in cholinergic synaptosomes. The other receptor is a low affinity, Na^+-independent receptor which transports choline into non-cholinergic cells. This receptor is responsible for transporting choline for subsequent phosphorylation by choline kinase and phosphatidylcholine synthesis. As ethanolamine and choline are structurally similar and both compounds are present in the plasma, we investigated the competitive effect of ethanolamine on phosphatidylcholine biosynthesis, and conversely the effect of choline on phosphatidylethanolamine biosynthesis [40,56]. In the isolated hamster heart, ethanolamine was found to inhibit the incorporation of labelled choline into phosphatidylcholine [40]. Ethanolamine competitively inhibited choline uptake, while conversion of choline into phosphocholine or the biosynthesis of phosphatidylcholine was unaffected. Similar results were observed in experiments using baby hamster kidney-21 cells [57]. In experiments aimed at investigating the effect of choline on phosphatidylethanolamine biosynthesis, choline was unable to inhibit ethanolamine uptake [58]. These observations provided evidence that the uptake of choline and ethanolamine by hamster cells occur via at least two distinct mechanisms. Entry of choline into the cell via a choline receptor may be competitively inhibited by ethanolamine, but the entry of ethanolamine through its cognate receptor is unaffected by choline. Although choline and ethanolamine are structurally similar, their differences are evidently sufficient to confer the observed specific inhibition. We hypothesise that steric hindrance due to the larger size and shape of choline does not allow it to associate with the ethanolamine binding site.

The influence of L-lysine on phosphatidylcholine biosynthesis in renal cortical slices has been reported [59]. Hence, the effect of amino acids, including L-lysine, on choline uptake and phosphatidylcholine biosynthesis in the isolated heart was investigated in our laboratory. Glycine induced a 30% increase in choline uptake, but interestingly did not affect the phosphorylation of choline or the rate of phosphatidylcholine biosynthesis [60]. In a subsequent study, the effect of neutral, basic and acidic amino acids on choline uptake was assessed. In the presence of L-alanine, L-serine or L-phenylalanine, choline uptake was enhanced 20–38%, and again, the enhancement of choline uptake did not affect the rate of phosphatidylcholine formation [61]. Basic and acidic amino acids had no effect on choline uptake. The enhancement of choline uptake by neutral amino acids was not additive or dose dependent, but required a concentration threshold. The effect of amino acids on choline uptake appears to be species- and tissue-specific, as amino acids did not affect choline uptake in guinea pig, rat or rabbit hearts [62], whereas glycine, L-alanine, L-serine, L-leucine, L-aspartate and L-arginine inhibited choline uptake in baby hamster kidney cells [57]. Since exogenous choline has no effect on the uptake of amino acids, we postulate that choline and the neutral amino acids are not co-transported by the same mechanism. The modulation of choline uptake may be facilitated by the direct interaction of neutral amino acids with the choline transport system.

3.3.2. Choline kinase

Choline kinase catalyses the phosphorylation of choline to phosphocholine in the presence of Mg^{2+}, with ATP as the phosphate donor. Choline kinase has been purified from a number of sources, including African green monkey lung [63], rat kidney [64], rat brain [65] and rat liver [66]. Choline kinase exists in at least two isoforms (alpha and beta) in mammalian tissues [67,68] which catalyse the phosphorylation of both choline and ethanolamine [69]. A cDNA for choline kinase has been cloned from rat liver [70]. The cDNA sequence predicted a 435 amino acid protein, and yielded a protein of approximately 49.7 kDa when expressed in *E. coli*. The recombinant kinase catalysed the phosphorylation of choline, ethanolamine, *N*-monomethylethanolamine and *N,N*-dimethylethanolamine. A human choline kinase cDNA has been cloned from a human glioblastoma expression library [71]. The cDNA sequence predicted a protein of 456 amino acids and molecular weight 52 kDa, and exhibited a high degree of sequence similarity to the rat cDNA. Choline kinase activity may be regulated by Ras-dependent activation [72].

Upon entry into the heart cell, choline is immediately phosphorylated to form phosphocholine [40]. Choline kinase activity has been detected in both the cytosolic and membrane fractions of various tissues [55]. Subcellular fractionation by differential centrifugation, however, does not provide definitive evidence as to the native subcellular location of the enzyme, as redistribution may have occurred during the preparation procedure. One possibility is that choline kinase is a loosely associated membrane protein adjacent to and/or near the choline uptake system. The close proximity of choline kinase to the choline uptake system would facilitate immediate choline phosphorylation. The disruption of the cell would lead to the presence of choline kinase activity in both cytosolic and membrane fractions.

3.3.3. Energy status

The possible rate-limiting role of the reaction catalysed by choline kinase was suggested in a study in which rat hearts were perfused under ischemic conditions [73]. Ischemia severely lowered the ATP and CTP levels in the hearts, with a concomitant decrease in the synthesis of phosphatidylcholine. The conversion of choline to phosphocholine was reduced, resulting in an accumulation of choline. The synthesis of CDP-choline was unaffected. This study demonstrated that the reaction catalysed by choline kinase can become rate-limiting in the event of a greatly diminished ATP pool.

Both ATP and CTP are required cofactors in the CDP-choline pathway and the availability of these high energy compounds may affect the rate of phosphatidylcholine biosynthesis [45,73–76]. An increase in cytoplasmic CTP in polio-infected HeLa cells caused an enhancement of phosphatidylcholine biosynthesis in these cells, but the increase in ATP level did not affect the phosphorylation of choline [77]. The effects of ATP and CTP levels on phosphatidylcholine biosynthesis in the heart were examined using a strain of cardiomyopathic hamsters as an experimental model [74]. Through autosomal recessive inheritance, cardiomyopathy develops spontaneously in a strain of inbred Syrian hamsters (BIO 14.6 strain) in which the myocardium exhibits degenerative lesions with 100% incidence [78]. Severe decreases in the ATP and CTP concentrations in the hearts of myopathic animals were observed [74]. Incorporation of radioactive choline into phosphocholine, and the pool size of phosphocholine, was unaltered despite the depressed ATP levels. In contrast, the labelling and pool size of CDP-choline were decreased. The net rate of phosphatidylcholine synthesis, however, was maintained in the myopathic hearts, and this phenomenon was attributed to an observed increase in cytidylyltransferase activity. This activation of cytidylyltransferase was regarded as a compensatory mechanism to maintain a minimum CDP-choline level in order to prevent a net reduction of phosphatidylcholine biosynthesis. This study further substantiated the rate-limiting role of cytidylyltransferase in the CDP-choline pathway.

The effect of an acute reduction in the energy status of the heart was examined using hypoxic and ischemic isolated heart models [73,75]. Hypoxic conditions were produced by perfusing the hearts with Krebs-Henseleit buffer saturated with N_2 rather than O_2, while ischemia was induced by perfusing with oxygenated buffer at 10% of the normal flow rate. Severe decreases in the levels of ATP and CTP were observed in hypoxia and ischemia, with resulting decreases in phosphatidylcholine biosynthesis. Examination of choline-containing metabolites revealed differences between the reaction of the hearts to hypoxia or ischemia. In ischemia, a decrease in the conversion of choline to phosphocholine was observed, but CDP-choline labelling was unaffected [73]. In contrast, hypoxia did not alter the conversion of choline to phosphocholine, but the conversion of phosphocholine to CDP-choline was decreased. As was the case in the myopathic hamster model, a concomitant increase in CTP:phosphocholine cytidylyltransferase activity was observed in hypoxia [75]. This stimulation of cytidylyltransferase was attributed to an accumulation of fatty acids in the hypoxic heart (see below). The importance of the CTP supply to phosphatidylcholine synthesis was corroborated in a study in which rat heart myoblastic (H9c2) cells were incubated with cyclopentenylcytosine, whose metabolite, cyclopentenylcytosinetriphosphate, is a potent and specific inhibitor of CTP synthetase [76]. CTP was reduced to 10% of control levels, while other nucleotides were unaffected. The synthesis of phosphatidylcholine and all other phospholipids was reduced by treatment with cyclopentenylcytosine. These findings indicate that the intracellular ATP and CTP concentrations are important factors for the maintenance of phosphatidylcholine and phospholipid biosynthesis in the heart.

3.3.4. Modulation of the key enzyme

CTP:phosphocholine cytidylyltransferase has been extensively reviewed elsewhere [12,41,79–83]. In this section, we will highlight some recent developments and discuss the regulation of this enzyme as it relates to phosphatidylcholine metabolism in the heart. The cytidylyltransferase has been purified to homogeneity from rat liver [84,85] and has been cloned from a rat liver cDNA library [86]. The cytidylyltransferase protein has a molecular weight of approximately 42 kDa [84–86], and the cDNA predicts a protein of 367 amino acids [86]. At least four isoforms of the cytidylyltransferase (alpha, beta-1, beta-2 and beta-3) exist in mammalian tissues [87]. One plausible mechanism for the posttranslational regulation of cytidylyltransferase activity is summarised in the translocation hypothesis which postulates that the enzyme exists in both soluble and membrane-bound forms [12,41,80–83]. The soluble form is relatively inactive, and its activity is enhanced by translocation to intracellular membranes where it is activated by association with certain phospholipids. Cytidylyltransferase has been shown to exist in both the cytoplasm and the nucleus, and it can be associated with the endoplasmic reticulum or the nuclear membrane [12,41,82,88]. However, cytidylyltransferase-alpha does not localise to the nucleus in pulmonary tissues indicating that nuclear localisation of cytidylyltransferase-alpha may not be a universal event [89]. The soluble form of the cytidylyltransferase is highly phosphorylated, while the membrane-associated form exists in a dephosphorylated state; it is thought that the lesser negative charge of the dephosphorylated state may allow the enzyme to associate more readily with membranes [90,91]. Translocation of the enzyme can be induced by fatty acids, diacylglycerol and acidic phospholipids [41].

Several studies have provided further details in the understanding of role of phosphorylation in the translocation and activation of cytidylyltransferase. The phosphorylation sites of the enzyme have been identified to be 16 Ser residues between Ser 315 and the carboxyl terminus [90]. Using site-directed mutagenesis, cytidylyltransferase mutants were created in which the serine residues in the phosphorylation region of the enzyme were replaced with alanine [91]. In cells transfected with the mutant enzyme, the cytidylyltransferase associated with membranes was found to be 10-fold greater than in cells transfected with wild type enzyme. This result is consistent with the notion that removal of negative charge from the phosphorylation region favours association of the enzyme with membranes. Despite the high degree of membrane-bound cytidylyltransferase in cells transfected with mutant enzyme, however, phosphatidylcholine synthesis increased by only 75% compared with cells transfected with the wild type enzyme, indicating that association with the membrane does not necessarily result in full activation of the enzyme. All mutant enzymes, including a mutant form in which the serine residues in the phosphorylation region were replaced with glutamate in order to impart a permanent negative charge, translocated to membranes upon treatment of cells with oleate. This observed translocation by all forms of the enzyme indicated that a negative charge does not prevent association of the enzyme with membranes, and that membrane composition also plays a role. In another study, dephosphorylation was shown not to be required before membrane association; rather, treatment of cells with oleic acid or phospholipase C induced the translocation of cytidylyltransferase to the membrane, where it was subsequently dephosphorylated [92]. Dephosphorylation of the enzyme, however, seems to favour, but is not a prerequisite for, association with the membrane. The dephosphorylation of the enzyme upon membrane association may stabilise that association, where the enzyme may be able to interact with lipid activators.

Amino acid residues 228–315 of the rat cytidylyltransferase sequence were predicted to form two α-helices interrupted by a turn between residues 294 and 297 [86]. The two helices have an asymmetric distribution of polar residues on one face and hydrophobic residues on the other, and it is thought that this domain was important for the enzyme to associate with membranes [86,93–95]. The helix appears to intercalate into membrane bilayers and that this interaction stabilises the helical structure [93,94]. Upon association with membranes, the cytidylyltransferase is able to interact with lipid activators. Cytidylyltransferase is activated by oleic acid and anionic phospholipids such as phosphatidylinositol, phosphatidylserine and phosphatidylglycerol [41,82,96] and by neutral lipids with small polar head groups such as diacylglycerol [41,82,97]. Zwitterionic phospholipids such as phosphatidylcholine and phosphatidylethanolamine are weak activators [41,82]. Lipid activators stimulate cytidylyltransferase activity by lowering the K_m for CTP [98]. The α-helical region is important for interaction of the enzyme with lipid activators, as its removal has been shown to abrogate the stimulation of enzymatic activity by lipid activators. Furthermore, the truncated enzyme was no longer inhibited by the negative lipid modulator 1-*O*-octadecyl-2-*O*-methyl-glycero-3-phosphocholine, an anti-neoplastic analogue of lysophosphatidylcholine. In another study, mutation of a phosphorylated Ser (Ser 315) to Ala lowered the K_m for activation by mixtures of phosphatidylcholine and oleic acid or diacylglycerol [99]. The complete deletion of the phosphorylation domain (residues 312–367) further lowered the K_m for lipid activation. The results of the latter study indicate that phosphorylation of the carboxyl terminal of the cytidylyltransferase interferes with the stimulation of the enzyme by lipid activators.

The regulation of phosphatidylcholine biosynthesis by fatty acids appears to be different between the liver and the heart. In hamster hearts perfused with fatty acids, only stearic acid was found to be effective in stimulating phosphatidylcholine synthesis [100] whereas in rat hepatocytes [101] stimulatory effects were produced by all fatty acids examined. The stimulation of phosphatidylcholine biosynthesis by stearic acid in the heart appears to be dependent on the intracellular accumulation of the fatty acid to a critical level in order to produce effective modulation of the cytidylyltransferase activity. In rat hepatocytes, activation of cytidylyltransferase can be duplicated in vitro by the addition of free fatty acids to the assay mixture [101]. Such direct activation of enzyme activity was not observed with the cytidylyltransferase from the hamster heart [100]. The inability of stearic acid to activate hamster heart cytidylyltransferase in vitro suggests that modulation of enzyme activity by fatty acid is not the same in the heart and in the liver.

The importance of the microsomal form of the cytidylyltransferase in the regulation of phosphatidylcholine biosynthesis in the heart has been demonstrated. When isolated hamster hearts were perfused with exogenous stearic acid (50 μM), phosphatidylcholine biosynthesis was stimulated, and this stimulation was attributed to a 2.3-fold increase in the activity of the microsomal cytidylyltransferase [100]. No redistribution of cytidylyltransferase activity, however, from the cytosolic to the microsomal fraction was detected. In the hypoxic hamster heart, a stimulation of the microsomal cytidylyltransferase activity and a corresponding decrease in the cytosolic activity were detected [75]. This redistribution of enzyme activity was attributed to an increase in cytosolic fatty acid levels in hypoxia. Rat heart myocytes incubated in the presence of 0.2 μM vasopressin exhibited enhanced cytidylyltransferase activity in both particulate and soluble fractions, with a corresponding increase in phosphatidylcholine synthesis [102]. A higher concentration of vasopressin (1.0 μM) resulted in decreased phosphatidylcholine synthesis, which was attributed to a redistribution of cytidylyltransferase activity from the membranes to the cytosolic fraction. Incubation of H9c2 cells with angiotensin II caused an increase in phosphatidylcholine synthesis [103]. The latter was due to a stimulation of cytidylyltransferase activity in both soluble and particulate fractions, resulting in a greater conversion of phosphocholine to CDP-choline. The stimulation of phosphatidylcholine synthesis by angiotensin II was not affected by an AT_1 receptor antagonist, losartan, at concentrations which block AT_1 receptors (1.0–100 μM) [104]. At higher concentrations (500 μM), however, losartan caused the inhibition of phosphatidylcholine synthesis, independent of angiotensin II. Losartan was found to inhibit choline uptake in a competitive manner. Interestingly, an increase in the particulate cytidylyltransferase activity and a concomitant decrease in the soluble activity were observed in the cells treated with 500 μM losartan. The translocation of cytidylyltransferase from soluble to particulate fractions was demonstrated by immunoblotting with an antibody against the cytidylyltransferase. This redistribution of enzyme activity to the membranes may be a compensatory mechanism in response to the losartan-mediated reduction in new phosphatidylcholine synthesis.

The level of phosphatidylcholine within cells may also regulate membrane cytidylyltransferase activity. In murine P19 teratocarcinona cells induced to undergo differentiation into cardiac myocytes, phosphatidylcholine levels were reduced during dimethylsulfoxide-induced differentiation [31]. This reduction in phosphatidylcholine was associated with a 3.6-fold elevation in membrane cytidylyltransferase activity may act as a compensatory mechanism for the reduction in cellular phosphatidylcholine levels that occurred during the differentiation process.

3.3.5. Cholinephosphotransferase

The purification of CDP-choline:1,2-diacylglycerol cholinephosphotransferase has proven to be difficult, due to the susceptibility of this enzyme to loss of activity upon removal from its native membrane environment [105]. To date, only partial purification of this enzyme has been achieved [106–108]. Using the yeast *Saccharomyces cerevisiae* as an experimental model, the *S. cerevisiae CPT1* gene was cloned by genetic complementation of a mutant strain deficient in cholinephosphotransferase activity [109]. The *CPT1* gene predicts a 407 amino acid protein with seven transmembrane helices [110]. In order to study functional domains with the cholinephosphotransferase enzyme, chimeric genes were constructed by splicing together segments of the *CPT1* and *EPT1* genes [111]. *EPT1* encodes a choline/ethanolamine phosphotransferase that can utilise CDP-choline or CDP-ethanolamine as substrates in the synthesis of phosphatidylcholine or phosphatidylethanolamine, respectively. Regions of cholinephosphotransferase responsible for substrate specificity were determined by expressing the chimeric gene constructs in a *cpt1 ept1* double null mutant background that lacked phosphoaminoalcohol transferase activity and analysing the substrate utilisation profiles of the chimeric enzymes. A region conferring CDP-aminoalcohol specificity was assigned to a predicted cytoplasmic domain comprising amino acids 79–186, while diacylglycerol acyl chain specificity was assigned to an internal 218 amino acid segment that was predicted to contain three transmembrane segments.

The human cDNA for cholinephosphotransferase (CPT1) has been cloned and characterised [112]. Expression of the mRNA varied greater than 100-fold between tissues being most abundant in testes followed by colon, small intestine, heart, prostate and spleen. The enzyme utilised exclusively CDP-choline as substrate in the phosphobase transfer to 1,2-diacyl-*sn*-glycerol. A second human cholinephosphotransferase (CEPT1) utilises both CDP-choline and CDP-ethanolamine for phosphobase transfer.

While it is clear that CTP:phosphocholine cytidylyltransferase plays a major role in the regulation of phosphatidylcholine biosynthesis, the step catalysed by CDP-choline:1,2-diacylglycerol cholinephosphotransferase may offer an additional point of control. The coordination of phosphatidylcholine biosynthesis with other major metabolic pathways in the liver, via regulation at the cholinephosphotransferase-catalysed step, was revealed in studies involving fasting hamsters [113]. In the livers of hamsters fasted for up to 48 h, phosphatidylcholine biosynthesis was reduced due to a number of factors, including a decreased rate of choline uptake; reductions in the pool sizes of ATP, CTP and metabolites in the CDP-choline pathway such as phosphocholine, CDP-choline and diacylglycerol; and decreased activity of cholinephosphotransferase. The inhibition of cholinephosphotransferase was found to be due to accumulation of an inhibitory cytosolic factor in the livers of fasted animals. This factor was identified as arginosuccinate, a metabolite in the urea cycle. In the fasting animal, an increased utilisation of amino acids for gluconeogenesis in the liver leads to activation of enzymes in the urea cycle [114–116]. The regulation of phosphatidylcholine synthesis via inhibition of cholinephosphotransferase by arginosuccinate may represent a novel mechanism for the coordination of phospholipid and protein metabolism in the liver during gluconeogenesis [113].

In the heart, evidence for the regulation of phosphatidylcholine biosynthesis at the level of cholinephosphotransferase was obtained in studies using hamster hearts perfused in the presence or absence of the antiarrhythmic drug lidocaine, under normoxic and hypoxic conditions [117,118]. In hearts perfused with lidocaine, using radiolabelled glycerol as a radioactive precursor, an increase in the synthesis of phospholipids such as phosphatidylethanolamine, phosphatidylinositol and phosphatidylserine, as well as the neutral lipid diacylglycerol was observed. In contrast, the synthesis of phosphatidylcholine was decreased by lidocaine [118]. In hypoxic hearts, the synthesis of all phospholipids was decreased [117]. Whereas lidocaine partially restored the synthesis of other phospholipids and diacylglycerol in hypoxia, the synthesis of phosphatidylcholine remained depressed. The distinct behaviour of phosphatidylcholine biosynthesis under lidocaine treatment was attributed to inhibition of cholinephosphotransferase by lidocaine.

Recent coimmunofluorescence confocal microscopy and subcellular fractionation studies have determined that CPT1 may be localised to the Golgi whereas CEPT1 may be localised to both nuclear and endoplasmic reticulum membranes [119]. Hence, substrate channelling from the nuclear cytidylyltransferase alpha to the nuclear CEPT1 could be a mechanism by which upregulation of the CDP-choline pathway increases new phosphatidylcholine biosynthesis.

4. The biosynthesis of phosphatidylethanolamine

The CDP-ethanolamine pathway (Fig. 2), which is corollary to the CDP-choline pathway, is the major route for phosphatidylethanolamine biosynthesis in most mammalian tissues [46,120]. The decarboxylation of phosphatidylserine, however, has been recognised as an important pathway for phosphatidylethanolamine biosynthesis in prokaryotes and some eukaryotic cells [121]. The decarboxylation of phosphatidylserine is catalysed by phosphatidylserine decarboxylase. The mammalian phosphatidylserine decarboxylase (PSD) is localised to the inner mitochondrial membrane and contains mitochondrial membrane targeting and sorting sequences [122]. A calcium mediated base-exchange pathway appears to be a minor route for the quantitative biosynthesis of phosphatidylethanolamine [46].

In the CDP-ethanolamine pathway, ethanolamine is phosphorylated to phosphoethanolamine which is then converted to CDP-ethanolamine and finally to phosphatidylethanolamine. Although the rate-limiting step of this pathway appears to be the conversion of phosphoethanolamine to CDP-ethanolamine [46,120], the phosphorylation of ethanolamine can also become rate limiting under certain conditions [123]. Phosphorylation of choline and ethanolamine appears to be catalysed by the same enzyme in all tissues (see Section 3.3.2) [107,124].

The conversion of phosphoethanolamine for the formation of CDP-ethanolamine is catalysed by CTP:phosphoethanolamine cytidylyltransferase. The purified enzyme from rat liver is localised to areas in the cytoplasm rich in rough endoplasmic reticulum and exhibits a bimodal distribution between the cisternae of the rough endoplasmic reticulum and the cytosolic space [125]. Although the enzyme was purified by Sundler [50] almost two decades ago, limited information is available concerning its regulation [126]. The enzyme utilises both CTP and deoxyCTP as substrates. The CTP:phosphoethanolamine cytidylyltransferase cDNA was cloned from rat liver and the cDNA codes for a protein of 404 amino acids and a molecular mass of 45.2 kDa [127]. A human CTP: phosphoethanolamine cytidylyltransferase cDNA clone was isolated from a human glioblastoma cDNA expression library [128]. Introduction of the gene into a yeast mutant in which the cytidylyltransferase gene was disrupted restored formation of CDP-ethanolamine and phosphatidylethanolamine in these cells.

The conversion of CDP-ethanolamine to phosphatidylethanolamine is catalysed by the CDP-ethanolamine:diacylglycerol ethanolaminephosphotransferase. As is the case for cholinephosphotransferase, this enzyme (CEPT-1) is also located in the endoplasmic reticulum and nucleus [119,129,130]. We have shown that the enzyme can be solubilised by Triton QS-15, and the cholinephosphotransferase activity can be partially separated from the ethanolaminephosphotransferase activity by ion exchange chromatography [131]. The two activities also display some differences in their characteristics and kinetic properties [108].

4.1. Regulation of phosphatidylethanolamine biosynthesis

Recently, we have demonstrated that phosphatidylethanolamine biosynthesis is elevated during murine P19 teratocarcinoma cell differentiation into cardiac myocytes [132]. The mechanism for the increase in phosphatidylethanolamine biosynthesis during differentiation was a 2.8-fold elevation in ethanolaminephosphotransferase activity. Interestingly, addition of the phosphatidylethanolamine biosynthesis inhibitor 8-(4-chlorophenylthio)-cAMP attenuated the differentiation-induced elevation in phosphatidylethanolamine mass but did not affect the expression of striated myosin indicating that elevation in phosphatidylethanolamine biosynthesis is an early but not essential event for cardiac cell differentiation. In addition, CTP:phosphoethanolamine cytidylyltransferase expression appears to be regulated at both the transcriptional as well as the translational level during development of the rat liver [127].

5. Biosyntheses of phosphatidylinositol and phosphatidylserine

These phosphoglycerides are present in significant quantities in mammalian tissues. They carry a net negative charge and are called acidic phospholipids. The key step for the synthesis of phosphatidylinositol is the conversion of phosphatidic acid to CDP-diacylglycerol by the action of CTP:phosphatidic acid cytidylyltransferase. The enzyme is located in both the microsomal and the mitochondrial fractions. The microsomal form of the enzyme is activated by GTP whereas the mitochondrial form is not, suggesting that these two forms of the enzyme are different. In support of this, we have observed a twofold elevation in mitochondrial cytidylyltransferase in a Chinese hamster lung fibroblast cell line (CCL16-B2) deficient in mitochondrial energy metabolism compared to the CCL16-B1 wild type [133]. Microsomal cytidylyltransferase activity was unaltered in CCL16-B2 cells. The mitochondrial enzyme has been purified from yeast [134]. It has a native mass of 114 kD and a subunit mass of 56 kD, and $MgCl_2$ is required for full enzymatic activity.

CTP:phosphatidic acid cytidylyltransferase has been cloned and characterised from rat brain [135]. The enzyme had a high degree of molecular species specificity for

1-stearoyl-2-arachidonoyl phosphatidic acid as substrate. Since phosphatidylinositol consists of predominantly 1-stearoyl-2-arachidonoyl species, these authors suggested that the enzyme may selectively participate in the phosphoinositide cycle but not in the synthesis of phosphatidylglycerol and cardiolipin. Northern blot analysis indicated the absence of the mRNA in rat heart, an organ that requires significant cardiolipin biosynthesis. It is possible that the gene undergoes alternative splicing, in which different sized transcripts of mRNA would be produced, or the enzyme is a member of a family of genes. In support of the latter, two human CTP:phosphatidic acid cytidylyltransferases genes, CDS1 and CDS2, have been cloned and sequenced and are localised to chromosomes 4q21 and 20p13 [136].

The CDP-diacylglycerol formed from the CTP:phosphatidic acid cytidylyltransferase reaction condenses with inositol for the formation of the phosphatidylinositol. The reaction is catalysed by phosphatidylinositol synthase (EC 2.7.8.11) which has been purified to near homogeneity [137]. Overexpression of the human cDNA coding for both CTP:phosphatidic acid cytidylyltransferase and phosphatidylinositol synthase in COS-7 cells leads to the overproduction of CTP:phosphatidic acid cytidylyltransferase, phosphatidylinositol synthase and phosphatidylinositol:inositol exchange reactions but did not enhance the rate of phosphatidylinositol biosynthesis [138]. These data indicate that the level of CTP:phosphatidic acid cytidylyltransferase and phosphatidylinositol synthase protein expression may not play a central role in the regulation of phosphatidylinositol biosynthesis. However, phosphatidylinositol biosynthesis may be regulated by the intracellular level of CTP [76] and at the level of inositol uptake in H9c2 cardiac myoblast cells [139].

In prokaryotes, phosphatidylserine is formed via the CDP-diacylglycerol pathway (similar to the formation of phosphatidylinositol). In mammalian tissues, phosphatidylserine is synthesised from a base-exchange reaction in which the base group of a pre-existing phospholipid is exchanged for serine. The base-exchange enzyme is located in the microsomal fraction and requires calcium for activity. The calcium requirement, however, can be circumvented by the presence of ATP. The purified enzyme from brain microsomes has both ethanolamine and serine base-exchange activity [140]. There are two isoforms of mammalian phosphatidylserine synthase (Pss1 and Pss2). They are structurally similar (32% amino acid identity) but differ in substrate specificities. Although in vitro phosphatidylserine synthase activity is reduced between 90 and 95% in Pss2 null mice, they do not show a deficiency in phospholipid content nor severe developmental abnormalities [141].

6. Biosyntheses of polyglycerophospholipids

Cardiolipin, the first polyglycerophospholipid discovered, was isolated from beef heart by Pangborn (reviewed in Ref. [142]). In rat liver, cardiolipin is localised to both inner and outer mitochondrial membranes and within contact sites in eukaryotic cells [143–145]. In the rat heart, cardiolipin is found exclusively in mitochondria and comprises approximately 12–15% of the total phospholipid mass [146]. Phosphatidylglycerol comprises 1–1.5% of the total phospholipid mass in mammalian tissues [142].

Phosphatidylglycerol is synthesised in mitochondria and microsomes and is localised not only to mitochondria but also in non-mitochondrial subcellular membranes.

Cardiolipin is required for the reconstituted activity of a number of key mitochondrial enzymes involved in cellular energy metabolism and may be the "glue" that holds the respiratory chain together [147]. In a temperature sensitive Chinese hamster ovary mutant cell line (PGP-S), phosphatidylglycerol and cardiolipin were shown to be essential for cell growth and function of the electron transport chain [148]. The temperature sensitive mutant, with thermolabile phosphatidylglycerol phosphate (PGP) synthase, was defective in phosphatidylglycerol and cardiolipin production at the non-permissive temperature. The mutant exhibited morphological and functional mitochondrial abnormalities including impairment of rotenone-sensitive NADH-ubiquinone reductase (Complex 1). When these mutant cells were placed on galactose medium, in which 98% of the cellular energy must be supplied by oxidative phosphorylation, growth was markedly attenuated. The role of cardiolipin in apoptosis is now well established. In staurosporine-treated granulosa cells undergoing apoptosis cardiolipin levels were reduced [149]. In addition, peroxidation of the fatty acyl chains of cardiolipin induced release of cytochrome c from mitochondria into the cytosol and this was associated with the induction of apoptosis [150–152]. Incubation of isolated rat heart cardiomyocytes with palmitate reduced cardiolipin levels and this was associated with an increase in cytochrome c release from the inner mitochondrial membrane [153]. In lower eukaryotes, phosphatidylglycerol and cardiolipin were shown to be essential for growth and survival of yeast under stress conditions [154,155], to improve the efficiency of oxidative phosphorylation [156], for maintenance of proper mitochondrial membrane potential and protein import [157] and translation of protein components of the electron transport chain [158]. In addition to its importance as a precursor for cardiolipin biosynthesis, phosphatidylglycerol is actively synthesised and secreted by alveolar type II cells and is a critical component of alveolar surfactant in the lung [159].

The de novo biosynthesis of mammalian phosphatidylglycerol via the CDP-diacylglycerol pathway was elucidated by Kennedy and co-workers in the 1960s (Fig. 3) [160]. In prokaryotes, cardiolipin is formed by the condensation of two phosphatidylglycerol molecules [161]. In eukaryotic cells, phosphatidylglycerol is converted to cardiolipin by condensation with CDP-diacylglycerol catalysed by CL synthase (Fig. 3) [162]. In the rat heart de novo biosynthesis of phosphatidylglycerol occurs via the CDP-diacylglycerol

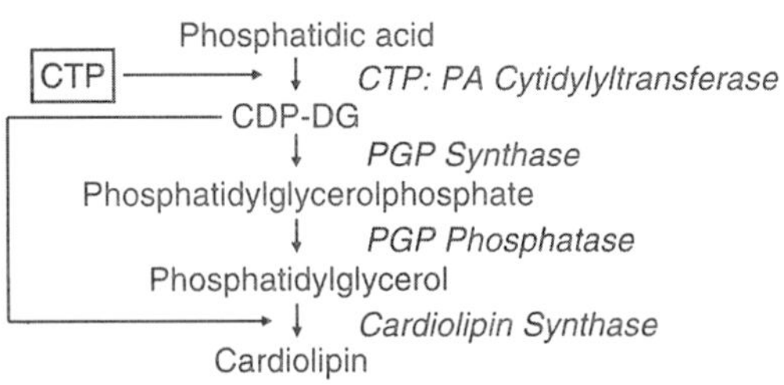

Fig. 3. The biosynthesis of phosphatidylglycerol and cardiolipin. Abbreviations: CDP-DG, CDP-diacylglycerol; PA, phosphatidic acid; PGP, phosphatidylglycerolphosphate.

pathway [146,163]. Using labelled glycerol, phosphatidylglycerol and cardiolipin are actively synthesised from the newly formed phosphatidic acid. Phosphatidic acid is converted to CDP-diacylglycerol by CTP:phosphatidic acid cytidylyltransferase. The mammalian enzyme has proven difficult to purify. The McMurray group characterised the cytidylyltransferase from rat liver mitochondrial and microsomal preparations [164]. The cytidylyltransferase was solubilised and partially purified from bovine brain [165]. Pulse-chase heart perfusion studies have indicated that one of the rate-limiting steps of phosphatidylglycerol and cardiolipin biosynthesis in the rat heart is the conversion of phosphatidic acid to CDP-diacylglycerol [146,166]. Evidence suggests that the cellular levels of CTP, lysophosphatidylcholine and 1,2-diacyl-*sn*-glycerol may control this reaction [76,167,168].

The CTP:phosphatidic acid cytidylyltransferase involved in polyglycerophospholipid biosynthesis was cloned from a human NT2 neuronal cell library [169]. The mRNA was expressed in all tissues examined including a predominant expression in heart. Furthermore, an additional mRNA band was observed in heart. Expression of this human cDNA in COS cells resulted in greater than a fourfold increase in phosphatidic acid:CTP cytidylyltransferase activity. The predicted reading frame encoded a protein of 444 amino acids with a molecular mass of 51.4 kDa. This was similar to the subunit molecular weight of 56 kDa of the purified protein from yeast as predicted by SDS-PAGE [134]. Another human cDNA clone for phosphatidic acid:CTP cytidylyltransferase was expressed in yeast [170]. Transfection of this cDNA into NCI-460 and ECV304 cells resulted in an approximate twofold increase in enzyme activity. Furthermore, the increase in phosphatidic acid:CTP cytidylyltransferase was associated with an increased secretion of tumour necrosis factor alpha and interleukin-6 from ECV304 cells upon stimulation with interleukin-1 beta. This finding led the authors of the study to suggest that phosphatidic acid:CTP cytidylyltransferase overexpression may amplify cellular signalling responses from cytokines.

In the second and third steps of the CDP-diacylglycerol pathway CDP-diacylglycerol condenses with *sn*-glycerol-3-phosphate to form phosphatidylglycerol catalysed by phosphatidylglycerol-phosphate synthase (PGS-S) and phosphatidylglycerolphosphate phosphatase. PGS-S was recently purified from Chinese hamster ovary cells [171]. The 60 kDa enzyme was devoid of phosphatidylglycerolphosphate phosphatase activity. The phosphatidylglycerolphosphate synthase and phosphatase have been characterised in the rat heart [172]. The rat heart phosphatidylglycerolphosphate phosphatase is a heat labile enzyme which may be stimulated in vitro by unsaturated fatty acids.

In the final step of the pathway cardiolipin is formed from the condensation of phosphatidylglycerol and CDP-diacylglycerol catalysed by cardiolipin synthase in eukaryotic cells [76,146,162,173]. The cardiolipin synthase is localised exclusively to the inner side of the inner mitochondrial membrane [174,175] and was purified to homogeneity from rat liver mitochondria [176]. The enzyme appears as a 50 kDa band in SDS-PAGE gel electrophoresis and requires cobalt for optimum activity. In prokaryotes the cardiolipin synthase is encoded by the *cls* gene [177]. The gene encoding cardiolipin synthase from yeast has been isolated and characterised [154]. Disruption of the gene in a haploid yeast strain resulted in loss of cardiolipin synthase activity and no detectable levels of cardiolipin but greatly elevated levels of phosphatidylglycerol.

Interestingly, although the absence of cardiolipin was not lethal, quantitatively the cells grew poorly on non-fermentable carbon sources in which oxygen-dependent energy metabolism is required.

6.1. Regulation of phosphatidylglycerol and cardiolipin biosynthesis

Continuous pulse and pulse-chase radiolabelling studies provide evidence that one of the rate-limiting steps of phosphatidylglycerol and cardiolipin biosynthesis in the rat heart is the conversion of phosphatidic acid to CDP-diacylglycerol [146]. In support of this is the observation that reduction of cellular CTP levels reduces de novo phosphatidylglycerol and cardiolipin biosynthesis from glycerol but not from linoleic acid [76]. Evidence suggests that exogenous CDP-diacylglycerol could regulate phosphatidylglycerol biosynthesis. In isolated digitonin permeabilised hepatocytes exogenous CDP-diacylglycerol was imported into mitochondria through mitochondrial inner and outer membrane contact sites and presented to the PGS-S for phosphatidylglycerol biosynthesis [178]. In H9c2 cardiac myoblast cells mitochondrial PGS-S was shown to be regulated by ceramide via a ceramide-activated and okadaic acid-sensitive protein phosphatase [179]. In addition, PGS-S activity was stimulated by treatment of H9c2 cells with CPT-cAMP. Mitochondrial PGS-S activity is increased 3.5-fold in thyroxine-treated animals and this accounts for the elevated levels of phosphatidylglycerol observed in that organ [180]. Evidence suggests that separate pools of phosphatidylglycerol may be utilised for cardiolipin biosynthesis in the heart [181] and that exogenous extra-mitochondrial phosphatidylglycerol may be utilised for cardiolipin biosynthesis [182]. Thyroxine treatment of rats stimulated the activity of liver and cardiac mitochondrial cardiolipin synthase 2.5-fold [180,183] and this may account for the elevated levels of cardiolipin observed in these organs [184].

Other polyglycerophospholipids found in mammalian tissues include bis(monoacylglycero)phosphate and acylphosphatidylglycerol which normally comprise less than 1% of the entire phospholipid mass in cells [142]. An exception is the alveolar macrophage in which bis(monoacylglycero)phosphate may comprise 14–18% of the entire phospholipid mass. The biosynthesis of bis(monoacylglycero)phosphate and acylphosphatidylglycerol occurs in the lysosome/endosome [185–187]. Synthesis of bis(monoacylglycero)phosphate may occur via deacylation of phosphatidylglycerol to form acyl-lysophosphatidylglycerol which is subsequently acylated to form bis(monoacylglycero)phosphate [188]. The transacylase and phospholipases required for the synthesis of bis(monoacylglycero) phosphate have been solubilised and characterised [188].

7. Resynthesis/remodelling of phospholipids

7.1. The deacylation–reacylation cycle

In general, the content and composition of phospholipids in a membrane system are well defined [1]. Changes in their content have been shown to alter membrane properties in tissues of normal and disease models [4,189]. In view of the importance of electrical signal generation and conduction to cardiac function and the central role of cellular membrane in those processes, alterations in these membrane phospholipids may affect membrane properties which have direct consequences to the maintenance of the cardiac rhythm [190]. In the biosynthesis of phosphatidylcholine, the CDP-choline:1,2-diacylglycerol cholinephosphotransferase has only limited specificity for diacylglycerol species with specific acyl chain compositions [191]. In the heart, a large proportion of phosphatidylcholine, and possibly phosphatidylethanolamine, undergoes remodelling in order to acquire the proper acyl groups [192]. The newly synthesised phospholipid is deacylated into the corresponding lysophospholipid which is then acylated back to the parent phospholipid by the action of the acyltransferase (Fig. 4). Impairment of the deacylation/reacylation process may result in the accumulation of lysophospholipids which are cytolytic at high concentrations [193]. A decrease in the activity of lysophosphatidylcholine:acyl-CoA acyltransferase has been shown in the ischemic pig heart [194,195]. Indeed, the elevated level of lysophospholipid during cardiac ischemia has been regarded as an important biochemical factor for the generation of cardiac arrhythmias [190]. Phospholipases and acyltransferases are generally regarded as the principal enzymes involved in phospholipid remodelling in mammalian tissue [146,196]. Phospholipids can be catabolised through the action of various phospholipases. The different types of phospholipases are categorised according to the hydrolysis of specific bonds in a phospholipid. Phospholipase A_1 specifically cleaves the acyl group at the *sn*-1 position, while cleavage of the acyl group at the *sn*-2 position is catalysed by phospholipase A_2. Phospholipases C and D are responsible for cleavage at the phosphate and phosphobase group. The phospholipases responsible for the hydrolysis of membrane phospholipids will be dealt with in other chapters within this series.

Structural studies reveal that acyl groups are distributed in an asymmetrical manner in most of the diacylphosphoglycerides. Saturated fatty acids are usually esterified at the *sn*-1 position whereas unsaturated acyl groups are located at the *sn*-2 position. The distribution of molecular species is different among animal species and tissues as well as at cellular and subcellular levels [192]. It is clear that cellular mechanisms exist to bring about the observed distinctive and non-random distribution of acyl groups in the

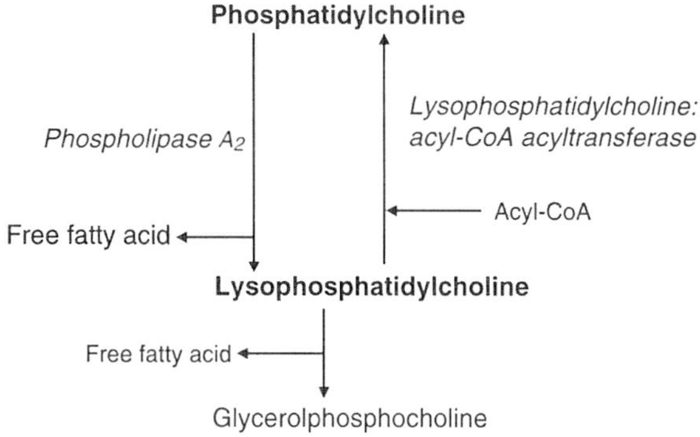

Fig. 4. The deacylation–reacylation of phosphatidylcholine. (*For a colored version of this figure, see plate section, page 610.*)

membrane phospholipids. The remodelling of phosphatidylcholine has been well studied in the heart [191]. Since limited selectivity for the molecular species of diacylglycerol has been displayed by cholinephosphotransferase phosphotransferase, the newly formed phospholipid must undergo extensive remodelling of the acyl group. At present, two mechanisms for the remodelling of the acyl groups in phosphatidylcholine have been identified. Lands [197,198] first introduced a scheme for the deacylation–reacylation of phospholipids which was based on the presence of both phospholipase A_2 and acyltransferase activities in mammalian tissues [199,200]. The formation of phosphatidylcholine from lysophosphatidylcholine by an energy-independent transfer of an acyl group from a phospholipid (or lysophospholipid) was subsequently demonstrated. We have characterised the in vitro and in vivo remodelling of phosphatidylglycerol in mammalian heart. Lysophosphatidylglycerol acyltransferase activity was observed in both mitochondrial and microsomal fractions [201]. The mitochondrial enzyme had a preference for unsaturated acyl-Coenzyme A substrates whereas the microsomal enzyme utilised both unsaturated as well as saturated acyl-Coenzyme A substrates to a similar extent. In addition, radiolabelled lysophosphatidylglycerol added to the perfusion medium of isolated rat hearts was rapidly converted to labelled phosphatidylglycerol indicating the ability of the heart to utilise exogenous lysophosphatidylglycerol for phosphatidylglycerol biosynthesis. The acylation of lysophosphatidylethanolamine to phosphatidylethanolamine by an acyl-CoA dependent process has been demonstrated in rat liver [202].

7.2. Acyl-CoA:lysophosphatidylcholine and acyl-CoA:lysophosphatidylethanolamine acyltransferases

The deacylation of diacyl phospholipids by the action of phospholipase A_1 or A_2 has been well documented [192]. In general, the reaction is catalysed by the Ca^{++}-dependent enzyme which is normally associated with the cellular membrane [203]. The lysophospholipid formed can be further deacylated by lysophospholipase or reacylated to the parent phospholipid (Fig. 3). The acylation of lysophosphatidylcholine to phosphatidylcholine by an acyl-CoA dependent process has been demonstrated in a large number of tissues. The transfer of acyl groups to 1-acyl-glycerophosphocholine is catalysed by acyl-CoA:1-acyl-glycerophosphocholine acyltransferase whereas the transfer of an acyl group to 2-acyl-glycerophosphocholine is catalysed by acyl-CoA:2-acyl-glycerophosphocholine acyltransferase [192]. Indirect evidence shows that these two activities are acyl specific and catalysed by separate enzymes [204]. In the last two decades, acyl-CoA:1-acyl-glycerophosphocholine acyltransferase has been studied extensively in the liver, lung and brain and the subject has been reviewed [205]. Acyl-CoA:1-acyl-glycerophosphocholine acyltransferase activity has been reported in cardiac microsomes [204,206,207], mitochondria [208,209] and cytosol [209,210].

We have characterised acyl-CoA:1-acyl-glycerophosphocholine activity in guinea pig heart microsomes, and found the enzyme to be more active with unsaturated acyl-CoAs such as arachidonoyl-CoA, oleoyl-CoA and linoleoyl-CoA, than with saturated species as acyl donors [191]. The purification of an acyl-CoA:1-acyl-glycerophosphocholine acyltransferase from bovine heart muscle microsomes was reported in 1988 [211]. The purified preparation exhibited high activity when assayed using 1-palmitoyl-glycerophosphocholine as the acyl acceptor and unsaturated acyl-CoAs as acyl donors. Substantial purification of an acyl-CoA:1-acyl-glycerophosphocholine acyltransferase from rat liver has been achieved by solubilising the enzyme with high concentrations of oleoyl-CoA and lysophosphatidylcholine [212]. The acyltransferase would convert the oleoyl-CoA/lysophosphatidylcholine micromicelles to phosphatidylcholine. After a 20 h incubation period, the phosphatidylcholine micelles containing the acyltransferase were separated from the acyl-CoA/lysophosphatidylcholine micromicelles by centrifugation. Higher acyltransferase activity was observed using unsaturated acyl-CoAs as substrates than when saturated acyl-CoAs were used [212]. The results of these studies are consistent with a role for acyl-CoA:l-acyl-glycerophosphocholine acyltransferase in maintaining an asymmetric acyl distribution in phosphatidylcholine molecules.

An acyl-CoA:l-acyl-glycerophosphocholine acyltransferase activity with the unusual property of utilising only linoleoyl-CoA as acyl donor has been reported in guinea pig heart mitochondria [208]. Although linoleate is the major acyl group at the *sn*-2 position of phosphatidylcholine in guinea pig heart mitochondria, significant quantities of other acyl groups are present [213]. The mitochondrion is incapable of de novo phosphatidylcholine synthesis and imports the phospholipid from the endoplasmic reticulum [214]. The mitochondrial acyltransferase may participate in the remodelling of the acyl groups of the imported phosphatidylcholine to yield the observed high linoleoyl content. In contrast, rabbit heart mitochondrial acyltransferase does not exhibit any selectivity for linoleoyl-CoA observed in the guinea pig heart [209]. In the latter study, the rabbit heart mitochondrial enzyme utilised a range of acyl-CoAs as substrates, and the rate of incorporation of the acyl groups was similar to that of the microsomal acyltransferase.

Although the purification of the enzymes from the brain [215] and bovine heart [211] have been reported, there has been no further study on the purified enzyme by these investigators. We have reported the partial solubilisation of an acyltransferase from rat liver microsomes [216]. Between 14 and 25% of the enzyme activity was solubilised by using the detergents sodium cholate, sodium deoxycholate and octylglucopyranoside. The solubilised and microsomal forms exhibited a similar pattern of acyltransferase activities toward palmitoyl-CoA, stearoyl-CoA, oleoyl-CoA, linoleoyl-CoA and arachidonoyl-CoA substrates. Storage of the enzyme preparations at -20 °C resulted in a gradual loss of activity, but no change in acyl specificities was observed, suggesting that the reduced enzyme activity was not caused by a selective denaturation of putative specific form(s) of the enzyme. Other attempts to purify or solubilise 1-acyl-glycerophosphocholine acyltransferase have often resulted in the loss of acyltransferase activity toward a particular species of acyl-CoA [192]. Selective solubilisation of acyltransferases with different acyl-CoA specificities may be explained by the existence of several forms of acyltransferase, each with a differing affinity toward specific acyl groups. The observation that ageing microsomal enzyme preparations at 4 °C results in a differential loss of specificity towards certain acyl-CoAs [217] is also consistent with the existence of multiple molecular forms of acyltransferases. Thus, the acyl substrate specificities previously determined in subcellular fractions would be a reflection of the quantitative distribution of the different acyltransferases in the various fractions.

7.3. Monolysocardiolipin acyltransferase

Mitochondrial cardiolipin is distinguished from other phospholipids by the great abundance of linoleoyl molecular species [173]. In rat liver, evidence was obtained for the presence of a deacylation–reacylation cycle by adding a labelled linoleoyl group to a mitochondrial preparation [218]. It was concluded that a cycle, comprising cardiolipin deacylation to monolysocardiolipin and its subsequent reacylation with linoleoyl-CoA, would provide a potential mechanism for the remodelling of molecular species of newly formed cardiolipin. We characterised this in vitro reacylation of exogenous monolysocardiolipin to cardiolipin with linoleoyl-Coenzyme A and oleoyl-Coenzyme A in rat heart mitochondrial fractions and other tissues [219]. More recently, we purified monolysocardiolipin acyltransferase from pig liver mitochondria to homogeneity, characterised its activity and raised a polyclonal antibody to the protein [220]. The 74 kDa monolysocardiolipin acyltransferase was purified by butanol extraction followed by hydroxyapatite and preparative SDS-PAGE. Purified monolysocardiolipin acyltransferase utilised exclusively monolysocardiolipin and unsaturated acyl-Coenzyme A's as substrates and exhibited a ping-pong reaction mechanism. Enzyme activity was optimum at a pH 7.0, which is in the range of the normal pH observed in respiring mitochondria. The enzyme was confirmed to be an acyltransferase by binding of the photoaffinity probe [^{125}I]12-[(4-Azidosalicyl)amino]dodecanoyl-Coenzyme A. In addition, photoaffinity labelling with 12-[(4-Azidosalicyl)amino]dodecanoyl-Coenzyme A inhibited enzyme activity in a concentration-dependent manner. The polyclonal antibody cross-reacted with the protein in crude pig and rat liver mitochondrial fractions.

7.4. The regulation of phospholipid remodelling

Are acyltransferases responsible for attaining and maintaining the observed molecular composition of phospholipids found in specific membranes? If so, one might expect subcellular and tissue differences in the properties of the acyltransferases to be reflected in the differences in the molecular species of phospholipids in the membrane. Although some correlation has been reported in a number of examples, a prediction of the acyl composition can rarely be obtained from the acyl specificity of the enzyme [192]. We feel that the acyl composition of any membrane is probably determined by several factors, including the selectivity of both phospholipase A and acyltransferase, and by the availability of acyl-CoAs. It is likely that the acyltransferases represent a superfamily of multiple molecular forms that may show specificity for selective lysophospholipid and acyl-CoA substrates. The lipid environment may also influence the affinity of acyl-CoA:1-acyl-glycerophosphocholine acyltransferase for its substrates [212].

Incorporation of exogenous linoleic acid into cardiolipin in H9c2 myoblast cells appears to be independent of de novo cardiolipin biosynthesis [76]. However, inhibition of fatty acid import into mitochondria by etomoxiryl-CoA channels exogenous fatty acid away from de novo cardiolipin biosynthesis in these cells [139]. The fatty acid composition of cardiolipin in mutant Chinese hamster lung fibroblast cells, CCL16-B2, has been examined [139]. These cells exhibit a futile elevation in mitochondrial metabolism but are deficient in oxidative energy production. The levels of unsaturated, but not saturated, fatty acids in cardiolipin were altered in B2 cells compared to wild type CCL16-B1 cells. Since the appropriate content and composition of cardiolipin are required for proper oxidative energy production, we hypothesise that the alteration in the fatty acid composition of cardiolipin in B2 cells, possibly mediated by altered cardiolipin remodelling, might be one of the mechanisms for the observed reduction in oxidative energy production in these cells.

Infection of the CCL16-B1 wild type cells with the intracellular bacterial parasite *C. trachomatis* increases host cell mitochondrial energy metabolism and results in an identical pattern of radioactive fatty acid incorporation into the mitochondrial phospholipids, phosphatidylglycerol and cardiolipin compared to uninfected mutant CCL16-B2 cells [221]. In addition, in *C. trachomatis*-infected Hela cells incorporation of radioactive unsaturated fatty acid into phosphatidylglycerol and cardiolipin is dramatically (7–10-fold) elevated relative to mock-infected cells yet phosphatidylglycerol and cardiolipin levels and their de novo biosynthesis are unaltered [222,223]. These data suggest that phosphatidylglycerol and cardiolipin molecular remodelling in eukaryotic cells infected with *C. trachomatis* may be linked to alterations in mitochondrial metabolism.

More recent evidence from our laboratory suggests that cardiac and liver monolysocardiolipin acyltransferase activity may be elevated in hyperthyroid rats when cardiolipin biosynthesis and cardiolipin levels are elevated and reduced in hypothyroid rats in which cardiolipin biosynthesis and cardiolipin levels are reduced [184,220,224, 225]. In hyperthyroid rats, the elevation in cardiac monolysocardiolipin acyltransferase activity and protein expression is likely required to maintain the appropriate fatty acid composition of cardiolipin as the cardiolipin levels increased within the heart. Hence, monolysocardiolipin acyltransferase activity may be rate limiting for the molecular remodelling of cardiolipin in the heart [220,225].

Microsomal lysophosphatidylethanolamine acyltransferase activity was elevated 2.4-fold and this corresponded with a 2.1-fold elevation in phosphatidylethanolamine in P19 teratocarcinoma cells induced to undergo differentiation into cardiac myocytes [31,132]. From these data it would be tempting to speculate that the level of acyltransferase activity for all glycerophospholipids may be directly coupled to de novo phospholipid biosynthesis. However, treatment of rats with thyroxine resulted in a 1.8-fold elevation in cardiac microsomal lysophosphatidylethanolamine acyltransferase activity without an alteration in the cardiac phosphatidylethanolamine pool size [226]. Thus, the regulation of specific lysophospholipid acyltransferase activities appears to be cell and likely tissue selective. It is clear, however, that the expression of selective acyltransferase activities is under thyroid hormone control in thyroid responsive tissues.

Barth Syndrome is a rare X-linked genetic disorder in young boys and is the only genetic disease in which the specific biochemical defect is a reduction in cardiolipin level [227]. The causative gene is localised to Xq28 and codes for a group of proteins known as tafazzins. Alternative splicing of the primary G4.5 transcript results in tafazzins that differ in the N-terminus and the central region introducing stop codons resulting in aberrant proteins [228]. Although more than 30 mutations in G4.5 have been described in patients with Barth Syndrome, no single mutation is concordant with the disease in each patient. In fibroblasts of patients with Barth Syndrome the ability to remodel phosphatidylglycerol and cardiolipin may be reduced [229] leading to a reduction in cardiolipin levels and an accumulation of monolysocardiolipin [230]. A reduced ability to remodel phosphatidylglycerol and cardiolipin, and possibly other phospholipids, could be one of or the underlying molecular mechanism(s) responsible for the Barth Syndrome and is an active area of investigation [229].

8. Concluding remarks

It is clear from the preceding discussion that there exist many mechanisms for the regulation of phospholipid biosynthesis and resynthesis in mammalian cells. The current challenge is to understand the coordination between the biosynthesis and resynthesis of phospholipids that should occur in order to maintain an appropriate molecular composition in the cell membrane. For example, vasopressin has been shown to stimulate the biosynthesis and hydrolysis of phosphatidylcholine in the mammalian heart [102,231]. Based on other studies, angiotensin II has been shown to stimulate the hydrolysis of phosphatidylcholine in smooth muscle cells [232] and in mesangial cells [233,234]. Several antiarrhythmic drugs have been shown to affect the biosynthesis, metabolism and remodelling of phospholipids in the mammalian heart [235,236]. The ability to coordinate the synthesis, catabolism and resynthesis of phospholipids should be of paramount importance for the cell to maintain its molecular composition in the membrane during agonist stimulation. The importance of coordinated regulation of phosphatidylcholine biosynthesis and catabolism has also been demonstrated in COS cells where the overexpression of CTP:phosphocholine cytidylyltransferase increased the synthesis of phosphatidylcholine, and also resulted in an accelerated rate of phosphatidylcholine degradation [237]. In addition, the coordination of monolysocardiolipin acyltransferase activity in concert with de novo cardiolipin biosynthesis in the heart has been demonstrated [224]. Clearly, an understanding of the functional roles of cellular phospholipids will require the characterisation of the molecular mechanisms for the coordination between synthesis and resynthesis of the cellular phospholipids. We anticipate that this will provide a fruitful area of research in the near future.

Acknowledgement

This work was supported by the Heart and Stroke Foundation of Manitoba and the Canadian Institutes of Health Research.

References

[1] White, D.A., 1973. In: Ansell, G.B., Hawthorne, J.N., Dawson, R.M.C., (Eds.), Form and Function of Phospholipids. Elsevier, Amsterdam, pp. 441–482.
[2] Hatch, G.M., 1998. Int. J. Mol. Med. 1, 33–41.
[3] Coleman, R., 1973. Biochim. Biophys. Acta 300, 1–30.
[4] Cullis, P.R., Hope, M.J., 1991. In: Vance, D.E., Vance, J.E. (Eds.), Biochemistry of Lipids, Lipoproteins and Membranes. Elsevier, Amsterdam, pp. 1–41.

[5] Fadok, W.A., Voelker, D.R., Campbell, P.A., Cohen, J.J., Bratten, D.L., Henson, P.M., 1992. J. Immunol. 148, 2207–2216.
[6] Fukami, K., Takenawa, T., 1992. J. Biol. Chem. 267, 10988–10993.
[7] Newmeyer, D.D., Ferguson-Miller, S., 2003. Cell 112, 481–490.
[8] Berridge, M.J., Irvine, R.F., 1989. Nature 341, 197–205.
[9] Exton, J.E., 1990. J. Biol. Chem. 265, 1–4.
[10] Kennedy, E.P., 1961. Fed. Proc. 20, 934–940.
[11] Kennedy, E.P., 1962. Harvey Lect. 57, 143–171.
[12] Kent, C., 1995. Annu. Rev. Biochem. 64, 315–343.
[13] van den Bosch, H., 1974. Annu. Rev. Biochem. 43, 243–277.
[14] Haldar, D., Vancura, A., 1992. Methods Enzymol. 209, 64–72.
[15] Coleman, R.A., Lewin, T., Muoio, D.M., 2000. Annu. Rev. Nutr. 20, 77–103.
[16] Paulauskis, J.D., Sul, H.S., 1988. J. Biol. Chem. 263, 7049–7054.
[17] Yet, S.F., Lee, S., Hahm, Y.T., Sul, H.S., 1993. Biochem. USA 32, 9486–9491.
[18] Ruan, H., Pownall, H.J., 2001. Diabetes 50, 233–240.
[19] Brindley, D.N., Waggoner, D., 1998. J. Biol. Chem. 273, 24281–24284.
[20] Kai, M., Wada, I., Imai, S., Sakane, F., Kanoh, H., 1996. J. Biol. Chem. 271, 18931–18938.
[21] Hooks, S.B., Ragan, S.P., Lynch, K.R., 1998. FEBS Lett. 427, 188–192.
[22] Waggoner, D.W., Martin, A., Dewals, J., Gomez-Munoz, A., Brindley, D.N., 1995. J. Biol. Chem. 270, 19422–19429.
[23] Carrasco, M.P., Segovia, J.L., Marco, C., 1998. Biochem. Pharmacol. 56, 1639–1644.
[24] Akesson, B., Elovson, J., Arvidson, G., 1970. Biochim. Biophys. Acta 210, 15–27.
[25] Wirthensohn, G., Vandewalle, A., Guder, W., 1981. Biochem. J. 198, 543–549.
[26] Kinsella, J.E., 1968. Biochim. Biophys. Acta 164, 540–549.
[27] Kasemsri, S., Bernardis, L.L., Chlouverakis, C., Schnatz, J.D., 1972. Proc. Soc. Exp. Biol. Med. 141, 38–42.
[28] Uhal, B.D., Longmore, W.J., 1988. Biochim. Biophys. Acta 961, 122–128.
[29] Chung, A.C., 1978. Atherosclerosis 29, 277–290.
[30] Wong, J.T., Man, R.Y.K., Choy, P.C., 1996. Lipids 31, 1059–1067.
[31] Fotheringham, J., Xu, F.Y., Nemer, M., Kardami, E., Choy, P.C., Hatch, G.M., 2000. Biochim. Biophys. Acta 1485, 1–10.
[32] Lee, D.P., Deonarine, A.S., Kienetz, M., Zhu, Q., Skrzypczak, M., Chan, M., Choy, P.C., 2001. J. Lipid Res. 42, 1979–1986.
[33] Guo, Z., Jensen, M.D., 1999. J. Biol. Chem. 274, 23702–23706.
[34] Lin, E.C., 1977. Annu. Rev. Biochem. 46, 765–795.
[35] Dominguez, M.C., Herrera, E., 1976. Biochem. J. 158, 183–190.
[36] Trach, V., Buschmans-Denkel, E., Schaper, W., 1986. Basic Res. Cardiol. 81, 454–464.
[37] Kohlschutter, A., Willig, H.P., Schlamp, D., Kruse, K., McCabe, E.R., Schafer, H.J., Beckenkamp, G., Rohkamm, R., 1987. Eur. J. Pediatr. 146, 575–581.
[38] King, R., Clements, J., 1972. Am. J. Physiol. 223, 715–726.
[39] Post, M., van Golde, L.M.G., 1988. Biochim. Biophys. Acta 947, 249–286.
[40] Zelinski, T.A., Savard, J.D., Man, R.Y.K., Choy, P.C., 1980. J. Biol. Chem. 255, 11423–11428.
[41] Vance, D.E., 1996. Eur. J. Med. Res. 1, 182–188.
[42] Bremer, J., Greenberg, J., 1961. Biochim. Biophys. Acta 46, 205–216.
[43] Vance, D.E., Ridgway, N.D., 1988. Prog. Lipid Res. 27, 61–79.
[44] Ridgway, N.D., Vance, D.E., 1992. Methods Enzymol. 209, 366–374.
[45] Vance, D.E., Choy, P.C., 1979. Trends Biochem. Sci. 4, 145–148.
[46] Vance, D.E., 2002. In: Vance, D.E., Vance, J.E. (Eds.), Biochemistry of Lipids, Lipoproteins and Membranes. 4th ed. Elsevier, Amsterdam, pp. 205–232.
[47] Dils, R.R., Hubscher, G., 1961. Biochim. Biophys. Acta 46, 205–216.
[48] Yavin, E., 1976. J. Biol. Chem. 251, 1392–1397.
[49] Pelech, S.L., Vance, D.E., 1984. Biochim. Biophys. Acta 779, 217–251.
[50] Sundler, R., Akesson, B., 1975. J. Biol. Chem. 250, 3359–3367.
[51] Haines, D.S.M., Rose, C.I., 1970. Can. J. Biochem. 48, 885–892.

[52] Blumenstein, J., 1964. Can. J. Biochem. 42, 1183–1194.
[53] Haeffner, E.W., 1975. Eur. J. Biochem. 51, 219–228.
[54] Plagemann, P.G.W., 1971. J. Lipid Res. 12, 715–724.
[55] Ishidate, K., 1989. In: Vance, D.E. (Ed.), Phosphatidylcholine Metabolism. CRC Press, Boca Raton, FL, pp. 9–32.
[56] Zelinski, T.A., Choy, P.C., 1984. Biochim. Biophys. Acta 794, 326–332.
[57] Zha, X., Jay, F.T., Choy, P.C., 1992. Biochem. Cell Biol. 70, 1319–1324.
[58] Zelinski, T.A., Choy, P.C., 1982. J. Biol. Chem. 257, 13201–13204.
[59] Havener, L.J., Toback, G., 1980. J. Clin. Invest. 65, 741–745.
[60] Hatch, G.M., Choy, P.C., 1986. Biochim. Biophys. Acta 884, 259–264.
[61] Hatch, G.M., Stevens, W.K., Choy, P.C., 1988. Biochem. Cell Biol. 66, 418–424.
[62] Chan, M., O, K., Man, A.S., Choy, P.C., 1992. Lipids 27, 466–469.
[63] Ulane, R.E., Stephenson, L.L., Farrell, P.M., 1978. Biochim. Biophys. Acta 531, 295–300.
[64] Ishidate, K., Nakagomi, K., Nakazawa, Y., 1984. J. Biol. Chem. 259, 14706–14710.
[65] Uchida, T., Yamashita, S., 1990. Biochim. Biophys. Acta 1043, 281–288.
[66] Porter, T.J., Kent, C., 1990. J. Biol. Chem. 265, 414–422.
[67] Cao, Z.M., Kanfer, J.N., 1995. Neurochem. Res. 20, 643–649.
[68] Uchida, T., Yamashita, S., 1992. J. Biol. Chem. 267, 10156–10162.
[69] Aoyama, C., Yamazaki, N., Terada, H., Ishidate, K., 2000. J. Lipid Res. 41, 452–464.
[70] Uchida, T., 1994. J. Biochem. Tokyo 116, 1241–1250.
[71] Hosaka, K., Tanaka, S., Nikawa, J., Yamashita, S., 1992. FEBS Lett. 304, 229–232.
[72] Ramirez de Molina, A., Penalva, V., Lucas, L., Lacal, J.C., 2002. Oncogene 21, 937–946.
[73] Choy, P.C., Chan, M., Hatch, G.M., Man, R.Y.K., 1992. Mol. Cell. Biochem. 116, 53–58.
[74] Choy, P.C., 1982. J. Biol. Chem. 257, 10928–10933.
[75] Hatch, G.M., Choy, P.C., 1990. Biochem. J. 268, 47–54.
[76] Hatch, G.M., McClarty, G., 1996. J. Biol. Chem. 271, 25810–25816.
[77] Choy, P.C., Paddon, H.P., Vance, D.E., 1980. J. Biol. Chem. 255, 1070–1073.
[78] Bajusz, E., 1971. Recent Adv. Stud. Card. Struct. Metab. 2, 291–292.
[79] Cornell, R.B., Kalmar, G.B., Kay, R.J., Johnson, M.A., Sanghera, J.S., Pelech, S.L., 1995. Biochem. J. 310, 699–708.
[80] Kent, C., Carman, G.M., Spence, M.W., Dowhan, W., 1991. FASEB J. 5, 2258–2266.
[81] Kent, C., 1990. Prog. Lipid Res. 29, 87–105.
[82] Tronchère, H., Record, M., Tercé, F., Chap, H., 1994. Biochim. Biophys. Acta 1212, 137–151.
[83] Vance, D.E., 1990. Biochem. Cell Biol. 68, 1151–1165.
[84] Weinhold, P.A., Rounsifer, M.E., Feldman, D.A., 1986. J. Biol. Chem. 261, 5104–5110.
[85] Feldman, D.A., Weinhold, P.A., 1987. J. Biol. Chem. 262, 9075–9081.
[86] Kalmar, G.B., Kay, R.J., Lachance, A., Aebersold, R., Cornell, R.B., 1990. Proc. Natl Acad. Sci. USA 87, 6029–6033.
[87] Lydikis, A., Baburina, I., Jackowski, S., 1999. J. Biol. Chem. 274, 26992–27001.
[88] Houweling, M., Tijburg, L.B., Vaartijes, W.J., Batenburg, J.J., Kalmar, G.B., Cornell, R.B., van Golde, L.M.G., 1993. Eur. J. Biochem. 214, 927–933.
[89] Ridsdale, R., Tseu, I., Wang, J., Post, M., 2001. J. Biol. Chem. 276, 49148–49155.
[90] MacDonald, J.I.S., Kent, C., 1994. J. Biol. Chem. 269, 10529–10537.
[91] Wang, Y., Kent, C., 1995. J. Biol. Chem. 270, 17843–17849.
[92] Houweling, M., Jamil, H., Hatch, G.M., Vance, D.E., 1994. J. Biol. Chem. 269, 7544–7551.
[93] Craig, L., Johnson, J.E., Cornell, R.B., 1994. J. Biol. Chem. 269, 3311–3317.
[94] Johnson, J.E., Cornell, R.B., 1994. Biochem. USA 33, 4327–4335.
[95] Wieder, T., Geilen, C.C., Wieprecht, M., Becker, A., Orfanos, C.E., 1994. FEBS Lett. 345, 207–210.
[96] Cornell, R.B., 1991. Biochem. USA 30, 5873–5880.
[97] Cornell, R.B., 1991. Biochem. USA 30, 5881–5888.
[98] Yang, W., Boggs, K.P., Jackowski, S., 1995. J. Biol. Chem. 270, 23951–23957.
[99] Yang, W., Jackowski, S., 1995. J. Biol. Chem. 270, 16503–16506.
[100] Mock, T., Slater, T.L., Arthur, G., Chan, A.C., Choy, P.C., 1986. Biochem. Cell Biol. 64, 413–417.
[101] Pelech, S.L., Pritchard, P.H., Brindley, D.N., Vance, D.E., 1983. J. Biol. Chem. 258, 6782–6788.

[102] Man, A.S., Lee, E., Choy, P.C., 1994. Lipids 29, 15–19.
[103] Tran, K., Man, R.Y.K., Choy, P.C., 1995. Biochim. Biophys. Acta 1259, 283–290.
[104] Hatch, G.M., Lee, D., Man, R.Y.K., Kroeger, E.A., Choy, P.C., 1997. Biochim. Biophys. Acta 1347, 183–190.
[105] Cornell, R.B., 1989. In: Vance, D.E. (Ed.), Phosphatidylcholine Metabolism. CRC Press, Boca Raton, FL, pp. 47–64.
[106] Bru, R., Blöchliger, E., Luisi, P.L., 1993. Arch. Biochem. Biophys. 307, 295–303.
[107] Ishidate, K., Nakazawa, Y., 1992. Methods Enzymol. 209, 121–134.
[108] O, K., Choy, P.C., 1990. Lipids 25, 122–124.
[109] Hjelmstad, R.H., Bell, R.M., 1987. J. Biol. Chem. 262, 3909–3917.
[110] Hjelmstad, R.H., Bell, R.M., 1990. J. Biol. Chem. 265, 1755–1764.
[111] Hjelmstad, R.H., Morash, S.C., McMaster, C.R., Bell, R.M., 1994. J. Biol. Chem. 269, 20995–21002.
[112] Henneberry, J., Wistow, G., McMaster, C.R., 2000. J. Biol. Chem. 275, 29808–29815.
[113] O, K., Choy, P.C., 1993. Biochem. J. 289, 727–733.
[114] Sakami, W., Harrington, H., 1963. Annu. Rev. Biochem. 32, 355–398.
[115] Schimke, R.T., 1962. J. Biol. Chem. 237, 1921–1924.
[116] Snodgrass, P.J., 1981. Pediatrics 68, 273–283.
[117] Wong, J.T., Man, R.Y.K., Choy, P.C., 1996. Lipids 31, 1059–1067.
[118] Wong, J.T., Man, R.Y.K., Choy, P.C., 1994. Lipids 29, 391–396.
[119] Henneberry, J., Wright, M.M., McMaster, C.R., 2002. Mol. Biol. Cell. 13, 3148–3161.
[120] McMaster, C.R., Choy, P.C., 1992. J. Biol. Chem. 267, 14586–14591.
[121] Voelker, D.R., 1984. Proc. Natl Acad. Sci. USA 81, 2669–2673.
[122] Voelker, D.R., 1997. Biochim. Biophys. Acta 1348, 236–244.
[123] Sundler, R., 1973. Biochim. Biophys. Acta 306, 218–226.
[124] Porter, T., Kent, C., 1992. Methods Enzymol. 209, 134–146.
[125] van Hellemond, J.J., Slot, J.W., Geelen, M.J., van Golde, L.M., Vermeulen, P.S., 1994. J. Biol. Chem. 269, 15415–15418.
[126] Tijburg, L.B.M., Vermeulen, P.S., van Golde, L.M.G., 1992. Methods Enzymol. 209, 258–263.
[127] Bladergroen, B.A., Houweling, M., Geelen, M.J., van Golde, L.M., 1999. Biochem. J. 343, 104–144.
[128] Nakashima, A., Hosaka, K., Nikawa, J., 1997. J. Biol. Chem. 272, 9567–9572.
[129] Cornell, R.B., 1992. Methods Enzymol. 209, 267–272.
[130] Kanoh, H., Ohno, K., 1975. Eur. J. Biochem. 66, 201–210.
[131] O, K., Siow, Y.L., Choy, P.C., 1989. Biochem. Cell Biol. 67, 680–686.
[132] Xu, F.Y., Kardami, E., Nemer, M., Choy, P.C., Hatch, G.M., 2000. Exp. Cell. Res. 256, 358–364.
[133] Rusnak, A., Mangat, R., Xu, F., McClarty, G., Hatch, G.M., 1997. J. Bioenerg. Biomembr. 29, 293–300.
[134] Carman, G.M., Kelley, M.J., 1992. Methods Enzymol. 209, 242–247.
[135] Saito, S., Goto, K., Tonosaki, A., Kondo, H., 1997. J. Biol. Chem. 272, 9503–9507.
[136] Halford, S., Dulai, K.S., Daw, S.C., Fitzgibbon, J., Hunt, D.M., 1998. Genomics 54, 140–144.
[137] Takenawa, T., Egawa, K., 1977. J. Biol. Chem. 252, 5419–5423.
[138] Lydikis, A., Jackson, P.D., Rock, C.O., Jackowski, S., 1997. J. Biol. Chem. 272, 33402–33409.
[139] Xu, F.Y., Taylor, W.A., Hurd, J.A., Hatch, G.M., 2003. J. Lipid Res. 44, 415–423.
[140] Kanfer, J.N., 1992. Methods Enzymol. 209, 341–348.
[141] Bergo, M.O., Gavino, B.J., Steenbergen, R., Sturbois, B., Parlow, A.F., Sanan, D.A., Skarnes, W.C., Vance, J.E., Young, S.G., 2002. J. Biol. Chem. 277, 47701–47708.
[142] Hostetler, K.Y., 1982. In: Hawthorne, J.N., Ansell, G.B. (Eds.), Phospholipids. Elsevier, Amsterdam, pp. 215–261.
[143] Hovius, R., Lambrechts, H., Nicolay, K., de Kruijff, B., 1990. Biochim. Biophys. Acta 1021, 217–226.
[144] Stoffel, W., Schiefer, H.-G., 1968. Hoppe-Seylers Z. Physiol. Chem. 349, 1017–1026.
[145] Hovius, R., Thijssen, J., van der Linden, P., de Kruijff, B., 1993. FEBS Lett. 330, 71–76.
[146] Hatch, G.M., 1994. Biochem. J. 297, 201–208.
[147] Zhang, M., Mileykovskya, E., Dowhan, W., 2002. J. Biol. Chem. 277, 43553–43556.
[148] Ohtsuka, T., Nishijima, M., Akamatsu, Y., 1993. J. Biol. Chem. 268, 22908–22913.
[149] Kahn, S.M., Dauffenbach, L.M., Yeh, J., 2000. Biochim. Biophys. Res. Commun. 269, 542–545.
[150] Poot, M., Pierce, R.H., 1999. Cytometry 35, 311–317.

[151] Shidoji, Y., Hayashi, K., Komura, S., Ohishi, N., Yagi, K., 1999. Biochim. Biophys. Res. Commun. 264, 343–347.
[152] Ushmorov, A., Ratter, F., Lehmann, V., Droge, W., Schirrmacher, V., Umansky, V., 1999. Blood 93, 2342–2347.
[153] Ostrander, D.B., Sparanga, G.C., Amoscato, A.A., McMillan, J.B., Dowhan, W., 2001. J. Biol. Chem. 276, 38061–38067.
[154] Pluschke, G., Hirota, Y., Overall, P., 1978. J. Biol. Chem. 253, 5048–5055.
[155] Jiang, F., Gu, Z., Granger, J.M., Greenberg, M., 1999. Mol. Microbiol. 31, 373–379.
[156] Koshkin, V., Greenberg, M.L., 2000. Biochem. J. 347, 687–691.
[157] Jiang, F., Ryan, M.T., Schlame, M., Zhao, M., Gu, Z., Klingenberg, M., Pfanner, N., Greenberg, M.L., 2000. J. Biol. Chem. 275, 22387–22394.
[158] Ostrander, D.B., Zhang, M., Mileykovskaya, E., Rho, M., Dowhan, W., 2001. J. Biol. Chem. 276, 25262–25272.
[159] Batenburg, J.J., 1992. Am. J. Physiol. 262, L357–L385.
[160] Kiyasu, J.Y., Pieringer, R.A., Paulus, H., Kennedy, E., 1963. J. Biol. Chem. 238, 2293–2298.
[161] Chang, S.-C., Heacock, P.N., Mileykovskaya, E., Voelker, D.R., Dowhan, W., 1998. J. Biol. Chem. 268, 14933–14941.
[162] Hostetler, K.Y., van den Bosch, H., van Deenen, L.L.M., 1971. Biochim. Biophys. Acta 239, 113–119.
[163] Hatch, G.M., 1996. Mol. Cell. Biochem. 159, 139–148.
[164] Mok, A., McDougall, G.E., McMurray, W.C., 1993. Biochem. Cell Biol. 71, 183–189.
[165] Lin, C.H., Lin, J., Strickland, K.P., 1991. Biochem. Int. 25, 299–306.
[166] Cheng, P., Hatch, G.M., 1995. Lipids 30, 513–519.
[167] Xu, F.Y., Taylor, W.A., Hatch, G.M., 1998. Arch. Biochem. Biophys. 349, 341–348.
[168] Xu, F.Y., Kelly, S.L., Taylor, W.A., Hatch, G.M., 1998. Mol. Cell. Biochem. 188, 217–223.
[169] Heacock, A.M., Uhler, M.D., Agranoff, B.W., 1996. J. Neurochem. 67, 2200–2203.
[170] Weeks, R., Dowhan, W., Shen, H., Balantac, N., Meengs, B., Nudelman, E., Leung, D.W., 1997. DNA Cell Biol. 16, 281–289.
[171] Kawasaki, K., Kuge, O., Yamakawa, Y., Nishijima, M., 2001. Biochem. J. 354, 9–15.
[172] Cao, S.G., Hatch, G.M., 1994. Lipids 29, 475–480.
[173] Schlame, M., Brody, S., Hostetler, K.Y., 1993. Eur. J. Biochem. 212, 727–735.
[174] Hostetler, K.Y., van den Bosch, H., 1972. Biochim. Biophys. Acta 260, 380–386.
[175] Schlame, M., Haldar, D., 1993. J. Biol. Chem. 268, 74–79.
[176] Schlame, M., Hostetler, K.Y., 1991. J. Biol. Chem. 266, 22398–22403.
[177] Shibuya, I., Hiraoka, S., 1992. Methods Enzymol. 209, 321–330.
[178] Xu, F.Y., Hatch, G.M., 1997. Biochem. Biophys. Res. Commun. 232, 261–265.
[179] Xu, F.Y., Kelly, S.L., Hatch, G.M., 1999. Biochem. J. 337, 483–490.
[180] Cao, S.G., Cheng, P., Angel, A., Hatch, G.M., 1995. Biochim. Biophys. Acta 1256, 241–244.
[181] Hatch, G.M., Cao, S.G., Angel, A., 1995. Biochem. J. 306, 759–764.
[182] Cao, S.G., Hatch, G.M., 1995. Biochim. Biophys. Acta 1256, 97–102.
[183] Hostetler, K.Y., 1991. Biochim. Biophys. Acta 1086, 139–140.
[184] Hoch, F.L., 1981. Prog. Lipid Res. 20, 225–228.
[185] Huterer, S.J., Wherrett, J.R., 1990. Biochem. Cell Biol. 68, 366–372.
[186] Huterer, S.J., Wherrett, J.R., 1989. Biochim. Biophys. Acta 1001, 68–75.
[187] Poorthuis, B.J., Hostetler, K.Y., 1976. J. Biol. Chem. 251, 4596–4602.
[188] Amidon, B., Brown, A., Waite, M., 1996. Biochem. USA 35, 13995–14002.
[189] Momchilova-Pankkova, A.B., Markovska, T.T., Koumanou, K.S., 1990. Biochimie 72, 863–866.
[190] Katz, A.M., Messineo, F.C., 1981. Circ. Res. 48, 1–16.
[191] Arthur, G., Choy, P.C., 1984. Biochim. Biophys. Acta 795, 221–229.
[192] Choy, P.C., Arthur, G., 1989. In: Vance, D.E. (Ed.), Phosphatidylcholine Metabolism. CRC Press, Boca Raton, FL, pp. 87–101.
[193] Mock, T., Man, R.Y.K., 1990. Lipids 25, 357–362.
[194] Das, D., Engelman, R., Rousou, R., Breyer, R., Lemeshow, S., 1986. Am. J. Physiol. 251, H71–H79.
[195] van der Vusse, G.J., Glatz, J.F.C., Stam, H.C.G., Reneman, R.S., 1992. Physiol. Rev. 72, 881–940.
[196] MacDonald, J.I.S., Sprecher, H., 1991. Biochim. Biophys. Acta 1084, 105–121.

[197] Lands, W.E.M., Hart, P., 1965. J. Biol. Chem. 240, 1905–1912.
[198] Lands, W.E.M., 1960. J. Biol. Chem. 235, 2233–2237.
[199] Eibl, H., Hill, E.E., Lands, W.E.M., 1969. Eur. J. Biochem. 9, 250–258.
[200] Eisenberg, S., Stein, Y., Stein, O., 1967. Biochim. Biophys. Acta 137, 115–120.
[201] Cheng, P., Dolinsky, V., Hatch, G.M., 1996. Biochim. Biophys. Acta 1302, 61–68.
[202] Akesson, B., Arner, A., Sundler, R., 1976. Biochim. Biophys. Acta 441, 453–464.
[203] Hatch, G.M., O, K., Choy, P.C., 1989. Biochem. Cell Biol. 67, 67–77.
[204] Arthur, G., 1989. Biochem. J. 261, 575–580.
[205] Choy, P.C., Tardi, P.G., Mukherjee, J.J., 1992. Methods Enzymol. 209, 80–86.
[206] Arthur, G., Choy, P.C., 1986. Biochem. J. 236, 481–487.
[207] Gross, R.W., Sobel, B.E., 1982. J. Biol. Chem. 257, 6702–6708.
[208] Arthur, G., Page, L., Zaborniak, C., Choy, P.C., 1987. Biochem. J. 242, 171–175.
[209] Needleman, P., Wyche, A., Sprecher, H., Elliot, W.J., Evers, A., 1985. Biochim. Biophys. Acta 836, 267–273.
[210] Severson, D.L., Fletcher, T., 1985. Can. J. Physiol. Pharmacol. 63, 944–951.
[211] Sanjanwala, M., Sun, G.Y., Cutrera, M.A., MacQuarrie, R.A., 1988. Arch. Biochem. Biophys. 265, 476–483.
[212] Fyrst, H., Pham, D.V., Lubin, B.H., Kuypers, F.A., 1996. Biochemistry 35, 2644–2650.
[213] Arthur, G., Mock, T., Zaborniak, C., Choy, P.C., 1985. Lipids 20, 693–698.
[214] McMurray, W.C., 1986. Biochem. Cell Biol. 64, 1115–1124.
[215] Deka, N., Sun, G.Y., MacQuarrie, R., 1986. Arch. Biochem. Biophys. 246, 554–563.
[216] Mukherjee, J.J., Tardi, P.G., Choy, P.C., 1992. Biochim. Biophys. Acta 1123, 27–32.
[217] Reitz, R.C., El-Sheikh, M., Lands, W.E.M., Isamail, I.A., Gunstone, F.D., 1969. Biochim. Biophys. Acta 176, 480–490.
[218] Schlame, M., Rustow, B., 1990. Biochem. J. 272, 589–595.
[219] Ma, B.J., Taylor, W.A., Dolinsky, V.W., Hatch, G.M., 1999. J. Lipid Res. 40, 1837–1845.
[220] Taylor, W.A., Hatch, G.M., 2003. J. Biol. Chem. 278, 12716–12721.
[221] Hatch, G.M., McClarty, G., 1998. Biochem. Biophys. Res. Commun. 243, 356–360.
[222] Hatch, G.M., McClarty, G., 1998. Infect. Immun. 66, 3727–3735.
[223] Wylie, J.L., Hatch, G.M., McClarty, G., 1997. J. Bacteriol. 179, 7233–7242.
[224] Taylor, W.A., Xu, F.Y., Ma, B.J., Mutter, T.C., Dolinsky, V.W., Hatch, G.M., 2002. BMC Biochem. 3, 9.
[225] Mutter, T., Dolinsky, V.W., Ma, B.J., Taylor, W.A., Hatch, G.M., 2000. Biochem. J. 346, 403–406.
[226] Dolinsky, V.W., Hatch, G.M., 1998. Biochim. Biophys. Acta 1391, 241–246.
[227] Barth, P.G., Wanders, R.J.A., Vreken, P., Janseen, E.A.M., Lam, J., Baas, F., 1999. J. Inherit. Metab. Dis. 22, 555–567.
[228] Bione, S., D'Adamo, P., Maestrini, E., Gedeon, A.K., Bolhuis, P.A., Torinolo, D., 1996. Nat. Genet. 12, 385–389.
[229] Vreken, P., Valinpour, F., Nijtmans, L.G., Grivell, L.A., Plecko, B., Wanders, R.J.A., Barth, P.G., 2000. Biochim. Biophys. Acta 279, 378–382.
[230] Schlame, M., Shankse, S., Doty, S., Konig, T., Sculo, T., DuMauro, S., Blank, T.J., 1999. J. Lipid Res. 37, 1585–1592.
[231] Tran, K., Zha, X., Chan, M., Choy, P.C., 1995. Mol. Cell. Biochem. 151, 69–76.
[232] Lassègue, B., Alexander, R.W., Clark, M., Akers, M., Griendling, K.K., 1993. Biochem. J. 292, 509–517.
[233] Andreatta van Leyen, S., Romero, M.F., Khosla, M.C., Douglas, J.G., 1993. Kidney Int. 44, 932–936.
[234] Barnett, R.L., Ruffini, L., Ramsammy, L., Pasmantier, R., Friedlaender, M.M., Nord, E.P., 1993. Am. J. Physiol. 265, C1100–C1108.
[235] Tardi, P.G., Man, R.Y.K., McMaster, C.R., Choy, P.C., 1990. Biochem. Cell Biol. 68, 745–750.
[236] Tardi, P.G., Man, R.Y.K., Choy, P.C., 1992. Biochem. J. 285, 161–166.
[237] Walkey, C.J., Kalmar, G.B., Cornell, R.B., 1994. J. Biol. Chem. 269, 5742–5749.

Membrane phospholipid asymmetry: biochemical and pathophysiological perspectives

Edouard M. Bevers, Paul Comfurius and Robert F.A. Zwaal*

Department of Biochemistry, Cardiovascular Research Institute Maastricht, Maastricht University, P.O. Box 616, 6200 MD Maastricht, The Netherlands
**Correspondence address: Cardiovascular Research Institute Maastricht, Maastricht University, P.O. Box 616, 6200 MD Maastricht, The Netherlands. Tel: # 31 (0)43 3881688; fax: # 31 (0)43 3884160*
E-mail: rfa.zwaal@bioch.unimaas.nl(R.F.A.Z.)

1. Membrane structure and asymmetry

Biological membranes are thin, sheet-like structures, composed of a great variety of different lipid and protein molecules, many of which bear carbohydrate residues. In water, the lipids assemble spontaneously to form a bimolecular leaflet with their polar headgroups on either surface and their apolar hydrocarbon tails pointing inwards. This leaflet acts as a permeability barrier and as a platform in or to which membrane proteins are embedded or attached. Three major classes of membrane proteins have been distinguished: (i) peripheral proteins that interact electrostatically with the polar lipid headgroups or with other membrane proteins but are positioned outside the lipid bilayer, (ii) integral proteins that insert their hydrophobic domains in the core of the bilayer, or span the membrane from one side to the other via one or multiple transmembrane segments, and (iii) anchor proteins that contain covalently attached lipid moieties (fatty acids, isoprenyl chains, or glycosylphosphatidylinositol), which are inserted in the bilayer and serve to anchor these proteins to the membrane.

The lipid bilayer is approximately 40 Å thick and behaves like a two-dimensional fluid in which the lipids and some, but not all membrane proteins, are constantly in rapid lateral motion. In this fashion, cell membranes are considered to form a two-dimensional fluid mosaic structure, in which proteins are floating in a sea of lipids [1]. Membrane fluidity, which is a.o. promoted by unsaturated fatty acid moieties of the lipid molecules, is imperative to the proper functioning of membrane proteins as they undergo conformational changes required to serve as selective channels and pumps, receptors to mediate signal transduction, energy transducers, generators of chemical and electrical impulses, and enzymes. Obviously, membrane protein content and composition vary

Advances in Molecular and Cell Biology, Vol. 33, pages 387–419

ISSN: 1569-2558 / DOI: 10.1016/S1569-2558(03)33019-X

widely among different biological membranes reflecting the diversity of functions that membranes perform.

The principal lipids of human cell membranes are phospholipids and cholesterol, with smaller amounts of glycolipids. Glycero-phospholipids contain a glycerol backbone esterified to two fatty acids and a phosphorylated alcohol that forms the polar headgroup. This group of phospholipids comprises phosphatidylcholine (PC), phosphatidyl-ethanolamine (PE), phosphatidylserine (PS), phosphatidylinositol (PI), and phosphatidyl-inositolphosphates. Sphingomyelin (Sph) is derived from sphingosine instead of glycerol, while its polar headgroup consists of phosphorylated choline, like PC. Glycolipids also contain a sphingosine backbone but differ from Sph in that they contain one or more sugar residues instead of phosphorylcholine. With the possible exception of cholesterol, spontaneous rotation or "flip-flop" of lipids from one side of the membrane to the other is extremely rare occurring only once a day, which is more than a billion times slower than the lateral movement over the same distance. These estimates are consistent with the asymmetric distribution of the different lipid classes between the two halves of the bilayer membrane [2]. Membrane asymmetry has been inferred from measurements of accessibility of membrane components to exogenous membrane-impermeable reagents (Fig. 1A). In plasma membranes, the choline-containing phospholipids PC and Sph are enriched in the outer membrane leaflet, whereas most of the PE and PI and virtually all PS occupy the inner leaflet. Glycolipids are only located in the outer half of the membrane

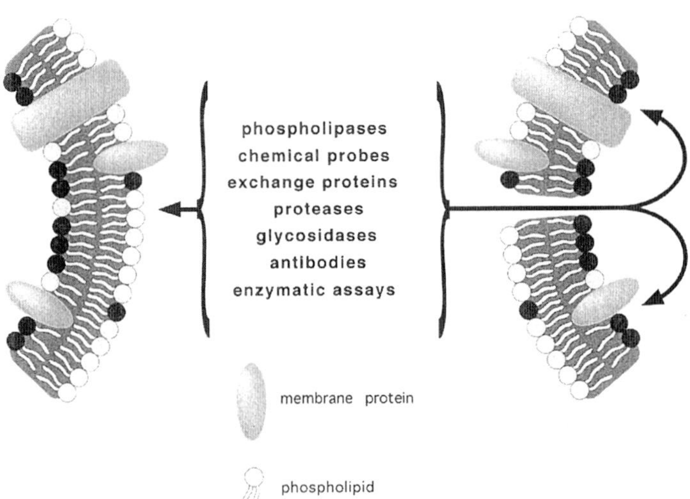

Fig. 1A. Measurement of membrane asymmetry. Membrane-impermeable reagents identify membrane constituents on either side of the membrane, by allowing them to interact with either intact cells to probe the outside of the membrane or with lysed cells to probe the inner membrane leaflet as well. Studies employing phospholipases or lipid transfer proteins as tools have been particularly useful in detecting the orientation of phospholipids. Lipid polar headgroups are represented as circles connected to two fatty acyl tails. Closed circles: aminophospholipids (PS and PE); open circles: choline-phospholipids (PC and Sph).

bilayer where they serve as antigenic determinants, e.g. of A, B, or O blood group specificity in erythrocytes. Whether cholesterol is also unequally distributed between both membrane leaflets is still controversial. Presumably, it is more abundant in the outer half due to its selective affinity for Sph with which it forms laterally phase-separated domains. These rafts play a role in lateral sorting of membrane proteins, for instance, by selective interactions with acylated and GPI anchor proteins while excluding proteins with an isoprenyl anchor [3,4]. While membrane lipid asymmetry is not absolute, membrane proteins have a unique asymmetric orientation, consistent with the fact that membrane proteins have never been observed to rotate from one face of the membrane to the other. Like glycolipids, glycosylated sites on membrane proteins are always exposed on the outer surface of the plasma membrane.

Membrane asymmetry is considered to be ubiquitous in eukaryotic plasma membranes. The origin of lipid sidedness lies in the vectorial biosynthesis of lipids in combination with lipid transporters that move lipids from one membrane face to the other. For example, most of the glycero-phospholipids are synthesized on the cytoplasmic surface of the endoplasmic reticulum, while sphingomyelin is synthesized on the luminal surface of the Golgi [4]. Newly synthesized lipids move to the plasma membrane via intracellular vesicular transport. In this way, sphingomyelin and the amino-phospholipids PE and PS are placed on the membrane surface that topologically corresponds to the outer and inner plasma membrane leaflet, respectively. However, since PC is synthesized on the cytoplasmic face of endoplasmic reticulum, proteins that transport lipids across membranes presumably move it to the opposite surface.

Most of the time, the non-random orientation of membrane lipids is preserved during the life span of the cell. However, in circumstances of cell activation or differentiation and during programmed cell death (apoptosis), bilayer asymmetry may readily collapse as a result of facilitated flip-flop of phospholipids. Also, perturbations of membrane phospholipid asymmetry are not uncommon in pathological cells. Since spontaneous transbilayer movement of phospholipids is rare, these phenomena dictate that biological membranes are assembled by specific mechanisms that control and maintain transbilayer lipid asymmetry, while harboring additional devices that can rapidly move phospholipids back and forth between the two membrane leaflets. Transbilayer migration of lipids can be inferred from detecting changes in accessibility of membrane lipids to exogenous reagents (Fig. 1B), or from real time experiments that rely on reporter lipids that contain a short chain fatty acid with a fluorescent tag or a spin-label (Fig. 1C). Indeed, the regulation of membrane lipid sidedness is controlled by specific membrane proteins, referred to as lipid transporters, which catalyze uni- or bidirectional transport of lipids from one membrane leaflet to the other. At least three protein-mediated activities can be distinguished: "flippase" that promotes inward-directed transport of lipids, "floppase" that promotes outward-directed lipid migration, and "scramblase" that mixes the lipids between the two layers (Fig. 2). While the first two activities primarily generate and maintain membrane lipid asymmetry, scramblase activity promotes its collapse.

This chapter will review the mechanisms and properties of these three activities, and describe some pathophysiological aspects associated with perturbations of membrane phospholipid asymmetry.

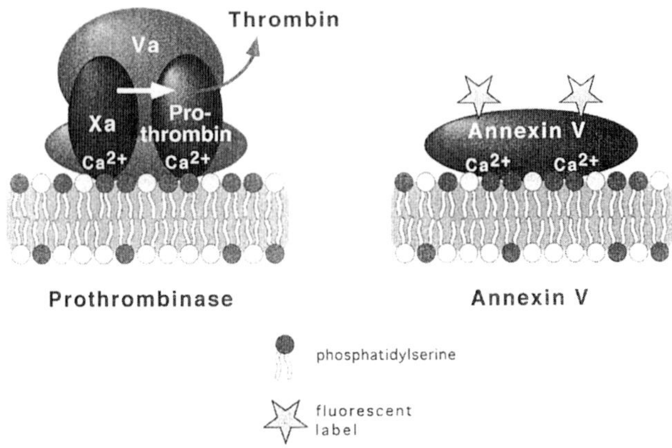

Fig. 1B. Detection of PS distribution over the membrane. Since PS is absent from the cell surface in quiescent cells, perturbations in membrane phospholipid asymmetry are often detected as appearance of PS in the outer membrane leaflet which can be measured by sensitive techniques. The prothrombinase complex (composed of coagulation factors Xa and Va) can be assembled on a membrane surface in such a way that the rate of prothrombin conversion into thrombin is a function of the mole fraction of PS in that surface. A more simple but somewhat less selective approach to detect surface-exposed PS uses fluorescent-labeled annexin V, a 35 kDa protein with a high affinity for anionic phospholipids.

2. Flippase: amino-phospholipid translocase

The discovery in 1984 of an ATP-dependent flippase activity in red blood cells has provided the first evidence for a role of distinct membrane proteins in the generation and maintenance of membrane asymmetry through the transport of specific lipids across the cell membrane [5,6]. This activity is characterized by its ability to catalyze vectorial

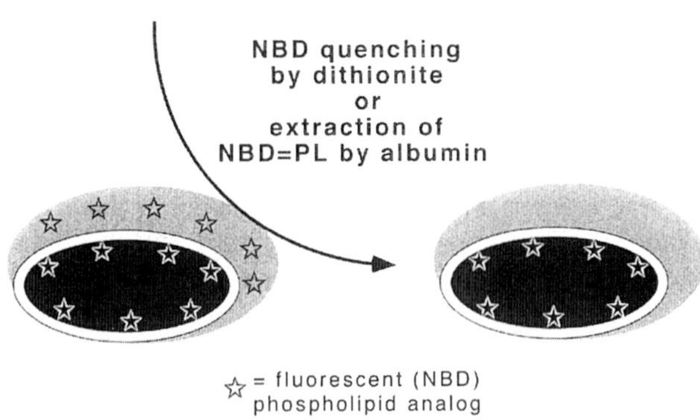

Fig. 1C. Lipid probes as reporters of endogenous phospholipids. Phospholipid analogs containing a fluorescent- or spin-labeled short chain fatty acid readily incorporate into the membrane bilayer. Lipid probes present in the outer half of the bilayer can be rapidly extracted with albumin or quenched with membrane-impermeable reducing agents (dithionite), providing information on the amount of probe present in the outer leaflet at any point in time. In general, the lipid analogs when used in trace amounts are reliable reporters of the behavior of the endogenous phospholipids.

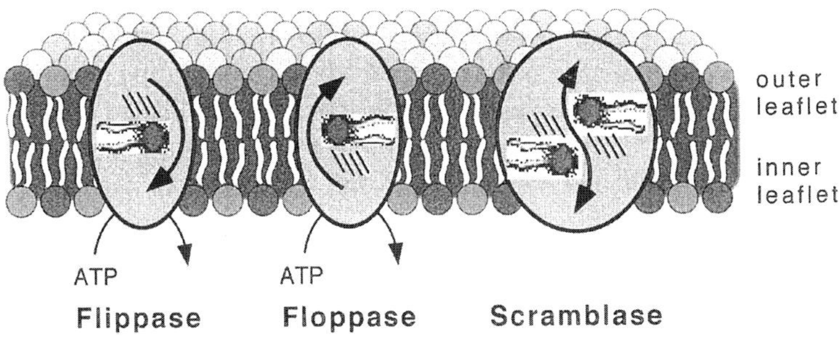

Fig. 2. Transmembrane lipid transporters. Transporter-controlled movement of (phospho)lipids between the two membrane leaflets of the lipid bilayer membrane. Unidirectional transport by flippase relates to inward-directed transport of phospholipids, while floppase catalyzes outward-directed transport of phospholipids. Since both type of transporters frequently move lipids against their concentration gradient, they are ATP dependent and promote generation of membrane lipid asymmetry. Bidirectional transport is catalyzed by a scramblase, activation of which promotes collapse of membrane lipid asymmetry.

transport of the amino-phospholipids PS and PE from the outer to inner leaflet of plasma membranes against the concentration gradient. Since other phospholipids are not moved, the activity is referred to as amino-phospholipid translocase. Competition experiments have revealed that the same protein transports both PS and PE, although PS is transported faster with half-times of 5–10 min [7,8]. Translocation of both amino-phospholipids requires a diacylglycerol backbone and is stereospecific for the naturally occurring L-isomers of the glycerol moiety [9]. While transport requires an amino group in the lipid polar headgroup, both L-serine and D-serine analogs of PS are transported equally well [10]. One molecule of ATP is hydrolyzed per molecule of lipid transported [11], and transport is inhibited by vanadate [5], indicating that it involves an ATP-hydrolyzing enzyme. Translocation is inhibited competitively by glycero-phosphoserine [10], and can be completely abolished by sulfhydryl- and histidine-reactive reagents [12–14], or when cytoplasmic Ca^{2+} levels reach micromolar concentrations [15].

The biochemical properties of amino-phospholipid translocase and its ubiquitous presence in cell membranes [10] underscores its importance in processes that assist in the maintenance of membrane phospholipid asymmetry, and in its regeneration once the transbilayer gradient of amino-phospholipids is lost. Although the observations clearly indicate that one or more membrane proteins catalyze lipid transport, its identity is still uncertain. A Mg^{2+}-ATPase that is stimulated by PS and inhibited by vanadate has been partially purified from human erythrocytes and reconstituted into artificial lipid vesicles with at least a fraction of its active center at the outer face [16]. These vesicles translocated a spin-labeled PS analog from the inner to outer leaflet upon the addition of Mg^{2+}-ATP, suggesting that this ATPase is responsible for amino-phospholipid translocation. However, the active fraction was not homogenous and contained several proteins ranging from 32 to 165 kDa, and efforts to further purify the enzyme have been unsuccessful [10]. Other strategies to identify the lipid transporter have focussed on a P-type ATPase of chromaffin granules, which are known to exhibit Mg^{2+}-ATP translocase activity [7]. The cDNA of this protein has been cloned [17] and found to be similar to a yeast gene *drs2*, originally discovered as a requisite for ribosome assembly. Although it has been claimed that a mutation in *drs2* generates a phenotype defective in transporting fluorescent-labeled PS [17,18], other groups have demonstrated normal transport of PS and PE analogs in this yeast mutant [19,20].

These observations preclude assignment of the transporter to a single protein and are not inconsistent with studies implicating the involvement of a 31 kDa polypeptide in amino-phospholipid transport [12]. This protein, which in erythrocytes may be complexed to Rhesus polypeptides, can be preferentially labeled with a photo-activatable PS analog only under conditions conducive to PS transport [21]. Although the labeled protein is neither an ATPase nor a member of the Rhesus family, it may be a regulatory component of a larger lipid-transporting complex together with a Mg^{2+}-ATPase [2]. Such a motif is not uncommon for P-type ATPases, which are transiently phosphorylated by ATP on an aspartate residue, and to which the amino-phospholipid translocase might belong [10].

3. Floppase: ATP-binding cassette (ABC) transporters

Studies with erythrocytes have demonstrated the existence of an ATP-dependent floppase activity, which facilitates lipid migration across the plasma membrane in a direction opposite to that of amino-phospholipid translocase [14,22]. This inward to outward movement appears to be less specific with respect to the lipid polar headgroup. Both choline- and amino-phospholipids are transported to the outer leaflet with half-times about 10 times slower than those of the translocase-mediated inward movement of PS and PE. Floppase activity is abrogated by ATP depletion, sulfhydryl oxidation, and histidine modification, similar to amino-phospholipid translocase activity. Yet, translocase and floppase operate independently, since a rapid inward movement of amino-phospholipids did not affect the rate of outward movement, suggesting that both processes are mediated by different lipid transporters [14]. Recently, the floppase in red blood cells has been shown to be identical to the ATP-binding cassette transporter encoded by the gene ABCC1 (also known as multidrug resistance protein MRP1) [23,24]. This is a member of the membrane protein family of ABC transporters best known to drive the transport of various molecules and hydrophobic drugs from the cytoplasmic leaflet to the outer layer or to an acceptor molecule [25,26]. MRP1 is a 180 kDa integral membrane protein with 17 putative membrane spanning domains, and contains a pair of ATP-binding sites at its cytoplasmic region. While inhibition of MRP1 results in a slow redistribution of endogenous PC to adopt a more random orientation, this does not affect the asymmetric distribution of PE and PS [27]. Thus, the concerted action of the ABCC1 (MRP1) and the amino-phospholipid translocase is thought to procure a dynamic asymmetric steady state in which all phospholipids are slowly but continuously moved to the outer membrane leaflet, whereas the amino-phospholipids are rapidly shuttled back to the inner leaflet. This equips the cell membrane with flexible machinery to correct for alterations in lipid asymmetry, for example, resulting from membrane fusion events during endo- or exocytosis [28].

A number of other members of the protein family of ABC transporters has been recognized or inferred to play a role in outward transport of phospholipids across the plasma membrane. For example, the gene product of ABCB4 (also known as human MDR3, or mouse mdr2) catalyzes unidirectional transport of PC in the canalicular domain of murine hepatocyte plasma membranes to provide PC for bile production [29,30]. However, given its predominant location in liver cells, this lipid transporter is unlikely to play a role in the maintenance of phospholipid asymmetry in plasma membranes. The closely related gene ABCB1 (also known as human MDR1) encodes for a protein that is capable of moving various short chain lipid analogs with a sphingosine backbone to the exterior leaflet of the cell membrane [30]. Also, selective inhibition of MDR1 in human leukemia cells has been shown to interfere with outward-directed transport of endogenous Sph leading to a near symmetric distribution of Sph over the bilayer [31]. This ABC transporter is also known to extrude a wide variety of hydrophobic drugs from the cell, and to confer multidrug resistance to tumor cells that overexpress this protein. Another ABC transporter that has been implicated in promoting lipid floppase activity is ABCA1 (also known as ABC1), which is one of the largest ABC transporters known (MW $\sim$ 260 kDa) and is expressed in nearly all mammalian cells [26]. Mutations in this gene lead to Tangier disease, an autosomal recessive disorder characterized by defective transfer of phospholipids and cholesterol from the cell to distinct serum apolipoprotein acceptors (apoA1 and apoE), resulting in an almost total absence of HDL cholesterol from the serum [32]. Moreover, ABCA1 has been implicated to promote transient and local PS exposure in apoptotic cells, suggesting that it may function as a PS floppase (a kind of reverse amino-phospholipid translocase), which selectively pumps this lipid from the cytoplasmic to the exofacial leaflet under conditions leading to programmed cell death [33].

4. Scramblase

Although assembly and maintenance of an asymmetric lipid membrane is an energy-consuming process, ATP depletion that results in inhibition of amino-phospholipid translocase will not readily lead to a loss of lipid asymmetry. However, as first shown for blood platelets in the early 1980s, a collapse of lipid asymmetry may occur rapidly upon particular conditions of cellular activation [34,35]. Platelet plasma membranes harbor a Ca^{2+}-dependent mechanism that can swiftly move phospholipids back and forth between the two membrane leaflets, leading within minutes to a loss of membrane phospholipid asymmetry. Because the influx of Ca^{2+} simultaneously abrogates amino-phospholipid translocase activity [36,37], Ca^{2+}-dependent loss of membrane phospholipid asymmetry is not corrected. Although several mechanisms have been postulated to explain Ca^{2+}-induced collapse of lipid asymmetry, lipid randomization is likely to be dependent on one or more membrane proteins with "lipid scramblase" activity [38]. Particularly, the discovery of an inherited bleeding disorder (Scott syndrome, see below), characterized by an impairment of scramblase activity, has strongly supported the notion that specific membrane proteins are involved in this process [39–42]. Ca^{2+}-induced scramblase activity has also been found in a wide variety of other cells, but its activity is usually lower than in blood platelets [15,28].

Scramblase activity requires the continuous presence of cytoplasmic calcium [43]. Extrusion of Ca^{2+} leads to restoration of lipid asymmetry, provided that the amino-phospholipid translocase is not irreversibly proteolyzed by intracellular calpain [36]. Lipid scrambling is bidirectional and involves all major phospholipid classes. In general, glycero-phospholipids move somewhat faster than sphingomyelin or other lipids with a ceramide backbone [44,45]. Using a variety of fluorescent lipid analogs with increasing polar headgroup size, it has been demonstrated that lipid movement comes to a stop when the polar headgroup is composed of a trisaccharide, possibly reflecting the upper size of a protein-mediated pore [45]. Pore-mediated flip-flop of phospholipids is thought to involve movement of lipid polar headgroups through a central aqueous channel while the fatty acid moieties diffuse along a hydrophobic interface between protein subunits [28]. Unlike amino-phospholipid translocase, lipid scrambling is not coupled to ATP hydrolysis. However, a gradual loss of scramblase activity occurs during prolonged ATP depletion [46] and can be restored by ATP repletion [15], suggesting that the scramblase transporter may be constitutively phosphorylated.

Reconstitution of proteins fractionated from platelet and erythrocyte membranes into artificial lipid vesicles can exhibit Ca^{2+}-dependent scramblase activity that is pronase-, heat-, and sulfhydryl-sensitive [47,48]. While these data strongly support the notion that one or more proteins are responsible for scramblase activity, its identity has not been unambiguously established. The active protein fraction of erythrocyte membranes has been further purified to homogeneity and appeared to be a 37 kDa protein. Using an internal peptide sequence of the purified protein, a cDNA has been cloned encoding a polypeptide of 318 amino acid residues, most likely a type II plasma membrane protein with a predicted single pass transmembrane domain near the exofacial C-terminus of the molecule and a putative calcium binding site at the cytoplasmic region (reviewed in Ref. [49]). Although a number of properties of this protein are not inconsistent with those alleged for a lipid scrambling transporter (reviewed in Refs. [15,49]), other data suggest that it may not function as a lipid scramblase. For example, B lymphocytes from a patient with Scott syndrome, despite of being deficient in scramblase activity, have normal levels of this membrane protein and its corresponding mRNA with nucleotide sequences identical to that of normal controls [50]. Also, the gene encoding this protein is under transcriptional control by interferon, but the resulting increase of the protein in the plasma membrane is not in any way accompanied by an increase in Ca^{2+}-dependent scramblase activity [49]. Knock-out mice, deficient in the gene encoding for the putative scramblase, have no hemostatic abnormality, and their erythrocytes and blood platelets show normal mobilization of PS to the cell surface upon cell stimulation [51].

Progressive loss of membrane lipid asymmetry is often accompanied by outward blebbing of the cell membrane and subsequent shedding of microvesicles, in which the phospholipids are completely randomized over the microvesicle membrane [37,52–54]. Shedding of microvesicles requires both lipid scrambling in the plasma membrane and activation of the intracellular Ca^{2+}-dependent protease calpain that degrades cytoskeletal membrane proteins. While the microvesicle membrane is lipid symmetric, randomization of lipids in the remnant plasma membrane is usually partial. Moreover, since Ca^{2+} influx inhibits amino-phospholipid translocase and activates lipid scramblase, intermediate Ca^{2+} levels could lead to a circumstance in which both mechanisms are active and oppose each other. Conceivably, these situations can accommodate a wide range of steady-state distributions of membrane phospholipids [55], commonly seen under a number of pathological conditions.

5. Membrane asymmetry and blood coagulation

Lipid–protein interactions play a pivotal role in blood coagulation [56,57]. Assembly of blood clotting enzyme complexes on appropriate phospholipid membranes leads to a profound increase of the reaction rate by which zymogens are converted into active serine proteases through limited proteolysis. To all intents and purposes, most coagulation enzymes in the absence of lipid membranes display negligible activity towards their respective substrates within a biologically relevant time span. It is now widely appreciated that anionic phospholipids, particularly PS (when mixed with a neutral phospholipid like PC), provide the most active catalytic surface. Activation of blood platelets, accompanied by an increase in cytoplasmic Ca^{2+} levels, may result in rapid activation of lipid scramblase while blocking amino-phospholipid translocase [37]. This leads to a collapse of platelet membrane phospholipid asymmetry and shedding of lipid-symmetric microvesicles, with concomitant surface exposure of PS to the coagulation proteins in plasma.

Although different blood coagulation pathways have been recognized, the most important one starts with tissue factor, an integral membrane protein expressed on the surface of activated or disrupted cells [58,59]. Tissue factor interacts with factor VII or VIIa, and this complex rapidly converts the zymogen factors IX, X, as well as factor VII itself, into their active forms (Fig. 3). Although assembly and catalytic activity of the tissue factor/factor VIIa complex are effective in the absence of anionic phospholipids, activity is increased by PS [60]. Surface exposure of PS on activated blood cells (e.g. platelets) promotes binding and catalysis of two subsequent coagulation factor complexes in the cascade that leads to thrombin formation [53,56]. The tenase complex is initiated by the interaction of factor VIIIa with a PS-containing membrane surface to create a high-affinity binding site for the enzyme factor IXa in the presence of Ca^{2+}. This complex rapidly activates the zymogen factor X into an active protease factor Xa. Likewise, in the prothrombinase complex, binding of factor Va to a membrane exposing PS promotes Ca^{2+}-dependent binding of factor Xa, which converts prothrombin to thrombin. This enzyme has multiple functions, among which its ability to promote aggregation of blood platelets and to catalyze the production of an insoluble fibrin gel. Together, these events secure effective hemostatic plug formation at the site of injury, or produce undesired thrombus formation as a response to internal damage of the vascular wall. PS is equally important in promoting the anticoagulant protein C pathway that provides feedback inhibition of thrombin formation [61,62]. Protein C, after being activated by

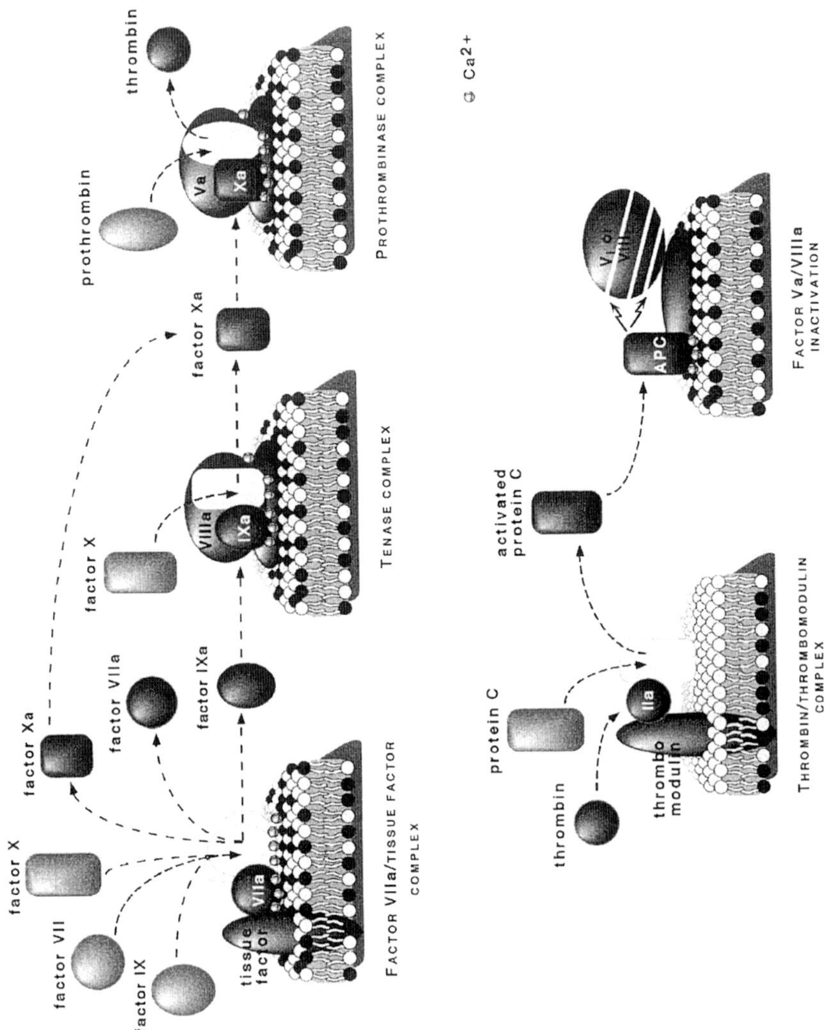
factor VII
factor IX
factor X
factor Xa
factor VIIa
factor IXa
tissue factor
VIIa
FACTOR VIIa/TISSUE FACTOR COMPLEX
factor X
VIIIa
IXa
TENASE COMPLEX
factor Xa
prothrombin
thrombin
Va
Xa
PROTHROMBINASE COMPLEX
Ca2+
thrombin
protein C
IIa
thrombomodulin
THROMBIN/THROMBOMODULIN COMPLEX
activated protein C
APC
FACTOR Va/VIIIa INACTIVATION

thrombin, effectively inactivates factors Va and VIIIa when bound to a PS-containing lipid surface such as platelet microvesicles, which leads to disassembly of the prothrombinase or tenase complex, respectively [63].

6. Membrane asymmetry and apoptosis

Besides promoting blood coagulation, another feature of PS-exposing cells is their propensity to be recognized by phagocytes [64,65]. In this way, activated blood platelets may be removed from the site of injury to promote tissue repair. Surface expression of PS is one of the hallmarks of cells undergoing programmed cell death or apoptosis [66,67], and targets the cell for uptake and removal by macrophages. Engulfment occurs before the cell undergoes more severe damage, particularly before it becomes necrotic and spills its aggressive contents into surrounding tissue [68]. Although the mechanisms of recognition and uptake of cells by macrophages presumably involve several distinct pathways, it has become clear that all types of phagocytes recognize PS on apoptotic cells [69]. Indeed, the PS-binding protein annexin V inhibits uptake [70], and apoptotic cells that fail to express PS externally are not engulfed by (stimulated) macrophages or fibroblasts [71].

Recent evidence indicates that phagocytes recognize PS-expressing cells via a distinct PS receptor that is stereospecifically inhibited by liposomes containing phosphatidyl-L-serine, and not by other (anionic) phospholipids including phosphatidyl-D-serine. The gene for the PS-specific receptor encodes a 48 kDa type II plasma membrane protein with runs of cationic amino acid residues near the extracellular C-terminal domain, which may provide a binding site for the anionic headgroup of PS [72]. The intracellular domain of the receptor contains a potential tyrosine phosphorylation site and multiple protein kinase C phosphorylation sites, which may provide signaling capabilities that promote engulfment after tethering the apoptotic cell to the phagocyte. In addition, clearance of activated and apoptotic blood cells can occur via more indirect pathways (Fig. 4). Because of its ability to bind PS, the serum protein β-2 glycoprotein I may interact with PS-expressing cells. A conformationally induced neo-epitope on this protein may be recognized by a distinct receptor on phagocytes, promoting engulfment of the cell [73]. Such a neo-epitope may also give rise to the generation of so-called antiphospholipid antibodies, that a.o. circulate in patients with lupus erythematosis and sickle cell disease (see below). Although these antibodies were originally thought to react with anionic phospholipids like cardiolipin or

---

Fig. 3. Membrane associated complexes in blood coagulation. The upper three complexes (factor VIIa/tissue factor-, tenase-, and prothrombinase complex, respectively) constitute the major blood coagulation pathway leading to thrombin (factor IIa) formation. The lower two complexes represent the anticoagulant protein C pathway which results in proteolytic degradation of the heavy chains of factors VIIIa and Va, thus leading to inactivation of the tenase and prothrombinase complex, respectively. This inactivation process is enhanced by protein S (not shown). A suffix "a" indicates the active form of the coagulation factors, while a suffix "i" indicates the inactive form. Factors Va and VIIIa are represented as dimers with a heavy and a light chain, the latter in interaction with phospholipids. Amino-phospholipids are shown with dark polar headgroups and choline-phospholipids with light polar headgroups. *Reprinted from R.F.A. Zwaal et al., Lipid–protein interactions in blood coagulation, Biochim. Biophys. Acta 1376, 433–453, Copyright 1998, with permission from Elsevier Science.*

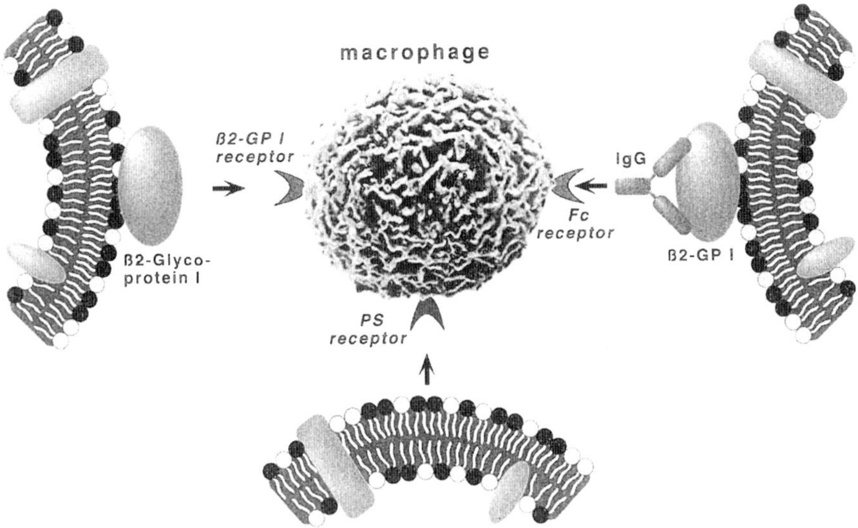

Fig. 4. PS-mediated recognition by macrophages. PS on the outer surface of activated and apoptotic cells marks the cell as a pathological target for elimination by macrophages. Recognition of the PS-expressing targets can occur via a PS-specific receptor that directly engages PS (below), via a β-2 glycoprotein I-specific receptor that interacts with the PS-bound glycoprotein (left), or via an Fc-receptor that recognizes IgG antibodies directed against lipid-bound β-2 glycoprotein I (right).

PS, it has been established that they are directed to lipid-bound plasma proteins [74–76]. The antibodies, which recognize PS-bound β-2 glycoprotein on the apoptotic cell surface, do promote elimination by macrophages through Fc-mediated phagocytosis [77].

While it has been proposed that surface exposure of PS during the process of apoptosis results from activation of lipid scramblase with concomitant inactivation of amino-phospholipid translocase [28,68], there are remarkable differences between PS exposure in activated cells and in apoptotic cells. For example, a rise in cytosolic Ca^{2+} activates scramblase during cell activation, but cytosolic Ca^{2+}-chelators do not prevent PS exposure in apoptotic cells, even though external presence of EDTA does [78]. Moreover, whereas Ca^{2+}-induced scramblase activity is clearly aberrant in B lymphocytes of Scott syndrome (see below), PS exposure in apoptotic Scott lymphocytes is indistinguishable from controls [79]. This raises the possibility that, during apoptosis, a Ca^{2+}-independent lipid transporter is activated in conjunction with inhibition of amino-phospholipid translocase. One possible candidate is the floppase ABCA1 (see above), which may promote a local high PS density on the surface of apoptotic cells [33], considered to be a requisite for efficient recognition by PS receptors on phagocytes [80].

7. Membrane asymmetry and cell development

While a collapse of membrane phospholipid asymmetry is a hallmark of cells undergoing apoptosis, surface exposure of PS in the absence of apoptosis has been recognized to play a distinct role during early cell development. For example, mammalian sperm cells have to go through a physiological maturation phase in order to become competent to fertilize an egg, a process called capacitation. When the sperm cell reaches the female reproductive tract, it encounters so-called capacitation factors (HCO_3^-, albumin, Ca^{2+}) that produce a subtle reorganization of membrane components. Bicarbonate, thought to be locally enriched in the upper region of the female genital tract, induces scrambling of sperm membrane phospholipids controlled by cAMP-dependent protein kinase A and tyrosine phosphorylation by an as yet unknown signaling pathway [81,82]. The response to bicarbonate occurs in 30–70% of the viable cells which rapidly expose PE followed by a slower expression of PS [83]. Exposure of amino-phospholipids is restricted to the apical region of the sperm head plasma membrane, and is accompanied by a lateral migration of cholesterol to the same region. Subsequent addition of albumin causes an efflux of cholesterol, but only in bicarbonate-responding cells. These bicarbonate- and albumin-mediated lipid rearrangements increase membrane fluidity and seem to be required for the initiation of the acrosome reaction, a Ca^{2+}-dependent exocytosis that involves fusion between the apical sperm plasma membrane and the underlying acrosomal membrane. This event releases proteins and hydrolytic enzymes from the acrosome that help the sperm to bind to and penetrate the egg's outer coat.

Another apoptosis unrelated event involves transient surface exposure of PS at distinct membrane regions of apparently viable myoblasts in the developing heart and skeletal muscle [84]. In vitro studies with differentiating skeletal myoblasts have shown that PS expression is restricted to cell–cell contact areas prior to actual fusion of individual cells to form multinucleate skeletal muscle cells, known as myotubes [85]. Myotube formation is inhibited by the PS-binding protein annexin V, indicating that PS exposure is a requisite for this process to occur. Whereas apoptotic myoblasts also expose PS, none of the other apoptotic characteristics (DNA fragmentation, loss of mitochondral membrane potential, caspase activation) is apparent in differentiating myoblasts, suggesting a different mechanism that regulates surface exposure of PS in these cells than occurs during apoptosis. Moreover, viable muscle cells lack chemotactic factors that attract phagocytes, which may explain why they avoid the attention of scavenger cells despite expressing PS as a signal for cell removal. Surface exposure of PS in differentiating muscle cells is transient and is followed by internalization before the fusion process is fully completed [85]. Whether or not this reflects a concerted regulation of scramblase and amino-phospholipid translocase activities remains to be resolved.

While differentiating muscle cells seem to be safeguarded against phagocytosis, erythropoiesis provides an instructive example in which a specialized phagocytotic event promotes cell development. During erythroblast proliferation, the cell extrudes its nucleus to become a reticulocyte that leaves the bone marrow and passes into the bloodstream. Erythrocyte clones develop in the bone marrow on the surface of a macrophage, which starts to engulf the membrane lobe surrounding the nucleus even before the segregation of the two bodies is complete [70,86]. Moreover, the portion of the plasma membrane that surrounds the lobe of the cell containing the extruding nucleus stains with the dye merocyanin 540, which reflects a looser packing of the lipids frequently associated with lipid scrambling [87]. The membrane surrounding the reticulocyte lobe of the erythroblast does not stain with merocyanin and is not recognized by phagocytes. Assuming that phagocytosis of the nucleus-containing membrane lobe is PS dependent, this may suggest that lateral rearrangement and localization of proteins that control lipid asymmetry occur in the plane of the erythroblast membrane, for example, resulting from compartmentalization of the spectrin-based cytoskeleton to the reticulocyte part of the enucleating cell [88].

8. Membrane asymmetry and disease

Although membrane lipid asymmetry is usually the rule for normal cells, loss of asymmetry, especially the appearance of PS at the cell surface, is associated with many pathological phenomena. The properties of PS-expressing cells to become procoagulant and to be marked for phagocytosis can be a major cause of the disorder or a matter of secondary importance. At present, no deficiencies or defects are known with respect to amino-phospholipid translocase or the less specific floppase, but a defective scramblase activity has been shown to lead to a bleeding disorder.

8.1. Scott syndrome, a bleeding disorder

The importance of a scramblase-induced loss of membrane phospholipid asymmetry can be illustrated in Scott syndrome, a rare, moderately severe, bleeding disorder characterized by a defect in platelet procoagulant activity that is not associated with decreased levels of coagulation factors [40]. Although stimulation of these platelets results in normal secretion and aggregation, these cells exhibit a decreased surface exposure of PS resulting in a reduced ability to promote both tenase and prothrombinase activity in response to agonists [39] (Fig. 5), and impaired capacity to shed membrane-derived microvesicles [52]. Family studies [42], and studies on dogs with Scott syndrome [89], have indicated that this bleeding disorder is transmitted as an autosomal recessive trait. The defect in Ca^{2+}-induced lipid scrambling is not restricted to platelets but can also be demonstrated in erythrocytes and erythrocyte ghosts [41], and in other peripheral blood cells including Epstein–Barr virus-transformed B lymphocytes [42,79]. While the studies on Scott syndrome suggest a deletion or mutation in multiple hematological lineages that either affects lipid scramblase directly or alters its Ca^{2+}-induced activation pathway, the molecular basis of this defect is still unresolved.

Whereas B lymphocytes from Scott syndrome do not expose PS following Ca^{2+}-influx, PS expression is normal in apoptotic lymphocytes from these patients [79]. Although induction of apoptosis-induced PS exposure occurs over a much longer time scale (hours) than the Ca^{2+}-induced scrambling process (minutes), the bidirectional and nonspecific characteristics of the apoptotic activity mirror those of the Ca^{2+}-induced activity. This may seem to rule out a role for the floppase ABCA1 (see above), but the question of whether this transporter could also promote bidirectional lipid movement has not been addressed so far. Once the lipid scrambling process is activated in apoptotic cells, no increase in rate of scrambling is seen when apoptotic cells are challenged with Ca^{2+}. These observations are not inconsistent with the view that a single scramblase can be activated via a Ca^{2+}-dependent and a Ca^{2+}-independent pathway, and that only the first route would be defective in Scott syndrome. However, the possibility of cells having different scramblases (with different activation pathways) cannot be ruled out.

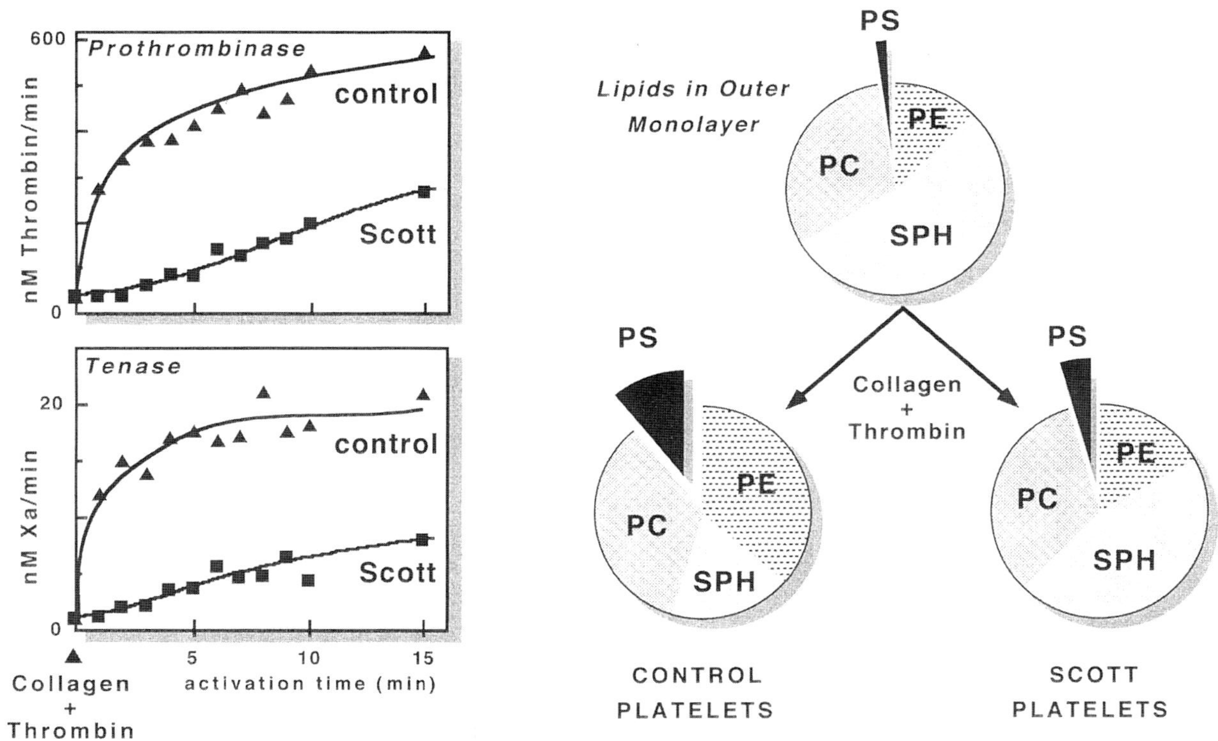

Fig. 5. Scott syndrome. Generation of prothombinase- and tenase activity of Scott and control platelets activated by collagen plus thrombin (left panel), and phospholipid composition of the outer leaflet of the platelet plasma membrane before and after activation (right panel). Scott platelets are characterized by a slower and diminished generation of platelet prothrombinase and tenase activity, and by a lower extent of PS exposure on the outer cell surface, as probed by phospholipases. (Unactivated platelets from Scott syndrome have a normal lipid composition and membrane phospholipid asymmetry.)

8.2. Antiphospholipid syndrome

The presence of circulating "antiphospholipid" antibodies in association with arterial and venous thrombosis, recurrent fetal loss, and thrombocytopenia defines the antiphospholipid syndrome [90]. Although these antibodies were first believed to recognize anionic phospholipids directly, it is now generally appreciated that the antibodies are directed against epitopes exposed by plasma proteins when they interact with anionic phospholipids [74–76]. Most often, the plasma proteins comprise β-2 glycoprotein I and prothrombin, and antibodies against the latter are frequently seen in patients with lupus erythematosis [91]. Both plasma proteins interact with PS-expressing membranes via electrostatic and hydrophobic interactions with K_d's in the micromolar range [56]. The antibodies potentiate this binding by two orders of magnitude, presumably resulting from the bivalent interaction of the IgG molecules with the lipid-bound proteins [92]. Since β-2 glycoprotein I and prothrombin undergo conformational changes upon binding to a lipid surface, it is possible that the antibodies recognize conformation-induced neo-epitopes [76].

The functional assembly of coagulation complexes on membrane surfaces (cf. Fig. 3) would predict that any protein with a high affinity for anionic phospholipids may interfere with the normal coagulation process. Indeed, in vitro, IgG-β-2 glycoprotein I complexes can restrict factor Xa binding to, and prothrombin activation on artificial membranes containing PS as well as on activated platelets or shed microvesicles [92–94]. Conceivably, IgG-prothrombin complexes may have a similar effect apart from a possible shielding of prothrombin from being activated by factor Xa. While these laboratory findings would predict that patients with antiphospholipid syndrome might have a bleeding tendency, these antibodies are associated with an increased risk for thrombosis. It has been suggested that, in view of the notion that patients with heterozygous protein C deficiency have thrombosis [61,62], interference of these antibodies with protein C-catalyzed inactivation of cofactors Va and VIIIa (cf. Fig. 3) on a membrane surface may underlie the thrombotic tendency of patients with anti-phospholipid syndrome [96]. Another possibility is that circulating antibodies would reflect a normal immune response towards persistent thrombogenic PS-expressing membrane surfaces rather than an aberrant auto-immune response [76,95]. Prolonged surface exposure of PS in these patients would be due to a disturbed balance between amino-phospholipid translocase and scramblase activity, or may be caused by a defective scavenging mechanism by which procoagulant cells or microvesicles are cleared from the circulation (Fig. 6). As outlined below, "antiphospholipid" antibodies are frequently observed in diseases that are compromised by increased cell surface exposure of PS, such as sickle cell anemia, thalassemia, malaria, uremia, diabetes, pre-eclampsia, and conditions associated with elevated levels of circulating microvesicles.

8.3. Sickle cell disease

Sickle cell anemia, which is caused by a point mutation in the β-chain of hemoglobin, is characterized by hemoglobin polymerization and sickling of erythrocytes under deoxygenated conditions that result in alterations in the plasma membrane architecture. A prominent change is a partial collapse of membrane phospholipid asymmetry with exposure of PS on the surface of a subpopulation of sickled cells [97–99], and on sickle cell-derived microvesicles that are produced during repeated cycles of oxygenation and de-oxygenation [100]. Flow-cytometric studies using fluorescent labeled annexin V have indicated that between 0.5 and 10% of the red blood cells of sickle cell patients contain a substantial amount of surface exposed PS, while normal donors show very little annexin-positive cells (less than 0.3%) [97,98]. While the small fraction of PS-exposing normal red cells presumably reflects the dense senescent cells known to expose PS [101], the PS-exposing subpopulation of sickle cells contains both the densest and the very light cells [99]. Apart from containing reticulocytes, the light fraction harbors those mature cells that have ion transport abnormalities which lead to high Na^+ and low K^+ content. All PS-exposing sickle cells are devoid of amino-phospholipid translocase activity, probably because of increased oxidation of membrane components in sickle cells. Deoxygenation-induced sickling is accompanied by transient periods of increased cytosolic Ca^{2+} particularly in cells with ion transport abnormalities [102], which may lead to temporary activation of the scramblase. In cells with an active amino-phospholipid translocase, PS exposure would be corrected. However, in the subpopulation of sickle cells with an inactivated translocase, this would lead to permanent PS exposure.

PS-expressing sickle cells can contribute to microvascular occlusion during sickle cell crisis in several ways. Although the size of the PS-exposing subpopulation may look rather small (average: 3%), numerically this equals to more than half the number of blood platelets in the circulation. Exposure of PS promotes blood coagulation, which may contribute to the thrombotic episodes during sickle cell crisis [103]. Moreover, thrombosis in sickle cell patients can be compromised by circulating "antiphospholipid" antibodies, which may be generated in response to cells with a sustained PS exposure [104]. Apart from thrombosis, microvascular occlusion may be promoted by the propensity of PS-exposing red cells to adhere to vascular endothelium and the endothelial matrix protein thrombospondin [105–107]. In addition to contributing to sickle cell crisis, the exposure of PS on sickle cells could be partially responsible for the decreased red blood cell survival and anemia, since these cells would be prone to clearance by macrophages [65,66,69–73].

8.4. *Thalassemia*

Thalassemia is a congenital hemolytic anemia caused by a partial or complete absence of the alpha- or beta-chain of hemoglobin. Homozygous carriers suffer from severe anemia and other serious complications from early childhood. Similar to sickle cells, subpopulations of thalassemic erythrocytes are present which expose PS on their external surface. The number of PS-expressing cells can vary considerably between different patients, from as low as those found in normal cells (less than 0.3%) to as high as 20% [108–110]. While the defect of the PS-expressing cells in thalassemia is unknown, it has been suggested that they may reflect cells with oxidized membranes resulting from the frequently observed iron overload in these patients [111], which for example could inactivate the oxidation-sensitive amino-phospholipid translocase [13,21].

Due to their procoagulant nature, PS-exposing red cells are thought to contribute to the profound thromboembolic complications in thalassemia, among which cerebral thrombosis, deep venous thrombosis, and pulmonary embolism are the most common [111]. Moreover, these cells may be more rapidly removed from the circulation by phagocytes that recognize PS, contributing to the anemia. Particularly, splenectomized patients lacking a significant reservoir of macrophages have the highest number of PS-expressing cells, although a direct correlation between the severity of the anemia and the proportion of these cells could not be identified [110]. It is evident that apart from early clearance from the circulation, ineffective erythropoiesis also exacerbates the anemia.

8.5. Renal insufficiency

Uremia, a toxic condition associated with accumulation of by-products of protein metabolism (e.g. urea) in the blood is a common feature of chronic renal failure. A subpopulation of about 3% of uremic erythrocytes has been found to expose PS at their outer surface [112], and to be preferentially removed from the circulation by phagocytosis of intact cells [113]. Phagocytosis of uremic erythrocytes is strongly inhibited in vitro when macrophages are pretreated with glycerophosphorylserine, a structural derivative of PS, whereas no inhibition is observed with the equivalent derivative of PE, glycerophosphorylethanolamine. Also, annexin V strongly hampers macrophage recognition of uremic erythrocytes. PS externalization promotes increased adhesion of uremic erythrocytes to endothelium, possibly via interaction with matrix thrombospondin [114]. Apart from depressed erythropoiesis, a shortened red blood cell life span has been found to contribute to anemia, a common feature in chronic renal failure. The form of dialytic treatment of uremic patients seems to influence the abnormal surface exposure of PS. Patients on continuous ambulatory peritoneal dialysis show a lower percentage of PS-expressing cells than patients on hemodialysis, which also correlates to the lower degree of anemia in the group of patients on peritoneal dialysis [112,115].

The toxic origin of shortened red cell survival in chronic renal failure is well known: red blood cells from uremic patients have a normal life span in healthy subjects, whereas red cells from healthy people have a reduced life span in uremic patients. Interestingly, incubation of uremic plasma with normal red blood cells promotes both PS exposure and erythrophagocytosis, the latter being independent of interaction between plasma and

---

Fig. 6. Possible associations between the presence of antiphospholipid antibodies and the increased risk of thrombosis in the antiphospholipid syndrome. Central to the hypothesis is a persistent cell surface exposure of anionic phospholipids (PS), which causes binding of distinct plasma proteins such as β-2 glycoprotein I and prothrombin. This initiates an immune response to neo-epitopes resulting in the formation of "antiphospholipid" antibodies. Enhanced binding of plasma proteins by antiphospholipid antibodies may cause a positive feedback in the maturation process of these antibodies. An increased and persistent exposure of PS may result from (i) defects in mechanisms that control transmembrane lipid asymmetry, (ii) increased cellular activation and microvesicle production, (iii) increased apoptosis or (iv) defective scavenging mechanisms of macrophages. Surface-exposed PS, however, also enhances coagulation leading to an increased risk for thrombosis. Since antiphospholipid antibodies interfere with lipid-dependent coagulation reactions, they may as well be considered as protective antibodies. On the other hand, but not necessarily in contrast, it has been reported that antiphospholipid antibodies may be causative to thrombosis.

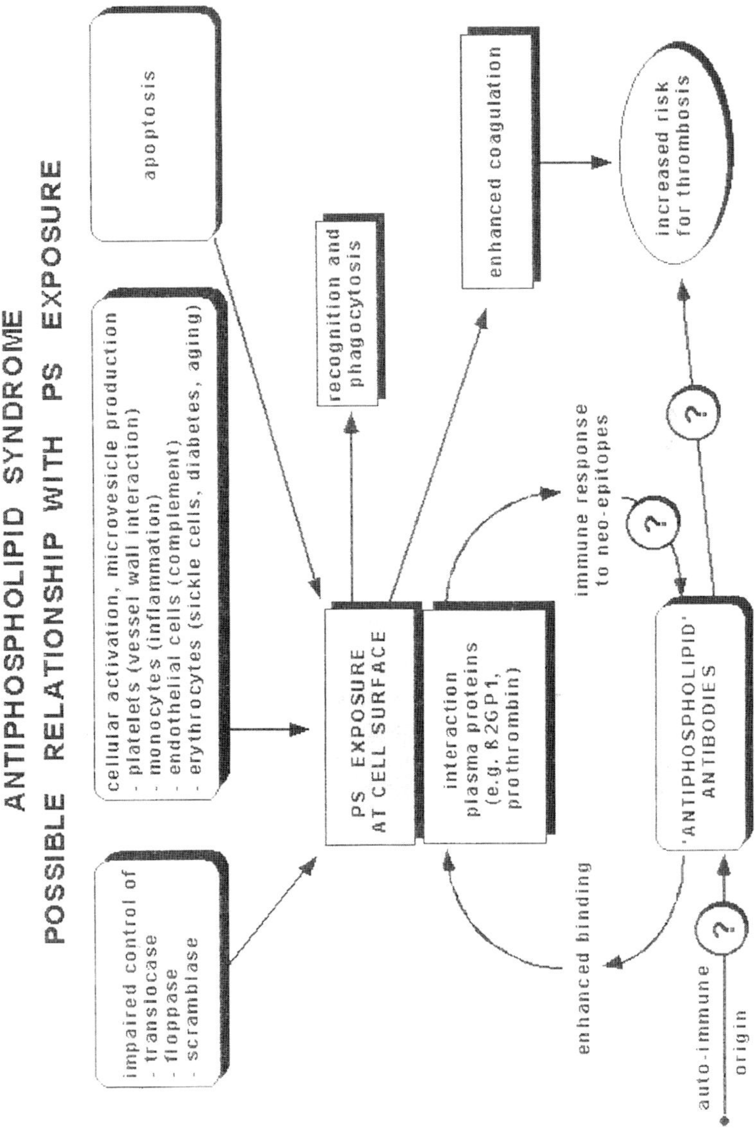
ANTIPHOSPHOLIPID SYNDROME
POSSIBLE RELATIONSHIP WITH PS EXPOSURE
impaired control of
- translocase
- floppase
- scramblase
cellular activation, microvesicle production
- platelets (vessel wall interaction)
- monocytes (inflammation)
- endothelial cells (complement)
- erythrocytes (sickle cells, diabetes, aging)
apoptosis
PS EXPOSURE AT CELL SURFACE
interaction plasma proteins (e.g. β2GP1, prothrombin)
recognition and phagocytosis
enhanced coagulation
increased risk for thrombosis
immune response to neo-epitopes
?
'ANTIPHOSPHOLIPID' ANTIBODIES
?
enhanced binding
auto-immune origin
?

macrophages [113]. Conversely, a marked decrease in PS-expressing uremic erythrocytes occurs following incubation in normal plasma, with a concurrent decrease in their propensity to be recognized by phagocytes. Although the mechanism of this toxicity is unknown, the ability of uremic plasma to promote surface exposure of PS as well as phagocytosis is associated with a plasma fraction of a molecular weight range between 10 and 20 kDa.

Surface exposure of PS in uremic patients is not restricted to red cells. Circulating platelet-derived microvesicles with procoagulant activity, originating from PS-exposing platelets, are found in these patients [116]. Clinical experience indicates, however, that bleeding and thrombotic tendencies co-exist in uremia. While functional platelet defects are thought to contribute to the bleeding tendency, the number of platelet-derived microvesicles was found to be significantly elevated in uremic patients with thrombotic events than in those without.

Abnormal distribution of PS in renal epithelial cell membranes has been suggested to play a role in kidney stone disease [117,118]. Exposure of cultured renal epithelial cells to oxalate produces surface exposure of PS, which in turn promotes binding of calcium oxalate crystals to the cell surface. This process may foster crystal retention and stone formation within the kidney. While the mechanism of oxalate-induced PS exposure is unclear, it may involve a direct physical interaction of oxalate with membrane lipids, rather than interfering with lipid transporters. This possibility is supported by the observation that calcium oxalate causes a redistribution of PS in artificial phospholipid vesicles that lack the biochemical machinery to maintain phospholipid asymmetry [118].

8.6. Hyperglycemia

Erythrocytes and platelets from patients with diabetes mellitus can exhibit a substantial loss of membrane phospholipid asymmetry with increased surface exposure of PS [119–121]. In a study on 25 diabetic patients, 12–18% of PS in the patients' erythrocytes has been found to be accessible to phospholipase A2 hydrolysis and chemical labeling by the non-permeant agent trinitrobenzene sulfonic acid [119]. Alterations in membrane surface characteristics may contribute to an increase in spontaneous aggregation of diabetic erythrocytes and platelets, as well as promoting microvascular occlusion by abnormal adherence of blood cells to vascular endothelium. Indeed, adhesion of red cells to human umbilical vein endothelial cells under flow conditions has been shown to be inhibited by PS liposomes and by annexin V, clearly indicating the PS dependence of these interactions [105]. In addition, diabetes is associated with several defects of coagulation and fibrinolysis, which together with PS-expressing blood cells predispose to a thrombogenic tendency. Abnormal surface exposure of PS is also found in obese mice that lack a functional receptor for leptin, the major regulator for fat storage in mammals [122]. These animals have an uncontrolled rise in blood sugar and display many of the characteristics of non-insulin-dependent diabetes, including an altered life span of erythrocytes.

Many of the alterations observed in diabetic red cells can be brought about by in vitro incubations of normal red cells in hyperglycemic buffers [123,124]. For example, incubating red cells for 18 h with 20 mM glucose induces an almost complete scrambling of all major phospholipid classes [123]. The observed loss of lipid asymmetry is not due to inhibition of amino-phospholipid translocase or to glucose-induced Ca^{2+}-influx. Incubation in hyperglycemic media causes depletion of vitamin E and accumulation of vitamin E–quinone and malondialdehyde, an end product of lipid peroxidation [124]. Pretreatment of cells with reducing agents like *N*-acetylcysteine prevents glucose-mediated lipid peroxidation and PS externalization. Presumably, hyperglycemia-induced loss of lipid asymmetry reflects an increased passive phospholipid flip-flop caused by lipid peroxidation or non-enzymatic glycosylation of membrane proteins, to an extent that it can no longer be corrected by amino-phospholipid translocase.

8.7. Infection

While many viruses induce apoptotic cell death accompanied by egress of PS to the outer surface of infected cells, the mechanisms involved are still obscure [125]. As first demonstrated for cytomegalovirus-infected endothelial cells [126], many virus-infected cells exhibit a procoagulant phenotype [127]. The best-documented example concerns influenza-infected HeLa cells [128]. Exposure of PS at the cell surface occurs several hours post-infection at about the time that efficient phagocytosis by peritoneal macrophages becomes detectable. Phagocytosis is largely inhibited by PS-containing liposomes, suggesting a role for a PS receptor in the uptake of virus-infected cells in addition to uptake via the asialo-receptor [129].

Several obligate and facultative intracellular bacteria are implicated in promoting procoagulant activity in vascular cells. Moreover, infection may be associated with atherosclerosis and has been considered a risk factor for myocardial infarction [127]. A bacterial agent that has attracted wide attention is *Chlamydia pneumoniae*, frequently found in atherosclerotic lesions [127,130]. Infection by this agent causes rapid (5 min post-infection) and Ca^{2+}-dependent externalization of PS in a wide variety of host cells [131]. PS exposure depends on the continuous presence of *Chlamydia* since the removal of inoculum leads to disappearance of PS from the surface. Also, *Chlamydia*-infected cells accelerate plasma clotting and are susceptible to PS-dependent uptake by phagocytes.

Protozoan parasites like malaria, which have elaborate life cycles within human erythrocytes and liver cells, have been suspected to disturb membrane phospholipid asymmetry. Flow-cytometric studies with human erythrocytes infected with *Plasmodium falciparum* – the most severe of the malaria-causing parasites – have shown that these cells bind annexin V provided that the extent of parasitemia is in excess of 25% with most of the parasites being multinucleate forms in order to reach statistical significance [132]. Previous reports disagree as to whether or not malaria parasites promote a collapse of lipid asymmetry [133–135], but this may depend on the different parasitic forms and extent of infection used in these studies. Considering the much lower extent of parasitemia in malaria patients compared to that used in the laboratory studies, it remains doubtful if PS exposure contributes to any of the clinical manifestations of *Plasmodium falciparum* infection. It should be mentioned, however, that the majority of falciparum malaria patients are positive for anti-phospholipid antibodies [136], which may reflect a response against PS-expressing cells (see above) and may underlie the frequently observed thrombocytopenia.

8.8. *Pre-eclampsia*

Pre-eclampsia is a serious multisystem disorder characterized by hypertension, proteinuria and a hypercoagulable state during the second half of pregnancy [137]. Red cells from patients with pre-eclampsia have been shown to display a twofold increase, relative to erythrocytes from normotensive pregnant women, in their ability to promote assembly and catalysis of the prothrombinase complex when used as a source of phospholipid [138]. Whilst this may reflect a slightly increased surface exposure of PS, its significance should not be underrated considering the abundance of circulating red cells, relative to other peripheral blood cells. Moreover, in some but not all cases, patients with pre-eclampsia are positive for antibodies directed against lipid-bound β-2 glycoprotein I [139,140]. Both conditions have been proposed to contribute to thrombosis in the intervillous spaces on the maternal side of the placenta, impeding placental perfusion.

8.9. *Hyperbilirubinemia*

Hyperbilirubinemia is a frequently observed complication in neonates during the first week of life, resulting from increased bilirubin production and decreased elimination. Unconjugated bilirubin binds to erythrocytes, particularly when the molar ratio of bilirubin to albumin exceeds unity. This leads to toxic manifestations, such as crenation of red cells, hemolysis, anemia, and release of phospholipids and cholesterol from the erythrocyte membrane [141–143]. Release from the cell of PC, PE, and Sph starts at bilirubin-to-albumin molar ratios of approximately 0.5, whereas release of PS occurs when this ratio becomes greater than unity. Incubation of red cells at a bilirubin-to-albumin ratio of 3 results in about 8% of the red cells to become positive for annexin V, suggesting transbilayer movement of PS from the cells' inner to outer membrane leaflet [139]. In samples pretreated with *N*-ethylmaleimide, to inhibit inward movement of PS by amino-phospholipid translocase, nearly 20% of the bilirubin-treated cells express PS at the outer surface, indicating that bilirubin does not inhibit translocase activity per se. Bilirubin-induced PS egress is observed in the presence of Ca^{2+}, which raises the possibility that it results from moderate Ca^{2+}-influx, reflecting a situation resembling that of senescent red cells [101] where both lipid scramblase and amino-phospholipid translocase are active but oppose each other. Irrespective of the molecular mechanism, lipid scrambling and release of lipids from the red cell may facilitate hemolysis and promote erythrophagocytosis, contributing to anemia during severe neonatal jaundice where bilirubin-to-albumin ratios higher than 1 are not uncommon.

8.10. *Cystic fibrosis and bronchiectasis*

Cystic fibrosis is a common autosomal recessive trait, characterized by a mutation of an ATP-dependent transmembrane protein that functions as part of a cyclic AMP-regulated chloride channel. This leads to a generalized dysfunction of the exocrine glands with formation of viscid mucus, which progressively plugs their ducts. Obstruction of the bronchi promotes persistent pulmonary infections by a.o. *Staphylococcus aureus* and *Pseudomonas aeruginosa*, which are also characteristic for bronchiectasis in which the lung periphery fails to develop resulting in focal bronchial dilation accompanied by inflammatory destruction of bronchial walls. These chronic bacterial infections evoke a sustained influx of polymorphonuclear neutrophils (PMNs) into the airways, where they die and release intracellular proteases that overwhelm antiprotease defenses thus producing a protease/antiprotease imbalance leaving proteases unimpeded to injure airways and impair host defense [144,145].

Resolution of inflammation is normally accomplished by phagocytosis of dying apoptotic inflammatory cells before disruption of the plasma membrane and leakage of potentially harmful intracellular components. However, airway fluid from patients with cystic fibrosis and bronchiectasis contains an abundance of apoptotic and necrotic cells, much more than seen for example in patients with chronic bronchitis or with acute respiratory distress syndrome [144–146]. Contrary to the latter disorders, there is a protease/antiprotease imbalance in cystic fibrosis and bronchiectasis airway fluid, with PMN-elastase in excess of their major inhibitors. In spite of prominent surface exposure of PS on the apoptotic PMNs, cystic fibrosis and bronchiectasis airway fluid inhibits removal of these cells by alveolar macrophages in a PMN-elastase-dependent manner [146]. Moreover, PMN-elastase cleaves the PS receptor on phagocytes in vitro, implying a potential mechanism for defective apoptotic cell removal in vivo, thus promoting ongoing airway inflammation. Thus, while collapse of membrane phospholipid asymmetry occurs normally in apoptotic inflammatory neutrophils of these patients, PS exposure is no longer recognized as a signal for cell removal due to elastase-mediated clipping of the PS receptor on macrophages.

8.11. Neoplasia

Despite the biological heterogeneity of tumor cells, cancer is presently understood as an improper control of the cell cycle associated with a loss of the cells' ability to steer into apoptosis [147]. Notwithstanding, many tumor cells have been shown to exhibit elevated expression of PS in the outer membrane leaflet [148–152]. This is particularly the case for undifferentiated, tumorigenic cells, which may express about five times as much PS as their differentiated, non-tumorigenic counterparts [149]. Tumor cells also release PS-expressing microvesicles, similar to phospholipid scrambling and microvesicle release in other cells [28]. Apart from the production of tissue factor, increased expression of PS in tumor cells and in their shed microvesicles may promote thrombin formation, and could be responsible for fibrin deposits often seen in solid tumors [150,151].

In vitro, surface exposure of PS in tumorigenic cells directly correlates with their ability to be recognized and bound by macrophages [148,149]. This is rather enigmatic considering that tumor cells often have faulty apoptotic pathways to escape from programmed cell death and subsequent elimination by phagocytes. Indeed, chemotherapy often activates the apoptotic program to dictate tumorigenic cells to commit suicide [153,154].

It has long been known, however, that tumor-associated mononuclear phagocytes accumulate in neoplastic tissues near the tumor–host tissue interface, and this may amount to half of the tumor's mass [155,156]. The question as to whether this inflammatory infiltrate helps or hinders tumor growth is still open to debate. Increased consensus exists, however, that phagocytes form part and parcel of the inflammatory responses that promote tumor growth and progression rather than mounting an effective antitumor response directed at elimination of neoplastic cells [157]. Conceivably, the characteristic surface exposure of PS by oncogenic cells facilitates recruitment of inflammatory phagocytes and cytokines to the benefit of tumor growth and progression.

8.12. Clinical aspects of microvesicles

As mentioned earlier, collapse of membrane phospholipid asymmetry is usually accompanied by shedding of lipid-symmetric microvesicles from the cell surface. Circulating PS-exposing microvesicles, mostly derived from activated platelets but sometimes also from other blood cells including endothelial cells, are elevated in a variety of clinical disorders [158–160]. The clinical relevance of circulating microvesicles may be best illustrated in immune thrombocytopenic purpura (ITP). In this bleeding disorder, interaction of autoimmune antiplatelet antibodies with platelets provokes microvesicle formation, the extent to which varies among patients. Interestingly, ITP patients with high levels of platelet-derived microvesicles do not bleed despite severe thrombocytopenia while others with higher platelet counts but lower levels of microvesicles bleed extensively [161]. Those with the highest levels of microvesicles often suffer from small vessel transient ischemic attacks.

Elevated platelet-derived microvesicles have been observed in many disorders associated with platelet activation. Apart from above-mentioned disorders like thrombocytopenia, diabetes, uremia, cancer, or antiphospholipid syndrome, it occurs in acute coronary syndromes, small vessel strokes, and during cardiopulmonary bypass surgery [159,160]. However, circulating microvesicles are not always associated with thrombotic tendencies. For example, in Stormorken syndrome (also referred to as inverse Scott syndrome) PS-expressing platelets are found without deliberate stimulation, and this condition is characterized by considerable spontaneous microvesiculation in the patients' blood [162]. This would be expected to result in a thrombotic disposition, but there is in fact a bleeding tendency. Although the reason for this is unclear, it should be recalled that PS is equally important in promoting the anticoagulant protein C pathway that inhibits thrombin formation through inactivation of cofactors Va and VIIIa [61–63]. The balance between the pro- and anticoagulant effect of the lipid surface may depend on the presence of other lipids such as PE. The presence of PE in PS-containing vesicles enhances their capacity to stimulate both protein C- and prothombinase activity, but this effect is much larger for protein C than for prothrombinase [56].

On the other hand, a clear correlation between bleeding tendency and inability to form platelet-derived microvesicles, despite normal PS expression on the platelet surface, has been observed in another inherited bleeding disorder known as Castaman's defect [163]. The question remains, however, to what extent microparticles are causative agents in pathology or merely an epiphenomenon. It should also be mentioned that microvesicles are not only composed of phospholipids, but also contain many membrane glycoproteins from the cell from which they are derived. Therefore, they may play additional roles in hemostasis, like promoting fibrinolysis, platelet–endothelium interactions or leukocyte adhesion [159]. However, irrespective of the role of platelet-derived microvesicles in hemostatic disorders, they may be regarded as a clinical marker of platelet activation.

9. Concluding remarks

Membrane phospholipid asymmetry was first appreciated in the early 1970s [164–166], and many studies since have led to the concept that it is a ubiquitous phenomenon of most if not all mammalian cells. Because cells invest energy to catalyze transbilayer lipid movement in order to generate and maintain a specific transmembrane phospholipid distribution, it is considered to be of major physiological importance. It is evident that lipid transporter-controlled emergence of PS at the cell's outer membrane leaflet results in the expression of altered surface properties that have their impact on the cell's interaction with its environment. PS clearly plays a pivotal role in promoting blood coagulation, and its overexpression may generate potentially dangerous thrombogenic surfaces. It is therefore crucial that distinct mechanisms exist for the recognition and ingestion of PS-expressing cells, which are equally important for the orderly removal of apoptotic cells. It is also increasingly appreciated that PS receptor-mediated uptake seems to play a key role in cell development by timely elimination of redundant intermediate parts. Defects in the cooperative mechanisms that regulate membrane lipid asymmetry can lead to surface expression of PS, which causes or compromises a wide variety of disorders. Understanding the mechanisms that generate and regulate lipid sidedness and those that promote its collapse may be crucial to assess the role of lipid transporters in disease.

References

[1] Singer, S.J., Nicolson, G.L., 1972. The fluid mosaic model of the structure of cell membranes. Science 175, 720–731.

[2] Schroit, A.J., Zwaal, R.F.A., 1991. Transbilayer movement of phospholipids in red cell and platelet membranes. Biochim. Biophys. Acta 1071, 313–329.

[3] Brown, D.A., London, E., 1998. Functions of lipid rafts in biological membranes. Annu. Rev. Cell Dev. Biol. 14, 111–136.

[4] Sprong, H., van der Sluijs, P., Meer, G.v., 2001. How proteins move lipids and lipids move proteins. Nat. Rev. 2, 504–513.

[5] Seigneuret, M., Devaux, P.F., 1984. Asymmetric distribution of spin-labeled phospholipids in the erythrocyte membrane: relation to shape changes. Proc. Natl Acad. Sci. USA 81, 3751–3755.

[6] Daleke, D.L., Huestis, W.H., 1985. Incorporation and translocation of aminophospholipids in human erythrocytes. Biochemistry 24, 5406–5416.

[7] Devaux, P.F., 1992. Protein involvement in transmembrane lipid asymmetry. Annu. Rev. Biophys. Biomol. Struct. 21, 417–439.

[8] Devaux, P.F., Zachowski, A., 1994. Maintenance and consequences of membrane phospholipid asymmetry. Chem. Phys. Lipids 73, 107–120.

[9] Martin, O.C., Pagano, R.E., 1987. Transbilayer movement of fluorescent analogs of phosphatidylserine and phosphatidylethanolamine at the plasma membrane of cultures cells. J. Biol. Chem. 262, 5890–5898.

[10] Daleke, D.L., Lyles, J.V., 2000. Identification and purification of aminophospholipid flippases. Biochim. Biophys. Acta 1486, 108–127.

[11] Beleznay, Z., Zachowski, A., Devaux, P.F., Navazo, M.P., Ott, P., 1993. ATP-dependent aminophospholipid translocation in erythrocyte vesicles – stoichiometry of transport. Biochemistry 32, 3146–3152.

[12] Connor, J., Schroit, A.J., 1988. Transbilayer movement of phosphatidylserine in erythrocytes. Inhibition of transport and preferential labeling of a 31 kDa protein by sulfhydryl reactive reagents. Biochemistry 27, 848–851.

[13] Connor, J., Schroit, A.J., 1990. Aminophospholipid translocation in erythrocytes: evidence for the involvement of a specific transporter and an endofacial protein. Biochemistry 29, 37–43.

[14] Connor, J., Pak, C.H., Zwaal, R.F.A., Schroit, A.J., 1992. Bidirectional transbilayer movement of phospholipid analogs in human red blood cells. J. Biol. Chem. 267, 19412–19417.

[15] Bevers, E.M., Comfurius, P., Dekkers, D.W.C., Zwaal, R.F.A., 1999. Lipid translocation across the plasma membrane of mammalian cells. Biochim. Biophys. Acta 1439, 317–330.

[16] Auland, M.E., Roufogalis, B.D., Devaux, P.F., Zachowski, A., 1994. Reconstitution of ATP-dependent aminophospholipid translocation in proteoliposomes. Proc. Natl Acad. Sci. USA 91, 10938–10942.

[17] Tang, X., Halleck, M.S., Schlegel, R.A., Williamson, P., 1996. A novel subfamily of P-type ATPases with aminophospholipid transporting activity. Science 272, 1495–1497.

[18] Gomès, E., Jakobsen, M.K., Axelson, K.B., Geisler, M., Palmgren, M.G., 2000. Chilling tolerance in Arabidopsis involves ALA1, a member of a new family of putative aminophospholipid translocases. Plant Cell 12, 2441–2453.

[19] Siegmund, A., Grant, A., Angeletti, C., Malone, L., Nichols, J.W., Rudolph, H.K., 1998. Loss of DRS2p does not abolish transfer of fluorescence-labeled phospholipids across the plasma membrane of *Saccharomyces cerevisiae*, L. Biol. Chem. 273, 34399–34405.

[20] Marx, U., Polakowski, T., Pomorski, T., Lang, C., Nelson, H., Nelson, N., Herrmann, A., 1999. Rapid transbilayer movement of fluorescent phospholipid analogues in the plasma membrane of endocytosis deficient yeast cells does not require the drs2 protein. Eur. J. Biochem. 263, 254–264.

[21] Connor, J., Schroit, A.J., 1991. Transbilayer movement of phosphatidylserine in erythrocytes: inhibitors of aminophospholipid transport block the association of photolabeled lipid to its transporter. Biochim. Biophys. Acta 1066, 37–42.

[22] Bitbol, M., Devaux, P.F., 1988. Measurement of outward translocation of phospholipids across human erythrocyte membrane. Proc. Natl Acad. Sci. USA 85, 6783–6787.

[23] Dekkers, D.W.C., Comfurius, P., Schroit, A.J., Bevers, E.M., Zwaal, R.F.A., 1998. Transbilayer movement of NBD-labeled phospholipids in red blood cell membranes; outward-directed transport by the multidrug resistance protein 1 (MRP1). Biochemistry 37, 14833–14837.

[24] Kamp, D., Haest, C.W.M., 1998. Evidence for a role of multidrug resistance protein (MRP) in outward translocation of NBD-phospholipids in the erythrocyte membrane. Biochim. Biophys. Acta 1372, 91–101.

[25] Borst, P., Zelcer, N., van Helvoort, A., 2000. ABC transporters in lipid transport. Biochim. Biophys. Acta 1486, 128–144.

[26] Dean, M., Hannon, Y., Chimini, G., 2001. The human ATP-binding cassette (ABC) transporter superfamily. J. Lipid Res. 42, 1007–1017.

[27] Dekkers, D.W.C., Comfurius, P., van Gool, R.G.J., Bevers, E.M., Zwaal, R.F.A., 2000. Multidrug resistance protein 1 (MRP1) regulates lipid asymmetry in red blood cell membranes. Biochem. J. 350, 531–535.

[28] Zwaal, R.F.A., Schroit, A.J., 1987. Pathophysiological implications of membrane phospholipid asymmetry in blood cells. Blood 89, 1121–1132.

[29] Smit, J.J.M., Schinkel, A.H., Oude Elferink, R.P.J., Groen, A.K., Wagenaar, E., Van Deemter, L., Mol, C.A.A.M., Ottenhoff, R., van der Lugt, N.M., van Roon, M.A., van der Vak, M.A., Offerhaus, G.J.A., Berns, A.J.M., Borst, P., 1993. Homozygous disruption of the murine mdr2 P-glycoprotein gene leads to a complete absence of phospholipids from bile and to liver disease. Cell 75, 451–459.

[30] van Helvoort, A., Smith, A.J., Sprong, H., Fritzsche, I., Schinkel, A.H., Borst, P., Van Meer, G., 1996. MDR1 P-glycoprotein is a lipid translocase of broad specificity, while MDR3 P-glycoprotein specifically translocates phosphatidylcholine. Cell 87, 507–517.
[31] Bezombes, C., Maestre, N., Laurent, G., Lavade, T., Bettabaïeb, A., Jaffrézou, J.P., 1998. Restoration of TNF-alpha-induced ceramide generation and apoptosis in resistant human leukemia KG1a cells by the P-glycoprotein blocker PSC833. FASEB J. 12, 101–109.
[32] Bodzioch, M., Orso, E., Klucken, J., Langmann, T., Bottcher, T., Diederich, W., Drobnik, W., Barlage, S., Buchler, C., Porsch-Ozcurumez, M., Kaminski, W.E., Hahmann, H.W., Oette, K., Rothe, G., Aslanidis, C., Lackner, K.J., Schmitz, G., 1999. The gene encoding ATP-binding cassette transporter 1 is mutated in Tangier disease. Nat. Genet. 22, 347–351.
[33] Hamon, Y., Broccardo, C., Chambenoit, O., Luciani, M.F., Toti, F., Chaslin, S., Freyssinet, J.M., Devaux, P.F., McNesh, J., Marguet, D., Chimini, G., 2000. ABC1 promotes engulfment of apoptotic cells and transbilayer redistribution of phosphatidylserine. Nat. Cell Biol. 2, 399–406.
[34] Bevers, E.M., Comfurius, P., van Rijn, J.L.M.L., Hemker, H.C., Zwaal, R.F.A., 1982. Generation of prothrombin-converting activity and the exposure of phosphatidylserine at the outer surface of platelets. Eur. J. Biochem. 122, 429–436.
[35] Bevers, E.M., Comfurius, P., Zwaal, R.F.A., 1983. Changes in membrane phospholipid distribution during platelet activation. Biochim. Biophys. Acta 736, 57–66.
[36] Bevers, E.M., Tilly, R.H.J., Senden, J.M.G., Comfurius, P., Zwaal, R.F.A., 1989. Exposure of endogenous phosphatidylserine at the outer surface of stimulated platelets is reversed by restoration of aminophospholipid translocase activity. Biochemistry 28, 2382–2387.
[37] Comfurius, P., Senden, J.M.G., Tilly, R.H.J., Schroit, A.J., Bevers, E.M., Zwaal, R.F.A., 1990. Loss of membrane phospholipid asymmetry in platelets and red cells may be associated with calcium-induced shedding of plasma membrane and inhibition of amino-phospholipid translocase. Biochim. Biophys. Acta 1026, 153–160.
[38] Zwaal, R.F.A., Comfurius, P., Bevers, E.M., 1993. Mechanism and function of changes in membrane phospholipid asymmetry in platelets and erythrocytes. Biochem. Soc. Trans. 21, 248–251.
[39] Rosing, J., Bevers, E.M., Comfurius, P., Hemker, H.C., van Dieijen, G., Weiss, H.J., Zwaal, R.F.A., 1985. Impaired factor X and prothrombin activation associated with decreased phospholipid exposure in platelets from a patient with a bleeding disorder. Blood 65, 1557–1561.
[40] Weiss, H.J., 1994. Scott syndrome: a disorder of platelet coagulant activity. Semin. Hematol. 31, 312–319.
[41] Bevers, E.M., Wiedmer, T., Comfurius, P., Shatill, S.J., Weiss, H.J., Zwaal, R.F.A., Sims, P.J., 1992. Defective Ca^{2+}-induced microvesiculation and deficient expression of procoagulant activity in erythrocytes from a patient with a bleeding disorder: a study of the red blood cells of Scott syndrome. Blood 79, 380–388.
[42] Toti, F., Satta, N., Fressinaud, E., Meyer, D., Freyssinet, J.M., 1996. Scott syndrome. Characterized by impaired transmembrane migration of procoagulant phosphatidylserine and hemorrhagic complications, is an inherited disorder. Blood 87, 1409–1415.
[43] Williamson, P., Bevers, E.M., Smeets, E.F., Comfurius, P., Schlegel, R.A., Zwaal, R.F.A., 1995. Continuous analysis of the mechanism of activated transbilayer lipid movement in platelets. Biochemistry 34, 10448–10455.
[44] Smeets, E.F., Comfurius, P., Bevers, E.M., Zwaal, R.F.A., 1994. Calcium-induced transbilayer scrambling of fluorescent phospholipid analogs in platelets and erythrocytes. Biochim. Biophys. Acta 1195, 281–286.
[45] Dekkers, D.W.C., Comfurius, P., Bevers, E.W.M., Zwaal, R.F.A., 2002. Comparison between Ca^{2+}-induced scrambling of various fluorescently labelled lipid analogues in red blood cells. Biochem. J. 362, 741–747.
[46] Martin, D.W., Jesty, J., 1995. Calcium stimulation of procoagulant activity in human erythrocytes – ATP dependence and the effects of modifiers of stimulation and recovery. J. Biol. Chem. 270, 10468–10474.
[47] Comfurius, P., Williamson, P., Smeets, E.F., Schlegel, R.A., Bevers, E.M., Zwaal, R.F., 1996. Reconstitution of phospholipid scramblase activity from human blood platelets. Biochemistry 35, 7631–7634.

[48] Basse, F., Stout, J.G., Sims, P.J., Wiedmer, T., 1996. Isolation of an erythrocyte membrane protein that mediates Ca^{2+}-dependent transbilayer movement of phospholipid. J. Biol. Chem. 271, 17205–17210.
[49] Sims, P.J., Wiedmer, T., 2001. Unravelling the mysteries of phospholipid scrambling. Thromb. Haemost. 86, 266–275.
[50] Zhou, Q., Sims, P.J., Wiedmer, T., 1998. Expression of proteins controlling transbilayer movement of plasma membrane phospholipids in the B lymphocytes from a patient with Scott syndrome. Blood 92, 1707–1712.
[51] Zhou, Q., Zhao, J., Wiedmer, T., Sims, P.J., 2002. Normal hemostasis but defective hematopoietic response to growth factors in mice deficient in phospholipid scramblase 1. Blood 99, 4030–4038.
[52] Sims, P.J., Wiedmer, T., Esmon, C.T., Weiss, H.J., Shattil, S.J., 1989. Assembly of the platelet prothrombinase complex is linked to vesiculation of the platelet plasma membrane. Studies in Scott syndrome: an isolated defect in platelet procoagulant activity. J. Biol. Chem. 264, 17049–17057.
[53] Zwaal, R.F.A., Comfurius, P., Bevers, E.M., 1992. Platelet procoagulant activity and microvesicle formation. Its putative role in hemostasis and thrombosis. Biochim. Biophys. Acta 1180, 1–8.
[54] Dachary-Prigent, J., Pasquet, J.M., Freyssinet, J.M., Nurden, A.T., 1995. Calcium involvement in aminophospholipid exposure and microparticle formation during platelet activation: a study using Ca^{2+}-ATPase inhibitors. Biochemistry 34, 11625–11634.
[55] Diaz, C., Schroit, A.J., 1996. Role of translocases in the generation of phosphatidylserine asymmetry. J. Membr. Biol. 151, 1–9.
[56] Zwaal, R.F.A., Comfurius, P., Bevers, E.M., 1998. Lipid–protein interactions in blood coagulation. Biochim. Biophys. Acta 1376, 433–453.
[57] Mann, K.G., 1999. Biochemistry and physiology of blood coagulation. Thromb. Haemost. 82, 165–174.
[58] Nemerson, Y., 1992. The tissue factor pathway of blood coagulation. Semin. Hematol. 29, 170–176.
[59] Ruf, W., Mueller, B.M., 1999. Tissue factor signalling. Thromb. Haemost. 82, 175–182.
[60] Bach, R., Gentry, R., Nemerson, Y., 1986. Factor VII binding to tissue factor in reconstituted phospholipid vesicles: induction of cooperativity by phosphatidylserine. Biochemistry 25, 4007–4020.
[61] Dahlbäck, B., 1995. The protein C anticoagulant system: inherited defects as basis for venous thrombosis. Thromb. Res. 77, 1–43.
[62] Esmon, C.T., Ding, W., Yasuhiro, K., Gu, J.M., Ferrell, G., Regan, L.M., Stearns-Kurosawa, D.J., Kurosawa, S., Mather, T., Laszik, Z., Esmon, N., 1997. The protein C pathway: new insights. Thromb. Haemost. 78, 70–74.
[63] Tans, G., Rosing, J., Thomassen, M.C.L.G.D., Heeb, M.J., Zwaal, R.F.A., Griffin, J.H., 1991. Comparison of anticoagulant and procoagulant activities of stimulated platelets and platelet-derived microvesicles. Blood 77, 2641–2648.
[64] Raz, A., Bugana, C., Fogler, W.E., Poste, G., Fidler, I.J., 1981. Biochemical, morphological, and ultrastructural studies on the uptake of liposomes by murine macrophages. Cancer Res. 41, 487–494.
[65] Schroit, A.J., Madsen, J., Tanaka, Y., 1985. In vivo recognition and clearance of red blood cells containing phosphatidylserine in their plasma membrane. J. Biol. Chem. 260, 5131–5138.
[66] Fadok, V.A., Voelker, D.R., Campbell, P.A., Cohen, J.J., Bratton, D.L., Henson, P.M., 1992. Exposure of phosphatidylserine on the surface of apoptotic lymphocytes triggers specific recognition and removal by macrophages. J. Immunol. 148, 2207–2216.
[67] Martin, S.J., Reutelingsperger, C.P.M., McGahon, A.J., Rader, J.A., VanSchie, R.C.A.A., Laface, D.M., Green, D.R., 1995. Early redistribution of plasma membrane phosphatidylserine is a general feature of apoptosis regardless of the initiating stimulus: inhibition by overexpression of bcl-2 and abl. J. Exp. Med. 182, 1545–1556.
[68] Henson, P.M., Bratton, D.L., Fadok, V.A., 2001. Apoptotic cell removal. Curr. Biol. 11, R795–R805.
[69] Fadok, V.A., Xue, D., Henson, P.M., 2001. If phosphatidylserine is the death knell, a new phosphatidylserine-specific receptor is the bellringer. Cell Death Differ. 8, 582–587.
[70] Schlegel, R.A., Williamson, P., 2001. Phosphatidylserine, a death knell. Cell Death Differ. 8, 551–563.
[71] Fadok, V.A., de Cathelineau, A., Daleke, D.L., Henson, P.M., Bratton, D.L., 2001. Loss of phospholipid asymmetry and surface exposure of phosphatidylserine is required for phagocytosis of apoptotic cells by macrophages and fibroblasts. J. Biol. Chem. 276, 1071–1077.

[72] Fadok, V.A., Bratton, D.L., Rose, D.M., Pearson, A., Ezekowitz, R.A., Henson, P.M., 2000. A receptor for phosphatidylserine-specific clearance of apoptotic cells. Nature 405, 85–90.
[73] Balasubramanian, K., Schroit, A.J., 1998. Characterization of phosphatidylserine-dependent beta(2)-glycoprotein I macrophage interactions – implications for apoptotic cell clearance by phagocytes. J. Biol. Chem. 273, 29272–29277.
[74] Galli, M., Comfurius, P., Maassen, C., Hemker, H.C., de Baets, M.H., van Breda-Vriesman, P.J., Barbui, T., Zwaal, R.F.A., Bevers, E.M., 1990. Anticardiolipin antibodies (ACA) are directed not to cardiolipin but to a plasma protein cofactor. Lancet 335, 1544–1547.
[75] McNeil, H.P., Simpson, R.J., Chesterman, C.N., Krilis, S.A., 1990. Antiphospholipid antibodies are directed against a complex antigen that includes a lipid-binding inhibitor of coagulation: beta 2-glycoprotein I (apolipoprotein H). Proc. Natl Acad. Sci. USA 87, 4120–4124.
[76] Bevers, E.M., Galli, M., Barbui, T., Comfurius, P., Zwaal, R.F.A., 1991. Lupus anticoagulant IgG's (LA) are not directed to phospholipids only, but to a complex of lipid-bound human prothrombin. Thromb. Haemost. 66, 629–632.
[77] Manfredi, A.A., Rovere, P., Heltai, S., Galati, G., Nebbia, G., Tincani, A., Balestrieri, G., Sabbadini, M.G., 1998. Apoptotic cell clearance in systemic lupus erythematosis. II. Role of beta(2)-glycoprotein I. Arthritis Rheum. 41, 215–223.
[78] Hampton, M.B., Vanags, D.M., Pomares, M.I., Orrenius, S., 1996. Involvement of extracellular calcium in phosphatidylserine exposure during apoptosis. FEBS Lett. 399, 277–282.
[79] Williamson, P., Christie, A., Kohlin, T., Schlegel, R.A., Comfurius, P., Harmsma, M., Zwaal, R.F.A., Bevers, E.M., 2001. Phospholipid scramblase activation pathways in lymphocytes. Biochemistry 40, 8065–8072.
[80] Henson, P.M., Bratton, D.L., Fadok, V.A., 2001. The phosphatidylserine receptor: a crucial molecular switch? Nat. Rev. 2, 627–633.
[81] Flesch, F.M., Brouwers, J.F.H.M., Nievelstein, P.F.E.M., Verkleij, A.J., van Golde, L.M.G., Colenbrander, B., Gadella, B.M., 2001. Bicarbonate stimulated phospholipid scrambling induces cholesterol redistribution and enables cholesterol depletion in the sperm plasma membrane. J. Cell Sci. 114, 3543–3555.
[82] Gadella, B.M., Rathi, R., Brouwers, J.F.H.M., Stout, T.A., Colenbrander, B., 2001. Capacitation and the acrosome reaction in equine sperm. Anim. Reprod. Sci. 68, 249–265.
[83] Gadella, B.M., Harrison, R.A., 2002. Capacitation induces cyclic adenosine 3′,5′-monophosphate-dependent, but apoptosis unrelated, exposure of aminophospholipids at the apical head plasma membrane of boar sperm cells. Biol. Reprod. 67, 340–350.
[84] van den Eijnde, S.M., Luijsterburg, A.J.M., Boshart, L., De Zeeuw, C.I., Reutelingsperger, C.P.M., Vermeij-Keers, C., 1997. In situ detection of apoptosis during embryogenesis with Annexin V: from whole mount to ultrastructure. Cytometry 29, 313–320.
[85] van den Eijnde, S.M., van den Hoff, M.J.B., Reutelingsperger, C.P.M., van Heerde, W.L., Henfling, M.E.R., Vermeij-Keers, C., Schutte, B., Borgers, M., Ramaekers, F.C.S., 2001. Transient expression of phosphatidylserine at cell–cell contact areas is required for myotube formation. J. Cell Sci. 114, 3631–3642.
[86] Skutelsky, E., Danon, D., 1967. An electron microscopic study of nuclear elimination from the late erythroblast. J. Cell Biol. 33, 625–635.
[87] Schlegel, R.A., Phelps, B.M., Waggoner, A., Terada, L., Williamson, P., 1980. Binding of merocyanine 540 to normal and leukemic erythroid cells. Cell 20, 321–328.
[88] Geiduschek, J.B., Singer, S.J., 1979. Molecular changes in the membranes of mouse erythroid cells accompanying differentiation. Cell 16, 149–163.
[89] Brooks, M.B., Catalfamo, J.L., Brown, H.A., Ivanova, P., Lovaglio, J., 2002. A hereditary bleeding disorder of dogs caused by a lack of platelet procoagulant activity. Blood 99, 2434–2441.
[90] Hughes, G.R.V., 1993. The antiphospholipid syndrome: ten years on. Lancet 342, 341–344.
[91] Galli, M., 1996. Non beta 2-glycoprotein I cofactors for antiphospholipid antibodies. Lupus 5, 388–392.
[92] Willems, G.M., Janssen, M.P., Pelsers, M.M.A.L., Comfurius, P., Galli, M., Zwaal, R.F.A., Bevers, E.M., 1996. Role of divalency in the high-affinity binding of anticardiolipin antibody-β2-glycoprotein I complexes to lipid membranes. Biochemistry 35, 13833–13842.

[93] Galli, M., Comfurius, P., Barbui, T., Zwaal, R.F.A., Bevers, E.M., 1992. Anticoagulant activity of β2-glycoprotein I is potentiated by a distinct subgroup of anticardiolipin antibodies. Thromb. Haemost. 68, 297–300.

[94] Galli, M., Bevers, E.M., Comfurius, P., Barbui, T., Zwaal, R.F.A., 1993. Effect of antiphospholipid antibodies on procoagulant activity of activated platelets and platelet-derived microvesicles. Br. J. Haematol. 83, 466–472.

[95] Casciola Rosen, L., Rosen, A., Petri, M., Schlissel, M., 1996. Surface blebs on apoptotic cells are sites of enhanced procoagulant activity: implications for coagulation events and antigenic spread in systemic lupus erythematosis. Proc. Natl Acad. Sci. USA 93, 1624–1629.

[96] Galli, M., Ruggeri, L., Barbui, T., 1998. Differential effects of anti-beta2-glycoprotein I and antiprothrombin antibodies on the anticoagulant activity of activated protein C. Blood 91, 1999–2004.

[97] Kuypers, F.A., Lewis, R.A., Hua, M., Schott, M.A., Discher, D., Ernst, J.D., Lubin, B.H., 1996. Detection of altered membrane phospholipid asymmetry in subpopulations of human red blood cells using fluorescently labeled annexin V. Blood 87, 1179–1187.

[98] Wood, B.L., Gibson, D.F., Tait, J.F., 1996. Increased erythrocyte phosphatidylserine exposure in sickle cell disease: flow-cytometric measurement and clinical associations. Blood 88, 1873–1880.

[99] de Jong, K., Larkin, S.K., Styles, L.A., Bookchin, R.M., Kuypers, F.A., 2001. Characterization of the phosphatidyl-exposing subpopulation of sickle cells. Blood 98, 860–867.

[100] Middelkoop, E., Lubin, B.H., Bevers, E.M., Op den Kamp, J.A.F., Comfurius, P., Chiu, D.T., Zwaal, R.F.A., van Deenen, L.L.M., Roelofsen, B., 1988. Studies on sickled erythrocytes provide evidence that the asymmetric distribution of phosphatidylserine in the red cell membrane is maintained by both ATP-dependent translocation and interaction with membrane skeletal proteins. Biochim. Biophys. Acta 937, 281–288.

[101] Connor, J., Pak, C.C., Schroit, A.J., 1994. Exposure of phosphatidylserine in the outer leaflet of human red blood cells: relationship to cell density, cell age, and clearance by mononuclear cells. J. Biol. Chem. 269, 2399–2404.

[102] Lew, V.L., Ortiz, O.E., Bookchin, R.M., 1997. Stochastic nature and red cell population distribution of the sickling-induced Ca^{2+} permeability. J. Clin. Invest. 99, 2727–2735.

[103] Setty, B.N.Y., Rao, A.K., Stuart, M.J., 2001. Thrombophilia in sickle cell disease: the red cell connection. Blood 98, 3228–3233.

[104] Westerman, M.P., Green, D., Gilman-Sachs, A., Beaman, K., Freels, S., Boggio, L., Allen, S., Zuckerman, L., Schlegel, R.A., Williamson, P., 1999. Antiphospholipid antibodies, proteins C and S, and coagulation changes in sickle cell disease. J. Lab. Clin. Med. 134, 352–362.

[105] Closse, C., Dachary-Prigent, J., Boisseau, M.R., 1999. Phosphatidylserine-related adhesion of human erythrocytes to vascular endothelium. Br. J. Haematol. 107, 300–302.

[106] Manodori, A.B., Barabino, G.A., Lubin, B.H., Kuypers, F.A., 1999. Adherence of phosphatidyl-exposing erythrocytes to endothelial matrix thrombospondin. Blood 95, 1293–1300.

[107] Setty, B.N.Y., Kulkami, S., Stuart, M.J., 2002. Role of erythrocyte phosphatidylserine in sickle red cell-endothelial adhesion. Blood 99, 1564–1571.

[108] Borenstein-Ben Yashar, V., Barenholz, Y., Hy-Am, E., Rachmilewitz, E.A., Eldor, A., 1993. Phosphatidylserine in the outer leaflet of red blood cells from beta-thalassemia patients may explain the chronic hypercoagulable state and thrombotic episodes. Am. J. Hematol. 44, 63–65.

[109] Ruf, A., Pick, M., Deutsch, V., Patscheke, H., Goldfarb, A., Rachmilewitz, E.A., Guillin, M.C., Eldor, A., 1997. In-vivo platelet activation correlates with red cell anionic phospholipid exposure in patients with beta-thalassaemia major. Br. J. Haematol. 98, 51–56.

[110] Kuypers, F.A., Yuan, J., Lewis, R.A., Snyder, L.M., Kiefer, C.R., Bunyaratvej, A., Fucharoen, S., Ma, L., Styles, L., De Jong, K., Schrier, S.L., 1998. Membrane phospholipid asymmetry in human thalassemia. Blood 91, 3044–3051.

[111] Eldor, A., Rachmilewitz, E.A., 2002. The hypercoagulable state in thalassemia. Blood 99, 36–43.

[112] Bonomini, M., Sirolli, V., Settefrati, N., Dottori, S., di Liberato, L., Arduini, A., 1999. Increased erythrocyte phosphatidylserine exposure in chronic renal failure. J. Am. Soc. Nephrol. 10, 1982–1990.

[113] Bonomini, M., Sirolli, V., Reale, M., Arduini, A., 2001. Involvement of phosphatidylserine exposure in the recognition and phagocytosis of uremic erythrocytes. Am. J. Kidney Dis. 37, 807–814.

[114] Bonomini, M., Sirolli, V., Gizzi, F., Di Stante, S., Grilli, A., Felaco, M., 2002. Enhanced adherence of human uremic erythrocytes to vascular endothelium: role of phosphatidylserine exposure. Kidney Int. 62, 1358–1363.

[115] Kong, Q.Y., Wu, X., Li, J., Peng, W.X., Ye, R., Lindholm, B., Wang, T., 2001. Loss of phospholipid asymmetry in red blood cells contributes to anemia in uremic patients. Adv. Perit. Dial. 17, 58–60.

[116] Ando, M., Iwata, A., Ozeki, T., Tsuchiya, K., Akiba, T., Nihei, H., 2002. Circulating platelet-derived microparticles with procoagulant activity may be a potential cause of thrombosis in uremic patients. Kidney Int. 62, 1757–1763.

[117] Wiessner, J.H., Hasegawa, A.T., Hung, L.Y., Mandel, N.S., 1999. Oxalate-induced exposure of phosphatidylserine on the surface of renal epithelial cells in culture. J. Am. Soc. Nephrol. 10, S441–S445.

[118] Cao, L.C., Jonassen, J., Honeyman, T.W., Scheid, C., 2001. Oxalate-induced redistribution of phosphatidylserine in renal epithelial cells; implications for kidney stone disease. Am. J. Nephrol. 21, 69–77.

[119] Wali, R.K., Jaffe, S., Kumar, D., Kalra, V.K., 1988. Alterations in organization of phospholipids in erythrocytes as factor in adherence to endothelial cells in diabetes mellitus. Diabetes 37, 104–111.

[120] Lupu, F., Calb, M., Fixman, A., 1988. Alterations of phospholipid asymmetry in the membrane of spontaneously aggregated platelets in diabetes. Thromb. Res. 50, 605–616.

[121] Wahid, S.T., Marshall, S.M., Thomas, T.H., 2001. Increased platelet and erythrocyte external cell membrane phosphatidylserine in type 1 diabetes and microalbuminuria. Diabetes Care 24, 2001–2003.

[122] Manadori, A.B., Kuypers, F.A., 2002. Altered red cell turnover in diabetic mice. J. Lab. Clin. Med. 140, 161–165.

[123] Wilson, M.J., Richter-Lowny, K., Daleke, D.L., 1993. Hyperglycemia induces a loss of phospholipid asymmetry in human erythrocytes. Biochemistry 32, 11302–11310.

[124] Jain, S.K., Palmer, M., Chen, Y., 1999. Effect of vitamin E and *N*-acetylcysteine on phosphatidylserine externalization and induction of coagulation by high-glucose-treated human erythrocytes. Metabolism 48, 957–959.

[125] Granville, D.J., Carthy, C.M., Yang, D., Hunt, D.W.C., McManus, B.M., 1998. Interaction of viral proteins with host cell death machinery. Cell Death Differ. 5, 653–659.

[126] van Dam-Mieras, M.C., Bruggeman, C.A., Muller, A.D., Debie, W.H., Zwaal, R.F.A., 1987. Induction of endothelial cell procoagulant activity by cytomegalovirus infection. Thromb. Res. 47, 69–75.

[127] Vercellotti, G.M., 2001. Overview of infections and cardiovascular diseases. J. Allergy Clin. Immunol. 108, S117–120.

[128] Shiratsuchi, A., Kaido, M., Takizawa, T., Nakanishi, Y., 2000. Phosphatidylserine-mediated phagocytosis of influenza A virus-infected cells by mouse peritoneal macrophages. J. Virol. 74, 9240–9244.

[129] Watanabe, Y., Shiratsuchi, A., Shimizu, K., Takizawa, T., Nakanishi, Y., 2002. Role of phosphatidylserine exposure and sugar chain desialylation at the surface of influenza virus-infected cells in efficient phagocytosis by macrophages. J. Biol. Chem. 277, 18222–18228.

[130] Kuo, C.C., Shor, A., Campbell, L.A., Fukushi, H., Patton, D.L., Grayston, J.T., 1993. Demonstration of *Chlamydia pneumoniae* in atherosclerotic lesions of coronary arteries. J. Infect. Dis. 167, 841–849.

[131] Goth, S.R., Stephens, R.S., 2001. Rapid, transient phosphatidylserine externalization induced in host cells by infection with *Chlamydia* spp. Infect. Immun. 69, 1109–1119.

[132] Sherman, I.W., Prudhomme, J., Tait, J.F., 1997. Altered membrane phospholipid asymmetry in *Plasmodium falciparum*-infected erythrocytes. Parasitol. Today 13, 242–243.

[133] Schwartz, R.S., Olson, J.A., Raventos-Suarez, C., Yee, M., Heath, R.N., Lubin, B., Nagel, R.L., 1987. Altered plasma membrane phospholipid organization in *Plasmodium falciparum*-infected human erythrocytes. Blood 69, 401–407.

[134] Moll, G.N., Vial, H.J., Bevers, E.M., Ancelin, M.L., Roelofsen, B., Comfurius, P., Slotboom, A.J., Zwaal, R.F.A., Op den Kamp, J.A.F., van Deenen, L.L.M., 1990. Phospholipid asymmetry in the plasma membrane of malaria infected erythrocytes. Biochem. Cell. Biol. 68, 579–585.

[135] Maquire, P.A., Prudhomme, J., Sherman, I.W., 1991. Alterations in erythrocyte membrane phospholipid organization due to the intracellular growth of the human malaria parasite, *Plasmodium falciparum*. Parasitology 102, 179–186.

[136] Facer, C.A., Agiostratidou, G., 1994. High levels of anti-phospholipid antibodies in uncomplicated and severe *Plasmodium falciparum* and in *P. vivax* malaria. Clin. Exp. Immunol. 95, 304–309.
[137] Walker, I.D., 2002. Prothrombotic genotypes and pre-eclampsia. Thromb. Haemost. 87, 777–778.
[138] Grisaru, D., Zwang, E., Peyser, M.R., Lessing, J.B., Eldor, A., 1997. The procoagulant activity of red blood cells from patients with severe preeclampsia. Am. J. Obstet. Gynecol. 177, 1513–1516.
[139] Allen, J.Y., Tapia-Santiago, C., Kutteh, W.H., 1996. Antiphospholipid antibodies in patients with preeclampsia. Am. J. Reprod. Immunol. 36, 81–85.
[140] Fialova, L., Mikulikova, L., Matous-Malbohan, I., Benesova, O., Zwinger, A., 2000. Prevalence of various antiphospholipid antibodies in pregnant women. Physiol. Res. 49, 299–305.
[141] Brito, M.A., Silva, R.M., Matos, D.C., da Silva, A.T., 1996. Alterations of erythrocyte morphology and lipid composition by hyperbilirubinemia. Clin. Chim. Acta 249, 149–165.
[142] Brites, D., Silva, R., Brito, A., 1997. Effect of bilirubin on erythrocyte shape and hemolysis, under hypotonic, aggregating or non-aggregating conditions, and correlation with cell age. Scand. J. Clin. Lab. Invest. 57, 337–350.
[143] Brito, M.A., Silva, R.F.M., Brites, D., 2002. Bilirubin induces loss of membrane lipids and exposure of phosphatidylserine in human erythrocytes. Cell Biol. Toxicol. 18, 181–192.
[144] Birrer, P., McElvaney, N.G., Rudeberg, A., Sommer, C.W., Liechti-Gallati, S., Kraemer, R., Hubbard, R., Crystal, R.G., 1994. Protease–antiprotease imbalance in the lungs of children with cystic fibrosis. Am. J. Respir. Crit. Care Med. 150, 207–213.
[145] Tsang, K.W., Chan, K., Ho, P., Zheng, L., Ooi, G.C., Ho, J.C., Lam, W., 2000. Sputum elastase in steady-state bronchiectasis. Chest 117, 420–426.
[146] Vandivier, R.W., Fadok, V.A., Hoffmann, P.R., Bratton, D.L., Penvari, C., Brown, K.K., Brain, J.D., Accurso, F.J., Henson, P.M., 2002. Elastase-mediated phosphatidylserine receptor cleavage impairs apoptotic cell clearance in cystic fibrosis and bronchiectasis. J. Clin. Invest. 109, 661–670.
[147] Igney, F.H., Krammer, P.H., 2002. Death and anti-death: tumor resistance to apoptosis. Nat. Rev. Cancer 2, 277–288.
[148] Connor, J., Bucana, C.D., Fidler, I.J., Schroit, A.J., 1989. Differentiation-dependent expression of phosphatidylserine in mammalian plasma membranes: quantitative assessment of outer-leaflet lipid by prothrombinase complex formation. Proc. Natl Acad. Sci. USA 86, 3184–3188.
[149] Utsugi, T., Schroit, A.J., Connor, J., Bucana, C.D., Fidler, I.J., 1991. Elevated expression of phosphatidylserine in the outer membrane leaflet of human tumor cells and recognition by activated human blood monocytes. Cancer Res. 15, 3062–3066.
[150] Rao, L.V., Tait, J.F., Hoang, A.D., 1992. Binding of annexin V to a human ovarian carcinoma cell line (OC-2008). Contrasting effects on cell surface factor VIIa/tissue factor activity and prothrombinase activity. Thromb. Res. 67, 517–531.
[151] Sigimura, M., Donato, R., Kakkar, V.V., Scully, M.F., 1994. Annexin V as a probe of the contribution of anionic phospholipids to the procoagulant activity of tumour cell surfaces. Blood Coagul. Fibrinolysis 5, 365–373.
[152] Ran, S., Downes, A., Thorpe, P.E., 2002. Increased exposure of anionic phospholipids on the surface of tumor blood vessels. Cancer Res. 62, 6132–6140.
[153] Wang, J., Weiss, I., Svoboda, K., Kwaan, H.C., 2001. Thrombogenic role of cells undergoing apoptosis. Br. J. Haematol. 115, 382–391.
[154] Green, A.M., Steinmetz, N.D., 2002. Monitoring apoptosis in real time. Cancer J. 8, 82–92.
[155] Sica, A., Saccani, A., Mantovani, A., 2002. Tumor-associated macrophages: a molecular perspective. Int. Immunopharmacol. 2, 1045–1054.
[156] Brigati, C., Noonan, D.M., Albini, A., Benelli, R., 2002. Tumors and inflammatory infiltrates: friends or foes? Clin. Exp. Metastasis 19, 247–258.
[157] Balkwill, F., Mantovani, A., 2001. Inflammation and cancer: back to Virchow? Lancet 357, 539–545.
[158] Solum, N.O., 1999. Procoagulant expression in platelets and defects leading to clinical disorders. Arterioscler. Thromb. Vasc. Biol. 19, 2841–2846.
[159] Horstman, L.L., Ahn, Y.S., 1999. Platelet microparticles: a wide-angle perspective. Crit. Rev. Oncol. Hematol. 30, 111–142.
[160] Nomura, S., 2001. Function and clinical significance of platelet-derived microparticles. Int. J. Hematol. 74, 397–404.

[161] Jy, W., Horstman, L.L., Arce, M., Ahn, Y.S., 1992. Clinical significance of platelet microparticles in autoimmune thrombocytopenias. J. Lab. Clin. Med. 119, 334–345.
[162] Stormorken, H., Holmsen, H., Sund, R., Sakariassen, K.S., Hovig, T., Jellum, E., 1995. Studies on the haemostatic defect in a complicated syndrome: an inverse Scott syndrome platelet membrane abnormality? Thromb. Haemost. 74, 1244–1251.
[163] Castaman, G., Yu-Feng, L., Battistin, E., Rodeghiero, F., 1997. Characterization of a novel bleeding disorder with isolated prolonged bleeding time and deficiency of platelet microvesicle generation. Br. J. Haematol. 96, 458–463.
[164] Bretscher, M.S., 1972. Asymmetrical lipid bilayer structure for biological membranes. Nat. New Biol. 236, 11–12.
[165] Zwaal, R.F.A., Roelofsen, B., Colley, C.M., 1973. Localization of red cell membrane constituents. Biochim. Biophys. Acta 300, 159–182.
[166] Zwaal, R.F.A., Roelofsen, B., Comfurius, P., van Deenen, L.L.M., 1975. Organization of phospholipids in human red cell membranes as detected by the action of various purified phospholipases. Biochim. Biophys. Acta 406, 83–96.

Regulation of $cPLA_2$ activity

Arie J. Verkleij* and Johannes Boonstra

Department of Molecular Cell Biology and Institute of Biomembranes, University of Utrecht, Padualaan 8, 3584 CH Utrecht, The Netherlands
**Correspondence address: E-mail: A.J. Verkleij@bio.uu.nl*

1. Introduction

Phospholipases A_2 (PLA_2) hydrolyse fatty acids from the *sn*-2 position of phospholipids resulting in the release of free fatty acids and lysophospholipids [1,2]. The *sn*-2 position of phospholipids in mammalian cells is enriched with arachidonic acid. Arachidonic acid is the substrate of cyclooxygenases, lipoxygenases and cytochrome P450s, and therefore an essential component in the synthesis of prostaglandins, leukotrienes and other eicosanoids. As a result, PLA_2s play a regulatory role in processes such as cell growth, inflammation, platelet activation and cytotoxicity [3–5].

The PLA_2 super family consists of different lipases such as the secretory PLA_2s ($sPLA_2$), the Ca^{2+}-independent PLA_2 ($iPLA_2$) and the cytosolic PLA_2 ($cPLA_2$) [6,7]. Within the PLA_2 family, $cPLA_2$ distinguishes itself by its arachidonic acid preference, and consequently $cPLA_2$ plays an important regulatory role in the biological processes described above. In this review, we will shortly describe the molecular properties of $cPLA_2$ and its activation mechanism. We will describe the interactions of $cPLA_2$ with membranes, the cellular localization and in vivo function.

2. Molecular properties of $cPLA_2$

The cDNA for human $cPLA_2$ was first cloned in 1991 [8,9]. The sequence encodes an 85 kDa protein consisting of 749 amino acids. Several domains have been identified on the protein, such as a N-terminal Ca^{2+}-dependent lipid-binding (CaLB) domain [8,10]. The CaLB domain preferentially binds to vesicles composed of phosphatidylcholine (PC) in response to physiological concentrations of Ca^{2+} [11]. The CaLB domain did not exhibit any preference for phospholipid vesicles composed of saturated, unsaturated *sn*-2 fatty acyl chains or the carbonyl oxygens at the *sn*-1 or *sn*-2 linkage [11], and therefore it was concluded that the CaLB domain interacts primarily with the headgroup of PC. Regarding the functional hydrolytic activity, a G-L-S^{228}-G-S sequence has been identified that closely resembles the lipase consensus motif G-X-S-X-G, which is present in many serine

Advances in Molecular and Cell Biology, Vol. 33, pages 421–430

ISSN: 1569-2558 / DOI: 10.1016/S1569-2558(03)33020-6

esterases and neutral lipases. Site-directed mutagenesis of Ser^{228} resulted in a complete loss of activity of $cPLA_2$ although the enzyme remained folded correctly [12,13]. Finally, $cPLA_2$ contains several consensus phosphorylation sites for serine/threonine kinases and tyrosine kinases [9]. Especially Ser^{505} and Ser^{727} have been demonstrated to have physiological relevance. Ser^{505} has been shown to represent the target of mitogen-activated protein kinase (MAPK) [14], while Ser^{727} represents the target for protein kinase C (PKC) or protein kinase A (PKA) [14].

Several isoforms of the 85 kDa $cPLA_2$ have been characterized, denoted as $cPLA_2\alpha$, $cPLA_2\beta$ and $cPLA_2\gamma$ [15–17]. $cPLA_2\alpha$ represents the regular 85 kDa $cPLA_2$, while $cPLA_2\beta$ and $cPLA_2\gamma$ are isoforms sharing approximately 30% sequence identity. $cPLA_2\beta$ is a 114 kDa protein, which is strongly expressed in pancreas, cerebellum, brain and liver [16,17]. $cPLA_2\gamma$ is a 61 kDa protein, which lacks the CaLB domain as compared to $cPLA_2\alpha$ and $cPLA_2\beta$ [15].

3. Regulation of $cPLA_2$ activity

$cPLA_2$ is constitutively expressed in most cell types. Extracellular stimuli, as interleukin-1, tumour necrosis factor-α (TNF-α), monocyte colony-stimulating factor and epidermal growth factor (EGF) have been demonstrated to induce prolonged protein expression in various cell lines [18–20]. $cPLA_2$ is modulated by phosphorylation on serine and threonine residues and requires Ca^{2+} to translocate from the cytosol to the membrane, as has been demonstrated in cells stimulated with agents that mobilize Ca^{2+}, including EGF and the Ca^{2+}-ionophore A23187, using cell fractionation and/or microscopical methods [21–24].

A wide variety of agents have been demonstrated to increase $cPLA_2$ phosphorylation and activity [21]. The phosphorylation of $cPLA_2$ was shown to be mediated by $p42/44^{MAPK}$ and to occur on Ser^{505} [25,26]. The importance of this phosphorylation site was demonstrated in Chinese hamster ovary (CHO) cells overexpressing mutant $cPLA_2$, in which Ser^{505} was substituted by Ala. This mutant $cPLA_2$ was not phosphorylated and evoked some agonist-induced arachidonic acid release [25]. However, $cPLA_2$ phosphorylation induced by different stimuli in various cells without a concomitant increase in intracellular Ca^{2+} did not result in arachidonic acid release, demonstrating that both Ca^{2+} signalling and phosphorylation are necessary for a full activation of $cPLA_2$ [27]. Of particular interest were the observations that phosphorylation of $cPLA_2$ has to precede an increase in intracellular Ca^{2+} to achieve maximal activity [28,29], implying that $cPLA_2$ is not available for phosphorylation after translocation. Recently it was reported that the signal transduction pathways leading to $p42/p44^{MAPK}$ and subsequent $cPLA_2$ activation were mediated through the raf-MEK pathway and also through PKC [30]. However, $p42/44^{MAPK}$ is not the only MAPK family member involved in $cPLA_2$ phosphorylation and activation. For example, in the human astrocytoma cell line 1321N1, stimulation with thrombin or TNF-α resulted in a phosphorylation and concomitant activation of $cPLA_2$, which is most likely mediated by the c-Jun NH_2-terminal kinase (JNK) [31–33]. Likewise, $cPLA_2$ activity was blocked in thrombin, collagen or stress-activated platelets that had been treated with an inhibitor for $p38^{MAPK}$ [34–37].

Ser^{727} lies in the consensus motif for basotrophic kinases, such as PKC. The involvement of PKC in $cPLA_2$ activation has been observed in a variety of cell types including macrophages, mesangial cells and thyroid cells [38–40]. Okadaic acid induces $cPLA_2$ phosphorylation of Ser^{727} and arachidonic acid release without elevated Ca^{2+} levels, and an inhibitor of $p38^{MAPK}$ resulted in a reduced Ser^{505} and Ser^{727} phosphorylation in thrombin-activated cells. The kinase responsible for Ser^{727} phosphorylation of $cPLA_2$ is a downstream substrate of $p38^{MAPK}$ and has recently been identified as the MAP kinase interaction protein kinase 1 (Mnk1) [41]. Furthermore, it was demonstrated that Ca^{2+}-/calmodulin-dependent kinase II binds directly to $cPLA_2$, resulting in $cPLA_2$ phosphorylation on Ser^{515} and an increase in $cPLA_2$ activity [42].

4. Translocation and binding of $cPLA_2$ to membrane lipids

$cPLA_2$ activity showed a striking Ca^{2+}-dependency in studies using purified enzyme preparations [43–45]. The $cPLA_2$ activity was increased within a Ca^{2+} concentration range of 100 nM to 1 μM, corresponding to the Ca^{2+} concentration range in cells upon activation. Subsequently, the association of $cPLA_2$ to natural membranes was shown to be a Ca^{2+}-dependent process [46]. Of particular interest were the observations that the Ca^{2+} dependency for membrane association was comparable to the Ca^{2+}-dependent activation of $cPLA_2$ [8,47], suggesting that the Ca^{2+}-dependent activation was due to the binding of $cPLA_2$ to the membrane. Molecular cloning of $cPLA_2$ revealed the presence of the CaLB domain responsible for the Ca^{2+}-dependent binding to the membrane as described above. This CaLB domain shares homology with the C2 domains first identified in the conventional isoforms of PKC [48]. Deletion of the CaLB domain prevented both the membrane binding and the Ca^{2+}-dependent activation of $cPLA_2$ [8]. Furthermore, a fragment of $cPLA_2$ containing the CaLB domain associated to the membrane at the same Ca^{2+} concentration as required for the intact protein [10]. These findings clearly demonstrate that the Ca^{2+}-dependent binding of $cPLA_2$ to the membrane by its CaLB domain plays a crucial role in the regulation of $cPLA_2$ activity. The binding of CaLB domain of $cPLA_2$ has a preference for head groups with hydrophobic features [49]. Indeed, both intact $cPLA_2$ and the $cPLA_2$ C2 domain show Ca^{2+} dependent, preferential binding to phospholipids with hydrophobic features of the head group, such as PC in preference to phosphatidylserine (PS), phosphatidylinositol (PI) or phosphatidylethanolamine (PE) [50–52]. However, addition of a small percentage of anionic lipids to PC vesicles leads to a tighter binding of $cPLA_2$ as compared to neutral vesicles [53–56]. Specifically the presence of 1 mol% PtdIns(4,5)P2 in PC vesicles resulted in a 20-fold increase in the binding affinity of $cPLA_2$ and a similar increase in substrate hydrolysis [57]. It has been shown that phosphatidylinositol-4,5-bisphosphate (PIP_2) mediates the binding of $cPLA_2$ to lipid vesicles, thereby increasing its activity in vitro in a calcium-independent manner [58–60]. However, from the crystal structure of $cPLA_2$ protein no clear PH domain was identified [61]. This suggests that $cPLA_2$ can interact directly with PIP_2, generating a binding site for $cPLA_2$ in PIP_2-enriched microdomains to which $cPLA_2$ might translocate upon cell stimulation.

Recently, it was demonstrated that ceramide may bind $cPLA_2$ directly via its CaLB domain, thereby targeting $cPLA_2$ to its substrate [62]. This binding occurred in a Ca^{2+}-dependent fashion and increased $cPLA_2$ activity towards ceramide-containing liposomes. Similar results were observed in Ca^{2+}-ionophore and epinephrine-stimulated CHO-2B cells. Besides an activation of $cPLA_2$ by ceramide, an inhibition in enzyme activity was measured in liposomal substrates containing sphingomyelin, which could subsequently be restored by addition of cholesterol or ceramide [63]. This is potentially of interest in view of the regulation and subcellular localization of $cPLA_2$ since ceramide, sphingomyelin and cholesterol are present in rafts and/or caveolea [64,65].

In addition, vimentin, an intermediate filament component, has been shown to act as a perinuclear adaptor for $cPLA_2$, in which the C2 domain associates with the head domain of vimentin in a Ca^{2+}-dependent manner [66].

PLA_2-interacting protein (PLIP) is a newly identified splice variant of TAT-interacting protein (Tip60) that can interact with the N-terminus of $cPLA_2$ [67] and colocalize within the nucleus of transfected COS cells and serum-deprived mesangial cells. Colocalization of PLIP/Tip60 with $cPLA_2$ was shown to temporally correlate with the induction of apoptosis. In addition, lipocortins may interact with $cPLA_2$ and thus regulate its activity, as discussed below.

5. Substrate availability

In model studies, it is shown that the binding of the enzyme to anionic vesicles can be sufficiently tight such that the enzyme hydrolyses many phospholipids without leaving the vesicle surface. This type of processive interfacial catalysis is termed scooting [68]. In this type of studies, mixtures of phospholipids containing 1-palmitoyl-2-arachidonoyl-*sn*-glycero-3-phosphocholine (PAPC) are used. The hydrolytic activity of $cPLA_2$ in such in vitro systems ceases prematurely. Only a small percentage of the available phospholipid substrate is hydrolysed. $cPLA_2$ phosphorylated on Ser^{505} by MAPK displays a 30% increase in the rate of *sn*-arachidonylphosphatidylcholine hydrolysis in the scooting mode compared to that of the non-phosphorylated enzyme. Using model membrane vesicles of dimyristoylphospholipid, the enzyme activity is at the phase transition [69], similar as was found with $sPLA_2$ in 1974 [70]. This latter can be interpreted as the enzymatic activity is higher at edges of solid and fluid lipid domains or in other words, at defects in the membranes. The activity of the enzyme or the premature cessation of the hydrolysis is dependent on the size of the vesicles, smaller vesicles allowed less enzyme activity. The presence of 30% glycerol or fusion conditions (2–10 mM Ca^{2+}) prevent premature cessation of enzyme activity [71].

Other studies [63] showed that sphingomyelin decreases and cholesterol and ceramides increase the efficiency of $cPLA_2$ activity on liposomes. These results are in line with data obtained in vivo with CHO cells using sphingomyelinase and cholesterol depletion. Recent work in our lab indicates that the presence of oxidized phospholipids can increase $cPLA_2$ activity [72,73].

In conclusion, $cPLA_2$ hydrolyses preferentially phospholipids containing the polyunsaturated fatty acid, arachidonic acid. The availability of the substrate is dependent on the curvature of the bilayer vesicle, membrane fluidity, phase separation, phospholipid composition i.e. the role of charge, presence of cholesterol and peroxidation of phospholipids. It is clear that all these parameters may play a role, but it is difficult to understand what really happens in vivo. Yet, $cPLA_2$ will be locally activated and will dock at places where locally Ca^{2+} was increased, where lipid domains may have a favourable composition and lipid fluidity.

6. Subcellular localisation of $cPLA_2$

The subcellular localisation of $cPLA_2$ in cells has been of particular interest in order to identify the specific target membranes from which $cPLA_2$ releases arachidonic acid, the precursor of eicosanoids. Using immunofluorescence microscopy, it was demonstrated that $cPLA_2$ translocates to the nuclear membrane upon Ca^{2+}-ionophore of IgE/antigen activation of rat basophilic leukaemia cells [23]. A Ca^{2+}-induced translocation of $cPLA_2$ to the nuclear envelope and ER has been observed in CHO cells overexpressing $cPLA_2$ [74]. In rat alveolar epithelial cells, this Ca^{2+}-ionophore-induced translocation of $cPLA_2$ to the nuclear envelope was associated with a preferential and localised release of arachidonic acid [75]. Similar observations were reported in glomerular epithelial cells in which $cPLA_2$ is associated with the plasma membrane, ER and nuclear membrane in both unstimulated and stimulated cells [76]. Recently, $cPLA_2$ has been fused to green fluorescent protein (GFP) [77,78] and using these constructs a Ca^{2+}-induced translocation of $cPLA_2$ to the perinuclear region, including the nuclear envelope, was observed [79]. $cPLA_2$-EGFP or $cPLA_2$-C2-EGFP constructs demonstrated in MDCK cells that both the $cPLA_2$ and its C2 domain were translocated to Golgi membranes in response to sustained $[Ca^{2+}]_i$ greater than 100–125 nM, and to Golgi, ER, and perinuclear membranes at $[Ca^{2+}]_i$ greater than 210–280 nM [80]. However, in ATP-depleted MDCK cells, $cPLA_2$ translocated from the cytosol to the nucleus [81]. In addition, Choukroun et al. [82] demonstrated, by immunofluorescence and subcellular fractionation, the presence of $cPLA_2$ in the Golgi fraction of rat kidney epithelial cells. Furthermore, these authors suggest that $cPLA_2$ plays a role in the regulation of Golgi structure and thereby modulates intracellular trafficking of membrane proteins. $cPLA_2$ localisation was shown to be cell density dependent in endothelial cells where $cPLA_2$ was present in the cytoplasm of confluent cells and in the cell nucleus in subconfluent cells [83], suggesting a role for $cPLA_2$ in angiogenesis.

Indirect immunofluorescence studies in fibroblasts demonstrated that endogenous $cPLA_2$ was present in punctate structures in the cytoplasm of untreated cells and of cells activated with EGF or A23187 [84]. This localisation was studied in detail using immunogold electron microscopy [84]. At this level, the punctate pattern was due to small clusters of $cPLA_2$ in the vicinity of membranes. These clusters were found in the cytosol in the vicinity of all organelles, except the Golgi system. The enzyme clusters showed no preference for the nuclear envelope, the endoplasmic reticulum or the plasma membrane (Fig. 1). Stimulation of cells with EGF or A23187 or both did not change the punctate immunofluorescence-labelling pattern. Furthermore, a similar pattern was observed by the introduction of extremely low or high intracellular calcium concentrations. Even by electron microscopy, immunogold labelling was not changed at all. No gold particles were found in membranes, and are still only visible in the clusters not associated to membranes. Thus no translocation of $cPLA_2$ from the clusters to membranes upon stimulation was detected with immunogold labelling and EM. The punctate labelling of $cPLA_2$ appears not to be restricted to fibroblasts. In addition, the human epidermal carcinoma A431 cells, neuroblastoma cells N2A, rat mesangial cells and CHO cells, showed punctate labelling of $cPLA_2$ with immunofluorescence. In addition in basophylic leukaemia cells, aggregates of $cPLA_2$ were occasionally found in stimulated cells on the electron microscopical level [23]. Our findings are surprising in light of the fractionation studies [24] showing a $cPLA_2$ translocation from the cytosol to membranes upon A23187 and EGF. From the biochemical data, it follows that $cPLA_2$ translocates from the cytosol to membranes. In cells, this translocation of $cPLA_2$ appears to take place over a small distance not visible at the ultrastructural level yet.

How to understand these data? We propose that these clusters reflect in the inactive pools of $cPLA_2$ in the vicinity of all membranes, from which monomers of $cPLA_2$ can be recruited by phosphorylation, to keep $cPLA_2$ activity under control at the subcellular level. Thus only a small fraction of $cPLA_2$ may be recruited as monomers to membranes where it can only bind if Ca^{2+} is increased locally to bind $cPLA_2$ and allow phospholipid hydrolysis. To investigate this hypothesis, $cPLA_2$ monomers obtained by gel filtration chromatography as described by Spaargaren [45] and clusters present in homogenates of HER 14 fibroblast were used in an electron microscopic in vitro approach. We show that only the $cPLA_2$ monomers, and not the clusters, bind to phospholipid vesicles composed of 1 stearoyl-2 arachidonoyl phosphatidylcholine, in a Ca^{2+}-dependent manner [85] (Fig. 2).

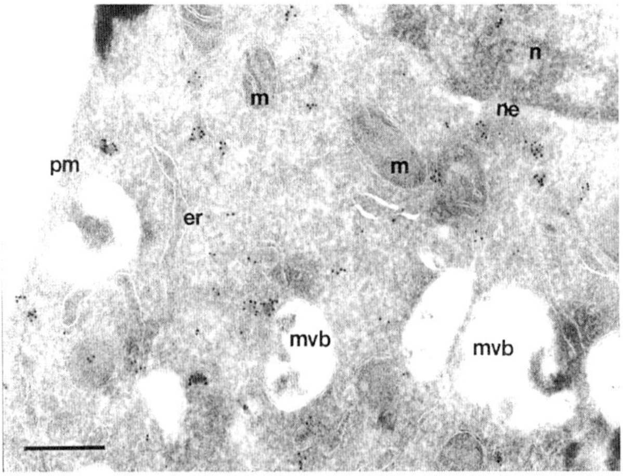

Fig. 1. Electron microscopic localization of $cPLA_2$ in an ultrathin cryosection of HER 14 fibroblasts. $cPLA_2$ immunoreactivity is detected by an antibody directed against $cPLA_2$ coupled to 10 nm gold as described previously [84]. $cPLA_2$ is present in particles in the proximity of membranes without preference for a particular membrane. g: Golgi; n: nucleus; ne: nuclear envelope; m: mitochondrion; mvb: multivesicular body; Bar: 300 nm.

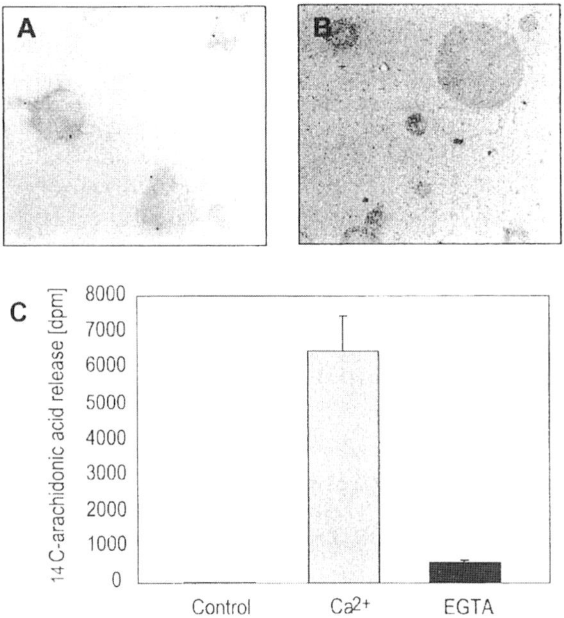

Fig. 2. Ca^{2+}-dependent membrane binding by cPLA$_2$ monomers as observed with the electron microscope is interrelated with hydrolytic activity. cPLA$_2$ monomers bind Ca^{2+} dependently to SAPC multilamellar vesicles (A: 1 mM Ca^{2+}; B: 1 mM EGTA) and exhibit Ca^{2+}-dependent arachidonic acid release activity towards S[^{14}C]APC MLVs under electron microscopic conditions (C). (*For a colored version of this figure, see plate section, page 611.*)

In addition, arachidonic acid release was detected with monomers only in the presence of Ca^{2+}. From these results we conclude that the clusters represent an inactive pool of cPLA$_2$ from which monomers can be recruited that are only active when bound to membranes. This is in line with the fact that phosphorylation has to precede an increase in intracellular calcium concentration for maximal activity [28]. All together, this has led us to the hypothesis that cPLA$_2$ is released from the inactive clusters where signal transduction cascades phosphorylate cPLA$_2$ and where Ca^{2+} concentration is increased as a result of the opening of Ca^{2+} channels nearby [85]. Thus, cPLA$_2$ activation by this model does not have to lead to a massive translocation and activation of cPLA$_2$, as is likely the case in inflammatory cells. In these cells, a sustained phosphorylation of cPLA$_2$ and sustained Ca^{2+} release at the endoplasmic reticulum and nuclear membrane is followed by a production of eicosanoids via very active lipoxygenase and cyclogenase present there. In the case of cell cycle dependent cPLA$_2$ activation [86,87], small production of arachidonic acid for eicosanoid production may be used in stimulated specific genes transcription essential in G1-S transition or mitosis.

Acknowledgements

We would like to thank Henk van den Bosch, Gerda van Rossum, Gertrude Bunt, and Casper Schalkwijk for their contributions in the cPLA$_2$ research described in this chapter.

References

[1] van den Bosch, H., 1980. Biochim. Biophys. Acta 604, 191–246.
[2] Dennis, E.A., 1997. Trends Biochem. Sci. 22, 1–2.
[3] Boonstra, J., van Rossum, G.S.A.T., 2003. Prog. Cell Cycle Res. 5, 181–190.
[4] Heller, A., Koch, T., Schmeck, J., van Ackern, K., 1998. Drugs 55, 487–496.
[5] Capdevila, J.H., Falck, J.R., Harris, R.C., 2000. J. Lipid Res. 41, 163–181.
[6] Six, D.A., Dennis, E.A., 2000. Biochim. Biophys. Acta 1488, 1–19.
[7] Murakami, M., Kudo, I., 2002. J. Biochem. (Tokyo) 131, 285–292.
[8] Clark, J.D., Lin, L.L., Kriz, R.W., Ramesha, C.S., Sultzman, L.A., Lin, A.Y., Milona, N., Knopf, J.L., 1991. Cell 65, 1043–1051.
[9] Sharp, J.D., White, D.L., Chiou, X.G., Goodson, T., Gamboa, G.C., McClure, D., Burgett, S., Hoskins, J., Skatrud, P.L., Sportsman, J.R., 1991. J. Biol. Chem. 266, 14850–14853.
[10] Nalefski, E.A., Sultzman, L.A., Martin, D.M., Kriz, R.W., Towler, P.S., Knopf, J.L., Clark, J.D., 1994. J. Biol. Chem. 269, 18239–18249.
[11] Nalefski, E.A., McDonagh, T., Somers, W., Seehra, J., Falke, J.J., Clark, J.D., 1998. J. Biol. Chem. 273, 1365–1372.
[12] Sharp, J.D., Pickard, R.T., Chiou, X.G., Manetta, J.V., Kovacevic, S., Miller, J.R., Varshavsky, A.D., Roberts, E.F., Strifler, B.A., Brems, D.N., 1994. J. Biol. Chem. 269, 23250–23254.
[13] Huang, Z., Payette, P., Abdullah, K., Cromlish, W.A., Kennedy, B.P., 1996. Biochemistry 35, 3712–3721.
[14] de Carvalho, M.G., McCormack, A.L., Olson, E., Ghomashchi, F., Gelb, M.H., Yates, J.R. III, Leslie, C.C., 1996. J. Biol. Chem. 271, 6987–6997.
[15] Underwood, K.W., Song, C., Kriz, R.W., Chang, X.J., Knopf, J.L., Lin, L.L., 1998. J. Biol. Chem. 273, 21926–21932.
[16] Pickard, R., Strifler, B.A., Kramer, R.M., Sharp, J.D., 1999. J. Biol. Chem. 274, 8823–8831.
[17] Song, C., Chang, X.J., Bean, K.M., Proia, M.S., Knopf, J.L., Kriz, R.W., 1999. J. Biol. Chem. 274, 17063–17067.
[18] Nakamura, T., Lin, L.L., Kharbanda, S., Knopf, J., Kufe, D., 1992. EMBO J. 11, 4917–4922.
[19] Schalkwijk, C.G., Vervoordeldonk, M., Pfeilschifter, J., van den Bosch, H., 1993. FEBS Lett. 333, 339–343.
[20] Chepenik, K.P., Diaz, A., Jimenez, S.A., 1994. J. Biol. Chem. 269, 21786–21792.
[21] Clark, J.D., Schievella, A.R., Nalefski, E.A., Lin, L.L., 1995. J. Lipid Mediat. Cell. Signal. 12, 83–117.
[22] Peters-Golden, M., McNish, R.W., 1993. Biochem. Biophys. Res. Commun. 196, 147–153.
[23] Glover, S., de Carvalho, M.S., Bayburt, T., Jonas, M., Chi, E., Leslie, C.C., Gelb, M.H., 1995. J. Biol. Chem. 270, 15359–15367.
[24] Schalkwijk, C.G., Spaargaren, M., Defize, L.H., Verkleij, A.J., van den Bosch, H., Boonstra, J., 1995. Eur. J. Biochem. 231, 593–601.
[25] Lin, L.L., Wartmann, M., Lin, A.Y., Knopf, J.L., Seth, A., Davis, R.J., 1993. Cell 72, 269–278.
[26] Nemenoff, R.A., Winitz, S., Qian, N.X., van Putten, V., Johnson, G.L., Heasley, L.E., 1993. J. Biol. Chem. 268, 1960–1964.
[27] Murakami, M., Nakatani, Y., Atsumi, G., Inoue, K., Kudo, I., 1997. Crit. Rev. Immunol. 17, 225–283.
[28] Schalkwijk, C.G., van der Heijden, M.A.G., Bunt, G., Maas, R., Tertoolen, L.G.J., van Bergen en Henegouwen, P.M.P., Verkleij, A.J., van den Bosch, H., Boonstra, J., 1996. Biochem. J. 313, 91–96.
[29] Abdullah, K., Cromlish, W.A., Payette, P., Laliberté, F., Huang, Z., Street, I., Kennedy, B.P., 1995. Biochim. Biophys. Acta 1244, 157–164.
[30] van Rossum, G.S.A.T., Klooster, R., van den Bosch, H., Verkleij, A.J., Boonstra, J., 2001. J. Biol. Chem. 276, 28976–28983.
[31] Hernández, M., Bayón, Y., Sánchez Crespo, M., Nieto, M.L., 1997. Biochem. J. 328, 263–269.
[32] Hernández, M., Bayón, Y., Sánchez Crespo, M., Nieto, M.L., 1999. J. Neurochem. 73, 1641–1649.
[33] van Putten, V., Refaat, Z., Dessev, C., Blaine, S., Wick, M., Butterfield, L., Han, S.Y., Heasley, L.E., Nemenoff, R.A., 2001. J. Biol. Chem. 276, 1226–1232.
[34] Waterman, W.H., Molski, T.F., Huang, C.K., Adams, J.L., Sha'afi, R.I., 1996. Biochem. J. 319, 17–20.

[35] Kramer, R.M., Roberts, E.F., Um, S.L., Börsch-Haubold, A.G., Watson, S.P., Fisher, M.J., Jakubowski, J.A., 1996. J. Biol. Chem. 271, 27723–27729.
[36] Börsch-Haubold, A.G., Kramer, R.M., Watson, S.P., 1997. Eur. J. Biochem. 245, 751–759.
[37] Buschbeck, M., Ghomashchi, F., Gelb, M.H., Watson, S.P., Börsch-Haubold, A.G., 1999. Biochem. J. 344, 359–366.
[38] Huwiler, A., Pfeilschifter, J., 1993. Eur. J. Biochem. 217, 69–75.
[39] Qiu, Z.H., Leslie, C.C., 1994. J. Biol. Chem. 269, 19480–19487.
[40] Ekokoski, E., Dugue, B., Vainio, M., Vainio, P.J., Tornquist, K., 2000. J. Cell. Physiol. 183, 155–162.
[41] Hefner, Y., Börsch-Haubold, A.G., Murakami, M., Wilde, J.I., Pasquet, S., Schieltz, D., Ghomashchi, F., Yates, J.R. III, Armstrong, C.G., Paterson, A., Cohen, P., Fukunaga, R., Hunter, T., Kudo, I., Watson, S.P., Gelb, M.H., 2000. J. Biol. Chem. 275, 37542–37551.
[42] Muthalif, M.M., Hefner, Y., Canaan, S., Harper, J., Zhou, H.L., Parmentier, J.H., Aebersold, R., Gelb, M.H., Malik, K.U., 2001. J. Biol. Chem. 276, 39653–39660.
[43] Clark, J.D., Milona, N., Knopf, J.L., 1990. Proc. Natl Acad. Sci. USA 87, 7708–7712.
[44] Kramer, R.M., Roberts, E.F., Manetta, J., Putnam, J.E., 1991. J. Biol. Chem. 266, 5268–5272.
[45] Spaargaren, M., Wissink, S., Defize, L.H.K., de Laat, S.W., Boonstra, J., 1991. Biochem. J. 287, 37–43.
[46] Yoshihara, Y., Watanabe, Y., 1990. Biochem. Biophys. Res. Commun. 170, 484–490.
[47] Channon, J.Y., Leslie, C.C., 1990. J. Biol. Chem. 265, 5409–5413.
[48] Coussens, L., Parker, P.J., Rhee, L., Yang-Feng, T.L., Chen, E., Waterfield, M.D., Francke, U., Ullrich, A., 1986. Science 233, 859–866.
[49] Perisic, O., Fong, S., Lynch, D.E., Bycroft, M., Williams, R.L., 1998. J. Biol. Chem. 273, 1596–1604.
[50] Nalefski, E.A., Sultzman, L.A., Martin, D.M., Kriz, R.W., Towler, P.S., Knopf, J.L., Clark, J.D., 1994. J. Biol. Chem. 269, 18239–18249.
[51] Wijkander, J., Sundler, R., 1992. Biochem. Biophys. Res. Commun. 184, 118–124.
[52] Nalefski, E.A., Falke, J.J., 1996. Biochemistry 37, 17642–17650.
[53] Diez, E., Louis-Flamberg, P., Hall, R.H., Mayer, R.J., 1992. J. Biol. Chem. 267, 18342–18348.
[54] Ghomashchi, F., Schüttel, S., Jain, M.K., Gelb, M.H., 1992. Biochemistry 31, 3814–3824.
[55] Hanel, A.M., Schüttel, S., Gelb, M.H., 1993. Biochemistry 32, 5949–5958.
[56] Witmer, M.R., Micanovic, R., Tredup, J., Lin, W., Hail, M., Villafranca, J.J., 1995. Arch. Biochem. Biophys. 318, 430–438.
[57] Dennis, E.A., 1994. J. Biol. Chem. 269, 13057–13060.
[58] Mosior, M., Six, D.A., Dennis, E.A., 1998. J. Biol. Chem. 273, 2184–2191.
[59] Leslie, C.C., Channon, J.Y., 1990. Biochim. Biophys. Acta 1045, 261–270.
[60] Buckland, A.G., Wilton, D.C., 1997. Biochem. Soc. Trans. 25, S599.
[61] Dessen, A., Tang, J., Schmidt, H., Stahl, M., Clark, J.D., Seehra, J., Somers, W.S., 1999. Cell 97, 349–360.
[62] Huwiler, A., Johansen, B., Skarstad, A., Pfeilschifter, J., 2001. FASEB J. 15, 7–9.
[63] Klapisz, E., Masliah, J., Béréziat, G., Wolf, C., Koumanov, K.S., 2000. J. Lipid Res. 41, 1680–1688.
[64] Brown, D.A., London, E., 1998. J. Membr. Biol. 164, 103–114.
[65] Dobrowsky, R.T., 2000. Cell. Signal. 12, 81–90.
[66] Murakami, M., Nakatani, Y., Kuwata, H., Kudo, I., 2000. Biochim. Biophys. Acta 1488, 159–166.
[67] Sheridan, A.M., Force, T., Yoon, H.-J., O'Leary, Y., Choukroun, G., Taheri, M.R., Bonventre, J.V., 2002. Mol. Cell. Biol. 21, 4470–4481.
[68] Jain, M.K., Rogers, J., Jahagirdar, D.V., Marecek, J.F., Ramirez, F., 1986. Biochim. Biophys. Acta 860, 435–447.
[69] Bayburt, T., Gelb, M.H., 1997. Biochemistry 36, 3216–3231.
[70] Op den Kamp, J.A., de Gier, J., van Deenen, L.L.M., 1974. Biochim. Biophys. Acta 29, 253–256.
[71] Burke, J.R., Witmer, M.R., Tredup, J.A., 1999. Arch. Biochem. Biophys. 365, 239–247.
[72] Van Rossum, G.S.A.T., Drummen, G.P.C., Verkleij, A.J., Post, J.A., Boonstra, J., 2003. Biochim. Biophys. Acta (submitted).
[73] Van Rossum, G.S.A.T., 2001. Thesis, University of Utrecht.
[74] Schievella, A.R., Rogier, M.K., Smith, W.L., Lin, L.-L., 1995. J. Biol. Chem. 270, 30749–30754.
[75] Peters-Golden, M., Song, K., Marshall, T., Brock, T., 1996. Biochem. J. 318, 797–803.
[76] Liu, J., Takano, T., Papillon, J., Khadir, A., Cybulsky, A.V., 2001. Biochem. J. 353, 79–90.

[77] Perisic, O., Paterson, H.F., Mosedale, G., Lara-Gonzalez, S., Williams, R.L., 1999. J. Biol. Chem. 274, 14979–14987.
[78] Gijon, M.A., Spencer, D.M., Kaiser, A.L., Leslie, C.C., 1999. J. Cell Biol. 145, 1219–1232.
[79] Hirabayashi, T., Kume, K., Hirose, K., Yokomizo, T., Iino, M., Itoh, H., Shimizu, T., 1999. J. Biol. Chem. 274, 5163–5169.
[80] Evans, J.H., Spencer, D.M., Zweifach, A., Leslie, C.C., 2001. J. Biol. Chem. 276, 30150–30160.
[81] Sheridan, A.M., Sapirstein, A., Lemieux, N., Martin, B.D., Kim, D.K., Bonventre, J.V., 2001. J. Biol. Chem. 276, 29899–29905.
[82] Choukroun, G.J., Marshansky, V., Gustafson, C.E., McKee, M., Hajjar, R.J., Rosenzweig, A., Brown, D., Bonventre, J.V., 2000. J. Clin. Invest. 106, 983–993.
[83] Sierra-Honigmann, M.R., Bradley, J.R., Pober, J.S., 1996. Lab. Invest. 74, 684–695.
[84] Bunt, G., de Wit, J., van den Bosch, H., Verkleij, A.J., Boonstra, J., 1997. J. Cell Sci. 110, 2449–2459.
[85] Bunt, G., van Rossum, G.S.A.T., Boonstra, J., van den Bosch, H., Verkleij, A.J., 2000. Biochemistry 39, 7847–7850.
[86] van Rossum, G.S.A.T., Vlug, A.S., van den Bosch, H., Verkleij, A.J., Boonstra, J., 2001. J. Cell. Physiol. 188, 321–328.
[87] van Rossum, G.S.A.T., Bijvelt, J.J.M., van den Bosch, H., Verkleij, A.J., Boonstra, J., 2002. Cell. Mol. Life Sci. 59, 181–188.

Mammalian phospholipase C

Martina Schmidt,* Paschal A. Oude Weernink, Frank vom Dorp, Matthias B. Stope and Karl H. Jakobs

Institut für Pharmakologie, Universitätsklinikum Essen, Hufelandstrasse 55, D-45122 Essen, Germany
**Correspondence address: Tel.: +49-201-723-3457; fax: +49-201-723-5968*
E-mail: martina.schmidt@uni-essen.de(M.S.)

Abbreviations

BCR: B cell receptor; DAG: diacylglycerol; EGF: epidermal growth factor; GAP: GTPase-activating protein; GEF: guanine nucleotide exchange factor; GPCR: G protein-coupled receptor; GTPγS: guanosine 5′-*O*-(3-thio)-triphosphate; IP_3: inositol-1,4,5-trisphosphate; PA: phosphatidic acid; PDGF: platelet-derived growth factor; PH: pleckstrin homology; PX: phox homology; PI-3-kinase: phosphatidylinositol-3-kinase; PIP_2: phosphatidylinositol-4,5-bisphosphate; PIP_3: phosphatidylinositol-3,4,5-trisphosphate; PKC: protein kinase C; PLC: phospholipase C; PLD: phospholipase D; PTX: pertussis toxin; RA: Ras-binding domain; RTK: receptor tyrosine kinase; SH: src homology; TCR: T cell receptor.

1. Introduction

The hydrolysis of cellular phospholipids leads to the formation of various bioactive lipid mediators, acting either as extracellular signaling molecules or as intracellular second messengers. A well-known second messenger-forming system involving phospholipid hydrolysis is the stimulation of phosphoinositide-specific phospholipase C (PLC) isoforms [1–3]. In the mid-1950s, Mabel and Lowell Hokin were the first to discover PLC as a distinct, phospholipid-specific phosphodiesterase activity in pancreas slices. Their pioneering work indicated that PLC plays a central role in agonist-stimulated phosphoinositide metabolism and calcium signaling, and has brought PLC signaling to the very forefront of current biological and biomedical research. Thus, PLC was classified as a bona fide signal-activated phospholipase, generating biologically active products. Upon activation by membrane receptors, PLC enzymes hydrolyze the membrane phospholipid, phosphatidylinositol-4,5-bisphosphate (PIP_2), and thereby produce the

Advances in Molecular and Cell Biology, Vol. 33, pages 431–450

ISSN: 1569-2558/DOI: 10.1016/S1569-2558(03)33021-8

two second messengers, diacylglycerol (DAG) and inositol-1,4,5-trisphosphate (IP_3), leading to the activation of several protein kinase C (PKC) isoforms and the release of Ca^{2+} from intracellular stores, respectively. It is now well established that PLC stimulation plays a major role in many early and late cellular responses to receptor activation, including smooth muscle contraction, secretion, neuronal signaling as well as cell growth and differentiation [4,5].

PIP_2 serves not only as a substrate for the PLC isoforms. Phosphorylation of PIP_2 at the 3 position of the inositol ring by PI-3-kinases generates phosphatidylinositol-3,4,5-trisphosphate (PIP_3) [6]. It is now generally assumed that PIP_2 and PIP_3 bind to pleckstrin homology (PH) and phox homology (PX) domain-bearing proteins and thereby modulate a remarkable number of cellular processes, such as clathrin-coated vesicle endocytosis, vesicular trafficking, membrane movement and actin cytoskeleton assembly. Modulation of intracellular localization and activity state upon binding of the phosphoinositides to PH/PX domains has been described for phospholipase A_2, phospholipase D (PLD), lipid kinases including PI-3-kinases as well as lipid phosphatases [7–14]. Thus, phosphoinositides control many aspects of signal transduction, and indeed the synthesis and degradation of PIP_2 and its derivatives are tightly regulated by phosphoinositide kinases and phosphatases [15–17]. In addition, hydrolytic cleavage of PIP_2 by PLC isoforms is compartmentalized in caveolae and lipid rafts [18–21]. These membrane-signaling microdomains most likely confer specificity into the complex mechanisms of signal transduction by phosphoinositides.

At present, 11 PLC isoforms have been identified in mammalian cells and are classified into 4 families: PLC-β, PLC-γ, PLC-δ and the newly identified PLC-ε. Consistent with their distinct structural organization, the PLC isoforms are susceptible to distinct modes of activation by membrane receptors. This review will focus on the recent insights into the regulation of PLC enzymes, and, in particular, into those mechanisms leading to activation of PLC-ε.

2. PLC isozymes: structure and organization

Eleven mammalian PLC isoforms (excluding alternatively spliced forms) have been identified and divided into four groups: PLC-β1-4 (130–155 kDa), PLC-γ1-2 (~145 kDa), PLC-δ1-4 (~85 kDa) and PLC-ε (230–260 kDa) [1–3]. PLC-ε is the homologue of PLC210 from *Caenorhabditis elegans*, probably acting as a downstream effector of Ras [22]. As yeast, slime molds and plants only contain PLC-δ enzymes, the other PLC isoforms in higher eukaryotes are likely to be derived from the archetypal PLC-δ (Fig. 1A). All PLC isoforms contain the X and Y domains that build the catalytic core, representing a highly conserved region (40–60% identity) in between the PLC enzyme family. At the NH_2-terminal region all PLC isoforms possess a PH domain, localizing the PLC enzymes to their substrate and other important signaling components. Furthermore, all PLC isoforms contain two EF-hands located between the PH domain and the X domain as well as a C2 domain, adjacent to the Y domain. In addition to this common principle, PLC-β, PLC-γ and PLC-ε isoforms contain specific regulatory domains (Fig. 1B).

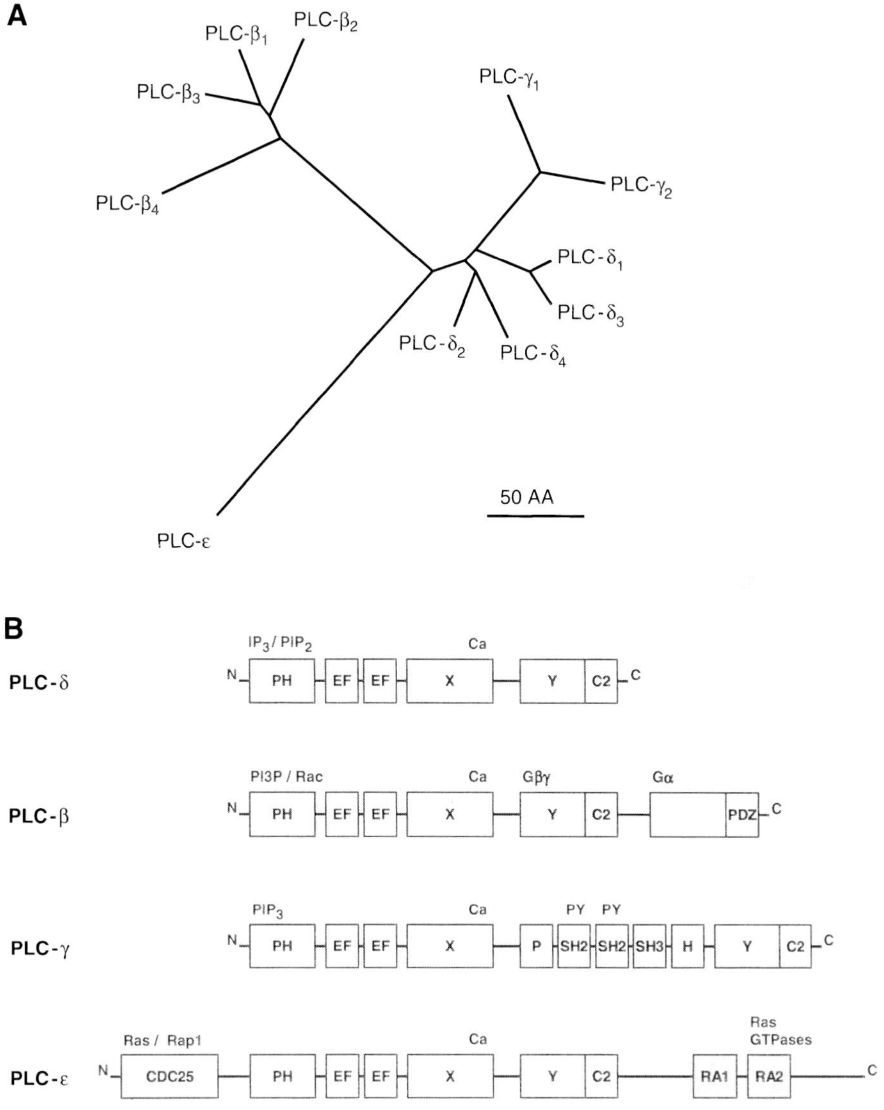

Fig. 1. A. Molecular phylogeny and domain organization of mammalian PLC isoforms. Phylogenetic analysis is based on the complete amino acid sequences of human PLC-β (1–4), PLC-γ (1,2), PLC-δ (1,3) and PLC-ε, as well as bovine PLC-δ2 and rat PLC-δ4 (using MegAlign software, DNASTAR Inc.; diagram kindly provided by Dr M. Schaefer). Please note the position of PLC-δ isoforms in the center of the phylogenetic tree. B. The molecular topology schematically shows the domains of the PLC isoforms. All the isoforms share the X and Y domains of the catalytic core, a PH domain, two EF-hands and a C2 domain. Isoform-specific domains are the PDZ-binding motif in the COOH-terminal tail of PLC-β, two SH2 and an SH3 domain within a split PH domain in PLC-γ, and a CDC25 domain and two RAs in PLC-ε. *(For a colored version of this figure, see plate section, page 612.)*

Based on the three-dimensional structure of PLC-δ1, it is now evident that the regulatory PH, EF and C2 domains are organized around a catalytic, irregular α/β barrel or TIM barrel (named after the structure in triose phosphate isomerase) formed by the X (147 residues) and Y (118 residues) domains [23,24]. PLC catalyzes the hydrolysis of the O–P bond between phosphoinositol and DAG resulting in the cyclic 1,2-phosphodiester intermediates being further processed to phosphomonoester [25]. The preference for PIP_2 as substrate is the result of a network of hydrogen bonds and salt-bridges involving Lys-438, Lys-440, Ser-522 and Arg-549, each located on distinct halves of the catalytic α/β barrel of PLC-δ1. These residues are conserved in between PLC isoforms, and indeed, mutation of Lys-440 abolished enzymatic activity of PLC-δ1 towards PIP_2 [26]. PLC-δ1-catalyzed PIP_2 hydrolysis requires calcium binding to the residues Asn-312, Glu-341, Asp-343 and Glu-390. Several catalytic residues are conserved in all PLC enzymes, in particular His-311 and His-356 of PLC-δ1. Consistently, mutation of His-311 and His-356 reduced the catalytic activity of PLC-δ1 [26].

PH domains are small protein modules of ~100–120 amino acids present in more than 100 different proteins involved in intracellular signaling, and are known to bind PIP_2 and PIP_3. In addition, PH domains can bind the phosphoinositides, PI3P and PI(3,4)P_2, the PLC product IP_3, as well as Gβγ subunits and GTP-bound Rac [7,27–29]. The PH domain of PLC-δ1 binds to both PIP_2 and IP_3, involving the residues Lys-30 and Lys-57 (interaction with 4- and 5-phosphoryl groups), Lys-32 (4-phosphoryl group) and Arg-40 (5-phosphoryl group) [23,30]. Binding to PIP_2 is essential for the localization of PLC-δ1 to the membrane as well as its catalytic activity, and the resulting formation of IP_3 may terminate this processive hydrolysis. Whereas the PH domain of PLC-γ isozymes specifically binds PIP_3, probably in concert with the COOH-terminal SH2 domain [31,32], thus linking PI-3-kinase activation and membrane recruitment of PLC-γ enzymes [33], it has recently been shown that the PH domain of PLC-β isoforms preferentially binds PI3P [34]. The PH domains of PLC-β1 and PLC-β2 are not simply tethering devices but mediate the regulation of the enzymes by Gβγ subunits and are even associated with the catalytic core [35–37]. Recently however, the functional significance of Gβγ binding to the PH domain of PLC-β2 has been questioned and evidence has been provided that the PH domain of PLC-β2 specifically interacts with GTP-loaded Rac proteins [28,29].

Although it was generally assumed that the EF-hands of PLC-δ1 solely serve as a flexible link between the PH domain and the rest of the enzyme [23], binding of calcium to the EF-hands was found to be necessary for proper interaction of the PH domain with PIP_2 [38]. The C2 domain of PLC enzymes consists of ~120 residues, and is a conserved membrane-targeting motif present in a multitude of calcium-regulated signaling proteins [39]. Binding of calcium to the catalytic core of PLC-δ1, but not to its C2 domain, is required for enzyme activity [40]. Recent studies indicated that PLC-δ1 and PLC-δ3, but not PLC-δ4, form a functional ternary complex with phosphatidylserine and calcium via their C2 domains, thereby enhancing their plasma membrane localization and facilitating processive substrate hydrolysis [41,42]. Recently, it has been shown that the C2 domain of PLC-β isozymes interacts with GTP-bound α subunits of G_q proteins [43].

PLC-β isoforms (except a splice variant of PLC-β4) are characterized by a long regulatory COOH-terminus, which upon dimerization is apparently responsible for the activation by GTP-bound α subunits of G_q proteins [44,45]. The COOH-terminus of PLC-β enzymes also bears a GTPase-activating protein (GAP) activity mimicking the $G\alpha_q$ interaction site of regulators of G protein signaling (RGS) proteins, a family of GAP proteins for heterotrimeric G proteins [44–46]. The COOH-terminus also triggers membrane association and localization of PLC-β1 into the nucleus [47]. In addition, PLC-β enzymes contain a COOH-terminal binding motif, X(Ser/Thr)-X(Val/Leu)-COOH, known to interact with PDZ domain-containing proteins that promote scaffolding of signaling molecules (reviewed in [48]).

PLC-γ isoforms bear an additional PH domain, which is split by two Src homology 2 (SH2) domains and one SH3 domain, known to mediate binding to phosphotyrosine residues and proline-rich sequences, respectively. Recently, it has been shown that the split PH domain of PLC-γ1 interacts with the elongation factor-1α (EF-1α), an activator of PI-4-kinases, and that the SH3 domain of PLC-γ1 exhibits guanine nucleotide exchange factor (GEF) activity for the nuclear GTPase PIKE, known to activate nuclear PI-3-kinase activity [49,50]. Both mechanisms may contribute to a rapid, highly dynamic phosphoinositide turnover in intact cells. Intriguingly, the SH3 domain of PLC-γ1 also interacts with PLD2, and may therefore directly link the two signaling phospholipase families [51].

The recently identified PLC-ε contains an NH_2-terminal CDC25 domain, possessing GEF activity for Ras and Ras-like GTPases, and two COOH-terminal Ras-binding (RA) domains, specifying upstream and downstream interactions with Ras-like GTPases [52–54].

3. Regulation of PLC-β isozymes

Heterotrimeric G proteins, coupling to the family of serpentine membrane receptors [55,56], activate PLC-β isoforms by two distinct mechanisms. Pertussis toxin (PTX)-insensitive heterotrimeric G proteins of the G_q family (G_q, G_{11}, G_{14}, G_{15} and G_{16}) activate PLC-β isoforms via GTP-loaded α subunits. PLC-γ, PLC-δ and PLC-ε isoforms do not respond to α subunits of the G_q family [1–3,53]. This mode of PLC-β activation includes the G protein-coupled receptors (GPCRs) for bradykinin, α_1-adrenergic agonists, angiotensin II, vasopressin, acetylcholine (muscarinic M_1 and M_3), thromboxane A2, bombesin and endothelin-1, to name but a few. Guanosine 5′-*O*-(3-thio)-triphosphate (GTγS)-bound α subunits of the G_q family, including $G\alpha_{16}$ being expressed only in hematopoietic cells, activate PLC-β isoforms, with the order PLC-β1 = PLC-β4 ≥ PLC-β3 > PLC-β2 [57]. Removal of the COOH-terminal region of PLC-β1 fully abrogated activation of the enzyme by $G\alpha_q$ leaving the catalytic activity of PLC-β1 unaffected [47]. Additional work demonstrated that the region required for PLC-β1 activation by $G\alpha_q$ is localized in the COOH-terminus comprising the residues 903–1143 [44] (see Fig. 2). The tertiary structure of turkey PLC-β (corresponding to the residues 904–1174 in the COOH-terminus of mammalian PLC-β1) revealed that the COOH-terminus of the enzyme composed almost entirely of three long helices forming a coiled-coil that dimerizes along its long axis in an antiparallel orientation, thereby generating an electrostatically positive dimer surface for the interaction with $G\alpha_q$ [45]. Consistent with the proposed extended coiled-coil structure of the PLC-β COOH-terminal dimer, Ross and colleagues reported

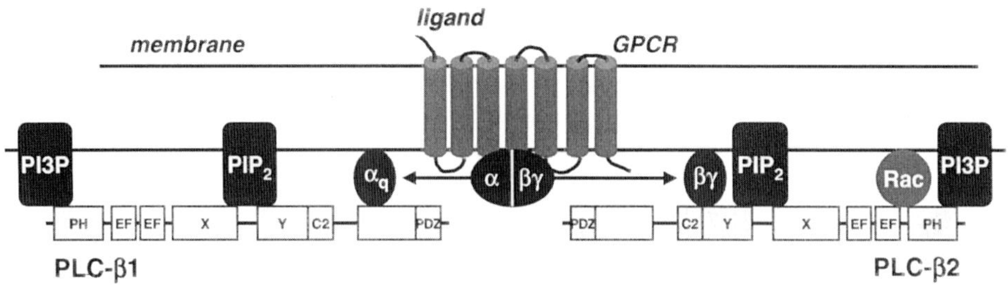

Fig. 2. Mechanism of stimulation of PLC-β by GPCRs. Stimulation of a GPCR by its appropriate ligand results in binding of GTP to the α subunit of the heterotrimeric G protein coupled to the receptor. After dissociation from the βγ dimer, activated $G\alpha_q$ tethers PLC-β1 to the membrane and stimulates its lipase activity by interaction with the COOH-terminal tail of the enzyme. Released βγ dimers of activated G_i are involved in the membrane localization and activation of PLC-β2. In addition, the monomeric GTPase Rac has been shown to modulate the activity of PLC-β2 via binding to the NH_2-terminal PH domain. Interaction of PI3P with the PH domains of PLC-β isoforms may contribute to the recruitment of the enzymes to the membrane. (*For a colored version of this figure, see plate section, page 613.*)

that the residues Lys-1058, Lys-1065, Lys-1069 and Tyr-1102, located at the interface of the dimer, probably stabilize the dimer, and that the residues Lys-921, Lys-925, Glu-1048 and Lys-1063, located at the outerface of two of the three helices, specifically interact with $G\alpha_q$ [44]. As it has been reported that GTPγS-activated $G\alpha_q$ additionally binds to the C2 domains of PLC-β1 and PLC-β2, allosteric activation of the enzymes may result due to interaction of the COOH-terminus with the catalytic core [43].

Gβγ subunits liberated from PTX-sensitive G proteins of the G_i family likewise activate PLC-β (except PLC-β4) isoforms. Whereas PLC-γ and PLC-δ isoforms did not respond to Gβγ subunits, activation of PLC-ε by Gβγ subunits has recently been reported [1–3,58] (see below). This mode of PLC-β activation includes the M_2 muscarinic receptors and receptors for chemoattractants. The affinity of PLC-β2 for Gβγ is at least 30-fold greater than that of PLC-β3 for Gβγ, and at least 200-fold greater than that of PLC-β1 for Gβγ. The activation of PLC-β isoforms by Gβγ subunits follow likewise the order PLC-β2 > PLC-β3 ≥ PLC-β1 [57] (see Fig. 2). The maximal extent of PLC-β activation by GTPγS-liganded $G\alpha_q$ and Gβγ subunits is comparable, however, the concentration of Gβγ subunits required for maximal PLC-β activation is much higher than the $G\alpha_q$ concentration [57]. Further work demonstrated that activation of PLC-β isozymes by Gβγ depends on their subunit composition. Whereas $G\beta_1\gamma_{1/2}$ activates PLC-β enzymes with the order PLC-β3 > PLC-β2 > PLC-β1, activation of PLC-β by $G\beta_5\gamma_2$ follows the order PLC-β2 > PLC-β1 ⋙ PLC-β3 [28,59]. It has been reported that Gβγ subunits bind to the isolated PH domain of PLC-β1 and PLC-β2, and microinjection of Gβγ subunits into NIH 3T3 cells induced membrane translocation of the PLC-β1 PH domain [27,34]. These data implicated that Gβγ binds to the PH domain of PLC-β-isoforms. However, the site of interaction of Gβγ with PLC-β2 was localized by others to the region between the catalytic subdomain Y residues Glu-574 and Lys-583 [60,61]. More recently, it has been reported that the PH domain of PLC-β2 specifically interacts with GTPγS-activated Rac proteins, and did not display significant affinity for Gβγ [28,29].

The distinct interaction sites of Gβγ subunits may promote membrane insertion of PLC-β enzymes. Although $G\alpha_q$ and Gβγ subunits interact with different regions in PLC-β isozymes, the heterotrimeric G protein subunits did not bind simultaneously to the enzymes and regulate enzyme activity independently [57]. However, activated members of the α_q subfamily of G proteins permit PLC-β stimulation by receptors acting through Gβγ subunits of PTX-sensitive G_i proteins [62,63]. Thus, $G\alpha_q$ and Gβγ-dependent signals may converge on the level of PLC-β isozymes to gain full activation in intact cells.

Very recently, evidence has been provided that PLC-β isozymes are also activated by members of the Rho family [28,29,64–66] (see Fig. 2). Furthermore, PLC-γ1 is apparently modulated by these monomeric GTPases, although by yet undefined mechanisms [67,68]. Initially, Gierschik and colleagues have demonstrated that in vitro PLC-β2 is activated by both Gβγ subunits and Rho family GTPases, in particular Cdc42Hs and Rac1, but not RhoA [64,65]. Additional work indicated that Rho GTPases stimulate PLC-β enzymes, with the rank order PLC-β2 > PLC-β3 ≥ PLC-β1. GTPγS-activated Rac1 and Rac2 enhanced PLC-β enzyme activity more potently than GTPγS-activated Cdc42Hs [65]. Regulation of PLC-β by Rho GTPases involved regions of PLC-β distinct from those required for the interaction with $G\alpha_q$ and Gβγ [28,65]. Stimulation of PLC-β2 by Gβγ and GTPγS-activated Rac proteins was additive, implicating independent stimulation of PLC-β2 by members of the Rho family and Gβγ subunits [28]. As a lipase-competent chimera of the PH domain of PLC-β1 with the remaining portion of PLC-β2 failed to respond to Rac proteins, it has been hypothesized that the PH domain of PLC-β2 is absolutely required for the stimulation by Rac proteins [28]. Meanwhile, Sondek and colleagues have shown that GTPγS-activated Rac proteins (Rac1, Rac2 and Rac3) bind to PLC-β2 (dissociation constant (K_D) ~5–10 μM) and PLC-β3 ($K_D > 25$ μM), but not to PLC-β1, probably via the PH domain of the PLC-β enzymes. Stimulation of PLC-β2 by Rac proteins was demonstrated both in vitro and in vivo [29]. Consistent with the fact that PLC-β2 is highly expressed in myeloid cells, Boulay and co-workers recently reported that Rac2 proteins are involved in the stimulation of the enzyme by chemoattractant receptor [69]. Intriguingly, it has been reported that constitutively active Rac2 [Rac2(12V)] not only stimulates PLC-β2, but also induced its membrane association in intact cells. The Rac2(12V)-induced membrane association of PLC-β2 absolutely required the NH_2-terminal PH domain [66]. As PLC-β2–membrane interaction by Rac2(12V) was also depending on the COOH-terminal region of PLC-β2, comprising the residues Phe-819–Glu-1166, these data point to a further function of this region in PLC-β2 to coordinate its association with the membrane compartment [66].

The distinct mechanisms of PLC-β regulation by heterotrimeric G protein subunits and monomeric GTPases are tightly controlled both in space and time. For example, PLC-β1 is a GAP protein for members of the $G\alpha_q$ family, and this mechanism, probably in concert with RGS proteins, may enhance temporal responsiveness to signaling processes mediated by $G\alpha_q$ and PLC-β1 to minimize spontaneous activation [46,70]. In addition, it has been reported that PKC and protein kinase A modulate PLC signaling; both inhibition and potentiation of PLC signals have been observed probably depending on the cellular setting (reviewed in [1,2]). Such mechanisms may contribute to oscillations in PLC-dependent second messenger generation by receptor activation, and may thereby control amplitude, frequency and duration of PLC signals [71,72]. Recruitment of PLC-β isozymes to caveolae and lipid rafts, compartmentalized membrane signaling microdomains bearing additional signaling components, probably controls the complex mechanisms of PLC-β signaling in space. The phosphorylation of phosphoinositides by PI-3-kinase contributes to the membrane recruitment of PLC-β enzymes, as specific interaction of the PH domain of PLC-β1 with PI3P triggered the membrane association of the enzyme, and this process was blocked by PI-3-kinase inhibitors [34]. In addition, Litosch has reported recently that the specific PLD reaction product, phosphatidic acid (PA) [13], enhances PLC-β1 activity, and renders PLC-β1 less sensitive to inhibition by PKC [73]. These data may even indicate that the products generated by PLC-β and PLD enzymes act in concert to regulate cellular signaling. In addition, PDZ domain-containing proteins induce clustering of PLC-β enzymes and other signaling components due to their interaction with the COOH-terminal binding motif, X(Ser/Thr)-X(Val/Leu)-COOH. For example, PLC-β1 and PLC-β2 interact with the PDZ-containing Na^+/K^+ exchanger factor (NHERF), which is coupled to Trp4, a calcium channel, and interaction of PLC-β3 with E3KARP (closely related to NHERF) enhances activation by muscarinic receptors (reviewed in [48]). Thus, the interaction of PLC-β enzymes with PDZ domain-containing proteins triggers scaffolding of signaling molecules. Scarlata recently reported that the presence of lipid rafts inhibits the interaction of PLC-β2 with Gβγ subunits, abolishing its stimulation by Gβγ, but not the access of PLC-β2 to its substrate PIP_2 [19]. It has been hypothesized that such a mechanism may interfere with the ability of Gβγ to activate PLC-β2, but may leave other binding sites available to alter protein–protein interactions. It would now be of particular interest to study, whether the Rac2(12V)-induced exchange of PLC-β2 between membrane-bound and cytoplasmic pools [66] is still operative in lipid rafts. Thus, interaction of PLC-β enzymes with the membrane compartment is a highly dynamic process depending on the cooperative signal integration from distinct regulatory domains.

4. Regulation of PLC-γ isozymes

Activation of PLC-γ enzymes is tightly regulated by receptor and nonreceptor tyrosine kinases (see Figs. 3 and 4). Whereas PLC-γ1 is expressed ubiquitously, PLC-γ2 is most abundant in hematopoietic cells. Binding of polypeptide growth factors, such as platelet-derived growth factor (PDGF) and epidermal growth factor (EGF), to their appropriate receptors induces receptor dimerization, followed by activation of their intrinsic tyrosine kinase activity and autophosphorylation of cytoplasmic tyrosine residues. RTK activities are strictly controlled by protein tyrosine phosphatases. As it has been reported that generation of reactive oxygen species by the EGF receptor suppresses protein tyrosine phosphatase activities, such a mechanism may be involved in sustaining the phosphorylation state of the EGF receptor [74]. Specific tyrosine residues now function as docking sites for SH2 domain-containing effector proteins, the binding specificity being determined by both the sequence surrounding the tyrosine residue of the receptor as well as by the structure of the SH2 domain of the effector. While PLC-γ1 exclusively binds at Tyr-1021 to the PDGF receptor [75], binding of PLC-γ1 to the EGF receptor can occur at five preferred autophosphorylation sites [76]. Association of PLC-γ1 to the receptors for

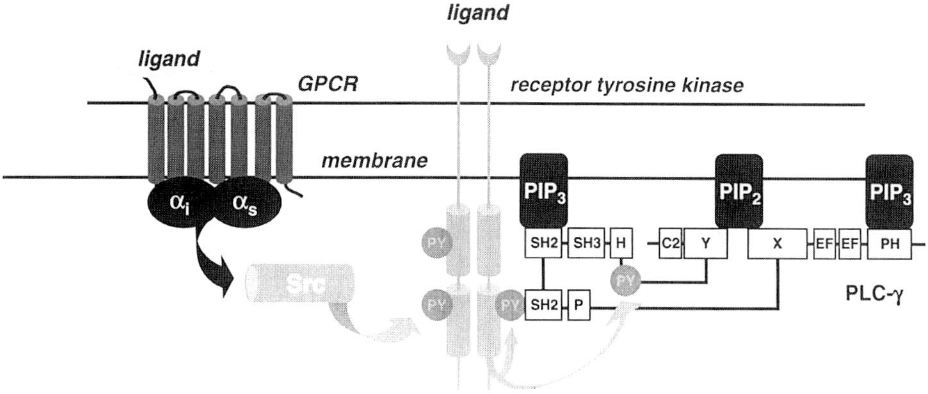

Fig. 3. Proposed activation of PLC-γ by tyrosine kinases. In most cells, PLC-γ is typically activated by RTKs. After binding the respective ligand, RTKs dimerize and become activated. Autophosphorylation of tyrosine residues in the cytoplasmic tail generates docking sites for target SH2 domain-containing proteins, including PLC-γ and PI3K. PLC-γ thereupon becomes phosphorylated and activated by the receptor kinase. Thus, PLC-γ interacts via its NH_2-terminal SH2 domain to the activated receptor and via its COOH-terminal SH2 domain as well as its PH domain to PIP_3 in the plasma membrane. Alternatively, phosphorylation of the RTK and, hence, recruitment and activation of PLC-γ, might be committed by the cytoplasmic tyrosine kinase Src. Src can become activated by stimulation of GPCRs, involving $G\alpha_i$ and $G\alpha_s$ subunits. (*For a colored version of this figure, see plate section, page 614.*)

PDGF and EGF requires both SH2 domains, however, the NH_2-terminal SH2 domain is apparently more important [76,77]. Tyrosine phosphorylation of PLC-γ1 occurs on the residues 771, 783 and 1254, but only Tyr-783 is required for activation of the enzyme [78]. As recently demonstrated by Pendergast and colleagues, regulation of PLC-γ1 by the PDGF receptor may also involve nonreceptor tyrosine kinases. Activation of the nonreceptor tyrosine kinase c-Abl requires functional PLC-γ1, acting probably in concert with a tyrosine kinase of the Src family to relieve c-Abl from inhibition by PIP_2. After activation, c-Abl negatively regulates PLC-γ1 by phosphorylation [79]. These findings identify a novel interdependence between the nonreceptor tyrosine kinase c-Abl and phosphoinositide signaling by PLC-γ1. In addition, it has been demonstrated that activated $G\alpha_s$ and $G\alpha_i$, but not $G\alpha_q$, $G\alpha_{12}$ or $G\beta\gamma$, can directly bind to c-Src and trigger its enzyme activity [80]. Such a mechanism may contribute to activation of PLC-γ1 by membrane receptors coupling to these heterotrimeric G proteins.

Tyrosine phosphorylation of PLC-γ1 is necessary but not sufficient for full activation. Indeed, recent studies have shown that activation of the PDGF receptor induces translocation and activation of PLC-γ1 by generating PIP_3 by the action of PI-3-kinase [31–33]. Membrane association of PLC-γ1 is mediated by specific interaction of PIP_3 with the PH domain and the COOH-terminal SH2 domain of PLC-γ1, acting probably in concert to enhance PIP_2 substrate availability of the enzyme. The PIP_3-dependent association of PLC-γ1 with membrane compartments is probably a highly dynamic process, as the cellular content of PIP_3 is tightly regulated by phosphoinositide phosphatases, in particular PTEN and SH2 domain-containing inositol-5-phosphatase [16,17]. In addition, PLC-γ1 binds via its split PH domain to EF-1α, an activator of

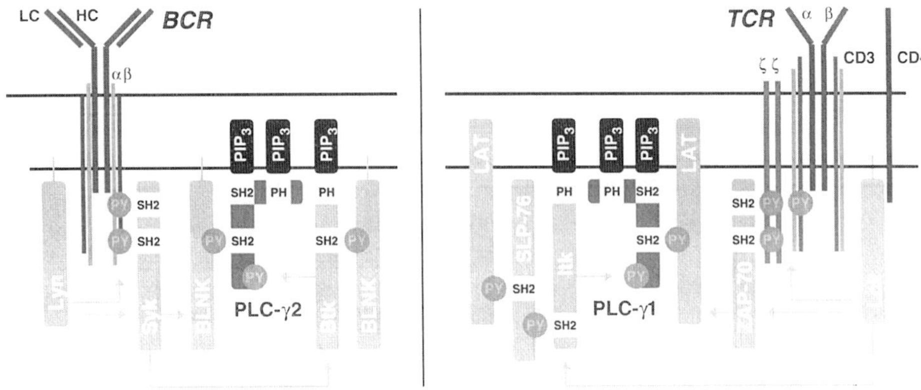

Fig. 4. Recruitment and activation of PLC-γ in lymphocytes by concerted tyrosine kinase cascades. In B cells (left panel), stimulation of the BCR complex results in the phosphorylation of tyrosine residues in the Igα and Igβ chains by activated Lyn. Syk docks with its tandem SH2 domains to the receptor and becomes phosphorylated (and activated) by Lyn as well. Activated Syk now phosphorylates the adaptor protein BLNK, which becomes associated with the membrane. BLNK serves as docking station for both PLC-γ2 and the tyrosine kinase Btk. Concomitant stimulation of PI3K (via phosphorylation of BCAP due to a concerted action of Syk and Btk, not shown) results in the production of PIP_3, which contributes to the recruitment of PLC-γ2 and Btk by binding their PH domains as well as the COOH-terminal SH2 domain of PLC-γ2. Thus brought together, Btk is phosphorylated and activated by Syk and now phosphorylates and activates PLC-γ2. In T cells (right panel), activation of PLC-γ1 is achieved by a very similar mechanism. After stimulation of the TCR complex, CD4-associated Lck phosphorylates the ITAM sequences located in CD3 and TCRζ, permitting the recruitment of ZAP-70. ZAP-70 becomes activated and phosphorylates the docking station LAT. PLC-γ1 binds with its NH2-terminal SH2 domain to LAT and, again, to locally produced PIP_3 in the membrane. Itk binds indirectly via SLP-76 (complexed with Gads, not shown) also to LAT, as well as to PIP_3. Lck activates Itk, which in turn phosphorylates and activates PLC-γ1. (*For a colored version of this figure, see plate section, page 615.*)

PI-4-kinases promoting the production of PIP and PIP_2. EF-1α also controls binding of PLC-γ1 to PIP_2 by this PH domain and thus may contribute to the membrane association and activation of PLC-γ1 [49]. Interestingly, Snyder and colleagues have reported that the SH3 domain of PLC-γ1 acts as a GEF for the nuclear GTPase PIKE, known to activate nuclear PI-3-kinase activity [50]. Thus, PLC-γ1 may positively control its PIP_3-dependent membrane association by an SH3 domain-mediated increase in the cellular content of PIP_3. Indeed, it has been demonstrated that a PLC-γ1 mutant lacking the SH3 domain exhibits reduced membrane association of the enzyme [81].

The generation of PIP_3 by PI-3-kinase is probably a general mechanism to induce sustained PLC-γ-dependent calcium signaling. In particular, in B and T lymphocytes, appropriate recruitment of nonreceptor tyrosine kinases and PLC-γ into signaling complexes at the plasma membrane requires activation of PI-3-kinase and thus, the generation of PIP_3 [82–84]. The B cell antigen receptor (BCR) and the T cell antigen receptor (TCR) belong to the family of multichain immune recognition receptors, consisting of distinct membrane-spanning subunits to recognize antigen/Ig and to recruit signaling components (see Fig. 4). Oligomerization of the BCR by multivalent ligand binding induces translocation of the BCR to lipid rafts, and the immune receptor tyrosine-based activation motifs (ITAMs) in the Igα and Igβ chains become phosphorylated by the Src family kinase

Lyn, thereby generating docking sites for the tandem SH2 domain of the tyrosine kinase Syk. Lyn also phosphorylates (activates) ITAM-bound Syk, and activated Syk now recruits the B-cell linker protein (BLNK) and the B cell adaptor for PI-3-kinase (BCAP), probably in concert with the Tec family tyrosine kinase Btk. PIP_3, produced by coincident activation of PI-3-kinase, binds to both Btk and PLC-γ2, thereby stabilizing their presence in the *signalosome*, and thus supports phosphorylation (activation) of PLC-γ2 due to a concerted action of Syk and Btk [83] (see Fig. 4, left panel). In T lymphocytes, activation of PLC-γ1 by TCR engagement occurs by a very similar scheme. The ITAM sequences of the TCR are phosphorylated by the Src family kinase Lck, followed by recruitment and phosphorylation (activation) of ZAP-70. The linker for activation of T cells (LAT) and the SH2-containing leukocyte-specific protein of 76 kDa (SLP-76) complexed with the adaptor protein Gads are phosphorylated by ZAP-70, and serve as docking stations for PLC-γ1 and the Tec family kinase Itk. TCR-induced activation of PI-3-kinase generates PIP_3, which detains PLC-γ1 and Itk in the lipid rafts and promotes phosphorylation (activation) of PLC-γ1 by Itk [84] (see Fig. 4, right panel). The detailed mechanisms of PLC-γ activation by BCR and TCR have recently been reviewed elsewhere [1,2,85].

Recent evidence indicates that PLC-γ enzymes are also activated by other lipid modulators, in particular PA (reviewed in [86]). Ryu and co-workers recently elucidated the molecular mechanisms of PLC-γ1 activation by PA [51]. PLC-γ1 binds specifically via its SH3 domain to the PX domain (comprising the residues Pro-145 and Pro-148) of PLD2 in an EGF-dependent manner. Binding of PLD2 to PIP_2 via its PH/PX domains induces translocation of the enzyme to PIP_2-enriched membrane compartments and activation of PLD2 [13]. Based on these findings it may be speculated that membrane recruitment and activation of PLD2 may result in membrane association of PLC-γ1 as well, accompanied by enhanced activation of PLC-γ1 due to PLD2-catalyzed PA generation.

5. Regulation of PLC-δ isozymes

The four PLC-δ enzymes are most sensitive to calcium. In PLC-δ1, binding of calcium to the EF-hands promotes the interaction of its PH domain with PIP_2 [38], and binding of calcium to the C2 domain enhances enzyme activity, due to increased substrate affinity after formation of the ternary complex enzyme–phosphatidylserine–calcium [41]. Further studies indicated that complex formation also promotes membrane targeting of PLC-δ1 and PLC-δ3, but not of PLC-δ4, and may thus be responsible for distinct subcellular distribution of PLC-δ enzymes [42]. PLC-δ1 contains both a nuclear localization signal (located at the COOH-terminus of the X domain) and a nuclear export signal (located at the residues 164–177 of the EF-hand domain), and thus PLC-δ1 may in concert with PLC-β1 generate second messengers in the nucleus to regulate nuclear shape and transcription [87,88]. Katan and co-workers have reported that changes in intracellular calcium concentration within the physiological range selectively stimulated the activity of PLC-δ1 [89]. Likewise, it has been demonstrated that PLC-δ1 is activated by capacitative calcium entry following the activation of PLC-β by the bradykinin receptor [90]. Recently, it has been reported that PLC-γ is required for agonist-induced calcium entry [91]. Thus, PLC-δ enzymes may act in concert with PLC-β and PLC-γ by activation of membrane receptors

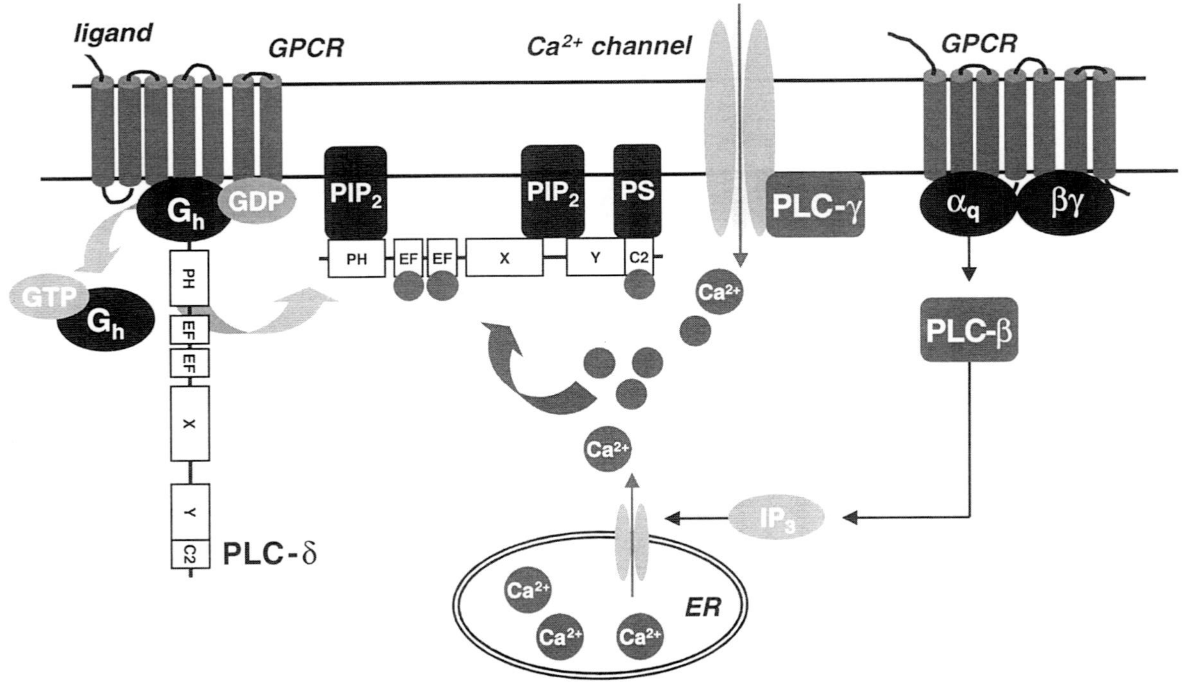

Fig. 5. Activation of PLC-δ by Ca^{2+}. Stimulation of PLC-β by GPCRs results in the production of IP_3 and hence, in the mobilization of Ca^{2+} from the endoplasmic reticulum. Capacitative entry of extracellular Ca^{2+} or opening of receptor- and second messenger-operated channels, processes recently shown to depend on PLC-γ, also contribute to the increase in intracellular Ca^{2+}. Binding of Ca^{2+} to the EF-hands and the C2 domain of PLC-δ is involved in the location and the activation of the enzyme. In resting cells, PLC-δ is held inactive by interaction with GDP-bound G_h proteins. Stimulation of receptors coupling to G_h induces the release of G_h from the membrane and dissociation of PLC-δ, which can now be relocated at the membrane by interaction with PIP_2, phosphatidylserine (PS) and Ca^{2+}. (*For a colored version of this figure, see plate section, page 616.*)

(see Fig. 5). Transglutaminase II (G_h), which has the ability to bind and hydrolyze GTP, directly interacts with and activates PLC-δ1 in vitro [92]. The interaction site of G_h in PLC-δ1 has been located to the C2 domain corresponding to the residues 721–736 [93]. In addition, evidence has been provided that binding of GDP-bound G_h to PLC-δ1 inhibits enzyme activity, and that GTP loading of G_h after stimulation of membrane receptors may thus release PLC-δ1 from its inhibitory constraint [94]. G_h probably mediates activation of PLC-δ1 by the oxytocin receptor, the thromboxane A_2 receptor and the α_{1B} adrenoceptor [95–97]. However, activation of PLC-δ isoforms by membrane receptors is only poorly understood (see Fig. 5).

6. Regulation of PLC-ε

Recently, PLC-ε has been identified as a novel mammalian PLC isoform with unique structural and regulatory features. PLC-ε is the mammalian homologue of *Caenorhabditis elegans* PLC210, known to bind to H-Ras in a GTP-dependent manner [22]. PLC-ε exists in two alternatively spliced forms with molecular weights of ~255 kDa (2281 residues) and ~230 kDa (1994 residues), and the messenger RNA encoding PLC-ε is most abundant in the heart, followed by the kidney, lung and brain [52–54]. The presence of a CDC25 domain possessing Ras-GEF activity at the NH_2-terminus and two RA domains (RA1 and RA2) at the COOH-terminus probably directly links PLC-ε with Ras signaling [98–100].

Initially, Kataoka and colleagues demonstrated by in vitro binding and yeast two-hybrid screening that PLC-ε specifically binds to GTγS-loaded H-Ras ($K_D \sim 40$ nM) and Rap1A, but not to R-Ras, RalA, RhoA, Rac1 and Cdc42, and that this binding required the RA domains of PLC-ε. The interaction with PLC-ε was abolished in effector mutants of Ras, suggesting that PLC-ε can act as a downstream effector of Ras proteins [54]. Binding of Ras proteins to PLC-ε did not affect its lipase activity, but profoundly changed its intracellular localization (see below). Smrcka and colleagues showed that H-Ras specifically binds to the RA2 domain of PLC-ε in a GTP-dependent manner, whereas low affinity binding of H-Ras to the RA1 domain of PLC-ε was GTP-independent [52]. Again, in vitro binding of GTPγS-loaded H-Ras did not directly affect PLC-ε enzyme activity, but cotransfection of constitutively active Q71L H-Ras and PLC-ε in COS-7 cells enhanced enzyme activity, and this effect required both RA domains of PLC-ε. The RA2 domain of PLC-ε seems to be a bona fide RA, as mutation of Lys-2150 to Ala or Lys-2152 to Glu, residues which are conserved in RA domains of other Ras effectors, largely reduced binding of Ras to PLC-ε. Likewise, mutation of Thr-35 to Ser or Tyr-40 to Cys in the effector domain of Ras fully inhibited binding to PLC-ε. Most importantly, mutational effects on binding directly correlated with the ability of Ras to enhance PLC-ε activity in intact cells, indicating that PLC-ε is indeed a Ras effector [52]. Although binding of H-Ras to PLC-ε did not require its post-translational modification, activation of PLC-ε in intact cells was dependent on lipid modification of H-Ras. Stimulation of PLC-ε by H-Ras seems to involve additional cellular components, which might be present at the plasma membranes or intracellular compartments.

Translocation of PLC-ε to the plasma membrane in response to EGF was Ras-dependent, whereas Rap1A induced translocation of PLC-ε to perinuclear Golgi regions [54]. Thus, RA domain-dependent signals by Ras and Rap1A probably regulate differential pathways at distinct subcellular locations. In further studies, it was demonstrated that PLC-ε exhibits significant GEF activity towards Rap1A, but not to H-Ras, R-Ras and Ral, and that this GEF activity was dependent on the CDC25 domain [101]. Furthermore, it was shown that the CDC25 domain is required to induce a long lasting translocation of PLC-ε to the Golgi compartment. Thus, PLC-ε specifically activates via the CDC25 domain Rap1A, which in turn stimulates PLC-ε via binding to the RA domains. By this mechanism, PLC-ε might function as amplifier for Rap1A signaling, leading to cellular proliferation, differentiation, lymphocyte aggregation, T-cell anergy and platelet activation [102]. The CDC25 domain of PLC-ε seems to have no GEF activity towards Rap2A and Rap2B, structurally close relatives of Rap1A [103]. However, both GTPγS-loaded Rap2A and Rap2B bind to PLC-ε in an RA domain-dependent manner. In addition, it has been shown that Rap2B, similar to H-Ras, TC21 and Rap1A, does enhance PLC-ε activity upon cotransfection in intact cells [103,104]. In contradiction to other studies, Lomasney and colleagues have found that PLC-ε exhibits GEF activity towards Ras, an effect depending on the CDC25 domain as well, but that Ras does not increase PLC-ε activity upon cotransfection [53]. Therefore, the ability of Ras to stimulate PLC-ε and the specificity of the CDC25 domain have not yet been unanimously answered. The data obtained so far indicate that two distinct Rap proteins, Rap1A and Rap2B, stimulate PLC-ε in an RA domain-dependent manner, but that PLC-ε only exhibits GEF activity towards Rap1A, but not Rap2B. Thus, a GEF distinct from the CDC25 domain probably mediates GTP loading of Rap2B (see below). Indeed, distinct Rap-specific GEFs have been identified, such as C3G, CalDAG-GEFs, PDZ-GEFs and Epac-GEFs, but their integration in specific receptor signaling pathways, and in particular their role in PLC-ε signaling, has not been analyzed so far [99,100].

PLC-ε contains a long regulatory COOH-terminal domain similar to the PLC-β enzymes, with which PLC-ε shares the closest evolutionary homology. The COOH-terminus of PLC-β enzymes are known to mediate their specific interaction with GTP-bound α subunits of G_q proteins [44,45]. Lomasney and colleagues demonstrated that the expression of constitutively active $G\alpha_{12}$, but not active mutants of $G\alpha_q$, $G\alpha_{13}$, Gαs and $G\alpha_i$, specifically increased PLC-ε activity, whereas Harden and colleagues found that both $G\alpha_{12}$ and $G\alpha_{13}$-stimulated PLC-ε activity in intact cells. Thus, stimulation of PLC-ε by G_{12} proteins may contribute to the mitogenic potential of the G_{12} family of heterotrimeric G proteins [105,106]. Harden and colleagues also studied the effects of Gβγ subunits on PLC-ε. Coexpression of $G\beta_1\gamma_2$ with PLC-ε markedly increased enzyme activity, an effect that was not additive with α subunits of G_{12} proteins and independent of the RA2 domain, thus Ras-independent [58]. It is presently unclear whether the effects of Gβγ are mediated via interaction with the NH_2-terminal PH domain of PLC-ε. However, as inhibition of PI-3-kinase did not abolish PLC-ε stimulation by Gβγ subunits, membrane targeting of PLC-ε by PIP_3 via its PH domain seems not to be involved in this process. $G\beta_1$ in combination with $G\gamma_1$, $G\gamma_2$, $G\gamma_3$ and $G\gamma_{13}$ all activated PLC-ε; $G\beta_2$ and $G\beta_4$ in combination with $G\gamma_2$ stimulated PLC-ε as the $G\beta_1$-containing dimers, whereas $G\beta_3$-containing dimers were less active and $G\beta_5$ was inactive [58]. Taken together, PLC-ε is obviously under control of α subunits of G_{12} type proteins, Gβγ subunits and Ras-like GTPases, in particular H-Ras, Rap1A and Rap2B, some of which may even act upstream

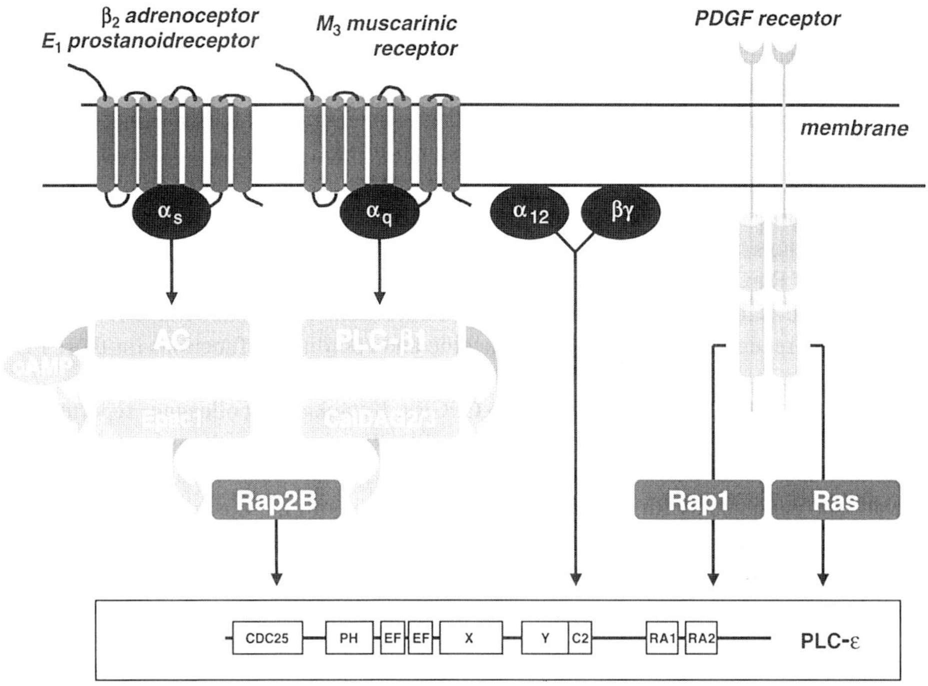

Fig. 6. Model of the regulation of PLC-ε by membrane receptors involving Ras-like GTPases. Rap2B is specifically involved in the stimulation of PLC-ε by GPCRs and likely converges signals of at least two different pathways. Stimulation of the β_2 adrenoceptor or the E_1 prostanoidreceptor coupling to $G\alpha_s$ leads to activation of adenylate cyclase (AC) and the formation of cAMP. As a result, the cAMP-activated GEF Epac1 facilitates GTP loading and hence, activation of Rap2B. Alternatively, activation of $G\alpha_q$ proteins, e.g. by the M_3 muscarinic receptor, directly stimulates PLC-β1, probably leading to an increase in the calcium- and DAG-regulated exchange activity of CalDAG2 and CalDAG 3 and, likewise, activation of Rap2B. Rap2B stimulates PLC-ε by a yet unknown mechanism. The M_3 muscarinic receptor can also contribute to the $G\alpha_s$-induced increase in cAMP. Furthermore, $G\alpha_{12}$ and βγ dimers may play a role in the regulation of PLC-ε. Activation of PLC-ε by the RTK for PDGF has been shown to involve Ras and Rap1. (*For a colored version of this figure, see plate section, page 617.*)

and downstream of PLC-ε. It should be emphasized, however, that none of these signaling molecules is sufficient to directly stimulate PLC-ε activity in vitro. Thus, a highly organized signaling complex is apparently required to achieve PLC-ε activation (see Fig. 6).

The regulation of PLC-ε by membrane receptors has been addressed in a few reports. We have found that activation of the β_2 adrenoceptor and the prostanoid receptor for prostaglandin E_1, both of which are typically G_s- and adenylyl cyclase-coupled GPCRs, can induce PLC-ε stimulation [107]. This novel PLC and calcium signaling pathway is dependent on cyclic AMP, but independent of protein kinase A. Stimulation of PLC-ε by these receptors was specifically mediated by Rap2B, but not by H-Ras, Rap1A and Rap2A. Activation of Rap2B involved the Rap-specific GEF, Epac. As reported for PLC-ε, Epac is abundantly expressed in the heart and exhibits a perinuclear localization [109,110]. Thus, integration of Epac and PLC-ε in β_2 adrenoceptor signaling may modulate heart physiology. Although protein kinase A-dependent phosphorylation inhibits some PLC isoforms, the increase of cellular cAMP triggered by activation of the β_2 adrenoceptor apparently activates PLC-ε via Epac. Cellular compartmentalization of cAMP and calcium signaling may allow coordinated integration of these distinct processes [111]. Also the M_3 muscarinic receptor can signal to PLC-ε via $G\alpha_s$, Epac and Rap2B [108]. However, an alternative pathway has become apparent. Stimulation of PLC-β by GPCRs coupling to $G\alpha_q$, including the M_3 muscarinic receptor, was also found to discharge in the activation of Rap2B and PLC-ε (own unpublished data). In this pathway, activation of Rap2B was achieved by the Ca^{2+} and DAG-sensitive exchange factors CalDAG2 and CalDAG3. This exciting cascade positions PLC-ε downstream of PLC-β during single receptor action, suggesting specific and coordinated functions for these PLC isoforms.

Regulation of PLC-ε is not restricted to GPCRs. Kataoka and colleagues reported that ectopically expressed PLC-ε is activated by a PDGF receptor deficient in its ability to phosphorylate and activate PLC-γ1. It was shown that the rapid and initial phase of PLC-ε activation is mediated by Ras, while prolonged activation of PLC-ε is mediated by Rap1A in a CDC25 domain-dependent manner [103]. In line with these findings, we have found recently that EGF receptor signaling to PLC-ε is mediated by Rap2B, but obviously involves a Rap-specific GEF distinct from Epac [112]. It has been reported that activation of the Raf-1 kinase/extracellular signal-regulated kinase pathway by PLC-ε is Ras-dependent, whereas activation of B-Raf kinase/extracellular signal-regulated kinase pathway by PLC-ε is Rap1A-dependent [53,54]. Thus, regulation of PLC-ε by Ras proteins probably modulates cellular growth and differentiation. Indeed, evidence was provided that activation of PLC-ε by Ras and Rap1A is required for cell proliferation [103]. As Ras proteins are products of genes mutated in ~15% of all human tumors, it will be of particular interest to understand the role played by PLC-ε in the regulation of cellular growth, differentiation and oncogenesis by Ras proteins.

7. Concluding remarks

We focus in the present overview on the remarkable progress in our knowledge about the mechanisms of signal integration from membrane receptors to the distinct PLC enzymes. It is evident that the PLC enzymes greatly differ in their responsiveness to receptor and nonreceptor tyrosine kinases, heterotrimeric G proteins coupling to the family of serpentine receptors, small GTPases and calcium, probably due to their structural organization. Besides distinct modes of activation by membrane receptors, agonist-induced recruitment of PLC enzymes to the plasma membrane, in particular to signaling microdomains such as caveolae and lipid rafts, represents the initial common regulatory principle. The generation of protein–protein and protein–lipid interactions due to the presence of distinct regulatory domains confer specificity into the highly dynamic organization of PLC signaling in space and time. Thus, the PI-3-kinase reaction product, PIP_3, tightly controls appropriate intracellular localization of PLC enzymes, thereby likely permitting sustained PLC signaling. Substantial progress has been made in recent years, particularly in our understanding of the regulation of PLC-β enzymes by heterotrimeric G protein subunits and Rho-like GTPases. The regulation of the newly identified PLC-ε exhibits most striking features. Activation of PLC-ε is apparently mediated by GPCRs and RTKs, involving Ras-like GTPases, the latter being activated by specific Ras-specific GEFs. Future studies will gain insights into the molecular mechanisms of signal integration leading to the regulation of PLC-ε.

Acknowledgements

The authors' work reported herein was supported by the Deutsche Forschungsgemeinschaft and the IFORES program of the Universitätsklinikum Essen.

References

[1] Rebecchi, M.J., Pentyala, S.N., 2000. Physiol. Rev. 80, 1291–1335.
[2] Rhee, S.G., 2001. Annu. Rev. Biochem. 70, 281–312.
[3] Fukami, K., 2002. J. Biochem. 131, 293–299.
[4] Berridge, M.J., Lipp, P., Bootman, M.D., 2000. Nat. Rev. Mol. Cell. Biol. 1, 11–21.
[5] Nishizuka, Y., 1995. FASEB J. 9, 484–496.
[6] Vanhaesebroeck, B., Leevers, S.J., Ahmadi, K., Timms, J., Katso, R., Driscoll, P.C., Woscholski, R., Parker, P.J., Waterfield, M.D., 2001. Annu. Rev. Biochem. 70, 535–602.
[7] Kavran, J.M., Klein, D.E., Lee, A., Falasca, M., Isakoff, S.J., Skolnik, E.Y., Lemmon, M.A., 1998. J. Biol. Chem. 273, 30497–30508.
[8] Mosior, M., Six, D.A., Dennis, E.A., 1998. J. Biol. Chem. 273, 2184–2191.
[9] Toker, A., 1998. Curr. Opin. Cell. Biol. 10, 254–261.
[10] Janmey, P.A., Xian, W., Flanagan, L.A., 1999. Chem. Phys. Lipids 101, 93–107.
[11] Czech, M.P., 2000. Cell 100, 603–606.
[12] Xu, Y., Seet, L.F., Hanson, B., Hong, W., 2001. Biochem. J. 360, 513–530.
[13] Exton, J.H., 2002. Rev. Physiol. Biochem. Pharmacol. 144, 1–94.
[14] Lemmon, M.A., Ferguson, K.M., Abrams, C.S., 2002. FEBS Lett. 513, 71–76.
[15] Anderson, R.A., Boronenkov, I.V., Doughman, S.D., Kunz, J., Loijens, J.C., 1999. J. Biol. Chem. 274, 9907–9910.
[16] Majerus, P.W., Kisseleva, M.V., Norris, F.A., 1999. J. Biol. Chem. 274, 10669–10672.
[17] Maehama, T., Taylor, G.S., Dixon, J.E., 2001. Annu. Rev. Biochem. 70, 247–279.
[18] Jang, I.H., Lee, B.D., Bae, S.S., Park, M.H., Suh, P.G., Ryu, S.H., 2001. FEBS Lett. 491, 4–8.
[19] Scarlata, S., 2002. Biochemistry 41, 7092–7099.
[20] Anderson, R.G.W., Jacobson, K., 2002. Science 296, 1821–1825.
[21] Zajchowski, L.D., Robbins, S.M., 2002. Eur. J. Biochem. 269, 737–752.
[22] Shibatohge, M., Kariya, K., Liao, Y., Hu, C.D., Watari, Y., Goshima, M., Shima, F., Kataoka, T., 1998. J. Biol. Chem. 273, 6218–6222.
[23] Ferguson, K.M., Lemmon, M.A., Schlessinger, J., Sigler, P.B., 1995. Cell 83, 1037–1046.
[24] Essen, L.O., Perisic, O., Cheung, R., Katan, M., Williams, R.L., 1996. Nature 380, 595–602.
[25] Hondal, R.J., Zhao, Z., Kravchuk, A.V., Liao, H., Riddle, S.R., Yue, X., Bruzik, K.S., Tsai, M.D., 1998. Biochemistry 37, 4568–4580.
[26] Ellis, M.V., James, S.R., Perisic, O., Downes, C.P., Williams, R.L., Katan, M., 1998. J. Biol. Chem. 273, 11650–11659.
[27] Wang, T., Pentyala, S., Rebecchi, M.J., Scarlata, S., 1999. Biochemistry 38, 1517–1524.
[28] Illenberger, D., Walliser, C., Nürnberg, B., Lorente, M.D., Gierschik, P., 2003. J. Biol. Chem. 278, 3006–3014.

[29] Snyder, J.T., Singer, A.U., Wing, M.R., Harden, T.K., Sondek, J., 2003. J. Biol. Chem. 278, 21099–21104.
[30] Yagisawa, H., Sakuma, K., Paterson, H.F., Cheung, R., Allen, V., Hirata, H., Watanabe, Y., Hirata, M., Williams, R.L., Katan, M., 1998. J. Biol. Chem. 273, 417–424.
[31] Bae, Y.S., Cantley, L.G., Chen, C.S., Kim, S.R., Kwon, K.S., Rhee, S.G., 1998. J. Biol. Chem. 273, 4465–4469.
[32] Rameh, L.E., Rhee, S.G., Spokes, K., Kazlauskas, A., Cantley, L.C., Cantley, L.G., 1998. J. Biol. Chem. 273, 23750–23757.
[33] Falasca, M., Logan, S.K., Lehto, V.P., Baccante, G., Lemmon, M.A., Schlessinger, J., 1998. EMBO J. 17, 414–422.
[34] Razzini, G., Brancaccio, A., Lemmon, M.A., Guarnieri, S., Falasca, M., 2000. J. Biol. Chem. 275, 14873–14881.
[35] Runnels, L.W., Jenco, J., Morris, A., Scarlata, S., 1996. Biochemistry 35, 16824–16832.
[36] Wang, T., Dowal, L., El-Maghrabi, M.R., Rebecchi, M., Scarlata, S., 2000. J. Biol. Chem. 275, 7466–7469.
[37] Philip, F., Guo, Y., Scarlata, S., 2002. FEBS Lett. 531, 28–32.
[38] Yamamoto, T., Takeuchi, H., Kanematsu, T., Allen, V., Yagisawa, H., Kikkawa, U., Watanabe, Y., Nakasima, A., Katan, M., Hirata, M., 1999. Eur. J. Biochem. 265, 481–490.
[39] Nalefski, E.A., Wisner, M.A., Chen, J.Z., Sprang, S.R., Fukuda, M., Mikoshiba, K., Falke, J.J., 2001. Biochemistry 40, 3089–3100.
[40] Grobler, J.A., Hurley, J.H., 1998. Biochemistry 37, 5020–5028.
[41] Lomasney, J.W., Cheng, H.F., Roffler, S.R., Kling, K., 1999. J. Biol. Chem. 274, 21995–22001.
[42] Ananthanarayanan, B., Das, S., Rhee, S.G., Murray, D., Cho, W., 2002. J. Biol. Chem. 277, 3568–3575.
[43] Wang, T., Pentyala, S., Elliott, J.T., Dowal, L., Gupta, E., Rebecchi, M.J., Scarlata, S., 1999. Proc. Natl Acad. Sci. USA 96, 7843–7846.
[44] Ilkaeva, O., Kinch, L.N., Paulssen, R.H., Ross, E.M., 2002. J. Biol. Chem. 277, 4294–4300.
[45] Singer, A.U., Waldo, G.L., Harden, T.K., Sondek, J., 2002. Nat. Struct. Biol. 9, 32–36.
[46] Ross, E.M., Wilkie, T.M., 2000. Ann. Rev. Biochem. 69, 795–827.
[47] Kim, G.G., Park, D., Rhee, S.G., 1996. J. Biol. Chem. 271, 21187–21192.
[48] Suh, P.G., Hwang, J.I., Ryu, S.H., Donowitz, M., Kim, J.H., 2001. Biochem. Biophys. Res. Commun. 288, 1–7.
[49] Chang, J.S., Seok, H., Kwon, T.K., Min, D.S., Ahn, B.H., Lee, Y.H., Suh, J.W., Kim, J.W., Iwashita, S., Omori, A., Ichinose, S., Numata, O., Seo, J.K., Oh, Y.S., Suh, P.G., 2002. J. Biol. Chem. 277, 19697–19702.
[50] Ye, K., Aghdasi, B., Luo, H.R., Moriarity, J.L., Wu, F.Y., Hong, J.J., Hurt, K.J., Bae, S.S., Suh, P.G., Snyder, S.H., 2002. Nature 415, 541–544.
[51] Jang, I.H., Lee, S., Park, J.B., Kim, J.H., Lee, C.S., Hur, E.M., Kim, I.S., Kim, K.T., Yagisawa, H., Sug, P.G., Ryu, S.H., 2003. J. Biol. Chem. 278, 18184–18190.
[52] Kelley, G.G., Reks, S.E., Ondrako, J.M., Smrcka, A.V., 2001. EMBO J. 20, 743–754.
[53] Lopez, I., Mak, E.C., Ding, J., Hamm, H.E., Lomasney, J.W., 2001. J. Biol. Chem. 276, 2758–2765.
[54] Song, C., Hu, C.D., Masago, M., Kariya, K., Yamawaki-Kataoka, Y., Shibatohge, M., Wu, D., Satoh, T., Kataoka, T., 2001. J. Biol. Chem. 276, 2752–2757.
[55] Lefkowitz, R.J., 2000. Nat. Cell. Biol. 2, E133–E136.
[56] Marinissen, M.J., Gutkind, J.S., 2001. Trends Pharmcol. Sci. 22, 368–376.
[57] Runnels, L.W., Scarlata, S.F., 1999. Biochemistry 38, 1488–1496.
[58] Wing, M.R., Houston, D., Kelley, G.G., Der, C.J., Siderovski, D.P., Harden, T.K., 2001. J. Biol. Chem. 276, 48257–48261.
[59] Maier, U., Babich, A., Macrez, N., Leopoldt, D., Gierschik, P., Illenberger, D., Nürnberg, B., 2000. J. Biol. Chem. 275, 13746–13754.
[60] Kuang, Y., Wu, Y., Smrcka, A., Jiang, H., Wu, D., 1996. Proc. Natl Acad. Sci. USA 93, 2964–2968.
[61] Sankaran, B., Osterhout, J., Wu, D., Smrcka, A.V., 1998. J. Biol. Chem. 273, 7148–7154.

[62] Offermanns, S., Toombs, C.F., Hu, Y.H., Simon, M.I., 1997. Nature 389, 183–186.
[63] Chan, J.S.C., Lee, J.W.M., Ho, M.K.C., Wong, Y.H., 2000. Mol. Pharmacol. 57, 700–708.
[64] Illenberger, D., Schwald, F., Gierschik, P., 1997. Eur. J. Biochem. 246, 71–77.
[65] Illenberger, D., Schwald, F., Pimmer, D., Binder, W., Maier, G., Dietrich, A., Gierschik, P., 1998. EMBO J. 17, 6241–6249.
[66] Illenberger, D., Walliser, C., Strobel, J., Gutman, O., Niv, H., Gaidzik, V., Kloog, Y., Gierschik, P., Henis, Y.I., 2003. J. Biol. Chem. 278, 8645–8652.
[67] Hong-Geller, E., Holowka, D., Siraganian, R.P., Baird, B., Cerione, R.A., 2001. Proc. Natl Acad. Sci. USA 98, 1154–1159.
[68] Thodeti, C.K., Massoumi, R., Bindslev, L., Sjölander, A., 2002. Biochem. J. 365, 157–163.
[69] Rabiet, M.J., Tardif, M., Braun, L., Boulay, F., 2002. Blood 100, 1835–1844.
[70] Mukhopadhyay, S., Ross, E.M., 1999. Proc. Natl Acad. Sci. USA 96, 9539–9544.
[71] Nash, M.S., Young, K.W., Challiss, R.A.J., Nahorski, S.R., 2001. Nature 413, 381–382.
[72] Young, K.W., Nash, M.S., Challiss, R.A.J., Nahorski, S.R., 2003. J. Biol. Chem. 278, 20753–20760.
[73] Litosch, I., 2003. Biochemistry 42, 1618–1623.
[74] Bae, Y.S., Kang, S.W., Seo, M.S., Baines, I.C., Tekle, E., Chock, P.B., Rhee, S.G., 1997. J. Biol. Chem. 272, 217–221.
[75] Claesson-Welsh, L., 1994. J. Biol. Chem. 269, 32023–32026.
[76] Chattopadhyay, A., Vecchi, M., Ji, Q.S., Mernaugh, R., Carpenter, G., 1999. J. Biol. Chem. 274, 26091–26097.
[77] Ji, Q.S., Chattopadhyay, A., Vecchi, M., Carpenter, G., 1999. Mol. Cell Biol. 19, 4961–4970.
[78] Kim, J.W., Sim, S.S., Kim, U.H., Nishibe, S., Wahl, M.I., Carpenter, G., Rhee, S.G., 1990. J. Biol. Chem. 265, 3940–3943.
[79] Plattner, R., Irvin, B.J., Guo, S., Blackburn, K., Kazlauskas, A., Abraham, R.T., York, J.D., Pendergast, A.M., 2003. Nat. Cell. Biol. 5, 309–319.
[80] Ma, Y.C., Huang, J., Ali, S., Lowry, W., Huang, X.Y., 2000. Cell 102, 635–646.
[81] DeBell, K.E., Stoica, B.A., Verí, M.C., Di Baldassarre, A., Miscia, S., Graham, L.J., Rellahan, B.L., Ishiai, M., Kurosaki, T., Bonvini, E., 1999. Mol. Cell. Biol. 19, 7388–7398.
[82] Scharenberg, A.M., Kinet, J.P., 1998. Cell 94, 5–8.
[83] Kurosaki, T., 2002. Nat. Rev. Immunol. 2, 354–363.
[84] Wange, R.L., 2000. Sci. STKE 63, 1–13.
[85] Wilde, J.I., Watson, S.P., 2001. Cell. Signal. 13, 691–701.
[86] Sekiya, F., Bae, Y.S., Rhee, S.G., 1999. Chem. Phys. Lipids 98, 3–11.
[87] Yamaga, M., Fujii, M., Kamata, H., Hirata, H., Yagisawa, H., 1999. J. Biol. Chem. 274, 28537–28541.
[88] Okada, M., Fujii, M., Yamaga, M., Sugimoto, H., Sadano, H., Osumi, T., Kamata, H., Hirata, H., Yagisawa, H., 2002. Genes Cells 7, 985–996.
[89] Allen, V., Swigart, P., Cheung, R., Cockcroft, S., Katan, M., 1997. Biochem. J. 327, 545–552.
[90] Kim, Y.H., Park, T.J., Lee, Y.H., Baek, K.J., Suh, P.G., Ryu, S.H., Kim, K.T., 1999. J. Biol. Chem. 274, 26127–26134.
[91] Patterson, R.L., van Rossum, D.B., Ford, D.L., Hurt, K.J., Bae, S.S., Suh, P.G., Gill, D.L., 2002. Cell 111, 529–541.
[92] Feng, J.F., Rhee, S.G., Im, M.J., 1996. J. Biol. Chem. 271, 16451–16454.
[93] Kang, S.K., Kim, D.K., Damron, D.S., Baek, K.J., Im, M. J., 2002. Biochem. Biophys. Res. Commun. 293, 383–390.
[94] Murthy, S.N., Lomasney, J.W., Mak, E.C., Lorand, L., 1999. Proc. Natl Acad. Sci. USA 96, 11815–11819.
[95] Vezza, R., Habib, A., Fitzgerald, G.A., 1999. J. Biol. Chem. 274, 12774–12779.
[96] Park, E.S., Won, J.H., Han, K.J., Suh, P.G., Lee, H.S., Yun, H.Y., Kwon, N.S., Baek, K.J., 1998. Biochem. J. 331, 283–289.
[97] Baek, K.J., Kang, S.K., Damron, D.S., Im, M.J., 2001. J. Biol. Chem. 276, 5591–5597.
[98] Cullen, P.J., 2001. Curr. Biol. 11, R342–R344.
[99] Cullen, P.J., Lockyer, P.J., 2002. Nat. Rev. 3, 339–348.
[100] Quilliam, L.A., Rebhuhn, J.F., Castro, A.F., 2002. Prog. Nucleic Acid. Res. 71, 391–444.
[101] Jin, T.G., Satoh, T., Liao, Y., Song, C., Gao, X., Kariya, K., Hu, C.D., Kataoka, T., 2001. J. Biol. Chem. 276, 30301–30307.

[102] Bos, J.L., de Rooij, J., Reedquist, K.A., 2001. Nat. Rev. 2, 369–377.

[103] Song, C., Satoh, T., Edamatsu, H., Wu, D., Tadano, M., Gao, X., Kataoka, T., 2002. Oncogene 21, 8105–8113.

[104] Murphy, G.A., Graham, S.M., Morita, S., Reks, S.E., Rogers-Graham, K., Voijtek, A., Kelley, G.G., Der, C.J., 2002. J. Biol. Chem. 277, 9966–9975.

[105] Dhanasekaran, N., Dermott, J.M., 1996. Cell. Signal. 8, 235–245.

[106] Gutkind, J.S., 1998. Oncogene 17, 1331–1342.

[107] Schmidt, M., Evellin, S., Oude Weernink, P.A., vom Dorp, F., Rehmann, H., Lomasney, J.W., Jakobs, K.H., 2001. Nat. Cell Biol. 3, 1020–1024.

[108] Evellin, S., Nolte, J., Tysack, K., vom Dorp, F., Thiel, M., Oude Weernink, P.A., Jakobs, K.H., Webb, E.J., Lomasney, J.W., Schmidt, M., 2002. J. Biol. Chem. 277, 16805–16813.

[109] de Rooij, J., Zwartkuis, F.J.T., Verheijen, M.H.G., Cool, R.H., Nijman, S.M.B., Wittinghofer, A., Bos, J.L., 1998. Nature 396, 474–477.

[110] Qiao, J., Mei, F.C., Popov, V.L., Vergara, L.A., Cheng, X., 2002. J. Biol. Chem. 277, 26581–26586.

[111] Zaccolo, M., Magalhaes, P., Pozzan, T., 2002. Curr. Opin. Cell. Biol. 14, 160–166.

[112] Evellin, S., Nolte, J., vom Dorp, F., Oude Weernink, P.A., Jakobs, K.H., Schmidt, M., von Eichel-Streiber, C., Lomasney, J.W., 2002. Naunyn Schmiedebergs Arch. Pharmacol. 354 (Suppl. R50).

Mammalian phospholipase D – properties and regulation

John H. Exton*

Howard Hughes Medical Institute, Department of Molecular Physiology and Biophysics, Pharmacology, Vanderbilt University School of Medicine, Nashville, TN 37232, USA
**Correspondence address: E-mail: john.exton@vanderbilt.edu(J.H.E.)*

1. Introduction

Phospholipase D (PLD) is an enzyme that hydrolyzes phospholipids to phosphatidic acid (PA) and the free head group. In mammalian cells, the substrate is almost exclusively phosphatidylcholine (PC) and the head group released is choline. Since choline levels in most mammalian cells are high, the signaling molecule produced by PLD is generally considered to be PA. PLD, widely distributed in prokaryotes and eukaryotes, is present in all mammalian tissues and almost all cell lines. In addition to catalyzing the hydrolysis of the phosphodiester bond of PC to release PA, PLD carries out the transphosphatidylation reaction. This is unique for this phospholipase and involves the transfer of the phosphatidyl moiety of PC to primary alcohols such as ethanol and 1-butanol. This reaction provides a specific assay for PLD and also a means for exploring the functions of PLD, because in the presence of high concentrations of primary alcohols, the production of PA is decreased.

All PLD isozymes share two conserved $HXKX_4DX_6GG/S$ sequences [1,2], designated HKD motifs, which mutational studies have shown are required for catalytic activity. These motifs are also present in other enzymes and proteins [1,3], which are said to comprise a PLD superfamily. Members of the PLD superfamily that exhibit PLD activity contain four conserved sequences (I–IV) of which sequences I and IV contain HKD motifs [4]. Two adjacent domains, Phox homology (PX) and pleckstrin homology (PH), are located in the N-terminal sequences of mammalian and certain other PLDs, but are absent from bacterial and most plant enzymes [5,6]. Mammalian, yeast and certain plant PLDs are dependent for activity on phosphatidylinositol 4,5-bisphosphate (PIP_2) [6–10]. Although the PH domain can bind PIP_2 [9], there is another PIP_2 binding site between conserved sequences II and III which is required for activity [8].

There are only two mammalian PLD genes whose protein products (PLD1 and PLD2) are alternatively spliced and show approximately 50% amino acid sequence identity [11,12].

Advances in Molecular and Cell Biology, Vol. 33, pages 451–462

ISSN: 1569-2558/DOI: 10.1016/S1569-2558(03)33022-X

PLD1 is regulated in vivo and in vitro by protein kinase C (PKC) and small G proteins of the Rho and Arf families. In contrast, PLD2 has a high basal activity and is not affected by PKC or Rho or Arf proteins in vitro. PLD1 and PLD2 are both expressed in most tissues and cell lines, but have different subcellular locations.

1.1. Structure and catalytic mechanism of phospholipase D

No mammalian PLD has been crystallized. However, the structure of a PLD from Streptomyces has been determined and that of the Nuc endonuclease from *S. typhimurium*, which is a low molecular mass member of the PLD superfamily [13,14]. The crystal structure of another member of the PLD superfamily, tyrosyl-DNA phosphodiesterase has been determined [15]. The Nuc protein crystallizes as a dimer, with the conserved HKD sequences opposed to produce a single active site [14]. The active site residues are held together by a network of H bonds that involve not only the conserved His, Ser and Asn residues, but also adjacent Glu residues. Substrate binding involves the His, Lys and Asn residues of the HKD motif [14].

Streptomyces PLD has a structure that is very similar to the Nuc dimer [13] and its catalytic center is very similar to that of the Nuc dimer. Conserved His, Lys and Asn residues are contributed by each HKD domain to form the active site. Thus, the structure and catalytic mechanism of this enzyme are very similar to those of other PLD superfamily members.

The catalytic mechanism involves a two-step (ping-pong) reaction with the formation of a covalent phospho-enzyme intermediate. One of the His residues in the HKD motif acts as the nucleophile, and the second His acts as a general acid to donate a H^+ to the O of the leaving group [14]. The mechanism is supported by experiments with hydroxylamine and the labeling of His with $^{32}P_i$ in phosphate–water exchange studies [16].

It is probable that the structure of the catalytic center of mammalian PLDs and the catalytic mechanism are very similar to those reported above. In fact, there is evidence that the two HDK motifs in mammalian PLDs must dimerize to form the catalytic center [17,18]. However, mutational studies indicate additional roles for conserved sequence III and also the extreme C-terminus [19,20]. Interestingly, deletion of the PH and/or PX domains does not reduce the catalytic activity of mammalian PLDs [17,21–23].

1.2. Cellular locations of PLD1 and PLD2

The cellular locations of the mammalian PLD isozymes are controversial. This is because immunological detection of the endogenous isozymes has been difficult, and most studies have utilized overexpression of epitope- or GFP-tagged PLDs. Subcellular fractionation studies have indicated that both PLD1 and PLD2 are membrane bound, except when the expression level is high [17,24,25]. Phosphorylation and palmitoylation of the enzyme can alter their membrane association [17,24–26] as considered in Section 1.3.

Studies of different subcellular fractions identified PLD activity in plasma membranes and intracellular membranes such as Golgi, nuclear membranes and secretory vesicles [27,28]. However, these did not distinguish between PLD1 and PLD2. There is one study of native PLD1 utilizing immunofluorescence, immunogold electron microscopy and cell fractionation [29]. This localized PLD1 predominantly to the Golgi apparatus, but there was some localization to lysosomes and late endosomes. In general, overexpressed tagged PLD1 exhibits a perinuclear and punctate cellular distribution and has been reported to be present in Golgi, endoplasmic reticulum, early endosomes, lysosomes, secretory granules and the plasma membrane [12,26,28,30–33]. However, there are large discrepancies between the localizations reported by different groups and it is unclear whether the differences are due to the extent of overexpression of PLD1 or result from the use of different methodologies and cell types. Overexpressed PLD2 appears to be localized to the plasma membrane [12,33–35], but there is evidence that the endogenous enzyme is associated with the Golgi [36]. PLD1 and PLD2 have been identified in caveolae [37–42] which may account for their presence at the plasma membrane. Co-localization of PLD1 and PLD2 with the actin cytoskeleton has also been reported [43].

1.3. Post-translational modification of PLD

Several studies have shown that PLD1 and yeast PLD (Spo14) are phosphorylated under basal conditions. The phosphorylation involves Ser and Thr residues and results in cellular relocalization of the enzymes [18,24,43,44]. The phosphorylation of PLD1 occurs predominantly in its N-terminal half and has little effect on the catalytic activity [18]. Cell fractionation studies have shown that the phosphorylated form of overexpressed PLD1 is present exclusively in the membrane fraction [18]. Treatment of cells with phorbol ester or overexpression of PKCα results in further phosphorylation of PLD1 and PLD2 and loss of catalytic activity [45].

Another post-translational modification of PLD1 and PLD2 is palmitoylation that occurs on two specific Cys residues in the PH domain [24–26]. The requirements for palmitoylation have been studied in detail for PLD1. Palmitoylation requires the presence of the N-terminal 168 amino acids of the enzyme, but does not depend on catalytic activity [24]. Mutation of the Cys residues that are the palmitoylation site(s) in PLD1 and PLD2 does not alter catalytic activity in vitro, but causes some reduction in vivo. However, the mutation reduces the association of PLD1 with membranes and decreases its Ser/Thr phosphorylation [24,25].

The determinants of membrane association of PLD1 and PLD2 are complex. As indicated above, phosphorylation and palmitoylation play a role. There is also evidence for involvement of the PH domain [9], but mutation of the PIP_2 binding site located between conserved sequences II and III does not alter membrane association [8].

1.4. Role of PIP_2

But both mammalian PLD isozymes are dependent on PIP_2 for activity, and no other phospholipid, except PIP_3, can substitute [8,11,12,46,47]. The PH domain can bind PIP_2 [9], but the relationship of this binding to the activity of the enzyme is unclear [8]. On the other hand, the PIP_2 binding site located between sequences II and III contains basic residues (Arg and Lys) that are required for the stimulation of the enzyme by PIP_2 [8]. When these residues are mutated, the enzymes lose catalytic activity, but are still membrane associated and show unchanged cellular localization [8]. Certain plant PLDs (β- and γ-isozymes) are dependent upon PIP_2 [6,48]. The PIP_2 binding site corresponds to the site identified in mammalian PLD [8,48]. Interestingly, PIP_2 causes a conformational change in PLDβ that promotes the binding of PC to the catalytic domain [48]. This binding is reduced in enzymes with deletions or mutations of conserved residues in the PIP_2 binding site.

Mammalian cells treated with neomycin or *C. difficile* Toxin B show reduce PIP_2 levels and inhibited PLD activity [49–51]. On the other hand, co-expression of PLD1 or PLD2 and PI 4-P 5-kinase, which synthesizes PIP_2, results in increased PLD activity [52]. This activity is much less when a kinase-dead mutant of PI 4-P 5-kinase is transfected in place of the wild type enzyme [52].

2. Regulation of PLD1 and PLD2

2.1. Role of PKC

Early studies showed that PKC was an important regulator of PLD. Thus, addition of phorbol esters to intact cells in vivo and of PKC to membranes or preparations of PLD in vitro caused a marked increase in PLD activity [53]. Surprisingly, the in vitro studies demonstrated that PKC could activate PLD in the absence of ATP and phosphorylation [11,47,54–57]. When the different isozymes of PLD became available, PLD1 was shown to be activated by PKCα in vitro whereas PLD2 was not [11,12,47]. Another unexpected finding was that the activation of PLD1 in vitro was observed with only the α- and β-isozymes of PKC [11,47]. Consistent with a direct interaction between PKCα and PLD1, in vitro binding of the two proteins was demonstrated and was shown to be enhanced by PMA [58,59].

The interaction between PKCα and PLD1 predominantly involves the N-terminus of the phospholipase, although there is some interaction with the C-terminus [21,23, 60]. Deletion of the N-terminus of PLD1 leads to a loss of activation by PKCα in vitro or by phorbol ester in vivo [17,21,23]. Studies of binding between PKCα and PLD1 also implicated the N-terminus of the phospholipase [23,59], although another site(s) might be involved [23]. The sites on PKCα involved in interaction with PLD1 are also multiple since neither the regulatory or catalytic domain alone can activate PLD1, and it appears that the PKC holoenzyme is required [45]. As alluded to above, PLD1 can be phosphorylated by PKCα in vivo and in vitro [41,45,47,51] causing inhibition of its catalytic activity [45,47]. In one study, three phosphorylation sites (Ser2, Thr147 and Ser561) were identified in PLD1 phosphorylated by PKCα in vitro by analysis of P-peptides by mass spectrometry [61]. Single or triple mutations of these residues caused a partial loss of PLD1 activation by PMA in COS7 cells. Although these data indicate that phosphorylation of these residues is associated with some activation of PLD1 by PKC in vivo, phosphorylation of more residues was observed in vivo, raising the possibility that the phosphorylation of other residues may be involved in the inhibition of PLD1.

Both PLD1 and PLD2 can be activated by phorbol ester in intact cells [62,63], whereas only PLD1 can be activated in vitro [11,12,47]. However, this may reflect differences in the in vitro assay conditions rather than an intrinsic difference between the two isozymes. The activation of PLD by phorbol esters in vivo has been demonstrated in many cell types [62–64] and the role of PKC isozymes in the effect has been shown by the actions of relatively selective PKC inhibitors [53]. Most of these inhibitors target the ATP binding site of PKC and block the phosphorylation of PLD [45]. However, as described above, studies of the phosphorylation of PLD1 by PKCα indicate that this causes inhibition rather than stimulation. Thus, the decrease of PLD activity induced by the inhibitors in vivo cannot be explained by reduced phosphorylation. On the other hand, there is increasing evidence that the inhibitors block the interaction between PKCα and PLD1.

One approach to exploring the role of phosphorylation in the activation of PLD by PKC has used kinase-dead PKC mutants which have lost their ability to phosphorylate PLD1 or PLD2. In one study, the kinase-dead D481E PKCα mutant was still able to activate PLD1 in COS-7 cells, but the subsequent decline in PLD activity was much less than seen with wild type PKCα [45]. These data indicate that phosphorylation is not required for the initial activation of PLD by PKC, but causes a later inhibition of PLD activity. In another report utilizing wild type and kinase-dead PKCα, evidence was obtained for PKC regulation of PLD1 by both protein–protein interaction and phosphorylation mechanisms [65]. However, differences in the expression of PLD1 were not controlled for. In a study of the effects of wild type and kinase-dead PKCδ on PLD2 activity in PC12 cells, the mutant kinase suppressed endogenous and PLD2 activity well below control levels [66], suggesting that it acted mainly as a dominant negative on endogenous PKC isozymes. In summary, the role of phosphorylation in the activation of PLD1 and PLD2 by PKC remains uncertain.

There is much evidence that PKC plays a major role in the activation of PLD by many agonists in vivo. Thus, PKC inhibitors often block the effects of agonists on PLD, although the extent of the inhibition may vary [67]. Down-regulation of PKC by prolonged treatment with phorbol esters usually causes partial or complete abrogation of the ability of growth factors and other agonists to activate PLD [67]. Growth factor receptors that are mutated so that they are unable to stimulate phosphoinositide phospholipase C (PLCγ) and activate PKC are incapable of activating PLD [68]. Likewise, in fibroblasts lacking PLCγ, the PLD response to platelet-derived growth factor (PDGF) is impaired [69]. In contrast, in fibroblasts overexpressing PLCγ, the PDGF response is enhanced [70]. Overexpression of PKCα or PKCβ also increases the response of PLD to PMA, PDGF and other agonists [71–73], while antisense depletion of PKCα decreases the activation [74]. In cells expressing a peptide that binds to RACK1, conventional PKC isozymes are inhibited and there is a loss of PMA-stimulated PLD activity [75].

Since both PLD1 and PLD2 are membrane associated, translocation of PKC isozymes to the relevant membrane loci is an essential component of the mechanism by which agonists activate PLD. Numerous studies have demonstrated that phorbol esters and agonists that activate PLC induce rapid membrane translocation of conventional (classical) and novel PKC isozymes [76] and that this occurs on a time scale consistent with the activation of PLD [31,45]. The association of PKCα with the membrane fraction and PLD1 occurs more rapidly than the phosphorylation of PLD1 [45] consistent with protein–protein interaction being the mechanism for the initial activation.

Marked synergism between PKC and Rho or Arf in the activation of PLD1 has been observed in vitro [11,56]. The molecular basis of this is unknown, but binding of these regulators presumably induces conformational changes in PLD1 that interact to produce enhanced activation. As indicated above, PKC interacts with the N-terminal region of PLD1, but probably has an additional site(s). Rho binds to a sequence in the C-terminus, but the site of interaction of Arf is unknown.

3. Role of Rho family GTPases

Early studies showed that PLD could be stimulated in vitro by Rho and Arf in the presence of GTPγS [64]. Subsequent studies showed that PLD1, but not PLD2 responded to these GTPases [11,12,22,46,47,77]. PLD1 was further shown to be stimulated by most members of the Rho family (RhoA, RhoB, Rac1, Rac2 and Cdc42Hs). However, Rho was more effective than Rac or Cdc42 [11,78,79] and geranylgeranylation of these GTPases greatly enhanced their effects [78].

Constitutively active V^{14}RhoA or V^{12}Rac1 was shown to enhance the activity of PLD in COS cells or fibroblasts [60,77,80,82], and expression of dominant negative forms of these GTPases suppressed the activation of PLD by epidermal growth factor (EGF), PMA and $G\alpha_{13}$ [82–84]. A role of Rho proteins in the activation of PLD in vivo is further supported by many studies that have shown that RhoA and Rac1 are activated by many agonists and are translocated to cell membranes [53].

Clostridial toxins (C3 exoenzyme, Toxin B) that inactivate Rho proteins attenuate agonist activation of PLD in several cell lines [82–89] and inhibit the activation of the enzyme by GTPγS in vitro [87,89,90]. Rho may play a role in the regulation of PLD in vivo by direct activation of the enzyme [11,47,78,79] or by indirect regulation of its activity through changes in PIP_2 [51]. The indirect mechanism arises from the fact that Rho activates PI 4-P 5-kinase, the enzyme that synthesizes PIP_2 [91–93].

Several studies have localized the interaction site for Rho proteins on PLD1 to a sequence in the C-terminus [4,80,81,94]. The residues in RhoA that are involved in the activation of PLD1 are located in the activation loop (Switch 1) of this GTPase, as expected [78]. A detailed study of the interaction of Cdc42 with PLD1 showed that a critical residue (Ser124) in the insert helix is involved in the PLD activation mechanism [79]. Interestingly, different mechanisms are involved in the activation by different Rho family members [79]. RhoA, Rac1 and Cdc42 bind initially to PLD1 through their Switch 1 regions. Interaction between the phospholipase and the insert helix of RhoA, but not Rac1, results in further activation. Stimulation of PLD1 by Cdc42 only occurs through interaction with the insert helix, with Ser124 playing a critical role [79]. All Rho proteins show a strong synergistic interaction with PKCα and Arf to activate PLD1 in vitro [11,56,95].

4. Role of Arf family GTPases

Arf was the first small GTPase shown to activate PLD [7,96]. Subsequently, it was found that all mammalian Arf isoforms (Arf 1–6) could stimulate the enzyme [97–99]. The potency of Arf to stimulate PLD was greatly increased by myristoylation of Gly2 at the N-terminus [97–100]. PLD1 shows a large response to Arf in vitro, whereas PLD2 shows little or no response [11,12,22,46,47,60,101,102]. As noted above, Arf and Rho or PKCα interact synergistically to activate PLD1 [11,56,95]. The interaction site on Arf for PLD1 is at the N-terminus [103,104], and differs from that for other Arf functions, e.g. cholera toxin activation or coatamer binding. On the other hand, the site on PLD1 at which Arf interacts is undefined, although it is not located at the N-terminus [23].

The role of Arfs 1–5 in the activation of PLD in vivo remains unclear, but there is more evidence for Arf6. Arfs 1–3 are associated with the Golgi, but the subcellular distribution of Arfs 4 and 5 is unclear. Arf6 is associated with the plasma membrane where it cycles between this membrane and the cytosol and a vesicular compartment [105–109]. Stimulation of chromaffin cells induces translocation of Arf6 to the plasma membrane resulting in an increase in PLD activity [109], and treatment of the cells with a myristoylated peptide corresponding to the N-terminal sequence of Arf6 inhibits PLD activation [109]. Similar results have been obtained with this peptide in myometrium [110].

In permeabilized HEK293 cells expressing the M3 muscarinic receptor, the stimulatory effect of carbachol was inhibited by brefeldin A, an inhibitor of some guanine exchange factors (GEFs) for Arf [111]. Other studies have reported inhibitory effects of brefeldin A on agonist-induced PLD activation [112–115], but this has not been uniformly observed [84,116] perhaps due to the involvement of different GEFs.

There have been other approaches to define a role for Arf in agonist activation of PLD. For example, PMA, GTPγS and certain agonists cause membrane translocation of Arf to sites of PLD activity [109,117–120]. In two studies, catalytically inactive mutants of Arf1 and Arf6 blocked agonist activation of PLD [112,113]. Involvement of the ARNO family GEFs in PLD activation has been indicated by the effects of overexpression of mSec7/ARNO on PLD activation [121]. ARP, an ARF-related protein that binds to ARNO, inhibited the stimulation of PLD induced by carbachol [121]. Another approach to defining a role for Arf in PLD activation has involved the use of the Arf inhibitor Arfaptin 1 [122], which impairs the activation of PLD in vitro and in vivo [99,123,124]. Like RhoA, Arf1 and Arf6 can activate PI 4-P 5-kinase [125,126] and may therefore also regulate PLD through changes in PIP_2.

5. Role of tyrosine kinase

Activation of growth factor receptors and receptors linked to the activation of soluble tyrosine kinases induces increased PLD activity [127]. However, the mechanisms appear to be indirect, i.e. involve the PKC and Rho pathways. In many cells, inhibitors of tyrosine kinases decrease agonist activation of PLD, but the specific kinases and mechanisms involved are unclear [67,128]. Vanadate, an inhibitor of protein tyrosine phosphatases, stimulates PLD activity either alone or in combination with H_2O_2 [67]. Although there is evidence that H_2O_2 and vanadate induce phosphorylation of PLD on Tyr residues, it is unclear this occurs with agonists [67,129,130]. Furthermore, it is unclear that the phosphorylation is responsible for the activation [129]. PLD2 is constitutively associated with the EGF receptor in transfected HEK293 cells [35]. The enzyme is phosphorylated on a specific Tyr residue (Tyr 11), but mutation of this residue does not alter its activation by EGF [35]. A recent report [131] has implicated Arf4 in the signaling of the EGF receptor to PLD2, but not PLD1.

6. Role of Ral

Recent evidence has implicated Ral, a member of the Ras subfamily of GTPases, in the activation of PLD by v-Src and v-Ras in vivo [132]. There is also a functional association between RalA and PLD1, although this does not lead to activation of the phospholipase [131]. Later work showed that Arf associated with Ral–PLD1 complexes [133,134] and that RalA enhanced the effect of Arf1 on PLD1 activity [134]. Clostridial toxins that inactivate members of the Ras subfamily [88,135] inhibited the activation of PLD induced by PMA in vivo or GTPγS in vitro [88,136]. Ral, but not other GTPases, restored the activation by PMA [88], and variants of one toxin supported a role for Ral in the GTPγS effect [136]. Expression of dominant negative or constitutively active forms of RalA in 3Y1 cells also implicated this GTPase in the regulation of PLD [137,138].

References

[1] Ponting, C.P., Kerr, I.D., 1996. Protein Sci. 5, 914–922.
[2] Koonin, E.V., 1996. Trends Biochem. Sci. 21, 242–243.
[3] Interthal, H., Pouliot, J.J., Champoux, J.J., 2001. Proc. Natl Acad. Sci. USA 98, 12009–12014.
[4] Sung, T.-C., Roper, R.I., Zhang, Y., Rudge, S.A., Temel, R., Hammond, S.M., Morris, A.J., Moss, B., Engebrecht, J.A., Frohman, M.A., 1997. EMBO J. 16, 4519–4539.
[5] Frohman, M., Sung, T.-C., Morris, A.J., 1999. Biochim. Biophys. Acta 1439, 175–186.
[6] Qin, C., Wang, X., 2002. Plant Physiol. 128, 1057–1068.
[7] Brown, H.A., Gutowski, S., Moomaw, C.R., Slaughter, C., Sternweis, P.C., 1993. Cell 75, 1137–1144.
[8] Sciorra, V.A., Rudge, S.A., Prestwich, G.D., Frohman, M.A., Engebrecht, J.A., Morris, A.J., 1999. EMBO J. 20, 5911–5921.
[9] Hodgkin, M.N., Masson, M.R., Powner, D., Saqib, K.M., Ponting, C.P., Wakelam, M.J.O., 1999. Curr. Biol. 10, 43–46.
[10] Wang, S., Banno, Y., Nakashima, S., Nozawa, Y., 2001. K. Eukaryot. Microbiol. 48, 194–201.
[11] Hammond, S.M., Jenco, J.M., Nakashima, S., Cadwallader, K., Gu, Q.-m., Cook, S., Nozawa, Y., Prestwich, G.D., Frohman, M.A., Morris, A.J., 1997. J. Biol. Chem. 272, 3860–3868.
[12] Colley, W.C., Sung, T.-C., Roll, R., Jenco, J., Hammond, S.M., Altshuller, Y., Bar-Sagi, D., Morris, A.J., Frohman, M.A., 1997. Curr. Biol. 7, 191–201.
[13] Leiros, I., Secundo, F., Zambonelli, C., Servi, S., Hough, E., 2000. Structure 8, 655–667.
[14] Stuckey, J.A., Dixon, J.E., 1999. Nat. Struct. Biol. 6, 278–284.
[15] Davaies, D.R., Interthal, H., Champoux, J.J., Hol, W.G.J., 2002. Structure 10, 237–248.
[16] Gottlin, E.B., Rudolph, A.E., Zhao, Y., Matthews, H.R., Dixon, J.E., 1998. Proc. Natl Acad. Sci. USA 95, 9202–9207.
[17] Xie, Z., Ho, W.-T., Exton, J.H., 1998. J. Biol. Chem. 273, 34679–34682.
[18] Xie, Z., Ho, W.-T., Exton, J.H., 2000. J. Biol. Chem. 275, 24962–24969.
[19] Xie, Z., Ho, W.-T., Exton, J.H., 2000. Eur. J. Biochem. 267, 7138–7146.

[20] Liu, M.Y., Gutowski, S., Sternweis, P.C., 2001. J. Biol. Chem. 270, 5556–5562.
[21] Park, S.-K., Min, D.S., Exton, J.H., 1998. Biochem. Biophys. Res. Commun. 244, 364–367.
[22] Sung, T.-C., Altshuller, Y.M., Morris, A.J., Frohman, M.A., 1999. J. Biol. Chem. 274, 494–502.
[23] Sung, T.-C., Zhang, Y., Morris, A.J., Frohman, M.A., 1999. J. Biol. Chem. 274, 3659–3666.
[24] Xie, Z., Ho, W.-T., Exton, J.H., 2001. J. Biol. Chem. 275, 24962–24969.
[25] Xie, Z., Ho, W.-T., Exton, J.H., 2002. Biochem. Biophys. Acta 1580, 9–21.
[26] Sugars, J.M., Cellek, S., Manifava, M., Coadwell, J., Ktistakis, N.T., 1999. J. Biol. Chem. 274, 30023–30027.
[27] Provost, J.J., Fudge, J., Israelit, S., Siddiqi, A.R., Exton, J.H., 1996. Biochem. J. 319, 285–291.
[28] Liscovitch, M., Czarny, M., Fiucci, G., Lavie, Y., Tang, X., 1999. Biochem. Biophys. Acta 1439, 245–263.
[29] Freyberg, Z., Sweeney, D., Siddhanta, A., Bourgoin, S., Frohman, M., Shields, D., 2001. Mol. Biol. Cell 12, 943–955.
[30] Brown, F.D., Thomas, N., Saqib, K.M., Clark, J.M., Powner, D., Thompson, N.T., Solari, R., Wakelam, M.J.O., 1998. Curr. Biol. 8, 835–838.
[31] Kim, Y., Kim, J.-E., Lee, S.D., Lee, T.G., Kim, J.H., Park, J.B., Han, J.M., Jang, S.K., Suh, P.-G., Ryu, S.H., 1999. Biochim. Biophys. Acta 1436, 319–330.
[32] Kam, Y., Exton, J.H., 2001. Mol. Cell. Biol. 21, 4055–4066.
[33] Emoto, M., Klarlund, J.K., Waters, S.B., Hu, V., Buxton, J.M., Chawla, A., Czech, M.P., 2000. J. Biol. Chem. 275, 7144–7151.
[34] Vitale, N., Caumont, A.-S., Golaz-Chasserot, S., Du, G., Wu, S., Sciorra, V.A., Morris, A.J., Frohman, M.A., Bader, M.-F., 2001. EMBO J. 20, 2424–2434.
[35] Park, J.B., Kim, J.H., Kim, Y., Ha, S.H., Kim, J.H., Yoo, J.-S., Du, G., Frohman, M.A., Suh, P.-G., Ryu, S.H., 2000. J. Biol. Chem. 275, 21295–21301.
[36] Freyberg, Z., Bourgoin, S., Shields, D., 2002. Mol. Biol. Cell 13, 3930–3942.
[37] Czarny, M., Lavie, Y., Fiucci, G., Liscovitch, M., 1999. J. Biol. Chem. 274, 2717–2724.
[38] Czarny, M., Fiucci, G., Lavaie, Y., Banno, Y., Nozawa, Y., Liscovitch, M., 2000. FEBS Lett. 467, 326–332.
[39] Iyer, S.S., Kusner, D.J., 1999. J. Biol. Chem. 274, 2350–2359.
[40] Sciorra, V.A., Morris, A.J., 1999. Mol. Biol. Cell 10, 3863–3876.
[41] Kim, J.H., Han, J.M., Lee, S., Kim, Y., Lee, T.G., Park, J.B., Lee, S.D., Suy, P.-G., Ryu, S.H., 1999. Biochemistry 38, 3763–3769.
[42] Kim, Y., Han, J.M., Han, B.R., Lee, K.-A., Kim, J.H., Lee, B.D., Jang, I.-H., Suy, P.-G., Ryu, S.H., 2000. J. Biol. Chem. 275, 13621–13627.
[43] Lee, S., Park, J.B., Kim, J.H., Kim, Y., Kim, J.H., Shin, K.-J., Lee, J.S., Ha, S.H., Suh, P.-G., Ryu, S.H., 2001. J. Biol. Chem. 276, 28252–28260.
[44] Rudge, S.A., Morris, A.J., Engebrecht, J.A., 1998. J. Cell Biol. 140, 81–90.
[45] Hu, T., Exton, J.H., 2003. J. Biol. Chem. 278, 2348–2355.
[46] Kodaki, T., Yamashita, S., 1997. J. Biol. Chem. 272, 11408–11413.
[47] Min, D.S., Park, S.-K., Exton, J.H., 1998. J. Biol. Chem. 273, 7044–7051.
[48] Zheng, L., Shan, J., Krishnamoorthi, R., Wang, X., 2002. Biochemistry 41, 4546–4553.
[49] Liscovitch, M., Chalifa, V., Pertile, P., Chen, C.-S., Cantley, L.C., 1994. J. Biol. Chem. 269, 21403–21496.
[50] Pertile, P., Liscovitch, M., Chalifa, V., Cantley, L.C., 1995. J. Biol. Chem. 270, 5130–5135.
[51] Schmidt, M., Rümenapp, M., Nehls, C., Ott, S., Keller, J., von Eichel-Streiber, C., Jacobs, K.H., 1996. Eur. J. Biochem. 240, 707–712.
[52] Divecha, N., Roefs, M., Halstead, J.R., D'Andrea, S., Fernandez-Borga, M., Oomen, I., Saqib, K.M., Wakelam, M.J.P., D'Santos, C., 2000. EMBO J. 19, 5440–5449.
[53] Exton, J.H., 2002. Rev. Physiol. Biochem. Pharmacol. 144, 1–94.
[54] Conricode, K.M., Brewer, K.A., Exton, J.H., 1992. J. Biol. Chem. 267, 7199–7202.
[55] Conricode, K.M., Smith, J.L., Burns, D.J., Exton, J.H., 1994. FEBS Lett. 342, 149–153.
[56] Singer, W.D., Brown, H.A., Jiang, X., Sternweis, P.C., 1996. J. Biol. Chem. 271, 4504–4510.
[57] Sciorra, V.A., Hammond, S.M., Morris, A.J., 2001. Biochemistry 40, 2640–2646.
[58] Min, D.S., Exton, J.H., 1998. Biochem. Biophys. Res. Commun. 248, 533–537.

[59] Lee, T.G., Park, J.B., Lee, S.D., Hong, S., Kim, J.H., Kim, Y., Yi, K.S., Bae, S., Hannun, Y.A., Obeid, L.M., Suh, P.-G., Ryu, S.H., 1997. Biochim. Biophys. Acta 1347, 199–204.
[60] Zhang, Y., Altshuller, Y.M., Hammond, S.M., Morris, A.J., Frohman, M.A., 1999. EMBO J. 18, 6339–6348.
[61] Kim, Y., Han, J.M., Park, J.B., Lee, S.D., Oh, Y.S., Chung, C., Lee, T.G., Kim, J.H., Park, S.-K., Yoo, J.-S., Suh, P.-G., Ryu, S.H., 1999. Biochemistry 38, 10344–10351.
[62] Siddiqi, A.R., Srajer, G.E., Leslie, C.C., 2000. Biochim. Biophys. Acta 1497, 103–114.
[63] Xie, Z., Ho, W.-T., Spellman, R., Cai, S., Exton, J.H., 2002. J. Biol. Chem. 277, 11979–11986.
[64] Exton, J.H., 1999. Biochim. Biophys. Acta 1439, 121–133.
[65] Oka, M., Hitomi, T., Okada, T., Nakamura, S.-I., Nagai, H., Ohba, M., Kuroki, T., Kikkawa, U., Ichihashi, M., 2002. Biochem. Biophys. Res. Commun. 294, 1109–1113.
[66] Han, J.M., Kim, J.H., Lee, B.D., Lee, S.D., Kim, Y., Jung, Y.W., Lee, S., Cho, W., Ohba, M., Kuroki, T., Suh, P.-G., Ryu, S.H., 2002. J. Biol. Chem. 277, 8290–8297.
[67] Exton, J.H., 1999. Physiol. Rev. 77, 303–320.
[68] Yeo, E.-J., Kazlauskas, A., Exton, J.H., 1994. J. Biol. Chem. 269, 27823–27826.
[69] Hess, J.A., Ji, Q.-s., Carpenter, G., Exton, J.H., 1998. J. Biol. Chem. 273, 20517–20524.
[70] Lee, Y.H., Kim, H.S., Pai, J.-K., Ryu, S.H., Suh, P.-G., 1994. J. Biol. Chem. 269, 26842–26847.
[71] Pai, J.-K., Dobek, E.A., Bishop, W.R., 1991. Cell Regul. 2, 897–903.
[72] Pachter, J.A., Pai, J.-K., Mayer-Ezell, R., Petrin, J.M., Dobek, E., Bishop, W.R., 1988. J. Biol. Chem. 267, 9826–9830.
[73] Eldar, H., Ben-AV, P., Schmidt, P.U.-S., Livneh, E., Liscovitch, M., 1993. J. Biol. Chem. 268, 12560–12564.
[74] Balboa, M.A., Firestein, B.L., Godson, C., Bekk, K.S., Insel, P.A., 1994. J. Biol. Chem. 269, 10511–10516.
[75] Thorsen, V.A.T., Bjorndal, B., Nolan, G., Fukami, M.H., Bruland, O., Lillehaug, J.R., Holmsen, H., 2000. Biochim. Biophys. Acta 1487, 163–176.
[76] Jaken, S., 1997. Molecular Biology Intelligence Unit. R.G. Landes Company, Austin, TX.
[77] Park, S.-K., Provost, J.J., Bae, C.D., Hos, W.-T., Exton, J.H., 1997. J. Biol. Chem. 272, 29263–29271.
[78] Bae, C.D., Min, D.S., Fleming, I.N., Exton, J.H., 1998. J. Biol. Chem. 273, 11596–11604.
[79] Walker, S.J., Brown, H.A., 2002. J. Biol. Chem. 277, 26260–26267.
[80] Du, G., Altshuller, Y.M., Kim, Y., Han, J.M., Ryu, S.H., Morris, A.J., Frohman, M.A., 2000. Mol. Biol. Cell 11, 4359–4368.
[81] Cai, S., Exton, J.H., 2001. Biochem. J. 355, 779–785.
[82] Hess, J.A., Ross, A.H., Qiu, R.-G., Symons, M., Exton, J.H., 1997. J. Biol. Chem. 272, 1615–1620.
[83] Plonk, S.G., Park, S.-K., Exton, J.H., 1998. J. Biol. Chem. 273, 4823–4826.
[84] Meacci, E., Vasta, V., Moorman, J.P., Bobak, D.A., Bruni, P., Moss, J., Vaughan, M., 1999. J. Biol. Chem. 274, 18605–18612.
[85] Senogles, S.E., 2000. Mol. Pharmacol. 58, 455–462.
[86] Murthy, K.S., Zhou, H., Grider, J.R., Makhlouf, G.M., 2001. Am. J. Physiol. 280, G381–G388.
[87] Schmidt, M., Rümenapp, U., Bienek, C., Keller, J., von Eichel-Streiber, C., Jakobs, K.H., 1996. J. Biol. Chem. 271, 2422–2426.
[88] Schmidt, M., Voss, M., Thiel, M., Baur, B., Grannas, A., Tapp, E., Cool, R.H., de Gunzburg, J., von Eichel-Streiber, C., Jakobs, K.H., 1998. J. Biol. Chem. 273, 7413–7422.
[89] Ojio, K., Banno, Y., Nakashima, S., Kato, N., Watanabe, K., Lyerly, D.M., Miyata, H., Nozawa, Y., 1996. Biochem. Biophys. Res. Commun. 224, 591–596.
[90] Kuribara, H., Tago, K., Yokozeki, T., Sasaki, T., Takai, Y., Morii, N., Narumiya, S., Katadda, T., Kanaho, Y., 1995. J. Biol. Chem. 270, 25667–25671.
[91] Chong, L.D., Traynor-Kaplan, A., Bokoch, G.M., Schwartz, M.A., 1994. Cell 79, 507–513.
[92] Ren, X.-D., Bokoch, G.M., Traynor-Kaplan, A., Jenins, G.H., Anderson, R.A., Schwartz, M.A., 1996. Mol. Biol. Cell 7, 435–442.
[93] Oude Weernink, P.A., Schulte, P., Guo, Y., Wetzel, J., Amano, M., Kaibuchi, K., Haverland, S., VoB, M., Schmidt, M., Mayr, G.W., Jakobs, K.H., 2000. J. Biol. Chem. 275, 10168–10174.
[94] Yamazaki, M., Zhang, Y., Watanabe, H., Yokozeki, T., Ohno, S., Kaibuchi, K., Shibata, H., Mukai, H., Ono, Y., Frohman, M.A., Kanaho, Y., 1999. J. Biol. Chem. 274, 6035–6038.

[95] Hodgkin, M.N., Clark, J.M., Rose, S., Saqib, K., Wakelam, M.J.O., 1999. Biochem. J. 339, 87–93.
[96] Cockcroft, S., Thomas, G.M.H., Fensome, A., Geny, B., Cunningham, E., Gout, I., Hiles, I., Totty, N.F., Truong, O., Hsuan, J.J., 1994. Science 263, 523–526.
[97] Massenburg, D., Han, J.-S., Liyanage, M., Patton, W.A., Rhee, S.G., Moss, J., Vaughan, M., 1994. Proc. Natl Acad. Sci. USA 91, 11718–11722.
[98] Brown, H.A., Gutowski, S., Kahn, R.A., Sternweis, P.C., 1995. J. Biol. Chem. 270, 14935–14943.
[99] Tsai, S.-C., Adamik, R., Hong, J.-X., Moss, J., Vaughan, M., Kanoh, H., Exton, J.H., 1998. J. Biol. Chem. 273, 20697–20701.
[100] Fensome, A., Cunningham, E., Prosser, S., Tan, S.K., Swigart, P., Thomas, G., Hsuan, J., Cockcroft, S., 1996. Curr. Biol. 6, 730–738.
[101] Lopez, I., Arnold, R.S., Lambeth, J.D., 1998. J. Biol. Chem. 273, 12846–12852.
[102] Slaaby, R., Du, G., Altshuller, Y.M., Frohman, M.A., Seedorf, K., 2000. Biochem. J. 351, 613–619.
[103] Zhang, G.-F., Patton, W.A., Lee, F.-J.L., Lliyanage, M., Han, J.-S., Rhee, S.G., Moss, J., Vaughan, M., 1995. J. Biol. Chem. 270, 21–24.
[104] Jones, D.H., Bax, B., Fensome, A., Cockcroft, S., 1999. Biochem. J. 341, 183–190.
[105] Cavenagh, M.M., Whitney, J.A., Carroll, K., Zhang, C.-j., Bowman, A.L., Rosenwald, A.G., Mellman, I., Kahn, R.A., 1996. J. Biol. Chem. 271, 21767–21774.
[106] Al-Awar, O., Radhakrishna, H., Powell, N.N., Donaldson, J.G., 2000. Mol. Cell. Biol. 20, 5998–6007.
[107] Gaschet, J., Hsu, V.W., 1999. J. Biol. Chem.(274), 20040–20045.
[108] Radhakrishna, H., Donaldson, J.G., 1997. J. Cell Biol. 139, 49–61.
[109] Caumont, A.-S., Galas, M.-C., Vitale, N., Aunis, D., Bader, M.-F., 1998. J. Biol. Chem. 273, 1373–1379.
[110] Le Stunff, H.L., Dokhac, L., Bourgoin, S., Bader, M.-F., Harbon, S., 2000. Biochem. J. 352, 491–499.
[111] Rümenapp, U., Geiszt, M., Wahn, F., Schmidt, M., Jakobs, K.H., 1995. Eur. J. Biochem. 234, 240–244.
[112] Shome, K., Nie, Y., Romero, G., 1998. J. Biol. Chem. 273, 30836–30841.
[113] Shome, K., Rizzo, M.A., Vasudevan, C., Andresen, B., Romero, G., 1999. Endocrinology 141, 2200–2208.
[114] Andresen, B.T., Jackson, E.K., Romero, G.G., 2001. Hypertension 37, 635–639.
[115] Mitchell, R., McCulloch, D., Lutz, E., Johnson, M., MacKenzie, C., Fennell, M., Fink, G., Zhou, W., Sealfon, S.C., 1998. Nature 392, 411–414.
[116] Guillemain, I., Exton, J.H., 1997. Eur. J. Biochem. 249, 812–819.
[117] Houle, M.G., Kahn, R.A., Naccache, P.H., Bourgoin, S., 1995. J. Biol. Chem. 270, 22795–22800.
[118] Houle, M.G., Naccache, P.H., Bourgoin, S., 1999. J. Leukoc. Biol. 66, 1021–1030.
[119] Whatmore, J., Morgan, C.P., Cunningham, E., Collison, K.S., Willison, K.R., Cockcroft, S., 1996. Biochem. J. 320, 785–794.
[120] Fensome, A., Whatmore, J., Morgan, C., Jones, D., Cockcroft, S., 1998. J. Biol. Chem. 273, 13157–13164.
[121] Schürmann, A., Schmidt, M., Asmus, M., Bayer, S., Fliegert, F., Koling, S., Massmann, S., Schilf, B., Subauste, M.C., Voss, M., Jakobs, K.H., Joost, H.-G., 1999. J. Biol. Chem. 274, 9744–9751.
[122] Kanoh, H., Williger, B.-T., Exton, J.H., 1997. J. Biol. Chem. 272, 5421–5429.
[123] Williger, B.-T., Provost, J.J., Ho, W.-T., Milstine, J., Exton, J.H., 1999. FEBS Lett. 454, 85–89.
[124] Williger, B.-T., Ostermann, J., Exton, J.H., 1999. FEBS Lett. 443, 197–200.
[125] Jones, D.H., Morris, J.B., Morgan, C.P., Kondo, H., Irvine, R.F., Cockcroft, S., 2000. J. Biol. Chem. 275, 13962–13966.
[126] Honda, A., Nogami, M., Yokozeki, T., Yamazaki, M., Nakamura, H., Watanabe, H., Kawamoto, K., Nakayama, K., Morris, A.J., Frohman, M.A., Kanaho, Y., 1999. Cell 99, 521–532.
[127] Natarajan, V., Scribner, W.M., Vepa, S., 1996. Chem. Phys. Lipids 80, 103–116.
[128] Houle, M.G., Bourgoin, S., 1999. Biochim. Biophys. Acta 1439, 135–150.
[129] Min, D.S., Kim, E.-G., Exton, J.H., 1998. J. Biol. Chem. 273, 29986–29994.
[130] Gomez-Cambronero, J., 1995. J. Inter. Cyto. Res. 15, 877–885.
[131] Kim, S.-W., Hayashi, M., Lo, J.-F., Yang, Y., Yoo, J.-S., Lee, J.-D., 2003. J. Biol. Chem. 278, 2661–2668.
[132] Jiang, H., Luo, J.-Q., Urano, T., Frankel, P., Lu, Z., Foster, D.A., Feig, L.A., 1996. Eur. J. Pharmacol. 302, 183–190.
[133] Luo, J.-Q., Liu, X., Frankel, P., Rotunda, T., Ramos, M., Flom, J., Jiang, H., Feig, L.A., Morris, A.J., Kahn, R.A., Foster, D.A., 1998. Proc. Natl Acad. Sci. USA 95, 3632–3637.

[134] Kim, J.H., Lee, S.D., Ham, J.M., Lee, T.G., Kim, Y., Park, J.B., Lambeth, J.D., Suh, P.-G., Ryu, S.H., 1998. FEBS Lett. 430, 231–235.

[135] Popoff, M.R., Chaves-Olarate, E., Lemichez, E., von Eichel-Streiber, C., Thelestam, M., Chardin, P., Cussac, D., Antonny, B., Chavrier, P., Flatau, G., Giry, M., de Gunzburg, J., Boquet, P., 1996. J. Biol. Chem. 271, 10217–10224.

[136] El Hadj, N.B., Popoff, M.R., Marvaud, J.-C., Payrastre, B., Boquet, P., Geny, B., 1999. J. Biol. Chem. 274, 14021–14031.

[137] Lu, Z., Hornia, A., Joseph, T., Sukezane, T., Frankel, P., Zhong, M., Bychenok, S., Xu, L., Feig, L.A., Foster, D.A., 2000. Mol. Cell. Biol. 20, 462–467.

[138] Voss, M., Weernink Oude, P.A., Haupenthal, S., Moller, U., Cool, R.H., Bauer, B., Camonis, J.H., Jakobs, K.H., Schmidt, M., 2000. J. Biol. Chem. 274, 34691–34698.

Metabolism and physiological functions of sphingolipids

Jacqueline Ohanian* and Vasken Ohanian

University Department of Medicine, University of Manchester, Manchester Royal Infirmary, Oxford Road, Manchester M13 9WL, UK
**Correspondence address: Tel.: +44-161-276-8670; fax: +44-161-274-4833*
E-mail: johanian@man.ac.uk(J.O.)

1. Introduction

Sphingolipids, through their biophysical properties and as reservoirs of bioactive molecules are involved in many cellular processes. They can form specialised structures within cell membranes which are linked to immune cell, G protein coupled receptor and growth factor receptor responses. They mediate cell–cell and cell–intracellular matrix interactions, modulate receptor and protein function and are involved in signal transduction. There are over 300 sphingolipid species in mammalian cells ranging from the small sphingoid bases, sphingosine and sphinganine to complex highly glycosylated cerebrosides. They are present in all eukaryotic membranes not only within the plasma membrane but also in membranes of subcellular organelles. However, despite their ubiquitous nature sphingolipid synthesis, degradation and distribution are tightly regulated reflecting the important physiological functions of this lipid class. Bioactive molecules derived from sphingolipids include, sphingosine, sphingosine-1-phosphate (S1P), ceramide and glucosylceramide (GlcCer) implicated in mitogenesis, angiogenesis, apoptosis, stress responses and multidrug resistant cancers. The generation of these lipids is through de novo biosynthesis and/or degradation of more complex sphingolipids. Furthermore, their site of generation within the cell and their restriction to the membrane in which they are formed acts to spatially regulate signals. Clearly the ways in which cells use sphingolipids to regulate physiological functions are multifaceted and complex. However, recent studies of the enzymes involved in synthesis, degradation and interconversion of sphingolipids have led to a greater understanding of the role of this fascinating class of lipid.

2. Sphingolipid chemistry

In order to appreciate the functions of sphingolipids it is necessary first to understand their basic structure, biosynthesis and metabolism.

Advances in Molecular and Cell Biology, Vol. 33, pages 463–502

ISSN: 1569-2558/DOI: 10.1016/S1569-2558(03)33023-1

2.1. Structure

Sphingolipids are made up of a lipophilic backbone (ceramide) conjugated to a hydrophilic headgroup (Fig. 1). Ceramide is composed of a long chain sphingoid base, predominantly D-erythrosphingosine in mammalian cells and an amide-linked fatty acid, which may vary in length from long (22:0, 24:0 and 24:1$^{\Delta 15(c)}$) to intermediate (16:0 and 18:00) [1]. The sphingoid base may also vary in chain length although C18-sphinganine is the most common in mammals. The sphingolipid species is defined by the hydrophilic headgroup. For instance, sphingomyelin is ceramide linked to phosphocholine whereas cerebrosides are formed by the addition of glucose or galactose. More complex glycolipids and gangliosides may then be formed by the further addition of complex sugars. These differences in headgroup and backbone (fatty acid chain length and unsaturation and sphingoid base heterogeneity) make sphingolipids the most complex lipid class, with over 300 species identified so far, in mammalian tissues [2].

2.2. Biosynthesis

In mammalian cells sphingolipid biosynthesis begins in the endoplasmic reticulum with the condensation of serine and palmitoylCoA by serine palmitoyl transferase (SPT) forming 3-ketosphinganine which in turn undergoes reduction to sphinganine catalysed by an NADPH dependent reductase (Fig. 2). Ceramide synthase then acts adding a fatty acyl group by an amide linkage to form dihydroceramide which is converted directly to ceramide, the precursor of all complex sphingolipids, by the introduction of a *trans* double bond between carbons 4 and 5 of the sphingoid base catalysed by dihydroceramide desaturase [3]. The formation of more complex sphingolipids derived from ceramide then occurs in the Golgi apparatus, the simplest of which is ceramide-1-phosphate formed by ceramide kinase (CERK). The transfer of phosphocholine from phosphatidylcholine by sphingomyelin synthase (SMS) forms sphingomyelin. The initial step in the formation of complex glycolipids is linking of glucose to ceramide forming GlcCer by glucosylceramide synthase (GCS) (Fig. 3). More complex headgroups may then be added to GlcCer

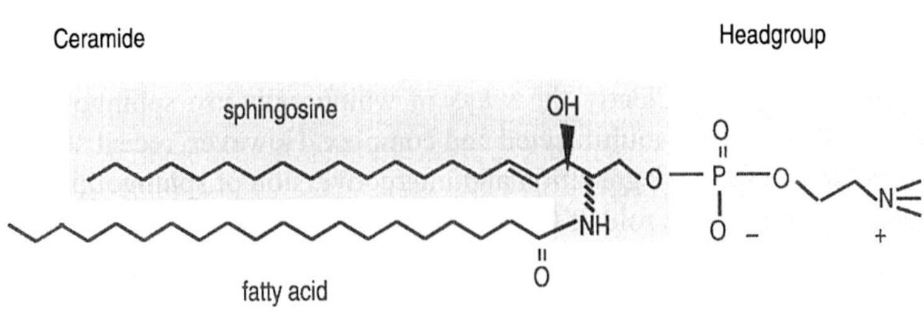

Fig. 1. Structure of sphingomyelin.

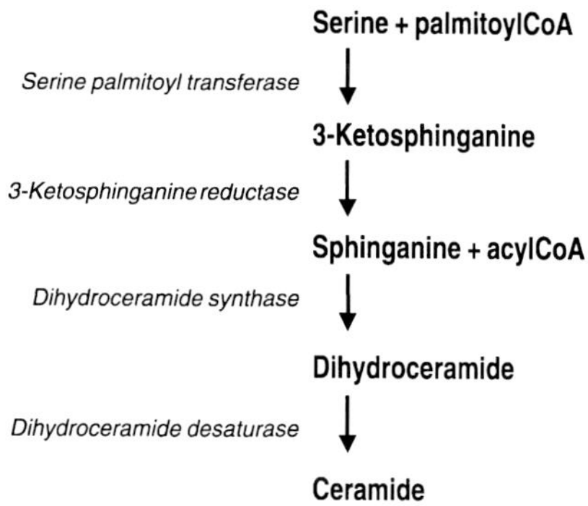

Fig. 2. De novo biosynthesis of ceramide at the endoplasmic reticulum.

to form cerebrosides, sulfatides and glycosphingolipids (GSLs). Gangliosides are a subclass of GSLs identified by the presence of sialic acid in the carbohydrate headgroup [4].

2.3. *Metabolism*

Complex GSLs may be stepwise degraded by a series of hydrolases to reform ceramide. In addition, sphingomyelin can be hydrolysed by sphingomyelinases to ceramide and several bioactive lipids may then be derived by catabolism of ceramide (Fig. 4). Ceramidases deacylate ceramide to form sphingosine which may then be phosphorylated to S1P. The reverse reaction can also occur such that S1P is dephosphorylated to sphingosine the sphingosine may then be converted to ceramide or it can be cleared by a lyase resulting in the formation of a fatty aldehyde and ethanolamine phosphate which may be used in glycolipid synthesis. Lyso-sphingolipids, *N*-deacylated derivatives such as 1-galactosylsphingosine, glucosylsphingosine and sphingosylphosphorylcholine (SPC) are also found. These sphingolipids are present in very low concentrations but may have important signalling effects either as second messengers or through their lytic and membrane destabilising effects [5].

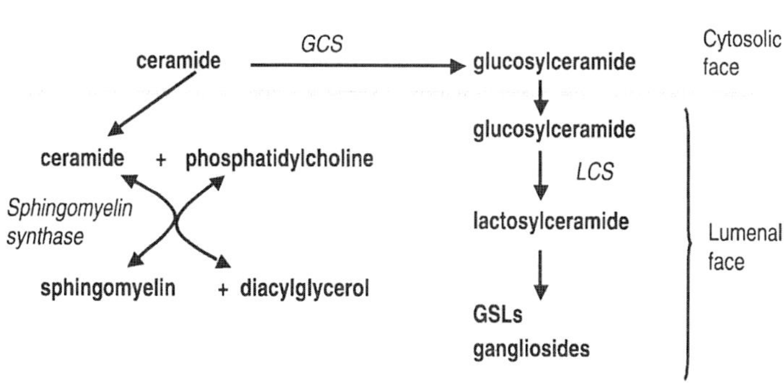

Fig. 3. De novo biosynthesis of complex sphingolipids at the Golgi apparatus.

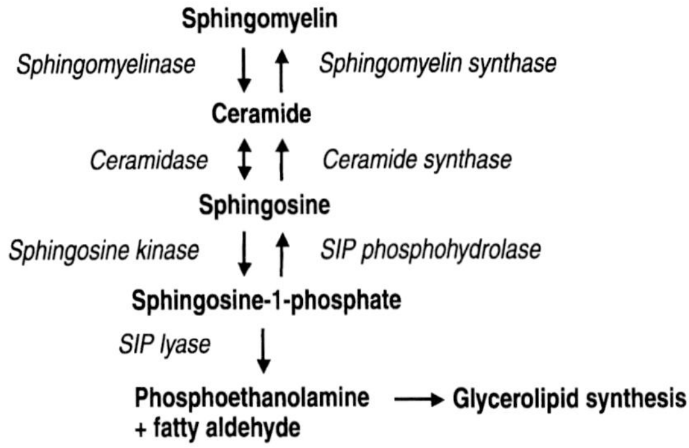

Fig. 4. Degradation of sphingolipids.

3. De novo synthesis of ceramide

Because of the important role of bioactive lipid species, their generation by de novo synthesis is known to be tightly regulated. However, due to their refractory nature, little is known of the enzymes involved in sphingolipid biosynthesis. They are known to be predominantly transmembrane proteins and have proven to be difficult to isolate and purify, although with the recent identification of genes coding for these enzymes their full characterisation should be feasible.

3.1. Serine palmitoyl transferase

SPT is the key enzyme in de novo sphingolipid biosynthesis. It initiates sphingolipid synthesis by catalysing the pyridoxal 5′-phosphate dependent condensation of L-serine and palmitoylCoA to 3-ketosphinganine (Fig. 2). Biochemical studies showed that SPT activity was predominantly associated with microsomes, was present in all tissues studied, required pyridoxal 5′-phosphate as cofactor and utilised predominantly fatty acyl-CoA with 16 ± 1 carbon atoms so explaining the preponderance of 18 carbon chain length sphingoid bases [6–8]. Further study localised SPT activity to the cytosolic face of the endoplasmic reticulum [9]. In mammals and yeast SPT is composed of two subunits LCB1 and LCB2 (long chain base) products of genes SPTLC1 and SPTLC2 (also called SPT1 and SPT2), respectively. The first clone, LCB1, was isolated from *Saccharomyces cerevisiae* and shown to encode a protein required for SPT activity [10]. A second clone, LCB2, a subunit of SPT was isolated from *S. cerevisiae* in 1994 [11]. This was followed by isolation of mammalian cDNAs for LCB1 and LCB2 from mouse and human sources [12–14] which were shown to encode 53 and 63 kDa proteins, respectively [13]. Studies utilising Chinese hamster ovary (CHO) cells deficient in LCB1 (strain LY-B) which have no SPT activity and are incapable of de novo sphingolipid synthesis have shown that transfection of LCB1 cDNA restored SPT activity and that endogenous LCB2 co-immunoprecipitated with expressed LCB1 [15] with an apparent stoichiometry of 1:1 [16]. Additionally, LCB1 subunit localised to the ER when expressed in LY-B cells and appeared to span the membrane with the N-terminal at the lumenal side and the C-terminal at the cytosolic face. Expression of LCB1 also increased LCB2 mRNA and protein levels which are normally low in LY-B cells demonstrating that LCB1 regulates LCB2 expression [17]. Further insights into the roles of the two SPT subunits have come from studies of hereditary sensory neuropathy type 1 (HSN1) a degenerative disorder of peripheral sensory neurons caused by mutations in SPTLC1, the gene encoding the LCB1 subunit [18,19]. In lymphoblasts from HSN1 patients SPT activity and sphingolipid synthesis are reduced but protein expression of LCB1 and LCB2 is normal suggesting that the LCB1 mutations affect SPT activity [20]. Mutating LCB1 in *S. cerevisiae* at the same sites as occurring in HSN1 reduces SPT activity by approximately 50% but does not affect binding to LCB2. Modelling studies suggested that the LCB1 mutations could affect the binding of pyridoxal 5′-phosphate to LCB2 so reducing SPT activity [21]. These studies demonstrate that SPT is composed of two subunits but that the regulation of catalytic activity is complex and requires the co-ordinate interaction of both subunits and cofactor. In yeast a third protein, Tsc3p is required for optimal SPT activity [22] whether a similar protein regulates mammalian SPT function is not known.

Regulation of SPT activity has been proposed as a major mechanism in cellular responses induced by sphingolipids. This pathway is implicated in both cell growth and death. For instance, inhibition of SPT with ISP1/myriocin – a fungal toxin, prevents T-cell proliferation [23] and CHO cell mutants (strain SPB-1) which have a thermolabile SPT will only grow at the permissive temperature unless supplemented with sphingolipids [24]. In the vasculature both SPT subunits were upregulated in balloon-injured rat carotid artery and the enhanced expression was localised to de-differentiating fibroblasts and smooth muscle cells [25]. Regulation of SPT at the transcriptional level has also been seen with UVB irradiation [26], cytokines [27] and retinoic acid [28]. Post-translational activation of SPT has also been reported with etoposide [29], cannabinoids [30] and heat shock in yeast [31] and mammalian cells [32]. De novo synthesis of ceramide is implicated in apoptotic cell death and recent studies suggest that activation of SPT is involved in this response [33]. One mechanism by which de novo synthesised ceramide may act is by regulating the alternative splicing of apoptosis regulators such as caspase 9 and Bcl-x. ISP1/myriocin inhibits de novo ceramide production in response to the pro-apoptotic agent gemcitabine implicating SPT in the response [34]. Whether activation of SPT leads to cell growth or cell death is most probably dependent on regulation of other enzymes involved in sphingolipid synthesis and the bioactive products that accumulate. For instance, ceramide and sphingosine are implicated in growth arrest and cytotoxicity whereas S1P is associated with growth stimulation and inhibition of apoptosis [35].

3.2. Dihydroceramide synthase

Dihydroceramide synthase (fatty acyl-CoA:sphingosine acyltransferase) catalyses the acylation of sphinganine to dihydroceramide (Fig. 2) utilising a range of fatty acyl-CoAs from C16:0 to C26:0. Dihydroceramide synthase activity is mainly localised in microsomal and mitochondrial membranes [36], it is dependent on acyl-CoA and inhibited by fungal toxins such as Fumonisin B1 (FB1) [37]. Partially purified dihydroceramide synthase from bovine liver mitochondria appeared as two protein bands of 62 and 72 kDa suggesting that the enzyme consists of at least two subunits [38]. Recently, two genes LAG1 and LAC1 have been identified in *S. cerevisiae* that are required for dihydroceramide synthase activity in yeast. They encode Lag1p and Lac1p, homologous transmembrane proteins of the ER [39]. Homologous genes are found in eukaryotes which should lead to the cloning of mammalian dihydroceramide synthases. A second route for the formation of ceramide is by the reverse action of ceramidases [40–42]. This ceramide synthase activity is acyl-CoA independent and is not inhibited by FB1 [43]. The enzyme has been cloned from yeast [41] and human [43] and when expressed in HEK293 cells it localised to mitochondria [43]. Biochemical studies have shown the enzyme has preference for sphingosine with low affinity for sphinganine (the main substrate for acyl-CoA dependent dihydroceramide synthase) suggesting that these two pathways utilise different sphingoid bases in vivo [42]. Possible regulators of acyl-CoA independent ceramide synthase activity are intracellular pH and phosphatidic acid [42] demonstrating potential for agonist induced ceramide synthesis and sphingosine removal by this pathway. The chemotherapeutic agent daunorubicin has been shown to increase acyl-CoA dependent dihydroceramide synthase activity [44] suggesting that this enzyme may also be regulated. Recently, coordinate activation of dihydroceramide synthase and SPT has been reported in fenritidine-induced apoptosis [45], suggesting that removal of sphingoid bases by dihydroceramide synthase may act as a positive feedback control stimulating SPT to increase their synthesis.

3.3. Dihydroceramide desaturase

Dihydroceramide desaturase inserts a 4,5-*trans*-double bond into dihydroceramide (Fig. 2) so producing ceramide [46,47]. This reaction is important as it converts a relatively inactive sphingolipid, dihydroceramide into the bioactive lipid ceramide [48]. Characterisation of dihydroceramide desaturase activity has shown it to be dependent on NADPH [47,49,50] and localised to ER membranes with desaturase activity oriented at the cytosolic face [46,50]. Sphingolipid desaturases have been cloned from plants [51] and delta4-desaturase (dihydroceramide desaturase) genes identified from human, mouse, fly and fungi [52]. A regulatory role for dihydroceramide desaturase is supported by the observation that enzyme activity is altered in tumours [53].

4. Synthesis of complex sphingolipids

The synthesis of complex sphingolipids from ceramide takes place at the Golgi in mammalian cells [54]. Ceramide is transferred from its site of synthesis, in the ER, to Golgi by vesicular traffic and also by direct transfer due to close apposition of ER and *trans*-Golgi membranes [55]. At the Golgi, ceramide is converted on the lumenal side to sphingomyelin by SMS [56] or at the cytosolic face to GlcCer by GlcCer synthase [57] from which complex GSLs are formed. GSLs, ceramide and sphingomyelin synthesised within the Golgi are delivered to other cellular membranes by transport vesicles [55].

4.1. Sphingomyelin synthase

Sphingomyelin synthase (SMS, phosphatidylcholine:ceramide phosphocholine transferase) transfers a phosphocholine group from phosphatidylcholine to ceramide forming sphingomyelin and releasing diacylglycerol (DAG) (Fig. 3). The reverse action has also been observed, i.e. conversion of sphingomyelin to phosphatidylcholine consuming DAG and releasing ceramide, at the basolateral surface of epithelial cells [58]. The potential of SMS to regulate levels of two lipid second messengers, ceramide and DAG with opposing cellular effects (cell death versus cell survival) has led to the suggestion that it may act as a "switch" in signalling pathways [59,60]. Recently, it has been suggested that regulation of the direction of SMS activity may be important for controlling DAG levels and vesicle transport at the *trans*-Golgi network [61]. SMS activity has been found on the lumenal face of *cis*-Golgi [56] and *trans*-Golgi [62]. Although Golgi appears to be the main site of SMS activity and sphingomyelin synthesis [63], activity has been found at plasma membrane [59,64], mitochondria [43] and nuclei [65,66]. However, the mammalian enzyme has neither been purified nor cloned. SMS activity is stimulated by thioacetamide [67] and basic fibroblast growth factor [68] and increased in SV40 transformed cells [59] and apoptosis [69]. Changes in SMS activity have been implicated in NFκB signalling [60], growth suppression [70] and vesicle trafficking [61].

4.2. Glucosylceramide synthase

Glucosylceramide synthase (GCS, ceramide glucosyl-transferase) catalyses the initial step in GSL synthesis, the transfer of glucose from UDP-glucose to ceramide forming GlcCer (Fig. 3) – the precursor for over 300 species of GSL. Accordingly GCS is a key regulator of intracellular levels of ceramide and GSLs. GSLs are found in the extracellular leaflet of the plasma membrane and play an important role in many cellular processes, including cell recognition, growth, development and differentiation [57,71]. GCS is found at the cytosolic face of the Golgi [57] and utilises de novo synthesised ceramide to form GlcCer [72], which then transfers to the lumenal side of the Golgi for further processing to complex GSLs [55]. GCS purified from rat liver Golgi membranes preferentially utilised dioleoylphosphatidylcholine and UDP-glucose as substrates and migrated as two proteins at 60–70 kDa [73]. GCS has been cloned from human and mouse sources and the cDNA encodes a 44.9 kDa protein with a membrane-association domain and a long cytoplasmic tail [74,75]. Antibodies raised against this protein detected GCS as a 38 kDa protein in rat liver membranes, crosslinking studies suggested that GCS forms oligomers possibly with other Golgi membrane proteins [76], which may explain the higher molecular mass obtained by the sedimentation of purified rat liver protein [73]. Targeted disruption of the gene encoding GCS was embryonic lethal showing that GSLs are required for embryonic development and normal tissue differentiation [77].

In addition to its role as the initiator of GSL biosynthesis GCS has also been proposed as a key regulator of chemotherapy drug resistance possibly through modulation of cellular ceramide levels [78]. Apoptotic stimuli such as TNFα and chemotherapeutic agents are thought to induce cell death through accumulation of ceramide. TNFα has been shown to inhibit the activity of GCS and SMS, which would contribute to the increase in ceramide [70]. Conversely, overexpression of GCS reduced ceramide accumulation in response to chemotherapeutic agents and protected against cell death induced by these drugs [79]. Furthermore, inhibition of GCS with specific inhibitors such as PDMP or P4, or downregulation of GCS with antisense increased ceramide levels restoring sensitivity to chemotherapeutic drugs [80]. However, GCS may only play a role in regulating the levels of de novo synthesised ceramide, as it does not have access to ceramide produced by sphingomyelinases at the plasma membrane [72,81]. Moreover, some of the effects of GCS inhibition in chemotherapeutic drug resistance may be due to changes in vesicle trafficking rather than increased ceramide levels [82].

4.3. Galactosylceramide synthase

Galactosylceramide synthase (GalCer synthase) catalyses the transfer of galactose from UDP-galactose to ceramide. This is the first step in the synthesis of galactosphingolipids (Fig. 3) major components of the myelin sheath surrounding the axons of neuronal cells [83]. GalCer synthase is a 61 kDa protein located in the ER and despite catalysing a similar reaction shows no homology to GCS [74]. GalCer synthase mRNA is expressed mainly in kidney and brain [84] and synthase knock out mice exhibit severe generalised tremor and ataxia [85,86].

4.4. Ceramide kinase

CERK catalyses the phosphorylation of ceramide producing ceramide-1-phosphate. CERK activity was first shown in brain synaptosomes as a calcium dependent enzyme activity specific for ceramide [87]. Further studies in HL-60 cells showed that CERK was distinct from diacylglycerol kinase and phosphorylated ceramide produced by sphingomyelinase-mediated hydrolysis of sphingomyelin at the plasma membrane [88]. The recently cloned Human CERK (hCERK) encodes a protein with similarities to sphingosine and diacylglycerol kinases but in a phylogenetically distinct class [89]. Ceramide-1-phosphate has been implicated in fusion of brain synaptic vesicles [87], neutrophil phagolysosome formation [90] and mitogenesis [91,92].

4.5. Lactosylceramide synthase

Lactosylceramide synthase (GalT-2) catalyses the second step of GSL synthesis by transferring galactose from UDP-galactose to GlcCer (Fig. 3). GalT-2 has been cloned from the rat and shown to be a highly glycosylated protein of 61 kDa [93]. GalT-2 activity is found at the lumenal side of the Golgi [94] and its activity may be associated with proliferation [95] and with endothelial cell responses to shear stress [96] suggesting that GSLs may play a role in atherosclerosis [97].

5. Metabolism of sphingolipids

Once formed by de novo synthesis sphingolipids may be broken down and/or modified to form additional bioactive lipids such as ceramide, sphingosine, S1P and SPC (lyso-sphingomyelin) (Fig. 4). The major enzymes involved are sphingomyelinases, ceramidases and sphingosine kinases (SPHKs), all of which may be regulated by diverse stimuli. Turnover of sphingolipids occurs mainly after endocytosis of plasma membranes. The internalised lipid may be recycled to the Golgi or returned to the plasma membrane, although most of the lipid is degraded within lysosomes through hydrolytic cleavage. Complex GSLs are broken down by a series of glycosidases that sequentially hydrolyse the glycosyl headgroups eventually releasing ceramide. Sphingomyelin is degraded within lysosomes by acidic sphingomyelinase releasing phosphocholine and ceramide. Ceramide may then be deacylated by lysosomal ceramidases producing sphingosine, which presumably crosses the lysosomal membrane and transfers to the endoplasmic reticulum where it is used for resynthesis of ceramide. In addition, degradation of sphingomyelin can occur outside of the well-characterised lysosomal pathway by membrane-associated sphingomyelinases and ceramidases with activity at neutral or alkaline pH. Activation of these enzymes and generation of sphingolipid metabolites at cellular membranes are thought to be involved in cell signalling. In addition to being used for ceramide resynthesis sphingosine may be phosphorylated by SPHK to S1P which in turn is dephosphorylated to sphingosine or degraded by a lyase located at the endoplasmic reticulum [98]. The mechanisms by which sphingosine traffics from the plasma membrane or lysosomes to the endoplasmic reticulum are unknown, however, cellular levels of this lipid are low suggesting that it is rapidly converted to either ceramide or S1P a reaction which is thought to act as a "switch" between apoptotic and proliferative signals [99].

5.1. Sphingomyelinases

Sphingomyelinases remove the phosphocholine headgroup from sphingomyelin releasing ceramide. There are three main types of sphingomyelinase; acid (aSMase), neutral (nSMase) and alkaline that have been separated into five classes according to their pH optima, cofactor requirements and subcellular localisation.

5.1.1. Acid sphingomyelinase

Acid sphingomyelinase (aSMase) is a soluble glycoprotein that is localised to lysosomes and exhibits maximal activity at ~pH 5 [100]. It has been purified and cloned [101,102]. Expression of the human cDNA in sf21 insect cells led to secretion of the recombinant protein with a molecular mass of 72 kDa [103]. aSMase deficiency is the cause of Niemann-Pick disease, a lysosomal lipid storage disorder [104,105]. A secreted form of aSMase has also been identified [106]. It is a product of the same gene as aSMase but it is differentially glycosylated and processed at the NH_2-terminus. Secretory SMase is dependent on exogenous Zn^{2+} for activity and targeted to the Golgi secretory pathway unlike lysosomal aSMase which appears to be constitutively exposed to endogenous Zn^{2+}. Recently, a third form of aSMase has been identified that is present on the surface of cells exposed to CD95 or CD40 [107,108]. Whether this is a third aSMase isoform or secretory SMase is not known.

aSMase is activated by many apoptotic stimuli for instance pro-apoptotic receptors (Fas and TNF) and stress (UV irradiation and hydrogen peroxide) and has been proposed as an important modulator of ceramide levels and apoptosis [109], although activation of aSMase is not always obligatory for, nor does it always induce apoptosis. For instance, stimuli that do not induce apoptosis such as CD40 and interleukin-5 also activate aSMase and some apoptotic stimuli (serum starvation) do not require aSMase (reviewed in Ref. [109] and references therein). However, studies using aSMase deficient hepatocytes from Niemann-Pick patients or mouse embryonic fibroblasts lacking aSMase have shown that they are resistant to apoptosis induced by CD95 (Fas receptor ligand) and that sensitivity is restored by the addition of C16-ceramide [110,111] suggesting that generation of ceramide is the implicit signal. The precise role of aSMase in apoptotic responses remains unclear, as Bezombes et al. [112] were unable to detect differences between aSMase deficient cells derived from Niemann-Pick patients and normal cells when stimulated with pro-apoptotic stimuli, anthracyclines, ionising radiation and Fas ligation leading these authors to suggest a role for neutral sphingomyelinase. A role for secretory SMase in atherosclerosis has been proposed. Serum low-density lipoprotein (LDL) when treated with bacterial sphingomyelinase forms small aggregates that adhere to extracellular matrix and stimulate macrophage foam cell formation [113]. More recently, it has been shown that inflammatory cytokines stimulate secretion of sSMase from endothelial cells [114] and that sSMase could hydrolyse and aggregate oxidised LDL [115] a key event in atherogenesis.

The mechanisms linking receptor stimulation to aSMase activation are poorly understood. Phosphoinositide 3 kinase (PI3K) may negatively regulate aSMase [116] and tyrosine phosphorylation of aSMase following CD95 stimulation has been observed [109]. However, as aSMase is predominantly located in vesicles [107] and tyrosine kinases are cytosolic the site of interaction with aSMase is unclear. Caspases are also implicated in linking the TNF receptor to aSMase possibly by activating the transport of aSMase containing vesicles to the membrane where aSMase is secreted to act at the cell surface (reviewed in Ref. [109]). Although it has also been suggested that activated aSMase may hydrolyse sphingomyelin at the lumenal face of the vesicle, fusion of the vesicle with the plasma membrane would then result in presentation of ceramide at the cell surface [72]. This would overcome the problem of aSMase acting at suboptimal pH on the extracellular cell surface.

5.1.2. Neutral sphingomyelinase

Two classes of neutral sphingomyelinase have been identified a Mg^{2+}-dependent enzyme (N-SMase) and a Mg^{2+} -independent enzyme. Little is known concerning the Mg^{2+}-independent neutral sphingomyelinase which has been found in the cytosol of HL60 cells [117] and in rabbit skeletal muscle [118].

5.1.2.1. Neutral, Mg^{2+}-dependent sphingomyelinases. Mammalian neutral Mg^{2+}-dependent sphingomyelinases (N-SMase) are integral membrane proteins with pH optima of approximately pH 7 and a requirement for Mg^{2+} or Mn^{2+} in the millimolar range. N-SMase activities have been purified from many sources, however the differences in molecular weights of these enzymes suggest that there are multiple mammalian isoforms. A 60 kDa N-Smase has been purified from rat brain membranes [119] that was highly dependent on phosphatidylserine and inhibited by glutathione. A similar activity was isolated from bovine brain but consisted of two closely related polypeptides of 97 and 46 kDa [120]. In rabbit skeletal muscle two isoforms were found of 92 and 53 kDa, although the smaller enzyme was Mg^{2+}-independent. These N-SMases were present in the T-tubule system suggesting a role in sphingosine-mediated calcium regulation [118]. Two microsomal N-SMases have been characterised from immature rat seminiferous tubules that differed in their sensitivity to Mg^{2+} and Mn^{2+} [121]. Three proteins have been cloned that show neutral sphingomyelinase activity. The first report was of the cloning, overexpression and characterisation of human and mouse Mg^{2+}-dependent N-SMase (N-SMase1) [122]. However, cells overexpressing this protein do not show altered sphingolipid metabolism [123,124] and it has been suggested that the natural in vivo substrate is not sphingomyelin [123]. Moreover, in Jurkat cells overexpression of N-SMase1 did not alter the ceramide response and apoptosis induced by Fas ligand [124] suggesting that this enzyme is not involved in death receptor induced cell death. N-SMase1 has been localised to the endoplasmic reticulum [125,126] and antibodies raised against this protein did not inhibit brain membrane Mg^{2+}-dependent N-SMase activity [127]. Furthermore, N-SMase1-deficient mice show no changed metabolism of sphingomyelin or other lipids, despite reduced N-SMase activity in all tissues except the brain [128] indicating that N-SMase1 is unlikely to be the membrane-associated brain activity previously identified [120,129]. The second clone identified was termed N-SMase2, this enzyme is expressed predominantly in brain tissue, has high selectivity for sphingomyelin and is dependent on Mg^{2+} ions [127]. Immunofluorescence studies showed that N-SMase2 co-localised with Golgi markers and was restricted to neurons in brain sections [127]. The restriction of N-SMase2 to brain and localisation in the Golgi suggest that this is not the plasma membrane activity identified in other tissues [100]. Finally, a third clone has been reported that encodes a protein with Mg^{2+}-dependent neutral sphingomyelinase activity, which when expressed in bacterial cells generates a recombinant protein of 45 kDa, however expression in a mammalian system (COS-7 cells) produced a protein of ~90 kDa [130]. Full characterisation of this N-SMase awaits further study.

N-SMase is implicated in TNF-induced cell death although the exact mechanism of its activation is not fully understood. The TNF receptor contains a cytosolic domain termed NSD (neutral sphingomyelinase domain) that associates with FAN (factor-associated with neutral sphingomyelinase) when bound to TNF. FAN then associates with and activates N-SMase [131]. FAN regulated activation of N-SMase is involved in CD40-induced apoptosis [132] and fibroblasts from FAN deficient mice or fibroblasts from normal mice transfected with a dominant negative FAN are resistant to TNF-induced cell death [133]. The mechanism by which FAN activates N-SMase is not known, although in a recent study the scaffold protein RACK1 (receptor for activated C kinase1) was shown to interact with FAN and enhance TNF-induced N-SMase activity [134] suggesting that assembly of a multi-component signalling complex occurs. Phospholipase A_2 (PLA_2) has also been implicated as a regulator of N-SMase. In HL60 cells activation of cPLA2 and accumulation of arachidonic acid preceded N-SMase activation following TNF stimulation [135,136] and cells lacking cPLA2 were resistant to TNF-induced N-SMase activation and apoptosis [137]. Arachidonic acid has been shown to activate N-SMase in vitro [119], although whether direct activation occurs in vivo is not known.

In many cell types N-SMase is activated by intracellular oxidation in response to TNF, hydrogen peroxide and hypoxia/reoxygenation leading to the proposal that it is a sensor of oxidative stress [99]. Mg^{2+}-dependent N-SMase is inhibited by reduced glutathione [129] and intracellular oxidation causes a drop in the levels of reduced glutathione and increase in oxidised glutathione. However, the effect of glutathione within the cell does not appear to be direct as Bcl-xl may act in between glutathione and N-SMase [138]. Furthermore, PKCζ may also be involved as a negative regulator of oxidative activation of N-SMase [139].

N-SMase activity has been found in caveolae, flask shaped invaginations of the plasma membrane involved in signal transduction [140]. In human skin fibroblasts N-SMase was present in caveolae and inhibited by a peptide corresponding to the caveolin scaffolding domain [141]. Caveolin is the major coat protein of caveolae and may act to inhibit signalling molecules through interaction with its scaffolding domain [142]. Stimulation with TNF increased caveolin scaffolding-domain sensitive N-SMase activity in non-caveolar regions of the plasma membrane, suggesting that dissociation of N-SMase from caveolae may be one mechanism of TNF-induced activation of this enzyme [141]. In endothelial cells activation of caveolar-associated N-SMase has been implicated in shear stress-induced ERK activity [143].

Recently, an inhibitor of Mg^{2+}-dependent N-SMase was reported that blocked TNF-induced N-SMase activation and cell death in breast cancer cells [144]. Further studies with such agents should help to unravel the role of N-SMase in cellular responses.

5.1.3. Alkaline sphingomyelinase

Dietary sphingomyelin is hydrolysed by an alkaline sphingomyelinase present in the gut. The enzyme has been purified from rat intestine and shown to be a 58 kDa protein with pH optima of 9–9.5 [145]. In vitro, purified alkaline sphingomyelinase is inhibited by glycerolipids, cholesterol and ceramide and activated by fatty acids [146] suggesting dietary fats would affect its activity. A role for alkaline sphingomyelinase in colonic carcinogenesis [147] and chronic colitis [148] has been proposed.

5.2. Ceramidases

Ceramidases hydrolyse the *N*-acyl linkage between the acyl side chain and sphingoid base of ceramide producing sphingosine and a fatty acid. They catalyse the rate-limiting step in the production of sphingosine which is not generated by de novo synthesis [149] and have the potential to regulate the relative levels of ceramide and sphingosine within cells. There are three classes of ceramidase, neutral, alkaline and acidic based on their pH optima. Neutral and alkaline (non-lysosomal) ceramidases are members of a gene family conserved from bacteria to mammals but phylogenetically distinct from acidic (lysosomal) ceramidases [150].

5.2.1. Neutral ceramidases

Neutral ceramidases (N-CDase) have been cloned from mouse [150], rat [151], human [43] and Drosophila [152]. The human N-CDase has a predicted molecular mass of 84 kDa and when overexpressed as a GFP-tagged protein in HEK293 cells was localised to mitochondria [43], suggesting that this N-CDase may regulate ceramide levels during apoptosis. The mouse liver N-CDase is a 94 kDa protein that is localised to late endosomes/lysosomes [40], in contrast the rat kidney enzyme is found in lipid rafts of the plasma membrane [151]. Overexpression of the rat kidney isoform in HEK293 cells resulted in production of two highly glycosylated forms which were differentially localised, a Golgi form of 133 kDa and an endoplasmic reticulum form of 113 kDa. Inhibition of N-glycosylation by tunicamycin resulted in both proteins being expressed as an 87 kDa form with low CDase activity [151]. Further studies have shown that *O*-glycosylation of N-CDases is important for localisation to the plasma membrane [153]. Therefore, post-translational modification of N-CDases appears to be important for both catalytic activity and cellular localisation. Both the mouse liver N-CDase and the human mitochondrial CDase catalyse the reverse reaction, condensation of sphingosine and a fatty acid to form ceramide in an acyl-CoA independent reaction [40,43] and Section 3.2 "Dihydroceramide Synthase".

There is evidence that N-CDases could be involved in signal transduction pathways to generate sphingosine for regulation of cell proliferation. Growth factors have been shown to activate membrane-associated N-CDase in rat glomerular mesangial cells [154] and interleukin-1β (Il-1β) activated N-CDase in rat hepatocytes [155]. In both studies tyrosine kinases appeared to regulate CDase activity. A tyrosine phosphorylation site has been identified in the deduced amino acid sequence of mouse N-CDase [150]. There is also evidence that cytokines regulate CDase at the translational level as Il-1β stimulation of renal mesangial cells increased N-CDase mRNA through a p38MAPK pathway resulting in increased CDase activity [156]. In contrast, stimulation of rat mesangial cells with nitric oxide decreases N-CDase activity [157] by downregulation of the enzyme through a proteosome pathway, an effect that is inhibited by activators of protein kinase C [158]. Inhibition of CDase activity is cardioprotective in animal models of coronary microembolisation [159] and ischaemia reperfusion [160] possibly due to a decrease in sphingosine production (see Section 6.4.2). In addition, in endothelial cells N-CDase activity was present in caveolin-enriched membranes, plasma membrane domains involved in cell signalling [140], treatment with β-cyclodextrin an agent that depletes cholesterol and interferes with caveolae increased N-CDase activity [161], suggesting that subcellular localisation regulates activity. Taken together these recent studies demonstrate that regulation of CDase activity is an important regulatory step in many cellular responses.

5.2.2. Alkaline ceramidase

Alkaline CDases have been cloned from yeast [41,162], Pseudomonas [163] and human [164]. The yeast enzymes are products of two genes YDC1 and YPC1, the protein product of YDC1 hydrolyses dihydroceramide [162] the precursor of ceramide, whilst that of YPC1 also catalyses a CoA-independent synthase activity [41] as does the enzyme from Pseudomonas [163]. The human alkaline CDase is ubiquitously expressed, demonstrated by Northern blot analysis, with the highest expression in placenta. The GFP-tagged protein localised to Golgi and endoplasmic reticulum, however the enzyme had no reverse synthase activity but did show preference for phytoceramide (a yeast sphingolipid) as substrate [164]. In endothelial cells bradykinin stimulated the secretion of both N-CDase and alkaline CDase activities [165] suggesting alkaline CDases could be involved in agonist responses. Moreover, alkaline CDases have broad pH optima therefore it is possible that they contribute to the agonist-induced effects attributed to N-CDases discussed above.

5.2.3. Acid ceramidase

Human acid CDase has been purified and characterised [166], and homologous genes encoding acid CDase have been characterised in mice and humans [167,168]. Recently, the full-length human acid CDase gene was isolated and characterised [169]. Human acid CDase is produced as a single precursor polypeptide (53–55 kDa) that is processed to a heterodimeric enzyme of 45 and 13 kDa. Glycosylation is important for both the processing of the enzyme and targeting to lysosomes [170]. Mutations affecting the activity of human acid CDase are the cause of a lipid storage disorder, Farbers disease, characterised by ceramide accumulation and lipogranulomatosis [168]. Acid CDase deficient mice die at embryonic day 8.5, whilst heterozygote mice survive but have elevated ceramide levels in several tissues especially liver as well as a lipid storage disorder similar to Farbers disease [171]. An inhibitory role for acid CDase in apoptosis has been suggested as its overexpression prevented TNFα-induced apoptosis and inhibiting its activity restored sensitivity [172]. Also, in HaCaT keratinocytes and human melanoma cells inhibition of acid CDase activity increased ceramide levels and inhibited proliferation [173].

5.3. Sphingosine kinase

SPHK phosphorylates sphingosine to form S1P. S1P acts as an agonist at specific G protein coupled receptors and as an intracellular second messenger to regulate many cellular functions, in particular, cell motility and development [98]. Seven clones of SPHK have been reported and appear to represent a gene family conserved from plants to mammals [174]. From yeast two genes Lcb4 and Lcb5 have been shown to encode proteins with SPHK activity [175], from plants a single clone has been reported [176] and from mammals two clones have been identified from mouse – mSPHK1 and mSPHK2 and human – hSPHK1 and hSPHK2 [177,178]. The yeast SPHKs show considerable homology to the mouse and human SPHK1 and mSPHK1 and hSPHK1 have 92% similarity at the amino acid level, demonstrating a highly conserved kinase finally. SPHK1 has been purified from rat kidney and shown to be a 49 kDa protein that is active as a monomer and preferentially phosphorylates D-erythrosphingosine [179]. Northern analysis showed hSPHK1 is widely expressed with abundant levels in liver, kidney, heart and skeletal muscle [180,181]. Antibody staining has shown expression of SPHK1 in brain, kidney, vascular endothelial cells, megakaryocytes and platelets [182]. Over-expression of SPHK1 induced increased cell proliferation in NIH 3T3 and HEK293 cells [183] and in a recent study NIH3T3 cells overexpressing SPHK1 acquired a transformed phenotype that involved Ras activation [184]. Additionally, expression of SPHK1 in human breast cancer MCF-7 cells increased S1P production, blocked cell death induced by anti-cancer drugs and increased cell proliferation and growth [185]. These studies suggest that SPHK1 may act as an oncogene. SPHK1 is also implicated in angiogenesis. Endothelial cells constitutively secrete SPHK by a non-classical vesicular pathway that requires an intact actin cytoskeleton and overexpression of SPHK1 in endothelial cells increased SPHK release and induced angiogenesis and vascular maturation [186]. mSPHK2 and hSPHK2 encode proteins with a molecular mass of 66 kDa that share five conserved domains with SPHK1 [178,180]. SPHK2 is most abundantly expressed in liver and heart and its transcript appears later during development than SPHK1 [178]. Differences also exist between the two isoforms in their substrate specificity and sensitivity to salt and detergents [174]. hSPHK2 contains four putative transmembrane domains however, only 20% of the cellular activity is membrane associated [178] raising the possibility that translocation to cellular membranes may be necessary for enzyme activity. Whether SPHK2 is involved in the same cellular responses as SPHK1 has not yet been investigated.

That SPHK activity is increased by a number of diverse stimuli in many cell types is well documented (for recent reviews see Ref. [98,187]). However, the molecular mechanisms of activation are only just beginning to be unravelled. Activation of SPHK by growth factors in TRMP cells expressing different mutants of the tyrosine kinase receptor for PDGF was dependent on the binding of PLCγ and subsequent release of calcium from intracellular stores [188]. Stimulation of SPHK by P2Y2 receptor agonists in HL60 cells also required calcium mobilisation [189] suggesting that calcium could be the common signal for SPHK activation by tyrosine kinase and G protein coupled receptors. Indeed, SPHK binds with high affinity to calmodulin in the presence of calcium [179] and in human SH-SY5Y cells muscarinic stimulation induced a calcium/calmodulin-dependent translocation of GFP-tagged SPHK1 to the plasma membrane [190]. However, only subcellular distribution and not activation appeared to be dependent on calcium/calmodulin in this cell type. In HEK293 cells, expressed GFP-tagged SPHK1 and endogenous SPHK translocated to the membrane in response to calcium ionophore possibly due to activation of protein kinase C as a similar translocation was observed in response to phorbol esters, which could be blocked by PKC inhibitors [191]. SPHK interacting proteins have also been identified that may regulate enzyme activity. Yeast two hybrid screens have identified two novel binding partners, SPHK interacting protein (SKIP) that reduced SPHK1 activity in an expression system [192] and RPK118 which colocalised with SPHK1 at the endoplasmic reticulum when the recombinant proteins were expressed [193], whether RPK118 affected SPHK1 activity was not reported. SPHK has also been shown to interact with TNFα receptor-associated factor 2 (TRAF2) through a specific binding motif. TRAF2 binding increased SPHK activity and was required for TRAF2-mediated activation of NF-κB. Moreover, the interaction of SPHK with TRAF2 was required to prevent TNFα-induced apoptosis [194].

5.4. Sphingosine phosphate phosphohydrolase and lyase

S1P can be degraded by general lipid phosphohydrolases (LPP), specific S1P phosphohydrolases (SPP) and S1P lyase. LPPs belong to a phosphohydrolase superfamily [195], they have broad substrate specificity readily dephosphorylating phosphatidic acid, lysophosphatidic acid and ceramide-1-phosphate in vitro, in addition to S1P [196,197]. Three mammalian LPPs have been characterised, they are membrane associated, share sequence homology and are insensitive to inhibition by *N*-ethylmaleimide [198,199]. Specific S1P phosphohydrolases have been identified also from yeast and mammalian sources. Two genes encoding sphingoid base phosphate phosphohydrolases have been cloned from yeast, LBP1 and LBP2 [200–202]. Despite sharing 53% sequence homology LBP2 cannot complement LBP1 function [200–202], suggesting that the two enzymes have separate physiological functions. LBP1 and LBP2 are important for heat stress responses in yeast. Two mammalian genes SPP1 and SPP2 have been cloned [203,204] that encode phosphatases with specificity for sphingoid base-1-phosphates. Overexpression of SPP1 in HEK293 cells increased membrane-associated S1P phosphohydrolase activity without increasing cytosolic activity, consistent with the presence of 8–10 putative transmembrane domains in SPP1. The levels of ceramide were increased and S1P decreased in cells expressing SPP1, elevated levels of sphingosine were not detected indicating rapid conversion to ceramide [203]. Consistent with an increase in ceramide, apoptosis was increased in NIH3T3 cells overexpressing SPP1 [203]. The SPP2 sequence has 70% similarity to SPP1, with high hydrophobicity and 9 putative transmembrane segments indicating an integral membrane protein. Expression of SPP2 in HEK293 cells substantially increased particulate phosphohydrolase activity and immunohistochemistry showed SPP2 was localised to endoplasmic reticulum consistent with a role in regulation of S1P signalling [204]. Northern blot analysis showed that in contrast to SPP1, which is ubiquitously expressed, SPP2 was restricted to brain, heart, colon, kidney, small intestine and lung [204], suggesting that the two isoforms may have different functions.

The final step in the degradation of S1P is catalysed by an endoplasmic reticulum pyridoxal phosphate-dependent SIP lyase (SPL). This lyase degrades S1P to form fatty aldehydes and phosphoethanolamine [205,206], which can be further metabolised to form glycerophospholipids [206]. S1P lyase activity is localised to the cytosolic face of the endoplasmic reticulum in rat liver [207] and is ubiquitous with regard to species and mammalian tissues, with the notable exception of platelets where it is absent [208] which may account for the high levels of S1P in platelets. SPL has been cloned from yeast, mouse and human sources. The first SPL gene, BST1, was cloned from yeast. BST1 conferred resistance to sphingosine in yeast and deletion of the gene resulted in accumulation of S1P in response to exogenous sphingosine treatment [209]. Mammalian homologues of the yeast gene have been identified [208,210]. The mouse and human clones share 84% identity at the amino acid level as well as a similar tissue distribution of mRNA, both being expressed ubiquitously but with high levels in liver and kidney [208] a similar tissue distribution was shown for mouse SPL [205]. Whether SPL activity is regulated in response to agonist stimulation is not known.

6. Physiological functions of sphingolipids

Sphingolipids have been implicated in many cellular responses such as regulation of apoptosis, proliferation, migration, adhesion and contractility. However, dissecting their physiological roles is complex as there are many sphingolipid-metabolising enzymes and generation of sphingolipids can occur in many different cellular compartments with potentially distinct targets. Furthermore, signalling sphingolipids may be formed by more than one pathway, for instance ceramide is formed by de novo synthesis and by hydrolysis of sphingomyelin and these two routes occur at different locations within the cell, Golgi and endoplasmic reticulum, respectively. Similarly the levels of S1P are regulated by three separate enzymes SPHK, SPP and SP lyase. Further complexity is apparent in sphingolipid biology as many of the sphingolipid-generating enzymes are present as multiple isoforms which have different cofactor requirements and pH optima and that are tightly regulated, for example acidic and neutral SMases (see Section 5.1). However, consensus within the literature is emerging for important roles of sphingolipids as mediators of apoptotic cell death, regulators of cell growth, components of plasma membrane microdomains essential for cell signalling and for their involvement in physiological and pathological function of cardiovascular cells.

6.1. Sphingolipids in apoptosis

Early studies demonstrated that extracellular stimuli could induce sphingomyelin hydrolysis leading to an increase in intracellular ceramide [211]. Furthermore, addition of ceramides to cells was a potent stimulus for cytotoxicity through apoptotic programmed cell death [212]. It is now clear that many inducers of apoptosis such as TNFα, Fas ligand, chemotherapeutic agents and ischaemia/reperfusion regulate enzymes involved in sphingolipid biosynthesis and metabolism leading to an increase in ceramide levels [213]. In addition, intracellular targets of ceramide such as kinases, phosphatases and proteases have been identified which appear to play a role in ceramide-mediated apoptosis [48]. However, recent evidence suggests that other sphingolipids such as sphingosine, S1P and GSLs are also involved in apoptosis. Recently, a detailed series of review articles were published [109,187,213–219] concerning the role of lipids in apoptosis accordingly only the key issues will be addressed in this section.

6.1.2. Role of ceramide in apoptosis

Despite controversy over the methodologies used to measure cellular ceramide levels [220], there is now considerable evidence showing that ceramide accumulates during the initiation phase of apoptosis (reviewed in Ref. [213]). Furthermore, the effect of ceramide is specific as dihydroceramide, the precursor of ceramide does not induce cell death [221] and detailed analyses using mass spectrometry techniques [222–224] suggest that the nature and species of ceramide generated confers specificity on the response. All three main routes for generation of ceramide, sphingomyelin hydrolysis by acidic or neutral SMases or de novo synthesis can be regulated by apoptotic signals (for reviews see Refs. [109,213,215]). For instance, A-SMase is activated by cytokines and stress [109], cytokines, vitamin D3, chemotherapeutic agents, stress and ischaemia/reperfusion [213] can activate N-SMase. Cytokines and chemotherapeutic agents increase de novo synthesis of ceramide resulting in apoptosis and inhibition of enzymes within the de novo synthesis pathway is cytoprotective [214,215]. Moreover, activation of enzymes that clear ceramide such as CDase and GCS can confer resistance to TNF or chemotherapeutic drug induced apoptosis [79,157,225]. Additionally, recent studies have shown that inhibition of N-SMase blocks TNFα-induced ceramide accumulation and cell death in MCF7 breast cancer cells [226] and A-SMase deficient cells are resistant to Fas-activated apoptosis an effect rescued by addition of long chain ceramides [110,227] demonstrating the importance of sphingomyelin hydrolysis and ceramide production. However, the molecular mechanisms involved in ceramide-induced apoptosis are still unclear.

Studies into the topology of sphingolipids and the localisation of metabolic enzymes to specific intracellular sites have shown that the site of production of sphingolipids determines their signalling functions (reviewed in Refs. [35,72]). Fractionation studies have shown that sphingomyelin levels are highest in the plasma membrane, predominantly in the outer leaflet, high in mitochondria and Golgi and low in nuclear and ER membranes [228]. Using fluorescent tagging and time lapse confocal microscopy ceramide was found in the perinuclear region, ER and mitochondria following which it rapidly accumulated in the Golgi and was metabolised to GluCer [214]. There is evidence that the site of ceramide generation in response to an agonist is important, as addition of exogenous bacterial SMase which would generate ceramide in the outer leaflet of the plasma membrane did not induce apoptosis whereas expression of bacterial SMase, which would act at intracellular sites did [229,230]. However, ceramide generated in the outer plasma membrane leaflet by A-SMase is important for Fas signalling and facilitates receptor clustering and capping in T cells, an event necessary for optimal Fas-induced cell death in some cell types [107,231]. Furthermore, there is evidence that endogenous ceramide generated by N-SMase at the plasma membrane is then restricted to that membrane as it is not available for metabolism by GCS at the Golgi. This is in contrast to metabolism of cell permeable fluorescent analogues of ceramide that appear to diffuse freely within the cell and are metabolised by GCS [81]. Reports of ceramide production at nuclear and mitochondrial membranes may have direct relevance to apoptosis. Radiation treatment of fractionated nuclei resulted in increases of SMase-derived ceramide [232] and nuclear SMase activity has been implicated in hepatocyte apoptosis [233]. Although, in epithelial cells nuclear ceramide production was not necessary for hydrogen peroxide-induced cell death [234].

Mitochondria play a key role in the control of cell survival and cell death and most apoptotic signals converge on the mitochondrion inducing release of cytochrome c and other proteins which lead to activation of effector caspases and commitment to cell death [235]. Understanding the upstream events that control mitochondrial responses is therefore crucial to understanding the initiation of apoptosis. Recent evidence suggests that sphingolipids at the mitochondria play an important role in regulating the apoptotic response. Enzymes of ceramide metabolism have been localised to mitochondria [43] and sphingomyelin is present in mitochondrial membranes [228]. The evidence that mitochondria are involved in ceramide-induced apoptosis has been reviewed recently [235]. The exact mechanism by which mitochondrial ceramides regulate the apoptotic response is still uncertain. However, ceramide can inhibit mitochondrial respiratory chain complex III, induce generation of reactive oxygen species in mitochondria and disrupt mitochondrial function. In addition, ceramide can be formed at mitochondrial membranes by the action of CDase and ceramides and neutral GSLs are increased in mitochondria by TNF (for references see [214,235]). Recently, expression of GFP-tagged bacterial SMase targeted to different subcellular locations showed that only when targeted to mitochondria did the cells undergo apoptosis. Targeting to the inner plasma membrane, cytoplasm, ER, Golgi or nucleus was ineffective despite generation of ceramide at these locations [236]. The effect was inhibited by Bcl 2, an anti-apoptotic protein, and mediated by the release of cytochrome c, demonstrating involvement of mitochondria. Ceramide may also affect mitochondrial function indirectly by conversion to ganglioside GD3 a potent stimulator of release of apoptotic factors from mitochondria [237] and reviewed in Ref. [218]. GD3 is implicated in Fas-induced apoptosis in lymphoid and myeloid cells [238] and inhibition of GD3 synthase by antisense protected HT-29 cells against TNFα-induced cell death [239]. However, no differences in Fas-mediated apoptosis were observed in lymphocytes from mice deficient in GD3 synthase when compared to cells from wild type mice [240], demonstrating that GD3 formation is not essential for Fas-stimulated apoptosis.

Several effectors of ceramide action in apoptosis have been identified including ceramide activated protein phosphatases (CAPP), CAP kinases and cathepsin D (reviewed in Refs. [48,214]). CAPK was first reported in A431 cells [241] and subsequently identified as KSR (kinase suppressor of Ras) [242]. KSR is implicated in TNF-induced inflammation by activation of Raf kinase and the subsequent activation of stress pathways leading to apoptosis [243,244]. Ceramide has also been shown to bind directly to Raf kinase possibly increasing its activity and activating the mitogen activated protein kinase (MAPK) pathway [245,246]. De novo synthesised ceramide has also been shown to activate the Raf/MAPK pathway [247]. Ceramide also interacts directly with PKCζ and this effect has been implicated in the regulation of alternative splicing of genes [248]. Whether KSR or PKCζ are direct effectors of ceramide actions in vivo is still not certain. Ceramide generated by A-SMase in lysosomes/endosomes has been shown to activate cathepsin D [249]. Cathepsin D has been implicated in mediating apoptosis in response to cytokines and chemotherapeutic agents [214] and lysosomal cathepsin D binds ceramide and translocates to the mitochondria following agonist-stimulation resulting in cytochrome c release [250]. In vitro studies have shown that ceramide can stereo-specifically activate protein phosphatases PP1 and PP2A [251]. Several pathways involved in apoptosis are regulated by CAPPs such as; c-jun, Bcl 2, Akt/PKB, Rb and SR proteins (reviewed in Ref. [214]). PP2A is implicated in TNFα and ceramide-induced dephosphorylation of c-jun [252], mitochondrial Bcl 2, possibly through inactivation of PKCα – a Bcl 2 kinase [253,254] and Akt/PKB [214]. However, whether endogenous ceramide inactivates these anti-apoptotic pathways through CAPP in vivo is not established. PP1 is implicated in ceramide-induced dephosphorylation of the retinoblastoma gene product Rb, involved in cell senescence [212,255] and in dephosphorylation of SR proteins in response to Fas-induced de novo synthesis of ceramide [256]. SR proteins are regulators of constitutive and alternative gene splicing, and may regulate alternative splicing of apoptotic mediators caspase 9 and Bcl-x [34] possibly linking agonist-induced de novo ceramide synthesis to apoptosis.

In summary, strong evidence is emerging for a role of ceramide in programmed cell death. Apoptotic stimuli increase sphingomyelin hydrolysis resulting in ceramide accumulation in the initiation phase of apoptosis and inhibitors of ceramide production or stimuli of its metabolism are cytoprotective. The mitochondrion is emerging as a key component of ceramide-induced apoptosis with several downstream effectors of ceramide converging at this organelle and with evidence accumulating for sphingolipid metabolism at mitochondrial membranes. However, little is known concerning the mechanisms regulating the enzymes involved in ceramide generation following agonist stimulation and few direct effectors of endogenous ceramide have been established in vivo.

6.1.3. Sphingosine in apoptosis

Many studies have shown that the hydrolysis of sphingomyelin in response to apoptotic stimuli is greater than the corresponding accumulation of ceramide (reviewed in Ref. [213]), suggesting rapid metabolism of ceramide. One such metabolite is sphingosine formed by the action of CDases on ceramide (Fig. 4). In many cell types sphingosine levels increase in response to apoptotic stimuli and their increase generally lags behind ceramide but precedes morphological changes associated with apoptosis leading to the suggestion that sphingosine is involved, either in its own right or cooperatively with ceramide, in the apoptotic response (reviewed in Ref. [216]). Addition of exogenous sphingosine causes apoptosis in a variety of leukaemic cells or solid cancer cell lines (reviewed in Ref. [216]) and overexpression of S1P phosphohydrolase decreased S1P levels, increased sphingosine and ceramide and promoted apoptosis [257] further supporting a role for sphingosine in programmed cell death. Treatment of cells with cell permeable ceramide or bacterial SMase leads to sphingosine production during apoptosis [258–260] implicating CDases, and indeed activation of CDases has been reported in phorbol ester treated HL-60 cells during apoptosis [261] and in Jurkat T cells after Fas ligation [258]. Furthermore in the latter study sphingosine production correlated with CDase activation. However, interpretation of these data is complicated as such apoptotic effects observed with exogenous sphingosine may be due to its conversion to ceramide by Cer synthase. Although, FB1 an inhibitor of acyl-CoA dependent ceramide synthase does not affect sphingosine-induced apoptosis in many cell types [216]. Nor did FB1 affect cell permeable ceramide and Fas-induced apoptosis, which both led to an increase in sphingosine levels during the apoptotic program [258]. However, recently an acyl-CoA independent Cer synthase that is not inhibited by FB1 was cloned and localised to mitochondria [42,43]. Therefore, it is possible that some of the effects of sphingosine could be attributed to conversion to ceramide at mitochondria. Furthermore, in MCF7 breast cancer cells FB1 and D-MAPP an alkaline CDase inhibitor both blocked sphingosine-induced apoptosis [260], leading the authors to suggest that exogenously added sphingosine could, in MCF7 cells, be converted to ceramide which in turn is metabolised to sphingosine, in a different cellular compartment. Clearly, further studies are required to elucidate the precise contributions of ceramide and sphingosine to agonist-induced apoptosis.

Sphingosine has been shown to activate many signalling pathways implicated in apoptosis. In HL60 cells sphingosine activated JNK [262,263] and in U937 cells it activated JNK and p38MAPK [262]. JNKs and p38MAPKs are activated in response to stress stimuli and are thought to play an important role in cell death [264]. However, inhibition of JNK and p38MAPK did not block sphingosine induced cell death [262]. Caspases are implicated in sphingosine-induced apoptosis as are mitochondria [216]. Recently, sphingosine was shown to inhibit basal and serum-induced Akt kinase activity in hepatoma cells and overexpression of activated Akt partially overcame sphingosine-induced cell death [265]. Sphingosine is a potent inhibitor of PKC [266] and because activation of PKC is known to promote cell survival [267] this is another pathway by which sphingosine could promote cell death. Sphingosine can also activate Rb protein, implicated in cell senescence [255]. However, the only known targets of sphingosine are PKC and cathepsin D both of which bind sphingosine, although in a lipid environment sphingosine did not activate cathepsin D whilst ceramide did [249]. Furthermore, many of the proposed pathways by which sphingosine could regulate apoptosis are also implicated in ceramide induced responses.

6.1.4. Sphingosine-1-phosphate in apoptosis

S1P is formed by the phosphorylation of sphingosine by SPHKs. S1P is unique amongst the bioactive sphingolipids in that it can act as an intracellular second messenger releasing calcium from inositol trisphosphate-independent calcium stores or as a first messenger acting on specific G protein coupled receptors [35,98,268]. Unlike ceramide and sphingosine, S1P promotes cell growth and inhibits apoptosis. Activation of SPHK and production of S1P levels in response to a number of growth and survival factors has been reported and S1P is implicated in proliferation, differentiation, cell survival, angiogenesis, cell migration and regulation of immune function (reviewed in Refs. [35,98,268]). With regard to apoptosis there is evidence that SPHK is anti-apoptotic. Overexpression of SPHK1 protects cells from apoptosis induced by ceramide or serum withdrawal [183,269]. SPHK1 expression reduced ceramide and increased S1P levels in NIH3T3, HEK293 and Jurkat T cells [183] suggesting that the anti-apototic effects may be due to reduction of ceramide. However, exogenous S1P mimicked the effects of SPHK1 in PC12 cells restoring resistance to apoptosis [270], indicating that S1P was indeed an anti-apoptotic signal.

S1P can act both as an intracellular second messenger and as an extracellular agonist through the endothelial differentiation gene-1 (EDG-1) family of G protein coupled receptors [98,187,268,271]. There is evidence that S1P can protect cells from apoptosis through both routes (reviewed in Ref. [187]). S1P activates ERKs [272,273] and inhibits JNK [267] so antagonising a stress-induced apoptotic pathway [274]. In hepatoma cells, TNFα activated SPHK, increased S1P and activated Akt/PKB protecting the cells from apoptosis. Inhibition of SPHK activity blocked Akt/PKB activation and induced apoptosis whilst exogenous S1P restored the response [275]. This study suggests that S1P could exert some of its effects through activation of the cell survival Akt/PKB pathway. S1P can also activate transcription factors such as AP-1 and NF-κB [276,277] which are implicated in protection against apoptosis. However, the mechanisms by which intracellular S1P regulates these cell survival pathways are unclear and direct targets for S1P are yet to be identified. S1P may also act extracellularly to regulate apoptosis. Effects of exogenously added S1P normally require micromolar concentrations to exert an intracellular effect, however in a T lymphoblastoma cell line nanomolar S1P was protective against ceramide and Fas-induced apoptosis [278]. The affinity of S1P receptors for S1P are in the nanomolar range [187] suggesting that these effects were mediated through a receptor coupled mechanism. Studies using S1P receptor antisense have shown attenuation of S1P inhibition of ceramide-induced apoptosis [279] possibly coupled to activation of calcium sensitive eNOS [280]. Additionally, S1P protected cells stably expressing S1P receptors from apoptosis induced by serum withdrawal possibly through ERK activation [281].

The cell proliferative anti-apoptotic effects of S1P in contrast to the pro-apoptotic effects of sphingosine and ceramide has led to the proposal that the relative levels of ceramide/sphingosine to S1P can determine the fate of a cell with SPHK acting as a regulatory switch [267]. CDases are also important in regulating the balance of ceramide and S1P as they are the rate-limiting step in the supply of sphingosine for conversion to S1P and recently it has been shown that overexpression of acid CDase inhibited TNFα-induced apoptosis [172] whereas inhibition of acid CDase activity inhibited proliferation [172,173] and increased ceramide levels [173]. Therefore, the cellular response to agonist stimulation of sphingolipid turnover will reflect the specific enzymes activated and the relative levels of the lipids that accumulate. For instance, a pro-apoptotic signal will require increased ceramide levels by de novo synthesis and/or activation of SMases. A parallel activation of CDases resulting in increased sphingosine levels will reinforce the apoptotic signal. In contrast, increased S1P by SPHK activation will result in reduced levels of ceramide and sphingosine and favour an anti-apoptotic signal.

6.1.5. Glucosylceramide synthase in apoptosis

Inhibition of GCS potentiated cell permeable ceramide-induced apoptosis in neuro-epithelial cells [282], suggesting that this pathway may allow cells to escape from agonist-induced apoptosis. However, in a recent study GCS did not have access to endogenous ceramide generated at the plasma membrane [81] indicating that this route may only influence apoptotic signals mediated through de novo synthesis of ceramide. Functionally GCS is implicated in multidrug resistant cancers (reviewed in Ref. [217]). Many chemotherapeutic agents stimulate ceramide production inducing apoptosis. GCS is thought to act as a mechanism for clearing the ceramide, protecting the cell from commitment to programmed cell death [217]. Expression of GCS in drug sensitive cells protected them from the cytotoxic agent Adriamycin and decreased the levels of ceramide. GluCer did not accumulate suggesting that the protective effect of GCS was due to ceramide removal rather than GluCer production [79]. In an additional study transfection of GCS into multidrug resistant MCF7 breast cancer cells increased their drug resistance whilst GCS antisense reversed the effect [225] further demonstrating the role of GCS in modulating cell survival. However, although removal of ceramide by GCS may be anti-apoptotic its conversion to GluCer is the first step in the synthesis of GSLs such as GD3 ganglioside which can release cytochrome c from intact mitochondria a pro-apoptotic signal ([283] and see above). Again, illustrating the complexity of sphingolipid signalling in determining cell fate.

6.2. Sphingolipids in cell proliferation

Mitogenic stimuli such as growth factors and GPCR agonists stimulate sphingolipid turnover and in 1991, S1P was shown to play an important role in cell growth regulation [284]. Agonists stimulate the production of S1P by increasing production of its precursors ceramide and sphingosine and by activating SPHK. Activation of SMases has been observed in response to mitogens such as growth factors, cytokines and arachidonic acid [100] and SPHK activation occurs following stimulation with growth factors, GPCR agonists, cytokines, phorbol esters, vitamin D3 and antigen (reviewed in Ref. [98]). Unravelling the mechanisms by which S1P induces proliferation is complicated because S1P can act as an intracellular second messenger, extracellularly through specific GPCRs and also in a paracrine/autocrine fashion following secretion into the extracellular space (reviewed in Refs. [268,271]). The extracellular mediated effects of S1P appear to be mainly regulation of cell migration and angiogenesis (reviewed in Ref. [268] and see below), whilst action as an intracellular second messenger is implicated in cell proliferation (reviewed in Ref. [271]). Increasing cellular S1P by overexpression of SPHK1 is mitogenic [272], whilst SPHK inhibitors block cell proliferation [98]. However, the molecular mechanisms by which S1P regulates proliferation are still unclear because intracellular targets of S1P are proving difficult to identify. Microinjection of S1P into fibroblasts increased DNA synthesis and released calcium from intracellular stores [285,286] and platelet derived growth factor (PDGF) induces SPHK translocation to the nucleus with an associated increase in nuclear SPHK activity [287], suggesting a direct action of S1P within the nucleus. S1P can activate the pro-mitogenic MAPKs–ERK and inhibit the pro-apoptotic MAPKs–JNKs [267,288] switching the balance towards a mitogenic signal. On the other hand ceramide opposes the effects of S1P on these pathways favouring an apoptotic response. As discussed above regulation of cell proliferation and cell death pathways by sphingolipids most probably reflects the relative levels of the bioactive lipids present during agonist stimulation.

The ganglioside GD3 is also implicated in proliferation. In most normal tissues GD3 is a minor ganglioside, however it is highly expressed during development and in pathological conditions such as atherosclerosis and cancers particularly of neuroectodermal origin (for a recent review see Ref. [218]). GD3 synthase overexpression in PC12 cells increased proliferation [289] and monoclonal antibodies against GD3 increased T cell proliferation [290,291] suggesting activation of signalling pathways involved in T cell growth. In support of this, GD3 has been shown to associate with Src tyrosine kinases in rat cerebellar granule cells [292]. However, in a variety of other cell types GD3 synthase expression was not associated with increased proliferation [218]. Given the evidence that GD3 can also promote apoptosis (see above) these differences may reflect a balance between these two actions.

6.3. Sphingolipids in membrane structure, raft formation and cell signalling

There is a growing body of evidence pointing to a signalling role of sphingolipids as structural components of cell membranes. Recently, the perception of the cell membrane structure and dynamics has changed from the "fluid mosaic" model of Singer and

Nicolson [293] – in which a homogeneous phospholipid bilayer acts mainly as a solvent for integral proteins and as a permeability barrier for the cell – to a membrane bilayer which has asymmetrical distribution of lipids between the inner and outer leaflet and lateral heterogeneity with lipids partitioning into ordered (rigid) and disordered (fluid) domains (reviewed in Refs. [35,294–296]). The ordered domains are composed predominantly of sphingomyelin and cholesterol and are termed lipid rafts [297]. Lipid rafts are small in size and estimated to cover ~50% of the cell surface [298] and to contain 70% of cellular sphingomyelin [299] in the presence of caveolins, palmitoylated hairpin-like proteins [142] they form flask-like invaginations known as caveolae [140]. Because of their small size each raft is thought to contain a subset of proteins and upon activation, e.g. receptor aggregation, it is proposed that rafts come together with their respective proteins to facilitate signalling by bringing together signalling kinases or phosphatases with their respective substrates [300]. Many signalling proteins such as Src family tyrosine kinases, Ras, G protein coupled receptors, ERKs, eNOS, PKC and integrins are either associated with or following stimulation move in and out of lipid rafts/caveolae suggesting that these membrane microdomains are important for cell signalling (reviewed in Refs. [35,300–302]).

Recently, it has been suggested that ceramide might signal through inducing changes in membrane structure, and that this effect may be physiologically more important than direct interaction with signalling targets [72]. Ceramides are amongst the least polar, most hydrophobic lipids in nature, and these biophysical properties restrict their localisation such that once generated they remain and act within the cellular membrane. For instance, in model membranes the estimated time for ceramide to flip/flop from one leaflet of the bilayer to the other is greater than 22 min [303]. Accordingly, ceramide generated following agonist stimulation; presumably in the outer membrane leaflet as 70% of sphingomyelin is located there, would be unlikely to rapidly activate a cytosolic target protein.

Ceramide because of its "cone" shape and tight packing within membranes induces membrane curvature and is implicated in vesicle formation and budding in model membranes (reviewed in Refs. [72,304]). Also in ATP-depleted macrophages treatment with exogenous bacterial SMase or ceramide led to vesicle formation [305]. Physiologically, local generation of ceramide and vesicle formation may be necessary for infection by *Neisseria gonorrhoeae* which require aSMase on the cell surface for uptake into mucosal epithelial cells [306]. Vesicular trafficking (both from the plasma membrane to internal organelles and vice versa) is an essential element in many signalling pathways. Ceramide appears to be involved in vesicle formation and trafficking at different stages. For instance, in domain coalescence (see below) and membrane curvature [72,107,231], internalisation of lipid rafts/caveolae and trafficking of ganglioside GD3 and possibly other pro-apoptotic signals to mitochondria [239,307,308] and vesicle biogenesis at the *trans*-Golgi network through diacylglycerol formation and recruitment of protein kinase D [61,309,310].

In model membranes ceramide partitions into gel-like (ordered) domains, excluding phospholipids (reviewed in Refs. [72,109,304]) suggesting that when generated by stress or by an agonist, ceramide could alter the fluidity of the plasma membrane and induce coalescence of lipid rafts. This effect of ceramide has been studied most extensively in immune cell responses. Capping is a process during which cell surface receptors or proteins aggregate on one pole of the cell after binding their cognate ligands or agonistic antibodies. Disruption of lipid rafts by removal of cholesterol prevents clustering of many receptors [311] and downstream signalling in T cells [312]. Recently, generation of ceramide by aSMase was shown to be necessary for capping and initiation of the full response to CD95 receptor activation [107]. Rafts have been most extensively studied in T cell responses and a model has been developed for the role of ceramide in the response to activation of the TNF receptor superfamily. Engagement of Fas by ligand or antibody induces an immediate secretion of aSMase to the outer surface of the plasma membrane where it hydrolyses sphingomyelin on the cell surface. Although it has also been suggested that aSMase is activated in lysosomes where it generates ceramide on the inner membrane and when the vesicles fuse with the plasma membrane the ceramide appears at the outer surface and aSMase is released [72], overcoming the need for aSMase to work at neutral pH. The ceramide formed causes coalescence of sphingolipid microdomains into ceramide enriched macrodomains facilitating capping and clustering of activated receptors with recruitment and activation of downstream signalling molecules (reviewed in Refs. [109, 304]). The development of a ceramide antibody has allowed visualisation, by confocal microscopy of this process [231]. Furthermore, aSMase deficient cells are defective in capping and Fas-induced apoptosis a response which can be restored by addition of ceramide [110,227]. In addition to involvement in apoptosis signalling, rafts are also implicated in cell migration. There is evidence that rafts associate with the underlying actin cytoskeleton (reviewed in Ref. [296]) and in polarised neutrophils CD44, which can interact with the actin cytoskeleton, is localised to lipid rafts and is involved in cell motility [313,314] is found in large detergent insoluble membrane areas at the rear of the cell during migration [315,316]. It is possible that formation of sphingolipid macrodomains may be involved in organisation of signalling molecules and signalling pathways in chemotaxis [317].

Finally, in addition to having an effect on membrane curvature ceramide can induce pore formation in model membranes with the loss of proteins up to 20 kDa in size [318], leading to local ion fluxes and changes in enzyme activity. In mitochondria C16-ceramide or SMase generated endogenous ceramide formed pores in the outer membrane with the release of cytochrome c [319], suggesting a physical/structural role for ceramide formed at the mitochondrial membrane in apoptosis.

6.4. Sphingolipids in the cardiovascular system

Recently, sphingolipids have emerged as important regulators of cardiovascular function; they are implicated in smooth muscle and endothelial cell responses such as contractility, nitric oxide production and proliferation, in vasculogenesis and atherogenesis. The role of sphingolipids in the cardiovascular system has been reviewed in depth recently, see Refs. [97,320,321], accordingly in this section we shall briefly describe their major functions and highlight new insights gained from recently published studies.

6.4.1. Contractility

Sphingolipids are potent mobilisers of intracellular calcium [322], an effect that is associated with mitogenesis, cardiac contractility and vascular smooth muscle tone [320]. Ghosh et al. [323] were the first to report that sphingosine could mediate a rapid and large release of calcium from intracellular stores, possibly through S1P formation. It is now clear that all three lysosphingolipids, sphingosine, S1P and SPC induce calcium mobilisation from intracellular stores [321]. S1P and SPC mediate calcium release through an endoplasmic reticulum calcium channel, sphingolipid calcium release-mediating protein of the endoplasmic reticulum (SCaMPER) [324]. Recently, SCaMPER was identified in cardiac tissue localised to the sarcotubular junction. Stimulation of cardiomyocytes with SPC initiated sarcoplasmic reticulum calcium release that was substantially inhibited by antisense knockdown of SCaMPER mRNA [325], suggesting that SCaMPER is a potentially important calcium channel in cardiomyocytes.

The effects of sphingolipids on vascular tone are still not clear. Addition of ceramide to pre-constricted smooth muscle induces relaxation correlated with a decrease in intracellular calcium [326,327], and ceramide is implicated in the vasodilator response to TNFα [328]. Conversely, when ceramide is added to non-treated smooth muscle contraction is observed [327] accompanied by a rise in intracellular calcium and activation of Src tyrosine kinases [329,330]. Similarly, S1P and SPC induce contraction in non-treated arterial rings [331,332] but relaxation of pre-constricted arteries was observed with SPC [333]. These results suggest that the contractile response to sphingolipids is dependent on the underlying level of tone and may reflect effects on nitric oxide (NO) production. For instance, the vasoconstriction induced by S1P and SPC was unaffected by NO synthase (NOS) inhibition [331], but SPC-induced relaxation of pre-constricted arteries was sensitive to NOS inhibition [333]. In endothelial cells S1P can regulate NO production through activation of eNOS [334] and NO can increase ceramide production by activation of SMase and downregulation of CDase in non-vascular cells [157,158]. Therefore, some of the conflicting effects of sphingolipids on contractility may reflect effects on NO production within the vascular wall. Moreover, these studies used exogenously added sphingolipids or non-specific generation of ceramide by bacterial SMases, therefore it is not clear exactly which lipid is exerting the contractile or relaxant effect due to the rapid metabolism of sphingolipids by tissues. More detailed studies of the sphingolipids produced by arteries and by vasoconstrictors and vasodilators are necessary to establish the role of sphingolipids in regulation of vascular tone. Recently, S1P was shown to constrict isolated rat cerebral arteries but not aorta. In smooth muscle cells isolated from cerebral artery S1P increased calcium but little change was observed in aorta-derived cells. Furthermore, using receptor subtype specific antibodies it was shown that cerebral artery smooth muscle cells expressed similar S1P1 but fourfold higher S1P2 and S1P3 receptors when compared to aorta, leading the authors to suggest that the contractile response may be regulated by receptor subtype expression [335]. Finally, SPC has been reported to increase the sensitivity of the contractile myofilaments to calcium so increasing force, through Src tyrosine kinase dependent activation of Rho kinase [332,336,337].

6.4.2. Myocardial ischaemia/reperfusion

Myocardial ischaemia results in tissue damage due to infarction, apoptosis and necrosis. Subsequent reperfusion of the myocardium whilst alleviating the ischaemia can induce further tissue injury through induction of apoptosis [338]. Antioxidants can attenuate the tissue damage caused by reperfusion in animal models [339] suggesting that redox stress and the production of free radicals are in part responsible for reperfusion injury. N-SMase is activated by intracellular oxidation and has been proposed as a sensor of redox stress [99] suggesting that the sphingomyelin/ceramide signalling pathway could be involved in reperfusion injury. In both cardiomyocyte and intact heart models of ischaemia/reperfusion ceramide levels increase following reperfusion of the ischaemic cells or tissues [340–342]. However, the precise mechanisms underlying ceramide production in reperfused myocardium are still unclear. Certainly, in cardiomyocytes N-SMase activity increased rapidly and ceramide accumulated following reoxygenation of hypoxic cells and the response was inhibited by antioxidants [342]. Furthermore, a recent study showed that cardiomyocytes express FAN (factor associated with N-SMase activation) which mediates activation of N-SMase and subsequent apoptosis. Cells expressing dominant negative FAN were resistant to hypoxia/reoxygenation injury and the protective effect was overcome by treatment with cell permeable ceramide suggesting that DN-FAN exerts its effect by preventing N-SMase activation [343]. However, in a rat heart model, although ceramide accumulated in the reperfused tissue this was due to a decrease in CDase activity rather than a change in SMase activity [342]. The differences in pathways for the production of ceramide in reperfused cells and tissue may reflect production of the inflammatory cytokine TNFα in response to ischaemia. TNFα is implicated in the pathogenesis of heart failure, ischaemia–reperfusion injury and cardiac dysfunction observed after coronary microembolisation [344–346]. There is evidence that some of the adverse effects of TNFα are mediated by sphingosine production. For instance, in perfused rat hearts bacterial endotoxins depress intracellular calcium cycling and myocardial contractility. Inhibition of TNFα or CDase activity prevented the cardiac dysfunction induced by endotoxin [347]. Additionally, inhibition of CDase activity inhibited coronary microembolisation induced sphingosine production and progressive contractile dysfunction [159] and improved the post-ischaemic recovery of myocardial function in hypertrophied hearts [160], suggesting that inhibition of sphingosine production and/or accumulation of ceramide may be beneficial for maintaining contractile function.

6.4.3. Sphingolipids in vascular maturation and angiogenesis

S1P stimulates migration of endothelial [279,348] and smooth muscle cells [349,350] critical events in the formation of blood vessels that requires migration of smooth muscle cells and pericytes around the initial tube formed by endothelial cells. In human vascular endothelial cells S1P induced cell migration and tube formation were inhibited by pertussis toxin [279] indicating that the effect was receptor mediated. Five S1P receptors (S1P1-5) members of the EDG family of GPCR have been identified which show different tissue expression and appear to couple to different intracellular signalling cascades so resulting in different cellular responses (reviewed in Ref. [351]). Disruption of the S1P1 receptor gene in mice is embryonic lethal and the mice die from haemorrhage due to incomplete maturation of the vascular system [352]. A similar failure to form mature blood vessels was also observed in mice in which PDGF-BB or PDGF-β genes were disrupted [353,354] suggesting a mechanistic link between the two phenotypes. Studies with fibroblasts from S1P1 null embryos showed that they failed to migrate or activate the GTPase Rac, a migratory signal, in response to S1P or PDGF [350] further suggesting a link between S1P and PDGF in vascular development. A model for the regulation of cell migration by S1P and PDGF has now emerged [271]. Binding of PDGF to its receptor activates and recruits SPHK to the leading edge of the cell [355], the resultant local increase in S1P stimulates the S1P1 receptor which activates downstream signals such as FAK and Src, regulators of cytoskeletal organisation and Rac GTPases involved in lamellipodia formation and forward movement [350,355]. Recently, it was reported that the PDGF receptor is tethered to the S1P1 receptor [356] providing a platform for their integrated signalling. There is also evidence that the S1P1 receptor can activate Rac and cell migration in the absence of S1P through transactivation by the insulin growth factor receptor mediated by Akt-dependent phosphorylation of S1P1 [357]. This observation raises the possibility that it is the S1P1 receptor per se that is required for cell migration rather than its ligand S1P. Further studies are required to validate these models.

Gangliosides are also implicated in angiogenesis in pathological conditions such as tumour growth. Tumour cells express many classes of ganglioside on their cell surface and the ratios present can regulate tumour growth through suppression and stimulation of angiogenesis for instance, GD3 is pro-angiogenic and GM3 anti-angiogenic [358–361]. Gangliosides can regulate VEGF (vascular endothelial growth factor) an important angiogenic factor. Suppression of GD3 synthase gene, downregulated VEGF expression and reduced angiogenesis and tumour growth [362], and GD1a pre-incubation enhanced VEGF-induced cell proliferation and migration [363]. Together these studies suggest that gangliosides may regulate angiogenesis through modulation of VEGF activity and expression.

The S1P2 receptor is thought to play a role in cardiac development. Kupperman et al. [364] showed that a gene called “miles apart” is essential for initial formation of the heart tube in zebra fish. “Miles apart” encodes a S1P receptor. However, disruption of the equivalent gene, S1P2 receptor in mice did not cause a similar cardiac defect and the mice appeared physiologically normal [365].

6.4.4. Sphingolipids in atherogenesis

GSLs, secretory SMase and ceramide are all present in atherosclerotic lesions [97] and there is a positive correlation between the plasma concentration of sphingomyelin and the severity of coronary heart disease in humans [366], implicating sphingolipids in atherogenesis. Secretory SMase, which is produced by cells in the vascular wall [114,367] can induce low density lipoprotein (LDL) aggregation [113] a primary event in atherogenesis [368] and LDL itself has SMase activity [369] which may facilitate its entry into cells [370]. The accumulation of aggregated LDL within the vessel wall and the uptake of LDL-cholesterol by macrophages, leading to foam cell formation, is a central event in the initiation of atherosclerosis [368]. Oxidation of LDL (oxLDL) also causes it to become atherogenic [371]. LDL can be oxidised in vitro by both non-enzymatic, induced by copper ions, and enzymatic mechanisms. Incubation of LDL in vitro with smooth muscle cells, endothelial cells or monocytes, all cell types found within atherosclerotic lesions causes it to become oxidised suggesting that oxidation would occur within the vascular wall. However, the mechanisms by which LDL is oxidised in vivo are still unclear. Although there is evidence that oxLDL accumulates in atherosclerotic lesions in humans and in animal models of the disease [371]. OxLDL appears to have a dual effect on smooth muscle cells as low concentrations stimulated proliferation but higher concentrations induced cell death [372]. Further studies showed that cell death was caused by oxLDL induced activation of N-SMase and ceramide production [373]. However, oxLDL has also been shown to increase S1P production in smooth muscle cells which would favour cell proliferation and migration [320]. Therefore, through activation of the sphingomyelin pathway oxLDL may play a dual role in atherogenesis, stimulation of smooth muscle cell proliferation and migration during the early stages of plaque formation and later when oxLDL levels rise within the lesion stimulation of apoptosis leading to plaque rupture and thrombus formation. In addition, S1P released from activated platelets has been implicated in proliferation and migration of smooth muscle cells in atherosclerosis [374,375]. Inflammatory responses occur at the site of atherosclerotic lesions and the initiating cytokines through generation of ceramide, lactosylceramide or S1P increase expression of cell surface adhesion molecules, attracting immune cells that migrate into the vessel wall another key event in the atherogenic process (for detailed reviews see Refs. [97,320]). Sphingolipids may also be atherogenic through effects on free radical production. Cell permeable ceramide (C6-ceramide) reduced nitric oxide production in endothelial cells reducing their ability to scavenge superoxide radicals [376] and ox-LDL stimulates the production of LacCer which in turn activated NADPH oxidase to produce superoxides. Superoxides are potent stimulators of the mitogenic MAPK signalling pathway and may contribute to LDL oxidation within the vessel wall (reviewed in Refs. [97,320,371]). Ganglioside GD3 can also activate NADPH oxidase stimulating superoxide production, activating MAPK and increasing smooth muscle cell proliferation. However, at higher concentrations superoxide production was inhibited and NO was produced triggering an apoptotic response [377]. Therefore, the concentration of ganglioside within an atherosclerotic lesion may regulate cell proliferation or cell death and so contribute to plaque stabilisation or rupture.

Recently, ceramide has been shown to reduce restenosis following angioplasty. Coating of catheters with C6-ceramide markedly reduced smooth muscle cell proliferation following balloon angioplasty in rabbit carotid arteries by inactivating ERK and Akt signalling mitogenic and survival pathways, respectively [378], suggesting that targeting of the sphingomyelin pathway may have therapeutic benefit in coronary artery restenosis.

7. Summary

Recent advances in the cloning and characterisation of many of the enzymes involved in sphingolipid metabolism have led to a greater understanding of the role of these lipids in cellular function. Specifically, how regulation of these enzymes by extra- and intracellular stimuli can modulate the production of "signalling" sphingolipids such as ceramide, sphingosine and S1P which in turn act as "molecular switches" in cellular responses. Such advances have also highlighted the complexity of this system where a particular sphingolipid can be generated from multiple sources and can be rapidly converted to other sphingolipid molecules, which may retain similar or antagonistic effects to the original product. Although ceramide is well established as a messenger in apoptosis it is only recently that a central role for sphingolipid signalling at the mitochondria has emerged and the importance of ceramide-induced membrane structural changes appreciated. Detailed molecular studies have unravelled the role of the S1P1 receptor in angiogenesis, however the function of the remaining four S1P receptors is still unclear. That sphingolipids form an integral part of lipid rafts and the signalling role associated with them is now established. However, future advances in imaging techniques will undoubtedly provide a greater understanding of the function of such specialised microdomains once it is possible to "visualise" these structures and observe changes induced by agonists. However, there is considerable variation between effects of sphingolipids observed in cells and those in tissues and animal models. Consequently, the physiological role of sphingolipids in tissues and organisms still remains elusive and presents a substantial challenge for future research.

References

[1] Slotte, J.P., 1999. Chem. Phys. Lipids 102, 13–27.
[2] Merrill, A.H., Schmelz, E.M., Dillehay, D.L., Spiegel, S., Shayman, J.A., Schroeder, J.J., Riley, R.T., Voss, K.A., Wang, E., 1997. Toxicol. Appl. Pharmacol. 142, 208–225.
[3] Merrill, A.H. Jr., 2002. J. Biol. Chem. 277, 25843–25846.
[4] Huwiler, A., Kolter, T., Pfeilschifter, J., Sandhoff, K., 2000. Biochim. Biophys. Acta 1485, 63–99.
[5] Iwabuchi, K., Zhang, Y., Handa, K., Withers, D.A., Sinay, P., Hakomori, S., 2000. J. Biol. Chem. 275, 15174–15181.
[6] Merrill, A.H. Jr., 1983. Biochim. Biophys. Acta 754, 284–291.
[7] Merrill, A.H. Jr., Williams, R.D., 1984. J. Lipid Res. 25, 185–188.
[8] Merrill, A.H. Jr., Nixon, D.W., Williams, R.D., 1985. J. Lipid Res. 26, 617–622.
[9] Mandon, E.C., Ehses, I., Rother, J., van Echten, G., Sandhoff, K., 1992. J. Biol. Chem. 267, 11144–11148.
[10] Buede, R., Rinker-Schaffer, C., Pinto, W.J., Lester, R.L., Dickson, R.C., 1991. J. Bacteriol. 173, 4325–4332.
[11] Nagiec, M.M., Baltisberger, J.A., Wells, G.B., Lester, R.L., Dickson, R.C., 1994. Proc. Natl Acad. Sci. USA 91, 7899–7902.
[12] Nagiec, M.M., Lester, R.L., Dickson, R.C., 1996. Gene 177, 237–241.
[13] Weiss, B., Stoffel, W., 1997. Eur. J. Biochem. 249, 239–247.
[14] Hanada, K., Hara, T., Nishijima, M., Kuge, O., Dickson, R.C., Nagiec, M.M., 1997. J. Biol. Chem. 272, 32108–32114.
[15] Hanada, K., Hara, T., Fukasawa, M., Yamaji, A., Umeda, M., Nishijima, M., 1998. J. Biol. Chem. 273, 33787–33794.
[16] Hanada, K., Hara, T., Nishijima, M., 2000. J. Biol. Chem. 275, 8409–8415.
[17] Yasuda, S., Nishijima, M., Hanada, K., 2003. J. Biol. Chem. 278, 4176–4183.
[18] Bejaoui, K., Wu, C., Scheffler, M.D., Haan, G., Ashby, P., Wu, L., de Jong, P., Brown, R.H. Jr., 2001. Nat. Genet. 27, 261–262.
[19] Dawkins, J.L., Hulme, D.J., Brahmbhatt, S.B., Auer-Grumbach, M., Nicholson, G.A., 2001. Nat. Genet. 27, 309–312.
[20] Bejaoui, K., Uchida, Y., Yasuda, S., Ho, M., Nishijima, M., Brown, R.H. Jr., Holleran, W.M., Hanada, K., 2002. J. Clin. Invest. 110, 1301–1308.

[21] Gable, K., Han, G., Monaghan, E., Bacikova, D., Natarajan, M., Williams, R., Dunn, T.M., 2002. J. Biol. Chem. 277, 10194–10200.
[22] Gable, K., Slife, H., Bacikova, D., Monaghan, E., Dunn, T.M., 2000. J. Biol. Chem. 275, 7597–7603.
[23] Miyake, Y., Kozutsumi, Y., Nakamura, S., Fujita, T., Kawasaki, T., 1995. Biochem. Biophys. Res. Commun. 211, 396–403.
[24] Hanada, K., Nishijima, M., Kiso, M., Hasegawa, A., Fujita, S., Ogawa, T., Akamatsu, Y., 1992. J. Biol. Chem. 267, 23527–23533.
[25] Uhlinger, D.J., Carton, J.M., Argentieri, D.C., Damiano, B.P., Dandrea, M.R., 2001. Thromb. Haemostasis 86, 1320–1326.
[26] Farrell, A.M., Uchida, Y., Nagiec, M.M., Harris, I.R., Dickson, R.C., Elias, P.M., Holleran, W.M., 1998. J. Lipid Res. 39, 2031–2038.
[27] Memon, R.A., Holleran, W.M., Moser, A.H., Seki, T., Uchida, Y., Fuller, J., Shigenaga, J.K., Grunfeld, C., Feingold, K.R., 1998. Arterioscler. Thromb. Vasc. Biol. 18, 1257–1265.
[28] Herget, T., Esdar, C., Oehrlein, S.A., Heinrich, M., Schutze, S., Maelicke, A., Echten-Deckert, G., 2000. J. Biol. Chem. 275, 30344–30354.
[29] Perry, D.K., Carton, J., Shah, A.K., Meredith, F., Uhlinger, D.J., Hannun, Y.A., 2000. J. Biol. Chem. 275, 9078–9084.
[30] Gomez, d.P., Velasco, G., Sanchez, C., Haro, A., Guzman, M., 2002. Biochem. J. 363, 183–188.
[31] Jenkins, G.M., Richards, A., Wahl, T., Mao, C., Obeid, L., Hannun, Y., 1997. J. Biol. Chem. 272, 32566–32572.
[32] Jenkins, G.M., Cowart, L.A., Signorelli, P., Pettus, B.J., Chalfant, C.E., Hannun, Y.A., 2002. J. Biol. Chem. 277, 42572–42578.
[33] Perry, D.K., 2000. Ann. NY Acad. Sci. 905, 91–96.
[34] Chalfant, C.E., Rathman, K., Pinkerman, R.L., Wood, R.E., Obeid, L.M., Ogretmen, B., Hannun, Y.A., 2002. J. Biol. Chem. 277, 12587–12595.
[35] Ohanian, J., Ohanian, V., 2001. Cell Mol. Life Sci. 58, 2053–2068.
[36] Shimeno, H., Soeda, S., Yasukouchi, M., Okamura, N., Nagamatsu, A., 1995. Biol. Pharm. Bull. 18, 1335–1339.
[37] Merrill, A.H. Jr., van Echten, G., Wang, E., Sandhoff, K., 1993. J. Biol. Chem. 268, 27299–27306.
[38] Shimeno, H., Soeda, S., Sakamoto, M., Kouchi, T., Kowakame, T., Kihara, T., 1998. Lipids 33, 601–605.
[39] Schorling, S., Vallee, B., Barz, W.P., Riezman, H., Oesterhelt, D., 2001. Mol. Biol. Cell 12, 3417–3427.
[40] Tani, M., Okino, N., Mitsutake, S., Tanigawa, T., Izu, H., Ito, M., 2000. J. Biol. Chem. 275, 3462–3468.
[41] Mao, C., Xu, R., Bielawska, A., Obeid, L.M., 2000. J. Biol. Chem. 275, 6876–6884.
[42] El Bawab, S., Birbes, H., Roddy, P., Szulc, Z.M., Bielawska, A., Hannun, Y.A., 2001. J. Biol. Chem. 276, 16758–16766.
[43] El Bawab, S., Roddy, P., Qian, T., Bielawska, A., Lemasters, J.J., Hannun, Y.A., 2000. J. Biol. Chem. 275, 21508–21513.
[44] Bose, R., Verheij, M., HaimovitzFriedman, A., Scotto, A., Fuks, Z., Kolesnick, R., 1995. Cell 82, 405–414.
[45] Wang, H., Maurer, B.J., Reynolds, C.P., Cabot, M.C., 2001. Cancer Res. 61, 5102–5105.
[46] Michel, C., Echten-Deckert, G., 1997. FEBS Lett. 416, 153–155.
[47] Geeraert, L., Mannaerts, G.P., Van Veldhoven, P.P., 1997. Biochem. J. 327(Pt 1), 125–132.
[48] Hannun, Y.A., Obeid, L.M., 2002. J. Biol. Chem. 277, 25847–25850.
[49] Mikami, T., Kashiwagi, M., Tsuchihashi, K., Akino, T., Gasa, S., 1998. J. Biochem. (Tokyo) 123, 906–911.
[50] Causeret, C., Geeraert, L., Van der, H.G., Mannaerts, G.P., Van Veldhoven, P.P., 2000. Lipids 35, 1117–1125.
[51] Sperling, P., Libisch, B., Zahringer, U., Napier, J.A., Heinz, E., 2001. Arch. Biochem. Biophys. 388, 293–298.
[52] Ternes, P., Franke, S., Zahringer, U., Sperling, P., Heinz, E., 2002. J. Biol. Chem. 277, 25512–25518.
[53] Kandyba, A.G., Kobliakov, V.A., Kozlov, A.M., Nagaev, I.Y., Shevchenko, V.P., Dyatlovitskaya, E.V., 2002. Biochemistry (Mosc.) 67, 597–599.
[54] van Meer, G., Holthuis, J.C., 2000. Biochim. Biophys. Acta 1486, 145–170.

[55] van Meer, G., Lisman, Q., 2002. J. Biol. Chem. 277, 25855–25858.
[56] Jeckel, D., Karrenbauer, A., Birk, R., Schmidt, R.R., Wieland, F., 1990. FEBS Lett. 261, 155–157.
[57] Ichikawa, S., Hirabayashi, Y., 1998. Trends Cell Biol. 8, 198–202.
[58] van Helvoort, A., van't Hof, W., Ritsema, T., Sandra, A., Van Meer, G., 1994. J. Biol. Chem. 269, 1763–1769.
[59] Luberto, C., Hannun, Y.A., 1998. J. Biol. Chem. 273, 14550–14559.
[60] Luberto, C., Yoo, D.S., Suidan, H.S., Bartoli, G.M., Hannun, Y.A., 2000. J. Biol. Chem. 275, 14760–14766.
[61] Baron, C.L., Malhotra, V., 2002. Science 295, 325–328.
[62] Allan, D., Obradors, M.J., 1999. Biochim. Biophys. Acta 1450, 277–287.
[63] van Helvoort, A., Stoorvogel, W., van Meer, G., Burger, N.J., 1997. J. Cell Sci. 110(Pt 6), 781–788.
[64] Vos, J.P., Giudici, M.L., van der, B.P., Magni, P., Marchesini, S., Van Golde, L.M., Lopes-Cardozo, M., 1995. FEBS Lett. 368, 393–396.
[65] Micheli, M., Albi, E., Leray, C., Magni, M.V., 1998. FEBS Lett. 431, 443–447.
[66] Albi, E., Magni, M.V., 1999. FEBS Lett. 460, 369–372.
[67] Miro-Obradors, M.J., Osada, J., Aylagas, H., Sanchez-Vegazo, I., Palacios-Alaiz, E., 1993. Carcinogenesis 14, 941–946.
[68] Riboni, L., Viani, P., Bassi, R., Giussani, P., Tettamanti, G., 2001. J. Biol. Chem. 276, 12797–12804.
[69] Flores, I., Jones, D.R., Merida, I., 2000. FASEB J. 14, 1873–1875.
[70] Bourteele, S., Hausser, A., Doppler, H., Horn-Muller, J., Ropke, C., Schwarzmann, G., Pfizenmaier, K., Muller, G., 1998. J. Biol. Chem. 273, 31245–31251.
[71] Kolter, T., Proia, R.L., Sandhoff, K., 2002. J. Biol. Chem. 277, 25859–25862.
[72] van Blitterswijk, W.J., van der Luit, A.H., Veldman, R.J., Verheij, M., Borst, J., 2003. Biochem. J. 369, 199–211.
[73] Paul, P., Kamisaka, Y., Marks, D.L., Pagano, R.E., 1996. J. Biol. Chem. 271, 2287–2293.
[74] Ichikawa, S., Sakiyama, H., Suzuki, G., Hidari, K.I., Hirabayashi, Y., 1996. Proc. Natl Acad. Sci. USA 93, 4638–4643.
[75] Ichikawa, S., Ozawa, K., Hirabayashi, Y., 1998. Biochem. Biophys. Res. Commun. 253, 707–711.
[76] Marks, D.L., Wu, K., Paul, P., Kamisaka, Y., Watanabe, R., Pagano, R.E., 1999. J. Biol. Chem. 274, 451–456.
[77] Yamashita, T., Wada, R., Sasaki, T., Deng, C., Bierfreund, U., Sandhoff, K., Proia, R.L., 1999. Proc. Natl Acad. Sci. USA 96, 9142–9147.
[78] Lavie, Y., Cao, H., Volner, A., Lucci, A., Han, T.Y., Geffen, V., Giuliano, A.E., Cabot, M.C., 1997. J. Biol. Chem. 272, 1682–1687.
[79] Liu, Y.Y., Han, T.Y., Giuliano, A.E., Cabot, M.C., 1999. J. Biol. Chem. 274, 1140–1146.
[80] Liu, Y.Y., Han, T.Y., Giuliano, A.E., Hansen, N., Cabot, M.C., 2000. J. Biol. Chem. 275, 7138–7143.
[81] Tepper, A.D., Diks, S.H., van Blitterswijk, W.J., Borst, J., 2000. J. Biol. Chem. 275, 34810–34817.
[82] Shabbits, J.A., Mayer, L.D., 2002. Mol. Cancer Ther. 1, 205–213.
[83] Coetzee, T., Dupree, J.L., Popko, B., 1998. J. Neurosci. Res. 54, 613–622.
[84] Stahl, N., Jurevics, H., Morell, P., Suzuki, K., Popko, B., 1994. J. Neurosci. Res. 38, 234–242.
[85] Coetzee, T., Fujita, N., Dupree, J., Shi, R., Blight, A., Suzuki, K., Suzuki, K., Popko, B., 1996. Cell 86, 209–219.
[86] Bosio, A., Binczek, E., Stoffel, W., 1996. Proc. Natl Acad. Sci. USA 93, 13280–13285.
[87] Bajjalieh, S.M., Martin, T.F., Floor, E., 1989. J. Biol. Chem. 264, 14354–14360.
[88] Kolesnick, R.N., Hemer, M.R., 1990. J. Biol. Chem. 265, 18803–18808.
[89] Sugiura, M., Kono, K., Liu, H., Shimizugawa, T., Minekura, H., Spiegel, S., Kohama, T., 2002. J. Biol. Chem. 277, 23294–23300.
[90] Hinkovska-Galcheva, V.T., Boxer, L.A., Mansfield, P.J., Harsh, D., Blackwood, A., Shayman, J.A., 1998. J. Biol. Chem. 273, 33203–33209.
[91] Gomez-Munoz, A., Duffy, P.A., Martin, A., O'Brien, L., Byun, H.S., Bittman, R., Brindley, D.N., 1995. Mol. Pharmacol. 47, 833–839.
[92] Gomez-Munoz, A., Frago, L.M., Alvarez, L., Varela-Nieto, I., 1997. Biochem. J. 325(Pt 2), 435–440.

[93] Nomura, T., Takizawa, M., Aoki, J., Arai, H., Inoue, K., Wakisaka, E., Yoshizuka, N., Imokawa, G., Dohmae, N., Takio, K., Hattori, M., Matsuo, N., 1998. J. Biol. Chem. 273, 13570–13577.
[94] Lannert, H., Gorgas, K., Meissner, I., Wieland, F.T., Jeckel, D., 1998. J. Biol. Chem. 273, 2939–2946.
[95] Chatterjee, S., Shi, W.Y., Wilson, P., Mazumdar, A., 1996. J. Lipid Res. 37, 1334–1344.
[96] Yeh, L.H., Kinsey, A.M., Chatterjee, S., Alevriadou, B.R., 2001. J. Vasc. Res. 38, 551–559.
[97] Chatterjee, S., 1998. Arterioscler. Thromb. Vasc. Biol. 18, 1523–1533.
[98] Pyne, S., Pyne, N.J., 2000. Biochem. J. 349, 385–402.
[99] Hannun, Y.A., Luberto, C., Argraves, K.M., 2001. Biochemistry 40, 4893–4903.
[100] Levade, T., Jaffrezou, J.P., 1999. Biochim. Biophys. Acta 1438, 1–17.
[101] Quintern, L.E., Schuchman, E.H., Levran, O., Suchi, M., Ferlinz, K., Reinke, H., Sandhoff, K., Desnick, R.J., 1989. EMBO J. 8, 2469–2473.
[102] Schuchman, E.H., Suchi, M., Takahashi, T., Sandhoff, K., Desnick, R.J., 1991. J. Biol. Chem. 266, 8531–8539.
[103] Bartelsen, O., Lansmann, S., Nettersheim, M., Lemm, T., Ferlinz, K., Sandhoff, K., 1998. J. Biotechnol. 63, 29–40.
[104] Levade, T., Salvayre, R., Douste-Blazy, L., 1986. J. Clin. Chem. Clin. Biochem. 24, 205–220.
[105] Schuchman, E.H., 1995. Hum. Mutat. 6, 352–354.
[106] Schissel, S.L., Keesler, G.A., Schuchman, E.H., Williams, K.J., Tabas, I., 1998. J. Biol. Chem. 273, 18250–18259.
[107] Grassme, H., Jekle, A., Riehle, A., Schwarz, H., Berger, J., Sandhoff, K., Kolesnick, R., Gulbins, E., 2001. J. Biol. Chem. 276, 20589–20596.
[108] Grassme, H., Jendrossek, V., Bock, J., Riehle, A., Gulbins, E., 2002. J. Immunol. 168, 298–307.
[109] Gulbins, E., Grassme, H., 2002. Biochim. Biophys. Acta 1585, 139–145.
[110] Paris, F., Grassme, H., Cremesti, A., Zager, J., Fong, Y., Haimovitz-Friedman, A., Fuks, Z., Gulbins, E., Kolesnick, R., 2001. J. Biol. Chem. 276, 8297–8305.
[111] Lozano, J., Menendez, S., Morales, A., Ehleiter, D., Liao, W.C., Wagman, R., Haimovitz-Friedman, A., Fuks, Z., Kolesnick, R., 2001. J. Biol. Chem. 276, 442–448.
[112] Bezombes, C., Segui, B., Cuvillier, O., Bruno, A.P., Uro-Coste, E., Gouaze, V., Andrieu-Abadie, N., Carpentier, S., Laurent, G., Salvayre, R., Jaffrezou, J.P., Levade, T., 2001. FASEB J. 15, 297–299.
[113] Xu, X., Tabas, I., 1991. J. Biol. Chem. 266, 24849–24858.
[114] Marathe, S., Schissel, S.L., Yellin, M.J., Beatini, N., Mintzer, R., Williams, K.J., Tabas, I., 1998. J. Biol. Chem. 273, 4081–4088.
[115] Schissel, S.L., Jiang, X., Tweedie-Hardman, J., Jeong, T., Camejo, E.H., Najib, J., Rapp, J.H., Williams, K.J., Tabas, I., 1998. J. Biol. Chem. 273, 2738–2746.
[116] Bilderback, T.R., Gazula, V.R., Dobrowsky, R.T., 2001. J. Neurochem. 76, 1540–1551.
[117] Okazaki, T., Bielawska, A., Domae, N., Bell, R.M., Hannun, Y.A., 1994. J. Biol. Chem. 269, 4070–4077.
[118] Ghosh, N., Sabbadini, R., Chatterjee, S., 1998. Mol. Cell. Biochem. 189, 161–168.
[119] Liu, B., Hassler, D.F., Smith, G.K., Weaver, K., Hannun, Y.A., 1998. J. Biol. Chem. 273, 34472–34479.
[120] Bernardo, K., Krut, O., Wiegmann, K., Kreder, D., Micheli, M., Schafer, R., Sickman, A., Schmidt, W.E., Schroder, J.M., Meyer, H.E., Sandhoff, K., Kronke, M., 2000. J. Biol. Chem. 275, 7641–7647.
[121] Raimann, P.E., Custodio, d.S., Bernard, E.A., Guma, F.C., 1999. Mol. Cell. Biochem. 201, 125–129.
[122] Tomiuk, S., Hofmann, K., Nix, M., Zumbansen, M., Stoffel, W., 1998. Proc. Natl Acad. Sci. USA 95, 3638–3643.
[123] Sawai, H., Domae, N., Nagan, N., Hannun, Y.A., 1999. J. Biol. Chem. 274, 38131–38139.
[124] Tepper, A.D., Ruurs, P., Borst, J., van Blitterswijk, W.J., 2001. Biochem. Biophys. Res. Commun. 280, 634–639.
[125] Neuberger, Y., Shogomori, H., Levy, Z., Fainzilber, M., Futerman, A.H., 2000. FEBS Lett. 469, 44–46.
[126] Tomiuk, S., Zumbansen, M., Stoffel, W., 2000. J. Biol. Chem. 275, 5710–5717.
[127] Hofmann, K., Tomiuk, S., Wolff, G., Stoffel, W., 2000. Proc. Natl Acad. Sci. USA 97, 5895–5900.
[128] Zumbansen, M., Stoffel, W., 2002. Mol. Cell. Biol. 22, 3633–3638.
[129] Liu, B., Andrieu-Abadie, N., Levade, T., Zhang, P., Obeid, L.M., Hannun, Y.A., 1998. J. Biol. Chem. 273, 11313–11320.
[130] Chatterjee, S., Han, H., Rollins, S., Cleveland, T., 1999. J. Biol. Chem. 274, 37407–37412.

[131] Adam-Klages, S., Adam, D., Wiegmann, K., Struve, S., Kolanus, W., 1996. Cell 86, 937–947.
[132] Segui, B., Andrieu-Abadie, N., Adam-Klages, S., Meilhac, O., Kreder, D., Garcia, V., Bruno, A.P., Jaffrezou, J.P., Salvayre, R., Kronke, M., Levade, T., 1999. J. Biol. Chem. 274, 37251–37258.
[133] Segui, B., Cuvillier, O., Adam-Klages, S., Garcia, V., Malagarie-Cazenave, S., Leveque, S., Caspar-Bauguil, S., Coudert, J., Salvayre, R., Kronke, M., Levade, T., 2001. J. Clin. Invest. 108, 143–151.
[134] Tcherkasowa, A.E., Adam-Klages, S., Kruse, M.L., Wiegmann, K., Mathieu, S., Kolanus, W., Kronke, M., Adam, D., 2002. J. Immunol. 169, 5161–5170.
[135] Visnjic, D., Batinic, D., Banfic, H., 1997. Blood 89, 81–91.
[136] Jayadev, S., Linardic, C.M., Hannun, Y.A., 1994. J. Biol. Chem. 269, 5757–5763.
[137] Jayadev, S., Hayter, H.L., Andrieu, N., Gamard, C.J., Liu, B., Balu, R., Hayakawa, M., Ito, F., Hannun, Y.A., 1997. J. Biol. Chem. 272, 17196–17203.
[138] Okamoto, Y., Obeid, L.M., Hannun, Y.A., 2002. FEBS Lett. 530, 104–108.
[139] Bezombes, C., de Thonel, A., Apostolou, A., Louat, T., Jaffrezou, J.P., Laurent, G., Quillet-Mary, A., 2002. Mol. Pharmacol. 62, 1446–1455.
[140] Anderson, R.G.W., 1998. Annu. Rev. Biochem. 67, 199–225.
[141] Veldman, R.J., Maestre, N., Aduib, O.M., Medin, J.A., Salvayre, R., Levade, T., 2001. Biochem. J. 355, 859–868.
[142] Okamoto, T., Schlegel, A., Scherer, P.E., Lisanti, M.P., 1998. J. Biol. Chem. 273, 5419–5422.
[143] Czarny, M., Liu, J., Oh, P., Schnitzer, J.E., 2003. J. Biol. Chem. 278, 4424–4430.
[144] Luberto, C., Hassler, D.F., Signorelli, P., Okamoto, Y., Sawai, H., Boros, E., Hazen-Martin, D.J., Obeid, L.M., Hannun, Y.A., Smith, G.K., 2002. J. Biol. Chem. 277, 41128–41139.
[145] Cheng, Y., Nilsson, A., Tomquist, E., Duan, R.D., 2002. J. Lipid Res. 43, 316–324.
[146] Liu, J.J., Nilsson, A., Duan, R.D., 2002. Lipids 37, 469–474.
[147] Nilsson, A., Duan, R.D., 1999. Chem. Phys. Lipids 102, 97–105.
[148] Sjoqvist, U., Hertervig, E., Nilsson, A., Duan, R.D., Ost, A., Tribukait, B., Lofberg, R., 2002. Inflamm. Bowel. Dis. 8, 258–263.
[149] Wang, E., Norred, W.P., Bacon, C.W., Riley, R.T., Merrill, A.H. Jr., 1991. J. Biol. Chem. 266, 14486–14490.
[150] 3Tani, M., Okino, N., Mori, K., Tanigawa, T., Izu, H., Ito, M., 2000. J. Biol. Chem. 275, 11229–11234.
[151] Mitsutake, S., Tani, M., Okino, N., Mori, K., Ichinose, S., Omori, A., Iida, H., Nakamura, T., Ito, M., 2001. J. Biol. Chem. 276, 26249–26259.
[152] Yoshimura, Y., Okino, N., Tani, M., Ito, M., 2002. J. Biochem. (Tokyo) 132, 229–236.
[153] Tani, M., Iida, H., Ito, M., 2003. J. Biol. Chem. 278, 10523–10530.
[154] Coroneos, E., Martinez, M., Mckenna, S., Kester, M., 1995. J. Biol. Chem. 270, 23305–23309.
[155] Nikolova-Karakashian, M., Morgan, E.T., Alexander, C., Liotta, D.C., Merrill, A.H. Jr., 1997. J. Biol. Chem. 272, 18718–18724.
[156] Franzen, R., Pautz, A., Brautigam, L., Geisslinger, G., Pfeilschifter, J., Huwiler, A., 2001. J. Biol. Chem. 276, 35382–35389.
[157] Huwiler, A., Pfeilschifter, J., van den, B.H., 1999. J. Biol. Chem. 274, 7190–7195.
[158] Franzen, R., Pfeilschifter, J., Huwiler, A., 2002. FEBS Lett. 532, 441–444.
[159] Thielmann, M., Dorge, H., Martin, C., Belosjorow, S., Schwanke, U., van De, S.A., Konietzka, I., Buchert, A., Kruger, A., Schulz, R., Heusch, G., 2002. Circ. Res. 90, 807–813.
[160] Stamm, C., Friehs, I., Cowan, D.B., Moran, A.M., Cao-Danh, H., Duebener, L.F., del Nido, P.J., McGowan, F.X. Jr., 2001. Circulation 104, I350–I355.
[161] Romiti, E., Meacci, E., Tanzi, G., Becciolini, L., Mitsutake, S., Farnararo, M., Ito, M., Bruni, P., 2001. FEBS Lett. 506, 163–168.
[162] Mao, C., Xu, R., Bielawska, A., Szulc, Z.M., Obeid, L.M., 2000. J. Biol. Chem. 275, 31369–31378.
[163] Kita, K., Okino, N., Ito, M., 2000. Biochim. Biophys. Acta 1485, 111–120.
[164] Mao, C., Xu, R., Szulc, Z.M., Bielawska, A., Galadari, S.H., Obeid, L.M., 2001. J. Biol. Chem. 276, 26577–26588.
[165] Romiti, E., Meacci, E., Tani, M., Nuti, F., Farnararo, M., Ito, M., Bruni, P., 2000. Biochem. Biophys. Res. Commun. 275, 746–751.

[166] Bernardo, K., Hurwitz, R., Zenk, T., Desnick, R.J., Ferlinz, K., Schuchman, E.H., Sandhoff, K., 1995. J. Biol. Chem. 270, 11098–11102.
[167] Li, C.M., Hong, S.B., Kopal, G., He, X., Linke, T., Hou, W.S., Koch, J., Gatt, S., Sandhoff, K., Schuchman, E.H., 1998. Genomics 50, 267–274.
[168] Koch, J., Gartner, S., Li, C.M., Quintern, L.E., Bernardo, K., Levran, O., Schnabel, D., Desnick, R.J., Schuchman, E.H., Sandhoff, K., 1996. J. Biol. Chem. 271, 33110–33115.
[169] Zhang, Z., Mandal, A.K., Mital, A., Popescu, N., Zimonjic, D., Moser, A., Moser, H., Mukherjee, A.B., 2000. Mol. Genet. Metab. 70, 301–309.
[170] Ferlinz, K., Kopal, G., Bernardo, K., Linke, T., Bar, J., Breiden, B., Neumann, U., Lang, F., Schuchman, E.H., Sandhoff, K., 2001. J. Biol. Chem. 276, 35352–35360.
[171] Li, C.M., Park, J.H., Simonaro, C.M., He, X., Gordon, R.E., Friedman, A.H., Ehleiter, D., Paris, F., Manova, K., Hepbildikler, S., Fuks, Z., Sandhoff, K., Kolesnick, R., Schuchman, E.H., Hepbiloikler, S., 2002. Genomics 79, 218–224.
[172] Strelow, A., Bernardo, K., Adam-Klages, S., Linke, T., Sandhoff, K., Kronke, M., Adam, D., 2000. J. Exp. Med. 192, 601–612.
[173] Raisova, M., Goltz, G., Bektas, M., Bielawska, A., Riebeling, C., Hossini, A.M., Eberle, J., Hannun, Y.A., Orfanos, C.E., Geilen, C.C., 2002. FEBS Lett. 516, 47–52.
[174] Liu, H., Chakravarty, D., Maceyka, M., Milstien, S., Spiegel, S., 2002. Prog. Nucleic Acid Res. Mol. Biol. 71, 493–511.
[175] Nagiec, M.M., Skrzypek, M., Nagiec, E.E., Lester, R.L., Dickson, R.C., 1998. J. Biol. Chem. 273, 19437–19442.
[176] Nishiura, H., Tamura, K., Morimoto, Y., Imai, H., 2000. Biochem. Soc. Trans. 28, 747–748.
[177] Nava, V.E., Lacana, E., Poulton, S., Liu, H., Sugiura, M., Kono, K., Milstien, S., Kohama, T., Spiegel, S., 2000. FEBS Lett. 473, 81–84.
[178] Liu, H., Sugiura, M., Nava, V.E., Edsall, L.C., Kono, K., Poulton, S., Milstien, S., Kohama, T., Spiegel, S., 2000. J. Biol. Chem. 275, 19513–19520.
[179] Olivera, A., Kohama, T., Tu, Z., Milstien, S., Spiegel, S., 1998. J. Biol. Chem. 273, 12576–12583.
[180] Kohama, T., Olivera, A., Edsall, L., Nagiec, M.M., Dickson, R., Spiegel, S., 1998. J. Biol. Chem. 273, 23722–23728.
[181] Melendez, A.J., Carlos-Dias, E., Gosink, M., Allen, J.M., Takacs, L., 2000. Gene 251, 19–26.
[182] Murate, T., Banno, Y., Koizumi, K., Watanabe, K., Mori, N., Wada, A., Igarashi, Y., Takagi, A., Kojima, T., Asano, H., Akao, Y., Yoshida, S., Saito, H., Nozawa, Y., 2001. J. Histochem. Cytochem. 49, 845–855.
[183] Olivera, A., Kohama, T., Edsall, L., Nava, V., Cuvillier, O., Poulton, S., Spiegel, S., 1999. J. Cell Biol. 147, 545–558.
[184] Xia, P., Gamble, J.R., Wang, L., Pitson, S.M., Moretti, P.A., Wattenberg, B.W., D'Andrea, R.J., Vadas, M.A., 2000. Curr. Biol. 10, 1527–1530.
[185] Nava, V.E., Hobson, J.P., Murthy, S., Milstien, S., Spiegel, S., 2002. Exp. Cell Res. 281, 115–127.
[186] Ancellin, N., Colmont, C., Su, J., Li, Q., Mittereder, N., Chae, S.S., Stefansson, S., Liau, G., Hla, T., 2002. J. Biol. Chem. 277, 6667–6675.
[187] Maceyka, M., Payne, S.G., Milstien, S., Spiegel, S., 2002. Biochim. Biophys. Acta 1585, 193–201.
[188] Olivera, A., Edsall, L., Poulton, S., Kazlauskas, A., Spiegel, S., 1999. FASEB J. 13, 1593–1600.
[189] Alemany, R., Sichelschmidt, B., zu Heringdorf, D.M., Lass, H., van Koppen, C.J., Jakobs, K.H., 2000. Mol. Pharmacol. 58, 491–497.
[190] Young, K.W., Willets, J.M., Parkinson, M.J., Bartlett, P., Spiegel, S., Nahorski, S.R., Challiss, R.A., 2003. Cell Calcium 33, 119–128.
[191] Johnson, K.R., Becker, K.P., Facchinetti, M.M., Hannun, Y.A., Obeid, L.M., 2002. J. Biol. Chem. 277, 35257–35262.
[192] Lacana, E., Maceyka, M., Milstien, S., Spiegel, S., 2002. J. Biol. Chem. 277, 32947–32953.
[193] Hayashi, S., Okada, T., Igarashi, N., Fujita, T., Jahangeer, S., Nakamura, S., 2002. J. Biol. Chem. 277, 33319–33324.
[194] Xia, P., Wang, L., Moretti, P.A., Albanese, N., Chai, F., Pitson, S.M., D'Andrea, R.J., Gamble, J.R., Vadas, M.A., 2002. J. Biol. Chem. 277, 7996–8003.
[195] Stukey, J., Carman, G.M., 1997. Protein Sci. 6, 469–472.

[196] Roberts, R., Sciorra, V.A., Morris, A.J., 1998. J. Biol. Chem. 273, 22059–22067.
[197] Alderton, F., Darroch, P., Sambi, B., McKie, A., Ahmed, I.S., Pyne, N., Pyne, S., 2001. J. Biol. Chem. 276, 13452–13460.
[198] Brindley, D.N., Waggoner, D.W., 1998. J. Biol. Chem. 273, 24281–24284.
[199] Le Stunff, H., Peterson, C., Liu, H., Milstien, S., Spiegel, S., 2002. Biochim. Biophys. Acta 1582, 8–17.
[200] Mandala, S.M., Thornton, R., Tu, Z., Kurtz, M.B., Nickels, J., Broach, J., Menzeleev, R., Spiegel, S., 1998. Proc. Natl Acad. Sci. USA 95, 150–155.
[201] Mao, C., Wadleigh, M., Jenkins, G.M., Hannun, Y.A., Obeid, L.M., 1997. J. Biol. Chem. 272, 28690–28694.
[202] Qie, L., Nagiec, M.M., Baltisberger, J.A., Lester, R.L., Dickson, R.C., 1997. J. Biol. Chem. 272, 16110–16117.
[203] Mandala, S.M., Harris, G.H., 2000. Methods Enzymol. 311, 335–348.
[204] Ogawa, C., Kihara, A., Gokoh, M., Igarashi, Y., 2003. J. Biol. Chem. 278, 1268–1272.
[205] Zhou, J., Saba, J.D., 1998. Biochem. Biophys. Res. Commun. 242, 502–507.
[206] van Veldhoven, P.P., Mannaerts, G.P., 1994. Biochem. J. 299(Pt 3), 597–601.
[207] Van Veldhoven, P.P., Mannaerts, G.P., 1991. J. Biol. Chem. 266, 12502–12507.
[208] Van Veldhoven, P.P., Gijsbers, S., Mannaerts, G.P., Vermeesch, J.R., Brys, V., 2000. Biochim. Biophys. Acta 1487, 128–134.
[209] Saba, J.D., Nara, F., Bielawska, A., Garrett, S., Hannun, Y.A., 1997. J. Biol. Chem. 272, 26087–26090.
[210] Wang, W., Zhou, G., Hu, M., Yao, Z., Tan, T., 1997. J. Biol. Chem. 272, 22771–22775.
[211] Okazaki, T., Bell, R.M., Hannun, Y.A., 1989. J. Biol. Chem. 264, 19076–19080.
[212] Obeid, L.M., Linardic, C.M., Karolak, L.A., Hannun, Y.A., 1993. Science 259, 1769–1771.
[213] Andrieu-Abadie, N., Levade, T., 2002. Biochim. Biophys. Acta 1585, 126–134.
[214] Pettus, B.J., Chalfant, C.E., Hannun, Y.A., 2002. Biochim. Biophys. Acta (BBA) – Mol. Cell Biol. Lipids 1585, 114–125.
[215] Perry, D.K., 2002. Biochim. Biophys. Acta (BBA) – Mol. Cell Biol. Lipids 1585, 146–152.
[216] Cuvillier, O., 2002. Biochim. Biophys. Acta 1585, 153–162.
[217] Bleicher, R.J., Cabot, M.C., 2002. Biochim. Biophys. Acta (BBA) – Mol. Cell Biol. Lipids 1585, 172–178.
[218] Malisan, F., Testi, R., 2002. Biochim. Biophys. Acta (BBA) – Mol. Cell Biol. Lipids 1585, 179–187.
[219] Salvayre, R., Auge, N., Benoist, H., Negre-Salvayre, A., 2002. Biochim. Biophys. Acta 1585, 213–221.
[220] Perry, D.K., Bielawska, A., Hannun, Y.A., 2000. Methods Enzymol. 312, 22–31.
[221] Bielawska, A., Linardic, C.M., Hannun, Y.A., 1992. J. Biol. Chem. 267, 18493–18497.
[222] Thomas, R.L. Jr., Matsko, C.M., Lotze, M.T., Amoscato, A.A., 1999. J. Biol. Chem. 274, 30580–30588.
[223] Sullards, M.C., 2000. Methods Enzymol. 312, 32–45.
[224] Kroesen, B.J., Jacobs, S., Pettus, B.J., Sietsma, H., Kok, J.W., Hannun, Y.A., de Leij, L.F., 2003. J. Biol. Chem. 278, 14723–14731.
[225] Liu, Y.Y., Han, T.Y., Giuliano, A.E., Cabot, M.C., 2001. FASEB J. 15, 719–730.
[226] Luberto, C., Hassler, D.F., Signorelli, P., Okamoto, Y., Sawai, H., Boros, E., Hazen-Martin, D.J., Obeid, L.M., Hannun, Y.A., Smith, G.K., 2002. J. Biol. Chem. 277, 41128–41139.
[227] Lozano, J., Menendez, S., Morales, A., Ehleiter, D., Liao, W.C., Wagman, R., Haimovitz-Friedman, A., Fuks, Z., Kolesnick, R., 2001. J. Biol. Chem. 276, 442–448.
[228] Merrill, A.H., Jones, D.D., 1990. Biochim. Biophys. Acta 1044, 1–12.
[229] Veldman, R.J., Klappe, K., Hoekstra, D., Kok, J.W., 1998. Biochem. J. 331(Pt 2), 563–569.
[230] Zhang, P., Liu, B., Jenkins, G.M., Hannun, Y.A., Obeid, L.M., 1997. J. Biol. Chem. 272, 9609–9612.
[231] Cremesti, A., Paris, F., Grassme, H., Holler, N., Tschopp, J., Fuks, Z., Gulbins, E., Kolesnick, R., 2001. J. Biol. Chem. 276, 23954–23961.
[232] Jaffrezou, J.P., Bruno, A.P., Moisand, A., Levade, T., Laurent, G., 2001. FASEB J. 15, 123–133.
[233] Tsugane, K., Tamiya-Koizumi, K., Nagino, M., Nimura, Y., Yoshida, S., 1999. J. Hepatol. 31, 8–17.
[234] Goldkorn, T., Balaban, N., Shannon, M., Chea, V., Matsukuma, K., Gilchrist, D., Wang, H., Chan, C., 1998. J. Cell Sci. 111, 3209–3220.
[235] Birbes, H., Bawab, S.E., Obeid, L.M., Hannun, Y.A., 2002. Adv. Enzyme Regul. 42, 113–129.
[236] Birbes, H., El Bawab, S., Hannun, Y.A., Obeid, L.M., 2001. FASEB J. 15, 2669–2679.

[237] Rippo, M.R., Malisan, F., Ravagnan, L., Tomassini, B., Condo, I., Costantini, P., Susin, S.A., Rufini, A., Todaro, M., Kroemer, G., Testi, R., 2000. FASEB J. 14, 2047–2054.

[238] De Maria, R., Lenti, L., Malisan, F., d'Agostino, F., Tomassini, B., Zeuner, A., Rippo, M.R., Testi, R., 1997. Science 277, 1652–1655.

[239] Colell, A., Morales, A., Fernandez-Checa, J.C., Garcia-Ruiz, C., 2002. FEBS Lett. 526, 135–141.

[240] Okada, M., Itoh, M.M., Haraguchi, M., Okajima, T., Inoue, M., Oishi, H., Matsuda, Y., Iwamoto, T., Kawano, T., Fukumoto, S., Miyazaki, H., Furukawa, K., Aizawa, S., Furukawa, K., 2002. J. Biol. Chem. 277, 1633–1636.

[241] Mathias, S., Dressler, K., Kolesnick, R., 1991. Proc. Natl Acad. Sci. USA 88, 10009–10013.

[242] Zhang, Y., Yao, B., Delikat, S., Bayoumy, S., Lin, X.H., Basu, S., McGinley, M., Chan-Hui, P.Y., Lichenstein, H., Kolesnick, R., 1997. Cell 89, 63–72.

[243] Xing, H.R., Kolesnick, R., 2001. J. Biol. Chem. 276, 9733–9741.

[244] Basu, S., Kolesnick, R., 1998. Oncogene 17, 3277–3285.

[245] Muller, G., Storz, P., Bourteele, S., Doppler, H., Pfizenmaier, K., Mischak, H., Philipp, A., Kaiser, C., Kolch, W., 1998. EMBO J. 17, 732–742.

[246] Huwiler, A., Brunner, J., Hummel, R., Vervoordeldonk, M., Stabel, S., van den Bosch, H., Pfeilschifter, J., 1996. Proc. Natl Acad. Sci. USA 93, 6959–6963.

[247] Blazquez, C., Galve-Roperh, I., Guzman, M., 2000. FASEB J. 14, 2315–2322.

[248] Lozano, J., Berra, E., Municio, M.M., Diaz-Meco, M.T., Dominguez, I., Sanz, L., Moscat, J., 1994. J. Biol. Chem. 269, 19200–19202.

[249] Heinrich, M., Wickel, M., Schneider-Brachert, W., Sandberg, C., Gahr, J., Schwandner, R., Weber, T., Saftig, P., Peters, C., Brunner, J., Kronke, M., Schutze, S., 1999. EMBO J. 18, 5252–5263.

[250] Kagedal, K., Johansson, U., Ollinger, K., 2001. FASEB J. 15, 1592–1594.

[251] Chalfant, C.E., Kishikawa, K., Mumby, M.C., Kamibayashi, C., Bielawska, A., Hannun, Y.A., 1999. J. Biol. Chem. 274, 20313–20317.

[252] Reyes, J.G., Robayna, I.G., Delgado, P.S., Gonzalez, I.H., Aguiar, J.Q., Rosas, F.E., Fanjul, L.F., Degalarreta, C.R., 1996. J. Biol. Chem. 271, 21375–21380.

[253] Ruvolo, P.P., Deng, X., Ito, T., Carr, B.K., May, W.S., 1999. J. Biol. Chem. 274, 20296–20300.

[254] Ruvolo, P.P., Deng, X., Carr, B.K., May, W.S., 1998. J. Biol. Chem. 273, 25436–25442.

[255] Dbaibo, G.S., Pushkareva, M.Y., Jayadev, S., Schwarz, J.K., Horowitz, J.M., Obeid, L.M., Hannun, Y.A., 1995. Proc. Natl Acad. Sci. USA 92, 1347–1351.

[256] Chalfant, C.E., Ogretmen, B., Galadari, S., Kroesen, B.J., Pettus, B.J., Hannun, Y.A., 2001. J. Biol. Chem. 276, 44848–44855.

[257] Mandala, S.M., Thornton, R., Galve-Roperh, I., Poulton, S., Peterson, C., Olivera, A., Bergstrom, J., Kurtz, M.B., Spiegel, S., 2000. Proc. Natl Acad. Sci. USA 97, 7859–7864.

[258] Cuvillier, O., Edsall, L., Spiegel, S., 2000. J. Biol. Chem. 275, 15691–15700.

[259] Nava, V.E., Cuvillier, O., Edsall, L.C., Kimura, K., Milstien, S., Gelmann, E.P., Spiegel, S., 2000. Cancer Res. 60, 4468–4474.

[260] Cuvillier, O., Nava, V.E., Murthy, S.K., Edsall, L.C., Levade, T., Milstien, S., Spiegel, S., 2001. Cell Death Differ. 8, 162–171.

[261] Ohta, S., Nishihara, J., Oka, Y., Todo, H., Kumon, Y., Sakaki, S., 1995. Acta Neurochir. 137, 217–225.

[262] Jarvis, W.D., Fornari, F.A., Auer, K.L., Freemerman, A.J., Szabo, E., Birrer, M.J., Johnson, C.R., Barbour, S.E., Dent, P., Grant, S., 1997. Mol. Pharmacol. 52, 935–947.

[263] Jarvis, W.D., Grant, S., 1998. Curr. Opin. Oncol. 10, 552–559.

[264] Davis, R.J., 2000. Cell 103, 239–252.

[265] Chang, H.C., Tsai, L.H., Chuang, L.Y., Hung, W.C., 2001. J. Cell Physiol. 188, 188–193.

[266] Hannun, Y.A., Loomis, C.R., Merrill, A.H. Jr., Bell, R.M., 1986. J. Biol. Chem. 261, 12604–12609.

[267] Cuvillier, O., Pirianov, G.G., Kleuser, B., Vanek, P.G., Coso, O.A., Gutkind, J.S., Spiegel, S., 1996. Nature 381, 800–803.

[268] Payne, S.G., Milstien, S., Spiegel, S., 2002. FEBS Lett. 531, 54–57.

[269] Edsall, L.C., Cuvillier, O., Twitty, S., Spiegel, S., Milstien, S., 2001. J. Neurochem. 76, 1573–1584.

[270] Edsall, L.C., Pirianov, G.G., Spiegel, S., 1997. J. Neurosci. 17, 6952–6960.

[271] Spiegel, S., Milstien, S., 2002. J. Biol. Chem. 277, 25851–25854.

[272] Olivera, A., Romanowski, A., Rani, C.S., Spiegel, S., 1997. Biochim. Biophys. Acta 1348, 311–323.
[273] Xia, P., Gamble, J.R., Rye, K.A., Wang, L., Hii, C.S., Cockerill, P., Khew-Goodall, Y., Bert, A.G., Barter, P.J., Vadas, M.A., 1998. Proc. Natl Acad. Sci. USA 95, 14196–14201.
[274] Xia, Z., Dickens, M., Raingeaud, J., Davis, R.J., Greenberg, M.E., 1995. Science 270, 1326–1331.
[275] Osawa, Y., Banno, Y., Nagaki, M., Brenner, D.A., Naiki, T., Nozawa, Y., Nakashima, S., Moriwaki, H., 2001. J. Immunol. 167, 173–180.
[276] Karin, M., Liu, Z., Zandi, E., 1997. Curr. Opin. Cell Biol. 9, 240–246.
[277] May, M.J., Ghosh, S., 1999. Science 284, 271–273.
[278] Goetzl, E.J., An, S., 1999. Adv. Exp. Med. Biol. 469, 259–264.
[279] Lee, O.H., Kim, Y.M., Lee, Y.M., Moon, E.J., Lee, D.J., Kim, J.H., Kim, K.W., Kwon, Y.G., 1999. Biochem. Biophys. Res. Commun. 264, 743–750.
[280] Kwon, Y.G., Min, J.K., Kim, K.M., Lee, D.J., Billiar, T.R., Kim, Y.M., 2001. J. Biol. Chem. 276, 10627–10633.
[281] An, S., Zheng, Y., Bleu, T., 2000. J. Biol. Chem. 275, 288–296.
[282] Spinedi, A., Bartolomeo, S.D., Piacentini, M., 1998. Cell Death Differ. 5, 785–791.
[283] Garcia-Ruiz, C., Colell, A., Paris, R., Fernandez-Checa, J.C., 2000. FASEB J. 14, 847–858.
[284] Zhang, H., Desai, N.N., Olivera, A., Seki, T., Brooker, G., Spiegel, S., 1991. J. Cell Biol. 114, 155–167.
[285] Himmel, H.M., Meyer zu, H.D., Windorfer, B., van Koppen, C.J., Ravens, U., Jakobs, K.H., 1998. Mol. Pharmacol. 53, 862–869.
[286] Van Brocklyn, J.R., Lee, M.J., Menzeleev, R., Olivera, A., Edsall, L., Cuvillier, O., Thomas, D.M., Coopman, P.J., Thangada, S., Liu, C.H., Hla, T., Spiegel, S., 1998. J. Cell Biol. 142, 229–240.
[287] Kleuser, B., Maceyka, M., Milstien, S., Spiegel, S., 2001. FEBS Lett. 503, 85–90.
[288] Goodemote, K.A., Mattie, M.E., Berger, A., Spiegel, S., 1995. J. Biol. Chem. 270, 10272–10277.
[289] Fukumoto, S., Mutoh, T., Hasegawa, T., Miyazaki, H., Okada, M., Goto, G., Furukawa, K., Urano, T., 2000. J. Biol. Chem. 275, 5832–5838.
[290] Norihisa, Y., McVicar, D.W., Ghosh, P., Houghton, A.N., Longo, D.L., Creekmore, S.P., Blake, T., Ortaldo, J.R., Young, H.A., 1994. J. Immunol. 152, 485–495.
[291] Ortaldo, J.R., Mason, A.T., Longo, D.L., Beckwith, M., Creekmore, S.P., McVicar, D.W., 1996. J. Leukoc. Biol. 60, 533–539.
[292] Kasahara, K., Watanabe, Y., Yamamoto, T., Sanai, Y., 1997. J. Biol. Chem. 272, 29947–29953.
[293] Singer, S.J., Nicolson, G.L., 1972. Science 175, 720–731.
[294] Brown, D.A., London, E., 1998. Annu. Rev. Cell Dev. Biol. 14, 111–136.
[295] Brown, R.E., 1998. J. Cell Sci. 111, 1–9.
[296] Maxfield, F.R., 2002. Curr. Opin. Cell Biol. 14, 483–487.
[297] Simons, K., Ikonen, E., 1997. Nature 387, 569–572.
[298] Gidwani, A., Holowka, D., Baird, B., 2001. Biochemistry 40, 12422–12429.
[299] Smart, E.J., Graf, G.A., McNiven, M.A., Sessa, W.C., Engelman, J.A., Scherer, P.E., Okamoto, T., Lisanti, M.P., 1999. Mol. Cell. Biol. 19, 7289–7304.
[300] Simons, K., Toomre, D., 2000. Nat. Rev. Mol. Cell. Biol. 1, 31–39.
[301] Brown, D.A., London, E., 2000. J. Biol. Chem. 275, 17221–17224.
[302] Sowa, G., Pypaert, M., Sessa, W.C., 2001. PNAS 98, 14072–14077.
[303] Bai, J., Pagano, R.E., 1997. Biochemistry 36, 8840–8848.
[304] Cremesti, A.E., Goni, F.M., Kolesnick, R., 2002. FEBS Lett. 531, 47–53.
[305] Zhang, Q., Cox, D., Tseng, C.C., Donaldson, J.G., Greenberg, S., 1998. J. Biol. Chem. 273, 19977–19981.
[306] Grassme, H., Gulbins, E., Brenner, B., Ferlinz, K., Sandhoff, K., Harzer, K., Lang, F., Meyer, T.F., 1997. Cell 91, 605–615.
[307] Garcia-Ruiz, C., Colell, A., Morales, A., Calvo, M., Enrich, C., Fernandez-Checa, J.C., 2002. J. Biol. Chem. 277, 36443–36448.
[308] Pol, A., Luetterforst, R., Lindsay, M., Heino, S., Ikonen, E., Parton, R.G., 2001. J. Cell Biol. 152, 1057–1070.
[309] Bankaitis, V.A., 2002. Science 295, 290–291.
[310] Liljedahl, M., Maeda, Y., Colanzi, A., Ayala, I., Van Lint, J., Malhotra, V., 2001. Cell 104, 409–420.
[311] Janes, P.W., Ley, S.C., Magee, A.I., Kabouridis, P.S., 2000. Semin. Immunol. 12, 23–34.

[312] Xavier, R., Brennan, T., Li, Q., McCormack, C., Seed, B., 1998. Immunity 8, 723–732.
[313] Oliferenko, S., Paiha, K., Harder, T., Gerke, V., Schwarzler, C., Schwarz, H., Beug, H., Gunthert, U., Huber, L.A., 1999. J. Cell Biol. 146, 843–854.
[314] Perschl, A., Lesley, J., English, N., Hyman, R., Trowbridge, I.S., 1995. J. Cell Sci. 108(Pt 3), 1033–1041.
[315] Seveau, S., Eddy, R.J., Maxfield, F.R., Pierini, L.M., 2001. Mol. Biol. Cell 12, 3550–3562.
[316] Gomez-Mouton, C., Abad, J.L., Mira, E., Lacalle, R.A., Gallardo, E., Jimenez-Baranda, S., Illa, I., Bernad, A., Manes, S., Martinez, A., 2001. Proc. Natl Acad. Sci. USA 98, 9642–9647.
[317] Pierini, L.M., Maxfield, F.R., 2001. PNAS 98, 9471–9473.
[318] Siskind, L.J., Colombini, M., 2000. J. Biol. Chem. 275, 38640–38644.
[319] Siskind, L.J., Kolesnick, R.N., Colombini, M., 2002. J. Biol. Chem. 277, 26796–26803.
[320] Auge, N., Negre-Salvayre, A., Salvayre, R., Levade, T., 2000. Prog. Lipid Res. 39, 207–229.
[321] Levade, T., Auge, N., Veldman, R.J., Cuvillier, O., Negre-Salvayre, A., Salvayre, R., 2001. Circ. Res. 89, 957–968.
[322] Spiegel, S., Foster, D., Kolesnick, R., 1996. Curr. Opin. Cell Biol. 8, 159–167.
[323] Ghosh, S., Strum, J.C., Sciorra, V.A., Daniel, L., Bell, R.M., 1996. J. Biol. Chem. 271, 8472–8480.
[324] Kindman, L.A., Kim, S., McDonald, T.V., Gardner, P., 1994. J. Biol. Chem. 269, 13088–13091.
[325] Cavalli, A.L., O'Brien, N.W., Barlow, S.B., Betto, R., Glembotski, C.C., Palade, P.T., Sabbadini, R.A., 2003. Am. J. Physiol. Cell Physiol. 284, C780–C790.
[326] Johns, D.G., Osborn, H., Webb, R.C., 1997. Biochem. Biophys. Res. Commun. 237, 95–97.
[327] Zheng, T., Li, W., Wang, J., Altura, B.T., Altura, B.M., 2000. Eur. J. Pharmacol. 391, 127–135.
[328] Johns, D.G., Webb, R.C., 1998. Am. J. Physiol. 275, H1592–H1598.
[329] Ibitayo, A.I., Tsunoda, Y., Nozu, F., Owyang, C., Bitar, K.N., 1998. Am. J. Physiol. 275, G705–G711.
[330] Su, X., Wang, P., Ibitayo, A., Bitar, K.N., 1999. Am. J. Physiol. 276, G853–G861.
[331] Bischoff, A., Czyborra, P., Fetscher, C., Meyer zu Heringdorf, D., Jakobs, K.H., Michel, M.C., 2000. Br. J. Pharmacol. 130, 1871–1877.
[332] Todoroki-Ikeda, N., Mizukami, Y., Mogami, K., Kusuda, T., Yamamoto, K., Miyake, T., Sato, M., Suzuki, S., Yamagata, H., Hokazono, Y., Kobayashi, S., 2000. FEBS Lett. 482, 85–90.
[333] Mogami, K., Mizukami, Y., Todoroki-Ikeda, N., Ohmura, M., Yoshida, K., Miwa, S., Matsuzaki, M., Matsuda, M., Kobayashi, S., 1999. FEBS Lett. 457, 375–380.
[334] Igarashi, J., Bernier, S.G., Michel, T., 2001. J. Biol. Chem. 276, 12420–12426.
[335] Coussin, F., Scott, R.H., Wise, A., Nixon, G.F., 2002. Circ. Res. 91, 151–157.
[336] Shirao, S., Kashiwagi, S., Sato, M., Miwa, S., Nakao, F., Kurokawa, T., Todoroki-Ikeda, N., Mogami, K., Mizukami, Y., Kuriyama, S., Haze, K., Suzuki, M., Kobayashi, S., 2002. Circ. Res. 91, 112–119.
[337] Nakao, F., Kobayashi, S., Mogami, K., Mizukami, Y., Shirao, S., Miwa, S., Todoroki-Ikeda, N., Ito, M., Matsuzaki, M., 2002. Circ. Res. 91, 953–960.
[338] Buja, L.M., Entman, M.L., 1998. Circulation 98, 1355–1357.
[339] Horwitz, L.D., Fennessey, P.V., Shikes, R.H., Kong, Y., 1994. Circulation 89, 1792–1801.
[340] Bielawska, A.E., Shapiro, J.P., Jiang, L., Melkonyan, H.S., Piot, C., Wolfe, C.L., Tomei, L.D., Hannun, Y.A., Umansky, S.R., 1997. Am. J. Pathol. 151, 1257–1263.
[341] Hernandez, O.M., Discher, D.J., Bishopric, N.H., Webster, K.A., 2000. Circ. Res. 86, 198–204.
[342] Zhang, D.X., Fryer, R.M., Hsu, A.K., Zou, A.P., Gross, G.J., Campbell, W.B., Li, P.L., 2001. Basic Res. Cardiol. 96, 267–274.
[343] O'Brien, N.W., Gellings, N.M., Guo, M., Barlow, S.B., Glembotski, C.C., Sabbadini, R.A., 2003. Circ. Res. 92, 589–591.
[344] Levine, B., Kalman, J., Mayer, L., Fillit, H.M., Packer, M., 1990. N. Engl. J. Med. 323, 236–241.
[345] Bryant, D., Becker, L., Richardson, J., Shelton, J., Franco, F., Peshock, R., Thompson, M., Giroir, B., 1998. Circulation 97, 1375–1381.
[346] Dorge, H., Neumann, T., Behrends, M., Skyschally, A., Schulz, R., Kasper, C., Erbel, R., Heusch, G., 2000. Am. J. Physiol. Heart Circ. Physiol. 279, H2587–H2592.
[347] Stamm, C., Cowan, D.B., Friehs, I., Noria, S., del Nido, P.J., McGowan, F.X. Jr., 2001. Anesthesiology 95, 1396–1405.
[348] Wang, F., Van Brocklyn, J.R., Hobson, J.P., Movafagh, S., Zukowska-Grojec, Z., Milstien, S., Spiegel, S., 1999. J. Biol. Chem. 274, 35343–35350.

[349] Boguslawski, G., Grogg, J.R., Welch, Z., Ciechanowicz, S., Sliva, D., Kovala, A.T., McGlynn, P., Brindley, D.N., Rhoades, R.A., English, D., 2002. Exp. Cell Res. 274, 264–274.
[350] Hobson, J.P., Rosenfeldt, H.M., Barak, L.S., Olivera, A., Poulton, S., Caron, M.G., Milstien, S., Spiegel, S., 2001. Science 291, 1800–1803.
[351] Spiegel, S., Milstien, S., 2000. Biochim. Biophys. Acta 1484, 107–116.
[352] Liu, Y., Wada, R., Yamashita, T., Mi, Y., Deng, C.X., Hobson, J.P., Rosenfeldt, H.M., Nava, V.E., Chae, S.S., Lee, M.J., Liu, C.H., Hla, T., Spiegel, S., Proia, R.L., 2000. J. Clin. Invest. 106, 951–961.
[353] Lindahl, P., Johansson, B.R., Leveen, P., Betsholtz, C., 1997. Science 277, 242–245.
[354] Hellstrom, M., Kalen, M., Lindahl, P., Abramsson, A., Betsholtz, C., 1999. Development 126, 3047–3055.
[355] Rosenfeldt, H.M., Hobson, J.P., Maceyka, M., Olivera, A., Nava, V.E., Milstien, S., Spiegel, S., 2001. FASEB J. 15, 2649–2659.
[356] Alderton, F., Rakhit, S., Kong, K.C., Palmer, T., Sambi, B., Pyne, S., Pyne, N.J., 2001. J. Biol. Chem. 276, 28578–28585.
[357] Lee, M.J., Thangada, S., Paik, J.H., Sapkota, G.P., Ancellin, N., Chae, S.S., Wu, M., Morales-Ruiz, M., Sessa, W.C., Alessi, D.R., Hla, T., 2001. Mol. Cell 8, 693–704.
[358] Ziche, M., Morbidelli, L., Alessandri, G., Gullino, P.M., 1992. Lab. Invest. 67, 711–715.
[359] Koochekpour, S., Pilkington, G.J., 1996. Cancer Lett. 104, 97–102.
[360] Alessandri, G., Cornaglia-Ferraris, P., Gullino, P.M., 1997. Acta Oncol. 36, 383–387.
[361] Slevin, M., Kumar, S., He, X., Gaffney, J., 1999. Int. J. Cancer 82, 412–423.
[362] Zeng, G., Gao, L., Birkle, S., Yu, R.K., 2000. Cancer Res. 60, 6670–6676.
[363] Lang, Z., Guerrera, M., Li, R., Ladisch, S., 2001. Biochem. Biophys. Res. Commun. 282, 1031–1037.
[364] Kupperman, E., An, S., Osborne, N., Waldron, S., Stainier, D.Y., 2000. Nature 406, 192–195.
[365] MacLennan, A.J., Carney, P.R., Zhu, W.J., Chaves, A.H., Garcia, J., Grimes, J.R., Anderson, K.J., Roper, S.N., Lee, N., 2001. Eur. J. Neurosci. 14, 203–209.
[366] Jiang, X.C., Paultre, F., Pearson, T.A., Reed, R.G., Francis, C.K., Lin, M., Berglund, L., Tall, A.R., 2000. Arterioscler. Thromb. Vasc. Biol. 20, 2614–2618.
[367] Schissel, S.L., Tweedie-Hardman, J., Rapp, J.H., Graham, G., Williams, K.J., Tabas, I., 1996. J. Clin. Invest. 98, 1455–1464.
[368] Ross, R., 1993. Nature 362, 801–809.
[369] Holopainen, J.M., Medina, O.P., Metso, A.J., Kinnunen, P.K., 2000. J. Biol. Chem. 275, 16484–16489.
[370] Kinnunen, P.K., Holopainen, J.M., 2002. Trends Cardiovasc. Med. 12, 37–42.
[371] Steinberg, D., Witztum, J.L., 1999. Lipoproteins, lipoprotein oxidation and atherogenesis. In: Chien, K.R. (Ed.), Molecular Basis of Cardiovascular Disease. W.B. Saunders Company, pp. 458–476.
[372] Chatterjee, S., 1992. Mol. Cell. Biochem. 111, 143–147.
[373] Chatterjee, S., 1999. Chem. Phys. Lipids 102, 79–96.
[374] Yatomi, Y., Ohmori, T., Rile, G., Kazama, F., Okamoto, H., Sano, T., Satoh, K., Kume, S., Tigyi, G., Igarashi, Y., Ozaki, Y., 2000. Blood 96, 3431–3438.
[375] English, D., Welch, Z., Kovala, A.T., Harvey, K., Volpert, O.V., Brindley, D.N., Garcia, J.G., 2000. FASEB J. 14, 2255–2265.
[376] Li, H., Junk, P., Huwiler, A., Burkhardt, C., Wallerath, T., Pfeilschifter, J., Forstermann, U., 2002. Circulation 106, 2250–2256.
[377] Bhunia, A.K., Schwarzmann, G., Chatterjee, S., 2002. J. Biol. Chem. 277, 16396–16402.
[378] Charles, R., Sandirasegarane, L., Yun, J., Bourbon, N., Wilson, R., Rothstein, R.P., Levison, S.W., Kester, M., 2000. Circ. Res. 87, 282–288.

Essential fatty acid metabolism during pregnancy and early human development

Gerard Hornstra[a,*] and Stephanie R. De Vriese[b]

[a]*Maastricht University, Maastricht, and NUTRI-SEARCH, Gronsveld, The Netherlands*
[b]*Department of Internal Medicine, Division of Nutrition, Ghent University Hospital, Ghent, Belgium*
[*]*Correspondence address: NUTRI-SEARCH, Brikkenoven 14, 6247 BG Gronsveld, The Netherlands.*
Tel.: +31-43-356-0537; fax: +31-43-356-0535
E-mail: g.hornstra@nutrisearch.nl(G.H.)

1. Introduction

Essential fatty acids (EFA) and their longer chain, more unsaturated derivatives, the LCPUFA, are major structural and functional components of cell membranes, and LCPUFA are particularly important for optimal visual and nerve cell development and function. During pregnancy and foetal development, accretion of maternal, placental and foetal tissue occurs and, therefore, the EFA and LCPUFA requirements of pregnant women and the developing foetus are high. This particularly holds for certain LCPUFA during the last trimester of pregnancy, because of rapid synthesis of brain tissue. In this chapter, EFA and LCPUFA biochemistry and metabolism in the foetoplacental unit are briefly reviewed, followed by a summary of their functional importance during pregnancy and early human development.

2. EFA and their LCPUFA

Almost 75 years ago, George and Mildred Burr observed that rats fed a fat-free diet for several months developed symptoms like growth retardation, dermatitis and reproductive failure. Supplementation with vegetable oils cured these symptoms and from this observation they suggested that certain fatty acids might be essential dietary components [1]. One year later, the Burrs identified linoleic acid (LA) and α-linolenic acid (ALA) as essential for growth and reproduction [2].

LA and ALA are 18-carbon fatty acids with 2 (LA) or 3 (ALA) methylene-interrupted double bonds. The first of these double bonds is located between the 6th and 7th (LA) or the 3rd and 4th (ALA) carbon atom, counted from the methyl head group. Therefore, LA and ALA are called *n*-6 (or omega-6) and *n*-3 (omega-3) fatty acids, respectively. Because mammals do not have the enzymes required for the insertion of these double bonds,

Advances in Molecular and Cell Biology, Vol. 33, pages 503–529

ISSN: 1569-2558 / DOI: 10.1016/S1569-2558(03)33024-3

LA and ALA cannot be synthesized de novo. Therefore, and because they are needed for optimum development and health [3], these fatty acids are called "essential fatty acids" (EFA). LA and ALA are present in seed oils and green leaves (mainly ALA).

LA has a well-defined function in the skin [4], but it is unclear whether ALA itself is essential only as a precursor or also in its own right [5].

LA and ALA can be converted to LCPUFA by microsomal desaturation and elongation, ultimately followed by one cycle of peroxisomal β-oxidation [6] (see Fig. 1). These metabolic reactions take place between the carboxyl group and the nearest double bond and, consequently, do not affect the molecular structure at the methyl end of the fatty acid molecule. Therefore, all fatty acids derived from the "parent" EFA LA (18:2*n*-6) and ALA (18:3*n*-3) remain *n*-6 and *n*-3 fatty acids, respectively.

The conversions of LA and ALA into their respective LCPUFA are catalysed by rate-limiting activities of Δ6- and Δ5-desaturases. These enzymes are bound to the microsomal lipid bilayer and require zinc as co-factor [7]. The Δ6-desaturase converts LA to γ-linolenic acid (GLA, 18:3*n*-6) and ALA to stearidonic acid (18:4*n*-3). These fatty acids are elongated to dihomo-γ-linolenic acid (20:3*n*-6) and 20:4*n*-3, respectively, which are further desaturated by the Δ5-desaturase, resulting in the formation of arachidonic acid (AA, 20:4*n*-6) and eicosapentaenoic acid (EPA, 20:5*n*-3), respectively. AA and EPA are then elongated twice to form 24:4*n*-6 and 24:5*n*-3, which are converted by a Δ6-desaturase to 24:5*n*-6 and 24:6*n*-3, respectively. Finally, peroxisomal chain shortening by β-oxidation is responsible for the formation of Osbond acid (ObA, 22:5*n*-6) and docosahexaenoic acid (DHA, 22:6*n*-3).

Recently, the Δ6- and Δ5-desaturases have been cloned [8,9], and it has been demonstrated that LA inhibits expression of the gene coding for the Δ6-desaturase enzyme [8], which likely leads to the phenomenon called "substrate inhibition". Since LA and ALA compete for the same enzymes, they both inhibit each other's desaturation and elongation [10]. Although the Δ6-desaturase has a preference for ALA [11], the excessive availability of LA compared to ALA in the present Western diet can be expected to promote LA conversion at the expense of ALA metabolism. In addition, it should be realized that the endogenous LCPUFA synthesis from EFA is far less efficient than in many other mammalian species. Therefore, under the present dietary conditions an adequate DHA status is likely to require the consumption of preformed DHA. Whether or not the intake of preformed AA is needed to guarantee adequate AA levels has not been properly investigated so far. Although at the current habitual (high) LA intake this may seem rather unlikely, it should be realized that even a modest increase in the consumption of *n*-3 LCPUFA causes the AA status to reduce (see Section 9). Therefore, consumption of preformed AA may be important to stabilize AA levels under this condition.

AA is a very important *n*-6 LCPUFA because it is the direct precursor of an eicosanoid family, which contains powerful physiological regulators [12]. In addition, AA is abundantly present in the central nervous system [13] and affects growth-related early gene expression and cell growth [14]. The most important ALA-derived LCPUFA are EPA and DHA. EPA is the precursor of a different family of eicosanoids, whereas DHA is a very important structural and functional LCPUFA in the central nervous system [13] and, like AA, actively modulates gene expression [15,16].

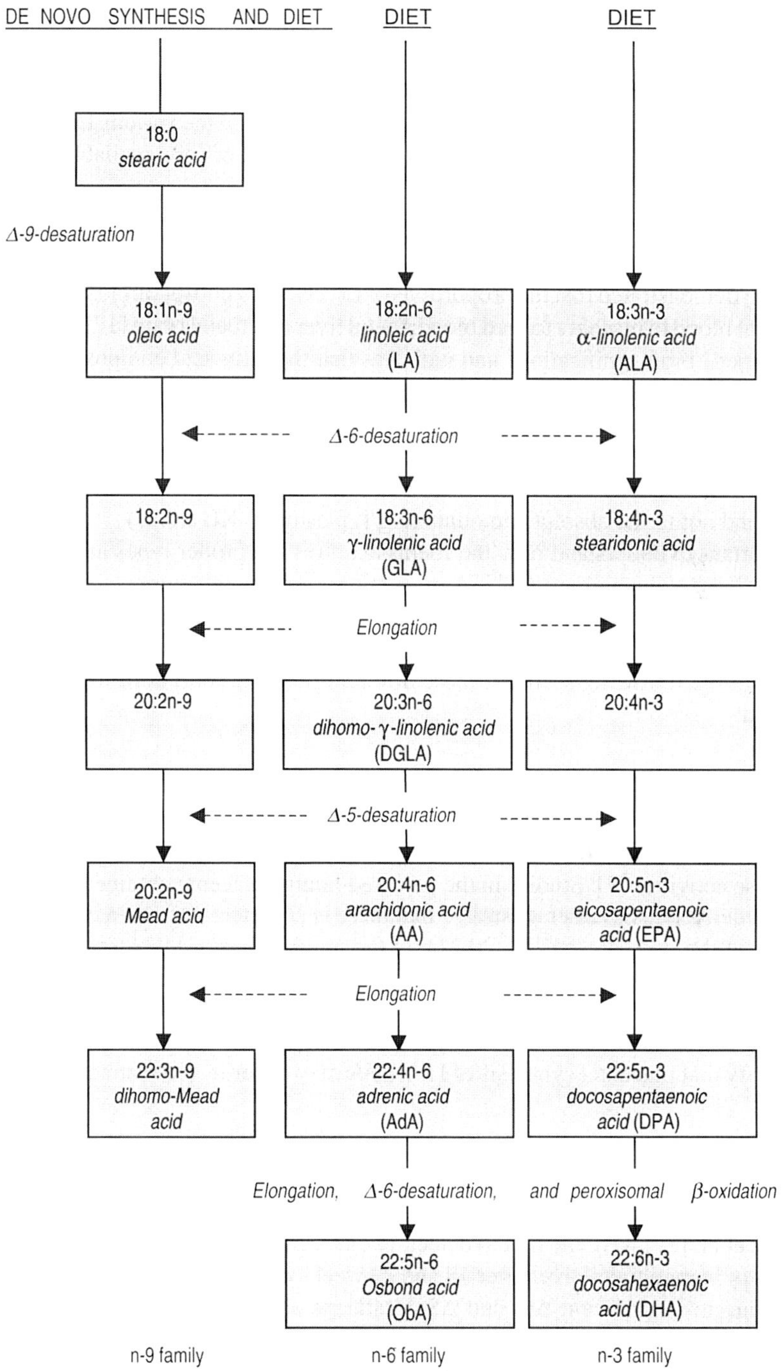

Fig. 1. Schematic representation of enzymatic fatty acid conversion.

Because EPA competes with AA for the enzymes involved in eicosanoid synthesis and the eicosanoids derived from EPA have different, often opposite, biological properties than those derived from AA, the biological response after eicosanoid release is at least partly dependent on the net balance of eicosanoids derived from the various LCPUFA and, consequently, from the ratio in which their LCPUFA precursors are available.

3. EFA and LCPUFA metabolism in the foetoplacental unit

It has been demonstrated that the proportions of LCPUFA in phospholipids (PL) increase from maternal blood to placenta to cord blood, foetal liver and foetal brain [17]. This process has been termed "biomagnification" and indicates that the fatty acid composition of PL in the foetoplacental unit is the result of alterations designed to achieve the high proportion of LCPUFA necessary for structural components of the developing brain. Fig. 2 illustrates the biomagnification process with regard to DHA [18]. As discussed by Hornstra [19], biomagnification only applies to the relative concentrations of LCPUFA (percentage of total fatty acids) and not to the absolute amounts (mg/L plasma or KG tissue).

It is important to understand how the foetus acquires the proper types and amounts of EFA and LCPUFA. The higher LCPUFA proportions in neonatal as compared to maternal blood could originate from preferential transfer of PUFA from the maternal circulation to the foetal circulation and/or the LCPUFA might be formed in either the placenta or the foetal liver. In the following sections, these different possible sources of foetal LCPUFA are discussed.

3.1. Desaturation enzyme system of the placenta

Different studies in vitro illustrated that the human placenta lacks both Δ6- and Δ5-desaturase activities [3]. Studies in the perfused human placenta obtained from normal or caesarean deliveries have demonstrated that there is no detectable chain elongation and desaturation of the two parent EFA [20,21]. In the microsomes of human placental tissue obtained at 18 and 22 weeks of gestation, no activity of the Δ6- or Δ5-desaturase enzymes has been detected either [22]. These data suggest that the long-chain derivatives of LA and ALA in the foetal circulation are synthesised by either the mother or the foetus. The possibility that placental synthesis of LCPUFA may occur in vivo cannot be ruled out, however.

3.2. Desaturation enzyme system of the foetal liver

Chambaz et al. [22] were the first to detect, in vitro, Δ5- and Δ6-desaturase activities in microsomes of human foetal liver after 18 and 22 weeks of gestation. Rodriguez et al. [23] established in vitro significant Δ6- and Δ5-desaturase activities in human foetal liver as early as the 17th week of gestation. The desaturation activities remained stable throughout the third trimester and the Δ6-desaturase activity was higher for *n*-3 than for *n*-6 fatty acids, regardless of gestational age. However, the desaturation capacity calculated from in vitro measurements appears to be lower than the foetal LCPUFA requirements [23].

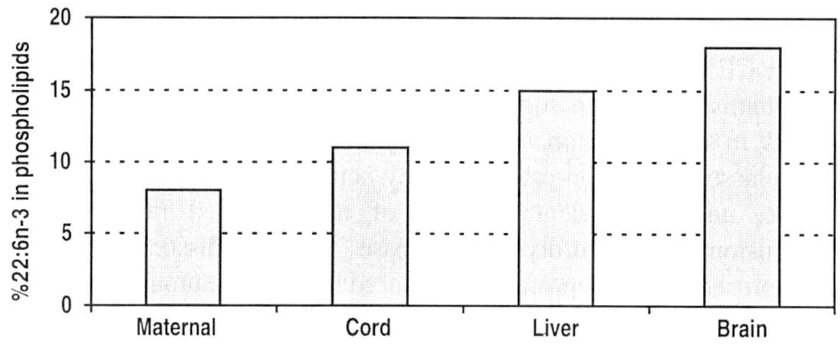

Fig. 2. "Biomagnification" of the relative DHA concentrations. Data from analysis of mid-term abortion material [18].

Stable isotope tracer studies demonstrated that human preterm infants are capable of synthesising DHA in vivo subsequent to enteral administration of [^{13}C]-labelled ALA [24,25]. Using [^{13}C]-tracers, Su et al. [26] recently showed that baboon foetuses have the capacity to convert ALA to DHA in vivo. In this study, [U-^{13}C]ALA or [U-^{13}C]DHA were administered as non-esterified fatty acids via the foetal jugular artery. The results demonstrated that in baboon foetuses, the liver is likely to be an important site for ALA to DHA conversion. In addition, foetal plasma DHA appeared about 8 times more effective as a substrate for brain DHA accretion than ALA.

Taken together, these data illustrate that PUFA can be synthesised in the human foetus during the second and third trimester. Despite this, desaturation enzyme systems in the human foetal liver seem to be immature and unable to supply enough LCPUFA to meet the high demands of rapidly growing tissues and organs [27]. Therefore, it is likely that foetal LCPUFA are primarily derived from maternal sources. Consequently, to obtain adequate amounts of the parent EFA and their LCPUFA, the developing foetus seems to depend mainly on the transport of these fatty acids from the maternal circulation across the placenta and thus, on the maternal essential PUFA status.

3.3. Fatty acid transport across the placenta

The ability of the placenta to extract LCPUFA from the maternal circulation and deliver them to the foetus is extremely important for neonatal development. Non-esterified fatty acids in the maternal compartment have been proposed as the major source of fatty acids for transport across the placenta, irrespective of the source from which they originate in the maternal circulation. Thus, Kuhn and Crawford [28] showed in the perfused human placenta that free fatty acids rather than triacylglycerols (TAG) or PL are taken up from the maternal circulation. In maternal plasma, however, LCPUFA are mainly esterified and associated with lipoproteins rather than in the form of free fatty acids [29,30]. Consequently, LCPUFA arrive at the microvillous membrane of the placenta (maternal side) as albumin-bound non-esterified fatty acids (minority of the fatty acids) or as TAG, PL and cholesteryl esters (CE) as components of lipoprotein particles (majority of the fatty acids). Endothelial-bound lipases are present that can hydrolyse these lipid classes and generate free fatty acids, which can be taken up [31–33]. Intact CE can be taken up by the placental tissue and hydrolysed there [34]. The latter mechanism has been suggested to provide the placenta with free cholesterol but also results in the liberation of free fatty acids. The relative contribution of the different lipid classes to the non-esterified fatty acid pool taken up by the placenta is, to our knowledge, unclear. Placental uptake of non-esterified fatty acids occurs by facilitated diffusion after hydrolysis by lipase or after dissociation from albumin [35–37]. The presence of lipoprotein lipase activity in the human placenta allows the utilization of maternal TAG. In guinea pigs, the placental lipoprotein lipase is present only at the microvillous membrane of the placental trophoblast [38]. This lipoprotein lipase hydrolyses TAG from maternal VLDL but not the TAG present in chylomicrons [32,37]. The preferential hydrolysis of posthepatic TAG (VLDL) by the placental lipoprotein lipase may result in an increased availability of LCPUFA for placental uptake and serve as a protection for the foetus from the immediate impact of an unusual fatty acid in a meal [35,39]. The TAG concentration in the maternal circulation increases much more with progressing gestation than the other lipid classes [40,41]. This suggests that maternal TAG are the major fatty acid sources for the placenta. Placental lipoprotein lipase preferentially hydrolyses fatty acids in the *sn*-2 position of the glycerol backbone and the *sn*-2 position generally is more unsaturated than the *sn*-1 or *sn*-3 positions. This implies selectivity by the placental lipases for the release of TAG-associated LCPUFA. Moreover, the preferential incorporation of DHA into the placental TAG fraction suggests that triglycerides may play an important role in the placental transport of DHA to the foetal circulation [35]. The presence of a phospholipase A_2 has been reported on the microvillous membrane for the hydrolysis of PL, but the hydrolysis of TAG by lipoprotein lipase is more pronounced [42]. More recently, an endothelial-derived lipase, which is expressed amongst other organs in the placenta, has been cloned [31]. This endothelial lipase has substantial phospholipase activity and less TAG lipase activity. Overexpression of this enzyme in mice reduced the plasma concentrations of HDL. As HDL is rich in PL and DHA is highly present in the PL fraction, this enzyme could be important for the hydrolysis of maternal PL and the supply of LCPUFA to the foetus.

Once maternal plasma TAG are hydrolysed, their components are taken up by the placenta, where re-esterification and intracellular hydrolysis facilitates diffusion of the released fatty acids to the foetus and their subsequent transport to the foetal liver [29].

Fatty acids can cross lipid bilayers (as in the syncytiotrophoblast) by simple diffusion. However, placental fatty acid-binding proteins in membranes and cytoplasm are thought to facilitate the transfer across membranes and intracellular channelling of fatty acids. These fatty acid-binding proteins are the fatty acid transfer proteins (FAT/CD36 and FATP) both on the microvillous and basal membranes, and a placenta-specific plasma membrane fatty acid-binding protein (p-FABPpm) located exclusively on the microvillous membrane [43,44]. Placental perfusion studies demonstrate that LA is more efficiently transferred from the maternal to the foetal circulation than AA [28]. But maternally derived LA was found mostly in the free fatty acid fraction of foetal circulating lipids whereas AA had been selectively incorporated in PL by the placenta and exported to the foetus in that form. It is known that free fatty acids may cross the placenta in either direction but the movement of PL and TAG may be restricted. Thus the selectivity with which the placenta distributes LCPUFA into a lipid fraction that does not re-cross the placental barrier may allow those fatty acids to be retained, relative to the parent EFA, in the foetoplacental unit [28].

Campbell et al. [45] investigated the binding characteristics of human placental membranes using four different radiolabelled fatty acids. Binding sites seem to have a strong preference for LCPUFA: the order of competition was AA $\gg$ LA $>$ ALA $\gg$ oleic acid. Haggarty et al. perfused placentas with a mixture of fatty acids and noted that the order of selectivity for uptake is AA $>$ DHA $>$ ALA $>$ LA [21]. But the placenta appears to retain AA in preference to the other fatty acids, resulting in a different order of selectivity for placental transfer to the foetal circulation: DHA $>$ ALA $>$ LA $>$ oleic acid $>$ AA. They, therefore, concluded that the human placenta has the capacity to selectively transfer individual fatty acids to the foetus with the greatest selectivity being shown for DHA.

Non-esterified fatty acids released at the foetal side of the placenta are transported in foetal blood bound to a specific foetal protein: the α-foetoprotein [29]. This protein has been shown in a number of studies to bind LCPUFA more strongly than does albumin [46]. The presence of this protein can account for the relatively high proportion of LCPUFA found in foetal plasma. The free fatty acids in foetal plasma are rapidly taken up by foetal liver, where they are esterified and released back into the foetal circulation as TAG. This may explain the significant linear correlation for certain LCPUFA between maternal and cord plasma TAG during gestation [47]. A strong correlation between the LCPUFA of maternal and umbilical plasma PL has also been observed (see Section 6).

To elucidate the role of p-FABPpm in the preferential transfer of LCPUFA from the maternal circulation across the placenta, the direct binding of various fatty acids to purified p-FABPpm has been determined. It was shown that p-FABPpm preferentially binds with LCPUFA despite a high concentration of non-essential fatty acids in the assay mixture [48]. This LCPUFA preference and the fact that this fatty acid-binding protein is exclusively located on the microvillous membrane, facing the maternal circulation [43], explains the earlier described unidirectional flow of maternal LCPUFA to the foetus [35].

The uptake of LA by brush-border (facing the maternal circulation) preparations from the human syncytiotrophoblast was greater than that for basal membrane (facing the foetal circulation) preparations [49]. It was suggested that this might be due to differences in the concentration of fatty acid-binding proteins between the brush-border and the basal membrane.

In summary, the observed trans-placental gradient of LCPUFA (known as biomagnification) suggests the presence of a highly active and specific trans-placental LCPUFA transport system. This system may be driven by (i) selectivity of the placental lipoprotein lipases for the release of LCPUFA from placental TAG; (ii) specific placental fatty acid-binding proteins; and (iii) foetal α-foetoprotein, which has a high affinity for LCPUFA.

4. Assessment of the biochemical EFA and LCPUFA status

4.1. EFA and LCPUFA status markers

For the assessment of the EFA or LCPUFA status of an individual, the total amounts of the various EFA and LCPUFA in blood or tissues give useful information [3,50]. It should be realized, however, that the presence of these fatty acids does not necessarily guarantee their proper use by cells and tissues. Therefore, additional "status markers" are required to estimate a more functional essential PUFA status of a given individual. In general, when insufficient LCPUFA are available to meet the physiological requirements, the body starts to synthesize certain fatty acids of comparable chain length and degree of unsaturation, which are hardly present under normal conditions and can, therefore, be used as LCPUFA "shortage markers". The best-known shortage marker is Mead acid (MA, 20:3*n*-9) that, in case of an overall shortage of essential PUFA, is synthesized by Δ6 desaturation, elongation and Δ5 desaturation of oleic acid (18:1*n*-9) [51]. The MA/AA ratio has been proposed as an index of the EFA status [52]. At a more severe essential LCPUFA shortage, even the MA elongation product, dihomo-Mead acid (22:3*n*-9), can be observed.

Another suitable indicator of the overall essential PUFA status is the "EFA status index", which is the ratio of the sum of all the essential *n*-3 and *n*-6 fatty acids to the sum of all the non-essential unsaturated *n*-7 and *n*-9 fatty acids. The higher this ratio, the better the essential PUFA status [53].

An isolated deficiency of DHA stimulates the synthesis of the most comparable LCPUFA of the *n*-6 family, ObA. Therefore, under steady state conditions, the ratio between DHA and ObA [i.e. the DHA sufficiency index (DHASI)] is considered a reliable biochemical indicator of the DHA status [54]. Since the synthesis of ObA also depends on the availability of its precursor 22:4*n*-6 [adrenic acid (AdA)], the ratio of ObA to AdA [i.e. the DHA deficiency index (DHADI) is also a reliable marker of the DHA status, this ratio being higher, the lower the DHA status [53,55].

Holman et al. [56] introduced the concept of the calculated mean melting point (MMP) and considered this parameter as the best current single measure of the overall EFA and LCPUFA status, although they prefer to use the entire fatty acid profile [57]. It should be pointed out, however, that all these "functional" status markers are based on biochemical pathways and measurements, and need validation with respect to physiological functions.

4.2. Determination of the biochemical EFA or LCPUFA status

In principle, the fatty acid profiles of the different lipid fractions (PL, CE, TAG and free fatty acids) extracted from plasma, red blood cell membranes or tissue can be used to document the biochemical EFA status of a given individual. Because of the different rates of turnover of fatty acids in these constituents, differences arise in the time required for a change in dietary fat type to be fully reflected in the fatty acid pattern [58]. Thus, the fatty acid composition of TAG, extracted from fasting serum or plasma, reflects the composition of the last few meals before blood sampling, while the fatty acid composition of CE and PL change more gradually during 2–3 weeks with a change in diet and reflect the average dietary composition during a longer time period.

In a group of institutionalised elderly subjects, a good relationship was found for LA between the four major serum fractions (PL, CE, TAG and free fatty acids). This implies that these four lipid fractions obtain fatty acids from a common precursor pool or that one lipid class provides fatty acids for synthesis of the other [59].

As compared to free fatty acids, TAG, CE and PL are structural lipids with a relatively slow turnover rate and the highest amounts of LCPUFA. Therefore, changes in LCPUFA profiles are most pronounced in PL and, consequently, the PL fatty acid profile provides the best reflection of the EFA or LCPUFA status [60]. The concentration of LCPUFA in plasma PL has been used as a marker of recent dietary intake of PUFA [61,62], whereas erythrocyte LCPUFA patterns have been more commonly used to indicate longer term dietary intake [63]. In a fish oil supplementation study, it was shown that the incorporation half lives of EPA in humans are about 5 days for serum CE, almost a month for erythrocytes and longer than a year for subcutaneous fat tissue [64]. Thus, determination of the fatty acid composition of CE gives a reflection of nutritional intakes over the past week or two, erythrocyte membranes over the past months or two and adipose tissue over a period of years.

The results discussed below are mainly based on PL, because their essential PUFA profile changes more gradually with changing conditions and, therefore, reflect the essential PUFA status over a longer period of time. The postprandial fatty acid profile of plasma triglycerides or free fatty acids is strongly influenced by the fatty acid composition of the last meal.

5. Maternal LCPUFA status during and after pregnancy

Pregnancy is associated with a generalized lipidaemia [40,41], and in a longitudinal study [65], we demonstrated that between early pregnancy (10th week) and delivery, the plasma amounts (mg/L) of the PL-associated essential PUFA increase by about 40%. For AA and DHA, these figures are 23 and 52%, respectively. Qualitatively similar pregnancy-associated fatty acid changes have been observed under highly different dietary and cultural conditions and, therefore, these changes seem a rather general phenomenon [66–70]. Although quantitatively different, comparable changes have been observed in maternal TAG [71,72] and CE [71–73].

From prepregnancy to 10 weeks of pregnancy, the absolute amount of DHA associated with maternal plasma PL increased by about 48% [74]. Daily supplementation of non-pregnant women with either 285 mg DHA as a microalgal oil or 266 mg DHA as tuna fish oil for a 4 week period increased plasma PL DHA by ca. 34 and 31%, respectively [75]. This demonstrates that the effect of pregnancy on the DHA increase in plasma PL is considerably stronger than that of the additional consumption of about 270 mg DHA/d. Many studies show that, in general, dietary habits remain unaltered during pregnancy. Neither the amount and type of fat nor the fatty acid composition of the maternal diet changes during pregnancy [74,76,77]. Because plasma PL mainly originate from hepatic synthesis, the increase in DHA concentrations during pregnancy might reflect a pregnancy-associated adjustment in the hepatic synthesis of DHA (increase in the activity of the desaturation and elongation system), with an enhanced or selective incorporation of this fatty acid into the PL. Other explanations for the improved absolute maternal DHA status could include LCPUFA mobilization from maternal stores or a metabolic LCPUFA shift from energy production to structural use.

The absolute amounts of the non-essential unsaturated fatty acids increase considerably stronger than those of the essential PUFA (65 versus 40%). Actually, the essential PUFA shortage marker MA and the specific DHA shortage marker ObA even increase by 92 and 125%, respectively [65] (for ObA see Fig. 3). This indicates that under the present dietary conditions, pregnancy is associated with a reduction of the functional (but still biochemically determined) PUFA status and the functional DHA status in particular. This is also suggested from the significant reductions in plasma PL of the relative concentrations (percentage of total PL-associated fatty acids) of AA, DHA and most other essential PUFA [78]. Interestingly, the relative DHA concentrations only decrease after an initial increase until week 18 of pregnancy and remains higher than prepregnancy levels throughout gestation [65,79]. Nonetheless, changes in the two DHA status indices indicate a continuous decrease of the functional DHA status during pregnancy. The same holds for the overall essential PUFA status, as reflected by the EFA status index [65].

After delivery, normalization of the essential PUFA status in maternal plasma PL takes place [65,80], but for most LCPUFA this is a relatively slow process [65], taking about 32 weeks [81]. In addition, significant differences were observed between lactating and non-lactating women for the postpartum changes of the relative concentrations of *n*-3 fatty acids, but not for the *n*-6 fatty acids [81]. Thus, the decline in plasma PL DHA values was enhanced in lactating women and reached values significantly lower than prepregnancy levels. As human milk contains significant amounts of DHA, the decrease of DHA in maternal plasma PL probably indicates the utilisation of DHA for breast milk. Moreover, the reductions in maternal DHA levels in plasma as well as erythrocyte PL became

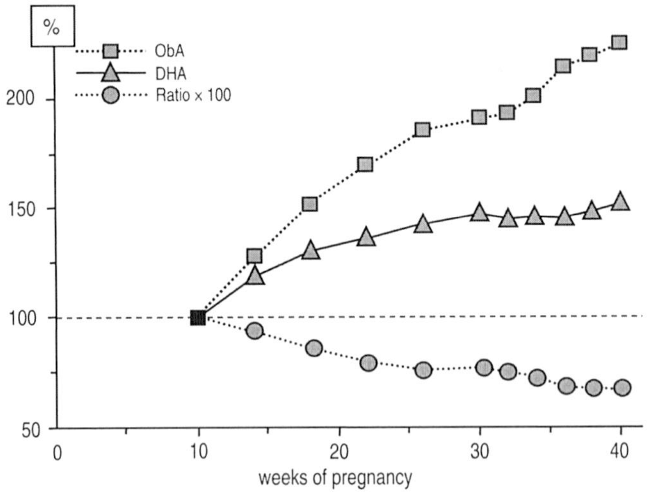

Fig. 3. Courses of maternal plasma PL DHA and ObA concentrations during pregnancy [%, value (mg/L) at 10th pregnancy week = 100]. Ratio represents DHA Sufficiency Index. Data from [65] and [79].

stronger, the longer the duration of breastfeeding. In other words, there is an extra drain on DHA stores in women who breastfeed their children and it will probably take them longer to replenish their DHA stores. After weaning of the infants, the maternal DHA values increase rapidly to values comparable to those of non-lactating women [81].

In a cross-sectional study, the absolute and relative amounts of DHA in maternal plasma PL of women who had been pregnant before (multigravidae) were significantly lower compared to women who were pregnant for the first time (primigravidae) [82]. Throughout pregnancy, the percentage of DHA in the plasma PL of multigravidae was on average ca. 10% lower than in primigravidae. In addition, a significant negative relation was observed between the gravida number and the percentage of DHA in the maternal plasma. These observations might indicate that under the usual dietary conditions, pregnancy is associated with mobilisation of DHA from a store that may not be readily replenished after the first pregnancy. Alternatively, DHA synthesis from precursor fatty acids may become diminished because of repeated pregnancies. This is suggested from the significant negative relationship between the *n*-6 LCPUFA/LA ratio of non-pregnant women (a proxy for the efficiency of the EFA–LCPUFA conversion) and the number of pregnancies completed by these women. Moreover, this ratio is significantly lower in mothers than in non-mothers [83]. In this non-pregnant population, nulligravidae and multigravidae did not differ significantly with respect to the relative amounts of DHA in their plasma PL. Moreover, no significant correlation was observed between parity and the percentage of DHA in plasma PL. Since the time lag between blood sampling and the last partus in the multigravidae was on average about 1 year, these observations indicate that the maternal DHA status after pregnancy (as reflected in plasma PL) normalises within 1 year [83]. However, the average DHA concentration in erythrocyte PL was significantly lower in mothers than in non-mothers and although no significant relationship was found with parity number, this observation nonetheless suggests that in domains with a slow fatty acid turnover, normalization of the DHA status after delivery may take longer than 1 year.

Whatever the reason, in pregnant women the plasma PL DHA content is lower, the higher their parity number. Since a highly significant and positive relationship exists between the LCPUFA status of the neonate and that of its mother (see Section 6), firstborn infants have a significantly higher DHA status than their later-born siblings [82].

6. The essential PUFA status during foetal development and at birth

As mentioned before, EFA and their LCPUFA cannot be synthesized de novo by humans and, therefore, the foetal essential PUFA supply will strongly depend on maternal essential PUFA consumption and metabolism, as well as on the placental transport of these fatty acids. This dependence is convincingly illustrated by the maternal–foetal correlations for most plasma EFA and LCPUFA [65,72]. Significant positive relationships between maternal plasma and umbilical vascular tissue fatty acid compositions were found for LA (vein, $r = 0.29$; $n = 337$; $p < 0.001$) and DHA (artery and vein, $r = 0.41$ and 0.53; $n = 334$; $p < 0.001$) only (G. Hornstra, unpublished observations). Data from an international comparative study involving differences in habitual diets [66] demonstrated that for plasma DHA the slope of the maternal–neonatal relationship is significantly affected by differences in maternal DHA consumption, as reflected by the DHA content of maternal plasma PL. No such association was observed for AA, suggesting a higher degree of foetal autonomy in establishing its AA as compared to its DHA status. This relative autonomy of the foetal *n*-6 LCPUFA status may be due to the fact that the habitual intake of *n*-6 PUFA is usually much higher than that of *n*-3 PUFA [66].

No matter the strong correlations between mothers and their term neonates with respect to the essential PUFA levels, plasma and erythrocyte phospholipid fatty acid profiles of neonates are very different from that of their mothers. In general, relative LCPUFA values (percentage of total PL-associated fatty acids) are considerably higher, whereas the concentrations of the parent EFA are greatly reduced in neonates as compared to their mothers [65,66,72,73,84,85]. When expressed in absolute figures, however (mg/L plasma), all fatty acid amounts are much lower in neonatal than in maternal plasma, which is due to considerably smaller neonatal plasma PL pools [19].

Preterm infants were shown to have an essential PUFA status significantly lower than that of term neonates [86]. However, the essential PUFA amounts in cord plasma of preterm infants at birth are not lower than that in cord plasma obtained by foetal blood sampling of ongoing pregnancies at a comparable gestational age (GA) [87]. Therefore, the low essential PUFA status of preterm infants is most probably a physiological situation and not a pathological condition. These comparative studies also demonstrate that the essential PUFA status of the foetus is not stable during its development, but changes with GA in a fatty acid-specific way. Thus, the foetal LA content strongly decreases during early gestation [88], after which it increases slightly during the second and third trimester [87]. Foetal AA levels, however, slowly decrease throughout gestation, whereas DHA concentrations rise strongly during the last 2 months of foetal development [87].

The usually observed declines of the maternal EFA and LCPUFA status occurring during pregnancy [65,66,73] may imply a suboptimal essential PUFA status of their newborn infants. This view is supported by the fact that the concentrations of most essential PUFA in the walls of the umbilical vein (the supplying blood vessel) are higher than those of the umbilical arteries (which carry the blood away from the foetus back to the placenta), whereas the opposite is true for the essential PUFA shortage markers MA and dihomo-Mead acid [66,89–93] (see also Fig. 4). Interestingly, the arterio-venous difference for DHA is not at all significant, and for ObA the concentration in the umbilical arteries is usually higher than in the umbilical vein. These observations suggest that foetal DHA synthesis is considerable, but still functionally insufficient, as it does not prevent an increased ObA production.

Although certain tissues may be preferred sites of essential PUFA uptake [94], the fatty acid profiles of umbilical venous and arterial walls likely reflect the PUFA status of "upstream" and "downstream" foetal tissue, respectively. Consequently, the typical fatty acid differences between umbilical veins and arteries indicate that the EFA status of the developing foetus is relatively low, and is lower in "downstream" as compared to "upstream" areas. A suboptimal neonatal EFA/LCPUFA status is also suggested from our observation that newborn singletons have a higher essential PUFA status than infants born after multiple pregnancies [95,96]. As mentioned before, the relative amounts of most essential PUFA in maternal plasma PL decrease during pregnancy. In a study comprising 627 mother–infant pairs, Rump et al. [97] observed that this decrease is more pronounced

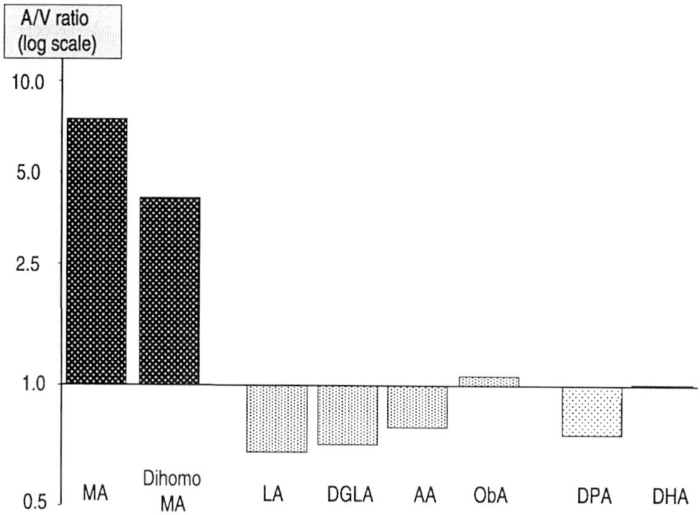

Fig. 4. PL-associated fatty acid concentrations in the walls of umbilical arteries relative to the concentrations in the walls of the umbilical veins (percentage, log scale). Results calculated from [93], which are representative for data presented in [66,89–92].

when the neonatal birth weight is higher. Nonetheless, in these term neonates the LCPUFA contents of umbilical plasma PL are negatively related to birth weight. This indicates that maternal-to-foetal LCPUFA transfer is limited.

Correlation studies in a population of over 300 mother–infant pairs revealed that the relative maternal plasma phospholipid LA contents are negatively related to DHA levels in PL isolated from umbilical plasma ($r = -0.36$) and arterial ($r = -0.31$) and venous ($r = -0.39$) vessel walls. In addition, the LA concentration (percentage of total fatty acids) in maternal plasma PL was negatively related to the neonatal concentration of total *n*-3 LCPUFA, not only in plasma PL ($r = -0.41$), but also in PL isolated from umbilical arteries ($r = -0.30$) and veins ($r = -0.33$; $p < 0.0001$ for all correlations) (G. Hornstra, unpublished observations). Since the plasma LA value is usually a reflection of the LA intake [58], it is of interest to study the relationship between maternal fatty acid intake during pregnancy and the maternal as well as neonatal LCPUFA levels.

7. Habitual fatty acid intake during pregnancy and maternal and neonatal LCPUFA levels

Humans are unable to synthesize EFA de novo, and LCPUFA synthesis from EFA precursors is inefficient in man. Therefore, the essential PUFA status of pregnant women is most likely determined by their intake of EFA and LCPUFA. Several investigators have now confirmed this suggestion.

7.1. Essential PUFA

As has been demonstrated before in non-pregnant subjects [58,61–64], a significant, positive correlation also exists between the dietary intake and the plasma PL contents of LA in pregnant women [76,77]. It is frequently thought that AA levels in plasma and tissue are directly dependent on the habitual LA intake. However, in a group of 288 pregnant women, the LA intake in mid-gestation was not significantly related to the AA content of maternal or neonatal plasma PL at delivery/birth [78]. Interestingly, a significant, negative relationship was observed between the maternal LA intake and the amounts of EPA, DPA and DHA in maternal as well as neonatal plasma PL. This may be due to an inhibitory effect of LA on the incorporation of *n*-3 PUFA in plasma and tissue PL, as has been demonstrated for DHA [98,99].

In the same 288 pregnant women, a significant, positive relationship was observed between the maternal ALA consumption and the ALA amounts in maternal plasma PL as well as the neonatal EPA concentrations. A higher maternal ALA consumption was not associated with a higher maternal or neonatal DHA status [78].

In non-pregnant subjects, the habitual intake of *n*-3 LCPUFA is reliably reflected by the *n*-3 LCPUFA content of plasma and erythrocyte PL [62]. This also holds for pregnant women [100,101].

7.2. *Trans fatty acids*

Trans isomers of unsaturated fatty acids are known to interfere with the conversion of parent EFA into their LCPUFA, especially when the parent EFA levels are low [102]. Studies by Koletzko and colleagues demonstrated that *trans* fatty acid can cross the placenta and may lower foetal LCPUFA concentrations [103,104], and we observed a highly significant positive correlation between the relative amounts of 18:1 *trans* in maternal plasma and foetal tissue [105]. In addition, we demonstrated that the presence of *trans* fatty acids in cord tissue is associated with proportionally lower amounts of essential PUFA, a reduced birth weight and a smaller head circumference. However, after correction for GA these latter two associations were no longer statistically significant. Most of these results were recently confirmed by Elias and Innis [106]. However, negative effects of *trans* fatty acids on foetal development have not yet been ascertained [107].

8. Functional implications of essential PUFA for mother and infant

8.1. Maternal LCPUFA status and pregnancy outcome

Although observational studies [108,109] suggest that a reduced *n*-3 LCPUFA status may contribute to pregnancy-induced hypertension (PIH), Al et al. observed in a prospective nested case control study that the *n*-3 LCPUFA status is slightly higher in women with PIH [91]. Therefore, a causal role of LCPUFA in the aetiology of PIH seems unlikely. This suggestion is supported by the results of another observational study [110]. In addition, supplementation during pregnancy with up to 6.1 g of *n*-3 LCPUFA/day did not lower PIH risk [111–113].

Hibbeln observed that a higher DHA content in mother's milk predicted a lower prevalence of postpartum depression [114]. This suggests that a low DHA status may be involved in the prevalence of postpartum depression. In support of this suggestion, Otto et al. [115] demonstrated in a prospective observational study that the postpartum increase of the functional DHA status, expressed as the ratio between DHA and ObA, was significantly lower in a "possibly depressed" group of young mothers compared to a non-depressed group. The DHA status at delivery was also lower in the group at risk of PPD, but this difference did not reach statistical significance. De Vriese, using a similar design, recently demonstrated that a lower DHA status at delivery is significantly associated with a higher risk of developing PPD symptoms indeed (S.R. De Vriese, Lower serum *n*-3 polyunsaturated fatty acids predict the occurrence of postpartum depression, submitted).

Olsen and his group extensively studied the relationship between the maternal *n*-3 LCPUFA intake and preterm delivery. Initially, their results were inconsistent, but their most recent prospective cohort study among almost 9000 pregnant women clearly demonstrated that the length of gestation is positively related to the intake of *n*-3 LCPUFA, and that low fish consumption is a risk factor for preterm delivery [116]. This finding is consistent with the modest but statistically significant positive results of intervention studies performed by the same group [112,117].

So far, no significant relationships have been reported between maternal EFA consumption or essential PUFA concentrations in maternal plasma and neonatal birth weight [97,118]. Relationships between maternal fish intake during pregnancy and infant birth weight have been found inconsistent, but tend to be positive [116,119,120]. Fish consumption during pregnancy has also been reported to reduce the risk of intra-uterine growth retardation [116]. However, fish oil supplementation did not reduce this risk [111].

Length of infants at birth appeared significantly and positively associated with maternal consumption of total PUFA minus LA [118]. However, birth length was not significantly associated with fish intake [119,120]. Placental weight and neonatal head circumference were shown to be associated with maternal fish consumption in a positive way [119].

8.2. Relationship between foetal essential PUFA status and pregnancy outcome

Al et al. [65] observed a highly significant, negative correlation between foetal LA availability (reflected by cord plasma phospholipid LA concentrations) and GA at birth, whereas the amounts and concentrations of DHA and the sum of all *n*-3 fatty acid were positively correlated with GA. In a cohort of 780 infants, these DHA findings were recently confirmed by Rump and Hornstra [78], who also observed a positive association between GA and the foetal availability of DPA (the precursor of DHA) and AdA (22:4*n*-6). Relationships with AA and ObA were not significant. MA concentrations, on the other hand, were negatively related with GA. In a recent study performed at the Faroer Islands (where the habitual intake of marine fatty acids is high), Grandjean et al. [121] observed a positive association between cord serum DHA concentrations and GA. The correlation with AdA was positive also.

Strong evidence has been presented that neonatal head circumference is negatively associated with maternal LA intake [65,118]. Since head circumference is an excellent predictor of brain mass [122], the negative relationship observed might result from the fact that under the present dietary conditions, maternal LA intake during pregnancy is negatively related to the neonatal availability of DHA [78], which is an important structural component of the brain. Further studies are warranted to investigate the functional relevance of this relationship.

As mentioned before, negative relationships have been reported for term neonates between birth weight and the concentrations of most LCPUFA in cord plasma and cord serum PL [97,121]. Correlations with the umbilical amounts of the essential PUFA shortage markers MA and ObA were positive and significant, suggesting that the maternal-to-foetal LCPUFA transfer is too limited to secure an adequate, birth weight independent neonatal LCPUFA status. In preterm and low birth weight neonates, the LCPUFA status has repeatedly been shown to be positively related to birth weight [18,123,124]. This discrepancy between term and preterm infants may be due to the differences in body composition, since preterm infants have much less adipose tissue than term infants.

8.3. *Early LCPUFA availability and later neurodevelopment*

Observational studies in infants reared on either mother's milk (contains LCPUFA) or formula without LCPUFA invariably show higher LCPUFA concentrations in blood of breastfed as compared to bottle-fed infants [125]. In addition, many observational studies demonstrated that breastfed infants have a long-term intellectual advantage over formula-fed infants [126], although results of higher quality studies were less convincing [127]. Intervention studies comparing developmental or visual outcomes of young infants given formula with and without added LCPUFA confirmed the importance of these fatty acids for short-term brain development and function of preterm infants [128–130]. At present, longer-term benefits have not yet been reported. Results for infants born at term are not conclusive, since only in a minority of the studies significant benefits have been observed [131–135] and longer-term benefits have not been reported at all. Therefore, further studies are certainly warranted [136]. Nonetheless, an expert panel already concluded that, although breastfeeding is the preferred option, formulas for term and preterm infants should contain DHA and AA [137].

Since the brain has its growth spurt during the third trimester of pregnancy and in the neonatal period, it seems feasible to suggest that the maternal LCPUFA status during pregnancy and lactation could affect infant cognitive development. Recently, the potential importance of maternal LCPUFA status during pregnancy for later infant development was investigated by relating the foetal availability of AA and DHA (as represented by their concentrations in umbilical plasma phospholipids) and cognitive, motor and visual functions at 7–8 years of age [138]. After appropriate correction for confounders, no significant associations were observed between either DHA or AA at birth and cognitive performance at 7 years of age [139]. Likewise, no significant relationship was observed between cognitive performance at 3.5 years of age and LCPUFA status of neonatal erythrocytes [140]. However, DHA status at birth was significantly and positively related to movement quality and visual acuity at 7–8 years of age [138]. In addition, speed of visual information processing, measured by visual evoked potentials and electroretinograms at follow-up, were also positively related to DHA levels at birth [138]. None of the functional outcome measures were significantly associated with DHA or AA levels at follow-up. These results indicate that a higher perinatal DHA availability may promote later neurodevelopment and function.

9. Effects of maternal essential PUFA supplementation on infant LCPUFA status and development

Since the essential PUFA status of pregnant women is determined by their intake of these fatty acids and the essential PUFA status of neonates is strongly correlated to that of their mothers, it is likely that the essential PUFA status of neonates can be influenced by nutritional intervention of their mothers during pregnancy. Intervention studies demonstrated that it is feasible indeed to increase the essential PUFA status of neonates or breastfed infants by dietary supplementation of their mothers. Thus, maternal supplementation with LA increased the neonatal *n*-6 PUFA concentrations, but reduced their *n*-3 PUFA status [92], whereas maternal supplementation with fish oil resulted in an increase of the neonatal *n*-3 PUFA status but was often associated with lower *n*-6 PUFA concentrations [141–143]. Therefore, it seems that an overall increase of the maternal and, consequently, neonatal LCPUFA status would require an increased maternal consumption of both *n*-6 and *n*-3 fatty acids. For the maternal LCPUFA status this has been confirmed by a series of studies with single-cell oils rich in DHA or AA [75,144].

It has amply been demonstrated that – maybe with the exception of AA [145] – the EFA or LCPUFA content of human milk can be influenced by dietary supplementation of lactating women with essential PUFA [146–153]. Therefore, it can also be expected that the essential PUFA status of breastfed infants can be influenced by modulating the fatty acid composition of breast milk via supplementation of lactating women with essential PUFA. Indeed, for DHA it was shown that the infant levels after 8–12 weeks of lactation were significantly and positively related to the DHA dose their mothers were supplemented with [149,151], independent of the source of DHA (fish oil, single-cell oil or DHA-enriched eggs). In one study [149], infant plasma and erythrocyte AA contents reduced significantly in the supplemented groups and in proportion to the maternal DHA supplementation doses. Despite these fatty acid changes, no neurodevelopment differences were observed between the various groups [154], but it should be realized that group sizes (8–12) might have been too small for a reliable assessment, considering the many potential sources of variability [155].

Effects of maternal fatty acid interventions during pregnancy and lactation on infant development have recently been reported. As compared to a placebo, maternal cod liver oil supplementation during pregnancy did not significantly affect mental development of the infants measured at 6 and 9 months of age. Brain maturity, as reflected by an electroencephalogram (EEG) at 3 months of age, was not significantly influenced either, although neonates with mature EEG scores had higher *n*-3 LCPUFA levels at birth than infants with immature EEG scores [142]. Interestingly, after 4 years of follow-up, infants born to the cod liver oil supplemented mothers had higher mental processing scores than the children born to the "control" mothers, suggesting that maternal intake of *n*-3

LCPUFA during pregnancy and lactation may be favourable for later mental development of children [156].

The use of cod liver oil supplements during pregnancy has also been reported to be associated with a lower risk of Type I (insulin-dependent) diabetes mellitus in the offspring, both unadjusted and after adjustment for age, sex, breastfeeding, maternal education and maternal use of "other supplements" [157].

10. Concluding remarks

From the results summarized above, it may be felt necessary to increase the dietary EFA and/or LCPUFA intake of pregnant women in order to prevent the decrease of their essential PUFA status during pregnancy and to optimise the foetal essential PUFA status. The latter may be of particular importance for infants born preterm, because they have a significantly lower LCPUFA status than term neonates [86]. In addition, their LCPUFA status drops considerably during the first postnatal weeks, even when given breast milk [158,159], whereas during intra-uterine life the foetal EFA status increases considerably during the same postconceptional period [87]. Consequently, during the growth spurt of the brain, the availability of LCPUFA is much lower for infants born preterm than for the intra-uterine foetus of comparable postconceptional age. Whether or not this contributes to the well-known developmental disadvantage of preterm versus term infants needs careful consideration.

As discussed in Section 5, the DHA content of maternal plasma PL is significantly lower in multiparous as compared to primiparous women, and infants born to multiparous women have significantly less DHA in umbilical tissue PL than infants born to primiparous women. Whether or not this has functional consequences for these infants is not known as yet. However, there is now good evidence that the pre- and early postnatal DHA status has important consequences for growth and function of the central nervous system and, consequently, for neurological and cognitive development (see Sections 8.3 and 9). Therefore, a lower pre- and perinatal DHA availability may – at least in part – present an explanation for observations that firstborn children, in general, do better than their younger siblings on several developmental, behavioural and intelligence tests [160,161].

If supplementation with essential PUFA during pregnancy is considered, it should be recalled that the two PUFA families compete for the same metabolic enzymes (see Section 2). Therefore, the supplement of choice should contain a mixture of *n*-6 and *n*-3 (LC)PUFA.

The major natural LCPUFA sources for human nutrition are fatty fish (mainly *n*-3 LCPUFA, but also some AA), egg yolk (mainly AA but also DHA in eggs from specially fed chickens), red meat and breast milk (DHA and AA). In addition, DHA and AA are available as dietary supplements, such as fish oil concentrates (mainly DHA and its precursors EPA and DPA), and single-cell oils (DHA and AA). In fatty fish, breast milk, fish oils and single-cell oils, the LCPUFA are mainly present as TAG; in lean fish, meat and egg yolk, PL are the major LCPUFA carriers, whereas in certain supplements *n*-3 LCPUFA are present as ethyl esters. Studies comparing the bioavailability of these various forms of LCPUFA indicated considerable differences [162–169], but are still incomplete. Therefore, further research on this issue is needed.

References

[1] Burr, G.O., Burr, M.M., 1929. A new deficiency disease produced by the rigid exclusion of fat from the diet. J. Biol. Chem. 82, 345–367.

[2] Burr, G.O., Burr, M.M., 1930. The nature and role of the fatty acids essential in nutrition. J. Biol. Chem. 86, 587–621.

[3] Innis, S.M., 1991. Essential fatty acids in growth and development. Progr. Lipid Res. 30, 39–103.

[4] Hansen, H.S., Jensen, B., 1985. Essential function of linoleic acid esterified in acylglucosylceramide and acylceramide in maintaining the epidermal water permeability barrier. Evidence from feeding studies with oleate, linoleate, arachidonate, columbinate and alpha-linolenate. Biochim. Biophys. Acta 834, 357–363.

[5] Cunnane, S.C., Chen, Z.Y., Yang, J., Liede, A.C., Hamadeh, M., Crawford, M.A., 1991. Alpha-linolenic acid in humans: direct functional role or dietary precursor? Nutrition 7, 437–439.

[6] Sprecher, H., Luthria, D.L., Mohammed, B.S., Baykousheva, S.P., 1995. Reevaluation of the pathways for the biosynthesis of polyunsaturated fatty acids. J. Lipid Res. 36, 2471–2477.

[7] Cunnane, S.C., 1988. Role of zinc in lipid and fatty acid metabolism and in membranes. Prog. Food Nutr. Sci. 12, 151–188.

[8] Cho, H.P., Nakamura, M.T., Clarke, S.D., 1999. Cloning, expression, and nutritional regulation of the mammalian Delta-6 desaturase. J. Biol. Chem. 274, 471–477.

[9] Cho, H.P., Nakamura, M., Clarke, S.D., 1999. Cloning, expression, and fatty acid regulation of the human D-5 desaturase. J. Biol. Chem. 274, 37335–37339.

[10] Emken, E.A., Adlof, R.O., Gulley, R.M., 1994. Dietary linoleic acid influences desaturation and acylation of deuterium-labeled linoleic and linolenic acids in young adult males. Biochim. Biophys. Acta 1213, 277–288.

[11] Brenner, R.R., Peluffo, R.O., 1969. Regulation of unsaturated fatty acid biosynthesis. I. Effect of unsaturated fatty acids of 18 carbons on the microsomal desaturation of linoleic acid into γ-linolenic acid. Biochim. Biophys. Acta 176, 471–479.

[12] Soberman, R.J., Christmas, P., 2003. The organization and consequences of eicosanoid signaling. J. Clin. Invest. 111, 1107–1113.

[13] Youdim, K.A., Martin, A., Joseph, J.A., 2000. Essential fatty acids and the brain: possible health implications. Int. J. Dev. Neurosci. 18, 383–399.

[14] Sellmayer, A., Danesch, U., Weber, P.C., 1996. Effects of different polyunsaturated fatty acids on growth-related early gene expression and cell growth. Lipids 31(Suppl.), S37–S40.

[15] Uauy, R., Mena, P., Rojas, C., 2000. Essential fatty acids in early life: structural and functional role. Proc. Nutr. Soc. 59, 3–15.

[16] Narayanan, B.A., Narayanan, N.K., Simi, B., Reddy, B.S., 2003. Modulation of inducible nitric oxide synthase and related proinflammatory genes by the omega-3 fatty acid docosahexaenoic acid in human colon cancer cells. Cancer Res. 63, 972–979.

[17] Crawford, M.A., Hassam, A.G., Stevens, P.A., 1981. Essential fatty acid requirements in pregnancy and lactation with special reference to brain development. Prog. Lipid Res. 20, 31–40.

[18] Crawford, M.A., Doyle, W., Drury, P., Lennon, A., Costeloe, K., Leighfield, M., 1989. n-6 and n-3 fatty acids during early human development. J. Intern. Med. Suppl. 225, 159–169.

[19] Hornstra, G., 1992. Essential fatty acids, pregnancy, and pregnancy complications: a roundtable discussion. In: Sinclair, A., Gibson, R. (Eds.), Essential Fatty Acids and Eicosanoids. American Oil Chemists' Society, Champaign, IL, pp. 177–182.

[20] Booth, C., Elphick, M.C., Hendrickse, W., Hull, D., 1981. Investigation of [14C] linoleic acid conversion into [14C] arachidonic acid and placental transfer of linoleic and palmitic acids across the perfused human placenta. J. Dev. Physiol. 3, 177–189.

[21] Haggarty, P., Page, K., Abramovich, D.R., Ashton, J., Brown, D., 1997. Long-chain polyunsaturated fatty acid transport across the perfused human placenta. Placenta 18, 635–642.
[22] Chambaz, J., Ravel, D., Manier, M.C., Pepin, D., Mulliez, N., Bereziat, G., 1985. Essential fatty acids interconversion in the human fetal liver. Biol. Neonate 47, 136–140.
[23] Rodriguez, A., Sarda, P., Nessmann, C., Boulot, P., Poisson, J.P., Leger, C.L., Descomps, B., 1998. Fatty acid desaturase activities and polyunsaturated fatty acid composition in human liver between the seventeenth and thirty-sixth gestational weeks. Am. J. Obstet. Gynecol. 179, 1063–1070.
[24] Salem, N.J., Wegher, B., Mena, P., Uauy, R., 1996. Arachidonic and docosahexaenoic acids are biosynthesized from their 18-carbon precursors in human infants. Proc. Natl Acad. Sci. USA 93, 49–54.
[25] Carnielli, V.P., Wattimena, D.J., Luijendijk, I.H., Boerlage, A., Degenhart, H.J., Sauer, P.J., 1996. The very low birth weight premature infant is capable of synthesizing arachidonic and docosahexaenoic acids from linoleic and linolenic acids. Pediatr. Res. 40, 169–174.
[26] Su, H.M., Huang, M.C., Saad, N.M., Nathanielsz, P.W., Brenna, J.T., 2001. Fetal baboons convert 18:3n-3 to 22:6n-3 in vivo. A stable isotope tracer study. J. Lipid Res. 42, 581–586.
[27] Makrides, M., Neumann, M., Simmer, K., Pater, J., Gibson, R., 1995. Are long-chain polyunsaturated fatty acids essential nutrients in infancy? Lancet 345, 1463–1468.
[28] Kuhn, D.C., Crawford, M., 1986. Placental essential fatty acid transport and prostaglandin synthesis. Prog. Lipid Res. 25, 345–353.
[29] Herrera, E., Amusquivar, E., 2000. Lipid metabolism in the fetus and the newborn. Diabetes Metab. Res. Rev. 16, 202–210.
[30] Herrera, E., 2002. Implications of dietary fatty acids during pregnancy on placental, fetal and postnatal development – a review. Placenta 23(Suppl. A), S9–S19.
[31] Jaye, M., Lynch, K.J., Krawiec, J., Marchadier, D., Maugeais, C., Doan, K., South, V., Amin, D., Perrone, M., Rader, D.J., 1999. A novel endothelial-derived lipase that modulates HDL metabolism. Nat. Genet. 21, 424–428.
[32] Hirata, K., Dichek, H.L., Cioffi, J.A., Choi, S.Y., Leeper, N.J., Quintana, L., Kronmal, G.S., Cooper, A.D., Quertermous, T., 1999. Cloning of a unique lipase from endothelial cells extends the lipase gene family. J. Biol. Chem. 274, 14170–14175.
[33] Waterman, I.J., Emmison, N., Sattar, N., Dutta-Roy, A.K., 2000. Further characterization of a novel triacylglycerol hydrolase activity (pH 6.0 optimum) from microvillous membranes from human term placenta. Placenta 21, 813–823.
[34] Waterman, I.J., Emmison, N., Dutta-Roy, A.K., 1998. Characterization of triacylglycerol hydrolase activities in human placenta. Biochim. Biophys. Acta 1394, 169–176.
[35] Kimura, R.E., 1989. Fatty acid metabolism in the fetus. Semin. Perinatol. 13, 202–210.
[36] Coleman, R.A., 1989. The role of the placenta in lipid metabolism and transport. Semin. Perinatol. 13, 180–191.
[37] Dutta-Roy, A.K., 2000. Transport mechanisms for long-chain polyunsaturated fatty acids in the human placenta. Am. J. Clin. Nutr. 71, 315S–322S.
[38] Thomas, C.R., Lowy, C., St-Hilliaire, R.J., Brunzell, J.D., 1984. Studies on the placental hydrolysis and transfer of lipids on the fetal guinea pig. In: Miller, R.K., Thiede, H.A. (Eds.), Fetal Nutrition, Metabolism and Immunology: The Role of the Placenta. Plenum Press, New York, pp. 135–146.
[39] Dutta-Roy, A.K., Campbell, F.M., Taffesse, S., Gordon, M.J., 1996. Transport of long-chain polyunsaturated fatty acids across the human placenta: role of fatty acid-binding proteins. In: Huang, Y.S., Mills, D.E. (Eds.), γ-Linoleic Acid: Metabolism and Its Role in Nutrition and Medicine. AOCS Press, Champaign, IL, pp. 42–53.
[40] Salameh, W.A., Mastrogiannis, D.S., 1994. Maternal hyperlipidemia in pregnancy. Clin. Obstet. Gynecol. 37, 66–77.
[41] Darmady, J.M., Postle, A.D., 1982. Lipid metabolism in pregnancy. Br. J. Obstet. Gynaecol. 89, 211–215.
[42] Rice, G.E., Wong, M.H., Farrugia, W., Scott, K.F., 1998. Contribution of type II phospholipase A2 to in vitro phospholipase A2 enzymatic activity in human term placenta. J. Endocrinol. 157, 25–31.
[43] Campbell, F.M., Dutta-Roy, A.K., 1995. Plasma membrane fatty acid-binding protein (FABPpm) is exclusively located in the maternal facing membranes of the human placenta. FEBS Lett. 375, 227–230.

[44] Haggarty, P., 2002. Placental regulation of fatty acid delivery and its effect on fetal growth – a review. Placenta 23(Suppl. A), S28–S38.
[45] Campbell, F.M., Gordon, M.J., Dutta-Roy, A.K., 1996. Preferential uptake of long chain polyunsaturated fatty acids by isolated human placental membranes. Mol. Cell. Biochem. 155, 77–83.
[46] Neville, M.C., 1999. Adaptation of maternal lipid flux to pregnancy: research needs. Eur. J. Clin. Nutr. 53(1), S120–S123.
[47] Berghaus, T.M., Demmelmair, H., Koletzko, B., 2000. Essential fatty acids and their long-chain polyunsaturated metabolites in maternal and cord plasma triglycerides during late gestation. Biol. Neonate 77, 96–100.
[48] Campbell, F.M., Gordon, M.J., Dutta-Roy, A.K., 1998. Placental membrane fatty acid-binding protein preferentially binds arachidonic and docosahexaenoic acids. Life Sci. 63, 235–240.
[49] Lafond, J., Simoneau, L., Savard, R., Gagnon, M.C., 1994. Linoleic acid transport by human placental syncytiotrophoblast membranes. Eur. J. Biochem. 226, 707–713.
[50] Innis, S.M., 2000. The role of dietary n-6 and n-3 fatty acids in the developing brain. Dev. Neurosci. 22, 474–480.
[51] Lundberg, W.O., 1980. The significance of cis, cis, cis 5,8,11 eicosatrienoic acid in essential fatty acid deficiency. Nutr. Rev. 38, 233–235.
[52] Holman, R.T., 1960. The ratio of trienoic:tetraenoic acids in tissue lipids as a measure of essential fatty acid requirement. J. Nutr. 70, 405–410.
[53] Hornstra, G., Al, M.D.M., van Houwelingen, A.C., Foreman-van Drongelen, M.M., 1995. Essential fatty acids in pregnancy and early human development. Eur. J. Obstet. Gynecol. Reprod. Biol. 61, 57–62.
[54] Hoffman, D.R., Uauy, R., 1992. Essentiality of dietary omega 3 fatty acids for premature infants: plasma and red blood cell fatty acid composition. Lipids 27, 886–895.
[55] Neuringer, M., Connor, W.E., Lin, D.S., Barstad, L., Luck, S.L., 1986. Biochemical and functional effects of prenatal and postnatal omega-3 fatty acid deficiency on retina and brain in rhesus monkeys. Proc. Natl Acad. Sci. USA 83, 4021–4025.
[56] Holman, R.T., Johnson, S.B., Kokmen, E., 1989. Deficiencies of polyunsaturated fatty acids and replacement by nonessential fatty acids in plasma lipids in multiple sclerosis. Proc. Natl Acad. Sci. USA 86, 4720–4724.
[57] Holman, R.T., Johnson, S.B., Ogburn, P.L., 1991. Deficiency of essential fatty acids and membrane fluidity during pregnancy and lactation. Proc. Natl Acad. Sci. USA 88, 4835–4839.
[58] Glatz, J.F., Soffers, A.E., Katan, M.B., 1989. Fatty acid composition of serum cholesteryl esters and erythrocyte membranes as indicators of linoleic acid intake in man. Am. J. Clin. Nutr. 49, 269–276.
[59] Asciutti-Moura, L.S., Guilland, J.C., Fuchs, F., Richard, D., Klepping, J., 1988. Fatty acid composition of serum lipids and its relation to diet in an elderly institutionalized population. Am. J. Clin. Nutr. 48, 980–987.
[60] Holman, R.T., 1986. Control of polyunsaturated fatty acids in tissue lipids. J. Am. Coll. Nutr. 5, 183–211.
[61] Lands, W.E., Libelt, B., Morris, A., Kramer, N.C., Prewitt, T.E., Bowen, P., Schmeisser, D., Davidson, M.H., Burns, J.H., 1992. Maintenance of lower proportions of (n-6) eicosanoid precursors in phospholipids of human plasma in response to added dietary (n-3) fatty acids. Biochim. Biophys. Acta 1180, 147–162.
[62] Bjerve, K.S., Brubakk, A.M., Fougner, K.J., Johnsen, H., Midthjell, K., Vik, T., 1993. Omega-3 fatty acids: essential fatty acids with important biological effects, and serum phospholipid fatty acids as markers of dietary omega 3-fatty acid intake. Am. J. Clin. Nutr. 57, 801S–805S.
[63] Brown, A.J., Pang, E., Roberts, D.C., 1991. Erythrocyte eicosapentaenoic acid versus docosahexaenoic acid as a marker for fish and fish oil consumption. Prostaglandins Leukot. Essent. Fatty Acids 44, 103–106.
[64] Katan, M.B., Deslypere, J.P., van Birgelen, A.P., Penders, M., Zegwaard, M., 1997. Kinetics of the incorporation of dietary fatty acids into serum cholesteryl esters, erythrocyte membranes, and adipose tissue: an 18-month controlled study. J. Lipid Res. 38, 2012–2022.
[65] Al, M.D.M., van Houwelingen, A.C., Kester, A.D.M., Hasaart, T.H.M., de Jong, A.E.P., Hornstra, G., 1995. Maternal essential fatty acid patterns during normal pregnancy and its relationship with the neonatal essential fatty acid status. Br. J. Nutr. 74, 55–68.

[66] Otto, S.J., van Houwelingen, A.C., Antal, M., Manninen, A., Godfrey, K., López-Jaramillo, P., Hornstra, G., 1997. Maternal and neonatal essential fatty acid status in phospholipids: an international comparative study. Eur. J. Clin. Nutr. 51, 232–342.
[67] De Vriese, S.R., van Houwelingen, A.C., Hornstra, G., Dhont, M., Christophe, A.B., 2001. The composition of saturated and monounsaturated fatty acids changes in a way to counteract changes in the mean melting point during pregnancy. Lipids 36, 15–20.
[68] Matorras, R., Ruiz, J.I., Perteagudo, L., Barbazan, M.J., Diaz, A., Valladolod, A., Sanjurjo, P., 2001. Longitudinal study of fatty acids in plasma and erythrocyte phospholipids during pregnancy. J. Perinat. Med. 29, 273–297.
[69] Ghebremeskel, K., Min, Y., Crawford, M.A., Nam, J.-H., Kim, A., Koo, J.-N., Suzuki, H., 2000. Blood fatty acid composition of pregnant and nonpregnant Korean women: red cells may act as a reservoir of arachidonic acid and docosahexaenoic acid for utilization by the developing fetus. Lipids 35, 567–574.
[70] Barros, M.F.A., Leger, C.L., Lira, P.I.C., Lima, M.C., Carbonneau, M.-A., Descomps, B., Alessio, M.L.M., 2001. Cord blood essential fatty acid and alpha-tocopherol in full-term newborns in a Northeast Brazil area. Int. J. Vitam. Res. 72, 155–160.
[71] van der Schouw, Y.T., Al, M.D.M., Hornstra, G., Bulstra-Ramakers, M.T.E.W., Huisjes, H.J., 1991. Fatty acid composition of serum lipids of mothers and their babies after normal and hypertensive pregnancies. Prostaglandins Leukot. Essent. Fatty Acids 44, 247–252.
[72] Berghaus, T.M., Demmelmair, H., Koletzko, B., 1998. Fatty acid composition of lipid classes in maternal and cord plasma at birth. Eur. J. Pediatr. 157, 763–768.
[73] De Vriese, S.R., Dhont, M., Christophe, A.B., 2003. Fatty acid composition of cholesteryl esters and phospholipids in maternal plasma during pregnancy and at delivery and in cord plasma at birth. Lipids 38, 1–7.
[74] Otto, S.J., van Houwelingen, A.C., Badart-Smook, A., Hornstra, G., 2001. Changes in the maternal essential fatty acid profile during early pregnancy and the relation of the profile to diet. Am. J. Clin. Nutr. 73, 302–307.
[75] Otto, S.J., van Houwelingen, A.C., Hornstra, G., 2000. The effect of different supplements containing docosahexaenoic acid on plasma and erythrocyte fatty acids of healthy non-pregnant women. Nutr. Res. 20, 917–927.
[76] Al, M.D., Badart-Smook, A., van Houwelingen, A.C., Hasaart, T.H., Hornstra, G., 1996. Fat intake of women during normal pregnancy: relationship with maternal and neonatal essential fatty acid status. J. Am. Coll. Nutr. 15, 49–55.
[77] De Vriese, S.R., Mattys, C., De Henauw, S., De Backer, G., Dhont, M., Christophe, A.B., 2002. Maternal and umbilical fatty acid status in relation to maternal diet. Prostaglandins Leukot. Essent. Fatty Acids 67, 389–396.
[78] Rump, P., Hornstra, G., 2002. The n-3 and n-6 polyunsaturated fatty acid compositions of plasma phospholipids in pregnant women and their infants. Relation with maternal linoleic acid intake. Clin. Chem. Lab. Med. 40, 32–39.
[79] Al, M.D.M., van Houwelingen, A.C., Hornstra, G., 2000. Long chain polyenes, pregnancy and pregnancy outcome. Am. J. Clin. Nutr. Suppl. 71, 285S–291S.
[80] De Vriese, S.R., Christophe, A.B., Maes, M., 2003. Fatty acid composition of phospholipids and cholestryl esters in maternal serum in the early puerperium. Prostaglandins Leukot. Essent. Fatty Acids 68, 331–335.
[81] Otto, S.J., van Houwelingen, A.C., Badart-Smook, A., Hornstra, G., 2001. Comparison of the peripartum and postpartum phospholipid polyunsaturated fatty acid profiles of lactating and nonlactating women. Am. J. Clin. Nutr. 73, 1074–1079.
[82] Al, M.D., van Houwelingen, A.C., Hornstra, G., 1997. Relation between birth order and the maternal and neonatal docosahexaenoic acid status. Eur. J. Clin. Nutr. 51, 548–553.
[83] van den Ham, E.C., van Houwelingen, A.C., Hornstra, G., 2001. Evaluation of the relation between n-3 and n-6 fatty acid status and parity in nonpregnant women from the Netherlands. Am. J. Clin. Nutr. 73, 622–627.
[84] Matorras, R., Perteagudo, L., Sanjurjo, P., Ruiz, J.I., 1999. Intake of long chain ω3 polyunsaturated fatty acids during pregnancy and the influence of levels in the mother on newborn levels. Eur. J. Obstet. Gynecol. Reprod. Biol. 83, 179–184.

[85] Min, Y., Ghebremeskel, K., Crawford, M.A., Nam, J.H., Kim, A., Le, I.S., Suzuki, H., 2001. Maternal–fetal n-6 and n-3 polyunsaturated fatty acids gradient in plasma and red cell phospholipids. Int. J. Vitam. Nutr. Res. 71, 286–292.
[86] Foreman-van Drongelen, M.M.P.H., Al, M.D.M., van Houwelingen, A.C., Blanco, C.E., Hornstra, G., 1995. Comparison between the essential fatty acid status of preterm and full-term infants, measured in umbilical vessels. Early Hum. Dev. 42, 241–251.
[87] van Houwelingen, A.C., Foreman-van Drongelen, M.M.H.P., Nicolini, U., Nicolaides, K.H., Al, M.D.M., Kester, A.D.M., Hornstra, G., 1996. Essential fatty acid status of fetal phospholipids: similar to postnatal values obtained at comparable gestational ages. Early Hum. Dev. 46, 141–152.
[88] van Houwelingen, A.C., Puls, J., Hornstra, G., 1992. Essential fatty acid status during early human development. Early Hum. Dev. 31, 97–111.
[89] Hornstra, G., van Houwelingen, A.C., Simonis, M., Gerrard, J.M., 1989. Fatty acid composition of umbilical arteries and veins: possible implications for the fetal EFA status. Lipids 24, 511–517.
[90] Otto, S.J., van Houwelingen, A.C., López-Jaramillo, P., Hornstra, G., 1999. Pregnancy induced hypertension has a different effect on the essential fatty acid status of Ecuadorian and Dutch women. Am. J. Obstet. Gynecol. 180, 1185–1190.
[91] Al, M.D.M., van Houwelingen, A.C., Badart-Smook, A., Hasaart, T.H., Roumen, F.J., Hornstra, G., 1995. The essential fatty acid status of mother and child in pregnancy-induced hypertension: a prospective longitudinal study. Am. J. Obstet. Gynecol. 172, 1605–1614.
[92] Al, M.D.M., van Houwelingen, A.C., Badart-Smook, A., Hornstra, G., 1995. Some aspects of neonatal essential fatty acid status are altered by linoleic acid supplementation of women during pregnancy. J. Nutr. 125, 2822–2830.
[93] De Vriese, S.R., 2003. Fatty acid composition of umbilical vessel walls. In: De Vriese, S. (Ed.), Essential Fatty Acids and Pregnancy, Thesis, Universiteit Gent.
[94] Makrides, M., Neumann, M.A., Byard, R.W., Simmer, K., Gibson, R.A., 1994. Fatty acid composition of brain, retina, and erythrocytes in breast- and formula-fed infants. Am. J. Clin. Nutr. 60, 189–194.
[95] Foreman-van Drongelen, M.M.H.P., Zeijdner, E.E., van Houwelingen, A.C., Kester, A.D.M., Al, M.D.M., Hasaart, T.H.M, Hornstra, G., 1996. Essential fatty acid status measured in umbilical vessel walls of infants born after a multiple pregnancy. Early Hum. Dev. 46, 205–215.
[96] Zeijdner, E.E., van Houwelingen, A.C., Kester, A.D.M., Hornstra, G., 1997. The essential fatty acid status in plasma phospholipids of mother and neonate after multiple pregnancy. Prostaglandins Leukot. Essent. Fatty Acids 56, 395–401.
[97] Rump, P., Mensink, R.P., Kester, A.D.M., Hornstra, G., 2001. Essential fatty acid composition of plasma phospholipids and weight at birth: a study in term neonates. Am. J. Clin. Nutr. 73, 797–806.
[98] Grønn, M., Gørbitz, C., Christensen, E., Levorsen, A., Ose, L., Hagve, T.-A., Christophersen, B.O., 1991. Dietary n-6 fatty acids inhibit the incorporation of dietary n-3 fatty acids in thrombocyte and serum phospholipids in humans: a controlled dietetic study. Scand. J. Clin. Lab. Invest. 51, 255–263.
[99] Cleland, L.G., James, M.J., Neumann, M.A., D'Angelo, M., Gibson, R.A., 1992. Linoleate inhibits EPA incorporation from dietary fish-oil supplements in human subjects. Am. J. Clin. Nutr. 55, 395–399.
[100] Olsen, S.F., Hansen, H.S., Jensen, B., Sandström, B., 1995. Erythrocyte levels compared with reported dietary intake of marine n-3 fatty acids in pregnant women. Br. J. Nutr. 73, 387–395.
[101] Matorras, R., Perteagudo, L., Sanjurjo, P., 1998. Biochemical markers of n-3 long chain polyunsaturated fatty acid intake during pregnancy. Clin. Exp. Obstet. Gynecol. 25, 135–138.
[102] Sugano, M., Ikeda, I., 1996. Metabolic interactions between essential and trans-fatty acids. Curr. Opin. Lipidol. 7, 38–42.
[103] Koletzko, B., 1992. Trans fatty acids may impair biosynthesis of long chain polyunsaturates and growth in man. Acta Paediatr. 81, 302–306.
[104] Koletzko, B., 1995. Potential adverse effects of trans fatty acids in infants and children. Eur. J. Med. Res. 1, 123–125.
[105] van Houwelingen, A.C., Hornstra, G., 1994. Trans fatty acids in early human development. In: Galli, C., Simopoulos, A.P., Tremoli, E., (Eds.), Fatty Acids and Lipids: Biological Aspects. World Review of Nutrition and Dietetics, Simopoulos, A.P. (Ed.), Vol. 75. Karger, Basel, pp. 175–178.

[106] Elias, S.L., Innis, S.M., 2001. Infant plasma trans, n-6, and n-3 fatty acids and conjugated linoleic acids are related to maternal plasma fatty acids, length of gestation, and birth weight and length. Am. J. Clin. Nutr. 73, 807–814.
[107] Carlson, S.E., Clandinin, M.T., Cook, H.W., Emken, E.A., Filer, L.J. Jr., 1997. *trans* Fatty acids: infant and fetal development. Am. J. Clin. Nutr. 66, 717S–736S.
[108] Popeski, D., Ebbeling, L.R., Brown, P.B., Hornstra, G., Gerrard, J.M., 1991. Blood pressure during pregnancy in Canadian Inuit: community differences related to diet. Can. Med. Assoc. J. 145, 445–454.
[109] Williams, M.A., Zingheim, R.W., King, I.B., Zebelman, A.M., 1995. Omega-3 fatty acids in maternal erythrocytes and risk of preeclampsia. Epidemiology 6, 232–237.
[110] Kesmodel, U., Olsen, S.F., Salvig, J.D., 1997. Marine n-3 fatty acid and calcium intake in relation to pregnancy induced hypertension, intrauterine growth retardation, and preterm delivery. Acta Obstet. Gynecol. Scand. 76, 38–44.
[111] Salvig, J.D., Olsen, S.F., Secher, N.J., 1996. Effects of fish oil supplementation in late pregnancy on blood pressure: randomised controlled trial. Br. J. Obstet. Gynecol. 103, 529–533.
[112] Olsen, S.F., Secher, N.J., Tabor, A., Weber, T., Walker, J.J., Gluud, C., 2000. Randomised clinical trials of fish oil supplementation in high risk pregnancies. Fish oil trials in pregnancy (FOTIP Team). Br. J. Obstet. Gynecol. 107, 382–395.
[113] Onwude, J.L., Lilford, R.J., Hjartardottir, H., Staines, A., Tuffnell, D., 1995. A randomized double blind placebo controlled trial of fish oil in high risk pregnancy. Br. J. Obstet. Gynaecol. 102, 95–100.
[114] Hibbeln, J.R., 2002. Seafood consumption, the DHA content of mothers' milk and prevalence rates of postpartum depression: a cross-national, ecological analysis. J. Affect. Disord. 69, 15–29.
[115] Otto, S.J., de Groot, R.H.M., Hornstra, G., 1995. Increased risk of postpartum depressive symptoms is associated with slower normalization after pregnancy of the functional docosahexaenoic acid status. Prostaglandins Leukot. Essent. Fatty Acids 69, 237–243.
[116] Olsen, S.F., Secher, N.J., 2002. Low consumption of seafood in early pregnancy as a risk factor for preterm delivery: prospective cohort study. Br. Med. J. 324, 1–5.
[117] Olsen, S.F., Sörensen, J.D., Secher, N.J., Hedegaard, M., Henriksen, T.B., Hansen, H.S., Grant, A., 1992. Randomised controlled trial of effect of fish-oil supplementation on pregnancy duration. Lancet 339, 1003–1007.
[118] Badart-Smook, A., van Houwelingen, A.C., Al, M.D.M., Kester, A.D.M., Hornstra, G., 1997. Fetal growth is associated positively with maternal intake of riboflavin and negatively with maternal intake of linoleic acid. J. Am. Diet. Assoc. 97, 867–870.
[119] Olsen, S.F., Olsen, J., Frische, G., 1990. Does fish consumption during pregnancy increase fetal growth? A study of the size of the newborn, placental weight and gestational age in relation to fish consumption during pregnancy. Int. J. Epidemiol. 19, 971–977.
[120] Olsen, S.F., Hansen, H.S., Secher, N.J., Jensen, B., Sandström, B., 1995. Gestation length and birth weight in relation to intake of marine n-3 fatty acids. Br. J. Nutr. 73, 397–404.
[121] Grandjean, P., Bjerve, K.S., Weihe, P., Steuerwald, U., 2001. Birthweight in a fishing community: significance of essential fatty acids and marine food contaminants. Int. J. Epidemiol. 30, 1272–1278.
[122] Cooke, R.W.I., Lucas, A., Yudkin, P.L.N., Pryse-Davies, J., 1977. Head circumference as an index of brain weight in the fetus and newborn. Early Hum. Dev. 1/2, 145–149.
[123] Koletzko, B., Braun, M., 1991. Arachidonic acid and early human growth: is there a relation? Ann. Nutr. Metab. 35, 128–131.
[124] Foreman-van Drongelen, M.M., van Houwelingen, A.C., Kester, A.D., Hasaart, T.H., Blanco, C.E., Hornstra, G., 1995. Long-chain polyunsaturated fatty acids in preterm infants: status at birth and its influence on postnatal levels. J. Pediatr. 126, 611–618.
[125] Lauritzen, L., Hansen, H.S., Jørgensen, M.H., Michaelsen, K., 2001. The essentiality of long-chain n-3 fatty acids in relation to development and function of the brain and retina. Progr. Lipid Res. 40, 1–94.
[126] Andersen, J.W., Johnstone, B.M., Remley, D.T., 1999. Breast-feeding and cognitive development: a meta-analysis. Am. J. Clin. Nutr. 70, 525–535.
[127] Jain, A., Concato, J., Leventhal, J.M., 2002. How good is the evidence linking breastfeeding and intelligence? Pediatrics 109, 1044–1053.

[128] Simmer, K., 2000. Longchain polyunsaturated fatty acid supplementation in preterm infants. Cochrane Database Syst. Rev. 2, CD000375.

[129] SanGiovanni, J.P., Parra-Cabrera, S., Colditz, G.A., Berkey, C.S., Dwyer, J.T., 2000. Meta-analysis of dietary essential fatty acids and long-chain polyunsaturated fatty acids as they relate to visual resolution acuity in healthy preterm infants. Pediatrics 105, 1292–1298.

[130] O'Connor, D.L., Hall, R., Adamkin, D., Auestad, N., Castillo, M., Connor, W.E., Connor, S.L., Fitzgerald, K., Groh-Wargo, S., Hartmann, E.E., Jacobs, J., Janowsky, J., Lucas, A., Margeson, D., Mena, P., Neuringer, M., Nesin, M., Singer, L., Stephenson, T., Szabo, J., Zemon, V., 2001. Ross preterm lipid study. Growth and development in preterm infants fed long-chain polyunsaturated fatty acids: a prospective, randomized controlled trial. Pediatrics 108, 359–371.

[131] Simmer, K., 2001. Longchain polyunsaturated fatty acid supplementation in infants born at term (Cochrane review). Cochrane Database Syst. Rev. 4, CD000376.

[132] SanGiovanni, J.P., Berkey, C.S., Dwyer, J.T., Colditz, G.A., 2000. Dietary essential fatty acids, long-chain polyunsaturated fatty acids, and visual resolution acuity in healthy fullterm infants: a systematic review. Early Hum. Dev. 57, 165–188.

[133] Agostoni, C., Trojan, S., Bellù, R., Riva, E., Giovannini, M., 1995. Neurodevelopmental quotient of healthy term infants at 4 months and feeding practice: the role of long-chain polyunsaturated fatty acids. Pediatr. Res. 38, 262–266.

[134] Willatts, P., Forsyth, J.S., DiModugno, M.K., Varma, S., Colvin, M., 1998. Effect of long chain polyunsaturated fatty acids in infant formula on problem solving at 10 months of age. Lancet 352, 688–691.

[135] Birch, E.E., Garfield, S., Hoffman, D.R., Uauy, R., Birch, D.G., 2000. A randomized controlled trial of early dietary supply of long-chain polyunsaturated fatty acids and mental development in term infants. Dev. Med. Child Neurol. 42, 174–181.

[136] Raiten, D.J., Talbot, J.M., Waters, J.H., 1998. Assessment of nutrient requirements for infant formulas. J. Nutr. 128, 2059S–2293S.

[137] Koletzko, B., Agostoni, C., Carlson, S.E., Clandinin, T., Hornstra, G., Neuringer, M., Uauy, R., Yamashiro, Y., Willatts, P., 2001. Long chain polyunsaturated fatty acids (LC-PUFA) and perinatal development. Acta Paediatr. 90, 460–464.

[138] Bakker, E.C., 2002. Long-chain polyunsaturated fatty acids and child development. PhD Thesis, Maastricht University, The Netherlands, 128p.

[139] Bakker, E.C., Ghys, A.J.A., Kester, A.D.M., Vles, J.S.H., Dubas, J.S., Blanco, C.E., Hornstra, G., 2003. Long-chain polyunsaturated fatty acids at birth and cognitive function at 7 years of age. Eur. J. Clin. Nutr. 57, 89–95.

[140] Ghys, A., Bakker, E., Hornstra, G., van den Hout, M., 2003. Red blood cell and plasma phospholipid arachidonic and docosahexaenoic acid levels at birth and cognitive development at 4 years of age. Early Hum. Dev. 69, 83–90.

[141] van Houwelingen, A.C., Sørensen, J.D., Hornstra, G., Simonis, M.M.G., Boris, J., Olsen, S.F., Secher, N.J., 1995. Essential fatty acid status in neonates after fish-oil supplementation during late pregnancy. Br. J. Nutr. 74, 723–731.

[142] Helland, I.B., Saugstad, O.D., Smith, L., Saarem, K., Solvoll, K., Ganes, T., Drevon, C.A., 2001. Similar effects on infants of n-3 and n-6 fatty acids supplementation to pregnant and lactating women. Pediatrics 108, e82.

[143] Velzing-Aarts, F.V., van der Klis, F.R., van der Dijs, F.P., van Beusekom, C.M., Landman, H., Capello, J.J., Muskiet, F.A., 2001. Effect of three low-dose fish oil supplements, administered during pregnancy, on neonatal long-chain polyunsaturated fatty acid status at birth. Prostaglandins Leukot. Essent. Fatty Acids 65, 51–57.

[144] Otto, S.J., van Houwelingen, A.C., Hornstra, G., 2000. The effect of supplementation with docosahexaenoic and arachidonic acid derived from single cell oils on plasma and erythrocyte fatty acids of pregnant women in the second trimester. Prostaglandins Leukot. Essent. Fatty Acids 63, 323–328.

[145] Smit, E.N., Koopmann, M., Boersma, E.R., Muskiet, F.A.J., 2000. Effect of supplementation of arachidonic acid (AA) or a combination of AA plus docosahexaenoic acid on breastmilk fatty acid composition. Prostaglandins Leukot. Essent. Fatty Acids 62, 335–340.

[146] Potter, J.M., Nestel, P.J., 1976. The effect of dietary fatty acids and cholesterol on the milk lipids of lactating women and the plasma cholesterol of breast-fed infants. Am. J. Clin. Nutr. 29, 54–60.

[147] Cant, A., Shay, J., Horrobin, D.F., 1991. The effect of supplementation with linoleic and γ-linolenic acids on the fat composition and content of human milk: a placebo-controlled trial. J. Nutr. Sci. Vitaminol. 37, 573–579.

[148] Jensen, R.G., Lammi-Keefe, C.J., Henderson, R.A., Bush, V., Ferris, J.A.M., 1992. Effect of dietary intake of n-6 and n-3 fatty acids on the fatty acid composition of human milk in North America. J. Pediatr. 120, S87–S93.

[149] Makrides, M., Neumann, M.A., Gibson, R.A., 1996. Effect of maternal docosahexaenoic acid (DHA) supplementation on breast milk composition. Eur. J. Clin. Nutr. 50, 352–357.

[150] Francois, C.A., Connor, S.L., Wander, R.C., Connor, W.E., 1998. Acute effects of dietary fatty acids on the fatty acids of human milk. Am. J. Clin. Nutr. 67, 301–308.

[151] Jensen, C.L., Maude, M., Andersen, R.E., Heird, W.C., 2000. Effect of docosahexaenoic acid supplementation of lactating women on the fatty acid composition of breast milk lipids and maternal and infant plasma phospholipids. Am. J. Clin. Nutr. 71, 292S–299S.

[152] Thijs, C., Houwelingen, A., Poorterman, I., Mordant, A., van den Brandt, P., 2000. Essential fatty acids in breast milk of atopic mothers: comparison with non-atopic mothers and effect of borage oil supplementation. Eur. J. Clin. Nutr. 54, 234–238.

[153] Francois, C.A., Connor, S.L., Bolewicz, L.C., Connor, W.E., 2003. Supplementing lactating women with flaxseed oil does not increase docosahexaenoic acid in their milk. Am. J. Clin. Nutr. 77, 226–233.

[154] Gibson, R.A., Neumann, M.A., Makrides, M., 1997. Effect of increasing breast milk docosahexaenoic acid on plasma and erythrocyte phospholipid fatty acids and neural indices of exclusively breast fed infants. Eur. J. Clin. Nutr. 51, 578–584.

[155] Tolley, E.A., Carlson, S.E., 2000. Considerations of statistical power in infant studies of visual acuity development and docosahexaenoic acid status. Am. J. Clin. Nutr. 71, 1–2.

[156] Helland, I.B., Smith, L., Saarem, K., Saugstad, O.D., Drevon, C.A., 2003. Maternal supplementation with very-long-chain n-3 fatty acids during pregnancy and lactation augments children's IQ at 4 years of age. Pediatrics 111, e39–e44.

[157] Stene, L.C., Ulriksen, J., Magnus, P., Joner, G., 2000. Use of cod liver oil during pregnancy associated with lower risk of type I diabetes in the offspring. Diabetologia 43, 1093–1098.

[158] Foreman-van Drongelen, M.M.H.P., van Houwelingen, A.C., Kester, A.D.M., de Jong, A.E.P., Blanco, C.E., Hasaart, T.H.M., Hornstra, G., 1995. Long-chain polyene status of preterm infants with regard to the fatty acid composition of their diet: comparison between absolute and relative fatty acid levels in plasma and erythrocyte phospholipids. Br. J. Nutr. 73, 405–422.

[159] Foreman-van Drongelen, M.M.H.P., van Houwelingen, A.C., Kester, A.D.M., Blanco, C.E., Hasaart, T.H.M., Hornstra, G., 1996. Influence of feeding artificial formulas containing docosahexaenoic and arachidonic acids on the postnatal, long-chain polyunsaturated fatty acid status of healthy preterm infants. Br. J. Nutr. 76, 649–667.

[160] Belmont, L., Marolla, F.A., 1973. Birth order, family size and intelligence. Science 182, 1096–1101.

[161] Gale, C.R., Martyn, C.N., 1996. Breast feeding, dummy use and intelligence. Lancet 347, 1072–1075.

[162] Lawson, L.D., Hughes, B.G., 1988. Human absorption of fish oil fatty acids as triacylglycerols, free fatty acids, or ethyl esters. Biochem. Biophys. Res. Commun. 152, 328–335.

[163] Nordøy, A., Barstad, L., Connor, W.E., Hatcher, L., 1991. Absorption of the n-3 eicosapentaenoic and docosahexaenoic acids as ethyl esters and triglycerides by humans. Am. J. Clin. Nutr. 53, 1185–1190.

[164] Krokan, H.E., Bjerve, K.S., Mork, E., 1993. The enteral bioavailability of eicosapentaenoic acid and docosahexaenoic acid is as good from ethyl esters as from glyceryl esters in spite of lower hydrolytic rates by pancreatic lipase in vitro. Biochim. Biophys. Acta 1168, 59–67.

[165] Hansen, J.B., Olsen, J.O., Wilsgard, L., Lyngmo, V., Svensson, B., 1993. Comparative effects of prolonged intake of highly purified fish oils as ethyl ester or triglyceride on lipids, haemostasis and platelet function in normilipaemic men. Eur. J. Clin. Nutr. 47, 497–507.

[166] Boehm, G., Müller, H., Kohn, G., Moro, G., Minoli, I., Böhles, H.J., 1997. Docosahexaenoic and arachidonic acid absorption in preterm infants fed LCP-free or LCP-supplemented formula in comparison to infants fed fortified breast milk. Ann. Nutr. Metab. 41, 235–241.

[167] Carnielli, V.P., Verlato, G., Pederzini, F., Luijendijk, I., Pendrotti, A., Boerlage, D., Sauer, P.J.J., 1998. Intestinal absorption of long-chain polyunsaturated fatty acids in preterm infants fed breast milk or formulas. Am. J. Clin. Nutr. 67, 97–103.
[168] Song, J.H., Inoue, Y., Miyazawa, T., 1997. Oxidative stability of docosahexaenoic acid-containing oils in the form of phospholipids, triacylglycerols, and ethyl esters. Biosci. Biotechnol. Biochem. 61, 2085–2088.
[169] Wijendran, V., Huang, M.-C., Diau, G.-Y., Boehm, G., Nathanielz, P.W., Brenna, T., 2002. Efficacy of dietary arachidonic acid provided as triglyceride or phospholipid as substrate for brain arachidonic acid accretion in baboon neonates. Pediatr. Res. 51, 265–272.

Phospholipid transfer protein and atherosclerosis

Rini de Crom[a,b,*] and Arie van Tol[c]

[a]*Department of Cell Biology and Genetics, Erasmus University Medical Center, P.O. Box 1738, 3000 DR Rotterdam, The Netherlands*
[b]*Department of Vascular Surgery, Erasmus University Medical Center, P.O. Box 1738, 3000 DR Rotterdam, The Netherlands*
[c]*Department of Biochemistry, Erasmus University Medical Center, P.O. Box 1738, 3000 DR Rotterdam, The Netherlands*
[*]*Correspondence address: Tel.: +31-10-4087459; fax: +31-4089468*
E-mail: m.decrom@erasmusmc.nl(R. de Crom)

1. Introduction

About two decades ago, two plasma proteins with lipid transfer activity were documented, which were designated as lipid transfer protein (LTP)-I and LTP-II [1]. LTP-I is nowadays known as cholesteryl ester transfer protein (CETP). LTP-II was identified by three different groups as a protein with the ability to transfer phospholipids between lipoproteins and/or vesicles [2–4]. Later, this protein was referred to as phospholipid transfer protein (PLTP) [5]. Following purification of the protein, a full-length cDNA encoding human PLTP was cloned and sequenced in 1994 [6]. It has an open reading frame of 1518 bases and encodes a mature protein of 476 amino acids. The gene has been mapped to human chromosome 20 [7] or mouse chromosome 2 [8]. It has 16 exons and spans approximately 13.3 kb [9]. It belongs to a gene family together with the genes for lipopolysaccharide binding protein (LBP), bactericidal permeability increasing protein (BPI) and CETP [10]. While the PLTP, LBP and BPI genes all map to human chromosome 20, CETP is on chromosome 16 [6].

In vitro studies have revealed two functions of PLTP: phospholipid transfer activity and conversion of high-density lipoproteins (HDL) [11]. Phospholipid transfer represents the activity by which the protein was initially identified, as mentioned above. PLTP can transfer all the common types of phosphatidylcholine as well as free cholesterol [12–14]. The assay of PLTP activity in plasma samples is usually performed by measuring the transfer of radiolabeled phospholipids between liposomes and HDL [15]. Mutation analysis revealed that the N-terminal

Advances in Molecular and Cell Biology, Vol. 33, pages 531–541

ISSN: 1569-2558/DOI: 10.1016/S1569-2558(03)33025-5

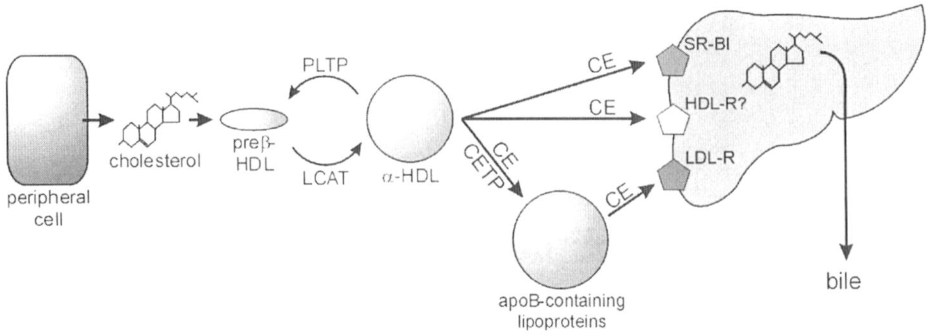

Fig. 1. Reverse cholesterol transport pathway. Excess of cholesterol from extrahepatic cells (macrophages or other peripheral cells) is taken up by an acceptor like preβ-HDL. Cholesterol is esterified by the plasma enzyme LCAT, causing the preβ-HDL particle to mature into spherical α-HDL. Cholesterylesters from α-HDL can be taken up selectively by hepatocytes via the scavenger receptor B type I (SR-BI). PLTP is involved in the regeneration of preβ-HDL, thus creating a PLTP–LCAT cycle. HDL can also be taken up directly by putative HDL receptors and HDL cholesterylesters can be transferred by CETP to apoB-containing lipoproteins that can be taken up by the liver via the LDLR. (*For a colored version of this figure, see plate section, page 618.*)

lipid binding pocket predicted by computer modeling of the protein is essential for the phospholipid transfer activity [16]. PLTP is also an HDL conversion factor [17,18]. In this process, HDL_3 is converted into small, lipid-poor apolipoprotein (apo)AI or preβ-HDL particles, while at the same time HDL_2 particles with increased size are formed [11]. The former type of particles are considered immature HDL particles, which can avidly take up cellular cholesterol [19,20]. Analysis of mutated recombinant PLTP demonstrated that phospholipid transfer activity is a prerequisite for HDL conversion by PLTP, including the formation of preβ-HDL particles [21].

Based on the involvement of PLTP in preβ-HDL formation, it has been proposed that PLTP has a physiological function in the reverse cholesterol transport pathway (Fig. 1), by which cholesterol is transported from extrahepatic tissues to the liver for excretion [20,22,23]. Preβ-HDL particles are thought to function as the primary acceptors of cholesterol entering the reverse cholesterol transport pathway following cellular efflux. The cholesterol molecules are consequently esterified by the plasma enzyme, lecithin cholesterol acyl transferase (LCAT), through which preβ-HDL matures into an α-HDL particle with a core consisting mainly of cholesterylesters. In order to study the function of PLTP in vivo, various models of genetically modified mice have been investigated.

2. Mouse models

2.1. Adenoviral transfer

Two research groups have investigated mice with a transient overexpression of human PLTP achieved by the use of adenoviral vectors.

Foger et al. [24] used two doses of recombinant PLTP adenovirus. Four days after injection, PLTP activity was increased 13- and 40-fold, respectively. Concomitant with this increase in PLTP activity, a decrease in plasma levels of HDL was found (−54% and −91%, respectively). This was attributed to an increased plasma clearance, probably by increased hepatic uptake of HDL. Interestingly, the formation of preβ-HDL was increased, as became evident after intravenous injection of radiolabeled HDL.

Ehnholm et al. [25] performed similar experiments with wild type, C57Bl6 mice or mice transgenic for human apoAI. They achieved a much more moderate level of overexpression in PLTP activity (4-fold in wild type mice, 2.5-fold in mice transgenic for human apoAI). Plasma cholesterol levels were decreased 70 and 45%, respectively. Density gradient analysis of lipoproteins showed that the decrease in plasma cholesterol occurs in HDL. In apoAI transgenic mice, HDL shifts to a lower density, representing larger HDL particles.

From these studies it could be concluded that increased PLTP activity resulted in HDL remodeling with accelerated HDL catabolism and increased formation of preβ-HDL. Besides, the presence of human apoAI in plasma affects HDL remodeling, which is induced by overexpression of human PLTP.

2.2. *First lines of transgenic mice*

In 1996, two independent laboratories generated transgenic mice with human PLTP.

Mice described by Albers et al. [26] expressed mRNA from the transgene in several organs. Minor differences were found in plasma lipoproteins. The significance of these differences is unclear however, because the transgenic mice showed no difference in plasma PLTP activity compared to wild type controls.

Jiang et al. [27] produced PLTP transgenic mice in which the plasma PLTP activity was increased to 29%. This moderate overexpression caused no effects on plasma lipids or lipoproteins. When the mice were cross-bred with human apoAI transgenic mice, a further increase in PLTP activity levels was found (47%). Double transgenic mice also had moderately increased HDL (25%) compared with mice transgenic for apoAI only.

Levels of mouse apoAI were unchanged in mice transgenic for human PLTP only when compared with wild type mice. When HDL subclass distribution was analyzed, it was found that apoAI either in preβ-HDL or in mature α-HDL particles was unchanged. Similar analyses were performed in double transgenic mice, expressing both human apoAI and human PLTP. In these animals, a marked increase in preβ-HDL was found, while α-HDL was only slightly increased (14%) when compared with mice transgenic for apoAI only.

From these studies, it was tentatively concluded that PLTP plays a role in reverse cholesterol transport by stimulating an increase in preβ-HDL levels, which might represent an anti-atherogenic property.

2.3. PLTP knockout mice

Jiang et al. continued their studies on the physiological function of PLTP by the generation of mice in which the PLTP gene was inactivated by gene targeting [14]. When compared with wild type littermates, PLTP knockout (KO) mice had a reduction in HDL-cholesterol (−65%) and apoAI (−85%), while VLDL + LDL cholesterol and apoB levels were unchanged. Preβ-HDL and α-HDL were decreased to the same extent. When the animals were fed a high cholesterol diet, resulting in an increase in plasma cholesterol of 50%, similar decreases were observed in HDL levels. However, VLDL + LDL cholesterol was increased in these animals (60%), while apoB levels were unchanged. Phospholipid transfer activity was measured in plasma samples, demonstrating a near complete absence of activity in PLTP KO mice. In vivo transfer was analyzed by intravenous injection of VLDL with radiolabeled phospholipids, demonstrating a rapid phospholipid transfer to HDL in wild type, but not in PLTP KO mice. After feeding the mice a high cholesterol diet, isolated plasma fractions of VLDL + LDL and HDL were inspected by negative stain electron microscopy. In PLTP KO mice, but not in wild type mice, vesicular stacks were seen next to normal shaped spherical particles. From these results, it was concluded that PLTP has an important physiological function in the transfer of phospholipids from VLDL to HDL. PLTP deficiency results in the accumulation of phospholipids in VLDL, particularly evident after an extreme diet, and in an impaired formation of HDL precursors and maturation of HDL, giving rise to enhanced HDL catabolism and consequently to lower plasma levels of HDL.

A few years later, the same investigators addressed the impact of PLTP deficiency on the development of atherosclerosis [28]. As normal mice, even the C57Bl6 strain, are very resistant to the development of diet-induced atherosclerosis, PLTP KO mice were cross-bred to three different hyperlipidemic strains: apoB transgenic mice, apoE KO mice and LDL-receptor (LDLR) KO mice. In all three backgrounds, a decline in plasma HDL was observed when the animals were deficient for PLTP, confirming the previous observations. PLTP deficiency resulted in a reduction of plasma apoB levels in apoB transgenic mice and in apoE KO mice, but not in LDLR KO mice. In cultured hepatocytes isolated from the various mouse lines, it was found that secretion of apoB was decreased by PLTP deficiency in cells from apoB transgenic mice or apoE KO mice, but not in cells from LDLR KO mice. This effect could be compensated for by reintroducing the PLTP gene in the hepatocytes by adenoviral vectors.

In each of the three backgrounds tested, (diet induced) atherosclerosis is decreased by PLTP deficiency, in spite of the reduced levels of plasma HDL. In the cases of apoB transgenic mice and apoE KO mice, this was explained by the observed novel effect of PLTP on apoB secretion, while in the LDLR KO mice the reduced atherosclerosis remained unexplained. Later it was shown that PLTP deficiency leads to elevated vitamin E levels in plasma LDL in each of these mouse models [29], as it was known already that PLTP is able to transfer vitamin E from lipoproteins to cells [30,31]. This could result in decreased levels of oxidized LDL and hence in less atherosclerosis. However, there is no direct proof for this mechanism. Supplementation of plasma with PLTP normalizes the vitamin E content in LDL from PLTP deficient mice, but has no additional effect on LDL from mice with normal PLTP levels. These results indicate that total PLTP deficiency specifically affects vitamin E bioavailability and therefore these findings cannot be extrapolated to situations of normal or elevated PLTP levels. Besides, in a previous study it was proposed that PLTP has anti-atherogenic potential because PLTP (at normal levels) transfers vitamin E to oxidized LDL and endothelial cells [31].

2.4. PLTP transgenic mice revisited

As the previously generated transgenic mice for PLTP showed virtually no or very modest elevations in plasma phospholipid transfer activity (see above), we set out to generate new PLTP transgenic mice. Results with the first line of PLTP transgenic mice were published in 2000 [32]. We used a cosmid containing the complete human PLTP gene, including its native promoter in approximately 15 kb of 5′ sequence, and 3.5 kb of 3′ sequence. Transgenic mice with this construct showed a 3–4-fold increase in PLTP activity in plasma, with a concomitant decrease in plasma HDL (− 30%). The formation of preβ-HDL in plasma from these animals was increased about 2–3-fold. Plasma from PLTP transgenic mice appeared to be more efficient in preventing the accumulation of cholesterol by macrophages in vitro than plasma from wild type littermates (Fig. 2). So, the increase in preβ-HDL outweighed the decrease in total HDL-cholesterol in this in vitro assay of reverse cholesterol transport. Therefore, we concluded that overexpression of PLTP might increase the anti-atherogenic potential of HDL.

The increase in the capacity for preβ-HDL formation was in line with findings in previous models of PLTP overexpression, either by adenoviral vectors or by transgenesis [24,25,27]. In addition, CETP is capable of the formation of preβ-HDL [33,34]. In double transgenic mice expressing both PLTP and CETP, we demonstrated that PLTP rather than CETP is responsible for the formation of preβ-HDL [35].

The decrease in plasma HDL was also found in mice with adenoviral-mediated overexpression of PLTP [24,25]. In an earlier model of PLTP transgenic mice [27], HDL

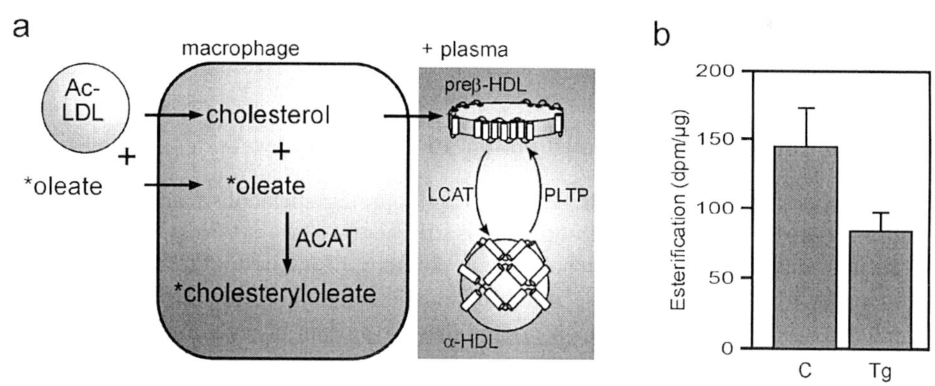

Fig. 2. Cholesterol accumulation in cultured macrophages. a. Peritoneal macrophages were incubated for 18 h in the presence of acetylated LDL (acLDL), radiolabeled oleate and mouse plasma. b. Intracellular cholesterol esterification, a measure for the cellular cholesterol content, was analyzed as described [32]. Adapted from Van Haperen et al., 2000. Arterioscler. Thromb. Vasc. Biol. 20, 1082–1088. (*For a colored version of this figure, see plate section, page 619.*)

levels were not affected, probably because of the very moderate level of PLTP overexpression (29%), or HDL was slightly increased upon cross-breeding with apoAI transgenic mice. When we cross-bred PLTP transgenic mice with apoAI transgenic mice, we found no difference in plasma levels of total cholesterol (unpublished results). Probably the increase in HDL caused by apoAI overexpression [36] balances the decrease in HDL caused by elevation of PLTP (see below). In the case of a combination of apoAI and a moderate PLTP overexpression, the balance shifts towards a slight increase in HDL [27].

In PLTP transgenic mice with a 15-fold increase in phospholipid transfer activity, we investigated the endpoint of the reverse cholesterol transport pathway, i.e. removal of cholesterol from the body [37]. In these mice, a strong reduction of plasma HDL was found concomitant with an almost 2-fold increase in cholesterol ester content of the liver and a 1.4-fold increase in bile acids in the bile compared with non-transgenic control mice. These results suggest a more efficient reverse cholesterol transport.

The increase in bile acid excretion and the stimulatory effect on preβ-HDL both suggest anti-atherogenic potential of elevated plasma PLTP activity levels. However, PLTP deficiency was shown to result in reduced atherosclerosis in different mouse models tested [28], as outlined above. Indeed, when we tested various lines of PLTP transgenic mice for susceptibility to diet-induced atherosclerosis, we found that overexpression of PLTP results in increased atherosclerosis [38]. The mouse lines tested showed a range of PLTP activities (Fig. 3a), with a PLTP dose-dependent decrease in plasma levels of HDL (Fig. 3b) and a PLTP dose-dependent increase in atherosclerosis (Figure 3c). Thus, we concluded that PLTP is an atherogenic protein because it lowers plasma levels of HDL.

In PLTP deficient mice, it was discovered that PLTP also plays a role in the secretion of apoB-containing lipoproteins by hepatocytes [28]. Therefore, we tested whether PLTP transgenic mice would have increased production of VLDL. This was indeed the case [38, 39], but there was no clear correlation with the level of PLTP activity [38]. Therefore, we conclude that the lowering effect on HDL is more important for the atherogenicity of PLTP than the stimulatory effect on VLDL secretion.

3. Human conditions with elevated plasma PLTP activity levels

Knowledge about the role of PLTP in human lipoprotein metabolism and development of atherosclerosis and coronary heart disease is still relatively sparse. Progress is hampered by the apparent absence of functional genetic polymorphisms in the PLTP gene [40]. In addition, PLTP deficiencies have not been documented in man until now. Another complication is that PLTP in human plasma may be present in both active and inactive forms, which probably causes the commonly observed lack of correlation between PLTP activity and mass [15].

Plasma PLTP activity is elevated in a variety of disease states with increased risk for atherosclerosis and coronary heart disease, e.g. diabetes mellitus type 1 and type 2 [41–44], alcohol abuse [45,46], obesity [47–49] and hypertriglyceridemia with insulin resistance [41,42,50]. With the exception of well-controlled type 1 diabetes patients, all

a

b

c

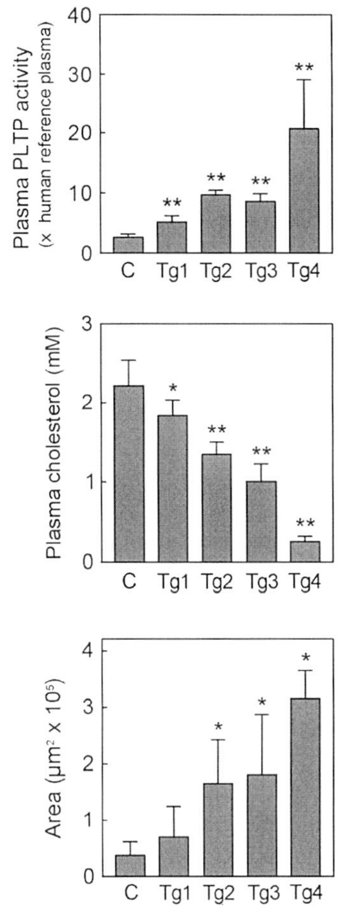

Fig. 3. PLTP activity, plasma cholesterol and atherosclerosis in PLTP transgenic mice. In non-transgenic, control mice (C) and in four lines of PLTP transgenic mice (Tg1–Tg4), the following parameters were measured: a. PLTP activity in plasma; b. total cholesterol levels in plasma; c. atherosclerotic lesion area in the aortic root. Adapted from Van Haperen et al., 2002. J. Biol. Chem. 277, 48938–48943.

these conditions coincide with elevated plasma triglycerides, in line with a role for PLTP in hepatic VLDL synthesis. When hepatic VLDL synthesis is decreased by interventions that lower plasma-free fatty acids by inhibiting lipolysis, a decrease in plasma PLTP activity levels is observed. These interventions include hyperglycemia-induced hyperinsulinemia [51], insulin infusion and treatment with the lipolysis inhibitor acipimox [52].

Drugs specifically affecting PLTP activity are not available at present. The chemical inhibitor (compound JTT-705) of CETP, another potentially atherogenic LTP, is available and dramatically increases the HDL-cholesterol in a dose-dependent fashion, with concomitant LDL lowering. JTT-705 did not affect PLTP activity in this human intervention study [53].

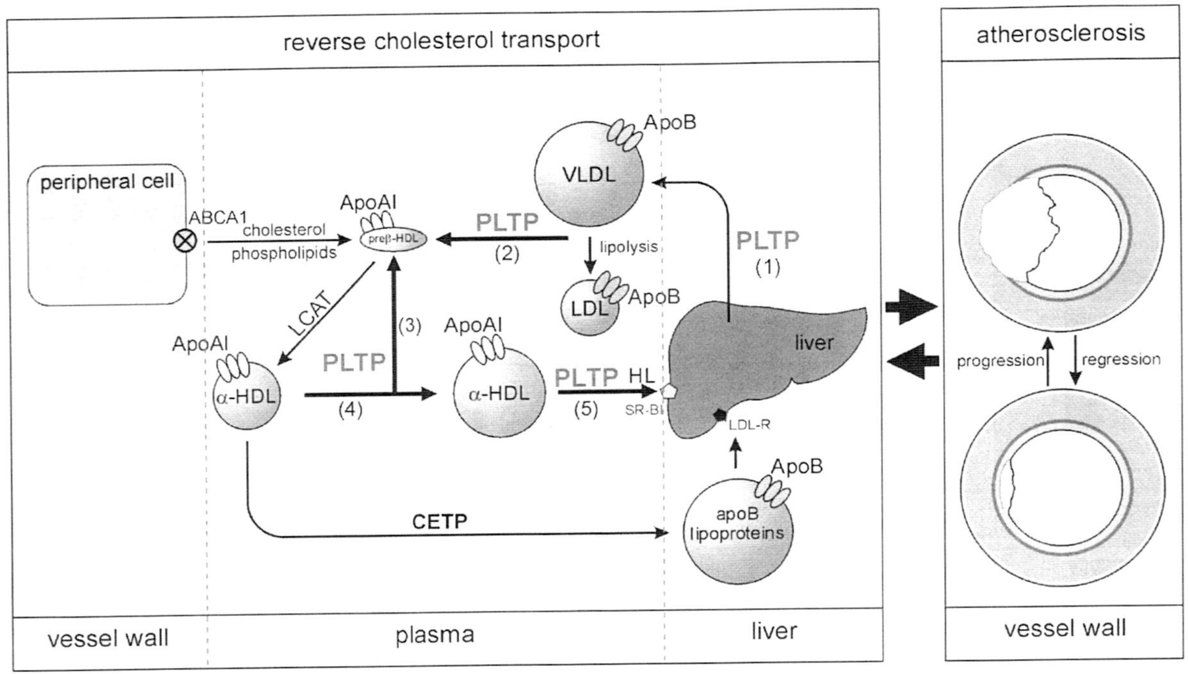

Fig. 4. Physiological functions of PLTP. Left panel: PLTP in reverse cholesterol transport: Cholesterol efflux from peripheral cells (including macrophages in an atherosclerotic lesion) is mediated by ATP-binding cassette A1 (ABCA1). Cholesterol is accepted by preβ-HDL, which rapidly matures to α-HDL by the action of plasma LCAT. Subsequently, HDL cholesterylesters can be taken up by the liver via SR-BI. Hepatic lipase (HL) activity may stimulate this uptake. Alternatively, HDL cholesterylesters can be transferred to apoB-containing lipoproteins via CETP. These lipoproteins are then taken up by the liver via the LDLR. PLTP has various functions in lipoprotein metabolism as it is involved in: 1) VLDL secretion; 2) phospholipid transfer from VLDL to HDL during lipolysis; 3) formation of preβ-HDL; 4) interconversion of α-HDL particles; 5) HDL catabolism (Modified from Van Tol and de Crom, 2002. Hart Bulletin, 33, 127–129). Right panel: Progression or regression of atherosclerosis is dependent on the activity of the reverse cholesterol transport pathway. (*For a colored version of this figure, see plate section, page 620.*)

4. Conclusion

From studies in genetically modified mice, the physiological functions of PLTP appear to include: 1) VLDL secretion; 2) phospholipid transfer from VLDL to HDL; 3) formation of preβ-HDL; 4) inter-conversion of α-HDL particles; 5) HDL catabolism (Fig. 4). The paradoxal situation that HDL levels are decreased in both PLTP KO mice and PLTP transgenic mice can be explained by different causes: In PLTP KO mice, HDL maturation is impaired, leading to accelerated breakdown of immature HDL particles; in PLTP transgenic mice, the catabolism of normal HDL is increased.

PLTP is atherogenic in mouse models of atherosclerosis, but the mechanism is complex. In PLTP KO mice, the decrease in VLDL secretion is at least part of the explanation for the lower levels of atherosclerosis that have been found. In mice with elevated PLTP activity levels, increased atherosclerosis is probably caused by the concomitant decline in HDL. Lower levels of HDL are thought to be atherogenic because of their role in reverse cholesterol transport, but other mechanisms have been proposed, including anti-oxidant and anti-inflammatory properties [54]. Therefore, the definitive answer to the question "why is PLTP atherogenic" is still uncertain.

Obviously, an important question remaining is whether PLTP is a causal factor for the development of atherosclerosis in humans. The same question can be asked for CETP, for which inhibitors have been developed and tested. More data are needed before such choices can be made with respect to PLTP. As mentioned above, increased plasma PLTP activity levels are related with human disease states, connected with coronary heart disease, as well as with the most important risk factor of all, age [40]. Future studies will have to show whether plasma PLTP activity or PLTP mass is an independent risk factor for vascular disease.

References

[1] Tollefson, J.H., Albers, J.J., 1986. Methods Enzymol. 129, 797–816.
[2] Damen, J., Regts, J., Scherphof, G., 1982. Biochim. Biophys. Acta 712, 444–452.
[3] Tall, A.R., Abreu, E., Shuman, J., 1983. J. Biol. Chem. 258, 2174–2180.
[4] Albers, J.J., Tollefson, J.H., Chen, C.H., Steinmetz, A., 1984. Arteriosclerosis 4, 49–58.
[5] Tall, A.R., Krumholz, S., Olivecrona, T., Deckelbaum, R.J., 1985. J. Lipid Res. 26, 842–851.
[6] Day, J.R., Albers, J.J., Lofton-Day, C.E., Gilbert, T.L., Ching, A.F., Grant, F.J., O'Hara, P.J., Marcovina, S.M., Adolphson, J.L., 1994. J. Biol. Chem. 269, 9388–9391.
[7] Whitmore, T.E., Day, J.R., Albers, J.J., 1995. Genomics 28, 599–600.
[8] LeBoeuf, R.C., Caldwell, M., Tu, A., Albers, J.J., 1996. Genomics 34, 259–260.
[9] Tu, A.Y., Deeb, S.S., Iwasaki, L., Day, J.R., Albers, J.J., 1995. Biochem. Biophys. Res. Commun. 207, 552–558.
[10] Hubacek, J.A., Buchler, C., Aslanidis, C., Schmitz, G., 1997. Biochem. Biophys. Res. Commun. 236, 427–430.
[11] Huuskonen, J., Olkkonen, V.M., Jauhiainen, M., Ehnholm, C., 2001. Atherosclerosis 155, 269–281.
[12] Nishida, H.I., Nishida, T., 1997. J. Biol. Chem. 272, 6959–6964.
[13] Rao, R., Albers, J.J., Wolfbauer, G., Pownall, H.J., 1997. Biochemistry 36, 3645–3653.
[14] Jiang, X.C., Bruce, C., Mar, J., Lin, M., Ji, Y., Francone, O.L., Tall, A.R., 1999. J. Clin. Invest. 103, 907–914.
[15] van Tol, A., 2002. Curr. Opin. Lipidol. 13, 135–139.

[16] Huuskonen, J., Wohlfahrt, G., Jauhiainen, M., Ehnholm, C., Teleman, O., Olkkonen, V.M., 1999. J. Lipid Res. 40, 1123–1130.
[17] Jauhiainen, M., Metso, J., Pahlman, R., Blomqvist, S., van Tol, A., Ehnholm, C., 1993. J. Biol. Chem. 268, 4032–4036.
[18] Tu, A.Y., Nishida, H.I., Nishida, T., 1993. J. Biol. Chem. 268, 23098–23105.
[19] Barrans, A., Jaspard, B., Barbaras, R., Chap, H., Perret, B., Collet, X., 1996. Biochim. Biophys. Acta 1300, 73–85.
[20] Fielding, C.J., Fielding, P.E., 2001. Biochim. Biophys. Acta 1533, 175–189.
[21] Huuskonen, J., Olkkonen, V.M., Ehnholm, C., Metso, J., Julkunen, I., Jauhiainen, M., 2000. Biochemistry 39, 16092–16098.
[22] von Eckardstein, A., Nofer, J.R., Assmann, G., 2001. Arterioscler. Thromb. Vasc. Biol. 21, 13–27.
[23] Sviridov, D., Nestel, P., 2002. Atherosclerosis 161, 245–254.
[24] Foger, B., Santamarina-Fojo, S., Shamburek, R.D., Parrot, C.L., Talley, G.D., Brewer, H.B., 1997. J. Biol. Chem. 272, 27393–27400.
[25] Ehnholm, S., van Dijk, K.W., van 't Hof, B., van der Zee, A., Olkkonen, V.M., Jauhiainen, M., Hofker, M., Havekes, L., Ehnholm, C., 1998. J. Lipid Res. 39, 1248–1253.
[26] Albers, J.J., Tu, A.Y., Paigen, B., Chen, H., Cheung, M.C., Marcovina, S.M., 1996. Int. J. Clin. Lab. Res. 26, 262–267.
[27] Jiang, X., Francone, O.L., Bruce, C., Milne, R., Mar, J., Walsh, A., Breslow, J.L., Tall, A.R., 1996. J. Clin. Invest. 98, 2373–2380.
[28] Jiang, X.C., Qin, S., Qiao, C., Kawano, K., Lin, M., Skold, A., Xiao, X., Tall, A.R., 2001. Nat. Med. 7, 847–852.
[29] Jiang, X.C., Tall, A.R., Qin, S., Lin, M., Schneider, M., Lalanne, F., Deckert, V., Desrumaux, C., Athias, A., Witztum, J.L., Lagrost, L., 2002. J. Biol. Chem. 277, 31850–31856.
[30] Kostner, G.M., Oettl, K., Jauhiainen, M., Ehnholm, C., Esterbauer, H., Dieplinger, H., 1995. Biochem. J. 305(Pt 2), 659–667.
[31] Desrumaux, C., Deckert, V., Athias, A., Masson, D., Lizard, G., Palleau, V., Gambert, P., Lagrost, L., 1999. FASEB J. 13, 883–892.
[32] van Haperen, R., van Tol, A., Vermeulen, P., Jauhiainen, M., van Gent, T., van den Berg, P., Ehnholm, S., Grosveld, F., van der Kamp, A., de Crom, R., 2000. Arterioscler. Thromb. Vasc. Biol. 20, 1082–1088.
[33] Francone, O.L., Royer, L., Haghpassand, M., 1996. J. Lipid Res. 37, 1268–1277.
[34] Rye, K.A., Hime, N.J., Barter, P.J., 1997. J. Biol. Chem. 272, 3953–3960.
[35] Lie, J., de Crom, R., Jauhiainen, M., van Gent, T., van Haperen, R., Scheek, L., Jansen, H., Ehnholm, C., van Tol, A., 2001. Biochem. J. 360, 379–385.
[36] Walsh, A., Ito, Y., Breslow, J.L., 1989. J. Biol. Chem. 264, 6488–6494.
[37] Post, S.M., De Crom, R., Van Haperen, R., Van Tol, A., Princen, H.M., 2003. Arterioscler. Thromb. Vasc. Biol. 23, 892–897.
[38] Van Haperen, R., Van Tol, A., Van Gent, T., Scheek, L., Visser, P., Van Der Kamp, A., Grosveld, F., De Crom, R., 2002. J. Biol. Chem. 277, 48938–48943.
[39] Lie, J., de Crom, R., van Gent, T., van Haperen, R., Scheek, L., Lankhuizen, I., van Tol, A., 2002. J. Lipid Res. 43, 1875–1880.
[40] Tahvanainen, E., Jauhiainen, M., Funke, H., Vartiainen, E., Sundvall, J., Ehnholm, C., 1999. Atherosclerosis 146, 107–115.
[41] Riemens, S., van Tol, A., Sluiter, W., Dullaart, R., 1998. Atherosclerosis 140, 71–79.
[42] Riemens, S.C., van Tol, A., Sluiter, W.J., Dullaart, R.P., 1998. Diabetologia 41, 929–934.
[43] Desrumaux, C., Athias, A., Bessede, G., Verges, B., Farnier, M., Persegol, L., Gambert, P., Lagrost, L., 1999. Arterioscler. Thromb. Vasc. Biol. 19, 266–275.
[44] Colhoun, H.M., Taskinen, M.R., Otvos, J.D., Van Den Berg, P., O'Connor, J., Van Tol, A., 2002. Diabetes 51, 3300–3305.
[45] Lagrost, L., Athias, A., Herbeth, B., Guyard-Dangremont, V., Artur, Y., Paille, F., Gambert, P., Lallemant, C., 1996. J. Biol. Chem. 271, 19058–19065.
[46] Liinamaa, M.J., Hannuksela, M.L., Kesaniemi, Y.A., Savolainen, M.J., 1997. Arterioscler. Thromb. Vasc. Biol. 17, 2940–2947.

[47] Dullaart, R.P., Sluiter, W.J., Dikkeschei, L.D., Hoogenberg, K., Van Tol, A., 1994. Eur. J. Clin. Invest. 24, 188–194.
[48] Kaser, S., Sandhofer, A., Foger, B., Ebenbichler, C.F., Igelseder, B., Malaimare, L., Paulweber, B., Patsch, J.R., 2001. Diabetologia 44, 1111–1117.
[49] Cheung, M.C., Knopp, R.H., Retzlaff, B., Kennedy, H., Wolfbauer, G., Albers, J.J., 2002. Biochim. Biophys. Acta 1587, 53–59.
[50] Jonkers, I.J.A.M., Smelt, A.H.M., Hattori, H., Scheek, L.M., van Gent, T., de Man, F.H.A.F., van der Laarse, A., van Tol, A., 2003. J. Lipid Res. 44, 1462–1469.
[51] Van Tol, A., Ligtenberg, J.J., Riemens, S.C., van Haeften, T.W., Reitsma, W.D., Dullaart, R.P., 1997. Scand. J. Clin. Lab. Invest. 57, 147–157.
[52] Riemens, S.C., van Tol, A., Sluiter, W.J., Dullaart, R.P., 1999. Diabetes 48, 1631–1637.
[53] de Grooth, G.J., Kuivenhoven, J.A., Stalenhoef, A.F., de Graaf, J., Zwinderman, A.H., Posma, J.L., van Tol, A., Kastelein, J.J., 2002. Circulation 105, 2159–2165.
[54] Rader, D.J., 2002. Am. J. Cardiol. 90, 62i–70i.

PPARs and atherosclerosis

Coralie Fontaine, Caroline Duval, Olivier Barbier, Giulia Chinetti, Jean-Charles Fruchart and Bart Staels*

UR545 INSERM, Faculté de Pharmacie, Département d'Athérosclérose, Institut Pasteur de Lille, Université de Lille II, 1 rue Calmette, 59019 Lille, France
**Correspondence address: Tel.: +33-3-20-87-73-88; fax: +33-3-20-87-71-98*
E-mail: bart.staels@pasteur-lille.fr(B.S.)

1. Introduction

Atherosclerosis is a chronic disease characterized by the accumulation of lipids and fibrous connective tissue in the large arteries, accompanied by a local inflammatory response [1]. Atherosclerosis is the main origin of cardiovascular diseases, such as myocardial infarction and stroke, the major causes of mortality and morbidity in industrialized countries. Epidemiological studies have revealed several genetic and environmental risk factors predisposing to atherosclerosis.

The metabolic syndrome, which is characterized by the simultaneous presence of one or more metabolic disorders, such as glucose intolerance, hyperinsulinemia, dyslipidemia, coagulation disturbances and hypertension, is defined as the clustering of cardiovascular risk factors with insulin resistance. Activators of peroxisome proliferator-activated receptors (PPARs), transcription factors, belonging to the superfamily of nuclear receptors, modulate several of these metabolic risk factors. Numerous studies have illustrated the role of PPARs in the control of glucose homeostasis, insulin resistance and hypertension (for review, see Refs. [2–4]).

The PPAR subfamily consists of three distinct subtypes termed α (NR1C1), β/δ (NR1C2) and γ (NR1C3), which display tissue-selective expression patterns reflecting their biological functions [4]. While PPARα is expressed preferentially in tissues where fatty acids are catabolized, PPARγ is mainly present in adipose tissue and PPARβ/δ is ubiquitously expressed (for review, see Refs. [4,5]). PPARs are expressed in most cell types of the vascular wall as well as in atherosclerotic lesions (for review, see Ref. [6]), where they affect atherogenic processes.

Most of the physiological functions of PPARs can be explained by their activity as transcription factors, modulating the expression of specific target genes (for reviews see Refs. [4,5]). Upon ligand activation, PPARs regulate gene transcription by dimerizing with the retinoid X receptor (RXR) and binding to PPAR response elements (PPREs)

Advances in Molecular and Cell Biology, Vol. 33, pages 543–560

ISSN: 1569-2558/DOI: 10.1016/S1569-2558(03)33026-7

within the regulatory regions of target genes [4]. These PPREs usually consist of a direct repeat of the hexanucleotide sequence, AGGTCA, separated by one or two nucleotides (DR1 or DR2) (Fig. 1). PPARs can also repress gene transcription in a DNA binding-independent manner by interfering with the nuclear factor-kappa B (NF-κB), signal transducer and activator of transcription (STAT), activator protein-1 (AP-1), nuclear factor of activated T cells (NFAT), CCAAT/enhancer-binding proteins (C/EBP) and Smad3 signaling pathways via protein–protein interactions and cofactor competition [6–9]. Such transrepression mechanism is likely to participate in the anti-inflammatory actions of PPARs (for review, see Ref. [6]).

Fatty acids (FA) and FA-derived compounds are natural ligands for PPARα and PPARγ. Similarly, PPARβ/δ is a receptor for unsaturated FA. Natural eicosanoids derived from arachidonic acid via the lipoxygenase pathway, such as 8-*S*-hydroxytetraenoic acid (8-*S*-HETE) and leukotrien B4 (LTB4), activate PPARα [4]. Oxidized phospholipids derived from oxidized lipoproteins are natural ligands for both PPARα and PPARγ [10,11]. In addition, PPARγ is a receptor for eicosanoid metabolites formed via the cycloxygenase pathway, e.g. prostaglandins (PG) such as PG-J2, PG-H1 and PG-H2, and also via the lipoxygenase pathway (15-HETE) [4] (Fig. 1).

Synthetic agonists of PPARs are used in the treatment of metabolic diseases, such as dyslipidemia and type 2 diabetes. The anti-diabetic glitazones, which are insulin sensitizers, are synthetic high-affinity ligands for PPARγ [12–14]. The hypolipidemic

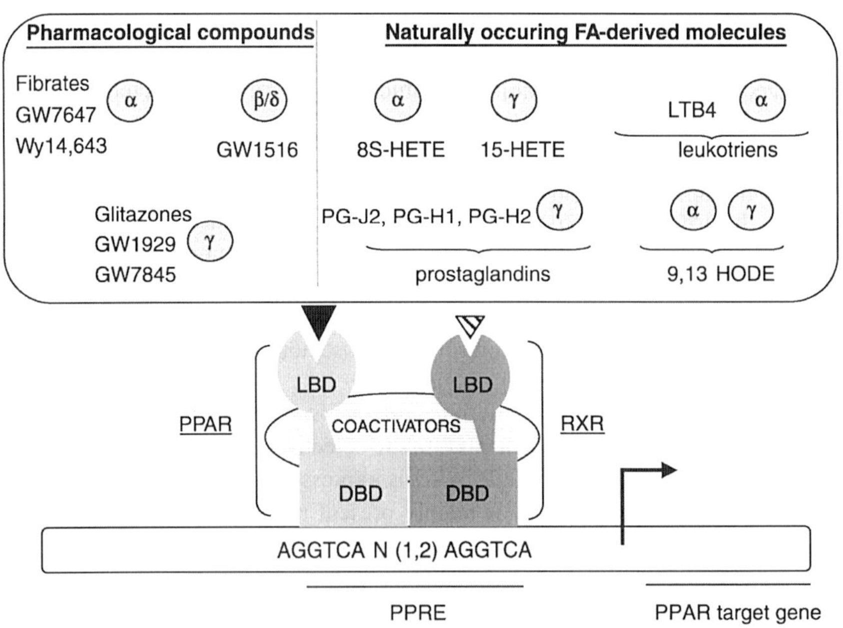

Fig. 1. Schematic representation of the mechanism of action and ligands of PPARs. PPARs act in a transcriptional complex as a heterodimer with RXR. The PPAR/RXR complex is activated by 9-*cis* retinoic acid, and either natural or synthetic agonists that bind to PPARs. Both RXR and PPAR possess a ligand-binding domain (LBD) and a DNA-binding domain (DBD).

fibrate drugs are PPARα ligands [15]. Recently, novel high-affinity subtype-specific PPAR agonists have been synthesized, including the human PPARα ligand GW7647 [16], the PPARγ activators GW1929 and GW7845 [17,18] and the PPARβ/δ-specific agonist GW501516 [19] (Fig. 1).

PPARs are exciting therapeutic targets for the treatment of disorders predisposing to atherosclerosis. We will focus on their role in the control of atherogenesis, by discussing their effects on each step of the atherogenic process.

2. PPARs modulate early stages of atherogenesis

Under basal conditions, the endothelium forms a relatively impermeable barrier between the circulating blood and the vessel wall. Endothelial injury is thought to be the primary event in atherosclerosis, which leads to the attraction, recruitment and activation of different cell types including monocytes/macrophages, T-lymphocytes, endothelial cells (ECs) and smooth muscle cells (SMCs) (Fig. 2).

The recruitment of monocytes to the intima requires the interaction of locally produced chemokines with specific cell surface receptors, such as the CCR2, the receptor for the monocyte chemoattractant protein-1 (MCP-1). MCP-1 expression is inhibited by PPARγ ligands [20] and glitazones decrease monocyte CCR2 expression [21,22], which demonstrates that the chemotaxis mediated by MCP-1 is blocked by PPARγ activation. In vivo, troglitazone reduces monocyte/macrophage homing to atherosclerotic plaques in apolipoprotein E (apoE) deficient mice [23]. Furthermore, glitazones also inhibit MCP-1 directed transendothelial migration of monocytes in low-density lipoproteins (LDL)-receptor deficient mice [24]. PPARα effects on monocyte recruitment are more controversial. Indeed, whereas both natural and synthetic PPARα ligands stimulate basal expression of MCP-1 in human aortic ECs [25,26], Pasceri et al. [26] have demonstrated that CRP-induced MCP-1 expression in human ECs is inhibited by synthetic PPARα ligands.

In addition T-cell recruitment to inflammation sites is modulated by PPARγ, which inhibits interferon gamma (IFNγ)-induced expression of certain CXC chemokines, such as IFNγ-inducible protein-10 (IP-10), monokine induced by IFNγ (Mig) and IFNγ-inducible T-cell α-chemoattractant (I-TAC) [27]. Furthermore, PPARγ agonists decrease EC expression of interleukin (IL)-8, a cytokine exerting chemotactic effects on T-lymphocytes, whereas PPARα ligands may induce its expression [25].

PPARα and PPARγ activators also decrease EC migration [28,29] by inhibiting vascular endothelial growth factor (VEGF)-induced Akt phosphorylation, one pathway required for the chemotactic signaling of ECs.

Rolling and adhesion of circulating monocytes to the ECs is the next critical early step in atherogenesis, and several cell adhesion molecules are involved in this process including intercellular adhesion molecule (ICAM-1) and vascular cell adhesion molecule (VCAM-1). PPARα and PPARγ activators inhibit cytokine-induced expression of VCAM-1 [30,31]. Furthermore, although expression of ICAM-1 is unaffected by PPARα activators, PPARγ agonists inhibit TNFα-induced ICAM-1 [23,32].

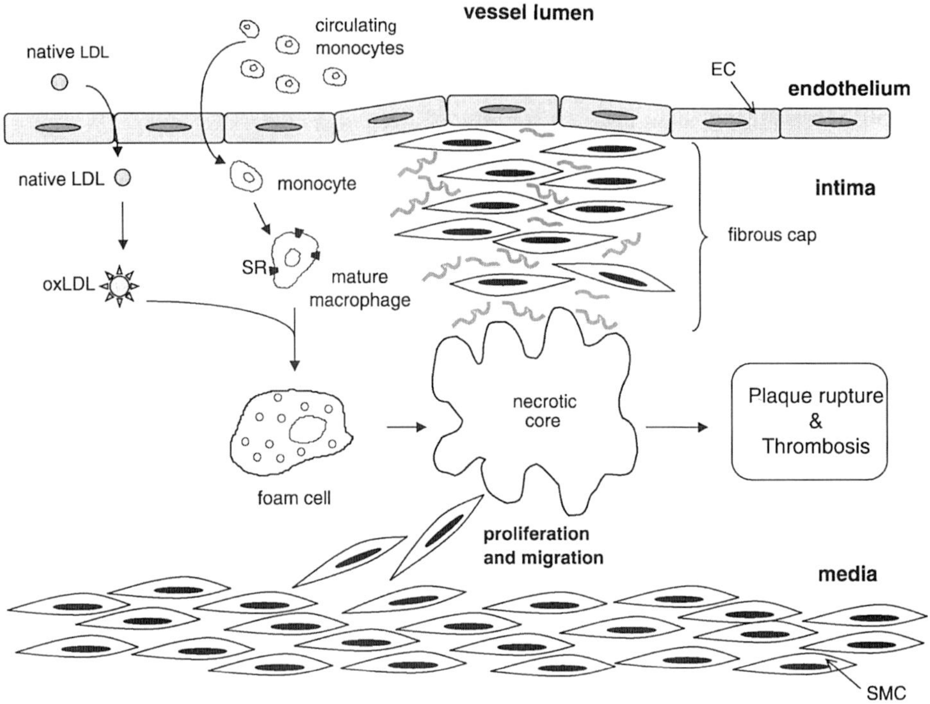

Fig. 2. Description of the atherosclerotic process. Atherosclerosis is a complex vascular disease initiated by oxidation and accumulation of low-density lipoproteins (LDL) in the subendothelial space of the vessels, followed by endothelial cell (EC) activation resulting in the recruitment of circulating monocytes. Trapped monocytes differentiate into macrophages, which take up oxidized LDL (OxLDL) through scavenger receptors (SRs), thus forming foam cells. Activated smooth muscle cells (SMCs) proliferate and migrate from the media, thus leading to fibrous cap formation. Activation of these cells leads to the release of pro-inflammatory cytokines, which combined with the secretion of metalloproteases and expression of pro-coagulant factors results in chronic inflammation and plaque instability. This can further evolve to plaque rupture and acute occlusion by thrombosis, resulting in myocardial infarction and stroke.

In addition, the expression [33] and the secretion [34–36] of endothelin-1 (ET-1), a vasoconstrictor peptide chemotactic to monocytes and a potent inducer of cell adhesion molecules (for review, see Ref. [37]), is repressed by both PPARα and PPARγ ligands in ECs. Since there is no PPRE in the ET-1 promoter, PPARs appear to act indirectly on the expression by blocking the AP-1 signaling pathway [33]. Iglarz et al. [38] have recently shown the beneficial effect of PPARα and PPARγ activators in vivo on vascular effects in deoxycorticosterone acetate (DOCA)-salt rats, a model of endothelin-dependent hypertension. This inhibition of ET-1 by PPARα and PPARγ activators has been previously demonstrated in rats with cardiac hypertrophy due to pressure overload provoked by abdominal aortic banding [39,40]. PPARγ agonists also enhance EC growth and secretion of C-type natriuretic peptide (CNP), an endothelium-derived relaxing peptide, in ECs [40]. Altogether, these effects might contribute to the slight decrease of blood pressure observed in diabetic patients treated with glitazones [41,42].

Under basal conditions, endothelium-derived nitric oxide (NO), produced by endothelial NO synthase (eNOS), inhibits leukocyte attachment and promotes vascular relaxation (for reviews, see Refs. [43,44]). However, NO may present pro-atherogenic effects, since it promotes oxidative stress and inflammation when produced at high concentration by inducible NO synthase (iNOS). PPARγ ligands increase NO release from ECs, thus inducing NO-dependent vasoprotective effects [45]. On the contrary, both PPARα and PPARγ ligands inhibit iNOS expression in macrophages and as such reduce inflammatory NO production [46,47].

Thus, PPAR-dependent regulation of chemokines, adhesion molecules and NO suggests that PPARα and PPARγ exert inhibitory effects on chemoattraction and cellular adhesion to ECs. PPARs may also have a protective action on the endothelium, thus decreasing endothelial injury.

3. PPARs control lipid accumulation and reverse cholesterol transport

After recruitment into the subendothelium, monocytes differentiate into macrophages. Accumulation of cholesterol in these cells leads to the formation of foam cells (Fig. 3).

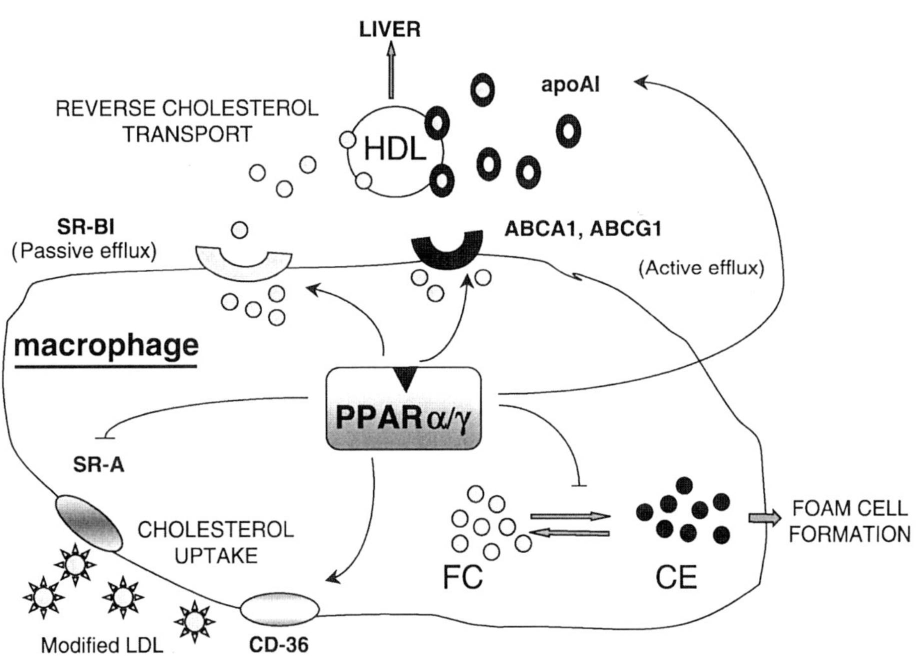

Fig. 3. Role of PPARα and PPARγ on cholesterol homeostasis in macrophages. Whereas PPARγ increases CD36 gene expression, SR-A expression is decreased. As net effect, no induction of foam cell formation is observed after treatment with PPARγ or PPARα. Moreover, both passive and active efflux are potentiated by PPARα and PPARγ agonists. In addition, the proportion of cholesteryl ester (CE) is decreased by PPARα and PPARγ ligands in macrophages, resulting in an enhanced availability of free cholesterol (FC) for efflux.

To be recognized by macrophages, LDL undergo modification, such as oxidation, glycosylation and aggregation. In ECs, PPARα and PPARγ modulate the generation of free radicals, which contributes to oxidized LDL (OxLDL) formation. Indeed, PPARα and PPARγ activators increase the expression of the Cu^{2+}, Zn^{2+} superoxide dismutase (CuZn-SOD), a superoxide scavenger enzyme, which protects arteries from the deleterious effects of reactive oxygen species (ROS) [48]. In addition, PPARα and PPARγ agonists decrease the NADPH oxidase expression, which contributes to superoxide formation [49]. Moreover, in macrophages, PPARγ activation attenuated ROS formation [49]. By contrast, in THP-1 macrophages, PPARα and not PPARγ activation induced generation of hydrogen peroxide, a marker of ROS [50]. Finally, PPARα activation by OxLDL leads to the downregulation of platelet-activating factor (PAF)-receptor gene expression in human monocytes and macrophages [51]. PAF is a potent pro-inflammatory substance playing an important role in LDL oxidation and whose effects are mediated by a specific cell receptor (PAF-receptor).

In addition, PPARs play an important role in regulating cholesterol uptake and homeostasis in macrophages. PPARα and PPARγ agonists upregulate human macrophage lipoprotein lipase (LPL) expression [52,53], but decrease LPL secretion and enzyme activity in differentiated macrophages [53]. Since LPL is required for the uptake of glycated LDL by macrophages [54], this downregulation may explain the decreased uptake of glycated LDL by macrophages after PPARα and PPARγ stimulation [53]. PPARγ ligands have been suggested to promote lipid accumulation in macrophages by inducing the receptor scavenger CD36 [55] and the adipocyte lipid-binding protein (ALBP/aP2) [56], both of which enhance cholesterol ester accumulation. However, PPARγ repression of the scavenger receptor A (SR-A) counterbalances CD36 induction [57]. Indeed, by acting through post-transcriptional mechanisms, PPARγ ligands decrease SR-A protein levels, which is involved in the uptake of modified LDL. Overall, there is no stimulatory effect of PPARγ or PPARα activation on LDL accumulation in macrophages [57,58]. Moreover, PPARα and PPARγ inhibit apoB48 receptor expression, which mediates lipid accumulation of triglyceride-rich lipoproteins in THP-1 cells [59]. On the contrary, Vosper et al. [60] have found that PPARβ/δ agonists promote lipid loading in human macrophages by oxLDL and that over-expression of PPARβ/δ in human differentiated THP-1 monocytes provokes a profound increase in lipid accumulation.

Finally, PPARγ activators inhibit TNFα-induced expression of the OxLDL receptor, lectin-like OxLDL receptor-1 (LOX-1) in ECs, an effect with potential beneficial consequences on cholesterol concentration in the endothelium [61].

Cholesterol efflux is the first step of the reverse cholesterol transport, which mediates the centripetal transport of cholesterol from peripheral cells back to the liver. Macrophage cholesterol efflux to high-density lipoprotein (HDL) occurs either via passive diffusion facilitated or not by protein, such as SR-B1/CLA-1, or via active efflux mediated by the ATP-binding cassette (ABC) proteins. In human macrophages, both PPARα and PPARγ activators induce protein levels of SR-B1/CLA-1 [62]. Moreover, the apoE expression is upregulated by PPARγ in macrophages [63], whereas PPARβ/δ downregulates genes involved in lipid efflux such as apoE and cholesterol 27-hydroxylase (CYP27) [60]. Interestingly, ligands of all three PPAR isotypes induce ABCA-1 expression [19,58] and as such stimulate cholesterol efflux from macrophages. PPARα and PPARγ induce

ABCA-1 by an indirect mechanism via induction of the liver X receptor α (LXRα) [58]. However, the molecular mechanism of ABCA1 induction by PPARβ/δ activators appears to be LXRα-independent [19]. Finally, PPARα and PPARγ-specific ligands decrease the cholesteryl ester (CE):free cholesterol (FC) ratio in macrophages resulting in an enhanced availability of FC for efflux through the ABCA-1 pathway [64,65].

In addition to the protective actions of HDL against atherosclerosis through the removal of excess cholesterol from the atherosclerotic lesion (for review, see Ref. [66]), HDL also exerts various other anti-atherogenic properties, such as the inhibition of monocyte chemotaxis, leukocyte adhesion, LDL oxidation and endothelial dysfunction (for review, see Ref. [67]).

PPARα and PPARγ (for review, see Refs. [68,69]) and, more recently, PPARβ/δ ([for review, see Refs. [70,71]) are known to influence HDL-cholesterol metabolism in humans. Fibrates influence the expression of HDL remodeling enzymes, such as the phospholipid transfer lipoprotein (PLTP) [72], an enzyme transferring phospholipids from VLDL/LDL to HDL, and lecithin:cholesterol acyl transferase (LCAT) [73], respectively, in mice and rats. PPARβ/δ-specific agonists also increase plasma HDL-cholesterol concentrations in insulin resistant mice and obese rhesus monkeys [19,74]. The molecular mechanisms behind this induction remain to be clarified.

ApoAI and apoAII are the major HDL apolipoproteins. In humans, PPARα activation increases the transcription of these two genes via binding to PPREs in their promoters [75, 76], an effect that contributes to the increase of HDL concentrations following fibrate treatment. In addition, fenofibrate increases apoAI plasma and hepatic expression levels in apoE deficient mice expressing a human apoAI transgene, which is associated with decreased atherosclerotic lesion formation [77]. Although the increase of plasma apoAI is undoubtedly beneficial, substantial controversy exists on the role of apoAII in atherosclerosis. Indeed, whereas transgenic mice over-expressing murine apoAII are more prone to develop atherosclerosis [78], over-expression of human apoAII protects against atherogenesis [79].

Thus, PPARα and PPARγ appear to be protective against accumulation of cholesterol formation into macrophages. Whereas they do not influence cholesterol accumulation, PPARα and PPARγ agonists promote cholesterol efflux, decrease cholesterol esterification and increase HDL concentrations. The role of PPARβ/δ in these processes is less clear, and more studies are required to specify its role in the control of atherogenesis.

4. PPARs, local immune and inflammatory responses

Atherosclerosis is characterized by a local immune response which occured via the recruitment, activation and proliferation of cells of the immune system. Dendritic cells (DCs) are critical initiators of the immune response by activating T-lymphocytes-presenting antigen. Activated T-lymphocytes differentiate into two subpopulations differing in their profiles of secreted cytokines. Th1 cells mediate the cellular immune response whereas Th2 cells potentiate the humoral response (Fig. 4).

In DCs, PPARγ activation inhibits the production of IL-12 [80,81], a pro-inflammatory cytokine known to play a role in the polarization of the acquired immune (Th1) response

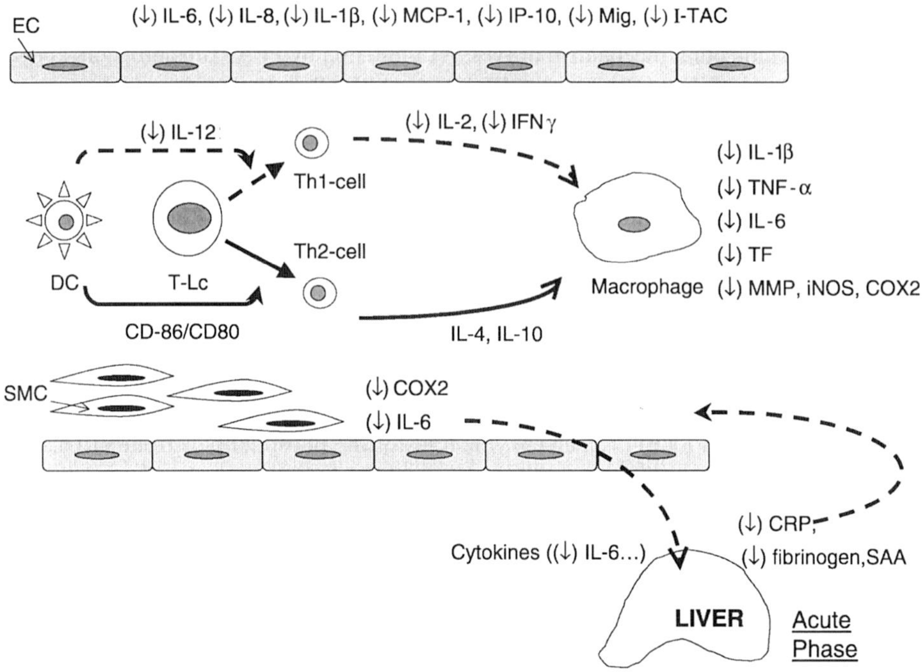

Fig. 4. Effects of PPARs on inflammation. PPARγ ligands deviate away from Th1 towards Th2 cytokine production. In addition, the anti-inflammatory activities of PPARs leads to the inhibition (↓) of pro-inflammatory molecule release by endothelial cells (ECs), macrophages, smooth muscle cells (SMCs), dendritic cells (DCs) and T-lymphocytes (T-Lc). Acute inflammation is neutralized by PPARs through the diminution of circulating interleukin (IL)-6, fibrinogen, C reactive protein (CRP) and serum amyloid A (SAA) plasma levels.

in mice and human. Moreover, PPARγ activation lowers the surface expression of the costimulatory molecules CD80, CD86, CD64, and the synthesis of chemokines involved in the Th1 immune response, such as RANTES, IP-10 and IL1-β [81–83].

In T-lymphocytes, activated PPARγ negatively interacts with T-cell-specific transcription factor NFAT-stimulated IL-2 transcription [7,83], and thus reduces the proliferative effect of this cytokine. The major histocompatibility complex class II, directly involved in T-lymphocyte activation, is inhibited by PPARγ ligands [84]. In addition, PPARγ increases T-lymphocyte apoptosis, thus decreasing their viability [85].

Altogether, the protective effects of PPARγ ligands on the immune response associated with atherogenesis may be due, on the one hand, to immune deviation from Th1 to Th2-type cytokine production and, on the other hand, to a global decrease of the immune response. The role of PPARα in the immune response is less documented. PPARα inhibits IL-2 secretion via NF-κB transrepression and also decreases IFNγ production in T-lymphocytes [83,86]. The PPARα-dependent suppression of adaptive immune responses was evidenced in vivo using the PPARα deficient mouse model [87].

The activation of ECs, macrophages, SMCs, T-lymphocytes leads to the release of pro-inflammatory molecules, such as cytokines, and the onset of a chronic inflammatory response (Fig. 4). In inflammation, the transcription factor NF-κB is activated and increases the expression of multiple pro-inflammatory genes. PPARα and PPARγ activation results in a negative cross-talk with inflammatory transcription factors, such as NF-κB, STAT-1 and AP-1, to block their downstream target genes (for review, see Ref. [6]). In human monocytes/macrophages, PPARγ activators inhibit the activation of inflammatory cytokines, such as IL-1β, IL-6, IL-8, IL-10 and TNFα (for review, see Refs. [6,83]). In addition, PPARγ activation induces the production of IL-1 receptor antagonist in THP-1 cells, providing a new anti-inflammatory action mechanism [88]. In macrophages, PG-J2 and thiazolidinediones (at high concentration) inhibit lipopolysaccharide (LPS) and IFNγ-induced macrophage activation [47,89]. The anti-inflammatory activity of PPARγ is also monitored by the downregulation of iNOS, COX2 and metalloproteinase 9 (MMP9) [47,90], three important actors of the inflammatory response. However, the most pronounced anti-inflammatory effects are observed with PG-J2, which is not selective for PPARγ and acts also via PPAR-independent mechanisms [91,92]. Nevertheless, PPARγ activation suppresses zinc finger transcription factor early growth response gene-1 (Egr-1) expression and, as such, inhibits expression of Egr-1-induced inflammatory target genes [93]. The interaction between activated PPARγ and C/EBPδ is another mechanism by which PPARγ may downregulate the production of inflammatory cytokines [94]. Thus, PPARγ ligands appear to exert anti-inflammatory activities via both PPARγ dependent and independent mechanisms. Activated PPARα inhibits inflammatory response markers, such as ET-1, VCAM-1, IL-6, TF, MMP9, COX2 and iNOS in ECs, SMCs and macrophages [25,30,31,33,34,95–99]. Zuckerman et al. [100] have recently shown that the synthesis of both NO and β-2 integrin CD11 induced by IFNγ is inhibited in peritoneal macrophages from apoE deficient mice treated with a PPARα/γ coagonist, thus leading to the inhibition of macrophage activation.

The anti-inflammatory activities of PPARα and PPARγ activators have been evidenced in humans. In patients with hyperlipidemia, fenofibrate treatment decreases circulating levels of IL-6 as well as fibrinogen and CRP, whose production is controlled by cytokines such as IL-1β and IL-6 [26,96]. PPARα activation reduces CRP expression in human umbilical vein ECs [26]. In addition, fibrates reduce nuclear CEBPβ-p50NF-κB complex formation, which enhances hepatic CRP promoter activity in response to IL-1β [101]. Moreover, fenofibrate treatment of mice abrogates the acute phase-induced elevation of plasma serum amyloid A (SAA) [102]. Ex vivo, isolated aortas from PPARα deficient mice show an exacerbated response to LPS, as measured by IL-6 secretion, thus demonstrating a role for PPARα in the control of vascular inflammation [103]. Similarly, rosiglitazone treatment of patients with type 2 diabetes significantly reduces plasma levels of IL-6, CRP, MMP9, NF-κB and SAA [90,104].

Thus, adaptive immune and chronic inflammatory responses involved in the process of atherogenesis are inhibited by both PPARα and PPARγ activation via direct and indirect mechanisms. The anti-inflammatory properties of PPARα and PPARγ may contribute to their protective role against atherogenesis.

5. Role of PPARs in plaque development and stability

Formation of the fibrous plaque is due to SMC migration and proliferation in the intima. These cells secrete an extracellular matrix (ECM) and give rise to a fibrous cap. The role of PPARγ in this stage of the atherosclerosis is well described. PPARγ ligands inhibit the proliferation and the migration of vascular SMCs (VSMCs), a phenomenon that may be of particular relevance for restenosis prevention [105–107]. PPARγ ligands inhibit migration of SMCs by inhibiting Ets-1 expression, a transcription factor required for matrix metalloprotease induction [108,109]. Inhibition of the ET-1 secretion, a potent inducer of SMC proliferation, by PPARα and PPARγ agonists [33–36,38] represents another way by which PPARα and PPARγ interfere with proliferation of SMCs. In addition, IL-1β-induced expression of the platelet-derived growth factor-α receptor (PDGFαR) – PDGFα being a potent mitogen for VSMCs – is suppressed by PPARγ activation via transrepression of CCAAT/enhancer-binding protein-δC/EBPδ) [110]. Furthermore, PPARγ inhibits the mitogenic induction of the cyclin-dependent kinase inhibitor p21 by modulating the protein kinase C delta (PKCδ) pathway in VSMCs [111]. Moreover, PPARγ ligands may inhibit angiotensin II-induced cell growth and hypertrophy in VSMCs by suppression of its receptor, the angiotensin II type 1 receptor (ATR1) [112, 113]. Recently, Abe et al. [114] have proposed another mechanism explaining the PPARγ-mediated growth inhibition of VSMCs through a GATA-6-dependent transcriptional mechanism. In addition, integrins play an important role in VSMC migration, and PPARα activation inhibits TGF-β-induced beta5 integrin expression [115]. Moreover, a PPARα/γ coagonist inhibits the beta2 integrin CD11 induced by IFNγ [100]. By contrast, PPARβ/δ induces post-confluent VSMC proliferation by increasing cyclinA and cyclin-dependent kinase CDK2, which are implicated in cell cycle regulation, as well as decreasing $p57^{kip2}$, a kinase activity inhibitor [116].

Plaque rupture is the end-stage of the atherogenic process, leading to thrombus formation, occlusion and the clinical sequels of atherosclerosis. Plaque instability is partly due to the degradation of the ECM in the fibrous cap. PPARα and PPARγ ligands inhibit gene expression [47], secretion and gelatinolytic activity [90,117,118] of MMP9, a matrix degrading protein secreted in response to inflammatory activation. In addition, PPARα agonists decrease expression of PAF receptor expression in macrophages [51,119]. PAF stimulates the secretion of elastase-type enzymes, which contribute to plaque stability and rupture. In addition, PPARγ activators reduce osteopontin [120] and osteoprotegerin [121] gene expression, respectively, in macrophages and VSMCs, suggesting that PPARγ could influence vascular calcification, since osteopontin and osteoprotegerin are described as inhibitors of arterial calcification [122,123]. Gemfibrozil, a PPARα agonist, significantly decreases vascular proteoglycan biosynthesis and glucosaminoglycan synthesis [124], two important components of the ECM.

Neovascularisation is also an important process that influences plaque stability. Even though VEGF appears protective against EC injury in the early stages of atherosclerosis by protecting ECs against oxLDL toxicity [125], this protein may exert pro-atherogenic activities, since VEGF induces angiogenesis leading to plaque destabilization and rupture. PPARγ induces VEGF production in VSMCs and macrophages, and thus may induce angiogenesis [126]. However, PPARγ ligands have been shown to inhibit tumor angiogenesis [127]. Thus, the role of PPARγ in angiogenesis remains to be clarified by further studies.

Finally, plaque stability is also influenced by apoptosis of SMCs and macrophages [128–130]. Apoptosis occurring in atherosclerotic areas is potentially involved in necrotic core formation and plaque rupture, which may trigger atherothrombotic events (for review, see Ref. [131]). PPARα and PPARγ control apoptosis via negative cross-talk with the anti-apoptotic NF-κB pathway in macrophages [132]. Rosiglitazone upregulates the expression of the tumor suppressor gene, phosphatase and tensin homolog (PTEN), which modulates, at least, cell survival and proliferation, in human macrophages via a PPRE in its promoter [133]. Furthermore, PG-J2 can also trigger the apoptosis of ECs via a PPARγ-dependent pathway [134]. The significance of apoptosis in atherosclerosis remains unclear. Although it has been proposed that apoptotic cell death contributes to plaque instability, rupture and thrombus formation, macrophage apoptosis also decreases inflammation of macrophage origin and avoids the destruction of ECM collagen, which maintains the elasticity of the plaque [135].

Thus, PPARs, and especially PPARγ, appear to reduce fibrous plaque formation by inhibiting SMC migration and proliferation. However, the role of PPARs in plaque stability or rupture and their consequences on atherogenesis are less understood.

6. PPARs and thrombosis

PPARs also modulate platelet aggregation. PPARγ inhibits thromboxane synthase (TXS) expression by a mechanism involving protein–protein interaction between PPARγ and the nuclear factor-E2-related factor 2 [136]. Moreover, the expression of the thromboxane receptor (TXR) is inhibited by PPARγ activators via an interaction with Sp1 in VSMCs [137]. Thus, PPARγ activation results in a decreased synthesis and action of thromboxane A2, a potent platelet aggregation inducer and vasoconstrictor. In addition, TF is a major factor in thrombus formation and blood coagulation. PPARα agonists inhibit TF expression in monocytes and macrophages [98,99]. Finally, glitazones and fibrates also modulate the secretion of the plasminogen activator inhibitor-1 (PAI-1) [138,139]. However, the exact mechanisms and effects are still unclear [139,140], and the role of PPAR in the regulation of PAI-1 and its consequences for atherothrombosis remain to be clarified.

On the whole, PPARα and PPARγ activation appears protective against atherothrombosis.

7. Anti-atherosclerotic effects of PPARs: results from in vivo studies

Compelling evidence for a regulatory role for PPARs on atherogenesis in vivo comes from studies in animal models of atherosclerosis and human clinical trials.

Various animals models presenting accelerated atherosclerosis have been developed to study the pathophysiology of the disease and/or the evaluation of potential therapeutic strategies (for review, see Ref. [141]). PPARγ ligands have been shown to prevent the progression of atherosclerotic lesions, with a concomitant decrease of macrophage accumulation in the plaque of both apoE and LDL receptor deficient mice [17,23,24,142]. Glitazones act by reducing monocyte/macrophage homing to atherosclerotic plaques, by inhibiting fatty streak formation, improving glucose homeostasis and enhancing reverse cholesterol transport [143]. Since rosiglitazone treatment of apoE deficient mice rendered diabetic by a low dose of streptozotocin leads to higher lipid levels and similar glucose levels but less atherosclerosis [144], the anti-atherogenic effect of glitazones seems to be mediated by direct vascular actions. In the same way, treatment of apoE deficient mice with PPARα/γ coagonist results in a reduction of atherosclerosis [100,142]. Moreover, fenofibrate also reduces lesion surface area in the aortic sinus of apoE deficient mice expressing a human apoAI transgene [77].

In addition, rosiglitazone treatment reduces myocardial infarction and limits post-ischemic injury in rats [145,146]. Moreover, protection against myocardial ischemic injury and improvement of endothelial vasodilatation by PPARα was demonstrated in mice [147]. However, surprisingly, PPARα deficiency in the apoE null background mice results in lowered atherosclerosis [148]. These double knockout mice are characterized by higher concentrations of atherogenic lipoproteins, but also higher insulin sensitivity, lower blood pressure and fewer intimal lesions.

A wealth of clinical studies have revealed that fibrates improve the cardiovascular risk profile. Several angiographic intervention trials, including the Lipid Coronary Angiographic Trial (LOCAT), the Diabetes Atherosclerosis Intervention Study (DAIS) and the Bezafibrate Coronary Atherosclerosis Intervention Trial (BECAIT), have demonstrated beneficial effects of fibrates on atherosclerotic lesion progression [149–151]. Furthermore, secondary prevention trials, such as the Veterans Administration-HDL-Cholesterol Intervention Trial (VA-HIT) [152] and the Helsinki Heart Study [153], demonstrated a decreased incidence of cardiovascular events following fibrate treatment. In patients with type 2 diabetes, who are characterized by moderate hypertriglyceridemia and low HDL-cholesterol concentrations, fibrates decrease the incidence of myocardial infarction, as observed in the Mary's Ealing, Northwick Park Diabetes (SENDCAP) study [154]. Moreover, adding fenofibrate to simvastatin increases significantly HDL cholesterol levels [155]. Thus, the development of combination drugs for atherosclerosis treatment appears to be a good way to increase their efficacy (for review, see Ref. [156]).

Some studies have been performed testing the effects of glitazone treatment on cardiovascular risks in various diabetic and insulin-resistant patient populations [157,158]. Glitazone therapy lowers both fasting and post-prandial glucose levels and improves insulin-stimulated glucose disposal. However, the effect of glitazones on the plasma lipid profile in humans is controversial and intervention trials assessing the influence of these compounds on the incidence of cardiovascular disease are still lacking. In patients with type 2 diabetes, pioglitazone and rosiglitazone appear to have distinct effects on the plasma levels of triglycerides and LDL. While pioglitazone treatment lowers serum concentrations of LDL and triglycerides, rosiglitazone treatment has no effect on these parameters [157,159]. Although the mechanistic basis for these differences is unclear, the fact that pioglitazone has, albeit limited, PPARα activity may be one possible explanation [160]. Nevertheless, troglitazone treatment of patients with type 2 diabetes [161] resulted in a significant reduction of intima:media thickness (IMT).

Running trials, such as the Fenofibrate Intervention and Event Lowering in Diabetes (FIELD) and PROACTIVE trials, will provide additional evidence for possible clinical cardiovascular benefits of both PPARα and PPARγ agonists in the diabetic population. PPARα/γ agonists are currently in phase 3 development for the treatment of patients with type 2 diabetes [162].

8. Conclusion

Considerable evidences indicate that PPARα and PPARγ have beneficial effects in inflammatory diseases, including atherosclerosis. Although molecular mechanisms are not yet fully established and the complexity of these systems appears important, PPARs interfere at different steps of atherogenesis by blocking vascular cell recruitment, modulating foam cell formation, interfering with the inflammatory response and inhibiting fibrous plaque development. Their implication in plaque stability and atherothrombosis is less clear, and its understanding requires further studies.

In conclusion, PPAR agonists represent pharmacological drugs with high potential. Combination treatment and development of coagonists appear to be promising future option for an optimal treatment of atherosclerosis.

Acknowledgements

Support by grants from ARC (Association pour la Recherche contre le Cancer) (to CD), the Fondation Lefoulon-Delalande, Institut de France (to OB), European Community (to GC) (grant QLRT-1999-01007), Fonds Européens de Développement Régional, Conseil Régional Région Nord/Pas-de-Calais "Genopole Project 01360124" and Leducq Foundation (to B.S. and J.C.F.) is kindly acknowledged.

References

[1] Lusis, A.J., 2000. Nature 407, 233–241.

[2] Torra, I.P., Chinetti, G., Duval, C., Fruchart, J.C., Staels, B., 2001. Curr. Opin. Lipidol. 12, 245–254.

[3] Willson, T.M., Lambert, M.H., Kliewer, S.A., 2001. Annu. Rev. Biochem. 70, 341–367.

[4] Barbier, O., Torra, I.P., Duguay, Y., Blanquart, C., Fruchart, J.C., Glineur, C., Staels, B., 2002. Arterioscler. Thromb. Vasc. Biol. 22, 717–726.

[5] Duval, C., Chinetti, G., Trottein, F., Fruchart, J.C., Staels, B., 2002. Trends. Mol. Med. 8, 422–430.

[6] Chinetti, G., Fruchart, J.C., Staels, B., 2000. Inflamm. Res. 49, 497–505.

[7] Yang, X.Y., Wang, L.H., Chen, T., Hodge, D.R., Resau, J.H., DaSilva, L., Farrar, W.L., 2000. J. Biol. Chem. 275, 4541–4544.

[8] Gervois, P., Vu-Dac, N., Kleemann, R., Kockx, M., Dubois, G., Laine, B., Kosykh, V., Fruchart, J.C., Kooistra, T., Staels, B., 2001. J. Biol. Chem. 276, 33471–33477.

[9] Fu, M., Zhang, J., Zhu, X., Myles, D.E., Willson, T.M., Liu, X., Chen, Y.E., 2001. J. Biol. Chem. 276, 45888–45894.

[10] Delerive, P., Furman, C., Teissier, E., Fruchart, J., Duriez, P., Staels, B., 2000. FEBS Lett. 471, 34–38.

[11] Davies, S.S., Pontsler, A.V., Marathe, G.K., Harrison, K.A., Murphy, R.C., Hinshaw, J.C., Prestwich, G.D., Hilaire, A.S., Prescott, S.M., Zimmerman, G.A., McIntyre, T.M., 2001. J. Biol. Chem. 276, 16015–16023.

[12] Lehmann, J.M., Moore, L.B., Smith-Oliver, T.A., Wilkison, W.O., Willson, T.M., Kliewer, S.A., 1995. J. Biol. Chem. 270, 12953–12956.
[13] Raji, A., Plutzky, J., 2002. Curr. Cardiol. Rep. 4, 514–521.
[14] Martens, F.M., Visseren, F.L., Lemay, J., de Koning, E.J., Rabelink, T.J., 2002. Drugs 62, 1463–1480.
[15] Willson, T.M., Brown, P.J., Sternbach, D.D., Henke, B.R., 2000. J. Med. Chem. 43, 527–550.
[16] Brown, P.J., Stuart, L.W., Hurley, K.P., Lewis, M.C., Winegar, D.A., Wilson, J.G., Wilkison, W.O., Ittoop, O.R., Willson, T.M., 2001. Bioorg. Med. Chem. Lett. 11, 1225–1227.
[17] Li, A.C., Brown, K.K., Silvestre, M.J., Willson, T.M., Palinski, W., Glass, C.K., 2000. J. Clin. Invest. 106, 523–531.
[18] Brown, K.K., Henke, B.R., Blanchard, S.G., Cobb, J.E., Mook, R., Kaldor, I., Kliewer, S.A., Lehmann, J.M., Lenhard, J.M., Harrington, W.W., Novak, P.J., Faison, W., Binz, J.G., Hashim, M.A., Oliver, W.O., Brown, H.R., Parks, D.J., Plunket, K.D., Tong, W.Q., Menius, J.A., Adkison, K., Noble, S.A., Willson, T.M., 1999. Diabetes 48, 1415–1424.
[19] Oliver, W.R. Jr., Shenk, J.L., Snaith, M.R., Russell, C.S., Plunket, K.D., Bodkin, N.L., Lewis, M.C., Winegar, D.A., Sznaidman, M.L., Lambert, M.H., Xu, H.E., Sternbach, D.D., Kliewer, S.A., Hansen, B.C., Willson, T.M., 2001. Proc. Natl Acad. Sci. USA 98, 5306–5311.
[20] Murao, K., Imachi, H., Momoi, A., Sayo, Y., Hosokawa, H., Sato, M., Ishida, T., Takahara, J., 1999. FEBS Lett. 454, 27–30.
[21] Ishibashi, M., Egashira, K., Hiasa, K., Inoue, S., Ni, W., Zhao, Q., Usui, M., Kitamoto, S., Ichiki, T., Takeshita, A., 2002. Hypertension 40, 687–693.
[22] Han, K.H., Quehenberger, O., 2000. Trends Cardiovasc. Med. 10, 209–216.
[23] Pasceri, V., Wu, H.D., Willerson, J.T., Yeh, E.T., 2000. Circulation 101, 235–238.
[24] Collins, A.R., Meehan, W.P., Kintscher, U., Jackson, S., Wakino, S., Noh, G., Palinski, W., Hsueh, W.A., Law, R.E., 2001. Arterioscler. Thromb. Vasc. Biol. 21, 365–371.
[25] Lee, H., Shi, W., Tontonoz, P., Wang, S., Subbanagounder, G., Hedrick, C.C., Hama, S., Borromeo, C., Evans, R.M., Berliner, J.A., Nagy, L., 2000. Circ. Res. 87, 516–521.
[26] Pasceri, V., Cheng, J.S., Willerson, J.T., Yeh, E.T., Chang, J., 2001. Circulation 103, 2531–2534.
[27] Marx, N., Mach, F., Sauty, A., Leung, J.H., Sarafi, M.N., Ransohoff, R.M., Libby, P., Plutzky, J., Luster, A.D., 2000. J. Immunol. 164, 6503–6508.
[28] Goetze, S., Bungenstock, A., Czupalla, C., Eilers, F., Stawowy, P., Kintscher, U., Spencer-Hansch, C., Graf, K., Nurnberg, B., Law, R.E., Fleck, E., Grafe, M., 2002. Hypertension 40, 748–754.
[29] Goetze, S., Eilers, F., Bungenstock, A., Kintscher, U., Stawowy, P., Blaschke, F., Graf, K., Law, R.E., Fleck, E., Grafe, M., 2002. Biochem. Biophys. Res. Commun. 293, 1431–1437.
[30] Jackson, S.M., Parhami, F., Xi, X.P., Berliner, J.A., Hsueh, W.A., Law, R.E., Demer, L.L., 1999. Arterioscler. Thromb. Vasc. Biol. 19, 2094–2104.
[31] Marx, N., Sukhova, G.K., Collins, T., Libby, P., Plutzky, J., 1999. Circulation 99, 3125–3131.
[32] Chen, N.G., Han, X., 2001. Biochem. Biophys. Res. Commun. 282, 717–722.
[33] Delerive, P., Martin-Nizard, F., Chinetti, G., Trottein, F., Fruchart, J.C., Najib, J., Duriez, P., Staels, B., 1999. Circ. Res. 85, 394–402.
[34] Satoh, H., Tsukamoto, K., Hashimoto, Y., Hashimoto, N., Togo, M., Hara, M., Maekawa, H., Isoo, N., Kimura, S., Watanabe, T., 1999. Biochem. Biophys. Res. Commun. 254, 757–763.
[35] Martin-Nizard, F., Furman, C., Delerive, P., Kandoussi, A., Fruchart, J.C., Staels, B., Duriez, P., 2002. J. Cardiovasc. Pharmacol. 40, 822–831.
[36] Kandoussi, A., Martin, F., Hazzan, M., Noel, C., Fruchart, J.C., Staels, B., Duriez, P., 2002. Clin. Sci. (Lond.) 103(Suppl. 48), 81S–83S.
[37] Agapitov, A.V., Haynes, W.G., 2002. J. Renin Angiotensin Aldosterone Syst. 3, 1–15.
[38] Iglarz, M., Touyz, R.M., Amiri, F., Lavoie, M.F., Diep, Q.N., Schiffrin, E.L., 2003. Arterioscler. Thromb. Vasc. Biol. 23, 45–51.
[39] Ogata, T., Miyauchi, T., Sakai, S., Irukayama-Tomobe, Y., Goto, K., Yamaguchi, I., 2002. Clin. Sci. (Lond.) 103(Suppl. 48), 284S–288S.
[40] Sakai, S., Miyauchi, T., Irukayama-Tomobe, Y., Ogata, T., Goto, K., Yamaguchi, I., 2002. Clin. Sci. (Lond.) 103(Suppl. 48), 16S–20S.
[41] Itoh, H., Nakao, K., 2000. Nippon Rinsho 58(Suppl. 1), 214–224.

[42] Parulkar, A.A., Pendergrass, M.L., Granda-Ayala, R., Lee, T.R., Fonseca, V.A., 2001. Ann. Intern. Med. 134, 61–71.
[43] Russo, G., Leopold, J.A., Loscalzo, J., 2002. Vascul. Pharmacol. 38, 259–269.
[44] Napoli, C., 2002. J. Card. Surg. 17, 355–362.
[45] Calnek, D.S., Mazzella, L., Roser, S., Roman, J., Hart, C.M., 2003. Arterioscler. Thromb. Vasc. Biol. 23, 52–57.
[46] Colville-Nash, P.R., Qureshi, S.S., Willis, D., Willoughby, D.A., 1998. J. Immunol. 161, 978–984.
[47] Ricote, M., Li, A.C., Willson, T.M., Kelly, C.J., Glass, C.K., 1998. Nature 391, 79–82.
[48] Inoue, I., Goto, S., Matsunaga, T., Nakajima, T., Awata, T., Hokari, S., Komoda, T., Katayama, S., 2001. Metabolism. 50, 3–11.
[49] Fischer, B., von Knethen, A., Brune, B., 2002. J. Immunol. 168, 2828–2834.
[50] Teissier, E., Chinetti, G., Staels, B., 2002. Circulation 106, II-261.
[51] Hourton, D., Delerive, P., Stankova, J., Staels, B., Chapman, M.J., Ninio, E., 2001. Biochem. J. 354, 225–232.
[52] Li, L., Beauchamp, M.C., Renier, G., 2002. Atherosclerosis. 165, 101–110.
[53] Gbaguidi, F.G., Chinetti, G., Milosavljevic, D., Teissier, E., Chapman, J., Olivecrona, G., Fruchart, J.C., Griglio, S., Fruchart-Najib, J., Staels, B., 2002. FEBS Lett. 512, 85–90.
[54] Zimmermann, R., Panzenbock, U., Wintersperger, A., Levak-Frank, S., Graier, W., Glatter, O., Fritz, G., Kostner, G.M., Zechner, R., 2001. Diabetes. 50, 1643–1653.
[55] Tontonoz, P., Nagy, L., Alvarez, J.G., Thomazy, V.A., Evans, R.M., 1998. Cell 93, 241–252.
[56] Fu, Y., Luo, N., Lopes-Virella, M.F., Garvey, W.T., 2002. Atherosclerosis 165, 259–269.
[57] Moore, K.J., Rosen, E.D., Fitzgerald, M.L., Randow, F., Andersson, L.P., Altshuler, D., Milstone, D.S., Mortensen, R.M., Spiegelman, B.M., Freeman, M.W., 2001. Nat. Med. 7, 41–47.
[58] Chinetti, G., Lestavel, S., Bocher, V., Remaley, A.T., Neve, B., Torra, I.P., Teissier, E., Minnich, A., Jaye, M., Duverger, N., Brewer, H.B., Fruchart, J.C., Clavey, V., Staels, B., 2001. Nat. Med. 7, 53–58.
[59] Haraguchi, G., Kobayashi, Y., Brown, M.L., Tanaka, A., Isobe, M., Gianturco, S.H., Bradley, W.A., 2003. J. Lipid Res. 16, 16.
[60] Vosper, H., Patel, L., Graham, T.L., Khoudoli, G.A., Hill, A., Macphee, C.H., Pinto, I., Smith, S.A., Suckling, K.E., Wolf, C.R., Palmer, C.N., 2001. J. Biol. Chem. 276, 44258–44265.
[61] Chiba, Y., Ogita, T., Ando, K., Fujita, T., 2001. Biochem. Biophys. Res. Commun. 286, 541–546.
[62] Chinetti, G., Gbaguidi, F.G., Griglio, S., Mallat, Z., Antonucci, M., Poulain, P., Chapman, J., Fruchart, J.C., Tedgui, A., Najib-Fruchart, J., Staels, B., 2000. Circulation 101, 2411–2417.
[63] Galetto, R., Albajar, M., Polanco, J.I., Zakin, M.M., Rodriguez-Rey, J.C., 2001. Biochem. J. 357, 521–527.
[64] Chinetti, G., Lestavel, S., Fruchart, J.C., Clavey, V., Staels, B., 2003. Circ. Res. 92, 212–217.
[65] Argmann, C.A., Sawyez, C.G., McNeil, C.J., Hegele, R.A., Huff, M.W., 2003. Arterioscler. Thromb. Vasc. Biol. 23, 475–482.
[66] von Eckardstein, A., Nofer, J.R., Assmann, G., 2001. Arterioscler. Thromb. Vasc. Biol. 21, 13–27.
[67] Nofer, J.R., Kehrel, B., Fobker, M., Levkau, B., Assmann, G., von Eckardstein, A., 2002. Atherosclerosis 161, 1–16.
[68] Xie, Y., Yang, Q., DePierre, J.W., 2002. Ann. N. Y. Acad. Sci. 973, 17–25.
[69] van Bilsen, M., van der Vusse, G.J., Gilde, A.J., Lindhout, M., van der Lee, K.A., 2002. Mol. Cell. Biochem. 239, 131–138.
[70] Michalik, L., Desvergne, B., Wahli, W., 2003. Curr. Opin. Lipidol. 14, 129–135.
[71] Skogsberg, J., Kannisto, K., Cassel, T.N., Hamsten, A., Eriksson, P., Ehrenborg, E., 2003. Arterioscler. Thromb. Vasc. Biol. 27, 27.
[72] Bouly, M., Masson, D., Gross, B., Jiang, X.C., Fievet, C., Castro, G., Tall, A.R., Fruchart, J.C., Staels, B., Lagrost, L., Luc, G., 2001. J. Biol. Chem. 276, 25841–25847.
[73] Staels, B., van Tol, A., Skretting, G., Auwerx, J., 1992. J. Lipid Res. 33, 727–735.
[74] Leibowitz, M.D., Fievet, C., Hennuyer, N., Peinado-Onsurbe, J., Duez, H., Bergera, J., Cullinan, C.A., Sparrow, C.P., Baffic, J., Berger, G.D., Santini, C., Marquis, R.W., Tolman, R.L., Smith, R.G., Moller, D.E., Auwerx, J., 2000. FEBS Lett. 473, 333–336.

[75] Vu-Dac, N., Schoonjans, K., Laine, B., Fruchart, J.C., Auwerx, J., Staels, B., 1994. J. Biol. Chem. 269, 31012–31018.
[76] Vu-Dac, N., Schoonjans, K., Kosykh, V., Dallongeville, J., Fruchart, J.C., Staels, B., Auwerx, J., 1995. J. Clin. Invest. 96, 741–750.
[77] Duez, H., Chao, Y.S., Hernandez, M., Torpier, G., Poulain, P., Mundt, S., Mallat, Z., Teissier, E., Burton, C.A., Tedgui, A., Fruchart, J.C., Fievet, C., Wright, S.D., Staels, B., 2002. J. Biol. Chem. 277, 48051–48057.
[78] Warden, C.H., Hedrick, C.C., Qiao, J.H., Castellani, L.W., Lusis, A.J., 1993. Science 261, 469–472.
[79] Tailleux, A., Bouly, M., Luc, G., Castro, G., Caillaud, J.M., Hennuyer, N., Poulain, P., Fruchart, J.C., Duverger, N., Fievet, C., 2000. Arterioscler. Thromb. Vasc. Biol. 20, 2453–2458.
[80] Faveeuw, C., Fougeray, S., Angeli, V., Fontaine, J., Chinetti, G., Gosset, P., Delerive, P., Maliszewski, C., Capron, M., Staels, B., Moser, M., Trottein, F., 2000. FEBS Lett. 486, 261–266.
[81] Gosset, P., Charbonnier, A.S., Delerive, P., Fontaine, J., Staels, B., Pestel, J., Tonnel, A.B., Trottein, F., 2001. Eur. J. Immunol. 31, 2857–2865.
[82] Hornung, D., Chao, V.A., Vigne, J.L., Wallwiener, D., Taylor, R.N., 2003. Gynecol. Obstet. Invest. 55, 20–24.
[83] Marx, N., Kehrle, B., Kohlhammer, K., Grub, M., Koenig, W., Hombach, V., Libby, P., Plutzky, J., 2002. Circ. Res. 90, 703–710.
[84] Kwak, B.R., Myit, S., Mulhaupt, F., Veillard, N., Rufer, N., Roosnek, E., Mach, F., 2002. Circ. Res. 90, 356–362.
[85] Harris, S.G., Phipps, R.P., 2001. Eur. J. Immunol. 31, 1098–1105.
[86] Jones, D.C., Ding, X., Daynes, R.A., 2002. J. Biol. Chem. 277, 6838–6845.
[87] Yang, Q., Xie, Y., Alexson, S.E., Nelson, B.D., DePierre, J.W., 2002. Biochem. Pharmacol. 63, 1893–1900.
[88] Meier, C.A., Chicheportiche, R., Juge-Aubry, C.E., Dreyer, M.G., Dayer, J.M., 2002. Cytokine 18, 320–328.
[89] Jiang, C., Ting, A.T., Seed, B., 1998. Nature 391, 82–86.
[90] Marx, N., Froehlich, J., Siam, L., Ittner, J., Wierse, G., Schmidt, A., Scharnagl, H., Hombach, V., Koenig, W., 2003. Arterioscler. Thromb. Vasc. Biol. 23, 283–288.
[91] Hinz, B., Brune, K., Pahl, A., 2003. Biochem. Biophys. Res. Commun. 302, 415–420.
[92] Park, E.J., Park, S.Y., Joe, E.H., Jou, I., 2003. J. Biol. Chem. 12, 12.
[93] Okada, M., Yan, S.F., Pinsky, D.J., 2002. FASEB J. 16, 1861–1868.
[94] Takata, Y., Kitami, Y., Yang, Z.H., Nakamura, M., Okura, T., Hiwada, K., 2002. Circ. Res. 91, 427–433.
[95] Xu, X., Otsuki, M., Saito, H., Sumitani, S., Yamamoto, H., Asanuma, N., Kouhara, H., Kasayama, S., 2001. Endocrinology 142, 3332–3339.
[96] Staels, B., Koenig, W., Habib, A., Merval, R., Lebret, M., Torra, I.P., Delerive, P., Fadel, A., Chinetti, G., Fruchart, J.C., Najib, J., Maclouf, J., Tedgui, A., 1998. Nature 393, 790–793.
[97] Delerive, P., Gervois, P., Fruchart, J.C., Staels, B., 2000. J. Biol. Chem. 275, 36703–36707.
[98] Marx, N., Mackman, N., Schonbeck, U., Yilmaz, N., Hombach, V.V., Libby, P., Plutzky, J., 2001. Circulation 103, 213–219.
[99] Neve, B.P., Corseaux, D., Chinetti, G., Zawadzki, C., Fruchart, J.C., Duriez, P., Staels, B., Jude, B., 2001. Circulation 103, 207–212.
[100] Zuckerman, S.H., Kauffman, R.F., Evans, G.F., 2002. Lipids 37, 487–494.
[101] Kleemann, R., Gervois, P.P., Verschuren, L., Staels, B., Princen, H.M., Kooistra, T., 2003. Blood 101, 545–551.
[102] Yamazaki, K., Kuromitsu, J., Tanaka, I., 2002. Biochem. Biophys. Res. Commun. 290, 1114–1122.
[103] Delerive, P., De Bosscher, K., Besnard, S., Vanden Berghe, W., Peters, J.M., Gonzalez, F.J., Fruchart, J.C., Tedgui, A., Haegeman, G., Staels, B., 1999. J. Biol. Chem. 274, 32048–32054.
[104] Fuell DL, F.M., Greenberg, A.S., Haffner, S., Chen, H., 2001. Diabetes 50, A435.
[105] Marx, N., Schonbeck, U., Lazar, M.A., Libby, P., Plutzky, J., 1998. Circ. Res. 83, 1097–1103.
[106] Law, R.E., Goetze, S., Xi, X.P., Jackson, S., Kawano, Y., Demer, L., Fishbein, M.C., Meehan, W.P., Hsueh, W.A., 2000. Circulation 101, 1311–1318.

[107] de Dios, S.T., Hannan, K.M., Dilley, R.J., Hill, M.A., Little, P.J., 2001. J. Diabetes Complications 15, 120–127.

[108] Goetze, S., Kim, S., Xi, X.P., Graf, K., Yang, D.C., Fleck, E., Meehan, W.P., Hsueh, W.A., Law, R.E., 2000. J. Cardiovasc. Pharmacol. 35, 749–757.

[109] Goetze, S., Kintscher, U., Kim, S., Meehan, W.P., Kaneshiro, K., Collins, A.R., Fleck, E., Hsueh, W.A., Law, R.E., 2001. J. Cardiovasc. Pharmacol. 38, 909–921.

[110] Takata, Y., Kitami, Y., Okura, T., Hiwada, K., 2001. J. Biol. Chem. 276, 12893–12897.

[111] Wakino, S., Kintscher, U., Liu, Z., Kim, S., Yin, F., Ohba, M., Kuroki, T., Schonthal, A.H., Hsueh, W.A., Law, R.E., 2001. J. Biol. Chem. 276, 47650–47657.

[112] Takeda, K., Ichiki, T., Tokunou, T., Funakoshi, Y., Iino, N., Hirano, K., Kanaide, H., Takeshita, A., 2000. Circulation 102, 1834–1839.

[113] Sugawara, A., Takeuchi, K., Uruno, A., Ikeda, Y., Arima, S., Sato, K., Kudo, M., Taniyama, Y., Ito, S., 2001. Hypertens. Res. 24, 229–233.

[114] Abe, M., Hasegawa, K., Wada, H., Morimoto, T., Yanazume, T., Kawamura, T., Hirai, M., Furukawa, Y., Kita, T., 2003. Arterioscler. Thromb. Vasc. Biol. 23, 404–410.

[115] Kintscher, U., Lyon, C., Wakino, S., Bruemmer, D., Feng, X., Goetze, S., Graf, K., Moustakas, A., Staels, B., Fleck, E., Hsueh, W.A., Law, R.E., 2002. Circ. Res. 91, e35–e44.

[116] Zhang, J., Fu, M., Zhu, X., Xiao, Y., Mou, Y., Zheng, H., Akinbami, M.A., Wang, Q., Chen, Y.E., 2002. J. Biol. Chem. 277, 11505–11512.

[117] Shu, H., Wong, B., Zhou, G., Li, Y., Berger, J., Woods, J.W., Wright, S.D., Cai, T.Q., 2000. Biochem. Biophys. Res. Commun. 267, 345–349.

[118] Marx, N., Sukhova, G., Murphy, C., Libby, P., Plutzky, J., 1998. Am. J. Pathol. 153, 17–23.

[119] Hourton, D., Stengel, D., Chapman, M.J., Ninio, E., 2001. Eur. J. Biochem. 268, 4489–4496.

[120] Oyama, Y., Kurabayashi, M., Akuzawa, N., Nagai, R., 2000. J. Atheroscler. Thromb. 7, 77–82.

[121] Fu, M., Zhang, J., Lin Yg, Y., Zhu, X., Willson, T.M., Chen, Y.E., 2002. Biochem. Biophys. Res. Commun. 294, 597–601.

[122] Browner, W.S., Lui, L.Y., Cummings, S.R., 2001. J. Clin. Endocrinol. Metab. 86, 631–637.

[123] Speer, M.Y., McKee, M.D., Guldberg, R.E., Liaw, L., Yang, H.Y., Tung, E., Karsenty, G., Giachelli, C.M., 2002. J. Exp. Med. 196, 1047–1055.

[124] Nigro, J., Dilley, R.J., Little, P.J., 2002. Atherosclerosis 162, 119–129.

[125] Kuzuya, M., Ramos, M.A., Kanda, S., Koike, T., Asai, T., Maeda, K., Shitara, K., Shibuya, M., Iguchi, A., 2001. Arterioscler. Thromb. Vasc. Biol. 21, 765–770.

[126] Jozkowicz, A., Dulak, J., Piatkowska, E., Placha, W., Dembinska-Kiec, A., 2000. Acta Biochim. Pol. 47, 1147–1157.

[127] Panigrahy, D., Singer, S., Shen, L.Q., Butterfield, C.E., Freedman, D.A., Chen, E.J., Moses, M.A., Kilroy, S., Duensing, S., Fletcher, C., Fletcher, J.A., Hlatky, L., Hahnfeldt, P., Folkman, J., Kaipainen, A., 2002. J. Clin. Invest. 110, 923–932.

[128] Kockx, M.M., De Meyer, G.R., Muhring, J., Jacob, W., Bult, H., Herman, A.G., 1998. Circulation 97, 2307–2315.

[129] Roberts, R.A., Chevalier, S., Hasmall, S.C., James, N.H., Cosulich, S.C., Macdonald, N., 2002. Toxicology 181–182, 167–170.

[130] Boitier, E., Gautier, J.C., Roberts, R., 2003. Comp. Hepatol. 2, 3.

[131] Geng, Y.J., Libby, P., 2002. Arterioscler. Thromb. Vasc. Biol. 22, 1370–1380.

[132] Chinetti, G., Griglio, S., Antonucci, M., Torra, I.P., Delerive, P., Majd, Z., Fruchart, J.C., Chapman, J., Najib, J., Staels, B., 1998. J. Biol. Chem. 273, 25573–25580.

[133] Patel, L., Pass, I., Coxon, P., Downes, C.P., Smith, S.A., Macphee, C.H., 2001. Curr. Biol. 11, 764–768.

[134] Bishop-Bailey, D., Hla, T., 1999. J. Biol. Chem. 274, 17042–17048.

[135] Martinet, W., Kockx, M.M., 2001. Curr. Opin. Lipidol. 12, 535–541.

[136] Ikeda, Y., Sugawara, A., Taniyama, Y., Uruno, A., Igarashi, K., Arima, S., Ito, S., Takeuchi, K., 2000. J. Biol. Chem. 275, 33142–33150.

[137] Sugawara, A., Uruno, A., Kudo, M., Ikeda, Y., Sato, K., Taniyama, Y., Ito, S., Takeuchi, K., 2002. J. Biol. Chem. 277, 9676–9683.

[138] Durrington, P.N., Mackness, M.I., Bhatnagar, D., Julier, K., Prais, H., Arrol, S., Morgan, J., Wood, G.N., 1998. Atherosclerosis 138, 217–225.

[139] Kato, K., Satoh, H., Endo, Y., Yamada, D., Midorikawa, S., Sato, W., Mizuno, K., Fujita, T., Tsukamoto, K., Watanabe, T., 1999. Biochem. Biophys. Res. Commun. 258, 431–435.

[140] Marx, N., Bourcier, T., Sukhova, G.K., Libby, P., Plutzky, J., 1999. Arterioscler. Thromb. Vasc. Biol. 19, 546–551.

[141] Mehta, D., Angelini, G.D., Bryan, A.J., 1996. Int. J. Cardiol. 56, 235–257.

[142] Claudel, T., Leibowitz, M.D., Fievet, C., Tailleux, A., Wagner, B., Repa, J.J., Torpier, G., Lobaccaro, J.M., Paterniti, J.R., Mangelsdorf, D.J., Heyman, R.A., Auwerx, J., 2001. Proc. Natl Acad. Sci. USA 98, 2610–2615.

[143] Chen, Z., Ishibashi, S., Perrey, S., Osuga, J., Gotoda, T., Kitamine, T., Tamura, Y., Okazaki, H., Yahagi, N., Iizuka, Y., Shionoiri, F., Ohashi, K., Harada, K., Shimano, H., Nagai, R., Yamada, N., 2001. Arterioscler. Thromb. Vasc. Biol. 21, 372–377.

[144] Levi, Z., Shaish, A., Yacov, N., Levkovitz, H., Trestman, S., Gerber, Y., Cohen, H., Dvir, A., Rhachmani, R., Ravid, M., Harats, D., 2003. Diabetes Obes. Metab. 5, 45–50.

[145] Yue Tl, T.L., Chen, J., Bao, W., Narayanan, P.K., Bril, A., Jiang, W., Lysko, P.G., Gu, J.L., Boyce, R., Zimmerman, D.M., Hart, T.K., Buckingham, R.E., Ohlstein, E.H., 2001. Circulation 104, 2588–2594.

[146] Khandoudi, N., Delerive, P., Berrebi-Bertrand, I., Buckingham, R.E., Staels, B., Bril, A., 2002. Diabetes 51, 1507–1514.

[147] Tabernero, A., Schoonjans, K., Jesel, L., Carpusca, I., Auwerx, J., Andriantsitohaina, R., 2002. BMC Pharmacol. 2, 10.

[148] Tordjman, K., Bernal-Mizrachi, C., Zemany, L., Weng, S., Feng, C., Zhang, F., Leone, T.C., Coleman, T., Kelly, D.P., Semenkovich, C.F., 2001. J. Clin. Invest. 107, 1025–1034.

[149] Steiner, G., 1996. Diabetologia 39, 1655–1661.

[150] Ericsson, C.G., Hamsten, A., Nilsson, J., Grip, L., Svane, B., de Faire, U., 1996. Lancet 347, 849–853.

[151] Frick, M.H., Syvanne, M., Nieminen, M.S., Kauma, H., Majahalme, S., Virtanen, V., Kesaniemi, Y.A., Pasternack, A., Taskinen, M.R., 1997. Circulation 96, 2137–2143.

[152] Rubins, H.B., Robins, S.J., Collins, D., Fye, C.L., Anderson, J.W., Elam, M.B., Faas, F.H., Linares, E., Schaefer, E.J., Schectman, G., Wilt, T.J., Wittes, J., 1999. N. Engl. J. Med. 341, 410–418.

[153] Frick, M.H., Elo, O., Haapa, K., Heinonen, O.P., Heinsalmi, P., Helo, P., Huttunen, J.K., Kaitaniemi, P., Koskinen, P., Manninen, V., Mäenpää, H., Mälkönen, M., Mänttäri, M., Norola, S., Pasternack, A., Pikkarainen, J., Romo, M., Jjöblom, T., Nikkilä, E.A., 1987. Helsinki Heart Study: primary-prevention trial with gemfibrozil in middle-aged men with dyslipidemia. Safety of treatment, changes in risk factors, and incidence of coronary heart disease. N. Engl. J. Med. 317, 1237–1245.

[154] Elkeles, R.S., Diamond, J.R., Poulter, C., Dhanjil, S., Nicolaides, A.N., Mahmood, S., Richmond, W., Mather, H., Sharp, P., Feher, M.D., 1998. Diabetes Care 21, 641–648.

[155] Vega, G.L., Ma, P.T., Cater, N.B., Filipchuk, N., Meguro, S., Garcia-Garcia, A.B., Grundy, S.M., 2003. Am. J. Cardiol. 91, 956–960.

[156] Black, D.M., 2003. Curr. Atheroscler. Rep. 5, 29–32.

[157] Kipnes, M.S., Krosnick, A., Rendell, M.S., Egan, J.W., Mathisen, A.L., Schneider, R.L., 2001. Am. J. Med. 111, 10–17.

[158] Kaplan, F., Al-Majali, K., Betteridge, D.J., 2001. J. Cardiovasc. Risk 8, 211–217.

[159] Gegick, C.G., Altheimer, M.D., 2001. Endocr. Pract. 7, 162–169.

[160] Smith, U., 2001. Int. J. Clin. Pract. Suppl., 121, 13–18.

[161] Minamikawa, J., Tanaka, S., Yamauchi, M., Inoue, D., Koshiyama, H., 1998. J. Clin. Endocrinol. Metab. 83, 1818–1820.

[162] Ebdrup, S., Pettersson, I., Rasmussen, H.B., Deussen, H.J., Frost Jensen, A., Mortensen, S.B., Fleckner, J., Pridal, L., Nygaard, L., Sauerberg, P., 2003. J. Med. Chem. 46, 1306–1317.

The role of the steroidogenic acute regulatory (StAR) protein in intramitochondrial cholesterol transfer and steroidogenesis

Douglas M. Stocco*

Department of Cell Biology and Biochemistry, Texas Tech University Health Sciences Center, Lubbock, TX 79430, USA

**Correspondence address: Tel.: + 1-806-743-2505; fax: + 1-806-743-2990*

E-mail: doug.stocco@ttmc.ttuhsc.edu(D.M.S.)

1. Background

In most cells, the intracellular distribution of cholesterol is carefully regulated, although the mechanisms involved in this regulation are not well understood. Much of the cholesterol found in cells can be found in the plasma membrane where it serves to affect membrane fluidity and other cellular membrane functions. Plasma membrane cholesterol, however, is not static and a constant cycling of this sterol occurs between the plasma membrane and other intracellular compartments such as the Golgi, lysosomes, peroxisomes, and the endoplasmic reticulum. In addition to the normal cycling of cholesterol between the plasma membrane and other cellular organelles that occurs to maintain cellular homeostasis in all mammalian cells, there exists a group of specialized cells, the steroidogenic cells, in which cholesterol transport has become highly evolved to perform a cell specific function.

In these cells cholesterol serves as the substrate for the synthesis of all steroid hormones. Steroid hormones make up a very important class of regulatory molecules that are synthesized mainly in the adrenal glands, the ovary, and the testis in response to steroidogenic stimuli. The importance of these compounds is readily seen when one considers that the adrenal gland produces glucocorticoids that serve to regulate carbohydrate metabolism and combat stress and also synthesize mineralocorticoids that regulate salt balance and maintain blood pressure. Further, in the female, estrogen and progesterone are synthesized in the ovary and placenta are essential for reproductive function and also function to maintain secondary sex characteristics. Lastly, in the male, testicular Leydig cells synthesize androgens that are responsible for maintaining reproductive function and secondary sex characteristics. These steroids are synthesized rapidly in the appropriate cell type in response to signals from the anterior pituitary gland.

Advances in Molecular and Cell Biology, Vol. 33, pages 561–577

ISSN: 1569-2558/DOI: 10.1016/S1569-2558(03)33027-9

The signals consist of the trophic hormones, namely adrenocorticotropic hormone (ACTH), luteinizing hormone (LH), or follicle stimulating hormone (FSH), and all transduce their effects through the cAMP second messenger pathway. In addition, the brain synthesizes neurosteroids from cholesterol and these compounds have been shown to have a number of different physiological functions. As stated above, while the steroid hormones can be distinguished from one another by their diverse physiological actions in the body they share a common feature in that they are all synthesized from cholesterol.

The biosynthesis of all steroids begins with the cleavage of a six carbon unit from the 27 carbon cholesterol molecule to form the first steroid synthesized, the 21 carbon-containing steroid, pregnenolone. This reaction is catalyzed by the cytochrome P450 side chain cleavage enzyme (P450*scc*), which is part of the cholesterol side chain cleavage enzyme system (CSCC) that is located on the matrix side of the inner mitochondrial membrane [1,2]. Thus, in steroidogenic cells, the mitochondrion represents an important delivery and metabolic site for intracellular cholesterol, and will be discussed in greater detail later.

Historically, the trophic hormone induced increase in the activity of the P450*scc* enzyme was long considered to be the rate-limiting step in steroidogenesis. However, with subsequent experimentation it became clear that the activity of the P450*scc* enzyme was not rate limiting in this process [3]. Rather, it was determined that to initiate and sustain steroidogenesis a constant supply of the substrate cholesterol for steroid biosynthesis must be delivered to the site of cleavage in the inner mitochondrial membrane where the P450*scc* enzyme resides. The stores of cholesterol found inside steroidogenic cells may be supplied from serum in the form of high-density lipoprotein or low-density lipoprotein depending on the species and cell type in question [4,5], or, from what appears to be a less important source of cholesterol in this type of cell, de novo synthesis from acetate. Given adequate intracellular cholesterol supplies, two equally important processes must occur. The first of these processes is the mobilization of cholesterol from cellular stores such as lipid droplets or other cellular membranes to the outer mitochondrial membrane. This constitutes an important process but is one that is not well understood at this time. The second process, the transfer of this cholesterol from the outer to the inner mitochondrial membrane and the P450*scc* enzyme [6,7], is in fact, the true rate-limiting step in steroidogenesis [8–12]. Proof of this was provided when hydroxylated analogs of cholesterol such as 22R-hydroxycholesterol, 20α-hydroxycholesterol, or 25-hydroxy-cholesterol, which can readily diffuse across the mitochondrial membranes to the P450*scc*, were placed on steroidogenic cells and resulted in high levels of steroids in the absence of hormone stimulation [13,14]. These observations indicated that the P450*scc* was fully active in unstimulated cells and that it was the lack of substrate cholesterol for cleavage that prevented the production of pregnenolone and eventually downstream steroids. As a result of observations such as these, it became clear that in the acute regulation of steroid hormone biosynthesis the major barrier to be overcome was the translocation of cholesterol to the P450*scc* through the aqueous space between the outer and inner mitochondrial membranes. Since cholesterol is a hydrophobic compound, its diffusion through this aqueous layer is very slow [15,16], and could not provide sufficient substrate to account for the rapid and large increase in steroid production observed. Therefore, it followed that stimulation of steroidogenesis required a mechanism that rapidly resulted in the transport of this steroid precursor across this aqueous barrier.

To summarize, the production of steroid hormones is controlled by events that facilitate the transport of cholesterol from cellular stores, first to the mitochondrial outer membrane and second, across the aqueous inter-membrane space of the mitochondria to the inner membrane. While both processes are necessary to ensure maximal rates of steroid production in response to stimulation, it is the second process that is known to be the rate-limiting step in hormone-regulated steroidogenesis. This review will attempt to focus on what is currently understood about the mechanism involved in the movement of cholesterol from the outer mitochondrial membrane to the inner mitochondrial membrane.

2. The acute regulation of steroidogenesis

The factors involved in and the mechanisms regulating the rapid production of steroids in response to trophic hormone stimulation have been the subject of intense investigation for over four decades. Regardless of the type of steroidogenic cell studied, the acute responses to trophic hormone stimulation usually share several of the same characteristics. Namely, the initial reaction in the steroid producing pathway occurs in an identical location, the inner mitochondrial membrane, and is catalyzed by the CSCC enzyme system in all steroidogenic tissues. In addition, steroidogenic responses are stimulated by trophic hormones in a dose- and time-dependent manner.

Early studies characterizing the regulation of steroid biosynthesis were performed in the adrenal gland where it had been observed that ACTH could stimulate the biosynthesis of steroids in vitro [17,18]. Using this in vitro system, one of the first and most fundamental observations concerning steroidogenesis was that acute steroid production in response to hormone stimulation had an absolute requirement for the synthesis of new proteins. This simple observation became pivotal to our understanding of the nature of the acute regulation of steroid hormone biosynthesis, and became one of the foundations for future studies designed to elucidate this regulation. The first of such studies were performed by Ferguson [19,20] who demonstrated that the acute stimulation of corticoid synthesis in adrenal glands by ACTH was sensitive to the protein synthesis inhibitor puromycin. At approximately the same time, Garren and co-workers also demonstrated that steroidogenesis in adrenal tissue was dependent upon the ACTH stimulated synthesis of new proteins [21,22]. Importantly, these studies also indicated that the regulated step was distal to cholesterol ester hydrolysis but proximal to cholesterol side chain cleavage, placing it precisely at the locus of the delivery of cholesterol to the P450*scc* enzyme [23]. Following these initial observations, many subsequent studies confirmed the need for de novo protein synthesis in the hormone regulated, acute production of steroids [8,11, 24–27]. Simpson and Boyd [1] made an important observation when they determined that the cycloheximide sensitive step was located in the mitochondria, and an important addition to this observation was made by Arthur and Boyd who noted that protein synthesis inhibitors had no effect on the activity of the P450*scc* itself [28]. The field progressed when it was demonstrated that inhibition of protein synthesis had no effect on the increased delivery of cellular cholesterol to the outer mitochondrial membrane, but that the delivery of this substrate from the outer membrane to the inner mitochondrial membrane was completely inhibited by cycloheximide [11,29]. As a result of these and many additional studies, the precise site of the cycloheximide-inhibited regulation had been pinpointed to the transfer of cholesterol to the P450*scc* enzyme in the inner mitochondrial membrane. Given the observations with protein synthesis inhibitors, it was assumed that a newly synthesized protein was responsible for transferring cholesterol to the inner mitochondrial membrane.

Summarizing many observations has resulted in the overall characterization of the acute regulation of steroidogenesis. Essentially, these characteristics indicated that the acute production of steroids was dependent upon a hormone stimulated, rapidly synthesized, cycloheximide sensitive protein whose function was to mediate the transfer of cholesterol from the outer mitochondrial membrane to the inner mitochondrial membrane and the P450*scc* enzyme. The effort to identify and characterize this acute regulatory protein(s) was continuous since the early observations of Ferguson and Garren and their colleagues. Several candidate proteins emerged from these efforts. A listing of these proteins and the data supporting their candidacies has been collectively reviewed [7], as has the characteristics for additional protein candidates [30]. This review will attempt to summarize what is known about the process of cholesterol transfer to the inner mitochondrial membrane in support of steroidogenesis and will focus mainly on the role of the StAR protein in cholesterol transfer.

3. The steroidogenic acute regulatory protein

The StAR protein has been proposed as the acute regulator of mediated cholesterol transfer to the inner mitochondrial membrane and hence, steroid biosynthesis. The StAR protein was initially described by Orme-Johnson and colleagues as an ACTH-induced 30 kDa phosphoprotein in hormone-treated rat and mouse adrenocortical cells, and as an LH-induced protein in rat corpus luteum cells and mouse Leydig cells [31–34]. Their carefully executed studies indicated that a close relationship between the appearance of the 30 kDa proteins and steroid hormone biosynthesis existed and that the synthesis of these proteins, as was steroidogenesis, was sensitive to cycloheximide. Proteins identical to those described by Orme-Johnson have also been characterized in hormone stimulated MA-10 mouse Leydig tumor cells by Stocco and colleagues [35–38]. In both laboratories, these proteins were localized to the mitochondria and consisted of several isoforms of a newly synthesized 30 kDa protein. In addition to the 30 kDa mitochondrial proteins, a 37 kDa precursor form of these proteins was also detected, a common observation with mitochondrial proteins [33,36]. The 37 kDa precursor contained N-terminal amino acid sequences that served to target the protein to the mitochondria and facilitate its import and processing by mitochondria. While many correlations between this protein and steroidogenesis existed, a direct cause-and-effect relationship between 37 kDa protein expression and steroidogenesis was lacking, and it became increasingly clear that it would be necessary to clone the cDNA for this protein to prove if it had a function in steroidogenesis. The cloning of the cDNA for the 37 kDa protein was successfully accomplished in 1994, when sufficient protein for sequence analysis was obtained from MA-10 cells [39]. The nucleic acid and protein sequences were found to be unique indicating the 37 kDa protein represented a novel protein. Transient transfection experiments demonstrated that expression of the cDNA-derived protein in MA-10 mouse Leydig tumor cells resulted in a significant increase in steroid production in the absence of hormone stimulation [39]. In addition, transient transfection of COS-1 cells with the cDNA for the 37 kDa protein resulted in a several-fold increase in the conversion of cholesterol to pregnenolone [8,40,41]. These results provided the cause-and-effect observation that had been previously lacking and indicated a direct role for the 37 kDa protein in hormone-regulated steroid production. As a result of these observations, the protein was named the Steroidogenic Acute Regulatory (StAR) protein [39]. Since the cDNA cloning of StAR in 1994, subsequent studies performed on the StAR protein strongly indicate that it is the best candidate for the rapidly synthesized factor that acutely regulates steroid hormone biosynthesis by mediating the delivery of cholesterol to the inner mitochondrial membrane.

4. StAR knockouts

Shortly after the cloning of the StAR cDNA, powerful corroborative proof for the essential role for StAR in regulated steroidogenesis was forthcoming. Congenital lipoid adrenal hyperplasia (lipoid CAH) is a lethal condition resulting from an almost complete inability of the newborn to synthesize steroids. This condition is inherited in an autosomal recessive pattern and is manifested by the presence of large adrenals containing very high levels of cholesterol and cholesterol esters and also by an increased amount of lipid accumulation in testicular Leydig cells. This disease was originally thought to be due to a mutation of the P450*scc* (20,22 desmolase) enzyme, which would necessarily result in a decreased capacity to convert cholesterol to downstream steroids [42], and this belief had persisted until relatively recent times [43]. However, with the cloning of the P450*scc* cDNA, it was determined that the gene for this enzyme was normal in several patients who had suffered from this disease [44]. Subsequently, it was deduced that the defect in lipoid CAH was upstream of P450*scc* at the point of cholesterol delivery to the enzyme, and StAR became an immediate candidate for scrutiny. In studies designed to determine if StAR was involved in lipoid CAH, Lin et al. [41] prepared StAR cDNA from testicular tissue of patients with lipoid CAH and identified three nonsense mutations and one deletion in the sequences of the first four patients analyzed. These mutations were confirmed in the genomic DNA and, importantly, while expression of the normal human StAR protein in COS-1 cells resulted in an eightfold increase in steroid production, expression of the StAR cDNA isolated from these patients indicated the proteins were completely inactive in promoting steroidogenesis. In addition to the first reports on StAR mutations causing lipoid CAH, many additional examples of mutations in StAR resulting in this disease have been and continue to be reported. At this time approximately 30 different mutations in the StAR gene resulting in lipoid CAH have been described.

That mutations in the StAR gene resulted in lipoid CAH in humans produced compelling evidence for the essential role of this protein in the acute regulation of steroidogenesis. Since the initial observation in humans, a knockout of the StAR gene has been produced in mice by Caron et al. [45]. Characterization of these mice indicated that regardless of genotype, all mice had female external genitalia, as is the case in the human condition. Following birth, all animals failed to grow normally and death within a short period of time occurred, presumably as a result of adrenocortical insufficiency. This was confirmed by the observation that serum levels of corticosterone and aldosterone were depressed while levels of ACTH and corticotropin releasing hormone (CRH) were elevated. These observations indicated an inhibition in the production of adrenal steroids with an accompanying loss of feedback regulation at the level of the hypothalamus or pituitary. Inspection of the adrenal gland revealed a normal medulla but an abnormal cortex, having a disrupted fascicular zone. Specific staining procedures revealed elevated lipid deposits in the adrenal cortex region of the StAR knockout mouse. While the StAR knockout mice were all phenotypically sex reversed, the testes of these animals appeared normal upon gross inspection. However, once again specific staining indicated the presence of elevated levels of lipid within this organ. In contrast, the ovaries of the StAR knockout mice were essentially indistinguishable from wild type animals, similar to the situation found with human StAR mutations [46–48]. When these animals are rescued and kept alive with appropriate mineralocorticoid therapy, huge deposits of lipid can be detected in the adrenals and testes of these animals indicating their inability to convert cholesterol to pregnenolone despite normal P450*scc* enzyme activity. More recent studies from the Parker laboratory have demonstrated that in StAR/gonadotropin deficient (hypogonadal, hpg) double mutant mice, gonadotropins are not only required for the expression of StAR, but also for the accumulation of gonadal lipids that will be utilized for steroid biosynthesis [49]. In this same study, using a StAR/apolipoprotein A-1 double mutant they further showed that the predominant source of steroidogenic lipid was from circulating high-density lipoprotein-derived cholesterol.

Clearly, the observations made in humans suffering from lipoid CAH and the subsequent production of StAR knockout mice all point to the indispensable role of StAR in regulated steroidogenesis.

5. Mechanism of action of StAR

One of the major questions being pursued in the area of regulated steroidogenesis is that of how StAR functions in mediating cholesterol transfer to the inner mitochondrial membrane. It is important to state at the outset, that while excellent work has been performed on this complex question, the entire answer has not yet been elucidated. There has been ample proof presented that StAR is required to mediate cholesterol transfer to the inner mitochondrial membrane and it is understandable that the question of how it does this would be of great interest and importance. Taking into consideration the observation that StAR is rapidly imported and processed by the mitochondria we originally proposed a model hypothesizing that during the course of import, contact sites were formed between the outer and inner mitochondrial membranes and, in some unknown manner, cholesterol could be transferred to the inner membrane during this process [7,39]. This model proposed that the formation of contact sites collapsed the inter-mitochondrial membrane aqueous space that served as a barrier to keep the hydrophobic cholesterol from crossing to the inner membrane and in this way allowed for its transfer via this newly formed lipid bridge. This model seemed reasonable in that other investigators had demonstrated the existence of contact sites at loci where proteins were imported into the mitochondria [50,51]. However, the validity of this model was severely questioned when it was found that when transfected into and expressed in intact cells, StAR proteins lacking the N-terminal 62 amino acids (and thus all of the signal sequence) supported steroid synthesis to the same extent as wild type StAR, yet never entered the mitochondria [52]. This indicated that StAR could act on the outside of the mitochondria to mediate cholesterol transfer to the inner membrane. Similarly, recombinant StAR lacking the N-terminal 62 amino acids could fully support steroid synthesis in isolated mitochondria, once again, without entering the organelle [53]. To further illustrate this point, Bose et al. utilized chimeric constructs of StAR bearing N-terminal sequences that targeted the protein to the outer mitochondrial membrane, the intramembranous mitochondrial space, the inner mitochondrial membrane or the matrix space and observed their capacities to stimulate steroidogenesis in transfected COS-1 cells [54]. Their data clearly showed that only StAR that was targeted to the outer mitochondrial membrane was capable of significantly stimulating steroid synthesis, thus corroborating earlier observations. However, this finding is in contrast to the studies of Artemenko and colleagues who utilized rat adrenal cells to demonstrate that it is the targeting of the phosphorylated form of StAR to the inner mitochondrial membrane that provides the dominant cholesterol transport mechanism in steroidogenic cells [55]. Clearly, this remains a controversial area.

Studies on the mechanism of cholesterol transport by StAR received a boost when experiments employing constructs truncated in their C-terminal end were performed. These experiments indicated that the cholesterol transferring capability of the StAR protein resided in the C-terminal portion of the molecule [41,46,56]. We now know that virtually all of the mutations causing lipoid CAH are found in the C-terminal region of the StAR protein [57], thus indicating a critical role for this segment of the protein in cholesterol transfer. The role of the C-terminal region of StAR became even more intensively investigated when it was found that a protein known as MLN64 contained an amino acid sequence which was highly homologous to the C-terminus of StAR [58], and could stimulate steroid synthesis when transfected into COS-1 cells [59]. MLN64 was subsequently found to be compartmentalized in late endosomes and a role in cytosolic cholesterol transport was hypothesized [60]. Interestingly, while MNL64 is found in many tissues, it is also found in the human placenta where it can be proteolytically cleaved to produce a fragment that has significant steroidogenic activity [61]. Conversely, StAR is not found in the human placenta and it has been speculated that MLN64 may play a role in steroidogenesis in this tissue [62]. In a recent and interesting study, Zhang et al. have demonstrated that MLN64 can also participate in intracellular cholesterol trafficking by mediating the transfer of lysosomal cholesterol to mitochondria in steroidogenic cells [63]. The potential mechanism of this transfer will be discussed in greater detail later when the function of START domains is discussed.

In the initial time period following the discovery of StAR, information on the mechanism by which StAR mediated cholesterol transfer to the inner mitochondrial membrane was indeed scarce. In one effort to explain StAR action, Kallen et al. [64], demonstrated that StAR can act as a sterol transfer protein and that the function of the StAR protein may be to enhance desorption of cholesterol from one sterol containing membrane to another. In this model, StAR is specifically directed to the mitochondria via its N-terminus and upon its arrival at the outer mitochondrial membrane the C-terminus produces alterations in this membrane that in some manner results in the transfer of cholesterol from the outer to the inner membrane. Interestingly, they observed that the transfer of cholesterol to both trypsin-treated and heat-treated mitochondria or to heat-treated microsomes by purified StAR protein was specific in that identical experiments employing phosphatidylcholine failed to show transfer of this phospholipid. These observations were particularly pertinent to the situation found in steroidogenic mitochondria in which desorption of cholesterol from the sterol-rich outer membrane to the sterol-poor inner membrane [65], would serve to enhance pregnenolone synthesis by the P450*scc* enzyme. While indicating that StAR had this capability, this study was not able to provide a molecular mechanism of how StAR could act as a sterol carrier protein. Recent studies by Petrescu et al. were highly consistent with the observation that StAR could function as a sterol transfer protein [66]. They demonstrated that cholesterol could interact with a hydrophobic region within the StAR molecule and predicted that each StAR molecule contained two cholesterol binding sites. They concluded from their studies that StAR was a cholesterol binding protein that could preferentially enhance the transfer of cholesterol when steroidogenic mitochondria were the donors and finally, that StAR could interact with mitochondrial membranes to alter the structure of the sterol domain within the membranes.

Is StAR action specific to steroidogenic tissues? In a recent study, Pandak et al. have been able to demonstrate that when transfected into primary rat hepatocytes, StAR can facilitate cholesterol transfer into the mitochondria to stimulate bile acid synthesis via the "alternative" pathway [67]. This pathway utilizes the sterol 27 hydroxylase enzyme (CYP27) which is located in the inner mitochondrial membrane of hepatocytes, similar to the situation found with the P450*scc* enzyme in steroidogenic tissues. In this study they concluded that the alternative pathway was the rate-limiting pathway in bile acid synthesis. However, more importantly, from the point of view of understanding StAR action, it demonstrated that StAR can transfer cholesterol across mitochondrial membranes in a nonsteroidogenic cell indicating that its sterol transferring capacity might be more universal than simply supporting steroidogenesis. Once again, when the function of the START domain is discussed, the importance of this observation will be re-visited in this review.

In order to determine structure–function relationships in the StAR protein, another approach was taken by Miller and colleagues. They attempted to determine the physical characteristics of the StAR protein under different physico-chemical conditions and then utilize these characteristics to provide insights into the mechanism of StAR action. In one study they subjected StAR to limited proteolysis at different pH values and found that the molecule behaves differently as the pH decreases [68]. They demonstrated that at pH values in the 3.5–4.0 range StAR undergoes conformational changes that result in a partial unfolding of the protein and a transition to a molten globule state. Molten globules are structures within proteins that have lost at least some of their tertiary structure but which have retained virtually all of their secondary structure. They speculated that if the pH microenvironment surrounding the mitochondria is acidic, possibly as a result of the expulsion of protons from the mitochondrial matrix by the proton pump and/or by the presence of the negatively charged head groups of the phospholipids in the outer mitochondrial membrane, the StAR molecule may undergo a conformational shift. They further hypothesized that as the transition to a molten globule occurs, this structural change could result in an opening of the StAR protein perhaps exposing a hydrophobic region or, it may prolong the interval with which StAR can reside on the outer membrane, thus allowing increased transfer of cholesterol during this period.

With the demonstration that StAR was fully active even without its signal sequence, that it could act as a sterol carrier protein and that it could form a molten globule while interacting with the mitochondrial outer membrane a scenario in which StAR might be causing perturbations of the outer membrane that resulted in cholesterol movement from the outer to the inner mitochondrial membrane emerged. An obvious possibility was that StAR was interacting with other mitochondrial outer membrane proteins and/or phospholipids to produce this effect. However, attempts to identify such binding partners using the yeast two-hybrid system, co-immunoprecipitation and binding assays utilizing radioactive StAR and isolated mitochondria have thus far failed to produce any components which specifically interact with StAR [69]. It should be noted that the methods utilized to identify StAR binding partners can be technically difficult and subject to artifacts but, nevertheless, have produced few positive results to date. In support of these findings it was demonstrated that StAR could promote cholesterol transfer into mitochondria in which the outer membrane proteins have been removed by partial proteolysis with trypsin [64]. Also, Tuckey et al. [70] have recently demonstrated that recombinant N-62 StAR was able to transfer cholesterol from donor phospholipid vesicles containing cholesterol to acceptor vesicles containing P450*scc* in the complete absence of other mitochondrial or cytosolic proteins. This would suggest that StAR does not require protein binding partners on the outer mitochondrial membrane to transfer cholesterol and can instead, interact directly with membrane phospholipids. However, the need for other mitochondrial proteins cannot yet be completely ruled out and in this light, a recent study using fluorescence energy transfer has demonstrated that StAR and the peripheral benzodiazepine receptor (PBR) are closely associated on the outer mitochondrial membrane, being less than 100 Å from each other [71]. PBR is a membrane protein found in high abundance in the outer mitochondrial membrane of steroidogenic cells and has been shown to be involved in cholesterol delivery to the inner mitochondrial membrane [72]. Based on this association, the authors proposed a model in which StAR targets cholesterol to the PBR that then facilitates its transfer to the inner mitochondrial membrane. The involvement of PBR in transferring cholesterol to the inner mitochondrial membranes of steroidogenic cells was also hypothesized in a recent review article by Jefcoate in which he envisioned PBR as generating cholesterol-rich domains in the outer mitochondrial membrane that would eventually be transferred to the inner mitochondrial membrane [73]. While PBR appears to be involved in cholesterol transfer to the inner mitochondrial membrane, little is known concerning the mechanism of its action in this process, and, in any event, it appears that StAR can transfer cholesterol into mitochondria and into phospholipid vesicles in the absence of other proteins [64,70]. Whether or not this transfer would be greatly escalated in the presence of PBR is a possibility that remains to be proven. The concept that StAR interacts with other mitochondrial proteins to effectively transfer cholesterol to the inner membrane clearly requires additional study.

6. START domains

As indicated earlier, the cholesterol-transferring region of the StAR protein appears to be located in the C-terminal region of the protein as demonstrated in studies with both N-terminally truncated StAR proteins and MLN64. A hint of how the C-terminus of these proteins might transfer cholesterol was put forth by Ponting and Aravind [74], who demonstrated that sequences in the C-terminus of StAR are homologous to sequences in several other proteins, including MNL64, which display diverse functions. They named these sequences START domains, for StAR Related lipid Transfer domains. START domains consist of approximately 200–210 amino acid stretches and the common characteristic of these domains is that they are capable of binding lipids. Thus, with this information, the possibility that StAR was a lipid binding carrier protein would have to be considered. This possibility received an exciting boost when Tsujishita and Hurley [75], succeeded in obtaining crystals and solving the structure for the START domain of the MNL64 protein. Because of problems in purifying and crystallizing the START domain from StAR they focused on the START domain of MLN64 that shows the highest degree of homology to the StAR–START domain (37%), and found that it readily crystallized. They demonstrated that both StAR–START and MLN64–START could bind cholesterol in essentially an identical manner and that binding occurred in a ratio of 1:1. These studies demonstrated for the first time that the START domains of both MLN64 and StAR behaved similarly and thus, was a confirmation of earlier studies that had been performed in vitro using transfected cells [56,59]. The crystal structure of the MLN64–START at 2.2 Å indicated that it consisted of an $\alpha + \beta$ fold built around a U-shaped incomplete β-barrel. MLN64–START contains a nine stranded antiparallel β-sheet, 4-α-helices, and 2-Ω-loops. Most importantly, the tertiary structure of MLN64–START revealed a hydrophobic tunnel that was $26 \times 12 \times 11$ Å in size and was large enough to bind a single molecule of cholesterol. This tunnel extended almost the entire length of the molecule. Interestingly, when three of the most common mutations resulting in lipoid CAH are projected onto the MLN64–START domain model they are all found to reside quite close to each other in the tertiary structure and two of these mutations reside within the cholesterol-binding hydrophobic tunnel. These mutations would be expected to disrupt the structure of the tunnel and quite likely result in a decrease in cholesterol binding. Based on these findings, the authors proposed that StAR functioned in transferring cholesterol to the inner mitochondrial membrane via its ability to bind and act as an inter-mitochondrial membrane cholesterol shuttling protein. However, several aspects of this model are in conflict with observations that have been made and with some hypotheses of other models that have been proposed. In support of the role of the START domain in sterol transfer, Strauss and colleagues have recently demonstrated that the MLN64–START domain is able to transfer cholesterol between intracellular membranes and can stimulate mitochondrial steroidogenesis in vitro [64]. They demonstrated that MLN64 can transfer lysosomal cholesterol to steroidogenic mitochondria and that this cholesterol can be utilized for steroid biosynthesis. They concluded from this study that the START domain of MLN64 plays an important role in the maintenance of intracellular cholesterol homeostasis.

The controversy over whether StAR can act as a cholesterol shuttling protein in the inter-membrane space, whether it can act on the outer mitochondrial membrane, or whether it can act only on the inner membrane to effect cholesterol transfer continues. The Miller laboratory have added structure–function studies to the body of information available, and in one such study they examined the structural properties of a bacterially produced segment of the StAR protein corresponding to amino acids 63–193, a protease resistant region [76]. They found that expression of the 63–193 domain in the absence of the molten globule 194–285 domain altered its structure rendering it more susceptible to protease digestion and devoid of tertiary structure. Treatment with detergents increased the secondary structure of this domain indicating that, like the 194–285 domain, the 63–193 domain could also form a molten globule. Most importantly, addition of 63–193 StAR to liposomes consisting of phosphatidylcholine or phosphatidylserine induced the formation of stable protein–liposome complexes. These data indicate the N-terminal region of the StAR protein can form a molten globule and that this structure can interact directly with membranes. This finding is important since when in the molten globule state, proteins lose tertiary structure, can become opened, thus exposing the interior portion of the molecule. If part of the exposed portion is hydrophobic in nature (as appears to be the case with StAR), this may allow an interaction with phospholipid membranes. The strong interaction of water-soluble proteins with phospholipid membranes following their transition to molten globule states has been well documented previously [77,78], and thus a case for StAR interaction with a phospholipid environment can be made. This observation has important implications in that StAR has been shown to closely interact with the outer mitochondrial membrane during the course of cholesterol transfer, and this interaction apparently does not require that it bind to other proteins [64].

These observations were followed by more extensive studies on StAR and its interactions with artificial membranes. Utilizing unilamellar artificial membranes composed of phosphatidylcholine or phosphatidylcholine:cholesterol [79], it was demonstrated that recombinant StAR can readily bind to these membranes in the complete absence of other proteins. This supported the hypothesis that StAR can interact directly with the outer mitochondrial membrane and does not require a receptor protein. Also, this binding occurred maximally at low pH, conditions favoring the formation of molten globule structures. While the degree of binding of StAR to these membranes varied with the heterogeneity of the membrane composition, a most interesting observation was that StAR was able to bind preferentially to the cholesterol-rich domains in cholesterol containing membranes. Cholesterol-rich domains have been previously demonstrated in biological membranes [80], and it is intriguing to speculate that StAR binds to such regions in the relatively cholesterol-rich mitochondrial outer membrane to more easily facilitate transfer of this substrate to the cholesterol poor inner membrane. Importantly, StAR proteins harboring mutations that cause lipoid CAH, and thus impaired steroidogenesis, did not bind to the artificial membranes as efficiently as did wild type StAR [79]. They also found that when StAR bound to artificial membranes containing cardiolipin in concentrations approximating that found in authentic mitochondrial outer membranes, it underwent a conformational change to a molten globule more readily than when cardiolipin-free membranes were used. These observations were strongly corroborated recently in studies demonstrating that in addition to being able to bind to artificial phospholipid membranes, StAR was able to transfer cholesterol from one phospholipid vesicle containing only cholesterol to another phospholipid vesicle containing the CSCC, and promote steroid formation [70].

It is clear at this point in time that we do not know precisely how StAR works to mediate cholesterol transfer to the inner mitochondrial membrane. Many of the observations that have been made to date cannot be disputed, but they also do not tell us how StAR works. For example, the data that StAR can act as a sterol carrier protein and promote desorption from one membrane to another is most convincing, but is not placed within the context of what is occurring in a steroidogenic cell. Also, the observation that StAR and PBR are in close proximity to each other on the outer mitochondrial membrane is very convincing but does not tell us if, in fact, they act together to promote cholesterol transfer. The discovery and characterization of the START domain that can bind and carry cholesterol have given investigators a new direction to pursue and will likely constitute a major part of the cholesterol transferring story, not only within the mitochondria, but within the entire cell. However, once again we do not precisely understand the function of the START domain in intramitochondrial cholesterol transfer. The two models that would appear to have the most credence at this time are the inter-membrane shuttle model and the molten globule model. In the former StAR acts as a carrier of cholesterol from the outer to the inner mitochondrial membrane and in the latter StAR acts to promote cholesterol transfer via changes in its conformation that might produce a hydrophobic tunnel or region through which cholesterol might pass. Each model has strong points and each model has facets that appear to be incompatible with the other. For example, the cholesterol shuttle model is inconsistent with the observation that StAR can act on the outer mitochondrial membrane and promote cholesterol transfer without ever entering the inter-membrane space or matrix. Also, the openings of the hydrophobic core of the START domain do not appear to be large enough to allow cholesterol molecules to enter or exit the pocket without some sort of conformational change in the protein. Perhaps the transformation of the START domain to something approximating a molten globule would allow for the opening of the hydrophobic core and thus cholesterol could enter and exit the tunnel more readily. As for the molten globule hypothesis, the data clearly indicates that StAR can form this structure at low pH, but it is unknown if the local pH in the vicinity of the outer mitochondrial membrane can reach the pH required (3.0–4.0) to form a molten globule. This model is, however, in agreement with observations that an active electrochemical force, in which protons would be actively extruded from the matrix, is required for the support of steroid hormone synthesis [81]. Also, the shuttle model argues that it is not reasonable to hypothesize that a molecule that has evolved such a highly ordered three-dimensional cholesterol binding pocket would function by eliminating the structure of this pocket through the formation of a molten globule. However, there is no data to support this conjecture. So, we are left at this time in the position that until more specific information regarding the interactions between StAR, the mitochondrial membranes and cholesterol are available, the mechanism of action of StAR in mediating cholesterol transfer in steroidogenic mitochondria remains one of the most intriguing questions to be answered in the field of steroidogenesis.

An interesting addition to this story has recently occurred with the finding of several new members of the StAR gene family [82,83]. These studies identified three

StAR-like genes in various tissues of the mouse. The expression of one of these genes, StarD4 was found to be decreased more than twofold in the livers of mice fed a high cholesterol diet. The closest homologs of this gene were two additional genes that the investigators named StarD5 and StarD6. The protein product of each gene consisted of 205–233 amino acids that almost entirely consisted of START domains. These genes constitute a subfamily that express START domain containing proteins that are 30% identical to one another and are approximately 20% identical to the START domains of StAR (StarD1) and MLN64 (StarD3). Interestingly, the tissue distribution of these genes is not restricted to steroidogenic tissues as seen with StAR. StarD4 is found in highest abundance in the liver and kidney with lesser amounts in the brain and testis. The distribution of StarD5 is similar to that of StarD4 while StarD6 is found almost exclusively in the testis. Crystallization of the START domain of StarD4 resulted in a structure that is highly similar to the START domain of MLN64, and presumably StAR [83]. Thus, it seems highly likely that the START domain containing proteins that have been uncovered to date are representative of a large family of proteins whose common trait is the binding and transfer of intracellular sterols. As such it is possible that the StAR protein represents one member of this family whose specific role is to transfer cholesterol to the inner mitochondrial membrane in steroidogenic cells only. The exact roles and functional significance of the remaining members of this interesting family remain to be determined. It is an exciting possibility that the START domain containing proteins are involved in a much more global intracellular cholesterol transporting process in both steroidogenic and nonsteroidogenic cells and as such may provide additional tools to study this little understood phenomenon.

Acknowledgements

The author would like to acknowledge the support of NIH grant HD17481 and funds from the Robert A. Welch Foundation. He would also like to acknowledge the help of Ms. Deborah Alberts in the preparation of this manuscript.

References

[1] Simpson, E.R., Boyd, G.S., 1966. The cholesterol side-chain cleavage system of the adrenal cortex: a mixed function oxidase. Biochem. Biophys. Res. Commun. 24, 10–17.

[2] Churchill, P.F., Kimura, T., 1979. Topological studies of cytochromes P450scc and P45011β in bovine adrenocortical inner mitochondrial membranes. J. Biol. Chem. 254, 10443–10448.

[3] Hanukoglu, I., Hanukoglu, Z., 1986. Stoichiometry of mitochondrials, cytochromes P450, adrenodoxin and adrenodoxin reductase in adrenal cortex and corpus luteum. Eur. J. Biochem. 157, 27–31.

[4] Kovanen, P.T., Goldstein, J.L., Chappell, D.A., Brown, M.S., 1980. Regulation of low density lipoprotein receptors by adrenocorticotropin in the adrenal gland of mice and rats in vivo. J. Biol. Chem. 255, 5591–5598.

[5] Gwynne, J.T., Mahaffee, D.D., 1989. Rat adrenal uptake and metabolism of high density lipoprotein cholesteryl ester. J. Biol. Chem. 264, 8141–8150.

[6] Jefcoate, C.R., McNamara, B.C., Artemenko, I., Yamazaki, T., 1992. Regulation of cholesterol movement to mitochondrial cytochrome P450scc in steroid hormone synthesis. J. Steroid Biochem. Mol. Biol. 43, 751–767.

[7] Stocco, D.M., Clark, B.J., 1996. Regulation of the acute production of steroids in steroidogenic cells. Endocr. Rev. 17, 221–244.
[8] Karaboyas, G.C., Koritz, S.B., 1965. Identity of the site of action of cAMP and ACTH in corticosteroidogenesis in rat adrenal and beef adrenal cortex slices. Biochemistry 4, 462–468.
[9] Crivello, J.F., Jefcoate, C.R., 1980. Intracellular movement of cholesterol in rat adrenal cells. J. Biol. Chem. 255, 8144–8151.
[10] Mori, M., Marsh, J.M., 1982. The site of luteinizing hormone stimulation of steroidogenesis in mitochondria of the rat corpus luteum. J. Biol. Chem. 257, 6178–6183.
[11] Privalle, C.T., Crivello, J., Jefcoate, C.R., 1983. Regulation of intramitochondrial cholesterol transfer to side-chain cleavage cytochrome P450scc in rat adrenal gland. Proc. Natl Acad. Sci. USA 80, 702–706.
[12] Jefcoate, C.R., DiBartolomeos, M.J., Williams, C.A., McNamara, B.C., 1987. ACTH regulation of cholesterol movement in isolated adrenal cells. J. Steroid Biochem. 27, 721–729.
[13] Lambeth, J.D., Kitchen, S.E., Farooqui, A.A., Tuckey, R., Kamin, H., 1982. Cytochrome P-450scc substrate interactions: studies of binding and catalytic activity using hydroxycholesterols. J. Biol. Chem. 257, 1876–1884.
[14] Tuckey, R.C., 1992. Cholesterol side-chain cleavage by mitochondria from the human placenta, studies using hydroxycholesterols as substrates. J. Steroid Biochem. Mol. Biol. 42, 883–890.
[15] Phillips, M.C., Johnson, W.J., Rothblat, G.H., 1987. Mechanisms and consequences of cellular cholesterol exchange and transfer. Biochim. Biophys. Acta 906, 223–276.
[16] Rennert, H., Chang, Y.J., Strauss, J.F. III, 1993. Intracellular cholesterol dynamics in steroidogenic cells: a contemporary view. In: Adashi, E.Y., Leung, P.C.K. (Eds.), The Ovary. Raven Press, New York, pp. 147–164.
[17] Stone, D., Hechter, O., 1954. Studies on ACTH action in perfused bovine adrenals: site of action of ACTH in corticosteroidogenesis. Arch. Biochem. Biophys. 51, 457–469.
[18] Hechter, O., Zaffaroni, A., Jacobsen, R.P., Levy, H., Jeanloz, R.W., Schenker, V., Pincus, G., 1951. The nature and the biogenesis of the adrenal secretory product. Recent Prog. Horm. Res. 6, 215–246.
[19] Ferguson, J.J., 1962. Puromycin and adrenal responsiveness to adrenocorticotropic hormone. Biochim. Biophys. Acta 57, 616–617.
[20] Ferguson, J.J., 1963. Protein synthesis and adrenocorticotropin responsiveness. J. Biol. Chem. 238, 2754–2759.
[21] Garren, L.D., Ney, R.L., Davis, W.W., 1965. Studies on the role of protein synthesis in the regulation of corticosterone production by ACTH in vivo. Proc. Natl Acad. Sci. USA 53, 1443–1450.
[22] Garren, L.D., Davis, W.W., Crocco, R.M., 1966. Puromycin analogs: action of adrenocorticotropic hormone and the role of glycogen. Science 152, 1386–1388.
[23] Davis, W.W., Garren, L.D., 1968. On the mechanism of action of adrenocorticotropic hormone. The inhibitory site of cycloheximide in the pathway of steroid biosynthesis. J. Biol. Chem. 243, 5153–5157.
[24] Cooke, B.A., Janszen, F.H.A., Clotscher, W.F., van der Molen, H.J., 1975. Effect of protein-synthesis inhibitors on testosterone production in rat testis interstitial tissue and Leydig-cell preparations. Biochem. J. 150, 413–418.
[25] Paul, D.P., Gallant, S., Orme-Johnson, N.R., Orme-Johnson, W.H., Brownie, A.C., 1976. Temperature dependence of cholesterol binding to cytochrome P-450scc of the rat adrenal. Effect of adrenocorticotropic hormone and cycloheximide. J. Biol. Chem. 251, 7120–7126.
[26] Farese, R.V., Prudente, W.J., 1977. On the requirement for protein synthesis during corticotropin-induced stimulation of cholesterol side chain cleavage in rat adrenal mitochondrial and solubilized desmolase preparations. Biochim. Biophys. Acta 496, 567–570.
[27] Crivello, J.F., Jefcoate, C.R., 1978. Mechanism of corticotropin action in rat adrenal cells. 1. Effect of inhibitors of protein synthesis and of microfilament formation on corticosterone synthesis. Biochim. Biophys. Acta 542, 315–329.
[28] Arthur, J.R., Boyd, G.S., 1976. The effect of inhibitors of protein synthesis on cholesterol side-chain cleavage in the mitochondria of luteinized rat ovaries. Eur. J. Biochem. 49, 117–127.
[29] Ohno, Y., Yanagibashi, K., Yonezawa, Y., Ishiwatari, S., Matsuba, M., 1983. A possible role of "steroidogenic factor" in the corticoidogenic response to ACTH; effect of ACTH, cycloheximide and

aminoglutethimide on the content of cholesterol in the outer and inner mitochondrial membrane of rat adrenal cortex. Endocrinology (Jpn) 30, 335–338.

[30] Papadopoulos, V., 1993. Peripheral-type benzodiazepine/diazepam binding inhibitor receptor: biological role in steroidogenic cell function. Endocr. Rev. 14, 222–240.

[31] Krueger, R.J., Orme-Johnson, N.R., 1983. Acute adrenocorticotropic hormone stimulation of adrenal corticosteroidogenesis. J. Biol. Chem. 258, 10159–10167.

[32] Epstein, L.F., Orme-Johnson, N.R., 1991. Acute action of luteinizing hormone on mouse Leydig cells: accumulation of mitochondrial phosphoproteins and stimulation of testosterone biosynthesis. Mol. Cell. Endocrinol. 81, 113–126.

[33] Epstein, L.F., Orme-Johnson, N.R., 1991. Regulation of steroid hormone biosynthesis: identification of precursors of a phosphoprotein targeted to the mitochondrion in stimulated rat adrenal cortex cells. J. Biol. Chem. 266, 19739–19745.

[34] Pon, L.A., Orme-Johnson, N.R., 1988. Acute stimulation of corpus luteum cells by gonadotropin or adenosine 3′5′-monophosphate causes accumulation of a phosphoprotein concurrent with acceleration of steroid synthesis. Endocrinology 123, 1942–1948.

[35] Stocco, D.M., Kilgore, M.W., 1988. Induction of mitochondrial proteins in MA-10 Leydig tumour cells with human choriogonadotropin. Biochem. J. 249, 95–103.

[36] Stocco, D.M., Sodeman, T.C., 1991. The 30-kDa mitochondrial proteins induced by hormone stimulation in MA-10 mouse Leydig tumor cells are processed from larger precursors. J. Biol. Chem. 266, 19731–19738.

[37] Stocco, D.M., 1992. Further evidence that the mitochondrial proteins induced by hormone stimulation in MA-10 mouse Leydig tumor cells are involved in the acute regulation of steroidogenesis. J. Steroid Biochem. Mol. Biol. 43, 319–333.

[38] Stocco, D.M., Ascoli, M., 1993. The use of genetic manipulation of MA-10 Leydig tumor cells to demonstrate the role of mitochondrial proteins in the acute regulation of steroidogenesis. Endocrinology 132, 959–967.

[39] Clark, B.J., Wells, J., King, S.R., Stocco, D.M., 1994. The purification, cloning, and expression of a novel LH-induced mitochondrial protein in MA-10 mouse Leydig tumor cells: characterization of the steroidogenic acute regulatory protein (StAR). J. Biol. Chem. 269, 28314–28322.

[40] Sugawara, T., Holt, J.A., Driscoll, D., Strauss, J.F. III, Lin, D., Miller, W.L., Patterson, D., Clancy, K.P., Hart, I.M., Clark, B.J., Stocco, D.M., 1995. Human steroidogenic acute regulatory protein: functional activity in COS-1 cells, tissue-specific expression, and mapping of the gene to 8p11.2 and a pseudogene to chromosome 13. Proc. Natl Acad. Sci. USA 92, 4778–4782.

[41] Lin, D., Sugawara, T., Strauss, J.F. III, Clark, B.J., Stocco, D.M., Saenger, P., Rogol, A., Miller, W.L., 1995. Role of steroidogenic acute regulatory protein in adrenal and gonadal steroidogenesis. Science 267, 1828–1831.

[42] Degenhart, H.J., Visser, H.K.A., Boon, H., O'Doherty, N.J., 1972. Evidence for deficient 20α-cholesterol-hydroxylase activity in adrenal tissue of a patient with lipoid adrenal hyperplasia. Acta Endocrinol. 71, 512–518.

[43] Muller, J., Torsson, A., Neilsen, M.D., Petersen, K.E., Christoffesen, J., Skakkeboek, N.E., 1991. Gonadal development and growth in 46, XX and 46 XY individuals with P450scc deficiency (congenital lipoid adrenal hyperplasia). Horm. Res. 36, 203–208.

[44] Lin, D., Gitelman, S.E., Saenger, P., Miller, W.L., 1991. Normal genes for the cholesterol side chain cleavage enzyme, P450scc, in congenital lipoid adrenal hyperplasia. J. Clin. Invest. 88, 1955–1962.

[45] Caron, K.M., Soo, S.C., Wetsel, W.C., Stocco, D.M., Clark, B.J., Parker, K.L., 1997. Targeted disruption of the mouse gene encoding steroidogenic acute regulatory protein provides insights into congenital lipoid adrenal hyperplasia. Proc. Natl Acad. Sci. USA 94, 11540–11545.

[46] Bose, H.S., Sugawara, T., Strauss, J.F. III, Miller, W.L., 1996. The pathophysiology and genetics of congenital lipoid adrenal hyperplasia. N. Engl. J. Med. 335, 1870–1878.

[47] Bose, H.S., Pescovitz, O.H., Miller, W.L., 1997. Spontaneous feminization in a 46,XX female patient with congenital lipoid adrenal hyperplasia due to a homozygous frameshift mutation in the steroidogenic acute regulatory protein. J. Clin. Endocrinol. Metab. 82, 1511–1515.

[48] Fujieda, K., Tajima, T., Nakae, J., Sageshima, S., Tachibana, K., Suwa, S., Sugawara, T., Strauss, J.F. III, 1997. Spontaneous puberty in 46,XX subjects with congenital lipoid adrenal hyperplasia. J. Clin. Invest. 99, 1265–1271.

[49] Ishii, T., Hasegawa, T., Pai, C., Yvgi-Ohana, N., Timberg, R., Zhao, L., Majdic, G., Chung, B., Orly, J., Parker, K.L., 2002. The roles of circulating high-density lipoproteins and trophic hormones in the phenotype of knockout mice lacking the steroidogenic acute regulatory protein. Mol. Endocrinol. 16, 2297–2309.

[50] Schwaiger, M., Herzog, V., Neupert, W., 1987. Characterization of translocation contact sites involved in the import of mitochondrial proteins. J. Cell Biol. 105, 235–246.

[51] Pon, L., Moll, T., Vestweber, D., Marshallsay, B., Schatz, G., 1989. Protein import into mitochondria: ATP-dependent protein translocation activity in a submitochondrial fraction enriched in membrane contact sites and specific proteins. J. Cell Biol. 109, 2603–2616.

[52] Arakane, F., Sugawara, T., Nishino, H., Liu, Z., Holt, J.A., Pain, D., Stocco, D.M., Miller, W.L., Strauss, J.F. III, 1996. Steroidogenic acute regulatory protein (StAR) retains activity in the absence of its mitochondrial import sequence: implications for the mechanism of StAR action. Proc. Natl Acad. Sci. USA 93, 13731–13736.

[53] Arakane, F., Kallen, C.B., Watari, H., Foster, J.A., Sepuri, N.B., Pain, D., Stayrook, S.E., Lewis, M., Gerton, G.L., Strauss, J. III, 1998. The mechanism of action of steroidogenic acute regulatory protein (StAR): StAR acts on the outside of mitochondria to stimulate steroidogenesis. J. Biol. Chem. 273, 16339–16345.

[54] Bose, H.S., Lingappa, V.R., Miller, W.L., 2002. Rapid regulation of steroidogenesis by mitochondrial protein import. Nature 417, 87–91.

[55] Artemenko, I.P., Zhao, D., Hales, D.B., Hales, K.H., Jefcoate, C.R., 2001. Mitochondrial processing of newly synthesized steroidogenic acute regulatory protein (StAR), but not total StAR, mediates cholesterol transfer to cytochrome P450 side chain cleavage enzyme in adrenal cells. J. Biol. Chem. 49, 46583–46596.

[56] Wang, X., Liu, Z., Eimerl, S., Timberg, R., Weiss, A.M., Orly, J., Stocco, D.M., 1998. Effect of truncated forms of the steroidogenic acute regulatory protein on intramitochondrial cholesterol transfer. Endocrinology 139, 3903–3912.

[57] Miller, W.L., 1997. Congenital lipoid adrenal hyperplasia: the human gene knockout of the steroidogenic acute regulatory protein. J. Mol. Endocrinol. 19, 227–240.

[58] Moog-Lutz, C., Tomasetto, C., Regnier, C.H., Wendling, C., Lutz, Y., Muller, D., Chenard, M.P., Basset, P., Rio, M.C., 1997. MLN64 exhibits homology with the steroidogenic acute regulatory protein [StAR] and is overexpressed in human breast carcinomas. Int. J. Cancer 7, 183–191.

[59] Watari, H., Arakane, F., Moog-Lutz, C., Kallen, C.B., Tomasetto, C., Gerton, G.L., Rio, M., Baker, M.E., Strauss, J.F. III, 1997. MLN64 contains a domain with homology to the steroidogenic acute regulatory protein (StAR) that stimulates steroidogenesis. Proc. Natl Acad. Sci. USA 94, 8462–8467.

[60] Alpy, F., Stoeckel, M.E., Dierich, A., Escola, J.M., Wendling, C., Chenard, M.P., Vanier, M.T., Gruenberg, J., Tomasetto, C., Rio, M.C., 2001. The steroidogenic acute regulatory protein homolog MLN64, a late endosomal cholesterol-binding protein. J. Biol. Chem. 276, 4261–4269.

[61] Bose, H.S., Whittal, R.M., Huang, M.C., Baldwin, M.A., Miller, W.L., 2000. N-218 MLN64, a protein with StAR-like steroidogenic activity, is folded and cleaved similarly to StAR. Biochemistry 39, 11722–11731.

[62] Christenson, L.K., Strauss, J.F. III, 2000. Steroidogenic acute regulatory protein (StAR) and the intramitochondrial translocation of cholesterol. Biochem. Biophys. Acta 1529, 175–187.

[63] Zhang, M., Liu, P., Dwyer, N.K., Christenson, L.K., Fujimoto, T., Martinez, F., Comly, M., Hanover, J.A., Blanchette-Mackie, E.J., Strauss, J.F. III, 2002. MLN64 mediates mobilization of lysosomal cholesterol to steroidogenic mitochondria. J. Biol. Chem. 277, 33300–33310.

[64] Kallen, C.B., Billheimer, J.T., Summers, S.A., Staybrook, S.E., Lewis, M., Strauss, J.F. III, 1998. Steroidogenic acute regulatory protein (StAR) is a sterol transfer protein. J. Biol. Chem. 273, 26285–26288.

[65] Martinez, F., Strauss, J.F. III, 1997. Regulation of mitochondrial cholesterol metabolism. In: Bittman, R. (Ed.), Subcellular Biochemistry, Vol. 28, Cholesterol: Its Functions and Metabolism in Biology and Medicine. Plenum Press, New York, pp. 205–234.

[66] Petrescu, A.D., Gallegos, A.M., Okamura, Y., Strauss, J.F. III, Schroeder, F., 2001. Steroidogenic acute regulatory protein binds cholesterol and modulates mitochondrial membrane sterol domain dynamics. J. Biol. Chem. 276, 36970–36982.

[67] Pandak, W.M., Ren, S., Marques, D., Hall, E., Redford, K., Mallonee, D., Bohdan, P., Heuman, D., Gil, G., Hylemon, P., 2002. Transport of cholesterol into mitochondria is rate limiting for bile acid synthesis via the alternative pathway in primary rat hepatocytes. J. Biol. Chem. 277, 48158–48164.

[68] Bose, H.S., Whittal, R.M., Baldwin, M.A., Miller, W.L., 1999. The active form of the steroidogenic acute regulatory protein, StAR, appears to be a molten globule. Proc. Natl Acad. Sci. USA 96, 7250–7255.
[69] Kallen, C.B., Arakane, F., Christenson, L.K., Watari, H., Devoto, L., Strauss, J.F. III, 1998. Unveiling the mechanism of action and regulation of the steroidogenic acute regulatory protein. Mol. Cell. Endocrinol. 145, 39–45.
[70] Tuckey, R.C., Headlam, M.J., Bose, H.S., Miller, W.L., 2002. Transfer of cholesterol between phospholipid vesicles mediated by the steroidogenic acute regulatory protein (StAR). J. Biol. Chem. 277, 47123–47128.
[71] West, A., Horvat, R.D., Roess, D.A., Barisas, B.G., Juengel, J.L., Niswender, G.D., 2001. Steroidogenic acute regulatory protein and peripheral-type benzodiazepine receptor associate at the mitochondrial membrane. Endocrinology 142, 502–505.
[72] Papadopoulos, V., 1993. Peripheral-type benzodiazepine/diazepam binding inhibitor receptor: biological role in steroidogenic cell function. Endocr. Rev. 14, 222–240.
[73] Jefcoate, C., 2002. High-flux mitochondrial cholesterol trafficking, a specialized function of the adrenal cortex. J. Clin. Invest. 110, 881–890.
[74] Ponting, C.P., Aravind, L., 1999. START: a lipid-binding domain in StAR, HD-ZIP and signalling proteins. Trends Biochem. Sci. 24, 130–132.
[75] Tsujishita, Y., Hurley, J.H., 2000. Structure and lipid transport mechanism of a StAR-related protein. Nat. Struct. Biol. 7, 408–414.
[76] Song, M., Shao, H., Mujeeb, A., James, T.L., Miller, W.L., 2001. Molten globule structure and membrane binding of the N-terminal protease-resistant domain (63–193) of the steroidogenic acute regulatory protein (StAR). Biochem. J. 356, 151–158.
[77] Kuwajima, K., Nitta, K., Sugai, S., 1975. Electrophoretic investigations of the acid conformational change of alpha-lactalbumin. J. Biochem. (Tokyo) 78, 205–211.
[78] Banuelos, S., Muga, A., 1995. Binding of molten globule-like conformations to lipid bilayers. Structure of native and partially folded alpha-lactalbumin bound to model membranes. J. Biol. Chem. 270, 29910–29915.
[79] Christensen, K., Bose, H.S., Harris, F.M., Miller, W.L., Bell, J.D., 2001. Binding of StAR to synthetic membranes suggests an active molten globule. J. Biol. Chem. 276, 17044–17051.
[80] Schroeder, F., Jefferson, J.R., Kier, A.B., Knittel, J., Scallen, T.J., Wood, W.G., Hapala, I., 1991. Membrane cholesterol dynamics: cholesterol domains and kinetic pools. Proc. Soc. Exp. Biol. Med. 196, 235–252.
[81] King, S.R., Liu, Z., Soh, J., Eimerl, S., Orly, J., Stocco, D.M., 1999. Effects of disruption of the mitochondrial electrochemical gradient on steroidogenesis and the steroidogenic acute regulatory (StAR) protein. J. Steroid Biochem. Mol. Biol. 69, 143–154.
[82] Soccio, R.E., Adams, R.M., Romanowski, M.J., Sehayek, E., Burley, S.K., Breslow, J.L., 2002. The cholesterol-regulated StarD4 gene encodes a StAR-related lipid transfer protein with two closely related homologues, StarD5 and StAR D6. Nature 99, 6943–6948.
[83] Romanowski, M.J., Soccio, R.E., Breslow, J.L., Burley, S.K., 2002. Crystal structure of the *Mus musculus* cholesterol-regulated START protein 4 (StarD4) containing a StAR-related lipid transfer domain. Nature 99, 6949–6954.

Colour Plates

Advances in Molecular and Cell Biology, Vol. 33, pages 579–620

ISSN: 1569-2558 / DOI: 10.1016/S1569-2558(03)33032-2

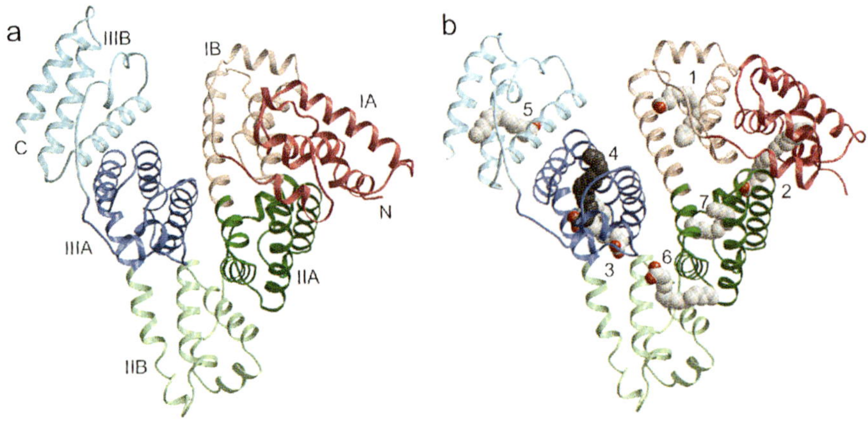

Fig. 1.

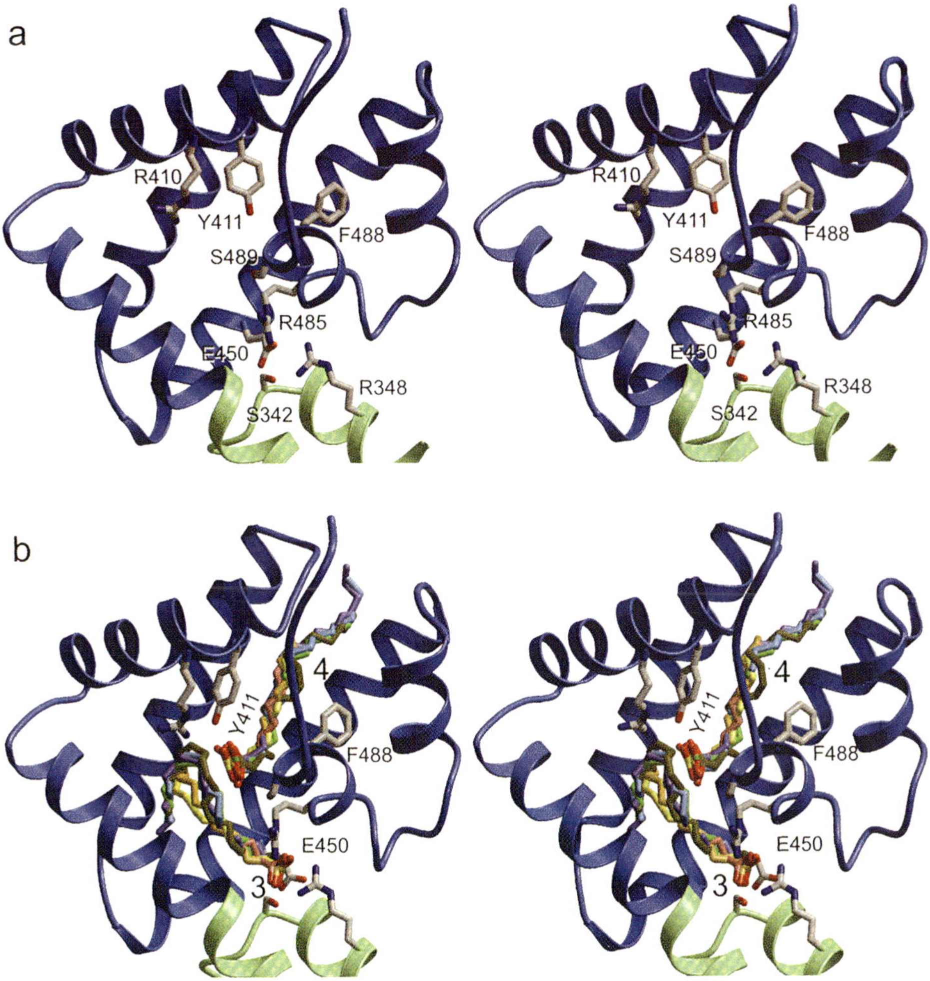
a
R410
Y411
F488
S489
R485
E450
R348
S342
R410
Y411
F488
S489
R485
E450
R348
S342
b
4
Y411
F488
E450
3
4
Y411
F488
E450
3

Fig. 3.

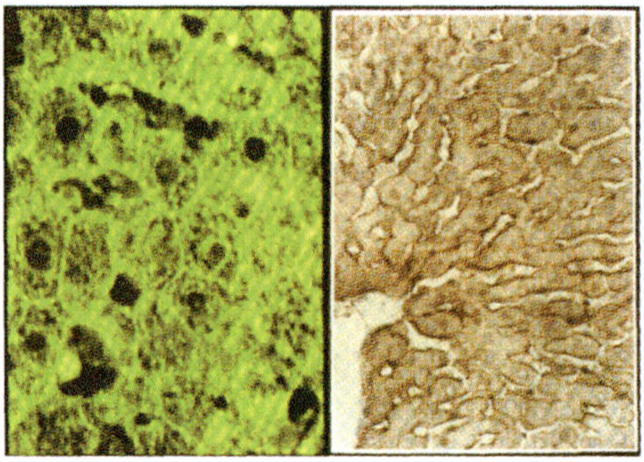

Fig. 4.

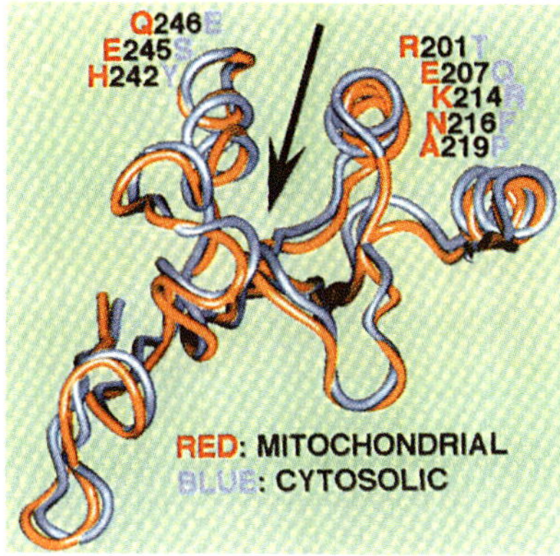
Q246E
E245S
H242Y
R201T
E207Q
K214R
N216F
A219P
RED: MITOCHONDRIAL
BLUE: CYTOSOLIC

Fig. 10.

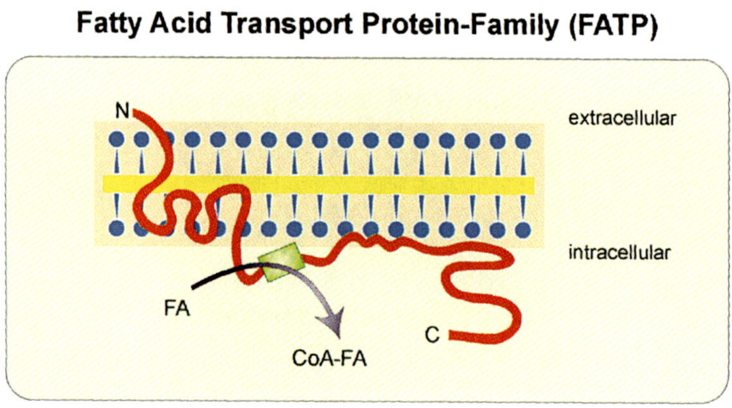
Fatty Acid Transport Protein-Family (FATP)
N
extracellular
intracellular
FA
CoA-FA
C

Fig. 1.

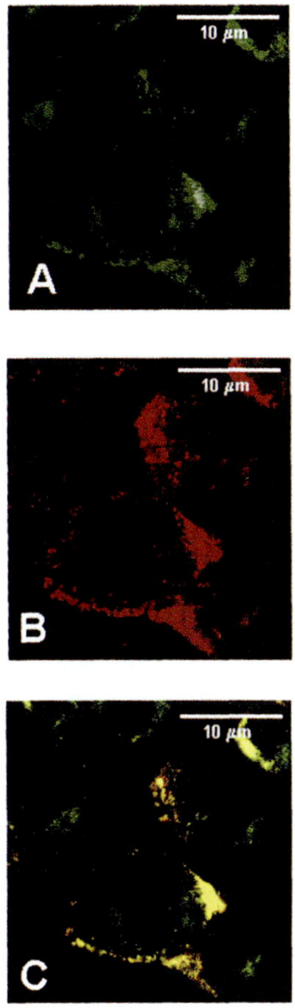

Fig. 2.

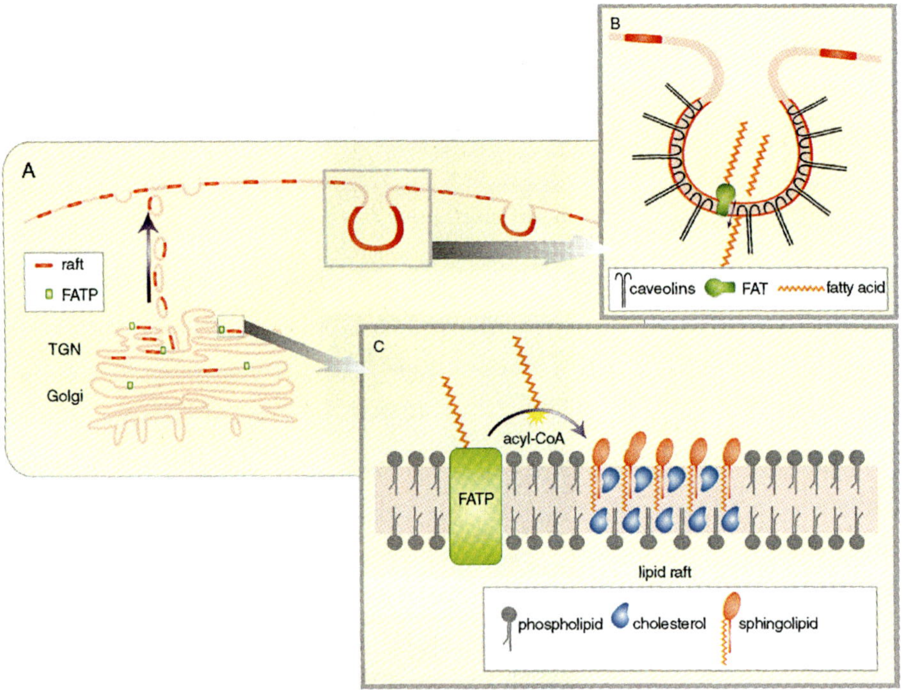
A
raft
FATP
TGN
Golgi
B
caveolins
FAT
fatty acid
C
acyl-CoA
FATP
lipid raft
phospholipid
cholesterol
sphingolipid

Fig. 3.

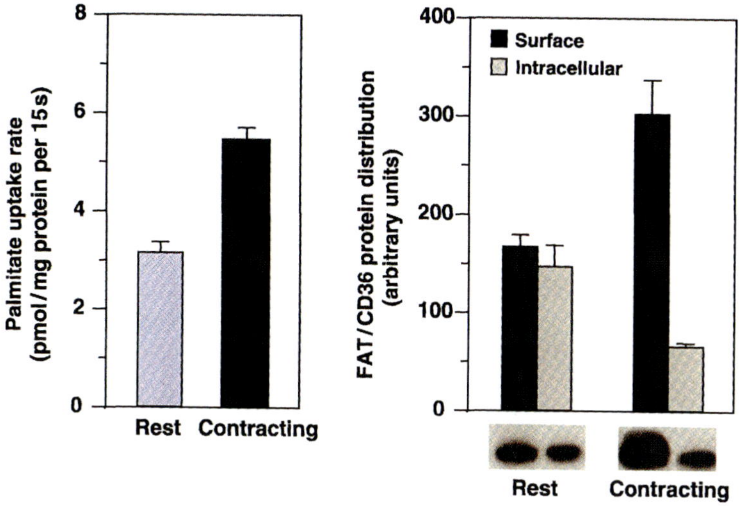
8
6
4
2
0
Palmitate uptake rate
(pmol/mg protein per 15 s)
Rest
Contracting
400
300
200
100
0
Surface
Intracellular
FAT/CD36 protein distribution
(arbitrary units)
Rest
Contracting

Fig. 3.

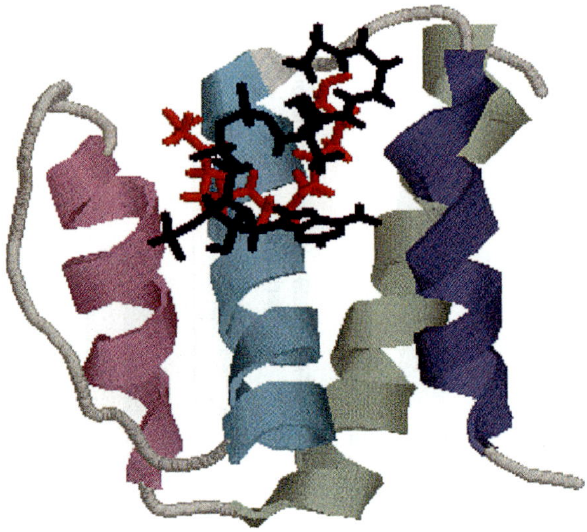

Fig. 2.

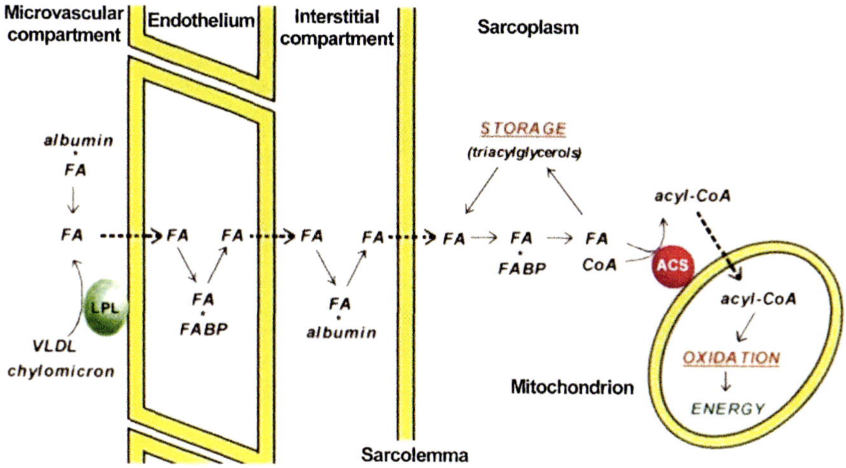
Microvascular compartment
Endothelium
Interstitial compartment
Sarcoplasm
albumin
FA
FA
FA
FA
FA
FA
FA
FA
FA
FA
FABP
FA
albumin
STORAGE
(triacylglycerols)
acyl-CoA
FA
FABP
CoA
ACS
acyl-CoA
OXIDATION
ENERGY
LPL
VLDL
chylomicron
Mitochondrion
Sarcolemma

Fig. 1.

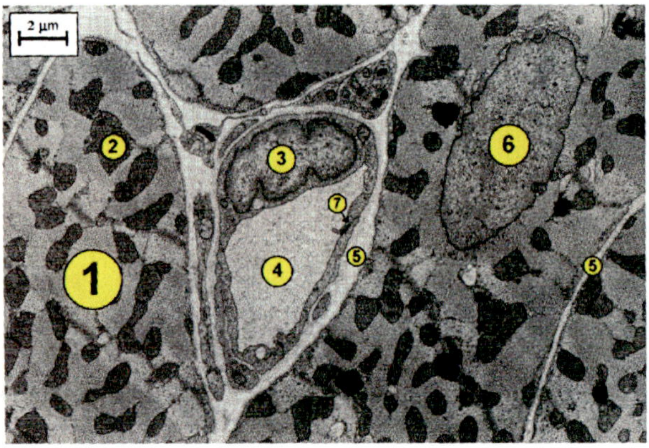

Fig. 2.

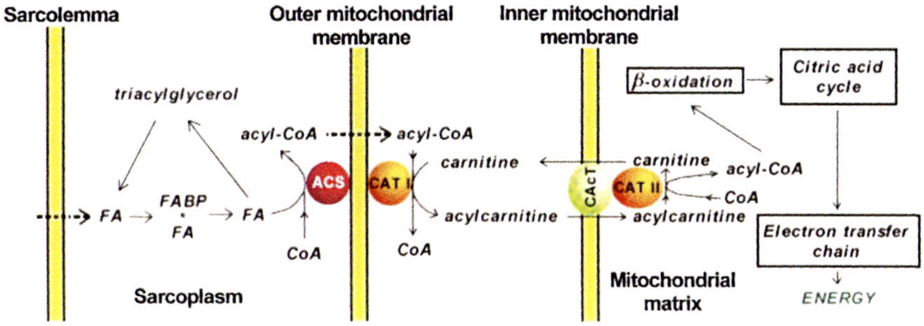
Sarcolemma
Outer mitochondrial membrane
Inner mitochondrial membrane
triacylglycerol
β-oxidation
Citric acid cycle
acyl-CoA
acyl-CoA
carnitine
carnitine
acyl-CoA
ACS
CAT I
CAcT
CAT II
CoA
FA
FABP
FA
FA
acylcarnitine
acylcarnitine
CoA
CoA
Electron transfer chain
Sarcoplasm
Mitochondrial matrix
ENERGY

Fig. 3.

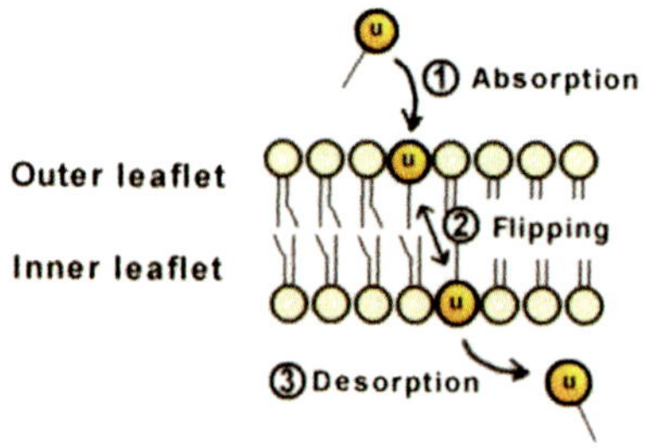
① Absorption
Outer leaflet
② Flipping
Inner leaflet
③ Desorption

Fig. 4.

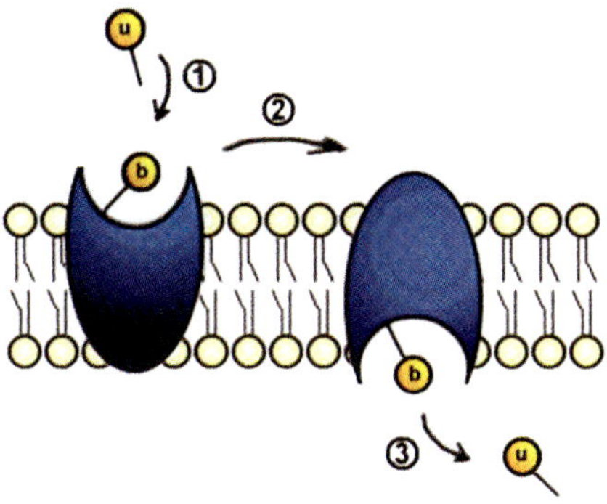

Fig. 5.

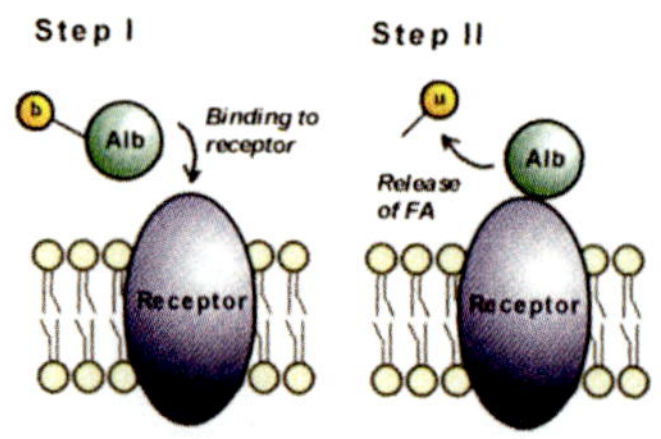
Step I
Step II
Alb
Binding to
receptor
Alb
Release
of FA
Receptor
Receptor

Fig. 6.

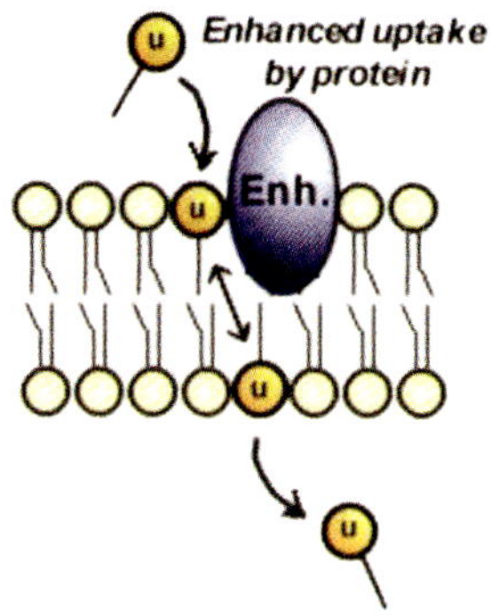

Fig. 7.

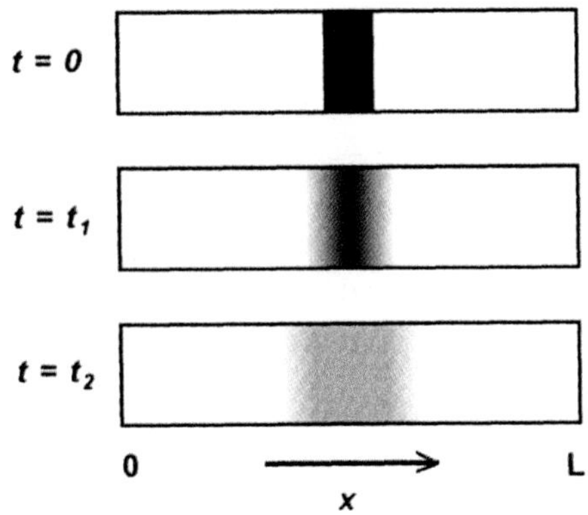

Fig. 8.

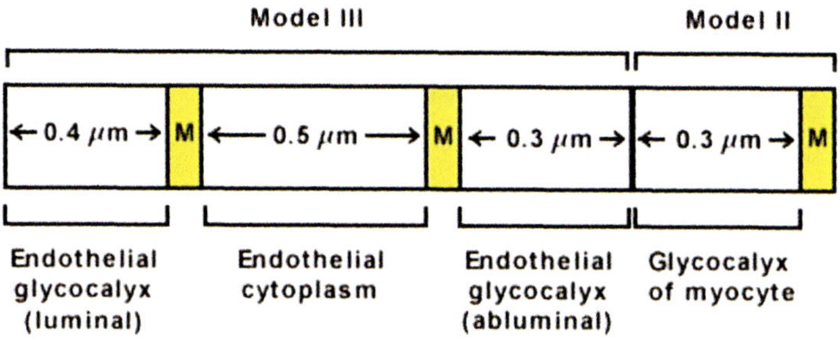
Model III
Model II
← 0.4 μm →
M
← 0.5 μm →
M
← 0.3 μm →
← 0.3 μm →
M
Endothelial glycocalyx (luminal)
Endothelial cytoplasm
Endothelial glycocalyx (abluminal)
Glycocalyx of myocyte

Fig. 10.

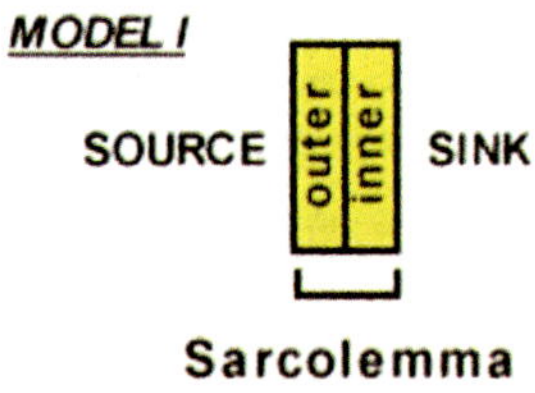

Fig. 11.

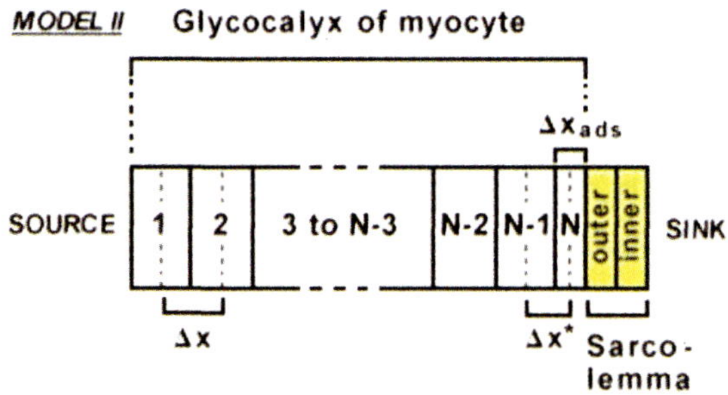
MODEL II
Glycocalyx of myocyte
Δxads
SOURCE
1
2
3 to N-3
N-2
N-1
N
outer
inner
SINK
Δx
Δx*
Sarco-
lemma

Fig. 12.

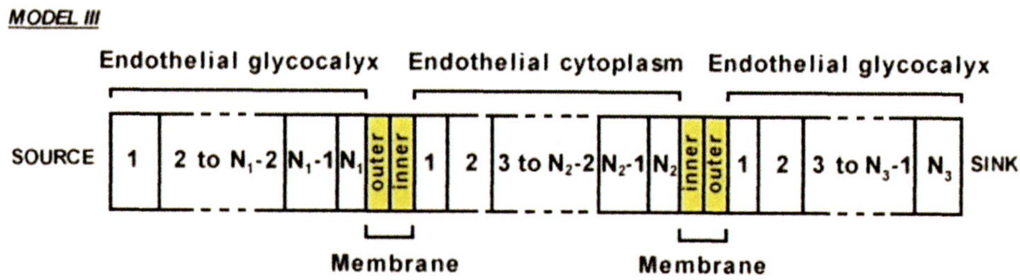
MODEL III
Endothelial glycocalyx
Endothelial cytoplasm
Endothelial glycocalyx
SOURCE
1
2 to N1-2
N1-1
N1
outer
inner
1
2
3 to N2-2
N2-1
N2
inner
outer
1
2
3 to N3-1
N3
SINK
Membrane
Membrane

Fig. 13.

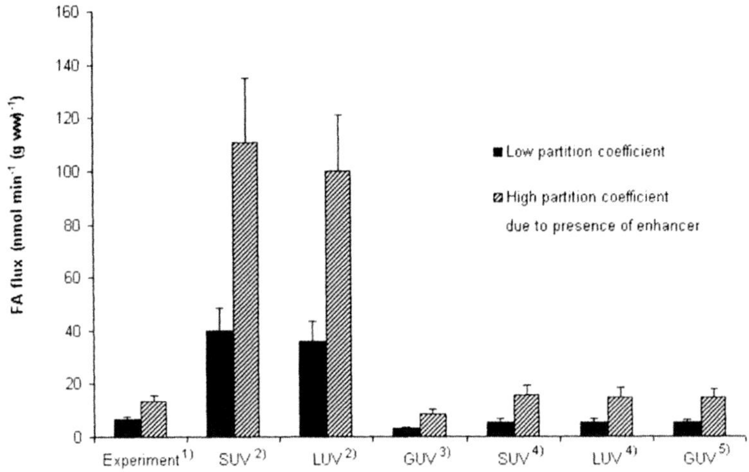
160
140
120
100
80
60
40
20
0
FA flux (nmol min-1 (g ww)-1)
Low partition coefficient
High partition coefficient
due to presence of enhancer
Experiment 1)
SUV 2)
LUV 2)
GUV 3)
SUV 4)
LUV 4)
GUV 5)

Fig. 14.

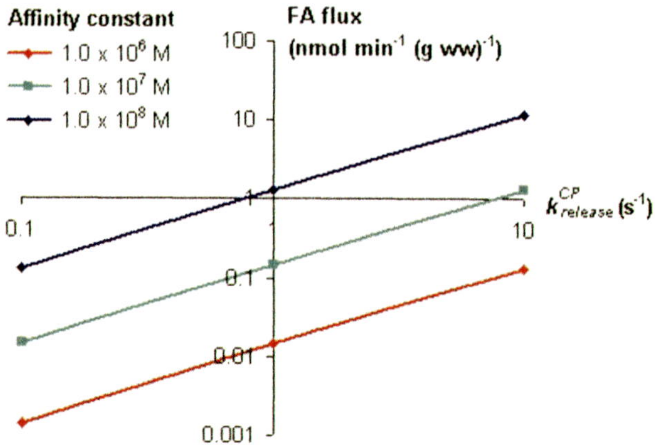
Affinity constant
1.0 x 10⁶ M
1.0 x 10⁷ M
1.0 x 10⁸ M
FA flux
(nmol min⁻¹ (g ww)⁻¹)
100
10
1
0.1
0.01
0.001
0.1
10
$k^{CP}_{release}$ (s⁻¹)

Fig. 16.

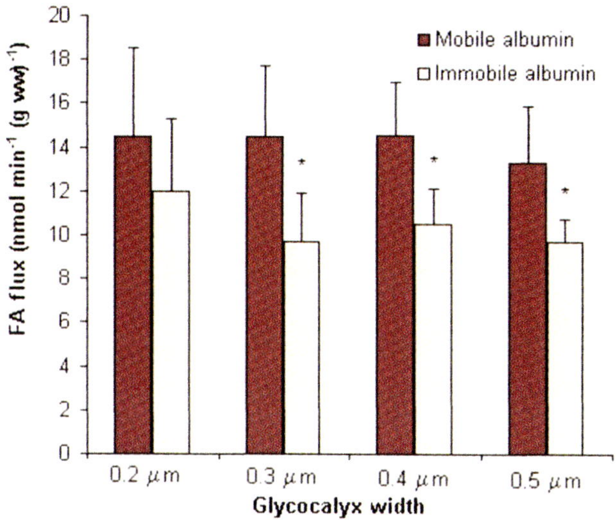
20
18
16
14
12
10
8
6
4
2
0
FA flux (nmol min⁻¹ (g ww)⁻¹)
Mobile albumin
Immobile albumin
*
*
*
0.2 μm
0.3 μm
0.4 μm
0.5 μm
Glycocalyx width

Fig. 17.

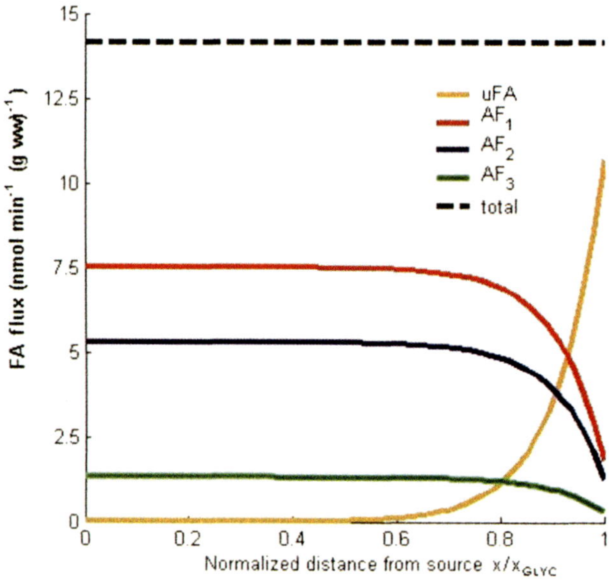
15
12.5
10
7.5
5
2.5
0
FA flux (nmol min^{-1} (g ww)$^{-1}$)
uFA
AF$_1$
AF$_2$
AF$_3$
total
0
0.2
0.4
0.6
0.8
1
Normalized distance from source x/x$_{GLYC}$

Fig. 18.

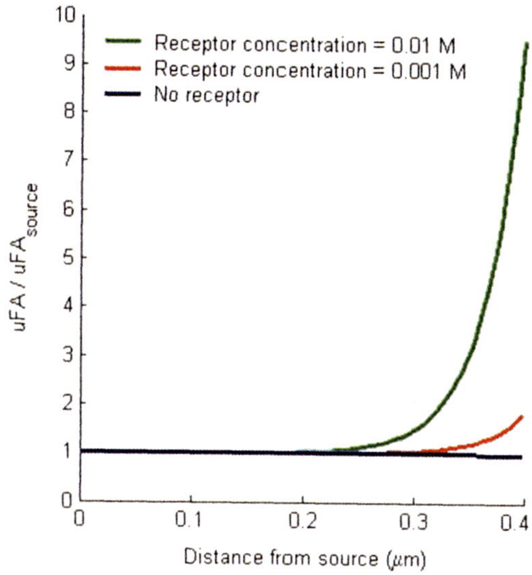

Receptor concentration = 0.01 M
Receptor concentration = 0.001 M
No receptor
uFA / uFA$_{source}$
Distance from source (μm)
0
1
2
3
4
5
6
7
8
9
10
0.1
0.2
0.3
0.4

Fig. 19.

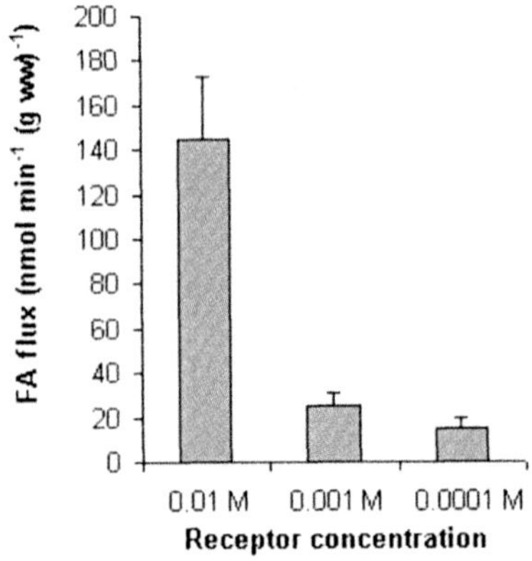
FA flux (nmol min-1 (g ww)-1)
200
180
160
140
120
100
80
60
40
20
0
0.01 M
0.001 M
0.0001 M
Receptor concentration

Fig. 20.

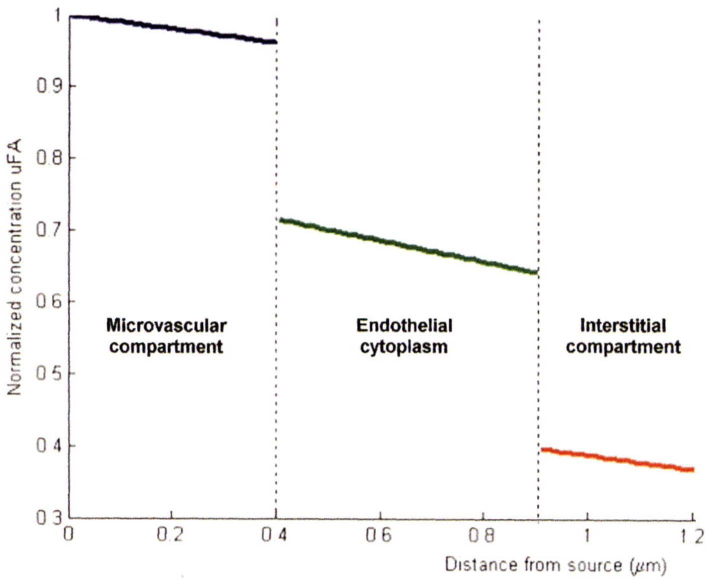
Normalized concentration uFA
Distance from source (μm)
Microvascular compartment
Endothelial cytoplasm
Interstitial compartment
0 0.2 0.4 0.6 0.8 1 1.2
0.3 0.4 0.5 0.6 0.7 0.8 0.9 1

Fig. 21.

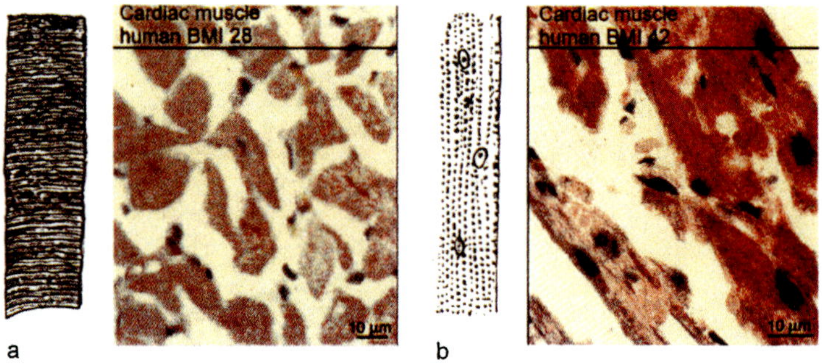
Cardiac muscle
human BMI 28
10 µm
a
Cardiac muscle
human BMI 42
10 µm
b

Fig. 3.

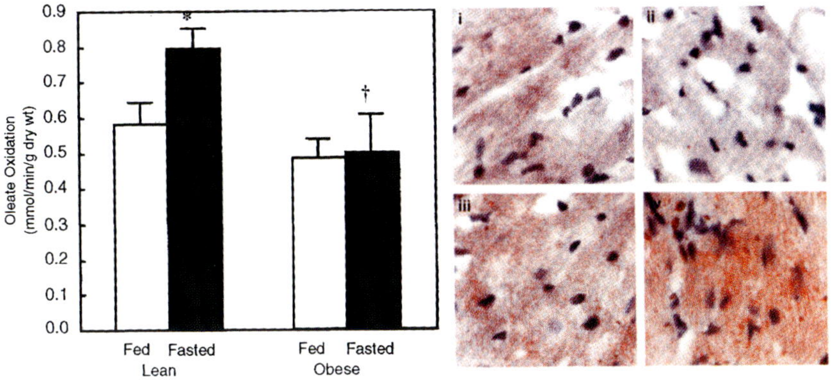
0.9
0.8
0.7
0.6
0.5
0.4
0.3
0.2
0.1
0.0
Oleate Oxidation
(mmol/min/g dry wt)
*
†
Fed
Fasted
Lean
Fed
Fasted
Obese
i
ii
iii

Fig. 4.

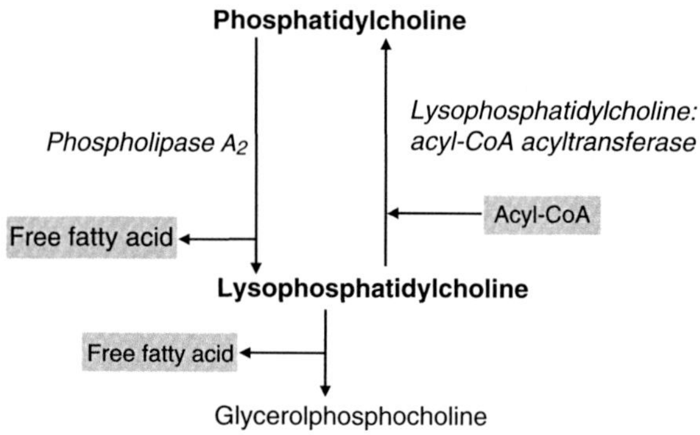
Phosphatidylcholine
Phospholipase A2
Lysophosphatidylcholine:
acyl-CoA acyltransferase
Acyl-CoA
Free fatty acid
Lysophosphatidylcholine
Free fatty acid
Glycerolphosphocholine

Fig. 4.

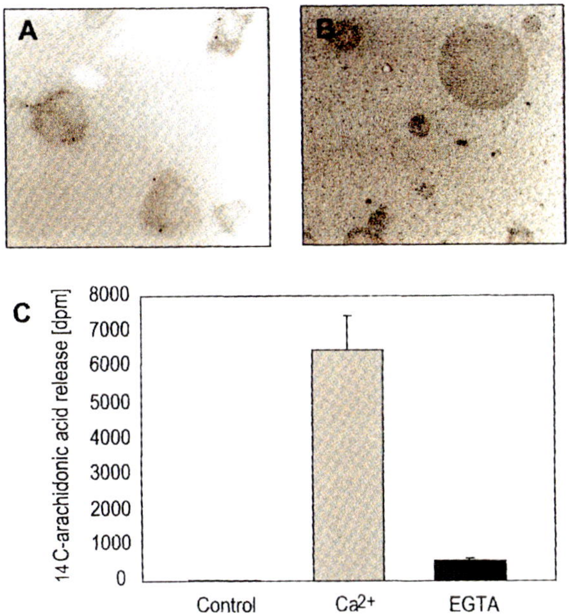
A
B
C
14 C-arachidonic acid release [dpm]
8000
7000
6000
5000
4000
3000
2000
1000
0
Control
Ca2+
EGTA

Fig. 2.

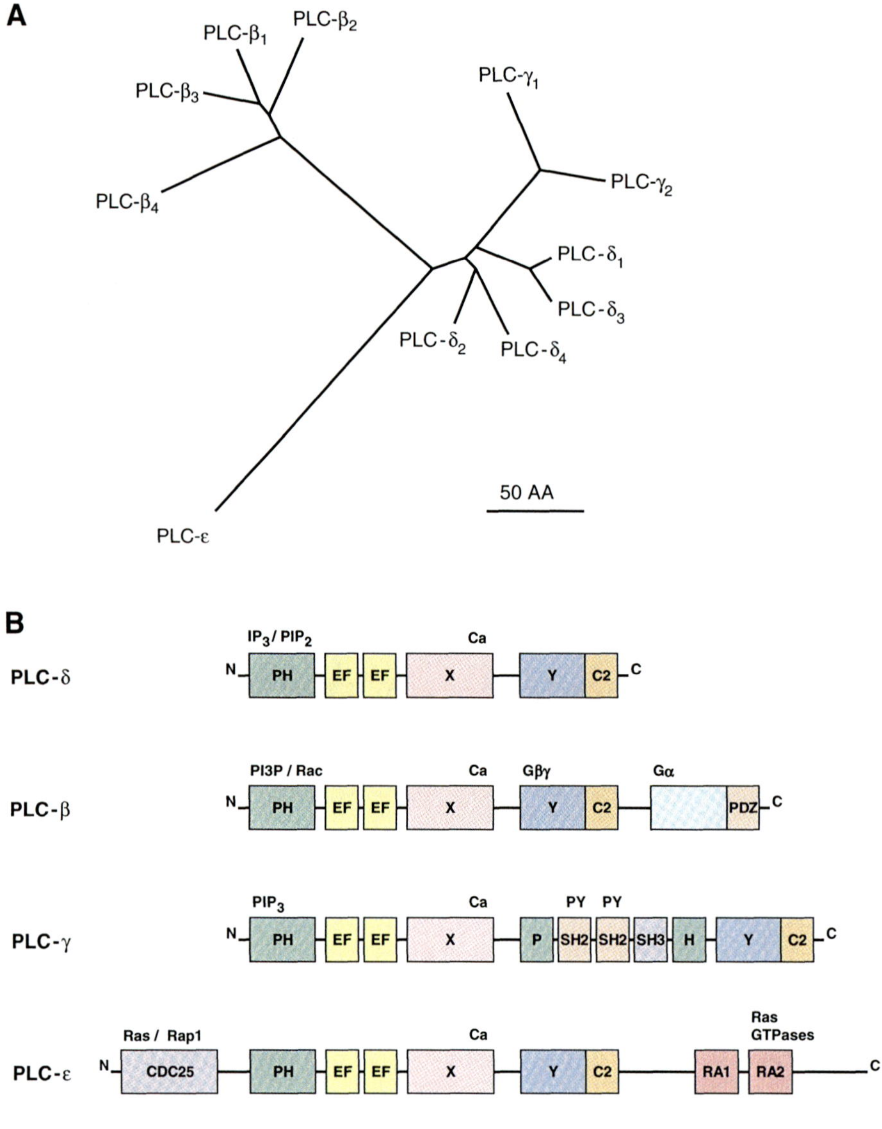
A
PLC-β1
PLC-β2
PLC-β3
PLC-β4
PLC-γ1
PLC-γ2
PLC-δ1
PLC-δ3
PLC-δ2
PLC-δ4
PLC-ε
50 AA
B
PLC-δ
IP3/ PIP2
Ca
N
PH
EF
EF
X
Y
C2
C
PLC-β
PI3P / Rac
Ca
Gβγ
Gα
PDZ
PLC-γ
PIP3
PY PY
P
SH2
SH2
SH3
H
PLC-ε
Ras / Rap1
CDC25
Ras
GTPases
RA1
RA2

Fig. 1.

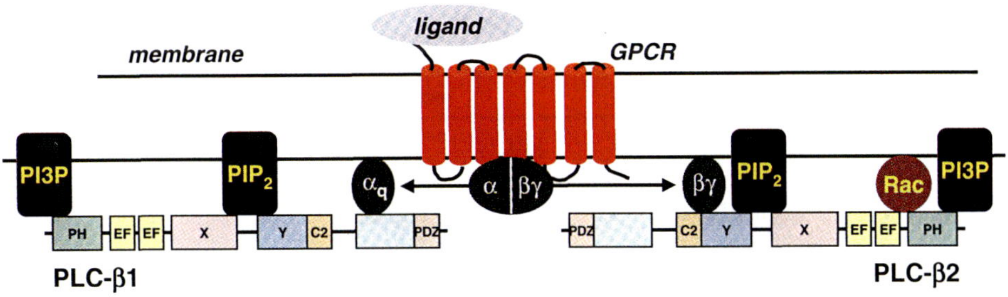
ligand
membrane
GPCR
PI3P
PIP2
αq
α
βγ
βγ
PIP2
Rac
PI3P
PH
EF
EF
X
Y
C2
PDZ
PDZ
C2
Y
X
EF
EF
PH
PLC-β1
PLC-β2

Fig. 2.

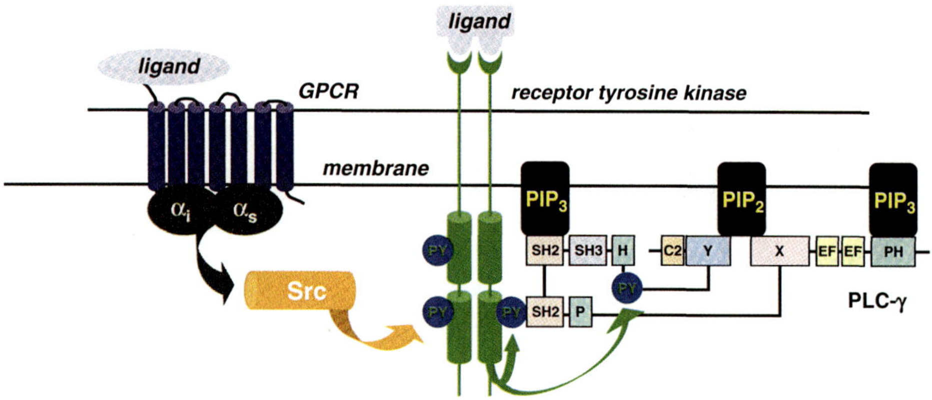
ligand
ligand
GPCR
receptor tyrosine kinase
membrane
αi
αs
Src
PY
PY
PY
PY
PIP3
SH2
SH3
H
SH2
P
C2
Y
PIP2
X
EF
EF
PIP3
PH
PLC-γ

Fig. 3.

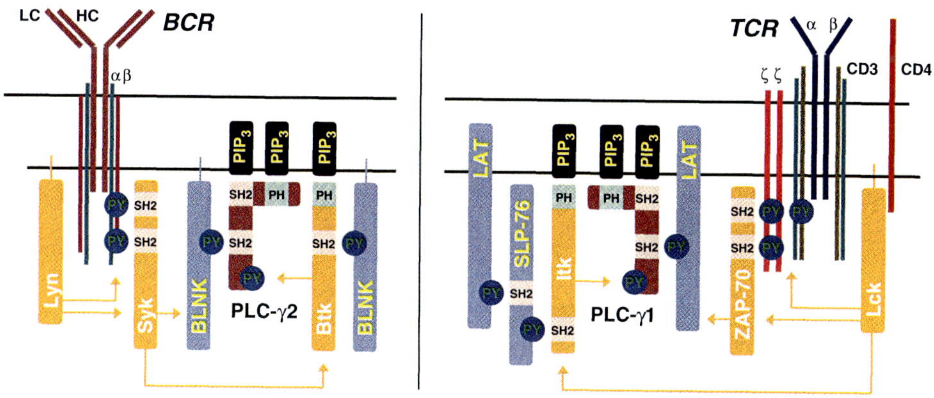

LC
HC
BCR
α β
PIP3
PIP3
PIP3
SH2
PH
PH
SH2
SH2
SH2
SH2
PY
Lyn
Syk
BLNK
PLC-γ2
Btk
BLNK
TCR
α β
ζ ζ
CD3
CD4
LAT
SLP-76
Itk
PLC-γ1
LAT
ZAP-70
Lck

Fig. 4.

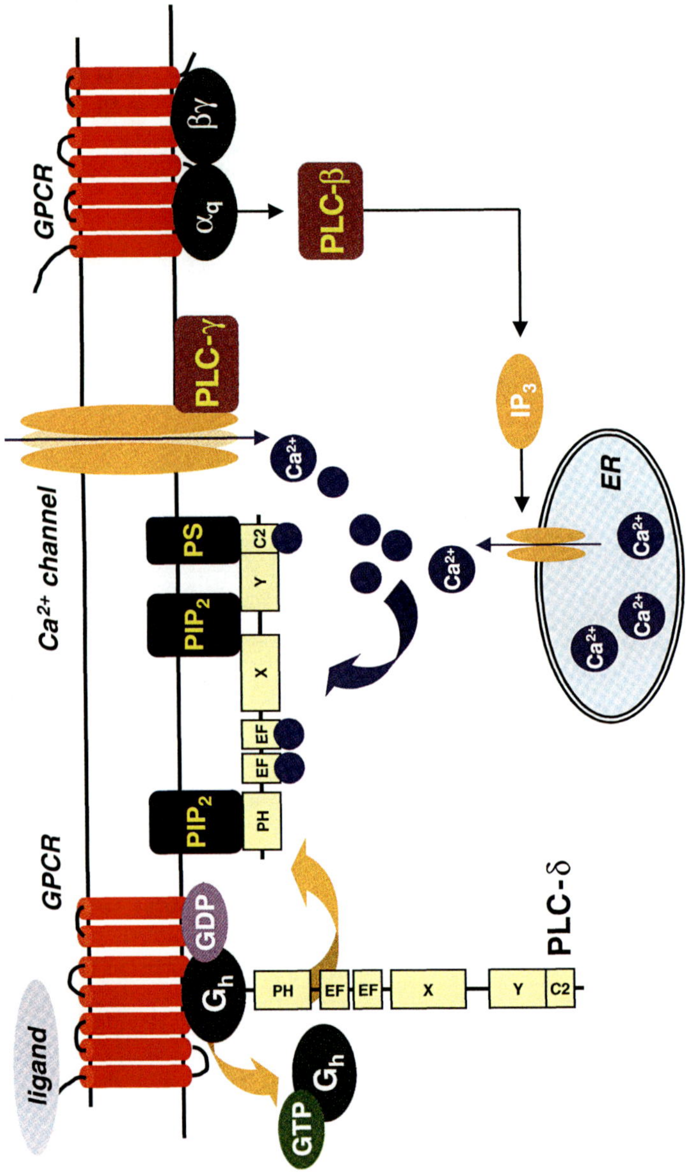
ligand
GPCR
GDP
G_h
GTP
G_h
PH
EF
EF
X
Y
C2
PLC-δ
PIP_2
PH
EF
EF
X
Y
C2
PIP_2
PS
Ca^{2+} channel
Ca^{2+}
PLC-γ
GPCR
α_q
βγ
PLC-β
IP_3
ER
Ca^{2+}
Ca^{2+}
Ca^{2+}
Ca^{2+}

Fig. 5.

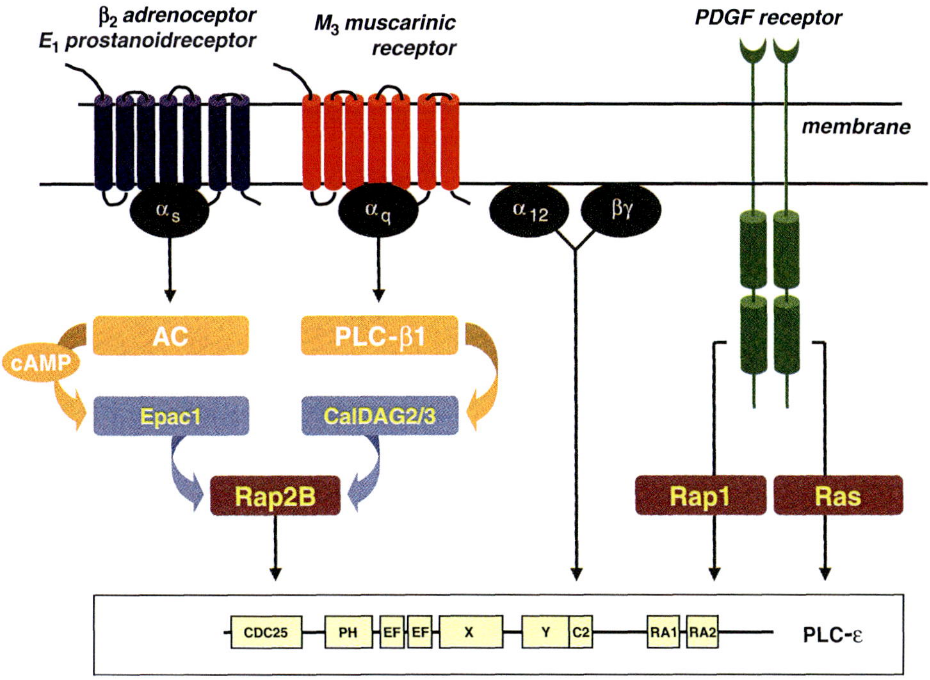

β2 adrenoceptor
E1 prostanoidreceptor
M3 muscarinic receptor
PDGF receptor
membrane
αs
αq
α12
βγ
AC
PLC-β1
cAMP
Epac1
CalDAG2/3
Rap2B
Rap1
Ras
CDC25
PH
EF
EF
X
Y
C2
RA1
RA2
PLC-ε

Fig. 6.

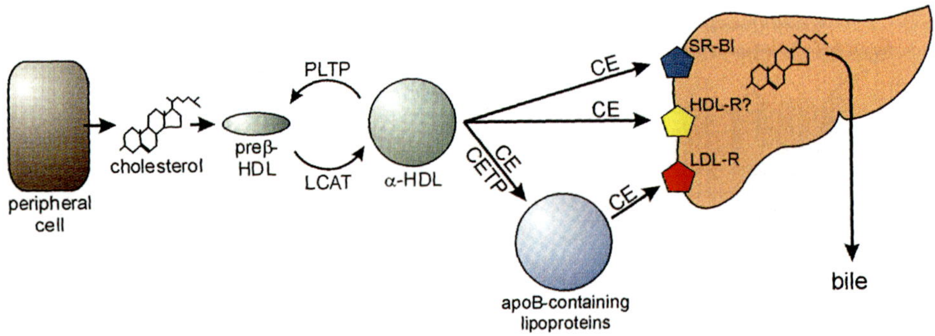

peripheral cell
cholesterol
preβ-HDL
PLTP
LCAT
α-HDL
CE
CE
CE
CETP
apoB-containing lipoproteins
CE
SR-BI
HDL-R?
LDL-R
bile

Fig. 1.

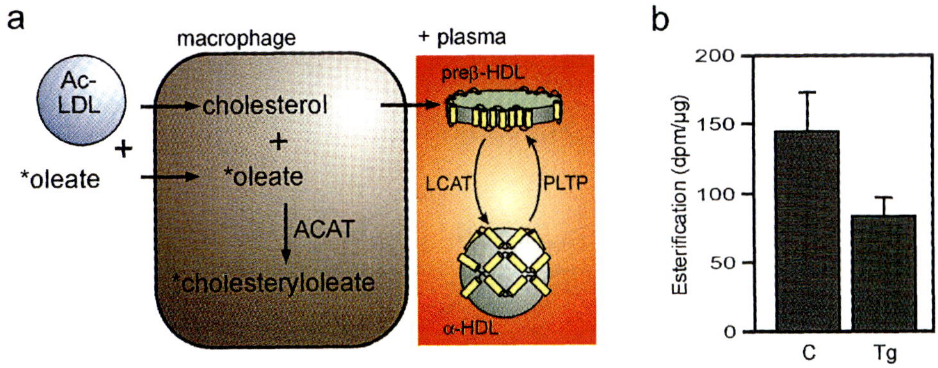
a
macrophage
+ plasma
Ac-
LDL
+
*oleate
cholesterol
+
*oleate
ACAT
*cholesteryloleate
preβ-HDL
LCAT
PLTP
α-HDL
b
Esterification (dpm/μg)
200
150
100
50
0
C
Tg

Fig. 2.

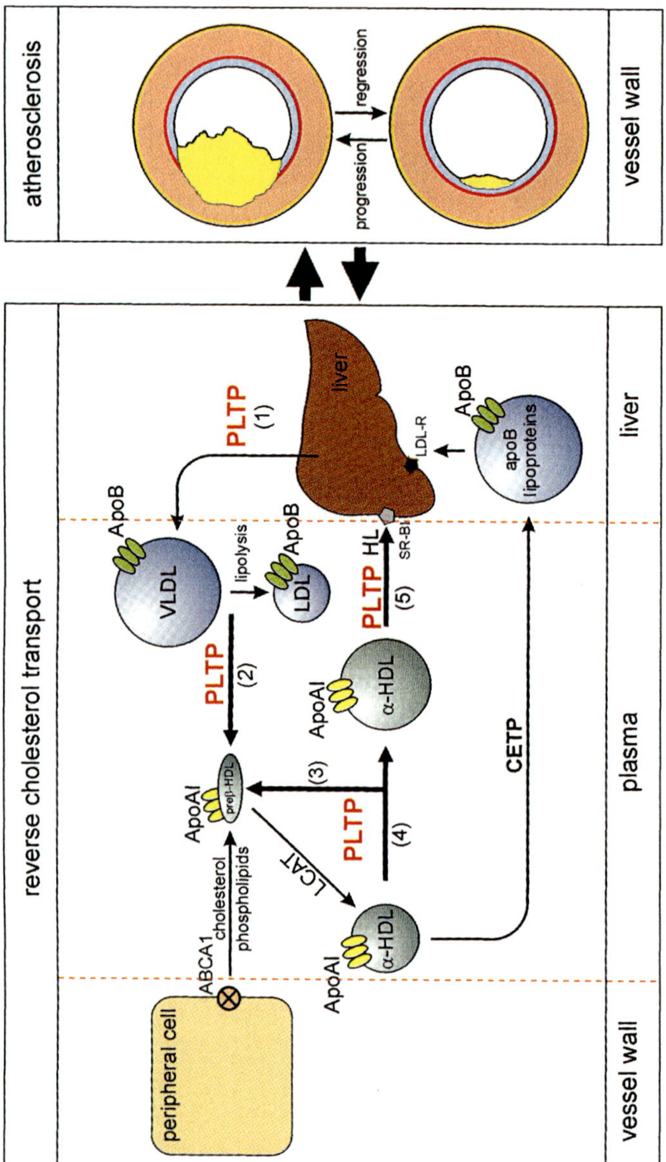
reverse cholesterol transport
vessel wall
plasma
liver
peripheral cell
ABCA1
cholesterol
phospholipids
ApoAI
preβ-HDL
LCAT
α-HDL
PLTP
(1)
(2)
(3)
(4)
(5)
HL
SR-BI
LDL-R
VLDL
LDL
ApoB
lipolysis
CETP
apoB lipoproteins
atherosclerosis
progression
regression

Fig. 4.

Edwards Brothers Malloy
Ann Arbor MI. USA
February 26, 2014